MODERN PRACTICE IN STRESS AND VIBRATION ANALYSIS

PROCEEDINGS OF THE 3RD INTERNATIONAL CONFERENCE ON MODERN PRACTICE IN STRESS AND VIBRATION ANALYSIS DUBLIN/IRELAND/3-5 SEPTEMBER 1997

MODERN PRACTICE IN STRESS AND VIBRATION ANALYSIS

Edited by

M.D.GILCHRIST

University College Dublin, Department of Mechanical Engineering, Ireland

A.A.BALKEMA/ROTTERDAM/BROOKFIELD/1997

The texts of the various papers in this volume were set individually by typists under the supervision of each of the authors concerned.

Published by
A.A. Balkema, P.O. Box 1675, 3000 BR Rotterdam, Netherlands (Fax: +31.10.413.5947)
A.A. Balkema Publishers, Old Post Road, Brookfield, VT 05036-9704, USA (Fax: 802.276.3837)

ISBN 90 5410 896 7

Printed in the Netherlands

Modern Practice in Stress and Vibration Analysis, Gilchrist (ed.) © 1997 Balkema, Rotterdam, ISBN 90 5410 896 7

Table of contents

Modern Practice in Stress and Vibration Analysis, Gilchrist (ed.) © *1997 Balkema, Rotterdam, ISBN 90 5410 896 7*

Preface

It is more important than ever before to use state-of-the-art stress and vibration analysis methods in the design of engineering components and structures. This demand is becoming increasingly urgent for scientists, researchers and engineers in today's environment given the demands for precision manufacturing, improved quality and reduced lead-times.

This 3rd international conference on Modern Practice in Stress and Vibration Analysis has attracted eighty papers on both analytical, computational and experimental techniques which reflect the state-of-the-art worldwide in stress and vibration analyses. The papers in this volume constitute the proceedings of the conference which was organised by the Stress & Vibration Group of the Institute of Physics at University College Dublin, 3-5 September 1997. These papers form a valuable contribution to the literature and it is hoped that they will provide an important source of information in these fields of research.

The Organising Committee would like to thank the authors, whose papers have been selected for presentation at the Conference, for the quality of their papers and for their co-operation in preparing the manuscripts of the papers.

M. D.Gilchrist
Dublin, September 1997

1 Finite element and boundary domain techniques

Modern Practice in Stress and Vibration Analysis, Gilchrist (ed.) © *1997 Balkema, Rotterdam, ISBN 90 5410 896 7*

Boundary element analysis of two-dimensional contact problems using a local coordinate system

R.S.Hack & A.A.Becker
Department of Mechanical Engineering, University of Nottingham, UK

ABSTRACT: In this paper, a local coordinate axes system is used to describe the contact conditions instead of the usual global Cartesian system used in most BE formulations. To demonstrate the accuracy of the algoritms, three contact examples covering stationary, receding and advancing contact are presented.

1. INTRODUCTION

The Boundary Element (BE) method is acknowledged as an accurate numerical tool for stress analysis. The prime advantage of this method is its surface-only modelling ability and high accuracy resolution of stresses making it very suitable for the study of practical contact problems. The BE method works on the premise of directly incorporating the contact surface relationships of equilibrium and compatibility into the characteristic system matrix prior to solution; as such unlike the Finite Element (FE) method there is no need for special elements, such as gap or interface elements, to represent the contact interface.

The analysis of theoretical contact problems is inherently non-linear, due to the number of unknowns, such as the size of the contact area after loading, and in the case of frictional contact problems the ultimate contact conditions. This is precisely the situation where a high accuracy numerical tool like the BE method becomes necessary. Currently the main alternative to BE solutions, and most widely used is the Finite Element (FE) method. In the BE modelling of elastic materials it is only necessary to model the surfaces of the components, hence the primary advantage of the BE method over the FE method, is the requirement for a reduced set of equations to represent the model which will carry through to faster solution times.

Although the BE method has been in use for the last two decades, it was Andersson (1982) who first applied the BEM to the solution of contact problems. Subsequent papers by Paris et al. (1995), Huesman and Kuhn (1995) and Zhu (1995) have elaborated on the more fundamental stick and slip contact boundary conditions; as did Olukoko et al. (1993), who proposed three benchmarks of frictional contact. From this literature it seems that all the authors have stated clearly that the contact algorithms used are based on global Cartesian coordinate systems, with contact boundary conditions to match.

The present authors are currently involved in the development of a more efficient BE algorithm for the solution of frictional contact problems, based on a local coordinate system, which are subject to combinations of normal and tangential loads. As a precursor to this it has been necessary to use three benchmark models to assess the performance of the new algorithm when the contacting surfaces are in stationary, receding and advancing contact and run comparisons with the previous algorithm which was based on a global Cartesian coordinate system, as well as the FE solutions derived from ABAQUS (HKS, 1996).

2. THE BE FORMULATION

As the name suggests, the BE method reduces a given structure into boundary elements, effectively avoiding to model the interior of the body; at least for the modelling of elastic bodies. It is however necessary to model the interior of a domain if the analysis involves elastoplastic materials, see, for example Huesman and Kuhn

(1995). For the purposes of this paper, its scope shall be restricted to the study of elastic materials. Boundary-only modelling presents a significant advantage, in that a smaller number of elements will be required to numerically represent a given domain.

The most improtant part of the BE formulation is the derivation of an integral equation which relates displacements and tractions on the boundary of the domain being modelled. The basis of this integral is the Somigliana Identity for Displacements. This is made up of traction and displacement kernels, which relate any node on the surface to all the remaining nodes which make up the entire domain, in terms of their relative positions and the material properties. At this point all displacements and tractions are unknown.

On the basis of this integral, it is then possible to introduce some boundary conditions, in the form of known displacements and tractions; at every node making up the domain. This step will create a distinct set of simultaneous equations, which can be solved using standard Gaussian elimination, and from this it is then possible to derive the stresses and strains. Further details on the BE formulations can be found in several textbooks, e.g. Becker (1992) and Banerjee (1994).

2.1 *Theory of contact analysis*

The primary classification will refer to conforming or non-conforming contacts where conforming contact implies that any two bodies in contact fit perfectly together without deformation, while non-conforming contact refer to bodies of dissimilar profiles. When brought together non-conforming bodies can make contact on two forms; either at a point or along a line.

In the theoretical modelling of line contact it is quite common to assume that the bodies have a unit finite length, hence they can be modelled as plane strain; this follows the reduction of the bodies in contact to a two-dimensional model. However for point contact it is common to assume that either one or both of the bodies have a spherical surface at the point of contact, hence an axisymmetric model can be used, thus reducing the analysis to a two-dimensional model.

The most significant theoretical models of contact analysis were established by Hertz (1896), but these were restricted to frictionless surfaces and perfectly elastic solids. More recently Gladwell (1980) and Johnson (1985), elaborated on the basic characteristic equations as proposed by Hertz, to include a wider array of bodies which did not fit the restricted descriptions of the Hertzian theories.

2.2 *Theory of BE contact analysis*

The approach of the BE analysis to contact problems is to define the interaction between any two bodies at the point of contact. This interaction will be defined in terms of mathematical relationships of the two surfaces relative to one another. The relationships will subdivide into two main categories; stick contact and frictional slip contact.

While the magnitude of any tractions or displacements at the contact interface may be unknown, the relationships will enable the evaluation of the system equations prior to solution for these tractions and displacements. However the actual formulation of these relationships will depend on the type of coordinate system being used. The purpose of this paper is to compare the BE contact formulations based on a global and a local coordinate system.

The main aspect of BE contact analysis is the fact that the contact variables of any two contacting pair nodes are coupled directly to make a set of linear simultaneous equations. This therefore allows the whole system to be solved using standard Gaussian elimination. Whether the contact analysis involves infinite friction (stick), frictionless or Coulomb frictional slip; the coupled equations must satisfy the two relationships of compatibility of displacements and equilibrium of tractions.

2.3 *Global contact coordinate system*

This coordinate system refers to a standard global Cartesian coordinate system, where the position of any given node is relative to the zero reference node. The contact conditions will be expressed by mathematical relationships, which take account of the geometrical position of the relevant node to the zero point.

Depending on whether the analysis is one of infinite friction or Coulomb friction, the compatibility and equilibrium relationships will vary accordingly. For infinite friction (stick) contact mode the equilibrium conditions are given by (see Becker, 1992):

$$T_{XA} = -T_{XB}$$
$$T_{YA} = -T_{YB} \qquad (1)$$

and the compatibility relationships are given by:

$$U_{XA} = U_{XB} + (x_B - x_A)$$
$$U_{YA} = U_{YB} + (y_B - y_A) \qquad (2)$$

where the subscripts A and B refer to the two bodies in contact, and T and U are the traction and displacement vectors respectively.

For Coulomb friction (slip) contact mode, the equilibrium equations are given by:

$$T_{XA} = -T_{XB}$$
$$T_{YA} = -T_{YB} \qquad (3)$$
$$T_{XA} = T_{YA}\left[\frac{\pm\mu\cos\theta + \sin\theta}{-\cos\theta + \mu\sin\theta}\right]$$

and the compatibility relationships are:

$$U_{YB} = U_{YA} + (y_A - y_B) + \left[(U_{XB} - U_{XA}) + (x_B - x_A)\right]\tan\theta \qquad (4)$$

Where θ is the angle of the tangent to the nodal point, as shown fig. 1. It important to note that when θ equals 90° ; tan θ becomes infinite. Hence when this occurs, it becomes necessary to rotate the axis of reference to overcome this problem, which makes it inconvenient to implement this formulation in a computer program.

2.4 *Local contact coordinate system*

A local coordinate system works on the same premise as the global system except that each node which makes up the domain will have an individual coordinate reference system, local to that node only. This system will have a normal and tangential direction, as shown in fig. 1, relative to the individual node and the orientation of the domain surface at the node. If two nodes from different domains are in contact, the two nodes will have a common but opposing normal and tangential directions.

However depending on whether the analysis is one of infinite friction (stick) or Coulomb friction (slip), the compatibility and equilibrium

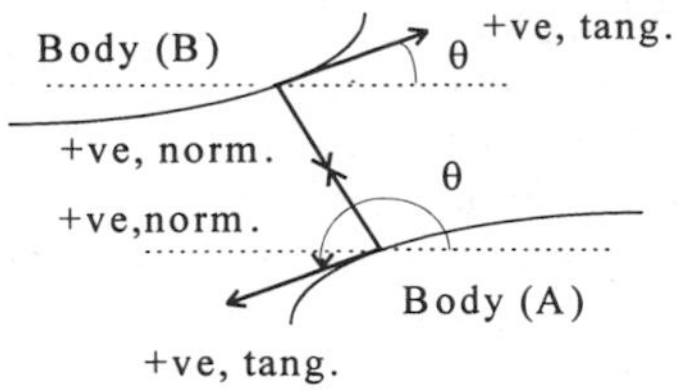

Figure 1. Local and tangential directions in contact problems.

relationships will also vary accordingly. For infinite friction (stick) contact mode the equilibrium equations are:

$$T_{NA} = T_{NB}$$
$$T_{NA} = T_{NB} \qquad (5)$$

and the compatibility equations are:

$$U_{NA} = -U_{NB} + \delta$$
$$U_{TA} = -U_{TB} \qquad (6)$$

Where δ is the gap between the two contacting nodes, given by:

$$\delta = \sqrt{(x_B - x_A) + (y_B - y_A)} \qquad (7)$$

For Coulomb friction (slip) contact mode, the equilibrium equations are given by:

$$T_{TA} = \pm\mu T_{NA}$$
$$T_{NB} = T_{NA} \qquad (8)$$
$$T_{TB} = T_{TA}$$

and the compatibility relationships are:

$$U_{NB} = -U_{NA} + \delta \qquad (9)$$

The sign of the traction must be chosen to oppose the direction of slipping. The sign convention used in the above equations closely follow that shown in fig. 1. From the above equations, it is clear that the local variable approach does not depend on the angle θ, hence it is applicable for any contact angle, including θ=90°.

3. EXAMPLES

Three examples are chosen to model

stationary, receding and advancing contact; comparing the BE global, BE local and FE solutions for frictional and frictionless contact problems. The first example is to test for stationary contact, and involves a two-dimensional punch on a foundation in which the extent of contact after loading is predetermined. The second example, is to test for receding contact, and involves a two-dimensional layer pressed against a foundation; where the extent of contact after loading is unknown, since a portion of the layer will lift away from the foundation surface with loading. The third example concerns a cylinder on a foundation, where the initial contact is at a point and once the loading is applied the extent of contact will increase hence advancing contact.

In this analysis a full set of results in terms of normal stress, shear stress and normal displacement at the contact interface are presented.The BE global solutions were obtained using the BEACON software (Becker, 1989). All the FE contact solutions were obtained analysis using the "contact pairs" function of the ABAQUS FE software (HKS, 1996), which allows elements to be brought into contact without using special gap or interface elements.

(a) BE Mesh

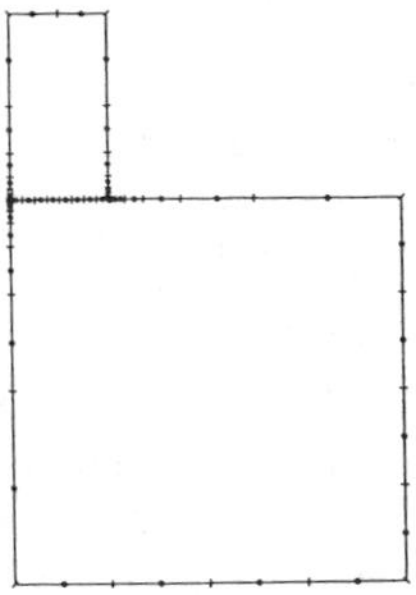

(b) FE Mesh

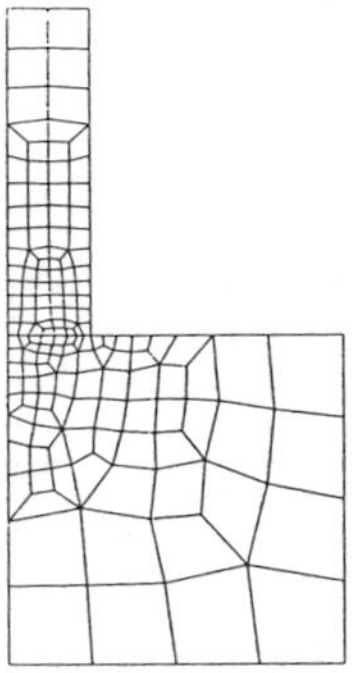

Figure 2: BE and FE meshes of a two-dimensional punch on a foundation

3.1 Example 1: Two-dimensional punch on a foundation

An elastic punch of height H_p and half-width W_p is in contact with an elastic foundation of height H_f and half-width W_f. Both components are made of the same material and a uniform pressure P is applied on the top of the punch. The two-dimensional plane-strain analysis is performed using the following data; $W_f/W_p=4$, $H_f/W_p=4$, $H_p/W_p\geq 2$, P=1, modulus of elasticity E=1, Poisson ratio $\nu=0.3$ and coefficients of friction $\mu=0.0$ and 0.2. Zero displacements are prescribed along the axis of symmetry in the x direction and in the y direction along the base of the foundation.

The BE mesh of this symmetric half model involves 72 isoparametric three-noded quadratic elements (38 elements for the punch and 34 elements for the foundation). The area of contact has been discretised by 8 elements of equal widths, as shown in fig. 2a. The FE mesh of this model is made up of 150 isoparametric quadratic quadrilateral eight-noded elements (72 elements for the punch and 78 elements for the foundation). In order to ensure comparability of FE and BE results, the discretization of the contact region is identical to the BE mesh with 8 equal elements, as shown in fig. 2b.

Both BE contact formulations and the FE, normal stress results are shown in fig. 3, at $\mu=0$ and 0.2 are all in close agreement with the results presented by Olukoko et al. (1993). If $\mu=0$ no shear stresses are recorded, however when $\mu=0.2$ the shear stresses are slightly different but are in the same

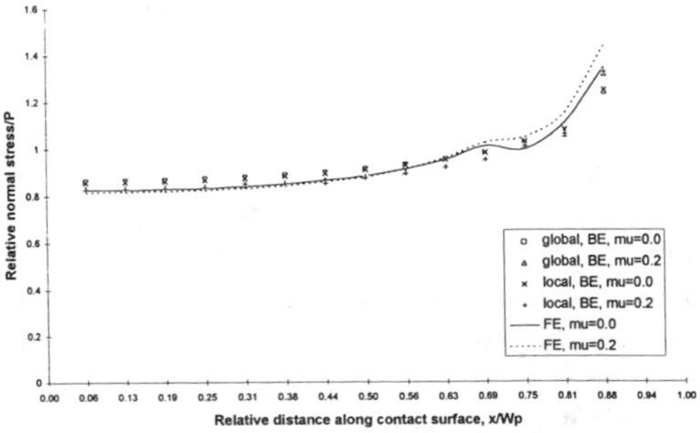

Figure 3. Normal contact stress for the punch and foundation

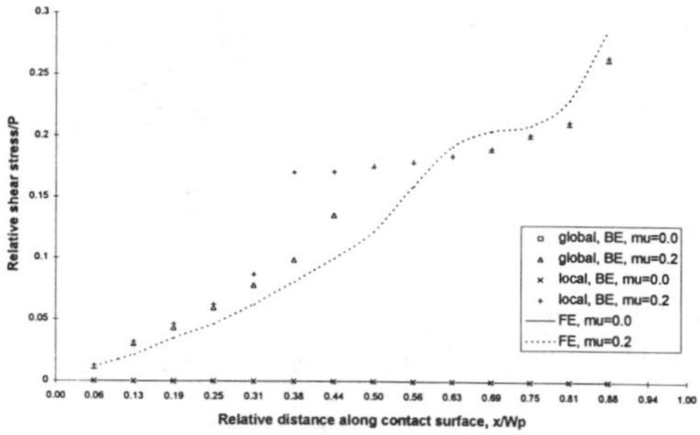

Figure 4. Shear stress stress for the punch and foundation

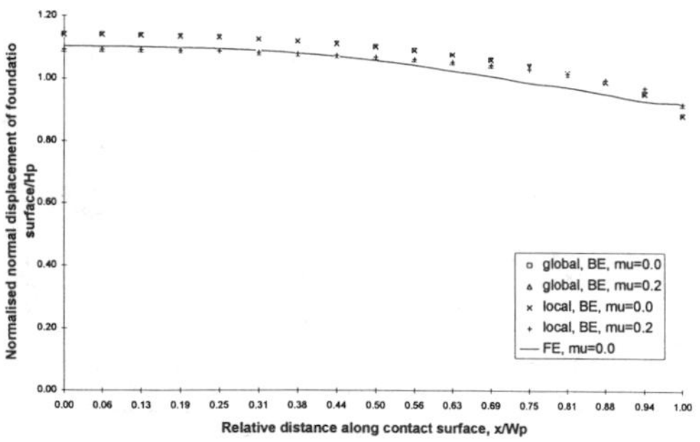

Figure 5. Normal displacement for the punch and foundation

(a) BE Mesh

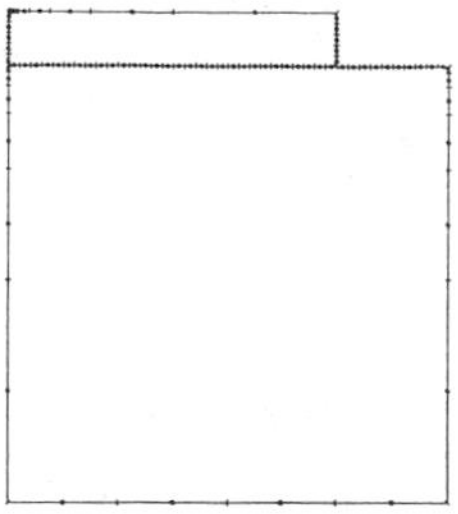

b) FE Mesh

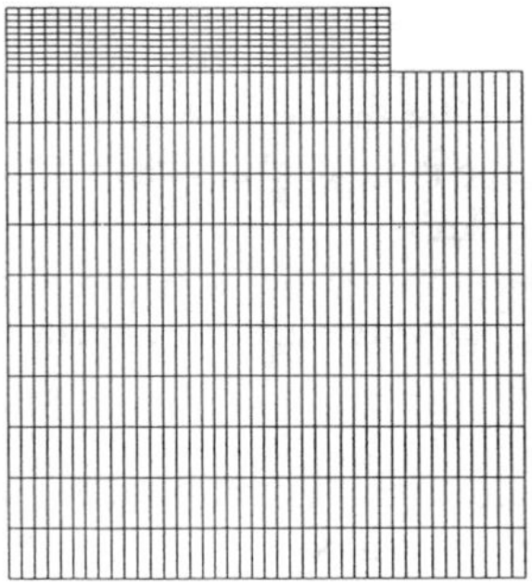

Figure 6: BE and FE meshes of a two-dimensional layer on a foundation

numerical range, as shown in fig. 4. As shear stress can be an indicator of where stick and slip contact occur, it is interesting to note that none of the methods display identical results; since the BE global formulation indicates that slip occurs over the range $0.5 \le x \le 1$; for the BEM local formulation this changes to $0.38 \le x \le 1$ and the FE solutions suggests that slip occurs in the range $0.69 \le x \le 1$. The normal displacement of the foundation surface in contact shows good agreement between all the methods and at $\mu=0$ and 0.2; where the displacement is greatest at the line of symmetry and decreases towards the edge of the contact zone, as shown on fig. 5.

3.2 Example 2: Two-dimensional layer on a foundation

An elastic layer of height H_p and half-width W_p is in contact with a foundation height of H_f and half-width W_f. Both components are made of the same material; and a point load P is applied along the line of symmetry on the layer. The two-dimensional plane-strain analysis is performed using the following dimensionless data; $H_p=0.5$, $W_p/H_p=6$, $H_f/H_p=8$, $W_f/H_p=8$, $P=1$, modulus of elasticity $E=1$, Poisson ration $\nu=0.3$, coefficients of friction $\mu=0.0$, 0.1 and 0.2. Zero displacements are prescribed in the x direction along the axis of symmetry and in the y direction along the base of the foundation.

The BE mesh of this symmetric half model involves 106 isoparametric three-noded quadratic elements (50 elements for the layer and 56 elements for the foundation). The area of contact has been divided into 30 elements of equal widths, as shown in fig. 6a. The FE mesh of this model is made-up of 700 isoparametric quadratic quadrilateral eight-noded elements (300 elements for the layer and 400 elements for the foundation). In order to ensure comparability of FE and BE results, the discretization of the contact region is identical to the BE mesh with 30 equal elements, as shown in fig. 6b.

The FEM and the two BE contact formulations show a good agreement as to the extent of the contact zone after the load is applied. In this instance the extent of contact before loading had a width of $6H_p$, after loading this had receded to a

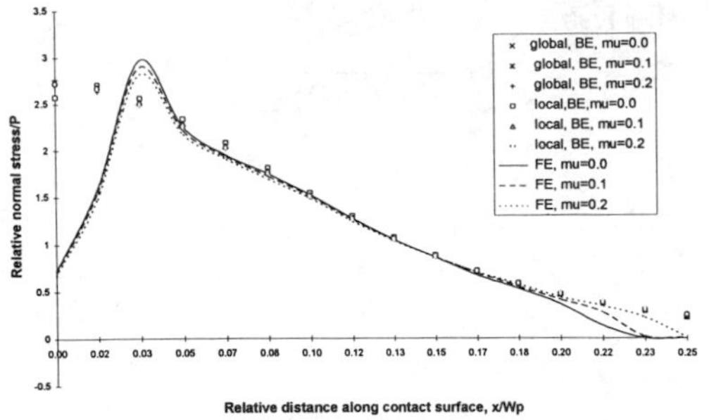

Figure 7. Normal contact pressure for the layer on a foundation

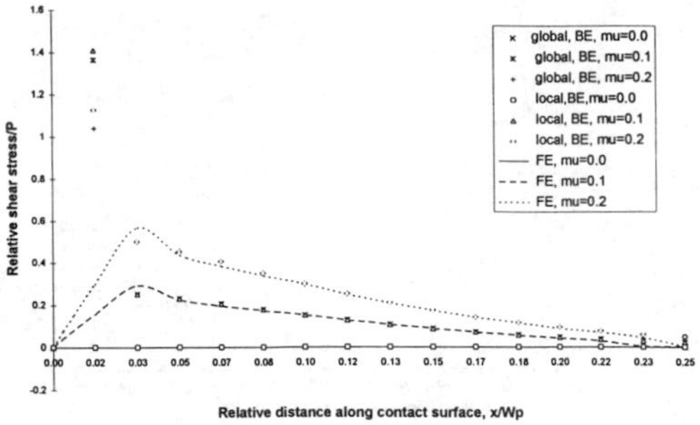

Figure 8. Contact shear stress for the layer on a foundation

width of 25% of the original contact zone. For most of this all the three methods show a gradual drop in the normal stress, from a high of approx. 2.7 at the line of symmetry, which decreases in a steady linear manner towards the edge of the contact zone.

3.3 Example 3: Axisymmetric sphere on a cylindrical foundation

A sphere of radius R_s is in contact with a cylindrical foundation of height H_f and radius R_f. Both components are made of the same material, and a point load P is applied on the sphere along the axis of rotation. The two-dimensional axisymmetric analysis is performed using the following dimensionless data; $R_s=1$, $H_f/R_s=2$, $R_f/R_s=2$, $P=1.172\times 10^9$, modulus of elasticity $E=200\times 10^9$, Poisson ratio $\nu=0.3$, coefficients of friction $\mu=0$ and 0.01. Zero displacements are prescribed in the axial Z direction along the base of the foundation.

However when the friction is varied from 0 to 0.1 to 0.2, there are small fluctuations at the edges of the contact zone, as shown in fig. 7. As for the shear stress data, as shown in fig. 8, there are clear variations when the friction varies from 0.1 to 0.2. Initially there is a sharp peak of the BE solutions only and in the range $0.03\leq x\leq 1$ all the methods show a very good agreement where increasing the friction from 0.1 to 0.2 will approximately double the shear stress values at the same positions along the contact interface. As expected the vertical displacement as shown in fig. 9, is not uniform across the contact region, indeed it is at its greatest at the line of symmetry and gradually decreases towards the edge of the contact zone.

The BE mesh of this symmetric half model involves 77 isoparametric three-noded quadratic elements (33

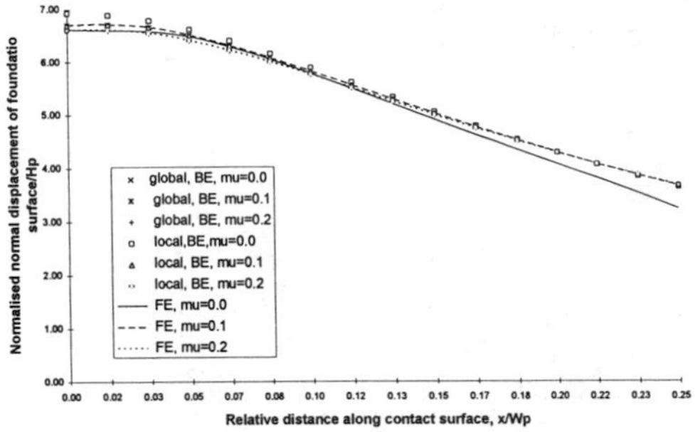

Figure 9. Normal displacement for the layer on a foundation

elements for the sphere and 29 elements for the foundation).The area of contact which is estimated at a width of $0.3R_s$, has been divided in 15 elements of equal widths, as shown in fig. 10a. The FE mesh of this model is made-up of 1050 isoparametric quadratic quadrilateral eight-noded elements (600 elements for the sphere and 450 elements for the foundation). In order to ensure comparability of FE and BE results, the discretization of the contact region is identical to the BE mesh with 15 equal elements, as shown in fig. 10b.

Both BE formulations and the FE showed very good agreement with the width contact as predicted by the Hertzian method (see Johnson, 1985). The resultant normal contact stress curves also show a very good agreement with the expected Hertzian normal stress curve for this problem, as shown on fig. 11. There does however seem to be some fluctuation closer to to the edge of contact for the FEM data. The normal displacement data for both the BEM and the FEM start from a maximum value along the axis of rotation and gradually decreases toward the edge of contact, as shown in fig. 12.

(a) BE Mesh

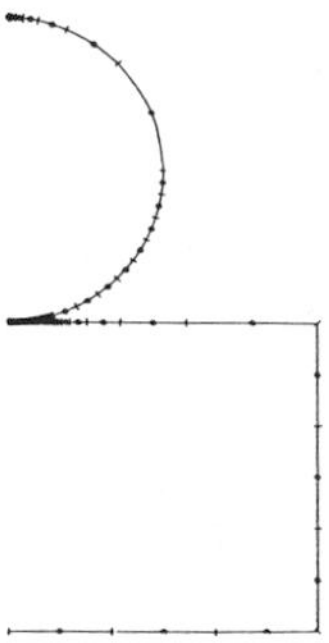

(b) Expanded view of contact area

(c) FE Mesh

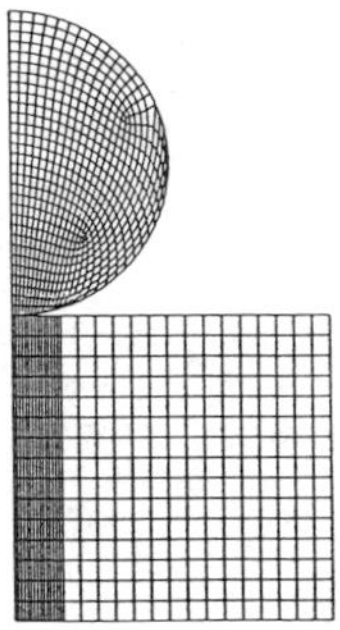

Figure 10: BE and FE meshes of a sphere on a cylindrical foundation

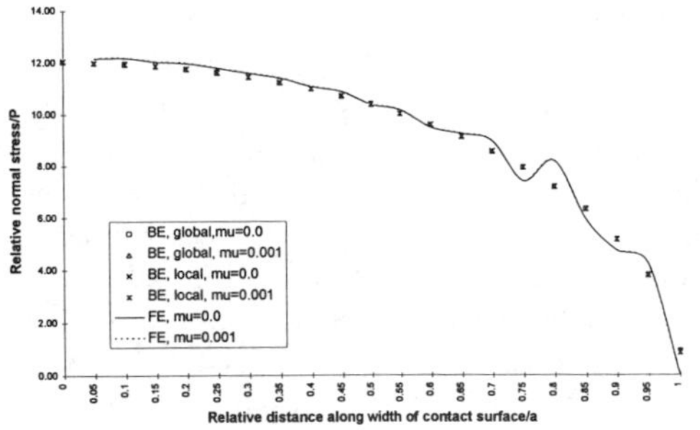

Figure 11. Normal contact pressure for a sphere on a foundation

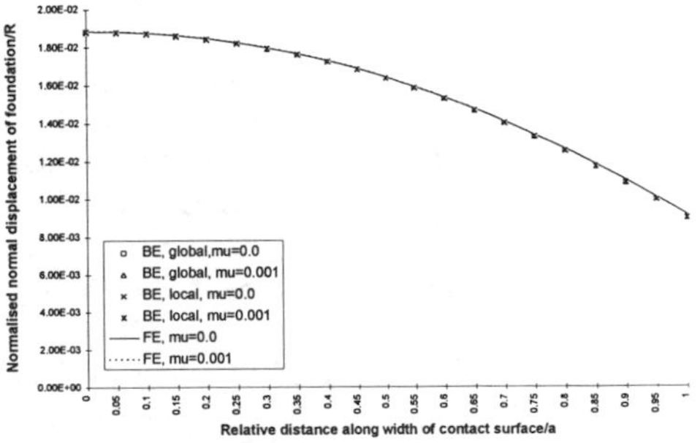

Figure 12. Normal displacement of a sphere on a foundation

4. DISCUSSION

Three two-dimensional examples involving stationary, receding and advancing contact, all with frictionless and frictional contact are presented. The intention of this exercise is to assess whether a BE contact formulation based on a local coordinate system would provide comparable accuracy to a BE formulation based on a global Cartesian coordinate system, and both these approaches are further compared to FE solutions. To facilitate the comparison the BE and FE solutions; discretizations of the contact surfaces were deliberately made identical between both the FE and BE models.

From straightforward observation, it can be seen that both BE contact formulations and the FE solutions generate broadly simillar results in terms of normal stress, shear stress and normal displacement. However there seems to be some small discrepencies at the edges of the various contact zones between the two BEM formulations, but most importantly it can be seen that the two formulations arrive at results which are in close agreement for most of the contact zones.

5. CONCLUSION

On the basis of the BE contact results it can be concluded that the BE contact algorithm based on a local coordinate system will produce sufficiently accurate results to compare favourably with alternative numerical methods, namely the FE method.

The main advantage of using local axes rather than global axes in coupling the contact variables is that any angle of contact can be accomodated including $\theta=90^{\circ}$. Furthermore the local axes approach can be easily extended into 3D contact problems, whereas the global axes approach presents significant problems.

6. REFERENCES

1. HKS, ABAQUS version 5.5, 1996, User's Manual, (HKS Inc, Rhode Island, USA)

2. Andersson T., 1982, "The second generation boundary element contact program", Proc. of 4th Int. Sem. on Recent Advances in BEM, (ed. C.A. Brebbia), Southampton, pp 409-427

3. Banerjee P.K., 1994, "The boundary element methods in engineering", McGraw-Hill, London

4. Becker A.A., 1989, "A boundary element computer program for practical contact problems", Modern Practise in Stress and Vibration Analysis, (ed. J.E. Mottershead), Pergamon Press, Oxford, pp 313-321

5. Becker A.A., 1992, "The boundary element method in engineering", McGraw-Hill, London

6. Gladwell G.M.L., 1980, "Contact problems in the classical theory of elasticity", Sijthoff & Noordhoff, Alphen ann den Rijn.

7. Hertz H., 1896, "On the contact of elastic solids", Miscellaneous

8. Huesman A. and Kuhn G., 1995, "Automatic load incrementation technique for plane elastoplastic frictional contact problems using boundary element method", Computers & Structures, vol. 56, pp 733-744

9. Johnson K.L., 1985, "Contact Mechanics", Cambridge University Press, Cambridge

10. Olukoko O.A., Becker A.A. and Fenner R.T., 1993, "Three benchmark examples for frictional contact modelling using finite element and boundary element methods", J. Strain Analysis, vol. 28, pp 293-301

11. Paris F., Blazquez A. and Canas J., 1995, "Contact problems with nonconforming discretization using boundary elemt method", Computers & Structures, vol. 57, pp 829-839

12. Zhu C., 1995, "A new boundary element/mathematical programming method for contact problems with friction", Comms. in Num. Mthds. in Eng., vol. 11, pp 683-690

Modern Practice in Stress and Vibration Analysis, Gilchrist (ed.) © 1997 Balkema, Rotterdam, ISBN 90 5410 896 7

A high-order element for linear and nonlinear stress and vibration analysis of composite layered plates and shells

O. Attia
School of Engineering, Oxford Brookes University, UK

A. El-Zafrany
Cranfield University, Bedford, UK

ABSTRACT: This paper introduces a family of high-order facetted shell elements for linear and nonlinear stress and vibration analysis of composite layered plate and shell structures. Engineering slope angles are employed in element equations, and the lateral deflection is modelled by conforming or non-conforming Hermitian shape functions. Nonlinear terms associated with geometrical nonlinearity are also derived using a practical approach based upon the actual components of strain. A programming package based on the developed elements was designed. Several case studies have been investigated and package results were compared with existing theoretical and/or experimental results. It has been proved that the developed elements can lead to accurate estimations of natural frequencies. The effect of fibre angles on natural frequencies has also been investigated with some case studies, and the results have proved that the package can be a useful tool for the design optimization of composite layered plate and shell structures.

1 INTRODUCTION

The use of composite materials in automotive and aircraft structures has increased in recent years, due to the high strength to weight ratio of composites and their many other advantages. Fibrous composites can optimally be designed with the tuning of fibre angles in different layers so as to obtain the required properties. Nevertheless, the sensitivity of composite layered plates to shear stresses necessitates the need to estimate transverse shear stresses as accurate as possible. Many industrial applications may also have a geometrical nonlinearity in the elastic range, which should be considered in the analysis.

There are many finite element publications for composite materials reported in the literature.

Reddy (1984, 1985) introduced a refined nonlinear theory of plates with transverse shear deformation, and a simple higher order theory for laminated composite plates. The deduction of two-dimensional theories from the three-dimensional theories in a transversely isotropic body was presented by Wang (1990).

A recent theory was introduced by Tessler (1993), who derived a two-dimensional laminated theory for linear elastic analysis of thick composite plates with the equivalent single-layer assumptions for the displacements, transverse shear strain, and transverse normal stress.

Cho and Parmerter (1994) devolved a three-node non-conforming triangular bending element, with five degrees of freedom, for symmetric laminated composites.

Many of the high-order element derivations presented in the literature are based on hypothetical nodal parameters, which are difficult to handle with different types of boundary conditions.

This paper introduces a family of high-order facetted shell elements for linear and nonlinear stress and vibration analysis of composite layered plate and shell structures, with transverse shear deformation being considered. Engineering slope angles are employed in element equations, and nonlinear terms associated with geometrical nonlinearity are derived using a practical approach based upon the actual components of strain.

2 DEFORMATION THEORY

2.1 *Transverse strain modelling*

Consider a composite layered plate, at an instant of time *t*, consisting of a number of orthotropic layers N_l as shown in Figure 1.

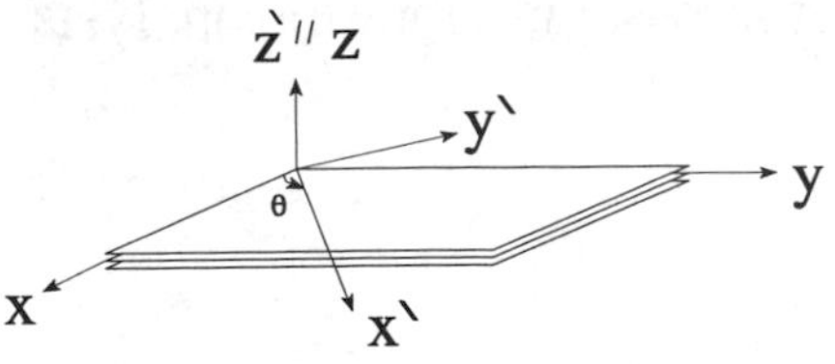

Figure 1. Composite layered plate

Let the midplane of the plate be the local cartesian *x-y* plane, and the total thickness at any point (*x,y*) on the midplane is $h(x,y)$. If the values of $\phi(x,y)$ and $\psi(x,y)$ represent the transverse shear strains at the midplane, then from the boundary and continuity conditions the distributions of the transverse shear strains across the plate thickness can be deduced by applying Lagrangian interpolation, leading to:

$$\gamma_{xz}(x,y,z,t) \equiv \phi\,(1 - 4z^2/h^2) \qquad (1)$$

$$\gamma_{yz}(x,y,z,t) \equiv \psi\,(1 - 4z^2/h^2) \qquad (2)$$

2.2 *Displacement and velocity components*

Assuming that the lateral deflection w is independent of z, and the transverse shear strains are infinitesimal, the displacement components in the x and y directions can be obtained as follows:

$$u(x,y,z,t) = u_o - z\frac{\partial w}{\partial x} + f_1(z)\,\phi(x,y,t) \qquad (3)$$

$$v(x,y,z,t) = v_o - z\frac{\partial w}{\partial y} + f_1(z)\,\psi(x,y,t) \qquad (4)$$

where $f_1(z) = z - 4z^3/(3h^2)$, and u_o, v_o represent the values of u and v at $z = 0$.

The corresponding velocity components can be obtained by differentiating displacement equations with respect to time.

2.3 *Strain components*

The *x-y* strain components will be considered finite and by using a Lagrangian frame of reference, and defining the following strain vector:

$$\boldsymbol{\epsilon}_{xy} = \{\epsilon_x \;\; \epsilon_y \;\; \gamma_{xy}\} \qquad (5)$$

then it can be deduced from Cauchy's strain-displacement equations that:

$$\boldsymbol{\epsilon}_{xy} = \boldsymbol{\epsilon}_o + \boldsymbol{\epsilon}_L \qquad (6)$$

where $\boldsymbol{\epsilon}_o$ represents the infinitesimal linear terms, and $\boldsymbol{\epsilon}_L$ represents the nonlinear terms of finite strains. Using Equations (3) and (4), the infinitesimal strain vector can be expressed as follows:

$$\boldsymbol{\epsilon}_o(x,y,z,t) = \boldsymbol{\epsilon}_m(x,y,t) - z\,\boldsymbol{\epsilon}_b(x,y,t) + f_1(z)\,\boldsymbol{\epsilon}_s(x,y,t) \qquad (7)$$

where

$$\boldsymbol{\epsilon}_m(x,y,t) = \begin{bmatrix} \dfrac{\partial u_o}{\partial x} \\ \dfrac{\partial v_o}{\partial y} \\ \dfrac{\partial u_o}{\partial y} + \dfrac{\partial v_o}{\partial x} \end{bmatrix} \qquad (8)$$

$$\boldsymbol{\epsilon}_b(x,y,t) = \begin{bmatrix} \dfrac{\partial^2 w}{\partial x^2} \\ \dfrac{\partial^2 w}{\partial y^2} \\ 2\dfrac{\partial^2 w}{\partial x\,\partial y} \end{bmatrix} \qquad (9)$$

$$\boldsymbol{\epsilon}_s(x,y,t) = \begin{bmatrix} \dfrac{\partial \phi}{\partial x} \\ \dfrac{\partial \psi}{\partial y} \\ \dfrac{\partial \phi}{\partial y} + \dfrac{\partial \psi}{\partial x} \end{bmatrix} \qquad (10)$$

Due to the non-linearity of terms in $\boldsymbol{\epsilon}_L$ equation, it will be represented as an integration of its differential, and it can similarly be deduced that:

$$\begin{aligned} d\boldsymbol{\epsilon}_L &= \boldsymbol{A}_m d\boldsymbol{\theta}_m + \boldsymbol{A}_w d\boldsymbol{\theta}_w \\ &+ f_1^2(z)\boldsymbol{A}_s d\boldsymbol{\theta}_s + z^2\boldsymbol{A}_b d\boldsymbol{\theta}_b \\ &- z f_1(z)\left[\boldsymbol{A}_b d\boldsymbol{\theta}_s + \boldsymbol{A}_s d\boldsymbol{\theta}_b\right] \\ &- z\left[\boldsymbol{A}_b d\boldsymbol{\theta}_m + \boldsymbol{A}_m d\boldsymbol{\theta}_b\right] \\ &+ f_1(z)\left[\boldsymbol{A}_m d\boldsymbol{\theta}_s + \boldsymbol{A}_s d\boldsymbol{\theta}_m\right] \end{aligned} \qquad (11)$$

where

$$\boldsymbol{\theta}_m = \left\{ \frac{\partial u_o}{\partial x} \quad \frac{\partial v_o}{\partial x} \quad \frac{\partial u_o}{\partial y} \quad \frac{\partial u_o}{\partial y} \right\} \tag{12}$$

$$\boldsymbol{\theta}_s = \left\{ \frac{\partial \phi}{\partial x} \quad \frac{\partial \psi}{\partial x} \quad \frac{\partial \phi}{\partial y} \quad \frac{\partial \psi}{\partial y} \right\} \tag{13}$$

$$\boldsymbol{\theta}_b = \left\{ \frac{\partial^2 w}{\partial x^2} \quad \frac{\partial^2 w}{\partial x \partial y} \quad \frac{\partial^2 w}{\partial y \partial x} \quad \frac{\partial^2 w}{\partial y^2} \right\} \tag{14}$$

$$\boldsymbol{\theta}_w = \left\{ \frac{\partial w}{\partial x} \quad \frac{\partial w}{\partial y} \right\} \tag{15}$$

$$\boldsymbol{A}_m = \begin{bmatrix} \frac{\partial u_o}{\partial x} & \frac{\partial v_o}{\partial x} & 0 & 0 \\ 0 & 0 & \frac{\partial u_o}{\partial y} & \frac{\partial v_o}{\partial y} \\ \frac{\partial u_o}{\partial y} & \frac{\partial v_o}{\partial y} & \frac{\partial u_o}{\partial x} & \frac{\partial v_o}{\partial x} \end{bmatrix} \tag{16}$$

$$\boldsymbol{A}_s = \begin{bmatrix} \frac{\partial \phi}{\partial x} & \frac{\partial \psi}{\partial x} & 0 & 0 \\ 0 & 0 & \frac{\partial \phi}{\partial y} & \frac{\partial \psi}{\partial y} \\ \frac{\partial \phi}{\partial y} & \frac{\partial \psi}{\partial y} & \frac{\partial \phi}{\partial x} & \frac{\partial \psi}{\partial x} \end{bmatrix} \tag{17}$$

$$\boldsymbol{A}_b = \begin{bmatrix} \frac{\partial^2 w}{\partial x^2} & \frac{\partial^2 w}{\partial x \partial y} & 0 & 0 \\ 0 & 0 & \frac{\partial^2 w}{\partial y \partial x} & \frac{\partial^2 w}{\partial y^2} \\ \frac{\partial^2 w}{\partial y \partial x} & \frac{\partial^2 w}{\partial y^2} & \frac{\partial^2 w}{\partial x^2} & \frac{\partial^2 w}{\partial x \partial y} \end{bmatrix} \tag{18}$$

$$\boldsymbol{A}_w = \begin{bmatrix} \frac{\partial w}{\partial x} & 0 \\ 0 & \frac{\partial w}{\partial y} \\ \frac{\partial w}{\partial y} & \frac{\partial w}{\partial x} \end{bmatrix} \tag{19}$$

3 FINITE ELEMENT DERIVATION

3.1 *Displacement interpolation*

Physically, the element is constructed of a number of layers (N_l), similar to that shown in Figure 1. Each layer has its own material axes, and the stress-strain relationship are defined for the lth layer with respect to the element local x, y, z axes, as follows:

$$\boldsymbol{\sigma}_{xy}^{(l)} \equiv \begin{bmatrix} \boldsymbol{\sigma}_x^{(l)} \\ \boldsymbol{\sigma}_y^{(l)} \\ \boldsymbol{\tau}_{xy}^{(l)} \end{bmatrix} = \boldsymbol{D}^{(l)} \; \boldsymbol{\epsilon}_{xy} \tag{20}$$

$$\boldsymbol{\tau}^{(l)} \equiv \begin{bmatrix} \tau_{xz)}^{(l)} \\ \tau_{yz)}^{(l)} \end{bmatrix} = \boldsymbol{\mu}^{(l)} \begin{bmatrix} \gamma_{xy} \\ \gamma_{yz} \end{bmatrix} \tag{21}$$

The matrices $\boldsymbol{D}^{(l)}$ and $\boldsymbol{\mu}^{(l)}$ are stress-strain matrices rotated from material axes to element axes. The midplane displacement components in the x and y directions and the transverse shear strains are modelled with Lagrangian interpolation as follows:

$$u_o(x,y,t) = \sum_{i=1}^{n} u_i(t)\, N_i(x,y) \tag{22}$$

$$v_o(x,y,t) = \sum_{i=1}^{n} v_i(t)\, N_i(x,y) \tag{23}$$

$$\gamma_{xz}^{o}(x,y,t) = \sum_{i=1}^{n} \phi_i(t)\; N_i(x,y) \tag{24}$$

$$\gamma_{yz}^{o}(x,y,t) = \sum_{i=1}^{n} \psi_i(t)\; N_i(x,y) \tag{25}$$

where u_i, v_i, ϕ_i, ψ_i are the nodal values defined at the midplane node i and time t, and $N_i(x,y)$ represents a Lagrangian shape function.

The lateral deflection w is interpolated by means of a Hermitian interpolation. Two types of interpolation, non-conforming and conforming, based upon El-Zafrany and Cookson (1986a, 1986b) are employed, where $w(x, y,t)$ is interpolated as follows:

$$w(x,y,t) = \sum_{i=1}^{n} \Big\{ w_i(t)\, F_i + w_{i,x}(t)\, G_i + w_{i,y}(t)\, H_i + w_{i,xy}(t)\, Q_i \Big\} \tag{26}$$

where w_i, $w_{i,x}$, $w_{i,y}$, and $w_{i,xy}$ represent the values of w and its partial derivatives at node i and time t, and F_i, G_i, H_i and Q_i are Hermitian shape functions, which are functions of (x, y). Notice also that Q_i dose not exist for the non-conforming Hermitian interpolation.

3.2 *Nodal displacement vector*

From the previous analysis it is clear that the displacement parameters required at each node i are u_i, v_i, ϕ_i, ψ_i, w_i, $w_{i,x}$, $w_{i,y}$, $w_{i,xy}$, and practical representation of nodal values will be described in section 3.6. To simplify the derivation, the local nodal displacement vector for an n-node element will be defined at time t as follows:

$$\boldsymbol{\delta}(t) = \{ \boldsymbol{\delta}_m(t) \ \ \boldsymbol{\delta}_b(t) \ \ \boldsymbol{\delta}_s(t) \} \tag{27}$$

where

$$\boldsymbol{\delta}_m(t) \equiv \{ u_1(t) \ v_1(t) \ \ldots \ u_n(t) \ v_n(t) \}$$

$$\boldsymbol{\delta}_b(t) \equiv \{ \ldots w_i(t) \ w_{i,x}(t) \ w_{i,y}(t) \ w_{i,xy}(t) \ldots \}$$

$$\boldsymbol{\delta}_s(t) \equiv \{ \phi_1(t) \ \psi_1(t) \ \ldots \ \phi_n(t) \ \psi_n(t) \}$$

3.3 *Velocity components*

The velocity components at any point (x, y, z) inside the plate at time t can be represented vectorially in terms of nodal values, and shape function as follows:

$$\dot{q}(x,y,z,t) \equiv \begin{bmatrix} \dot{q}_{xy}(x,y,z,t) \\ \dot{q}_z(x,y,z,t) \end{bmatrix} \tag{28}$$

where

$$\dot{q}_{xy} \equiv \begin{bmatrix} \dot{u} \\ \dot{v} \end{bmatrix} = \boldsymbol{N}_L(x,y)\dot{\boldsymbol{\delta}}_m(t) - z\,\boldsymbol{N}_b(x,y)\dot{\boldsymbol{\delta}}_b(t) + f_1(z)\,\boldsymbol{N}_L(x,y)\dot{\boldsymbol{\delta}}_s(t) \tag{29}$$

$$\dot{q}_z \equiv [\dot{w}] = \boldsymbol{N}_H(x,y)\,\dot{\boldsymbol{\delta}}_b(t) \tag{30}$$

and $\boldsymbol{N}_L$, $\boldsymbol{N}_b$, $\boldsymbol{N}_H$ are shape function matrices.

3.4 *Strain components*

Substituting from Equations (22) - (25) into (7), it can be deduced that:

$$\boldsymbol{\epsilon}_o = \boldsymbol{B}_m(x,y)\,\boldsymbol{\delta}_m(t) - z\,\boldsymbol{B}_b(x,y)\,\boldsymbol{\delta}_b(t) + f_1(z)\,\boldsymbol{B}_m(x,y)\,\boldsymbol{\delta}_s(t) \tag{31}$$

where the $\boldsymbol{B}$ matrices are functions of shape function derivatives.

Using the interpolation equations for displacement components, the following expressions can be obtained:

$$\mathrm{d}\boldsymbol{\theta}_m = \boldsymbol{G}_m\,\mathrm{d}\boldsymbol{\delta}_m \qquad \mathrm{d}\boldsymbol{\theta}_s = \boldsymbol{G}_m\,\mathrm{d}\boldsymbol{\delta}_s$$

$$\mathrm{d}\boldsymbol{\theta}_b = \boldsymbol{G}_b\,\mathrm{d}\boldsymbol{\delta}_b \qquad \mathrm{d}\boldsymbol{\theta}_w = \boldsymbol{G}_w\,\mathrm{d}\boldsymbol{\delta}_b$$

where the $\boldsymbol{G}$ matrices are functions of shape function derivatives. Hence, Equation (11) can be rewritten in terms of nodal values as follows:

$$\begin{aligned} \mathrm{d}\boldsymbol{\epsilon}_L = {} & \boldsymbol{A}_m\boldsymbol{G}_m\,\mathrm{d}\boldsymbol{\delta}_m + \boldsymbol{A}_w\boldsymbol{G}_w\,\mathrm{d}\boldsymbol{\delta}_w \\ & + f_1^2(z)\,\boldsymbol{A}_s\boldsymbol{G}_m\,\mathrm{d}\boldsymbol{\delta}_s + z^2\boldsymbol{A}_b\boldsymbol{G}_b\,\mathrm{d}\boldsymbol{\delta}_b \\ & - z f_1(z)\left[\boldsymbol{A}_b\boldsymbol{G}_m\,\mathrm{d}\boldsymbol{\delta}_s + \boldsymbol{A}_s\boldsymbol{G}_b\,\mathrm{d}\boldsymbol{\delta}_b\right] \\ & - z\left[\boldsymbol{A}_b\boldsymbol{G}_m\,\mathrm{d}\boldsymbol{\delta}_m + \boldsymbol{A}_m\boldsymbol{G}_b\,\mathrm{d}\boldsymbol{\delta}_b\right] \\ & + f_1(z)\left[\boldsymbol{A}_m\boldsymbol{G}_m\,\mathrm{d}\boldsymbol{\delta}_s + \boldsymbol{A}_s\boldsymbol{G}_m\,\mathrm{d}\boldsymbol{\delta}_m\right] \end{aligned} \tag{32}$$

3.5 *Element dynamic equations*

Applying the principle of minimum total potential energy at an instant of time t, then:

$$\mathrm{d}\chi(t) = \mathrm{d}KE(t) + \mathrm{d}U(t) - \mathrm{d}W(t) = 0 \tag{33}$$

where the terms are defined as follows:

(a) *Strain Energy*: $\mathrm{d}U$ represents the change in the strain energy of the element, *i.e.*

$$\mathrm{d}U(t) = \iiint_{element} (\mathrm{d}\boldsymbol{\epsilon}^t\,\boldsymbol{\sigma})\,\mathrm{d}x\,\mathrm{d}y\,\mathrm{d}z \tag{34}$$

and it can be shown that:

$$\mathrm{d}U(t) = \mathrm{d}\boldsymbol{\delta}^t(t)\left[\boldsymbol{K} + \boldsymbol{K}^\sigma\right]\boldsymbol{\delta}(t) \tag{35}$$

where $\boldsymbol{K}$, $\boldsymbol{K}^\sigma$ are the element stiffness matrices due to infinitesimal strains, and nonlinear strain terms, respectively.

(b) *Kinetic energy*: $\mathrm{d}KE$ represents the variation of the kinetic energy of the element, *i.e.*

$$dKE = \iiint_{element} \rho \left(d\dot{q}^t_{xy}\dot{q}_{xy} + d\dot{q}^t_z\dot{q}_z \right) dx\,dy\,dz \qquad (36)$$

and it can be proved that:

$$dKE(t) = d\boldsymbol{\delta}^t(t)\, \boldsymbol{M}\, \ddot{\boldsymbol{\delta}}(t) \qquad (37)$$

where $\boldsymbol{M}$ is the mass matrix of the element.

(c) *Work done by applied forces*: dW represents the variation of the work done by the applied forces at time t, and can be expressed as follows:

$$dW(t) = d\boldsymbol{\delta}^t(t)\, \boldsymbol{F}(t) \qquad (38)$$

where $\boldsymbol{F}(t)$ represents a nodal loading vector equivalent to the actual applied forces.

Substituting from Equations (35), (37), (38) into (33), then it can be deduced that:

$$\boldsymbol{M}\, \ddot{\boldsymbol{\delta}}(t) + \left[\boldsymbol{K} + \boldsymbol{K}^{\sigma}\right] \boldsymbol{\delta}(t) = \boldsymbol{F}(t) \qquad (39)$$

which represents the dynamic matrix equation of the element. The dynamic equations for the whole structure are assembled from the element equations using the usual assembly procedure.

3.6 *Practical nodal values and rotated matrices*

The use of derivatives of w as nodal values causes some difficulties with the specification of boundary conditions for thick plates. Engineering slope angles can be defined so as to lead to displacement components linearized over thickness (El-Zafrany et al. 1994, 1995), and they are defined as follows:

$$\theta_x = \frac{\partial w}{\partial y} - 0.8\,\phi\,, \quad \theta_y = -\frac{\partial w}{\partial x} + 0.8\,\psi \qquad (40)$$

The average values of transverse shear strains are used as nodal values, and can also be expressed as follows:

$$\bar{\gamma}_{xz} = \tfrac{2}{3}\phi \equiv \omega_y\,, \quad \bar{\gamma}_{yz} = \tfrac{2}{3}\psi \equiv -\omega_x \qquad (41)$$

Defining a new nodal displacement vector based on such practical values as follows:

$$\boldsymbol{\delta}_{new} = \{\ldots u_i\; v_i\; w_i \quad (\theta_x)_i\, (\theta_y)_i\, (\theta_z)_i \quad (\omega_x)_i\, (\omega_y)_i\, (\omega_z)_i \ldots\} \qquad (42)$$

then it can be shown that:

$$\boldsymbol{\delta} = \boldsymbol{P}\, \boldsymbol{\delta}_{new} \qquad (43)$$

where $\boldsymbol{P}$ is a transformation matrix defined from the previous relationships.

The element local axes are defined with respect to structural global axes by means of directional cosines, and it can be shown that:

$$\boldsymbol{\delta}_{new} = \boldsymbol{R}\, \boldsymbol{\delta}_g \qquad (44)$$

where $\boldsymbol{R}$ is the element rotation matrix, and $\boldsymbol{\delta}_g$ is the global nodal displacement vector. Hence, it can be deduced that:

$$\boldsymbol{\delta} = \boldsymbol{P}\boldsymbol{R}\, \boldsymbol{\delta}_g \equiv \boldsymbol{L}\, \boldsymbol{\delta}_g \qquad (45)$$

and the element matrices defined with respect to global axes can be obtained as follows:

$$\boldsymbol{K}_g = \boldsymbol{L}^t\boldsymbol{K}\boldsymbol{L}\,,\; \boldsymbol{K}^{\sigma}_g = \boldsymbol{L}^t\boldsymbol{K}^{\sigma}\boldsymbol{L}\,,\; \boldsymbol{M}_g = \boldsymbol{L}^t\boldsymbol{M}\boldsymbol{L} \qquad (46)$$

4 CASE STUDIES

A versatile finite element programming package for the static and dynamic analysis of composite layered plates and shells has been designed, based upon the elements developed in this work. An instantaneous-iteration frontal eigenvalue solver is used for the detection of relevant natural frequencies. A number of case studies, with published analytical and or experimental results, have been analysed, as summarized next.

4.1 *Composite layered square plate case*

This case has overall dimensions as shown in Figure 2, and it consists of 8 layers of a fibrous composite material. Each layer has a thickness of 0.13 mm, and made of a Hercules type AS/3501-6 graphite/epoxy composite material with properties as listed below: $E_{x'} = 1.28\times10^{11}$ Pa, $E_{y'} = 0.11\times10^{11}$ Pa, $\mu_{x'y'} = 4.48\times10^9$ Pa $\mu_{y'z'} = \mu_{z'x'} = 1.53\times10^9$ Pa, $\rho = 1.5\times10^3$ kg/m^3, $\nu_{y'x'} = 0.25$. The fibre angles θ, in the 8 layers are: 45°, -45°, -45°, 45°, 45°, -45°, -45°, 45°. A non-dimensional frequency parameter was employed for result presentations, and is defined as follows:

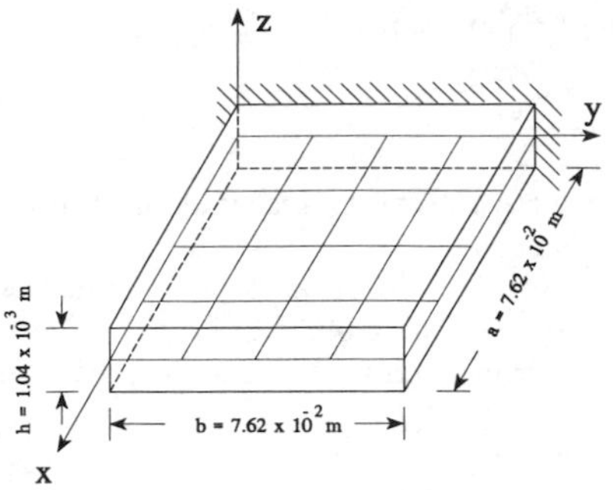

Figure 2. Composite layered square plate

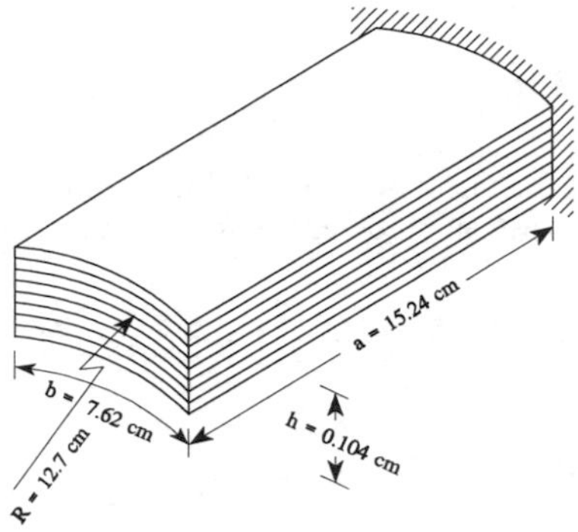

Figure 3. Composite layered cylindrical Shell

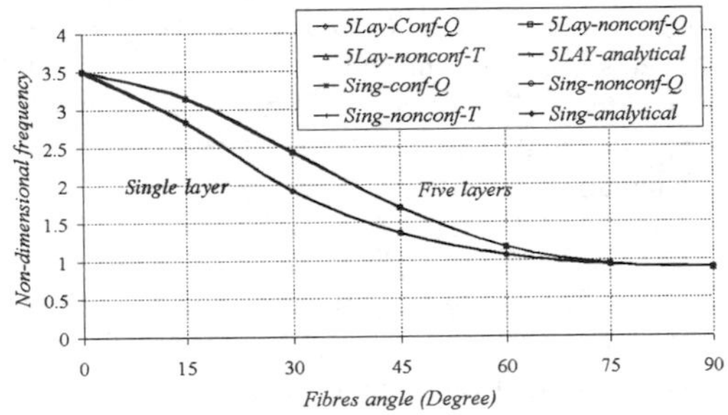

Figure 4. Effect of fibre angle on first natural frequency for single- and five-layer plates

$$\Phi = \omega a^2 \sqrt{\frac{\rho h}{D_o}}\,, \quad D_o = \frac{E_{x'} h^3}{12\left(1 - \nu_{x'y'} \nu_{y'x'}\right)}$$

and h represents the total thickness of the plate. The non-dimensional frequency parameters obtained with the developed package were compared with the analytical solution published by Narita and Leissa (1992), and also compared with an experimental work published by Crawley (1979), and his finite element results which were based on a moderately thick quadrilateral shallow shell element developed by Lee and Pian (1978), as shown in Table 1. The package results have generally proved to be closer to the analytical solution than those based on Lee and Pian's element. However, some deviation (≈ 8%) between all theoretical results and corresponding experimental results was observed. Crawley (1979)

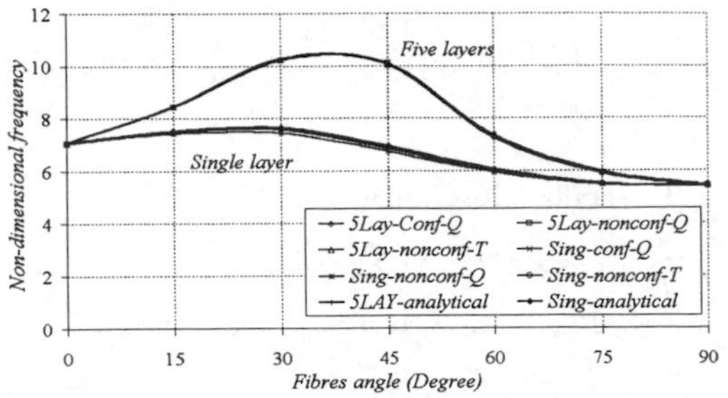

Figure 5. Effect of fibre angle on second natural frequecy of single- and five-layer plates

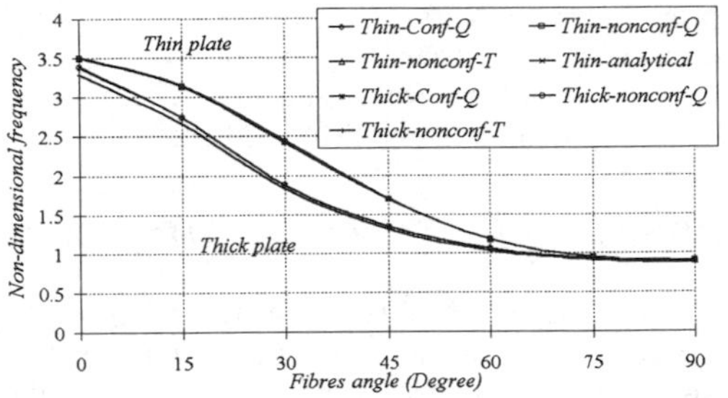

Figure 6. Effect of thickness on first natural frequecy for five-layer plate

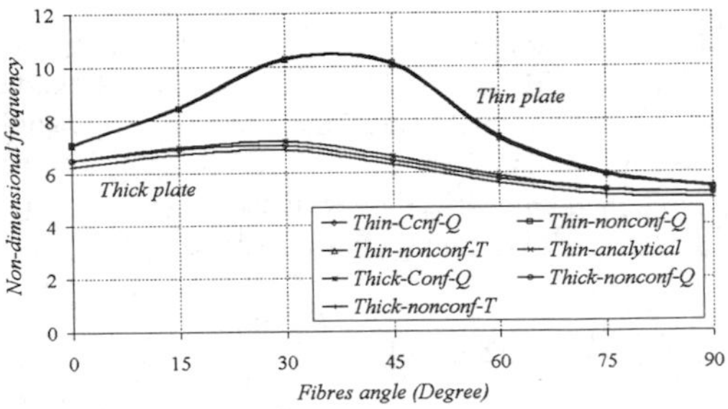

Figure 7. Effect of thickness on second natural frequecy for five-layer plate

Table 1: Non-dimensional frequency parameters for 8-layer composite square plate

Mode	1	2	3	4	5
Conforming-Q	1.812	6.521	10.44	17.11	21.23
Nonconf-Q	1.807	6.512	10.70	17.00	21.31
Nonconf-T	1.820	6.506	10.90	16.98	21.40
Analytical	1.813	6.553	10.48	17.29	21.49
Published FEM	1.792	6.443	10.38	17.11	21.26
Experimental	1.692	6.089	10.20	15.07	19.17

Table 2: Natural frequencies for 8-layer composite cylindrical shell

Mode	Experimental	Published FEM	Conf-Q	Nonconf-Q	Nonconf-T
1	161.0	165.7	168.72	126.34	141.27
2	245.1	289.6	295.09	141.74	195.37
3	555.6	597.1	606.06	452.03	539.76
4	670.0	718.5	713.01	793.62	982.72
5	794.0	833.3	816.94	964.10	1156.19

mentioned that the experimental results represented an average of the measured natural frequencies of nominally identical samples. Since there was a slight variation in the finished thickness from plate to plate, the measured frequencies were linearly corrected to a reference thickness before averaging.

4.2 *Composite layered cylindrical shell case*

The geometry of this case is shown in Figure 3, the boundary conditions are also illustrated in that figure, and the material properties are the same as in the previous case, with fibre angles: 0, 0, 30°, -30°, -30°, 30°, 0, 0. Package results were compared with published experimental and finite element results of Crawley (1979) as shown in Table 2. Observations similar to those seen with the previous case are found, emphasizing the accuracy of the developed elements for the dynamic analysis of composite layered shells.

4.3 *Case with variation of fibre angles*

This case represents a rectangular plate with the length along the x-direction being $a = 15.24\times10^{-2}$ m and the length in the y-direction $b = 7.62\times10^{-2}$ m. Two plates were considered: one with five layers and the thickness of each layer is 0.13×10^{-3} m, where the angles of fibres are arranged as θ, $-\theta$, θ, $-\theta$, θ, and the second is a single layered plate with a thickness equal to the total thickness of the five-layer plate, and the fibres at an angle θ. The natural frequency analysis has been carried out on plates with different values of θ, *i.e.* $\theta = 0$, 15°, 30°, 45°, 60°, 75°, 90°. The first two non-dimensional frequencies as obtained with the developed elements together with the corresponding analytical solution of Narita and Leissa (1992), were plotted against fibre angles as shown in Figures 4 and 5.

Results obtained using different finite elements for the two plates at different fibre angles are in a very close agreement to those obtained by the analytical solution of Narita and Leissa (1992). Increasing the fibre angle θ, changes the natural frequency, both for the single layer and the 5-layer plates, and the change in the fibre angle has a greater effect on the results of the 5-layer plate than on those of the single layer plate.

To investigate the thickness effect on the natural frequencies of a multilayered plate, the five-layer plate was employed with two values of plate thickness, the first is the same as before and the second thickness was ten times greater. The first two non-dimensional natural frequencies for the thin and thick plates were plotted against fibre angles, as shown in Figures 6 and 7. It is clear from those figures, that the transverse shear effect tends to reduce the non-dimensional frequency.

5 CONCLUSIONS

It is clear from the previous case studies that conforming and non-conforming elements developed in this work have led to accurate and reliable results. Although the developed elements consider transverse shear effects, they do not suffer from shear locking as do Mindlin-type elements, and reduced integration techniques are not required for the elements developed in this work.

The fibre angle θ can have a significant effect on the stiffness matrix of composite layered plates and shells, and this will result in a corresponding effect on their natural frequencies. This effect can be increased by increasing the number of layers, while keeping the same thickness. Such an effect may also provide a tool for designers to tune the natural frequencies away from resonance due to an excitation frequency, by simply changing the fibre angles without having to modify the mechanical design.

REFERENCES

Cho, M. & Parmerter, R. 1994. Finite element composite plate bending based on efficient higher order theory. *AIAA Journal,* 11: 2241-2248.

Crawley, E. F. 1979. The natural modes of Graphite/Epoxy cantilever plates and shells. *J. Composite Materials*, 13: 195-205.

El-Zafrany, A. M. & Cookson, R. A. 1986a. Derivation of Lagrangian and Hermitian shape functions for triangular elements. *Int. J. Numer. Meth. Engng.* 23: 275-285.

El-Zafrany, A. M. & Cookson, R. A. 1986b. Derivation of Lagrangian and Hermitian shape functions for quadrilateral elements. *Int. J. Numer. Meth. Engng.* 23: 1939-1958.

El-Zafrany, A., Debbih, M. & Fadhil, S. 1994. A modified Kirchhoff theory for boundary element bending analysis of thin plates. *Int. J. Solids & Structures,* 31: 2885-2899.

El-Zafrany, A., Fadhil, S. & Debbih, M. 1995. An efficient approach for boundary element bending analysis of thin and thick plates. *Int. Journal of Computers and Structures,* 56: 565-576.

Lee, S. W. & Pian, T. H. 1978. Improvements of plate and shell finite elements by mixed formulations. *AIAA Journal,*16.

Narita, Y. & Leissa, A. W. 1992. Frequencies and mode shapes of cantilevered laminated composite plates. *J. of Sound and Vibration,* 154: 161-172.

Reddy, J. N. 1984. A refined nonlinear theory of plates with transverse shear deformation. *Int. J. Solids & Structures,* 20: 881-896.

Reddy, J. N. 1985. A simple higher order theory for laminated composite plates. *J. Appl. Mech.* 51: 745-752.

Tessler, A. 1993. An improved plate theory of {1,2}-order for thick composite laminates. *Int. J. Solids & Structures,* 30: 981-1000.

Wang, F. Y. 1990. Two dimensional theories deduced from three-dimension theory for a transversely isotropic body, I. plate problems. *Int. J. Solids & Structures,* 26: 445-470.

Modern Practice in Stress and Vibration Analysis, Gilchrist (ed.) © *1997 Balkema, Rotterdam, ISBN 90 5410 896 7*

A 3-D variable order singularity finite element

M.M.Abdel Wahab & G.De Roeck
Department of Civil Engineering, Katholieke Universiteit, Leuven, Belgium

ABSTRACT: A three-dimensional six-noded prism (wedge) finite element containing a singularity of order λ is developed. The interpolation functions of the displacement allow for variations proportional to the power λ of the distance from the crack front along the crack surface and to the distance in the perpendicular direction. The element is compatible along one of its faces (which does not intersect the crack front) with standard wedge (6-noded) and/or brick (8-noded) isoparametric elements. The element is suitable to model the intersection points between a crack front and the free surface where the order of the singularity is different from 0.5 and depends on the material properties (mostly Poisson's ratio). Two problems are studied to examine the proposed element, namely, a penny shaped crack in a cylinder and an edge crack in an homogeneous material in which the crack front intersects with the free surface. The results prove the applicability of the element in 3-D crack problems with variable order singularity.

1 INTRODUCTION

In many engineering fracture mechanics applications, the stress singularity near a crack tip varies as $r^{\lambda-1}$, where r is the distance from the crack tip and λ is the order of singularity. For example, in case of a crack terminating at the interface of a bimaterial composite, the value of λ can be complex (this means oscillatory stresses and displacements) or real depending on the angle between the crack and the interface, and the material properties of the two materials (Bogy 1971, Cook & Erdogan 1972).

In case of a kinked crack, the order of the stress singularity arising at the knee is $-1/2+f(\theta)$ (Williams 1952), where $f(\theta)>0$. This singularity can interact with the singularity at the kink tip (which is of the square root type) as the kinking length becomes smaller. Therefore, a proper modelling of this singularity improves the accuracy of the results.

In a 3-D structure, when the crack front intersects the free surface, the order of the singularity (corner singularity) is different than 0.5 (Benthem 1977 and Bazant & Estenssoro 1979), and depends on the material properties (mostly on Poisson's ratio) and the loading conditions.

Due to the inherent mathematical difficulties, analytical solutions of those types of problems are limited to few cases. However, the finite element technique can be used widely for more complex crack problems.

In general, conventional finite elements cannot adequately represent the stress and the displacement singularities near a crack tip. Substantial mesh refinement may lead to accurate results. However, high computer time and data preparation effort are required for such refined meshes. Besides, there is no guarantee that the conventional elements are accurate enough near the crack tip (Hilton & Sih 1973). On the other hand, the inclusion of singularity elements in the region of the crack tip leads to highly accurate results even with moderate mesh refinement.

In some circumstances, a 3-D finite element analysis should be performed when neither plane strain (the interior point are far from the outer surfaces) nor plane stress (vanishing thickness) conditions prevail. One has also to realize that in a composite material, an out of plane deformation can take place due to in plane loading, requiring eventually a 3-D analysis.

Many 2-D λ singularity elements have been proposed throughout the literature (Akin 1976, Tracy & Cook 1977, Abdel Wahab & De Roeck 1995, 1996), while little attention has been payed to the 3-D case. For cracks in homogeneous materials, where λ is 0.5, the Barsoum quarter point element (Barsoum 1976) is widely used in 3-D analysis. Blackburn and Hellen (Blackburn & Hellen 1977) have used a special 3-D 15-noded element with displacement variations proportional to the square root of the distance from the tip. However, those elements are not suitable for non-homogeneous materials (unless an extremely fine mesh is used) where λ can take values other than 0.5.

We present here a 3-D element which has the capability of representing a displacement variation of r^{λ} along the crack face and hence a variation of the derivatives of $r^{\lambda-1}$. The element is compatible along one of its faces (which does not intersect the crack front) with standard wedge (6-noded) and/or brick (8-noded) isoparametric elements. Two examples are presented; the first deals with a penny shaped crack in a cylinder, while the second deals with an edge crack in an homogeneous material.

2 ELEMENT FORMULATION

Figure 1 shows the 3-D prism (wedge) element in the (ξ,η,ζ) coordinate system and the (ξ,η) plane. The crack front is located along the line 1-4. If we introduce a singularity λ along the ξ direction, the relation between ξ and r, where r is another local parameter measuring the distance from the crack front with its origin (r=0) at the line 1-4 and r=a at the surface 2-3-5-6, can be written as:

$$\xi=\left(\frac{r}{a}\right)^{\lambda} \tag{1}$$

Two different shape functions are assumed for the interpolation of the coordinates and the displacement; i.e.,

$$[X]_{3\times1}=[N']_{3\times18}\ [x_i]_{18\times1}$$
$$[U]_{3\times1}=[N]_{3\times18}\ [u_i]_{18\times1} \tag{2}$$

where [X] and [U] are the coordinate and displacement vectors respectively, i.e. $[X]^T=[x\ y\ z]$ and $[U]^T=[u_x\ u_y\ u_z]$, and consequently x_i and u_i are the corresponding nodal coordinates and displacements, respectively. The shape functions N' and N are given by:

$$\begin{aligned}
N'_1&=(1-\xi^{\frac{1}{\lambda}})\ \frac{(1-\zeta)}{2}\\
N'_2&=\xi^{\frac{1}{\lambda}}\ (1-\eta)\ \frac{(1-\zeta)}{2}\\
N'_3&=\xi^{\frac{1}{\lambda}}\ \eta\ \frac{(1-\zeta)}{2}\\
N'_4&=(1-\xi^{\frac{1}{\lambda}})\ \frac{(1+\zeta)}{2}\\
N'_5&=\xi^{\frac{1}{\lambda}}\ (1-\eta)\ \frac{(1+\zeta)}{2}\\
N'_6&=\xi^{\frac{1}{\lambda}}\ \eta\ \frac{(1+\zeta)}{2}\\
N_1&=(1-\xi)\ \frac{(1-\zeta)}{2}\\
N_2&=\xi\ (1-\eta)\ \frac{(1-\zeta)}{2}\\
N_3&=\xi\ \eta\ \frac{(1-\zeta)}{2}\\
N_4&=(1-\xi)\ \frac{(1+\zeta)}{2}\\
N_5&=\xi\ (1-\eta)\ \frac{(1+\zeta)}{2}\\
N_6&=\xi\ \eta\ \frac{(1+\zeta)}{2}
\end{aligned} \tag{3}$$

The Jacobian matrix [J] is evaluated as:

$$[J]_{3\times3}=\begin{bmatrix}
\sum\frac{\partial N'_i}{\partial\xi}x_i & \sum\frac{\partial N'_i}{\partial\xi}y_i & \sum\frac{\partial N'_i}{\partial\xi}z_i\\
\sum\frac{\partial N'_i}{\partial\eta}x_i & \sum\frac{\partial N'_i}{\partial\eta}y_i & \sum\frac{\partial N'_i}{\partial\eta}z_i\\
\sum\frac{\partial N'_i}{\partial\zeta}x_i & \sum\frac{\partial N'_i}{\partial\zeta}y_i & \sum\frac{\partial N'_i}{\partial\zeta}z_i
\end{bmatrix} \tag{4}$$

The global strain displacement matrix [B] can be calculated as follows:

$$[B]\ =\ [J]^{-1}\ [DN] \tag{5}$$

where the matrix [DN] contains the derivatives of the

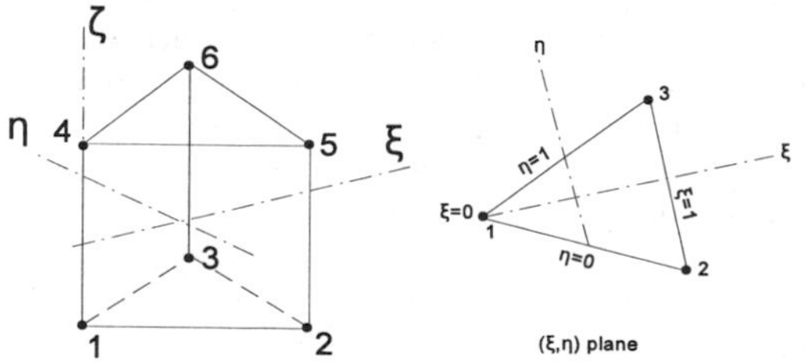

Figure 1. 3-D prism element.

shape functions N with respect to ξ,η,and ζ. The strain vector, [ϵ], is therefore given by:

$$[\epsilon]_{6\times1}=[B]_{6\times18}\ [u_i]_{18\times1} \tag{6}$$

where $[\epsilon]^T=[\epsilon_x\ \epsilon_y\ \epsilon_z\ \epsilon_{xy}\ \epsilon_{yz}\ \epsilon_{xz}]$. The inverse of the Jacobian,$[J]^{-1}$, contains terms as $\lambda\ r^{\lambda-1}$. Therefore, all strain components possess a λ-1 singularity.
Finally, the element stiffness matrix is obtained as:

$$[K]_{18\times18}=\int_V [B]^T_{18\times6}\ [D]_{6\times6}\ [B]_{6\times18} \tag{7}$$

where V is the volume of the element. [D] is the stress-strain matrix.

To integrate equation (7) numerically, the following quadrature rule is used (Cook et al. 1982):

$$\int_V \phi\ dV=\frac{1}{2}\sum w_{1i}\ w_{2i}\ J\ \phi_i \tag{8}$$

where w_{1i} and w_{2i} are the weight factors at the point where ϕ_i is evaluated. J is the determinant of the Jacobian matrix (eq.(4)).

The function ϕ in equation (8) is written in terms of the area coordinates L_1, L_2, L_3 and ζ. Therefore, the shape functions and in turn the matrix [B] are transformed from (ξ,η,ζ) to (L_1,L_2,L_3,ζ) using the following relations:

$$\xi=1-L_1 \qquad \eta=\frac{L_3}{1-L_1} \tag{9}$$

Figure 2 shows the different integration rules used in this analysis.

It can easily be shown that the derived element has a stress singularity at nodes 1 and 4 of order λ-1. It has continuity along the planes 1-2-4-5 and 1-3-4-6 with similar singular elements, and along the plane 2-3-5-6 with standard 6-noded wedge and/or 8-noded brick isoparametric elements.

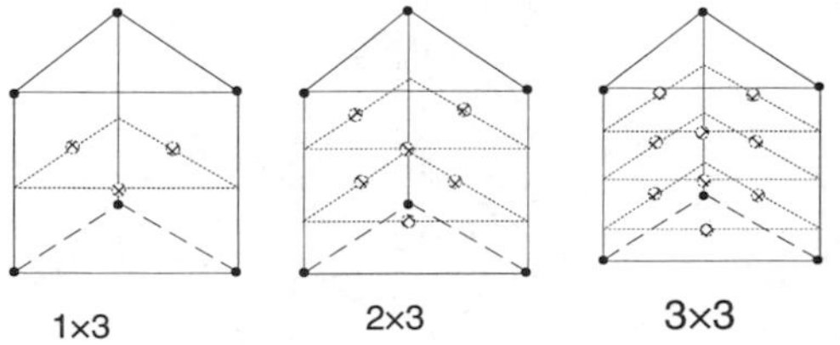

Figure 2. Integration rules.

The above technique can be easily extended to develop a 10-noded element (additional 4 mid-side nodes in the plane 2-3-5-6) which will be compatible with the 20-noded brick element.

3 INVESTIGATION OF ACCURACY

The present 3-D element has been introduced in the finite element program CALM (Geyskens et al. 1991). Two cases studies are given below to illustrate the effectiveness of the proposed element. In the first case study, a penny shaped crack in a cylinder (figure 3) is analysed.

Although this problem can be solved with 2-D axisymmetric assumption, it is solved here in three dimensions to illustrate the validity of the proposed 3-D singular element. Therefore, one-eighth of the body is modelled as shown in figure 4. The current singular element is used around the crack front, while standard 8-noded brick elements are used elsewhere. The model consists of 365 nodes and 272 elements.

The stress intensity factor is extrapolated from the opening displacement of the crack surface, and the normalized values of $(K/\sigma\ (\pi a)^{1/2})$ are given in table (1) for different integration rules.

The same problem has been solved by Banks-Sills and Sherman (Banks-Sills & Sherman 1992) using the stiffness derivative and the J-integral methods. The percentage differences between the current solution and the reference solution are in general less than 3%.

In the second case study, an edge crack in an homogeneous material is solved. The dimensions and

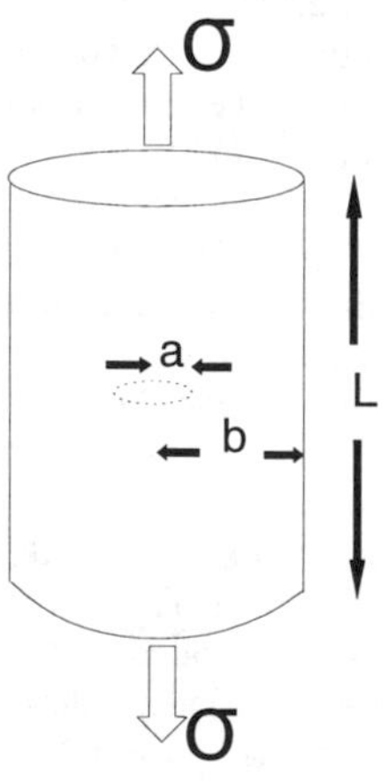

(L/b=2, a/b=0.5)

Figure 3. Penny shaped crack in a cylinder.

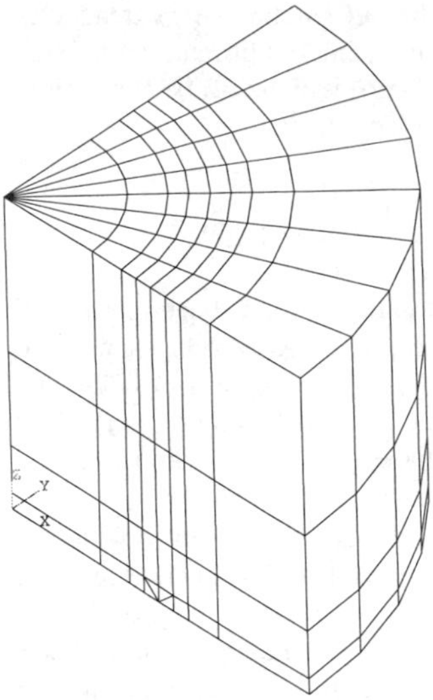

Figure 4. Finite element mesh for problem 1.

Table 1. Penny shaped crack

	Integration rules			Ref.
	1×3	2×3	3×3	
$K/\sigma(\pi a)^{1/2}$	0.7255	0.7267	0.7261	0.706

the configurations of the problem are shown in figure 5. Due to symmetry, only one-quarter of the edge crack need to be analysed (the shaded parts in figure 5). The finite element mesh used for this problem is given in figure 6.

The present element is used in the first row around the crack front with λ equal to 0.5 except the elements at the intersection between the crack front and the free surface where λ is put equal to 0.5477. These conclusions are drawn from the asymptotic analysis of Benthem (Benthem 1977) in which the corner singularity of a quarter infinite crack plane in a half space, (λ-1), is found equal to 0.4523 for Poisson's ratio of 0.3. Standard 8-noded brick elements are used away from the crack front. The crack tip element size to the crack length ratio is 1:16. Four elements are used through the half thickness of the sample. The model consists of 104 elements and 175 nodes. The stress intensity factor is calculated at different positions along the crack front. The results of $K/\sigma(\pi a)^{1/2}$ (where σ is the remote applied pressure) are given in table (2) for different integration orders. The ratio z/w=1.0 represents the centre of the specimen, while z/w=0.0 represents the free edge.

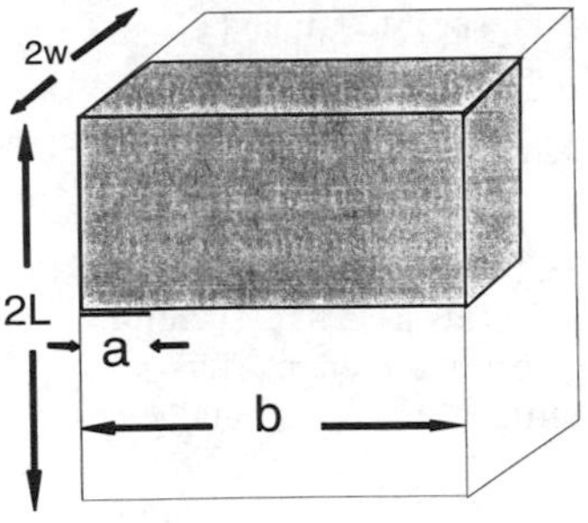

Figure 5. Dimensions of the edge crack.

The stress intensity factor distribution is nearly uniform through the thickness of the specimen except in the region near the crack front where the stress singularity is less severe. The integration rules 1×3, 2×3, and 3×3 give almost the same results.

The stress intensity factors for Poisson's ratio equal 0.0, 0.15 and 0.3 are plotted along the thickness z/w in figure 7. For ν=0.0, the singularity at the free surface is of the normal order (0.5). As expected, for this case, the stress intensity factor is constant through the thickness of the specimen and differs from the 2-D solution by 1.6%. For Poisson's ratio other than 0.0, the stress intensity factor decreases as the free surface is approached. These results are in agreement with

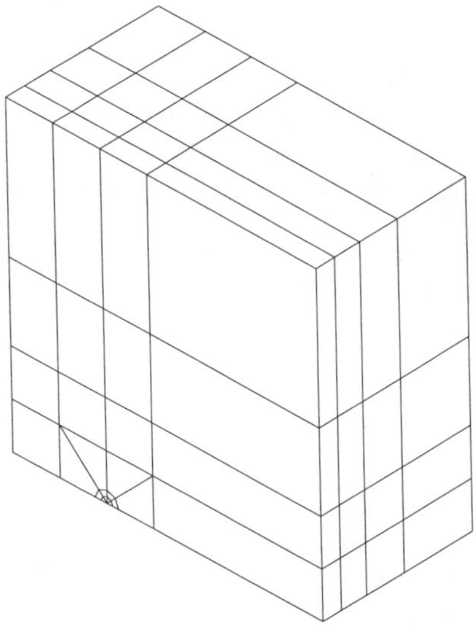

Figure 6. Finite element mesh for problem 2.

Table 2. Edge crack problem: values of $K/\sigma\sqrt{\pi a}$.

z/w	Integration orders		
	1×3	2×3	3×3
0.0	1.5451	1.535	1.5374
0.25	1.648	1.644	1.6413
0.5	1.676	1.679	1.6808
0.75	1.678	1.681	1.6877
1.0	1.687	1.689	1.691

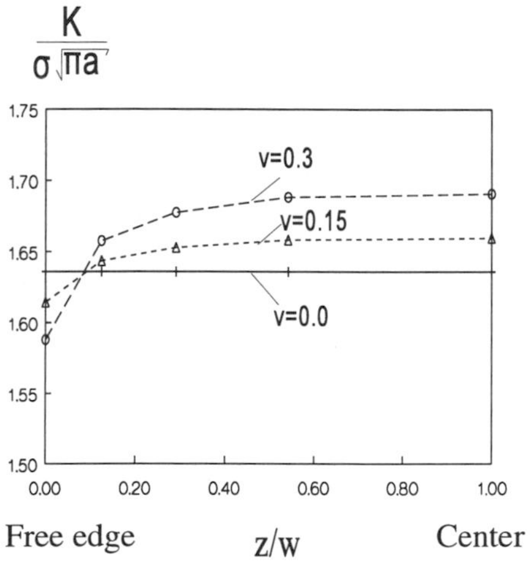

Figure 7. Effect of Poisson's ratio on the stress intensity factor.

those of other investigators (Burton et al. 1984 and Nakamura & Parks 1988).

4 SUMMARY AND CONCLUSIONS

A 3-D six-noded wedge element containing a singularity of order λ is developed. The interpolation functions of the displacements allow for variations of order r^{λ} along the crack surface and hence a variation of the stresses of $r^{\lambda-1}$. The element is compatible with standard (6-noded) wedge element and/or brick (8-noded) isoparametric finite elements. The element has been tested in two problems, namely: a penny shaped crack in a cylinder, and an edge crack in an homogeneous material.

REFERENCES

Abdel Wahab, M.M. & G. De Roeck 1995. A 2-D five noded finite element to model power singularity. *International Journal of Fracture*. 74: 89-97.

Abdel Wahab, M.M. & G. De Roeck 1996. A family of elements with λ singularity. *Advances in Engineering Software*: 23-28.

Akin, A.J. 1976. The generation of elements with singularities, *International Journal for Numerical Methods in Engineering*.10:1249-1259.

Banks-Sills, L. & D. Sherman 1992. On the computation of stress intensity factors for three dimensional geometries by means of stiffness derivatives and J-integral methods. *International Journal of Fracture*. 53: 1-20.

Barsoum, R.S. 1976. On the use of isoparametric finite elements in linear fracture mechanics, *International Journal for Numerical Methods in Engineering*. 10:25-37.

Bazant, Z.P. & L.F. Estenssoro 1979, Surface singularity and crack propagation. *International Journal of Solid and Structure*. 15:405-426.

Benthem, J.P. 1977. State of stress at the vertex of a quarter-infinite crack in a half space. *International Journal of solid and structure*. 13: 479-492.

Blackburn. W.S. & T.K. Helen 1977, calculation of stress intensity factors in three dimensions by finite element method, *International Journal for Numerical Methods in Engineering*. 11:211-229.

Bogy, D.B. 1971, On the plane elastostatic problem of a loaded crack terminating at a material interface, Journal of applied Mechanics. 12:911-918.

Burton, W. S., Sinclair, G. B., Solecki, J. S. and J.L. Swedlow 1984. On the implications for LEFM of the three-dimensional aspects in some crack/surface intersection problems. *International Journal of Fracture*. 25:3-32.

Cook, R.D., Malkus, D.S. & Phesha M.E. 1982, Concept and application of finite element analysis, John Wiley & Sons, Third edition.

Cook, T.S. & F. Erdogan 1972, Stresses in bonded materials with a crack perpendicular to the interface, *International Journal of Engineering Science*. 10:667-697.

Geyskens, P., Marien, W. & G. De Roeck 1991. Computer analysis language (modified version). Katholieke Universiteit Leuven.

Hilton, P.D. & G.C. Sih 1973, *in Mechanics of fracture* (edited by G.C. Sih). 1.1:426-483. Noordhohh, Amsterdam.

Nakamura, T. and D.M. Parks 1988. Three-dimensional stress field near the crack front of a thin elastic plate. *Journal of Applied Mechanics*. 55:805-813.

Tracy D.M. & S.T.Cook 1977 Analysis of power type singularities using finite elements, International *Journal for Numerical Methods in Engineering*. 11:1225-1233.

Williams, M.L. 1952. Stress singularities resulting from various boundary conditions in angular corners of plates in extension. Transactions, ASME, *Journal of Applied Mechanics*. 19:526-528.

Modern Practice in Stress and Vibration Analysis, Gilchrist (ed.) © *1997 Balkema, Rotterdam, ISBN 90 5410 896 7*

The use of J-integrals in the boundary element method for the fracture analysis of plates

J.L.Wearing & S.Y.Ahmadi-Brooghani
Department of Mechanical Engineering, University of Sheffield, UK

ABSTRACT: This paper discusses then application of the Boundary Element Method for the determination of the stress intensity factors in cracked plates. In the analysis the Dual Boundary Element Method has been used in conjunction with the J-integral method for the determination of the stress intensity factors. A number of case studies, illustrating the effectiveness of the technique for the determination of stress intensity factors are presented in the paper. The results from the case studies are compared with results from alternative approaches within the Boundary Element Method for the determination of the stress intensity factors and with analytical and finite elements solutions.

1 INTRODUCTION

The determination of the stress distribution in the vicinity of cracks, particularly in safety critical situations is of vital importance to engineers. The value of the stress in the vicinity of a crack is governed by the magnitude of the stress intensity factor at the crack tip. Due to the importance of stress intensity factors to the engineer, there is a considerable volume of published hand book data (Rooke and Cartwright, 1976) which gives detailed information of the stress intensity factors for a wide range of crack configurations, component geometries and loading conditions. This information has been obtained from a large number of sources and, although extensive, does not cover every possible situation. As structures, their loading conditions and crack configurations become more complex and with structural integrity being an important feature in their design, engineers have sought new approaches for determining the stress intensity factors and increasing their knowledge in this field of study. Attention has therefore been focused in the application of numerical methods like the Finite Element Method (FEM) and the Boundary Element Method (BEM) for the determination of the stress intensity factors in complex situations where existing hand book data has proved to be inadequate. Because the BEM only requires the boundary of the component to be modelled, it has many advantages over the FEM, particularly in the field of Linear Elastic Fracture Mechanics (LEFM). The BEM does, however, have one major drawback, particularly if the displacement boundary integral equations are used to model both faces of the crack. In such cases a system of singular algebraic equations is produced. This problem can be resolved by subdividing the component into subregions with each face of the crack being in different subregions. An alternative approach is to use the Dual Boundary Element Method (DBEM) in which one face of the crack is modelled using the displacement boundary integral equations and the other face is modelled using the traction boundary element equations, with the remainder of the boundary being modelled using the displacement boundary integral equations. The DBEM therefore eliminates the singular equations and avoids the necessity of using subregions. The DBEM has been used successfully for the determination of the stress intensity factors in two and three dimensional stress situations (Portela, Aliabadi and Rooke, 1992) and (Mi and Aliabadi, 1993) where LEFM applies. There is however little published work on the determination of the stress intensity factors in plate bending problems. The work in this paper therefore discusses the application of the DBEM for the fracture analysis of plate bending problems in LEFM with the J integral approach being used to determine the stress intensity factors. The J integral results are compared with Murakami's (1987) analytical results, Sosa and Eishen's (1986) finite element results and with Wearing and Ahmadi-Brooghani's (1996) BEM results, where alternative approaches have been used within the BEM to determine the stress intensity factors.

2 THE BOUNDARY ELEMENT METHOD FOR PLATE BENDING

The plate bending boundary element equations, which are used in this paper relate to Reissner's (1947) plate bending equations and are given by Karam and Telles (1985). The displacements at an internal point on the plate are obtained from the following equation

$$u_i(\xi) + \int_\Gamma T^*_{ij}(\xi,x)u_j(x)d\Gamma(x) = \int_\Gamma U^*_{ij}(\xi,x)t_j(x)d\Gamma(x) + \int_\Omega \left(U^*_{i3}(\xi,x) - \frac{\upsilon}{(1-\upsilon)\lambda^2} U^*_{i\alpha,\alpha}(\xi,x) \right) q(x)d\Gamma(x) \tag{1}$$

in which x is a point on the boundary of the plate and U^*_{ij} (ξ,x) and T^*_{ij} (ξ,x) are the fundamental solutions for displacement and traction respectively. When the distance, r, between ξ and x is not zero, the integrals in equation (1) are regular. The integrals become singular, however, when ξ is moved to the boundary of the plate and ξ coincides with x and in these circumstances equation (1) becomes

$$c_{ij}(\xi)u_j(\xi) + (CPV)\int_\Gamma T^*_{ij}(\xi,x)u_j(x)d\Gamma(x) = \int_\Gamma U^*_{ij}(\xi,x)t_j(x)d\Gamma(x) + \int_\Omega \left(U^*_{i3}(\xi,x) - \frac{\upsilon}{(1-\upsilon)\lambda^2} U^*_{i\alpha,\alpha}(\xi,x) \right) q(x)d\Gamma(x) \tag{2}$$

where (CPV)∫ denotes a Cauchy Principal Value Integral.

Stresses at internal points are obtained from integral equations for bending moment and shear force, which are derived from equations (1) and (2) as

$$M_{\alpha\beta}(\xi) = \int_\Gamma D^*_{\alpha\beta k}(\xi,x)t_k(x)d\Gamma(x) - \int_\Gamma S^*_{\alpha\beta k}(\xi,x)u_k(x)d\Gamma(x) + q\int_\Omega W^*_{\alpha\beta}(\xi,x)d\Gamma(x) + \frac{\upsilon}{(1-\upsilon)\lambda^2} q\delta_{\alpha\beta} \tag{3}$$

$$Q_\beta(\xi) = \int_\Gamma D^*_{3\beta k}(\xi,x)t_k(x)d\Gamma(x) - \int_\Gamma S^*_{3\beta k}(\xi,x)u_k(x)d\Gamma(x) + q\int_\Omega W^*_{3\beta}(\xi,x)d\Gamma(x) \tag{4}$$

Equations (3) and (4) also become singular when ξ is moved to the boundary and ξ and x coincide. In these circumstances, provided the boundary is smooth, these equations become

$$\frac{1}{2}M_{\alpha\beta}(\xi) = (CPV)\int_\Gamma D^*_{\alpha\beta k}(\xi,x)t_k(x)d\Gamma(x) - (HPV)\int_\Gamma S^*_{\alpha\beta k}(\xi,x)u_k(x)d\Gamma(x) + \left[(CPV)\int_\Gamma W^*_{\alpha\beta}(\xi,x)d\Gamma(x) + \frac{\upsilon}{(1-\upsilon)\lambda^2}\delta_{\alpha\beta} \right] q \tag{5}$$

$$\frac{1}{2}Q_\beta(\xi) = (CPV)\int_\Gamma D^*_{3\beta k}(\xi,x)t_k(x)d\Gamma(x) - (HPV)\int_\Gamma S^*_{3\beta k}(\xi,x)u_k(x)d\Gamma(x) + q\int_\Gamma W^*_{3\beta}(\xi,x)d\Gamma(x) \tag{6}$$

in which (HPV)∫ represents a Hadamard Principal Value Integral.

The components of traction, t_i, are obtained from

$$t_\alpha = M_{\alpha\beta}\, n_\beta \text{ and } t_3 = Q_\alpha\, n_\alpha \tag{7}$$

in which n represents the outward normal.

Equations (5), (6) and (7) can therefore be used to obtain the traction components on a smooth boundary as

$$\frac{1}{2}t_\alpha(\xi) = n_\beta(\xi)(CPV)\int_\Gamma D^*_{\alpha\beta k}(\xi,x)t_k(x)d\Gamma(x)$$
$$-n_\beta(\xi)(HPV)\int_\Gamma S^*_{\alpha\beta k}(\xi,x)u_k(x)d\Gamma(x)$$
$$+\left[n_\beta(\xi)(CPV)\int_\Gamma W^*_{\alpha\beta}(\xi,x)d\Gamma(x) + \frac{\upsilon}{(1-\upsilon)\lambda^2}n_\alpha(\xi)\right]q \qquad (8)$$

$$\frac{1}{2}t_3(\xi) = n_\beta(\xi)(CPV)\int_\Gamma D^*_{3\beta k}(\xi,x)t_k(x)d\Gamma(x)$$
$$-n_\beta(\xi)(HPV)\int_\Gamma S^*_{3\beta k}(\xi,x)u_k(x)d\Gamma(x)$$
$$+qn_\beta(\xi)\int_\Gamma W^*_{3\beta}(\xi,x)d\Gamma(x) \qquad (9)$$

In the case studies considered in this paper the domain integrals in equations (1) and (2) resulting from a uniformly distributed load are transferred to boundary integrals using the following equation

$$\int_\Omega\left[U^*_{i3}(\xi,x) - \frac{\nu}{(1-\nu)\lambda^2}U^*_{i\alpha,\alpha}(\xi,x)\right]d\Omega =$$
$$\int_\Gamma\left[v^*_{i,\alpha}(\xi,x) - \frac{\nu}{(1-\nu)\lambda^2}v^*_{i\alpha}(\xi,x)n_\alpha(x)\right]d\Gamma \qquad (10)$$

where v^*_i satisfies the equation $v^*_{i,\alpha\alpha} = v^*_{ie}$ and is given by

$$v^*_\alpha = \frac{1}{128\pi D\lambda^2}r_\alpha rz^2(4\ln z - 5) \qquad (11)$$

$$v^*_3 = \frac{-z^2}{256\pi D\lambda^4(1-\nu)}\left[64(\ln z - 1) - z^2(1-\nu)(2\ln z - 3)\right] \qquad (12)$$

In equations (1) to (12) the Roman subscripts i, j and k vary from 1 to 3 and the Greek subscripts α and β have values of 1 and 2.

Equations (2), (8) and (9) are the integral equations which constitute the Dual Boundary Element Method using Reissner's (1947) plate bending equations and are used in the analysis of the case studies discussed in this paper.

3 CASE STUDIES

Four case studies are presented in this paper. Two of these illustrate the use of symmetry in the fracture analysis of plate bending problems and the other two illustrate the use of the DBEM.

Case Study 1: A rectangular plate with a central crack, loaded by two equal, uniformly distributed moments on opposite edges, parallel to the crack, with the other two edges of the plate being free (Fig. 1).

Case Study 2: A rectangular plate loaded by two equal uniformly distributed moments on opposite edges and the other two boundaries free, with two equal opposite edge cracks on the free edges. (Fig. 2).

As the plates in the above two case studies are symmetrical only a quarter of the plate needs to be analysed. In each of these analyses the boundary adjacent to the crack tip was analysed using semi-continuous quarter point elements and the remainder of the boundary was analysed using continuous quadratic elements. For each of these case studies a convergence study was undertaken with 5, 10, 15, 20 and 22 elements. When 5, 10, 15 and 20 elements

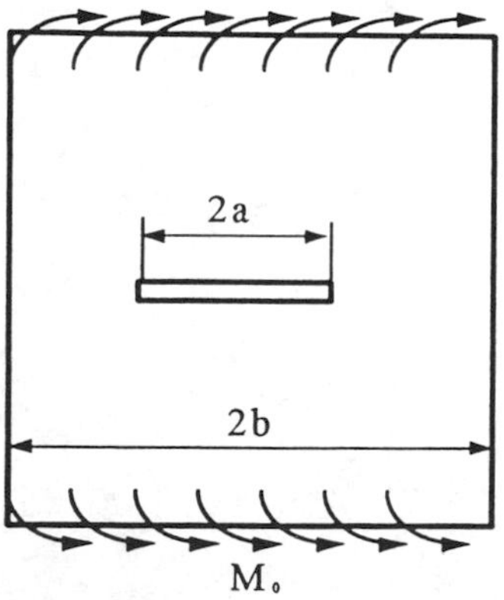

Figure 1. Plate with end moments and central crack.

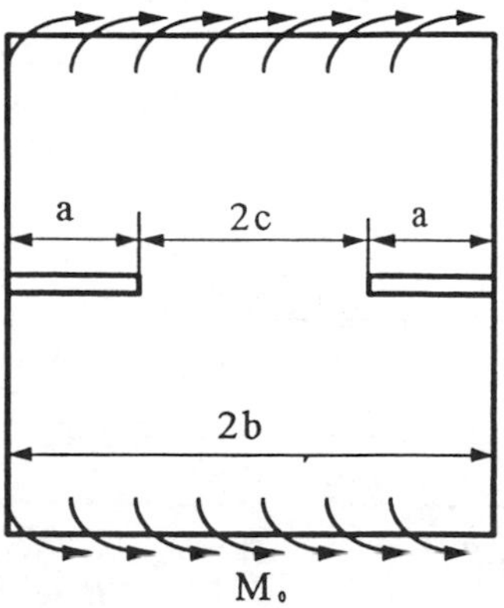

Figure 2. Plate with end moments and edge cracks.

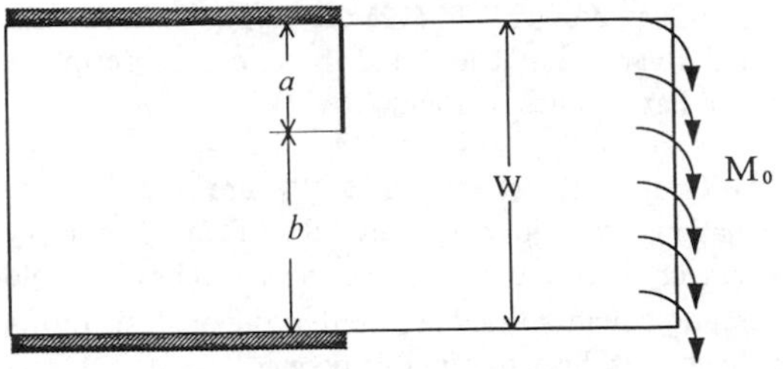

Figure 3. Rectangular plate with mixed boundary conditions and edge crack.

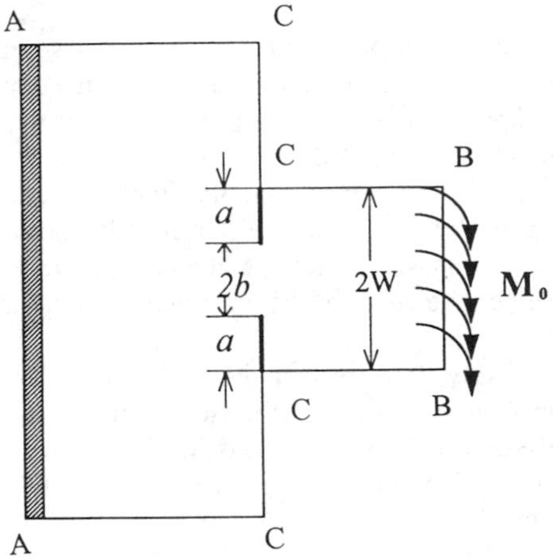

Figure 4. Cantilevered tee plate with edge cracks.

were used there was an equal number of elements on the crack face, the uncracked boundary collinear with the crack face and on each of the other boundaries. However when 22 elements are used there are 4 elements on each of the uncracked boundaries and 10 elements along the boundary containing the crack, with the number of elements being used to model the crack depending on the crack length.

Case Study 3: A rectangular plate with one half of each long boundary built in and the remainder of the boundary free, acted on by uniformly distributed moment along the free edge adjacent to the free sections of the long boundaries (Fig. 3). The plate has a single crack at the junction of the built in and the free portions on one of the long boundaries (Fig. 3). The boundary element model for this case study comprised thirty two elements. Each long external boundary was modelled using eight elements, each short external boundary was modelled using four elements and each crack face was modelled using four elements (Fig. 3).

Case Study 4: This case study comprises a cantilevered tee shaped plate acted on by a uniformly distributed moment along the free boundary opposite the built in boundary AA. The plate has two cracks at the tee junction as shown in figure 4. It was modelled using fifty two boundary elements. Boundary AA had eight elements,

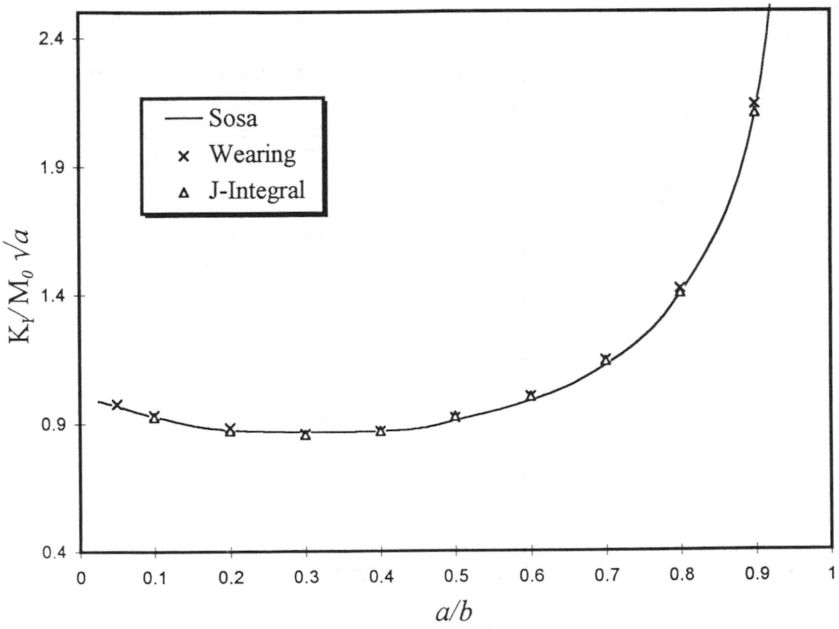

Figure 5. Variation of K_I with crack length for plate with central crack and end moments.

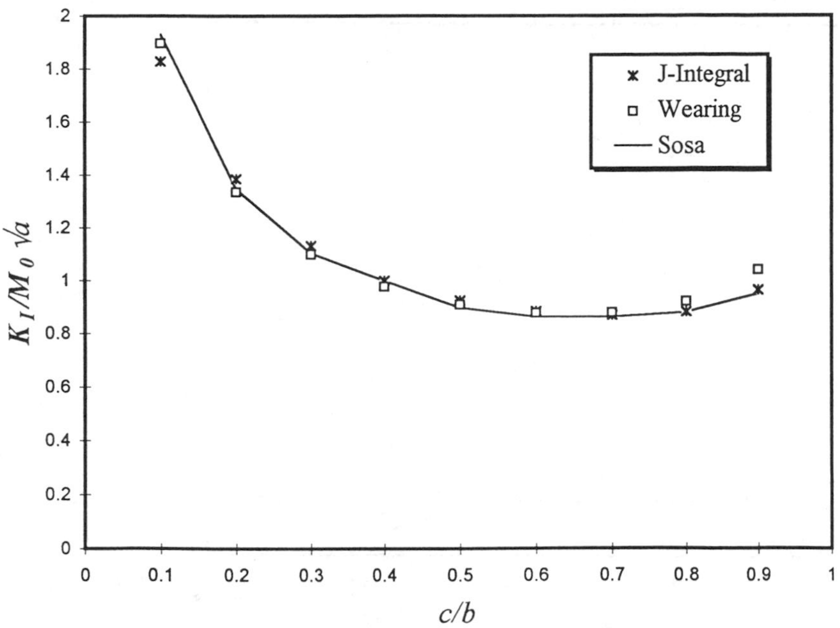

Figure 6. Variation of K_I with crack length for plate with edge cracks and end moments.

boundary BB had four elements and boundaries AC, CC and CB each had four elements. Each side of the crack face was modelled using four elements (Fig. 4).

In case studies three and four, the DBEM was used in the analyses, with equation (2) for boundary displacements being used to model one of the crack faces and equations (8) and (9) for boundary tractions being used to model the other crack face. The remainder of the boundary was modelled using the equation for boundary displacements. Discontinuous elements were used to model the crack faces, at the junction of the crack face and the external boundary semi-continuous elements were used and the remainder of the boundary was analysed using continuous elements.

4 RESULTS

In all case studies considered in this paper, the stress intensity factors at the crack tip were determined using the J integral method and the displacement extrapolation method. For case studies one and two a square plate having a breadth to depth ratio of two and a range of crack lengths was analysed for each case study. Results from the analyses are shown in figures 5 and 6 and are compared with finite element results obtained by Sosa and Eischen (1986), who used the J integral method to determine the stress intensity factors. The results are also compared with Wearing and Ahmadi-Brooghani's (1996) results, who used the BEM and the displacement extrapolation method to determine the K_I stress intensity factors. Figure 5 gives details of the results for the variation of normalised stress intensity factor at the crack tip with the ratio of crack length to plate breadth as shown in figure 1. For the second case study (Fig. 2) results are shown in figure 6 for the variation of normalised stress intensity factor at the crack tip with the ratio of the distance between the crack tips to plate breadth.

In case studies three and four the Dual Boundary Element Method was used in conjunction with the J integral method to calculate the stress intensity factors. For case study three (Fig. 3) results of the normalised stress intensity factors K_I, K_{II} and K_{III} were calculated from the following expressions

$$K_I = \frac{6\,M\omega}{t^2\sqrt{b}}F_I \tag{13}$$

$$K_{II} = \frac{6\,M\omega}{t^2\sqrt{b}}F_{II} \tag{14}$$

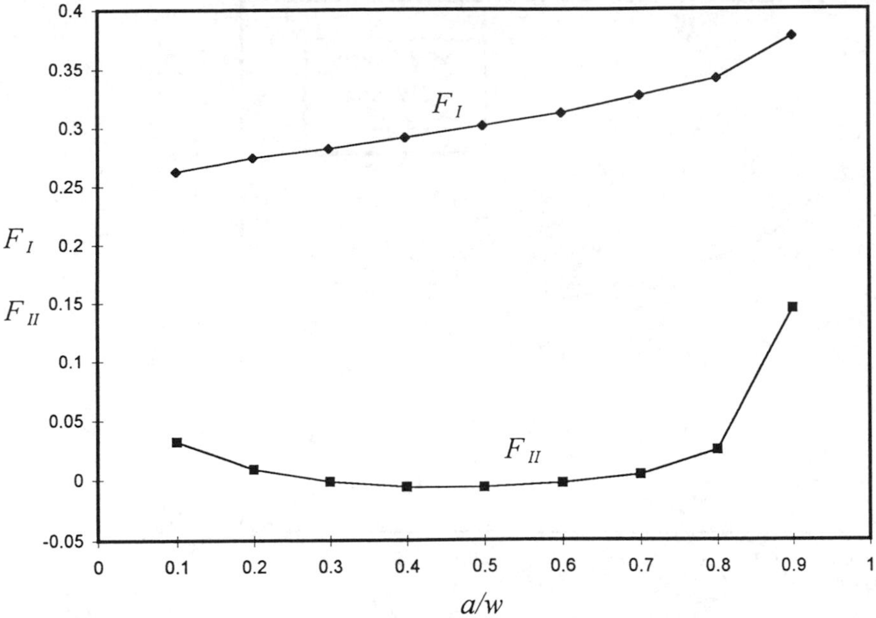

Figure 7. Variation of K_I and K_{II} with crack length for rectangular plate with mixed boundary conditions.

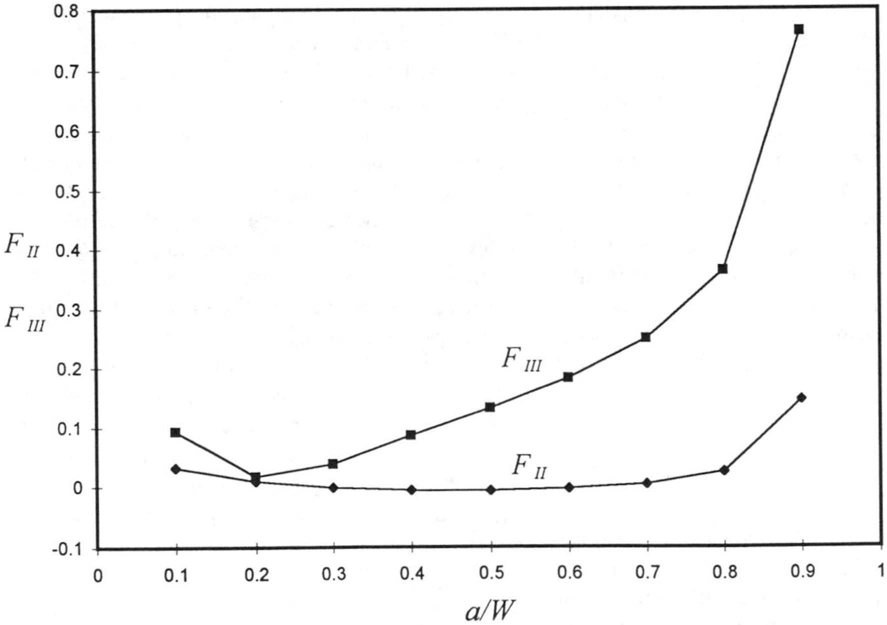

Figure 8. Variation of K_{III} with crack length for rectangular plate with mixed boundary conditions.

$$K_{III} = \frac{6\,M\omega}{t^2 a\sqrt{b}} F_{III} \qquad (15)$$

in which F_I, F_{II} and F_{III} are factors relating the ratio of crack length, a, to plate width, w, and t is the thickness of the plate (Hasabe, Miwa and Nakamuri, 1990). The results for a range of ratios of crack length, a, to plate width, w, for K_I and K_{II} are shown in figure 7 and results for K_{III} are given in figure 8. Results for the stress intensity factors using the displacement extrapolation method are also given in figures 7 and 8.

For case study four (Fig. 4) results of the normalised stress intensity factor K_I were calculated from the following expression

$$K_I = \frac{6\,M\omega}{t^2\sqrt{b}} F_I \qquad (16)$$

in which F_I is a factor relating the ratio of crack length, a, to distance, b, between the crack tips and t is the plate thickness (Hasabe, Miwa, and Nakamuri, 1990). Results are presented in figure 9 for a range of ratios of crack length, a, to distance, b, between the crack tips. Results for the stress intensity factors using both the J integral method and the displacement extrapolation method are shown in figure 9.

In all case studies the J integrals were calculated along a number of paths. Details of the J integral paths for the calculation of the stress intensity factors are shown in figure 10.

5 DISCUSSION OF RESULTS

The results presented in figures 5, 6, 7, 8 and 9 for the plates shown in figures 1, 2, 3 and 4 clearly indicate the effectiveness of the BEM, using the J integral method, to calculate the stress intensity factors in the linear elastic fracture mechanics analyses of plate bending problems. Plates with a number of loading conditions, boundary conditions and crack configurations have been analysed and two different approaches have been used in the analyses. In case studies one and two the symmetry of the plates was exploited and the displacement boundary integral equations were used to model the boundary of the quarter plate models. Case studies three and four did not have the advantage of symmetry and the DBEM was used, eliminating the singular equations and the requirement for subregioning.

Figures 5 and 6 give the results of the K_I stress intensity factors for case studies one and two (Figs. 1 and 2) for the models with twenty-two boundary elements. These models showed rapid convergence

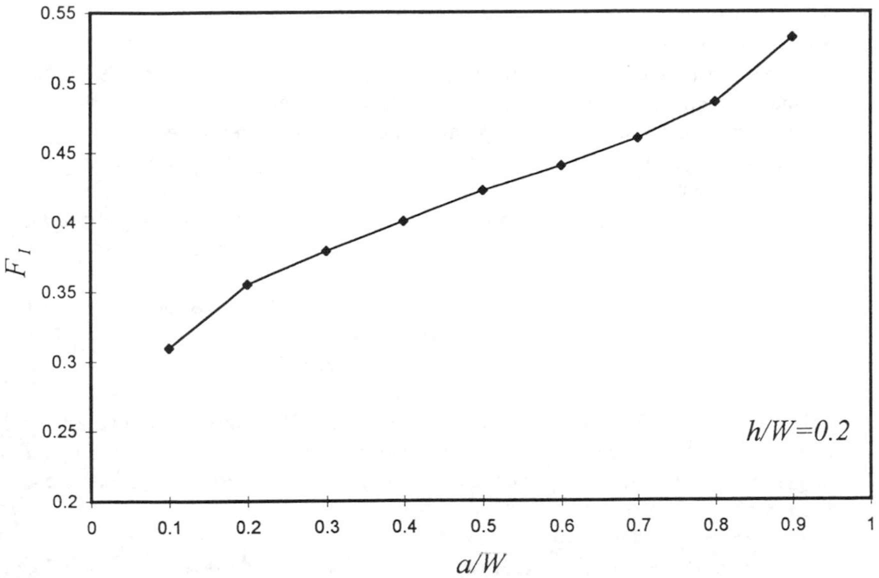

Figure 9. Variation of K_I with crack length for cantilevered tee plate.

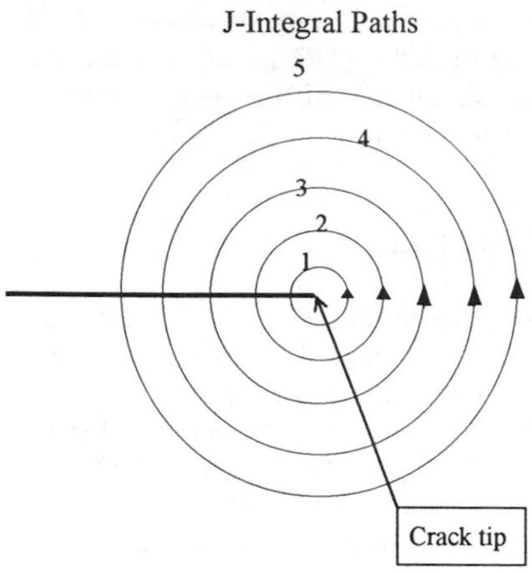

Figure 10. J integral paths.

from the five element models and are in close agreement with finite element results (Sosa and Eischen, 1986) and boundary element, displacement extrapolation results (Wearing and Ahmdadi-Brooghani, 1996). However the size of the quarter plate models in both boundary element analyses (J integral and displacement extrapolation), with the 22 elements and 44 noes is in sharp contrast to the finite element quarter plate models with their 140 elements and 469 nodes. The results for the three different approaches are in close agreement.

Figures 7 and 8 give the results of the K_I, K_{II} and K_{III} stress intensity factors for case study three (Fig. 3) for a range of crack lengths and figure 9 gives the results of the K_I stress intensity factors for case study four (Fig. 4). In each case study the results are compared with boundary element results using the displacement extrapolation technique to determine the stress intensity factors and in each case both sets of results are in close agreement.

6 CONCLUSIONS

The results presented in this paper confirm that the BEM can be used successfully for the determination of stress intensity factors in plate bending problems when LEFM conditions apply. Boundary element results have been presented for stress intensity factors using both the J integral approach and the displacement extrapolation approach. The BEM results for two case studies have been compared with finite element results and in each case the BEM results and in close agreement with the FEM results. The BEM also been shown to have a clear advantage over the FEM as results of comparable accuracy have been achieved using the BEM with considerably fewer elements and degrees of freedom, compared to the FEM. The work discussed in this paper also indicates that, where, appropriate, the DBEM can be used for the fracture analysis of plate bending problems.

REFERENCES

Hasabe N., Miwa M. & Nakamuri T. 1990. A mixed boundary value problem of a strip with a crack under concentrated bending and torsional moments. *Trans. Japan Soc. Civil Engg.*, 416: 395-401.

Karam V.J. & Telles J.C.F. 1985. On boundary elements for Reissner's plate theory. *Engg. Analysis with Boundary* Elements. 5: 21-27.

Mi Y. & Aliabadi M.H. 1993. Dual boundary element method for three dimensional crack growth analysis. In C.A. Brebbia & J.J. Rencis (eds), *Proc. 15th Int. Conf. on Boundary Element Methods, Worcester, U.S.A., August* 1993: 249-260. Southampton: Computational Mechanics Publications.

Murakami Y. 1987, *Stress intensity factors handbook.* Oxford: Pergamon Press.

Portela A., Aliabadi M.H. & Rooke D.P. 1992. The dual boundary element method : effective implementation for crack problems. *Int. J. for Numerical Methods in Engg.* 33: 1269-1287.

Reissner N. 1947. On the bending of elastic plates. *Quarterly of Applied* Mathematics. 5:55-68.

Rooke D.P. & Cartwright D.J. 1976. *A compendium of stress intensity factors.* London : Her Majesty's Stationery Office.

Sosa H.A. & Eischen J.W. 1986. Computation of stress intensity factors for plate bending via a path dependent integral. *Engg. Fracture* Mechanics, 25: 451-462.

Wearing J.L. & Ahmadi-Broghani S.Y. 1996. Quarter point boundary elements for the analysis of crack problems in plate bending. In C.A. Brebbia (ed), *Proc. 18th Int. Conf. on Boundary Element Methods, Braga, Portugal, September* 1993 : Southampton, Computational Mechanics Publications.

ACKNOWLEDGEMENT

Mr Ahmadi-Brooghani would like to thank the Iranian Research organisation for Science and Technology, Mashad Centre, for providing the financial support to enable him to undertake the work discussed in this paper.

2 System identification and parameter estimation

Modern Practice in Stress and Vibration Analysis, Gilchrist (ed.) © 1997 Balkema, Rotterdam, ISBN 90 5410 896 7

An inverse method to measure the vibration parameters of a liquid-filled shell

A.J. Hull
Submarine Sonar Department, Naval Undersea Warfare Center Division, Newport, R.I., USA

ABSTRACT: This paper develops an inverse method for determining the parameters of a model that describes a liquid-filled cylindrical shell under axial tension subjected to an external displacement. The model assumes that an extensional wave and a breathing wave each propagate in both the positive and negative axial directions of the shell, which results in six unknown parameters: two forced wavenumbers and four wave propagation coefficients. Based on experimental testing, the extensional and breathing wavenumbers are first estimated and then inserted as numerical values into the model at four spatial pressure field locations. The model equations are now rewritten as a four-by-four system of linear equations that are equated to the data. The solutions are the unknown wave propagation coefficients. The model and experimental results are discussed, and recommendations to improve the inverse method are presented.

1 INTRODUCTION

A comparison of experimental data with a model of the interior pressure field of a liquid-filled cylindrical shell often shows disagreement between measurement and prediction. One reason is that the models for this field (Morgan and Kiely 1954, Chow and Apter 1968, Jameson et al. 1974) typically do not take into account such unmodeled processes as internal hydrophone mount noise, support cable interaction, and other noise effects that cause deviations from the measurements. In this paper, a model of the pressure field in a liquid-filled cylindrical shell is developed based on experimental pressure measurements of the field. The underlying assumption is that two waves (an extensional wave and a breathing wave) are propagating in both axial directions of the shell. A four-term model that describes this motion is derived, and the six unknown model parameters are determined with an inverse method. The model makes no assumptions about the boundary conditions of the shell and the material properties of the shell or liquid. Concerning the shell/liquid interaction, the only assumption is that the dynamics of this process creates a breathing wave in the structure that originates at the bulkheads. This empirical model, developed as a "bridge" between theory and experiment, will identify the frequency ranges where disagreement typically occurs.

2 SYSTEM MODEL

The pressure field inside the liquid-filled shell is derived from two wave equations in the spatial domain, both of which use pressure as the independent variable. The first equation models the extensional wave contribution and is written as

$$\frac{\mathrm{d}^2 P_e(x,\omega)}{\mathrm{d}x^2} + k_e^2 P_e(x,\omega) = 0 \ , \tag{1}$$

where $P_e(x,\omega)$ is the temporal Fourier transform of the pressure that is generated by the extensional wave, x is the spatial location (m), ω is the excitation frequency (rad/s), and k_e is the complex extensional wavenumber, which is equal to ω / c_e, where c_e is the complex extensional wave speed (m/s). The second wave equation models the breathing wave contribution and is written as

$$\frac{\mathrm{d}^2 P_b(x,\omega)}{\mathrm{d}x^2} + k_b^2 P_b(x,\omega) = 0 \ , \tag{2}$$

where $P_b(x,\omega)$ is the temporal Fourier transform of the pressure that is generated by the breathing wave and k_b is the complex breathing wavenumber (rad/m), which is equal to ω/c_b, where c_b is the complex breathing wave speed (m/s). The solutions to equations (1) and (2) are complex exponential functions and can be added together using the principle of superposition, which yields the total pressure in the shell as

$$\begin{aligned} P(x,\omega) &= P_b(x,\omega) + P_e(x,\omega) \\ &= \overline{A}(\omega)\mathrm{e}^{\mathrm{i}k_b x} + \overline{B}(\omega)\mathrm{e}^{-\mathrm{i}k_b x} \\ &\quad + \overline{C}(\omega)\mathrm{e}^{\mathrm{i}k_e x} + \overline{D}(\omega)\mathrm{e}^{-\mathrm{i}k_e x} , \end{aligned} \tag{3}$$

where $P(x,\omega)$ is the temporal Fourier transform of the pressure that is generated by both the extensional and breathing waves; i is the square root of -1; and $\overline{A}(\omega)$, $\overline{B}(\omega)$, $\overline{C}(\omega)$, and $\overline{D}(\omega)$ are wave propagation coefficients determined by the boundary conditions. It is now noted that the pressure field at x divided by an accelerometer placed at the forward end of the shell is

$$\begin{aligned} \frac{P(x,\omega)}{\ddot{U}} &= A(\omega)\mathrm{e}^{\mathrm{i}k_b x} + B(\omega)\mathrm{e}^{-\mathrm{i}k_b x} \\ &\quad + C(\omega)\mathrm{e}^{\mathrm{i}k_e x} + D(\omega)\mathrm{e}^{-\mathrm{i}k_e x} , \end{aligned} \tag{4}$$

where $\ddot{U}$ is the temporal Fourier transform of the acceleration; $A(\omega)$, $B(\omega)$, $C(\omega)$, and $D(\omega)$ are wave propagation coefficients; and $P(x,\omega)/\ddot{U}$ has units of Pa//(m/s^2). Physically, the coefficient $A(\omega)$ corresponds to forward-traveling breathing wave energy, $B(\omega)$ corresponds to aft-traveling breathing wave energy, $C(\omega)$ corresponds to forward-traveling extensional wave energy, and $D(\omega)$ corresponds to aft-traveling extensional wave energy.

There are six unknowns in equation (4), which are the four wave propagation coefficients and the two forced wavenumbers (breathing and extensional). Using a laboratory configuration in which a longitudinal shaker is placed at the forward end of the shell allows all these unknowns to be estimated. Due to the nonlinear (and sinusoidal) nature of the parameters in equation (4), the most numerically stable method to determine the parameters is to first estimate the extensional wavenumber, then the breathing wavenumber, and finally the unknown wave propagation coefficients.

3 ESTIMATION OF FORCED WAVENUMBERS

In order to estimate the wave propagation coefficients and the breathing wavenumber, it is necessary to know the extensional wavenumber. This complex, frequency-dependent quantity is determined with measurements from forward and aft impedance heads that are attached to the shell ends in the laboratory test facility (Hull 1996a). Although these measurements contain a breathing wave contribution, it typically occurs only at low frequencies and can be easily discerned from the extensional wave effects. Such behavior is described in the experiment section.

The governing differential equation of the extensional wave is expressed in the spatial domain as a single wave equation with particle displacement as the independent variable:

$$\frac{\mathrm{d}^2 U(x,\omega)}{\mathrm{d}x^2} + k_e^2 U(x,\omega) = 0 , \tag{5}$$

where $U(x,\omega)$ is the temporal Fourier transform of the axial displacement. Equation (5) is a one-dimensional "lumped" approximation of the extensional wave motion in the system, combining the effect of the shell and the liquid together as a homogeneous medium that supports longitudinal wave motion. Although this approximation is not sufficient to model the radial motion of the shell, it is an accurate model of axial motion and the corresponding extensional wave propagation in the liquid-filled shell. The energy attenuation in the shell is defined with a structural damping law, and therefore the wave speed is a complex quantity. The real part of the wave speed corresponds to energy transmission and the imaginary part corresponds to energy attenuation.

The solution to equation (5) is

$$U(x,\omega) = G(\omega)\mathrm{e}^{\mathrm{i}k_e x} + H(\omega)\mathrm{e}^{-\mathrm{i}k_e x} , \tag{6}$$

where $G(\omega)$ and $H(\omega)$ are coefficients determined by the boundary conditions at the ends of the shell. The temporal Fourier transform of the axial force in the shell is

$$\begin{aligned} F(x,\omega) &= A_s E_x \frac{\mathrm{d}U(x,\omega)}{\mathrm{d}x} \\ &= A_s E_x \mathrm{i}k_e \left[G(\omega)\mathrm{e}^{\mathrm{i}k_e x} - H(\omega)\mathrm{e}^{-\mathrm{i}k_e x} \right], \end{aligned} \tag{7}$$

where A_s is the cross-sectional area of the shell (m^2) and E_x is the effective longitudinal modulus of the shell (N/m^2). The known parameters in equations (6) and (7) are the location of the accelerometers and force transducers (x) and the frequency of excitation (ω). Although the effective longitudinal modulus (E_x) is unknown, equation (7) will be rewritten as a ratio of forces permitting the cancellation of this term and the cross-sectional area (A_s). Additionally, the coefficients $G(\omega)$ and $H(\omega)$ are unknown; however, they will be condensed out of the mathematical relationships. The inversion of equations (6) and (7) at the sensor locations will allow for a measurement of the unknown extensional wavenumber k_e and extensional wave speed c_e. This technique is described next.

In the extensional wave measurement part of the experiment, the forward and aft pair of sensors (impedance heads) described earlier collect data that are in the form of transfer functions between each pair. The position of the forward pair of sensors is defined as $x = 0$ and of the aft pair as $x = L$, where L is the length of the shell (m). The two transfer function measurements used are the forward displacement divided by the aft displacement and the forward force divided by the aft force. Their theoretical form can be rewritten using equations (6) and (7) as

$$\frac{U(0,\omega)}{U(L,\omega)} = \frac{G+H}{Ge^{ik_eL} + He^{-ik_eL}} = R_1 \tag{8}$$

and

$$\frac{F(0,\omega)}{F(L,\omega)} = \frac{G-H}{Ge^{ik_eL} - He^{-ik_eL}} = R_2 \ , \tag{9}$$

where R_1 and R_2 are transfer function data from the experiment at a specific test frequency. Equations (8) and (9) are rewritten as functions of H divided by G and are set equal to each other, yielding

$$\cos(k_eL) = \frac{R_2R_1 + 1}{R_2 + R_1} = \phi \ , \tag{10}$$

where ϕ is a complex quantity. Using an angle-sum relationship on the complex cosine term in equation (10) and separating the equation into real and imaginary parts results in

$$\cosh[\mathrm{Im}(k_e)L] = \frac{\mathrm{Re}(\phi)}{\cos[\mathrm{Re}(k_e)L]} \tag{11}$$

and

$$\sinh[\mathrm{Im}(k_e)L] = \frac{-\mathrm{Im}(\phi)}{\sin[\mathrm{Re}(k_e)L]} \ , \tag{12}$$

where Re denotes the real part and Im denotes the imaginary part of the corresponding complex quantity.

Equation (12) is now squared and subtracted from the square of equation (11), yielding

$$\{\cosh[\mathrm{Im}(k_e)L]\}^2 - \{\sinh[\mathrm{Im}(k_e)L]\}^2 = \tag{13}$$

$$\frac{[\mathrm{Re}(\phi)]^2}{\{\cos[\mathrm{Re}(k_e)L]\}^2} - \frac{[\mathrm{Im}(\phi)]^2}{\{\sin[\mathrm{Re}(k_e)L]\}^2} = 1 \ ,$$

Equation (13) can be simplified with trigonometric power relationships to

$$\cos[2\,\mathrm{Re}(k_e)L] = s =$$

$$[\mathrm{Re}(\phi)]^2 + [\mathrm{Im}(\phi)]^2 - \left\{\left([\mathrm{Re}(\phi)]^2 + [\mathrm{Im}(\phi)]^2\right)^2 - 2[\mathrm{Re}(\phi)]^2 + 2[\mathrm{Im}(\phi)]^2 + 1\right\}^{1/2} \tag{14}$$

Note that only a negative sign in front of the square root is used. The real part of k_e in equation (14) is now solved for by

$$\mathrm{Re}(k_e) = \begin{cases} \dfrac{1}{2L}\mathrm{Arc}\cos(s) + \dfrac{n\pi}{2L} & n \text{ even} \\ \dfrac{1}{2L}\mathrm{Arc}\cos(-s) + \dfrac{n\pi}{2L} & n \text{ odd} \end{cases} \ , \tag{15}$$

where n is a nonnegative integer and capital A denotes the principal value of the inverse cosine function. The value of n is determined from the function s, which is a cosine function with respect to frequency. At zero frequency, n is 0. Every time s cycles through π radians, n is increased by 1. The imaginary part of k_e is determined by adding equations (11) and (12) together, resulting in

$$\mathrm{Im}(k_e) = \frac{1}{L}\log_e\left\{\frac{\mathrm{Re}(\phi)}{\cos[\mathrm{Re}(k_e)L]} - \frac{\mathrm{Im}(\phi)}{\sin[\mathrm{Re}(k_e)L]}\right\} \ . \tag{16}$$

Now that the real and imaginary parts of the wavenumber k_e are known, the complex-valued extensional wave speed can be determined at each frequency by

$$c_e = \mathrm{Re}(c_e) + \mathrm{i}\,\mathrm{Im}(c_e) = \frac{\omega}{k_e} \ . \tag{17}$$

Note that the extensional wave speed has been measured without knowing the boundary conditions at $x = 0$ and $x = L$.

Once the extensional wave speed is known, five independent, equally-spaced measurements of the spatial pressure field are needed to eliminate the wave propagation coefficients and solve for the breathing wave speed (Hull 1996b). Without loss of generality, the origin of the coordinate system is defined as $x = 0$ at the middle (third) pressure sensor (hydrophone). Equation (4) is written to correspond to the locations of the five pressure sensors as

$$\frac{P(-2\delta,\omega)}{\ddot{U}} = S_1 = \tag{18}$$

$$Ae^{-ik_e2\delta} + Be^{ik_e2\delta} + Ce^{-ik_b2\delta} + De^{ik_b2\delta} \ ,$$

$$\frac{P(-\delta,\omega)}{\ddot{U}} = S_2 = Ae^{-ik_e\delta} + Be^{ik_e\delta} + Ce^{-ik_b\delta} + De^{ik_b\delta} , \quad (19)$$

$$\frac{P(0,\omega)}{\ddot{U}} = S_3 = A + B + C + D , \quad (20)$$

$$\frac{P(\delta,\omega)}{\ddot{U}} = S_4 = Ae^{ik_e\delta} + Be^{-ik_e\delta} + Ce^{ik_b\delta} + De^{-ik_b\delta} , \quad (21)$$

and

$$\frac{P(2\delta,\omega)}{\ddot{U}} = S_5 = Ae^{ik_e2\delta} + Be^{-ik_e2\delta} + Ce^{ik_b2\delta} + De^{-ik_b2\delta} , \quad (22)$$

where δ is the sensor-to-sensor spacing (m) and S_1 through S_5 correspond to the measured transfer function data of the hydrophone divided by the forward accelerometer at a specific frequency.

Equations (19) and (21) are now added together to yield

$$(A+B)\cos(k_e\delta) + (C+D)\cos(k_b\delta) = (1/2)(S_2 + S_4) , \quad (23)$$

and equations (18) and (22) are added together to produce

$$(A+B)\cos(k_e2\delta) + (C+D)\cos(k_b2\delta) = (1/2)(S_1 + S_5) . \quad (24)$$

Equation (20) is then rewritten with the term $(A + B)$ on the left-hand side and is substituted into equations (23) and (24), yielding

$$C + D = \frac{(1/2)(S_2 + S_4) - S_3\cos(k_e\delta)}{\cos(k_b\delta) - \cos(k_e\delta)} \quad (25)$$

and

$$C + D = \frac{(1/2)(S_1 + S_5) - S_3\cos(k_e2\delta)}{\cos(k_b2\delta) - \cos(k_e2\delta)} , \quad (26)$$

respectively. Equations (25) and (26) are next set equal to each other. Applying a double-angle trigonometric relationship to the $\cos(k_b2\delta)$ term then produces

$$X\cos^2(k_b\delta) + Y\cos(k_b\delta) + Z = 0 , \quad (27a)$$

where

$$X = (S_2 + S_4) - 2S_3\cos(k_e\delta) , \quad (27b)$$

$$Y = S_3\cos(k_e\delta) - (1/2)(S_1 + S_5) , \quad (27c)$$

and

$$Z = [(1/2)(S_1 + S_5) + S_3]\cos(k_e\delta) - (1/2)(S_2 + S_4)\cos(k_e2\delta) - (1/2)(S_2 + S_4) . \quad (27d)$$

Equation (27) is a quadratic form with the solution

$$\cos(k_b\delta) = \frac{-Y \pm \sqrt{Y^2 - 4XZ}}{2X} = \psi , \quad (28)$$

where ψ is a complex quantity. Rewriting equation (28) as real and imaginary terms produces

$$\cos[2\,\mathrm{Re}(k_b)\delta] = r = [\mathrm{Re}(\psi)]^2 + [\mathrm{Im}(\psi)]^2 - \left\{ \left([\mathrm{Re}(\psi)]^2 + [\mathrm{Im}(\psi)]^2 \right)^2 - 2[\mathrm{Re}(\psi)]^2 + 2[\mathrm{Im}(\psi)]^2 + 1 \right\}^{1/2} . \quad (29)$$

Note that only a negative sign in front of the square root is used in equation (29). However, both the negative and the positive signs in front of the radical in equation (28) are needed. The real part of k_b in equation (29) is now solved for by

$$\mathrm{Re}(k_b) = \begin{cases} \dfrac{1}{2\delta}\mathrm{Arc}\cos(r) + \dfrac{m\pi}{2\delta} & m \text{ even} \\ \dfrac{1}{2\delta}\mathrm{Arc}\cos(-r) + \dfrac{m\pi}{2\delta} & m \text{ odd} \end{cases} , \quad (30)$$

where m is a nonnegative integer and capital A denotes the principal value of the inverse cosine function. The value of m is determined from the function r, which is a cosine function with respect to frequency. At zero frequency, m is 0. Every time r cycles through π radians, m is increased by 1. The imaginary part of k_b is determined from equation (28), resulting in

$$\mathrm{Im}(k_b) = \frac{1}{\delta}\log_e\left\{ \frac{\mathrm{Re}(\psi)}{\cos[\mathrm{Re}(k_b)\delta]} - \frac{\mathrm{Im}(\psi)}{\sin[\mathrm{Re}(k_b)\delta]} \right\} . \quad (31)$$

Now that the real and imaginary parts of the wavenumber k_b are known, the complex-valued breathing wave speed can be determined at each frequency with

$$c_b = \mathrm{Re}(c_b) + \mathrm{i}\,\mathrm{Im}(c_b) = \frac{\omega}{k_b} \; . \qquad (32)$$

Use of this method produces two wave speed measurements because of the retention of the positive and negative signs in equation (28). One of the wave speeds is the breathing wave speed, and the other is the extensional wave speed, which was previously known. The extensional wave speed is typically at least one order of magnitude greater than the breathing wave speed.

4 ESTIMATION OF WAVE PROPAGATION COEFFICIENTS

Four hydrophone measurements are now used to estimate the wave propagation coefficients. Equation (4) is rewritten at the different hydrophone measurement locations as

$$\frac{P(x_{1:4},\omega)}{\ddot{U}} = T_{1:4} = A\mathrm{e}^{\mathrm{i}k_b x_{1:4}} + B\mathrm{e}^{-\mathrm{i}k_b x_{1:4}} + C\mathrm{e}^{\mathrm{i}k_e x_{1:4}} + D\mathrm{e}^{-\mathrm{i}k_e x_{1:4}} \; , \qquad (33)$$

where $T_{1:4}$ are the transfer function data for a specific frequency at hydrophones 1, 2, 3, and 4 divided by the forward accelerometer. Unlike the estimation of the breathing wave speed, it is not necessary that the spatial distances x_1 through x_4 be equally spaced. Equation (33) is now written in matrix form as

$$\begin{bmatrix} \mathrm{e}^{\mathrm{i}k_b x_1} & \mathrm{e}^{-\mathrm{i}k_b x_1} & \mathrm{e}^{\mathrm{i}k_e x_1} & \mathrm{e}^{-\mathrm{i}k_e x_1} \\ \mathrm{e}^{\mathrm{i}k_b x_2} & \mathrm{e}^{-\mathrm{i}k_b x_2} & \mathrm{e}^{\mathrm{i}k_e x_2} & \mathrm{e}^{-\mathrm{i}k_e x_2} \\ \mathrm{e}^{\mathrm{i}k_b x_3} & \mathrm{e}^{-\mathrm{i}k_b x_3} & \mathrm{e}^{\mathrm{i}k_e x_3} & \mathrm{e}^{-\mathrm{i}k_e x_3} \\ \mathrm{e}^{\mathrm{i}k_b x_4} & \mathrm{e}^{-\mathrm{i}k_b x_4} & \mathrm{e}^{\mathrm{i}k_e x_4} & \mathrm{e}^{-\mathrm{i}k_e x_4} \end{bmatrix} \begin{Bmatrix} A \\ B \\ C \\ D \end{Bmatrix} = \begin{Bmatrix} T_1 \\ T_2 \\ T_3 \\ T_4 \end{Bmatrix} . \qquad (34)$$

The coefficients are then solved for by multiplication of each side of equation (34) by a matrix inverse, which results in

$$\begin{Bmatrix} A \\ B \\ C \\ D \end{Bmatrix} = \begin{bmatrix} \mathrm{e}^{\mathrm{i}k_b x_1} & \mathrm{e}^{-\mathrm{i}k_b x_1} & \mathrm{e}^{\mathrm{i}k_e x_1} & \mathrm{e}^{-\mathrm{i}k_e x_1} \\ \mathrm{e}^{\mathrm{i}k_b x_2} & \mathrm{e}^{-\mathrm{i}k_b x_2} & \mathrm{e}^{\mathrm{i}k_e x_2} & \mathrm{e}^{-\mathrm{i}k_e x_2} \\ \mathrm{e}^{\mathrm{i}k_b x_3} & \mathrm{e}^{-\mathrm{i}k_b x_3} & \mathrm{e}^{\mathrm{i}k_e x_3} & \mathrm{e}^{-\mathrm{i}k_e x_3} \\ \mathrm{e}^{\mathrm{i}k_b x_4} & \mathrm{e}^{-\mathrm{i}k_b x_4} & \mathrm{e}^{\mathrm{i}k_e x_4} & \mathrm{e}^{-\mathrm{i}k_e x_4} \end{bmatrix}^{-1} \begin{Bmatrix} T_1 \\ T_2 \\ T_3 \\ T_4 \end{Bmatrix} . \qquad (35)$$

5 EXPERIMENT

Use of this model with the inverse method corresponds to the physical testing configuration in the Axial Vibration Test Facility (AVTF) at the Naval Undersea Warfare Center Division, Newport, Rhode Island. The AVTF has been designed to provide a simple procedure for testing long structures under varying tensions and temperatures. The longitudinal shaker at the forward end of the structure provides axial excitation. A rope attached to the aft end and a winch allows the tension to be adjusted. A mass is inserted between the shell and the rope to increase the force levels and decrease the acceleration levels. This mass also produces an impedance change at the end of the cylinder that is sufficiently large to allow accurate modeling of the rope behavior by a spring and damper rather than by a continuous media expression. A rail from which thin Kevlar lines can be hung to provide lateral support to heavy or long test specimens runs the entire length of the facility. The unit is completely surrounded by an air-conditioned PVC duct to permit temperature-dependent testing. To collect data during a test, impedance heads are attached to the forward and aft ends of the structure (as described earlier). Each impedance head consists of a single axial force transducer and an accelerometer. Additional data are collected by hydrophones in the liquid-filled shell. A load cell that measures the tension on the structure is located between the rope and the mass.

In the first part of the experiment, which measures the extensional wave speed, a longitudinally-reinforced, liquid-filled urethane shell containing the five equally-spaced hydrophones used in this experiment was placed in tension. The shell had a mean radius of 0.015 m and a thickness of 0.0028 m; the internal liquid had a density of 760 kg/m^3. The axial tension on the shell was 890 N and the stressed length was 12.0 m. The point mass had a weight of 13.6 kg. The data from the force transducers, accelerometers, and hydrophones were acquired in the time domain with a Hewlett Packard (HP) 3562 dynamic signal analyzer. The analyzer then Fourier transformed the raw data to the frequency domain to obtain the desired transfer functions. The test was run with a frequency range between 3 and 100 Hz.

Equations (8)-(17) were applied to the experimental test data taken with the force transducers and accelerometers, and the resulting extensional wave speed of the structure was found. Figure 1 shows the calculated extensional wave speed versus frequency. The solid line corresponds to the computed wave speed and the dashed line to the ordinary least squares (OLS) straight-line fit. The OLS fit was applied to the data between 35 and 100 Hz to minimize the effect of the breathing wave interaction seen at lower frequencies. The resulting

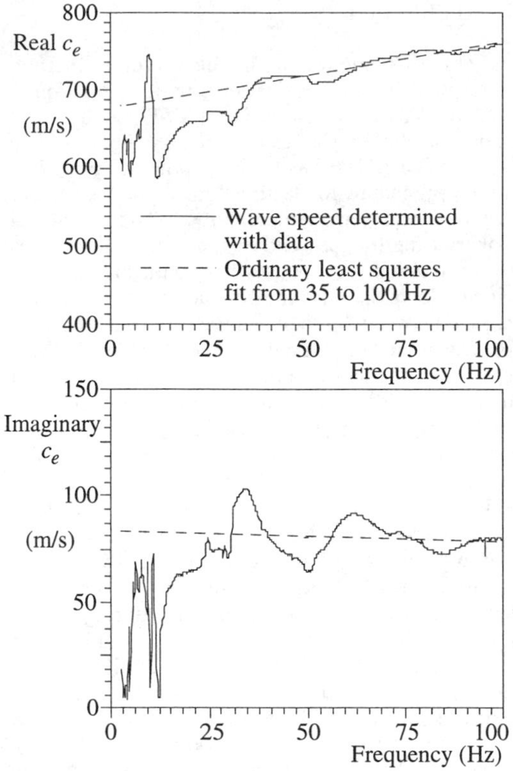

Figure 1. Extensional wave speed versus frequency

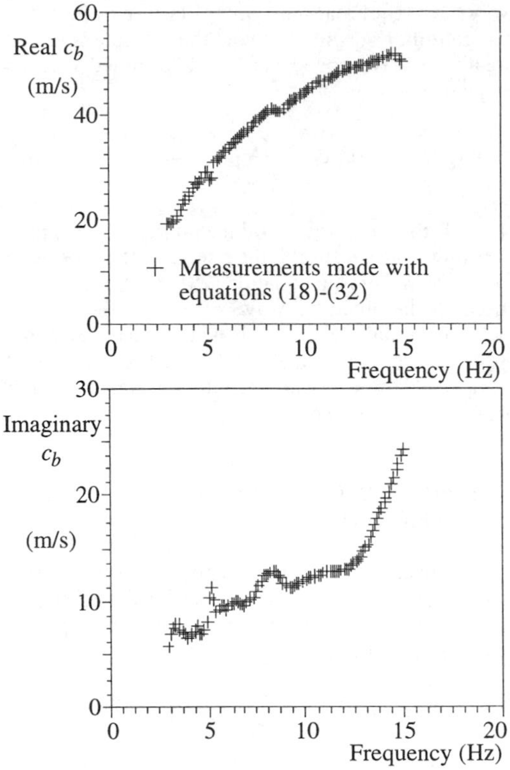

Figure 2. Breathing wave speed versus frequency

OLS fit was $c_e = 677.3 + 0.8f$ (m/s) for the real part and $c_e = 83.3 - 0.05f$ (m/s) for the imaginary part (where f is frequency in Hertz). These same values are used for the extensional wave speed in the following calculations in order to determine the breathing wave speed and estimate the wave propagation coefficients.

In the second part of the experiment, transfer functions of the hydrophone data divided by the forward accelerometer data were collected. The five hydrophones were spaced at intervals of 1.83 m, with a distance of 2.96 m from the forward end of the shell to the first hydrophone. Equations (18)-(32) were applied to the experimental test data, and the breathing wave speed of the structure was calculated. Figure 2 is a plot of the breathing wave speed versus frequency. The plus symbols mark the data from this experiment that were obtained with the method developed in section 3. In the figure, it is noted that the breathing wave for this specific structure is spatially coherent from 3 to 13 Hz. Above 13 Hz, the imaginary part of the measurement begins to diverge. It is possible that effects other than spatial incoherence are preventing the breathing wave from being identified at higher frequencies. Changing the extensional wave speed by ±20 percent and recalculating the breathing wave speed produced a change of less than 1 percent. Thus, the breathing wave speed measurement method is very insensitive to incorrect extensional wave speeds. Additionally, the breathing wave speed was insensitive to varying axial tensions.

Once the extensional and breathing wave speeds are measured, equations (33)-(35) can be applied to the hydrophone data. Although four hydrophone measurements are sufficient to define the empirical pressure field, they are insufficient to verify the accuracy of the model. This is because the model is defined by the four data points used to create it. To resolve the problem of determining the accuracy of the model, 12 hydrophones were placed in the shell and were spaced at intervals of 0.5 m, with a distance of 3.0 m from the forward end of the shell to the first hydrophone. The signal-to-noise ratio of each hydrophone and the forward accelerometer was approximately 60 dB.

Figures 3 and 4 are plots of the pressure field versus distance at 5.4 and 61 Hz, respectively, for both the experimental data and the model. The solid line represents the model of the pressure field and is

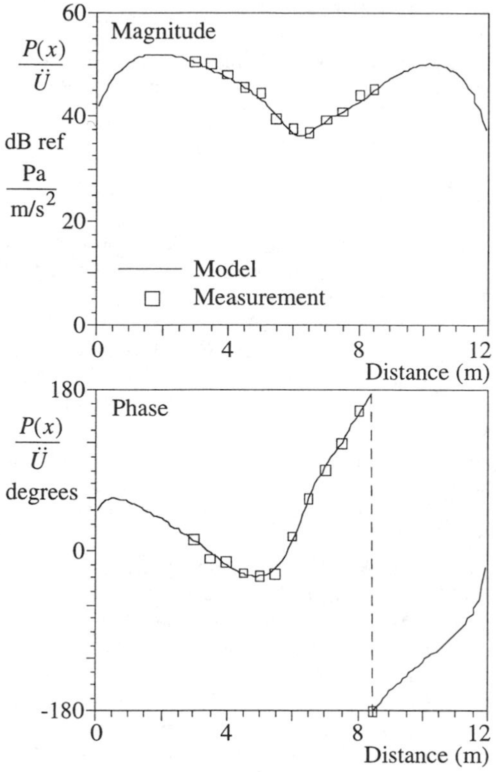

Figure 3. Pressure field versus distance at 5.4 Hz

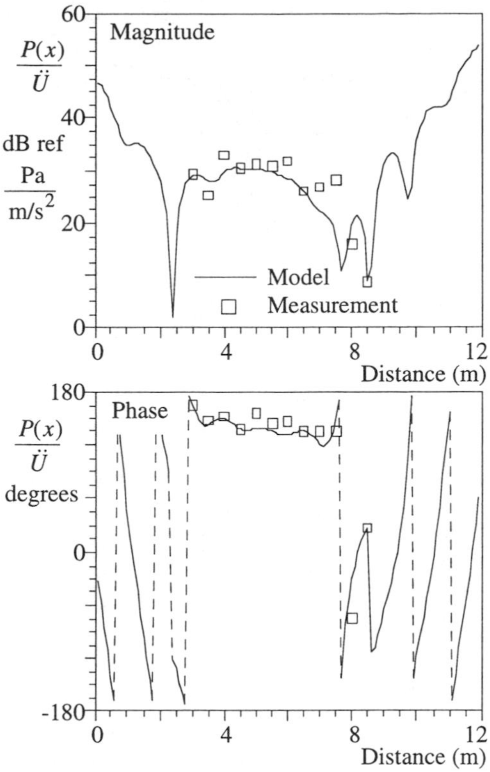

Figure 4. Pressure field versus distance at 61 Hz

based on the spatial locations corresponding to the first (x_1 = 3.0 m), fourth (x_2 = 4.5 m), eighth (x_3 = 6.5 m), and twelfth (x_4 = 8.5 m) hydrophone sensors. The extensional wave speeds used in deriving the model were determined from figure 1, and the breathing wave speeds from figure 2. For the region above 15 Hz, the breathing wave speed was extrapolated.

Predictions from the model were compared to the experimental data at various frequencies. The formula used for this comparison, in decibels, was

$$\beta(\omega) = 20 \log_{10}\left(\frac{1}{N}\right)\sum_{i=1}^{N}\left|\frac{|h_i(\omega)| - |y_i(\omega)|}{\bar{y}(\omega)}\right| , \tag{36}$$

where

$$\bar{y}(\omega) = \frac{1}{N}\sum_{i=1}^{N}|y_i(\omega)| . \tag{37}$$

The term i is the ith member of the corresponding vector and N is the total number of measurements (N = 12). The average difference between the models and the measurements (equation (36)) at various frequencies is listed in table 1.

The model is extremely accurate from 3 to approximately 13 Hz. The method is moderately accurate up to approximately 27 Hz. From 27 to 50 Hz, there is a difference of about 4 dB between the experiment and the model. Because the model is being defined by the experimental values and the coherence of the measurements are almost unity, this difference is larger than expected. There could be a number of reasons why the divergence occurs:

- A transverse wave is introducing unmodeled energy into the measurements.
- The wave speeds are not consistent across the length of the shell.
- The hydrophones themselves are creating significant breathing wave energy sources in this frequency range.
- The transverse support lines are a noise source.
- The spatial coherence length of the breathing wave generated at the aft bulkhead is causing a situation where some breathing wave energy is being input into the aft hydrophones but not into the forward ones.

Table 1. Difference between model and measurement in decibels

Frequency (Hz)	Difference (β) (dB)
3.0	0.6
7.9	0.4
12.7	0.6
17.5	1.2
22.4	0.9
27.2	1.2
32.1	3.6
37.0	5.1
41.8	4.8
46.7	3.3
51.5	2.9
56.3	2.7
61.2	2.3
66.1	2.4
70.9	2.0
75.8	1.9
80.6	1.7
85.4	1.4
90.3	1.8
95.2	2.2
100.0	1.8

- Nonaxisymmetric effects of wave motion are being detected by the hydrophones.
- The hydrophone mounts are transmitting unmodeled energy into the hydrophones.

The model above 50 Hz shows a difference from the measured values of about 2 dB, which is considered borderline for accuracy in this particular experiment. Additionally, the model is creating pressure levels that are too high in the spatial region around the aft bulkhead. The phase angle associated with the model at the aft bulkhead is incorrect in the 50- to 100-Hz frequency range.

6 CONCLUSIONS AND RECOMMENDATIONS

A two-wave empirical model of the pressure field in a liquid-filled cylindrical shell can be derived based on four hydrophone measurements in the liquid. An inverse method is used to derive this pressure field and the associated parameters. It is found that in the frequency range of 3 to 30 Hz and 50 to 100 Hz, the model was accurate. It is likely that an unmodeled noise source is contributing to the pressure field from 30 to 50 Hz. It is recommended that the experiment be rerun with shells of varying diameters, stiffness, and length to determine if the differences between the model and the experiment are shell dependent.

7 ACKNOWLEDGMENTS

The author would like to thank Karen A. Holt for her help with the paper. This work was sponsored by the Office of Naval Research.

REFERENCES

Chow, J.C.F. & J.T. Apter 1968. Wave propagation in a viscous incompressible fluid contained in flexible viscoelastic tubes. *Journal of the Acoustical Society of America.* 44(2): 437-443.

Hull, A.J. 1996a. An inverse method to measure the axial modulus of composite materials under tension. *Journal of Sound and Vibration.* 195(4): 545-551.

Hull, A.J. 1996b. An inverse method to measure the breathing wave speed in a liquid-filled cylindrical shell. Naval Undersea Warfare Center Technical Report 11,093.

Jameson, P.W., S.A. Africk, D.M. Chase, & E.C.H. Schmidt 1974. Investigation of towed-array self-noise mechanisms by analytical models and laboratory experiments. Bolt Beranek and Newman Inc. Report No. 2819.

Morgan, G.W. & J.P. Kiely 1954. Wave propagation in a viscous liquid contained in a flexible tube. *Journal of the Acoustical Society of America.* 26(3): 323-328.

Modern Practice in Stress and Vibration Analysis, Gilchrist (ed.) © *1997 Balkema, Rotterdam, ISBN 90 5410 896 7*

Detection of faults in carbon fibre reinforced plates using Lamb wave based novelty detection

K.Worden, W.J.Staszewski & G.R.Tomlinson
Department of Mechanical Engineering, University of Sheffield, UK
S.G.Pierce, W.R.Philp & B.Culshaw
Department of Electronic and Electrical Engineering, University of Strathclyde, Glasgow, UK

ABSTRACT: This paper describes the application of Lamb wave testing to the location of defects in a carbon fibre composite panel. Fundamental symmetric (S_0) Lamb waves were initiated in the sample using a perspex wedge coupling technique to match the incident compressional wavefield to the S_0 mode. A broadband optical fibre detector was used to monitor the Lamb waves in a geometry such that the outgoing, defect reflected and backwall echos of the S_0 mode could all be monitored. The sample plate contained centre line defects comprising a delamination and a resin rich area both of 20mm diameter. Data processing based on a novelty measure allowed both defects to be clearly resolved.

1 INTRODUCTION

Lamb wave testing of composite plates has been discussed by numerous authors (Alleyne & Cawley 1992a,b,1996, Guo & Cawley 1994, Jansen *et al* 1994, Pierce *et al* 1996a), describing the detection of various material failures including delaminations, fibre fracture and matrix cracking. The technique is complementary to the conventional C scan, as it divides the scanned area into a series of strips rather than performing a point by point measurement. There is therefore scope for reduction of the inspection time, albeit at the likely sacrifice of defect resolution, in comparison to C scanning. A significant practical complication of Lamb wave inspection lies in the propagation characteristics of the elastic waves. The propagation is typically characterised by the product of the wave frequency and the sample thickness, or *frequency-thickness product* (FT). For low values of this parameter (typically, FT $<$ 1 MHz.mm in Aluminium), only two modes can propagate, the fundamental symmetric (S_0) and the fundamental antisymmetric (A_0). As the frequency-thickness product increases, so does the number of allowed modes. In general, these modes will have widely different phase velocities and will often display considerable phase velocity dispersion as a function of frequency-thickness product. Clearly Lamb wave testing is considerably simplified if a known single mode is launched over a non-dispersive region of the FT product.

The information about defects in the plate is encoded in the waves scattered by the faults. Because the time-variation of the wave intensity at the fibre-optic is quite complex due to the presence of boundaries, low reflection coefficients from the fault and the uncertain coupling between the transducer and the plate, it is advantageous to use an automatic pattern recognition technique to signal anomalous (faulted) condition.

Neural networks have proved to be extremely powerful tools for pattern recognition (Bishop 1995) and they are adopted here for the fault diagnosis. There are essentially two methods of application. In the first case, if detailed models of the plate and defect are available, one can conceivably simulate the Lamb wave patterns corresponding to normal condition and various faults. If such *training data* are available, a neural network can be trained to provide a fault classification on the basis of experimentally measured data. This option was considered intractable due to the complex nature of Lamb wave interactions with say, delaminations in composite plates. The second method is based on the idea of *novelty* or anomaly detection; the network is trained only on normal condition data and is required to signal if there is a deviation from this condition. Such techniques can not provide classification information. The advantage of adopting

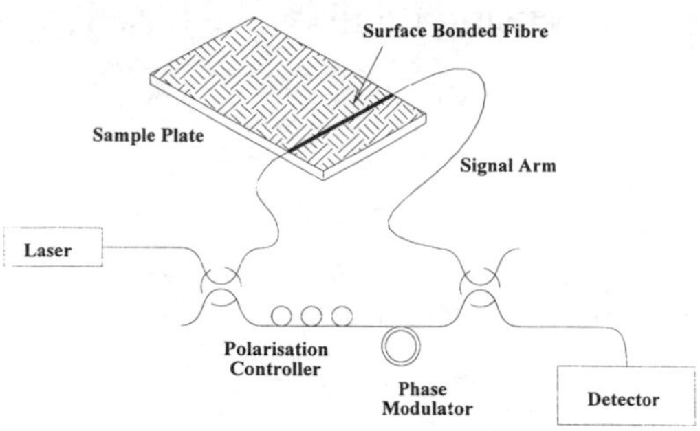

Figure 1. Optical fibre Mach-Zehnder interferometer.

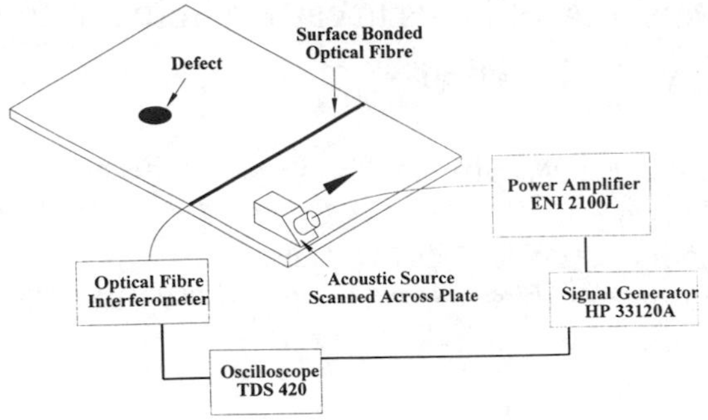

Figure 2. Experimental arrangement for Lamb wave scanning of sample plates.

the second approach is that the normal condition data can be taken from experiment and so the need for a detailed *a priori* model is removed. Examples of the use of novelty detection in Medical diagnosis can be found in (Tarassenko *et al*), case studies for Engineering fault detection can be found in (Worden 1997, Surace *et al* 1997).

The layout of the paper is as follows. The experimental set-up is described in the following section. Section Three introduces the neural network used to produce the novelty detector, and Section Four describes how the network is applied. Section Five discusses the results of the procedure.

2. EXPERIMENTAL ARRANGEMENT

The use of optical fibre interferometers for the monitoring of Lamb waves has been investigated in depth by the authors (Pierce *et al* 1996b). By subjecting the signal arm of a stabilised Mach-Zenhder interferometer to the ultrasonic field, the stress-induced modulation of the refractive index of the optical fibre, leads to intensity fluctuation at the interferometer output. A sensitive photo-receiver thus allows observation of the acoustic wavefield. Advantages of the optical fibre receiver are high fidelity, wide bandwidth response, high electromagnetic noise immunity and structural compatibility with composite materials especially when embedded (Pierce *et al* 1996b). The main drawback associated with the detection scheme is that since broadband detection is used, the sensitivity is lower than that associated with conventional narrowband piezoelectric transducers. It has been found that whilst gross defects like holes can be found with relative ease (Pierce *et al* 1996a), the identification of more realistic defects like delaminations can be difficult due to low reflection coefficients. Use of conventional signal processing techniques (such as correlation with the outgoing wave pulse) and consideration of acoustic wave mode conversion at the defect allowed some enhancement in delamination detection (Pierce *et al* 1996a), but with considerable scope for improvement.

Figure 1 shows details of the optical Mach-Zehnder interferometer. A Helium-Neon laser, launched into single-mode fibre was split at a coupler into signal and reference fibres. The signal arm was surface-bonded on the carbon fibre plate under test, the dimensions of which were 470mm wide by 430mm long. The reference arm contained polarisation control for maximising interference fringe visibility, and a phase modulator which was used to stabilise the interferometer against low-frequency drifts. Propagating Lamb waves in the sample plate caused a phase modulation in the signal arm which was demodulated by the Mach-Zehnder into an intensity change monitored by the detector. For defect detection, the arrangement of Figure 2 was adopted. Fundamental symmetric (S_0) Lamb waves were launched using a perspex wedge (Pierce *et al* 1996a), angled for appropriate phase matching into the S_0 mode. The source was driven by a 1 cycle toneburst from the signal generator subsequently amplified by the broadband RF amplifier (ENI 2100L). The signal fibre of the interferometer was bonded across the full plate width using a quick-drying varnish. (Figure 3). The fibre was positioned between the source and the known line of defects; in this way the outgoing S_0 wave could be monitored following reflections from the defects and the far edge of the plate. Signals from the interferometer were monitored by a digital storage oscilloscope from where they could be transferred to a PC for storage. In operation,

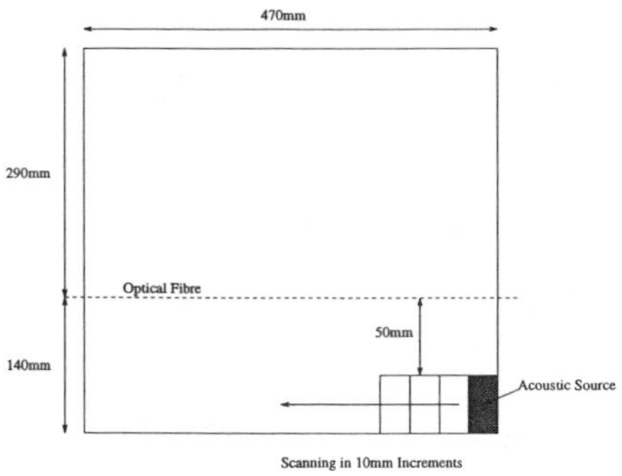

Figure 3. Details of Carbon fibre plate.

the acoustic source was scanned across the plate width (470mm) in 10mm increments and a waveform recorded from each source position. Reflections would vary as the source was scanned thus identifying the strips that contained the defects. The acoustic source was driven at a centre frequency of 250kHz, which combined with a plate thickness of nominally 3mm gave an FT product of less than 1 MHz.mm such that only the S_0 and A_0 modes could propagate.

The Carbon fibre test plate incorporated two defects. A thin layer of PTFE material of diameter 20mm simulated a delamination at a distance of 75mm from the left-hand plate edge, and a cut section of Carbon mat (again 20mm diameter) simulated a resin rich area at a distance of 225mm from the same edge. Both defects were inserted at the mid-plane of the sample plate. The plate lay-up was $[(0/90),(0,90),(+45/-45),(0/90),(0/90)]_S$, manufactured from T300 Carbon fibre using a Resin Transfer Moulding (RTM) process.

3. THE NEURAL NETWORK

For the sake of completeness, a brief description of the Multi-Layer Perceptron (MLP) follows; for a more detailed discussion, the reader is referred to the seminal work (Rumelhart & McClelland 1988).

The MLP is simply a collection of connected processing elements called nodes or neurons, arranged together in layers (Figure 4). Signals pass into the input layer nodes, progress forward through the network hidden layers and finally emerge from the output layer. Each node i is connected to each node j in its preceding layer through a connection of weight w_{ij}, and similarly to nodes in the following layer. Signals pass through the node as follows: a weighted sum is performed at i of all the

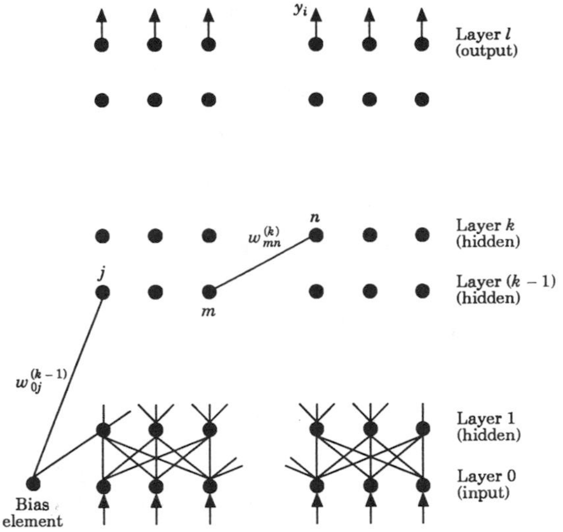

Figure 4. Multi-Layer Perceptron (MLP) neural network structure.

signals x_j from the preceding layer, giving the excitation z_i of the node; this is then passed through a nonlinear *activation function* f to emerge as the output of the node x_i to the next layer i.e.

$$x_i = f(z_i) = f(\sum_j w_{ij} x_j) \qquad (1)$$

Various choices for the function f are possible; the hyperbolic tangent function $f(x) = \tanh(x)$ was used here. One node of the network, the *bias* node is special in that it is connected to all other nodes in the hidden and output layers, the output of the bias node is held constant throughout, in order to allow constant offsets in the excitations z_i of each node.

The first stage of using a network to model an input-output system is to establish the appropriate values for the connection weights w_{ij}. This is the *training* or *learning* phase. Training is accomplished using a set of network inputs for which the desired network outputs are known. At each training step, a set of inputs are passed forward through the network yielding trial outputs which are then compared to the desired outputs. If the comparison error is considered small enough, the weights are not adjusted. If, however, a significant error is obtained, the error is passed *backwards* through the net and a *training algorithm* uses the error to adjust the connection weights. The algorithm used in this work is the *back-propagation* algorithm which uses the parameter update rule,

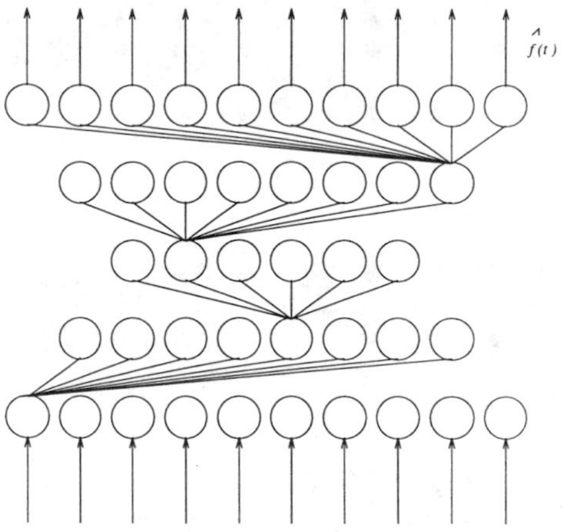

Figure 5. Auto-Associative Network (ANN) structure.

$$w_{ij}^{(m)}(t) = w_{ij}^{(m)}(t-1) + \eta\delta_i^{(m)}(t)x_j^{(m-1)}(t) \quad (2)$$

where $\delta_i^{(m)}$ is the error in the output of the ith node in layer m and t is the index for the iteration. This error is not known *a priori* but must be constructed from the known errors $\delta_i^{(l)} = y_i - \hat{y}_i$ between the network outputs $\hat{y}_i$ and the desired outputs y_i. This is the origin of the name back-propagation. The update used here is modified by the inclusion of an additional *momentum* term which allows previous updates to persist,

$$\triangle w_{ij}^{(m)}(t) = \eta\delta_i^{(m)}(t)x_j^{(m-1)}(t) + \alpha\triangle w_{ij}^{(m)}(t-1) \quad (3)$$

The effect of this extra term is to damp out oscillations in the weight estimates. The coefficients η and α determine the overall speed of learning; unfortunately, there are no hard and fast rules as to their optimum values for a given problem.

Once the comparison error is reduced to an acceptable level over the whole training set, the training phase ends and the network is established.

4. THE NOVELTY INDEX

As described above, the object of novelty detection is to establish if a new pattern differs from previously obtained patterns in some significant respect. The application to on-line damage detection is clear. It is assumed that damage will alter the measured patterns, so novelty will indicate a

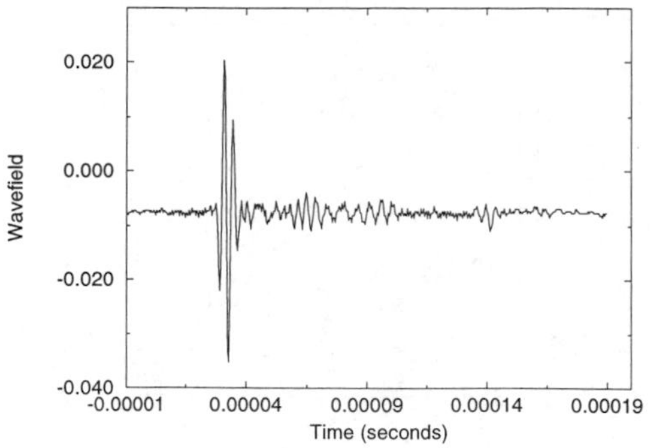

Figure 6. Measured Lamb wave record.

fault. The important point is to identify *significant* changes, i.e. those which can not be attributed to fluctuations in the measured patterns due to noise.

The approach taken here, based on (Pomerleau 1993), is simply to train an Auto-Associative Network (AAN) on the patterns. This means a feed-forward Multi-Layer Perceptron (MLP) network which is asked to reproduce at the output layer, those patterns which are presented at the input. This would be a trivial exercise except that the network structure has a 'bottleneck' i.e. the patterns are passed through hidden layers which have fewer nodes than the input layer (Figure 5). This forces the network to learn the significant features of the patterns; the activations of the smallest, central layer, correspond to a compressed representation of the input. Training proceeds by presenting the network with many versions of the pattern corresponding to normal condition corrupted by noise and requiring a copy at the output.

The novelty index $\nu(\underline{z})$ corresponding to a pattern vector $\underline{z} = z_i, i = 1, \ldots, N$ is then defined as the Euclidean distance between the pattern $\underline{z}$ and the result of presenting it to the network $\hat{\underline{z}}$,

$$\nu(\underline{z}) = ||\underline{z} - \hat{\underline{z}}|| \quad (4)$$

It is clear how this works. If learning has been successful, then $\underline{z} = \hat{\underline{z}}$ for all data in the training set so $\nu(\underline{z}) \approx 0$ if $\underline{z}$ represents normal condition. If $\underline{z}$ corresponds to damage, $\nu(\underline{z})$ is non-zero. Note that there is no guarantee that ν will increase monotonically with the level of damage. This is why novelty detection only gives a yes/no diagnostic.

5. RESULTS

The first stage of producing the novelty detector

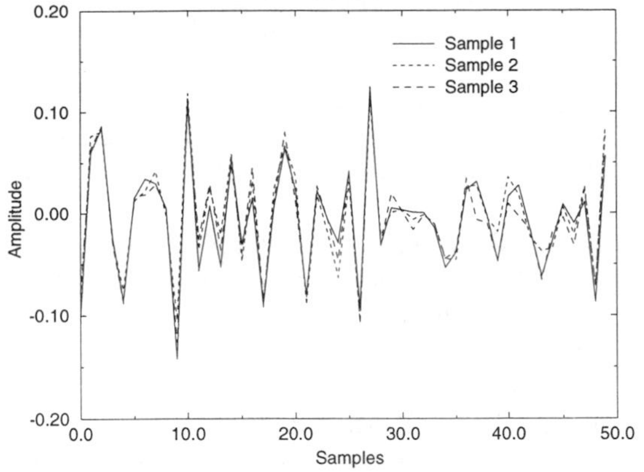

Figure 7. Samples of noise-corrupted training vectors.

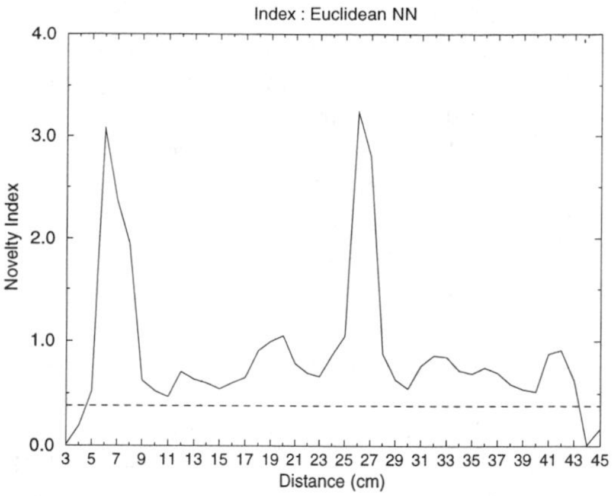

Figure 8. Novelty index for sample plate.

is to establish the pattern or set of measurements used for the diagnosis. In order to obtain the simplest possible diagnostic the raw time data were used with one or two modifications. Figure 6 shows the 500 point Lamb wave record for the leftmost source location. The first pulse shows the first passage of the wave under the fibre and contains no information regarding the defects. The other feature which is expected is the backwall reflection and this occurs at about 0.00014 seconds in Figure 6. The backwall reflection potentially contains useful information as the wave in question would pass through the defect twice. If any direct wave scattering from a defect is present it will occur between these features. In order to compensate for the effect of uncertain coupling between the source and plate, all the patterns - scanned at 10mm increments across the plate - were normalised by dividing by the initial pulse height. The means of the signals were removed in all cases. The final stage of preprocessing was to focus on the time interval of interest by taking only the points between the 125th and the 375th in the record. The resulting vector was decimated by a factor of 5, giving a 50-point feature vector for training the novelty detector.

The second stage of producing the novelty index is training the AA network. The two leftmost and two rightmost data vectors were taken as normal condition. The training set was obtained by making 250 copies of each of the patterns corresponding to normal condition and corrupting each copy with different Gaussian noise vectors (three sample patterns are shown in Figure 7). The object of this exercise is to produce a novelty detector which never fires purely because a measured pattern is noisy. In the absence of any prescription for the noise, the Gaussian process (with unit-proportional covariance matrix) was chosen; a minimal requirement for any pattern recognition system is that it should be transparent to normally distributed noise. In geometrical terms, the assumption here is that the normal condition set in pattern space is spherical.

After the neural network was trained, it was presented with the patterns measured from the plate sequentially from left to right. The resulting novelty index as a function of position is given in Figure 8. Two faults are clearly identified. The first is at 70 ± 10 mm (three index values are high). In excellent agreement with the delamination at 70 mm. One can infer that the delamination extends for 20mm as expected. The second fault is indicated at 260 ± 10mm which shows reasonable agreement with the expected 225 mm. Again a defect extent of 20mm is indicated.

The dotted line in the graph is the novelty threshold computed as in (Worden 1997). This fails to properly characterise the anomalous events although the values on the training data are subthreshold as required. The reason for this is that the probability distribution for novelty values is not unimodal, so the Gaussian statistics applied in (Worden 1997) do not apply.

As an attempted check, a correlation technique (Pierce *et al* 1996b) was applied; however it was unable to resolve the two defects. In fact the expected S_0 reflection coefficients from the mid-plane defects are very small due to the zero shear stress component of the S_0 mode at the mid-plane of the plate.

CONCLUSIONS

A novelty detection technique based on neural network analysis of Lamb wave reflection data has been presented which has proves considerably superior to conventional methods for the detection of certain defects in a particular Carbon fibre plate. The defects are both located and sized accurately by the new method.

ACKNOWLEDGEMENTS

The authors would like to acknowledge very useful conversations with Dr. Peter Cowley and Ms. Rachel Pierce of the Rolls-Royce Applied Science Laboratory at Derby. Also Dr. Lionel Tarassenko of the Department of Engineering Sciences, University of Oxford, helped with the literature.

REFERENCES

Alleyne, D.N. & Cawley, P. 1992a. The interaction of Lamb waves with defects. *IEEE Transactions UFFC.* 39:381-397.

Alleyne, D.N. & Cawley, P. 1992b. Optimisation of Lamb wave inspection techniques. *NDT&E International.* 25:11-22.

Bishop, C.M. 1995. *Neural Networks for Pattern Recognition.* Oxford University Press.

Cawley, P. & Alleyne, D.N. 1996. The use of Lamb waves for the long range inspection of large structures. *Ultrasonics.* 34:287-290.

Guo, N. & Cawley, P. 1994. lamb wave reflection for the quick non-destructive evaluation of large composite laminates. *Materials Evaluation.* 52:404-411.

Jansen, D.P., Hutchins, D.A. & Mottram, J.T. 1994. Lamb wave tomography of advanced composite laminates containing damage. *Ultrasonics.* 32:83-89.

Pierce, S.G., Philp, W.R., Culshaw, B., Gachagan, A., McNab, A., Hayward, G., & Lecuyer, F. 1996. Surface-bonded optical fibres for the inspection of CFRP plates using ultrasonic Lamb waves. *Smart Materials and Structures.* 5:776-787.

Pierce, S.G., Philp, W.R., Gachagan, A., McNab, A., Hayward, G., Culshaw, B. 1996b. Surface-bonded and embedded optical fibres as ultrasonic sensors. *Applied Optics.* 35:5191-5197.

Pomerleau, D. 1993. Input reconstruction reliability estimation. In S.J.Hanson, J.D.Cowan & C.L.Giles (eds), *Advances in Neural Information Processing Systems 5.* Morgan Kaufman Publishers.

Rumelhart, D.E. & McClelland, J.L. 1988 *Parallel Distributed Processing: Explorations in the Microstructure of Cognition (Two Volumes).* MIT press.

Surace, C., Worden, K. & Tomlinson G.R. 1997. A novelty detection approach to diagnose damage in a cracked beam. *Proceedings of the 15th International Conference on Modal Analysis, Orlando, Florida.* 947-953.

Tarassenko, L., Hayton, P., Cerneaz, N. & Brady, M. Novelty detection for the identification of masses in mammograms. *Preprint, Department of Engineering Sciences, University of Oxford.*

Worden, K. 1997. Structural fault detection using a novelty measure. *Journal of Sound and Vibration.* 201:85-101.

Modern Practice in Stress and Vibration Analysis, Gilchrist (ed.) © 1997 Balkema, Rotterdam, ISBN 90 5410 896 7

Parameter identification in a system with both linear and coulomb damping

C. Meskell & J.A. Fitzpatrick
Department of Mechanical and Manufacturing Engineering, Trinity College, Dublin, Ireland

ABSTRACT: A single degree of freedom system with both viscous and coulomb damping has been described and tested using band limited random excitation. The dynamic data obtained was used as the input to a non-linear parameter identification procedure. Values for the linear modal parameters and the coulomb damping parameter were estimated. These values were then used in a simulation of the system. The simulated response was found to be in good agreement with the measured response, thus validating both the identified model and the analysis technique.

1 INTRODUCTION

Coulomb damping (dry friction) is a common form of damping in engineering applications and on it's own is easily analysed; see for example Inman (1994). However, it often occurs with linear viscous damping as well. Den Hartog (1931) described an analytical solution for the steady state response with mixed damping for sinusoidal excitation. Ravindra & Mallik (1993) have extended this by considering a system with a cubic restoring force (*i.e.* a hardening spring).

If the normal force is unknown, as might be the case in a vibration isolator, the coulomb and viscous damping coefficients can be estimated by fitting the response to an analytical solution such as Den Hartog's. However, for continuous systems with several modes of interest or for systems with other non-linear terms, obtaining such a solution may be difficult. Alternatively, the equivalent linear damping can be estimated (*e.g.* Rice 1995, Tan & Rogers 1995). However, the viscous damping coefficient obtained in this manner will be dependent on the excitation amplitude.

Since the system is inherently non-linear, the system parameters could be estimated using one of the many techniques currently available for non-linear system identification, a summary of which can be found in Jezquel & Lamarque (1991). In this paper the procedure described by Rice & Fitzpatrick (1988) has been applied to a single degree of freedom linear system with coulomb damping added. This technique has been used because it is easily implemented with additional non-linearities and it can be readily extended to multi-degree of freedom systems (Rice & Fitzpatrick, 1991).

2 EXPERIMENTAL APPARATUS.

2.1 Test facility.

The test rig consists of a block of aluminium supported by two slender aluminium beams (3mm x 50mm x 500mm) set parallel to each other and rigidly fixed at the other end. This arrangement allows only translational motion perpendicular to the plates. An electromagnetic shaker was used to excite the system. A sketch of the mounting scheme can be seen in figure 1.

Acceleration was measured with a Bruel & Kajer 4370 accelerometer attached to the outside of one beam. A Dantec 55X laser vibrometer was focused on the accelerometer to monitor the velocity. The input voltage to the shaker was recorded as a measure of the excitation force. This was calibrated against a BK 8200 force transducer. All the signals were digitised and logged using a HP35650 data acquisition frame.

A series of random forced tests were conducted to ensure linearity and to locate the modes of the structure. The first natural frequency is at 11.25Hz while the next is at about 100Hz. Therefore, this part of the model can sensibly be regarded as a

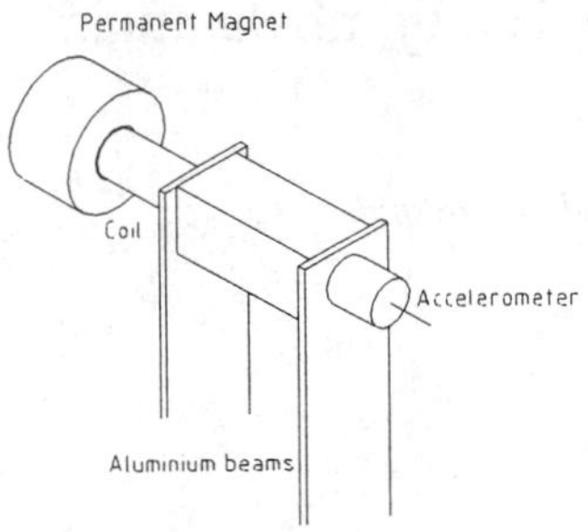

Figure 1 Sketch of linear part of the system.

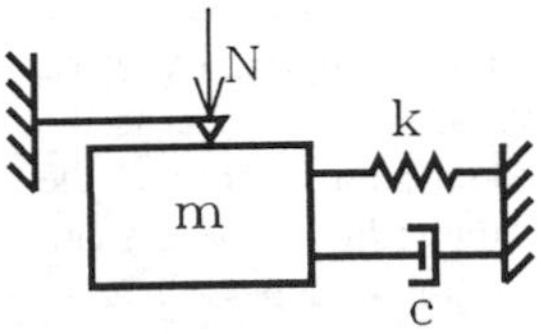

Figure 2 Schematic of system

single degree of freedom mass-spring-damper system.

A sheet of balsa wood was placed in contact with the top of the aluminium block, but fixed rigidly to ground. A mass of *30g* was secured to the top of the balsa to provide a normal downward force. In this way dry friction was added to the system.

2.2 Theoretical model

The system described above can be replaced with the lumped parameter model shown in fig. 2. The equation of motion for this system is

$$m\ddot{x}+c\dot{x}+kx+\mu N\,\mathrm{sgn}(\dot{x})=f \qquad (1)$$

This model excludes the possibility of sticking (i.e. the friction force being non-zero when the mass is at rest). This assumption is justified because for the displacements under consideration the spring force is large in comparison to the static friction.

3 IDENTIFICATION PROCEDURE

The frequency domain technique proposed by Rice & Fitzpatrick (1988) has been used successfully by Esmonde et al. (1992a, 1992b) to validate the model for and identify the elements of a squeeze film system. For identification of a non-linear single degree of freedom model, the procedure is based on reformulating the system as a multiple input single output (MISO) system. In general this is achieved by using the linear and non-linear terms as inputs with the excitation force as the output. In practice, there will always be additional inputs to represent extraneous excitation, line noise and non-linearities which have not been modelled. Once the auto and cross spectra associated with the inputs and the output have been estimated an uncorrelated model can be constructed in which the inputs and output have been conditioned to remove the mutually correlated components, so that the "optimised" transfer functions are calculated directly as:

$$L_{if}=S_{if\cdot(i-1)!}\,/\,S_{ii\cdot(i-1)!}$$

where the conditioned spectra have been computed iteratively from the relationship:

$$S_{ij\cdot r}=S_{ij\cdot(r-1)!}-L_{rj}S_{ir\cdot(r-1)!}$$

for $i,j=r+1,n+1;\ r=1,2..j$ and n is the number of inputs. The original transfer functions can now be computed from the optimised transfer functions using the relationship:

$$H_{if}=L_{if}-\sum_{j=i+1}^{n}L_{if}H_{jf} \quad \text{for } i=(n-1),(n-2)\ \ldots\ 1$$

This back substitution is initiated by setting $H_{nf}=L_{nf}$. The relevance of each successive conditioned input to the overall dynamics of the system can be assessed using the partial coherence function which can be expressed as:

$$\gamma_{if}^{2p}=\frac{\left|S_{if\cdot(i-1)!}\right|^{2}}{S_{ii\cdot(i-1)!}\,S_{ff\cdot(i-1)!}}$$

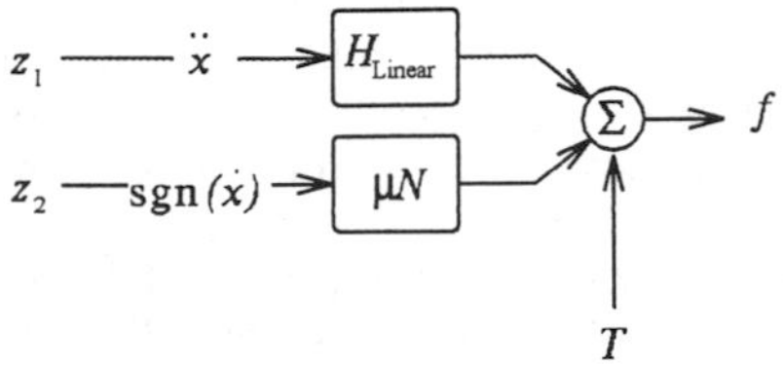

Figure 3 Block diagram of system

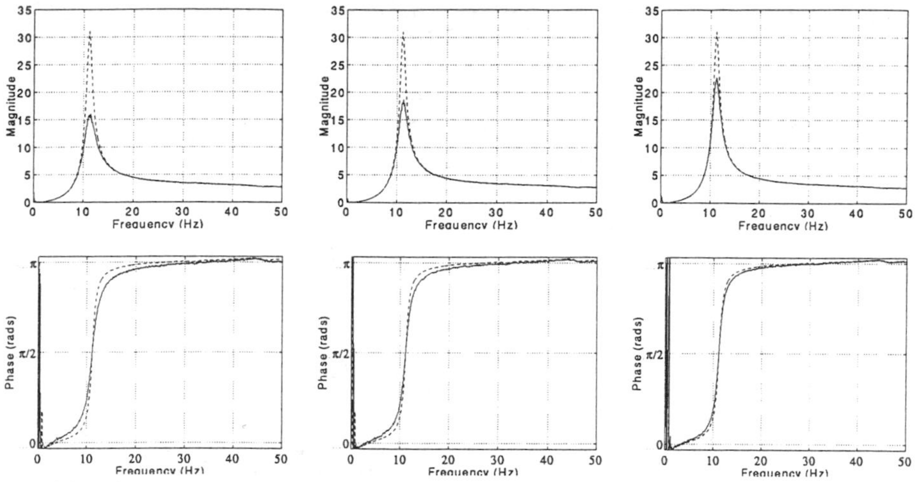

(a) Low excitation (b) Medium excitation (c) High excitation

Figure 4 Linear transfer function (ignoring friction): effective,—; actual, - -

This technique can be applied to the problem under consideration by first reformulating equation 1 as a two input, single output system as shown in figure 3. The inputs z_1 and z_2 are the acceleration and sign of the velocity respectively. The auto and cross spectra associated with z_1, z_2 and f are calculated using a FFT algorithm with 40 averages and 2048 points per average.

For the special case of a two input model the desired transfer functions can be written as:

$$H_1 = \frac{S_{1f}}{S_{11}[1-\gamma_{12}^2]}\left[1-\frac{S_{12}\,S_{2f}}{S_{22}\,S_{1f}}\right]$$

$$H_2 = \frac{S_{2f}}{S_{22}[1-\gamma_{12}^2]}\left[1-\frac{S_{21}\,S_{1f}}{S_{11}\,S_{2f}}\right] \qquad (2)$$

where the ordinary coherence between the inputs is given as

$$\gamma_{12}^2 = \frac{|S_{12}|^2}{S_{11}S_{22}}$$

Bendat and Peirsol (1986) give full details of the process.

4 RESULTS AND DISCUSSION

So as to provide a reference for comparison purposes, the system was tested with no coulomb damping present and the transfer function between force and acceleration was calculated.

Coulomb damping was introduced and the system was excited at three different levels of excitation. The excitation was band limited white noise between *5Hz* and *45Hz*. The RMS values for the three levels of excitation were *0.36N*, *0.45N* and *0.88N*.

4.1 Equivalent linear system.

Figure 4 shows the linear transfer function calculated if the friction forces are ignored. The actual linear transfer function is shown for comparison. It can be seen that the presence of Coulomb damping introduces a noticeable bias to both the magnitude and the phase in the region of the natural frequency. Using the half power point method the equivalent linear damping ratios can be estimated:

Excitation	Damping ratio	Damping coefficient (Ns/m)
Low	.092	4.4
Medium	.074	3.6
High	.062	3.0

It is apparent that the equivalent linear damping approaches the actual linear damping as the excitation level increases. This can be attributed to the fact that the friction force is not proportional to the velocity, whereas the viscous damping force is, and so, as the excitation level increases, the coulomb damping becomes less significant.

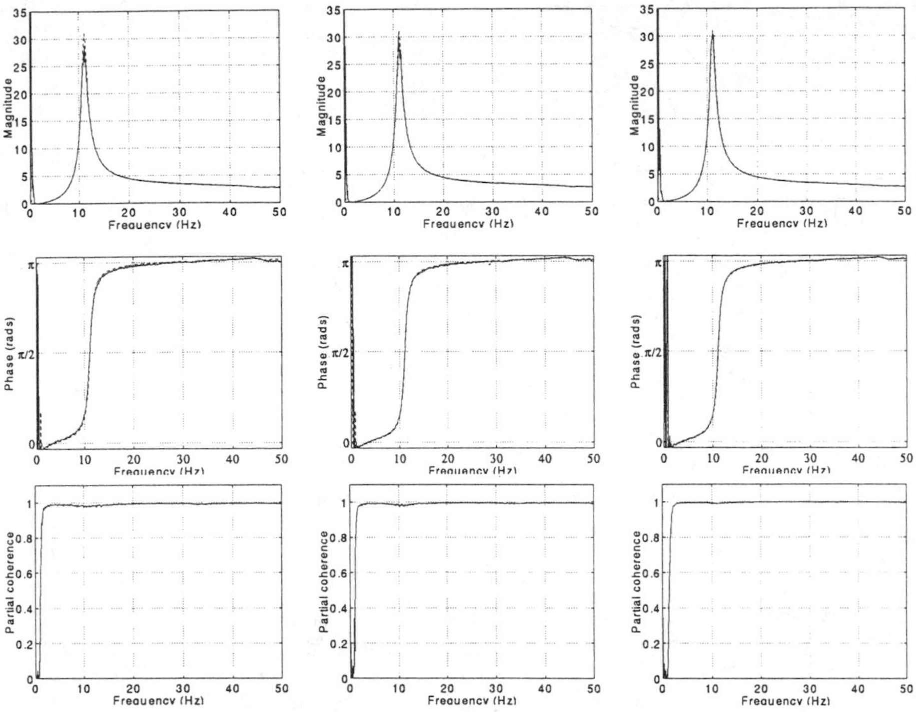

(a) Low excitation (b) Medium excitation (c) High excitation

Figure 5 linear path (using eqn 2) : effective, ___; actual, - -

4.2 Identified non-linear system.

Figure 5 shows the linear transfer function identified as part of a two input model using equations 2. The levels of excitation are as before and the actual linear path is again shown for comparison. In contrast to figure 4, the graphs are almost coincident. Close inspection of the partial coherence reveals that there is a small drop below 1 in the region of the natural frequency, and this drop out decreases as the excitation increases.

Figure 6 shows the estimate of the parameter associated with the friction force. In all three cases, the partial coherence in the region around the natural frequency is higher than elsewhere. Furthermore, the fluctuations of the phase in this region are small.

4.3 Model verification.

To verify the identified model, the measured velocity of the physical system has been compared with the response obtained from a numerical simulation of the experimental set-up with the measured force as input. The numerical model consisted of a 4th order Runge-Kutta integration of equation 1, using the values obtained above. As can be seen in figure 7, the simulated and the measured response to the medium level of excitation are in good agreement.

5 CONCLUSIONS

The frequency domain parameter estimation technique described by Rice & Fitzpatrick (1988) has been shown to be appropriate for a system with both coulomb and viscous damping. The derived parameters values are independent of excitation amplitude. This technique may be applied in the presence of other non-linearities or with a multi-degree of freedom system. In conclusion, the technique is computationally inexpensive allowing implementation on a desktop computer.

6 NOMENCLATURE

c viscous (linear) damping coefficient
f the excitation force

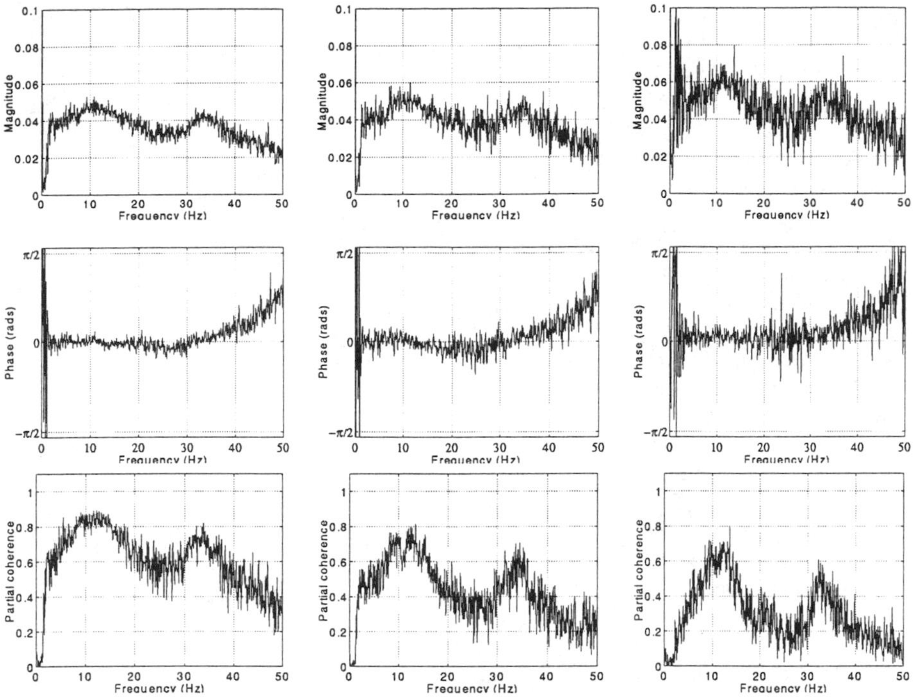

(a) Low excitation (b) Medium excitation (c) High excitation

Figure 6 Identified coulomb damping path (using eqn 2)

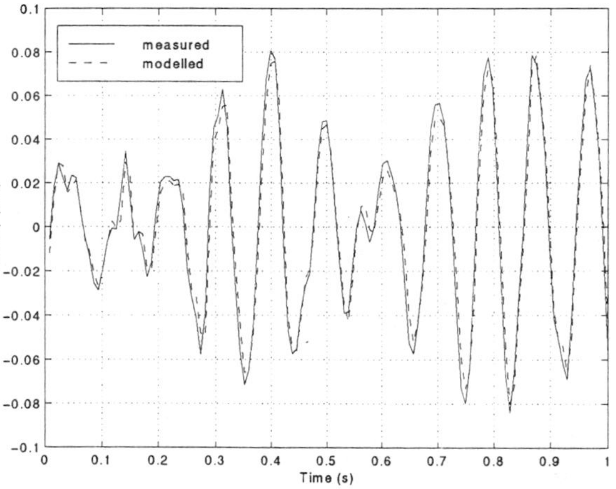

Figure 7 Comparison between the measured and modelled response to random excitation

k stiffness
H_{ij} identified transfer function
L_{ij} optimised transfer function between I and j
m mass
N normal force
S_{ij} spectral density function
x displacement
γ^2_{ij} ordinanry coherenece
γ^{2p}_{ij} partial coherence
μ coefficient of dynamic friction

7 REFERENCES

Bendat, J.S. & Piersol, J.S. (1986). *Random Data: Analysis and Measurement procedures.* Wiley Interscience.

Den Hartog, J.P. (1931) Forced vibrations with coulomb and viscous damping. *Trans. ASME, 53:107-11.*

Esmonde, H, J.A. Fitzpatrick, H.J. Rice & F. Axisa (1992a) Modelling and identification of non-linear squeeze film dynamics. *J. Fluids and Structures, 6:223-248.*

Esmonde, H, J.A. Fitzpatrick, H.J. Rice & F. Axisa (1992b) Reduced order modelling of non-linear squeeze film dynamics. *Proc. I. Mech. E., 206:225-238.*

Inman, D.J. (1994) Engineering Vibrations. *Prentice Hall.*

Jezequel, L. & C.H. Lamarque (Ed.s) (1991) *Proceedings of the International Symposium on Identification of Non-linear Mechanical Systems from Dynamic Tests, Euromech 280, Ecully, France.*

Ravindra, B. & A.K. Mallik (1993) Hard duffing-type vibration isolator with combined coulomb & viscous damping. *Int. J. Non-linear Mechanics, 28(4):427-440.*

Rice, H.J. (1995) Identification of weakly non-linear systems using equivalent linearization. *J. Sound and Vibration, 185(3):473-481.*

Tan, X. & R.J. Rogers (1995) Equivalent viscous damping models of coulomb friction in multi-degree vibration systems. *J. Sound and Vibration 185(1),33-50.*

Rice, H.J. & J.A. Fitzpatrick (1988) A generalised technique for spectral analysis of non-linear systems. *Mech, Systems and Signal Processing, 2(2)195-207.*

Rice, H.J. & J.A. Fitzpatrick (1991) A procedure for the identification of linear and non-linear multi-degree-of-freedom systems. *J. Sound and Vibration, 149(3),397-411.*

Modern Practice in Stress and Vibration Analysis, Gilchrist (ed.) © *1997 Balkema, Rotterdam, ISBN 90 5410 896 7*

Identifying noise modes in estimated turbo-generator foundation models

M.I. Friswell, A.W. Lees, M.G. Smart & U. Prells
Department of Mechanical Engineering, University of Wales Swansea, UK

ABSTRACT: This paper addresses the problem of identifying the foundation model of a turbo-generator using run-down data. Using response measurements at the pedestals, the force applied to the foundations during a run-down may be estimated from the rotor and bearing model. These force estimates, together with the response measurements, are used to identify the foundation model and this paper considers the case when there are fewer modes in the real foundation than in the model over the frequency range of interest. This produces an ill-conditioned estimation problem that may generate numerical or noise modes. The noise modes should be dealt with carefully, since different forcing regimes may excite these modes leading to inaccurate predictions. This paper considers strategies to identify the noise modes, and demonstrates these strategies on a simulated two bearing example. In practice these noise modes are easily identified and may be removed from the model.

1 INTRODUCTION

The analysis of the vibrational behaviour of turbo machinery is a topic of great importance in most process industries and particularly in power generation. Apart from the need to design machinery to operate within acceptable limits, dynamic models are now used to great effect in the diagnosis of operational difficulties. With the high returns from modern plant there is an increasing need to develop reliable plant models for fault diagnosis but the models are not yet developed to the stage of being applicable with confidence across a full range of plant problems (Lees & Simpson, 1983).

A number of authors (Lees, 1988, Zanetta, 1992, Feng & Hahn, 1995) have addressed this problem in recent years and the conclusion is that the supporting structure has a significant effect on the machine dynamics. Over the past 30 years there has been a strong trend towards the use of flexible steel supports for large turbines as this approach offers a number of practical advantages, not least the cost of fabrication. However this does highlight the importance of the supporting structure.

Lees and Simpson (1983) discussed the modelling of this type of structure: in principle it should be possible to develop a suitable model from finite element techniques, but there are a number of practical difficulties. It is often found that similar units, built to the same drawings, display substantially different vibrational behaviour. The different vibrational behaviour is mainly due to small changes in a huge number of joints, that combine to give a substantial change in the stiffness of the structure. With these difficulties it is unlikely that the techniques of finite element model updating (Friswell & Mottershead, 1995) could be used, as there are too many uncertain parameters in the joints. The most promising avenue is to identify a model of the foundations directly from acceleration measurements at the pedestals during run-down (Lees & Friswell, 1996).

2 MODELLING

The aim is to model the foundation of the machine based on response measurements. An initial assumption is that a good model of the rotor and an 'adequate' model of the bearings is available. By assuming knowledge of the rotor and bearing models, the number of parameters to identify may be reduced, and thus the quality of the foundation model is improved. Also, to enable the location of faults in the rotor it is desirable to separate the rotor/bearing model and the foundation model.

Assume that the full model of the rotor/bearing/foundation system may be partitioned using degrees of freedom (DoF) relating to the rotor and the foundation. Both sets of DoFs may be further

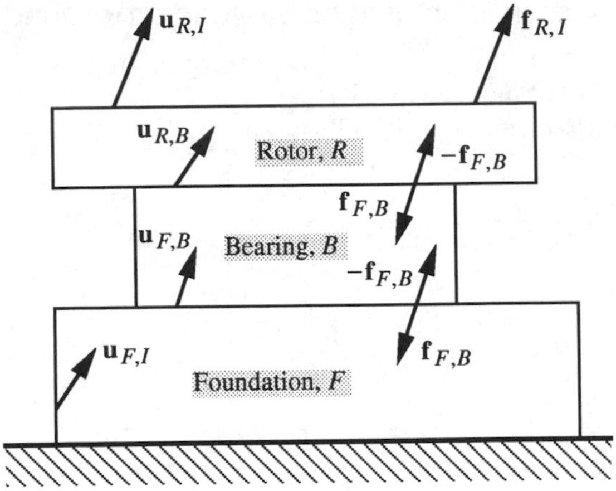

Figure 1. Representation of the Rotor, Bearing and Foundation Model as Three Substructures

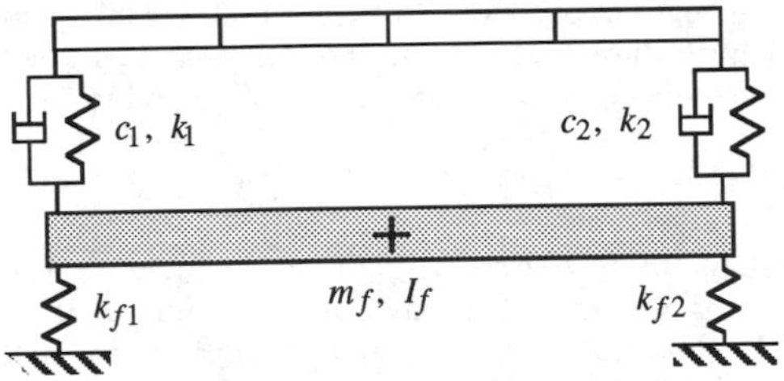

Figure 2. The Simple Two Bearing Example

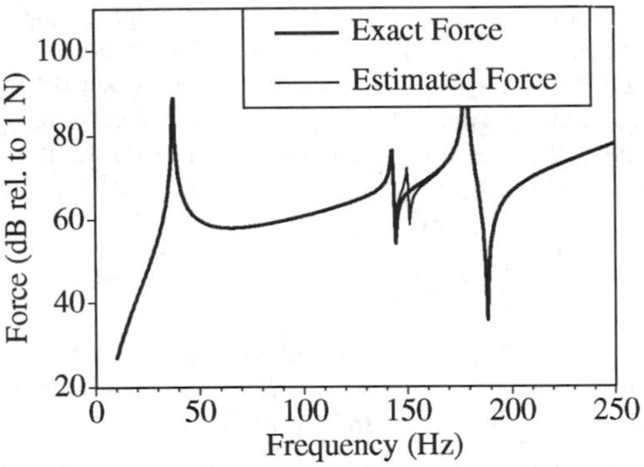

Figure 4. Exact and Estimated Force into the Foundations at Bearing 1. The Estimation had a 50% Error in the Bearing Model.

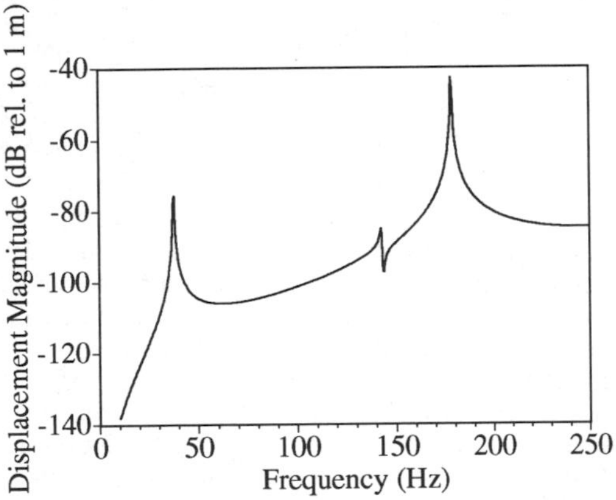

Figure 3. The Pedestal Response at Bearing 1

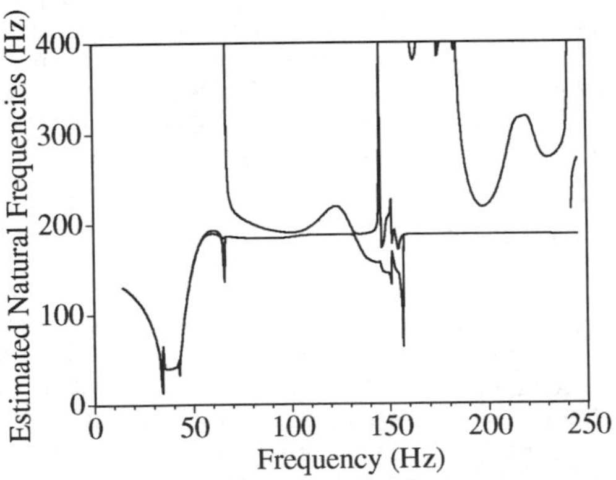

Figure 5. Case 1. Estimated Natural Frequencies of the Foundation Alone using a 9Hz Wide Sliding Frequency Window. Overdamped Modes are not Shown.

partitioned into internal DoFs and DoFs at the bearings, as shown in Figure 1. It is assumed that the bearing can be modelled using DoFs on the rotor and on the foundation only, so that the bearing model does not contain any internal DoFs. Thus the equations of motion of the system, in terms of dynamic stiffness matrices, may be written as

rotor, $$\begin{bmatrix} \mathbf{D}_{R,II} & \mathbf{D}_{R,IB} \\ \mathbf{D}_{R,BI} & \mathbf{D}_{R,BB} \end{bmatrix} \begin{Bmatrix} \mathbf{u}_{R,I} \\ \mathbf{u}_{R,B} \end{Bmatrix} = \begin{Bmatrix} \mathbf{f}_{R,I} \\ -\mathbf{f}_{F,B} \end{Bmatrix} \tag{1}$$

bearings, $$\begin{bmatrix} \mathbf{B} & -\mathbf{B} \\ -\mathbf{B} & \mathbf{B} \end{bmatrix} \begin{Bmatrix} \mathbf{u}_{R,B} \\ \mathbf{u}_{F,B} \end{Bmatrix} = \begin{Bmatrix} \mathbf{f}_{F,B} \\ -\mathbf{f}_{F,B} \end{Bmatrix} \tag{2}$$

foundation, $$\begin{bmatrix} \mathbf{D}_{F,BB} & \mathbf{D}_{F,BI} \\ \mathbf{D}_{F,IB} & \mathbf{D}_{F,II} \end{bmatrix} \begin{Bmatrix} \mathbf{u}_{F,B} \\ \mathbf{u}_{F,I} \end{Bmatrix} = \begin{Bmatrix} \mathbf{f}_{F,B} \\ \mathbf{0} \end{Bmatrix} \tag{3}$$

In equations (1) to (3), **D** represents a dynamic stiffness, **u** the response, and **f** a force. The first subscripts *R, B* and *F* represent the rotor, bearing and force models. The second subscripts represent the DoFs at the bearing (*B*) or at internal DoFs (*I*).

Note the special form of the bearing dynamic stiffness matrix arises from the assumption that the force required to produce a given relative bearing deflection only depends on the relative displacement, $\mathbf{u}_{R,B} - \mathbf{u}_{F,B}$. The only force into the foundations is assumed to be from the bearings. Any force excitation applied to the foundations will produce extra force terms in equation (3).

Equations (1) to (3) may be combined to give the full equations of motion

$$\begin{bmatrix} \mathbf{D}_{R,II} & \mathbf{D}_{R,IB} & \mathbf{0} & \mathbf{0} \\ \mathbf{D}_{R,BI} & \mathbf{D}_{R,BB}+\mathbf{B} & -\mathbf{B} & \mathbf{0} \\ \mathbf{0} & -\mathbf{B} & \mathbf{B}+\mathbf{D}_{F,BB} & \mathbf{D}_{F,BI} \\ \mathbf{0} & \mathbf{0} & \mathbf{D}_{F,IB} & \mathbf{D}_{F,II} \end{bmatrix} \begin{Bmatrix} \mathbf{u}_{R,I} \\ \mathbf{u}_{R,B} \\ \mathbf{u}_{F,B} \\ \mathbf{u}_{F,I} \end{Bmatrix} = \begin{Bmatrix} \mathbf{f}_{R,I} \\ \mathbf{0} \\ \mathbf{0} \\ \mathbf{0} \end{Bmatrix} \tag{4}$$

Normally internal DoFs of the foundation are of no interest and using the last equation of (4) $\mathbf{u}_{F,I}$ can be eliminated to give

$$\begin{bmatrix} \mathbf{D}_{R,II} & \mathbf{D}_{R,IB} & \mathbf{0} \\ \mathbf{D}_{R,BI} & \mathbf{D}_{R,BB}+\mathbf{B} & -\mathbf{B} \\ \mathbf{0} & -\mathbf{B} & \mathbf{B}+\mathbf{D}^{R}_{F,BB} \end{bmatrix} \begin{Bmatrix} \mathbf{u}_{R,I} \\ \mathbf{u}_{R,B} \\ \mathbf{u}_{F,B} \end{Bmatrix} = \begin{Bmatrix} \mathbf{f}_{R,I} \\ \mathbf{0} \\ \mathbf{0} \end{Bmatrix} \tag{5}$$

where $\mathbf{D}^{R}_{F,BB} = \mathbf{D}_{F,BB} + \mathbf{D}_{F,BI}\,\mathbf{D}^{-1}_{F,II}\,\mathbf{D}_{F,IB}$. All the elements of the dynamic stiffness matrix in equation (5) except $\mathbf{D}^{R}_{F,BB}$ are assumed known, and the object is to identify this matrix, or more generally a reduced order estimate of it, over the frequency range of interest.

2.1 *Rotor Model*

The rotor is relatively easy to model. Analysis packages are available to produce a model and to perform analysis, such as model reduction (Nuclear Electric, 1994, Genta, 1995). Experiments may be performed on the free-free rotor to validate the analytical model, or indeed provide an experimental modal model that may be used in the following analysis. Such experiments performed on real rotors have produced natural frequency results with a 1-2% difference to those produced by analysis. These considerations lead to the conclusion that the rotor may be assumed to be modelled accurately, particularly when compared to the models of the bearings and foundations.

2.2 *Bearing Model*

The models of the dynamics of journal bearing are much less accurate than the rotor models. Given the physical dimensions of the bearing, the properties of the oil and the load that the bearing carries, reasonable approximations for the linearised damping and stiffness coefficients may be calculated from the fluid dynamic equations (Smith, 1969). Because of the inherent non-linearities in the bearings these coefficients depend on the rotational speed. This may cause significant problems for some procedures that try to identify the bearings and the foundation at the same time. The foundation will be approximated closely by a linear, reduced order model, either in terms of mass, damping and stiffness matrices, or in terms of a modal model or an ARMA model. There is no guarantee that the bearing properties will be adequately represented by such a model. Given an estimate of the bearing properties it is possible to apply the identification techniques to the foundation alone, but the estimated parameters should be robust with respect to errors in the bearing parameters. This is most important, as the combination of a relatively inaccurate bearing model and an uncertain static load may lead to large errors in the assumed bearing properties.

3 FORCE ESTIMATION TECHNIQUES

The most attractive option to determine the foundation's dynamic properties is to estimate the total force that is applied to the foundation via the bearing. Assuming that the force into the foundation, $\mathbf{f}_{F,B}$, may be estimated then, from equation (3) on eliminating $\mathbf{u}_{F,I}$, we obtain

$$\mathbf{D}^{R}_{F,BB}\,\mathbf{u}_{F,B} = \mathbf{f}_{F,B}. \tag{6}$$

where $\mathbf{u}_{F,B}$ is measured. Equation (6) will be used to identify $\mathbf{D}_{F,BB}^{R}$.

Lees & Friswell (1996) have described a method for force evaluation in terms of the dynamic Greens function of the rotor in isolation, following the description outlined by Lees (1988). It is more convenient however to formulate the method using matrices, as described by Smart *et al.* (1996). In the analysis presented here, the basis of the measured data is the motion of the bearing pedestals. This is normally measured in only one direction: the analysis presented below is equally valid in either one or two dimensions. Combining the rotor and bearing models, equation (1) and (2) we have

$$\begin{bmatrix} \mathbf{D}_{R,II} & \mathbf{D}_{R,IB} & \mathbf{0} \\ \mathbf{D}_{R,BI} & \mathbf{D}_{R,BB}+\mathbf{B} & -\mathbf{B} \\ \mathbf{0} & -\mathbf{B} & \mathbf{B} \end{bmatrix} \begin{Bmatrix} \mathbf{u}_{R,I} \\ \mathbf{u}_{R,B} \\ \mathbf{u}_{F,B} \end{Bmatrix} = \begin{Bmatrix} \mathbf{f}_{R,I} \\ \mathbf{0} \\ -\mathbf{f}_{F,B} \end{Bmatrix} \tag{7}$$

Taking the upper 2 sets of equations gives

$$\begin{Bmatrix} \mathbf{u}_{R,I} \\ \mathbf{u}_{R,B} \end{Bmatrix} = \begin{bmatrix} \mathbf{D}_{R,II} & \mathbf{D}_{R,IB} \\ \mathbf{D}_{R,BI} & \mathbf{D}_{R,BB}+\mathbf{B} \end{bmatrix}^{-1} \begin{Bmatrix} \mathbf{f}_{R,I} \\ \mathbf{B}\mathbf{u}_{F,B} \end{Bmatrix} \tag{8}$$

Equation (8) estimates the response of the rotor based on the unbalance force contained in $\mathbf{f}_{R,I}$, which is assumed known, the measured response $\mathbf{u}_{F,B}$, and the rotor and bearing models. Having established the rotor motion the forces exerted on the foundation are readily calculated from the last set of equations in (7), assuming that there are no external forces on the foundation structure,

$$\mathbf{f}_{F,B} = -\mathbf{B}\mathbf{u}_{F,B} + [\mathbf{0}\ \ \mathbf{B}] \begin{bmatrix} \mathbf{D}_{R,II} & \mathbf{D}_{R,IB} \\ \mathbf{D}_{R,BI} & \mathbf{D}_{R,BB}+\mathbf{B} \end{bmatrix}^{-1} \begin{Bmatrix} \mathbf{f}_{R,I} \\ \mathbf{B}\mathbf{u}_{F,B} \end{Bmatrix} \tag{9}$$

Hence, given a model of the rotor and bearings, the force may be estimated given the set of measured pedestal responses. The sensitivity of this force estimate to uncertainties in the bearing stiffness has been studied by Lees & Friswell (1996), Smart *et al.* (1996) and Prells *et al.* (1997). In the case where the bearing is substantially stiffer than the foundation, errors in the force estimate are restricted to a limited and predictable part of the frequency range. The problem frequencies are where the inverse in equation (9) is ill-conditioned, which are the resonance frequencies of the rotor and bearing system mounted on rigid pedestals. Note that if the bearing parameters are speed dependent these resonance frequencies will change with speed, and these problems may not be encountered. The situations of greatest practical interest are those in which the foundation is very flexible compared to the bearing, thereby exerting a significant influence on the behaviour of the machine.

4 IDENTIFICATION TECHNIQUES

Once the force applied to the foundation through the bearings has been identified, a parameter based model of the foundations may be estimated. The dynamic stiffness may be represented as a frequency dependent matrix at the measured frequencies provided a complete set of independent forces are applied at the bearing pedestals. This may be possible for direct force application, but is rarely possible using data from run-downs. Such data would require many unbalance runs, with the unbalances chosen so that the force at the bearings are independent. Thus in general there are not enough independent forces to obtain the dynamic stiffness of the foundation directly. Indeed, with independent forcing the standard methods of experimental modal analysis could be used to identify a modal model of the foundation.

The alternative method to obtain a model is to assume a parameteric representation of the foundation model and to identify the unknown parameters. In the analysis presented here, the basis of the measured data is the motion of the bearing pedestals. Only the identification of foundation models based on physical matrices will be considered. Other models include the identification of an ARMA model (Lees *et al.,* 1996) or a modal model (Feng & Hahn, 1996, Vania, 1996).

4.1 *Identifying Physical Matrices*

Having established the forces acting on the structure, there is no further requirement to include the rotor in the model. Equation (6) represents the equations of motion of the foundation as measured at the bearing location. Lees (1988) presented a method for estimating the foundation model based on expressing the dynamic stiffness of the foundation $\mathbf{D}_{F,BB}^{R}$ in terms of a mass, damping and stiffness matrix, thus

$$\mathbf{D}_{F,BB}^{R}(\omega) = -\omega^2\mathbf{M}_F + j\omega\mathbf{C}_F + \mathbf{K}_F . \tag{10}$$

The unknown matrix elements are estimated by substituting equation (10) into equation (6) and finding the optimum values in the least squares sense. If the undetermined matrix coefficients are assembled into a vector θ then the equation may be rewritten in the form

$$\mathbf{P}(\omega)\theta = \mathbf{q}(\omega) \tag{11}$$

where the matrix $\mathbf{P}(\omega)$ contains the response, whilst the vector θ contains elements of the mass, stiffness and damping matrices (Lees & Friswell, 1996). The elements of the vector θ have now to be determined in the usual least squares sense, by minimising the residual in equation (11) over all the measured frequencies, with the constraint that θ is real.

This method of modelling the foundations is quite straight-forward and the parameters have some physical meaning. Unfortunately there are some significant disadvantages. The method requires a complete set of forces and responses so that a full set of equations may be generated for the unknown parameters. In some circumstances, it may be possible to estimate a sub-set of the parameters using incomplete data. Also the number of modes within the measured frequency range is constrained to equal the number of DoFs, given by the number of response measurements. In their example, Feng & Hahn (1995) split the frequency range and identified a separate foundation model within each section of the frequency spectrum. In large machines the problem may be different, in that the number of DoFs may be larger than the number of modes of the foundation that are excited, and this may lead to ill-conditioning problems in the estimation that are addressed next.

4.2 *Ill-Conditioning*

In Section 4.1, the regression equation obtained from equation (12) at all measured frequencies must be solved to evaluate the foundation properties. An important issue is the with respect to inversion of the coefficient or regression matrix. The application of the singular value decomposition leads to the consideration of the distribution of the singular values of the regression matrix. In particular, the number of zero or low singular values indicates the number of parameters that cannot be identified from the data. Other regularisation techniques may be used to obtain well conditioned parameter estimates, although their use is limited in this application because there are no constraints to apply (for example from an initial finite element model).

It is also possible to use physical insight to reduce the number of parameters. For example, in a large steam turbine, it is anticipated that the diagonal terms of the three structural matrices have the greatest influence on the rotor and that terms remote from the diagonal assume relatively less importance.

Note also that the number of foundation modes in the frequency range of interest cannot be chosen independently as this number will be equal to the size of the matrices. It is unlikely that the foundation will contain exactly this number of modes in the frequency range of interest. With measurement noise present the algorithm described above will identify a complete set of modes. The question of how to determine which modes are physical modes of the foundation, and which modes are due to measurement noise and errors is addressed next.

4.3 *Strategies to Identify Noise Modes*

There are number of features that identify spurious modes in the identified foundation model. Procedures similar to these are common in experimental modal analysis of static structures.

Heavily Damped Modes. Modes with a large damping ratio are likely to be noise or numerical modes. Any real mode of the structure is likely to have low damping.

Problem Frequencies. As already stated, there are frequencies where the force estimation is poor. Any modes estimated close to these frequencies must be treated with suspicion.

Natural Frequencies Outside the Measured Frequency Range. It may be that natural frequencies are identified that are outside the frequency range used for identification. Even if the natural frequencies relate to physical modes it is very unlikely that they will be identified very accurately, and they should be treated with suspicion.

The Run-Down Responses. Although the run-down responses may not be used directly, some idea of the number of foundation modes to be expected may be gauged from this information. Of course these responses combine the dynamics of the foundation with those of the rotor and bearing, and so care must be exercised.

One useful strategy in experimental modal analysis is to increase the number of modes identified and plot the natural frequencies against the number of modes identified. The physical modes are those that remain constant, whereas the noise modes change frequency with increase numbers of modes estimated. This concept is not easily transferred to the estimation of foundation models because the number of identified modes cannot be chosen arbitrarily. It is possible to change other properties of the identification procedure, for example,

i) the frequency range used for the identification
ii) the number of singular values retained in the parameter estimation
iii) a regularisation parameter incorporated into the estimation.

These strategies will be demonstrated in the numerical example.

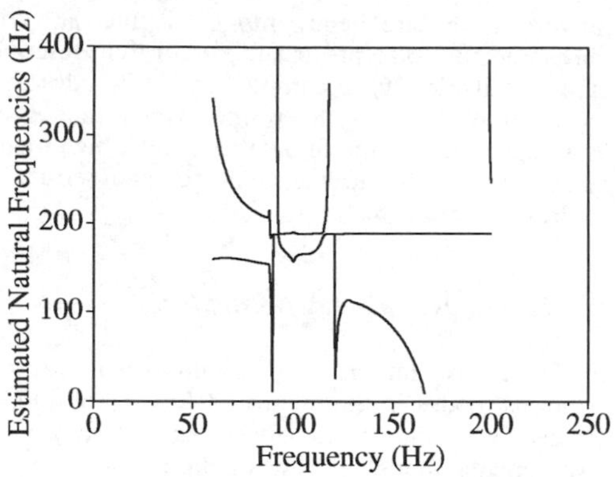

Figure 6. Case 2. Estimated Natural Frequencies of the Foundation Alone using a 100Hz Wide Sliding Frequency Window.

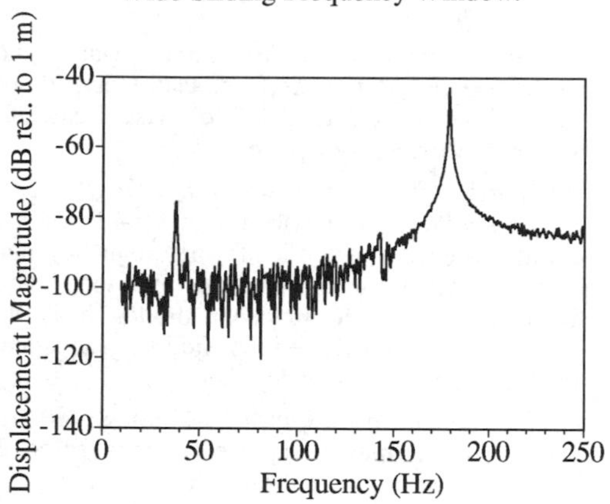

Figure 7. The Response at Bearing 1, Including Random Noise.

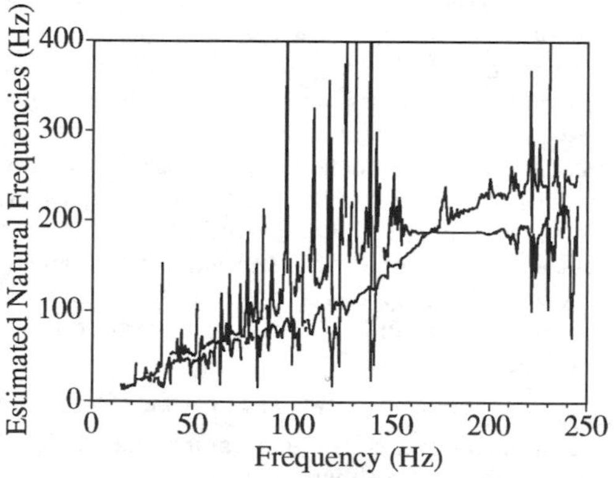

Figure 8. Case 3. Estimated Natural Frequencies of the Foundation Alone using a 9Hz Wide Sliding Frequency Window. Random Noise Included.

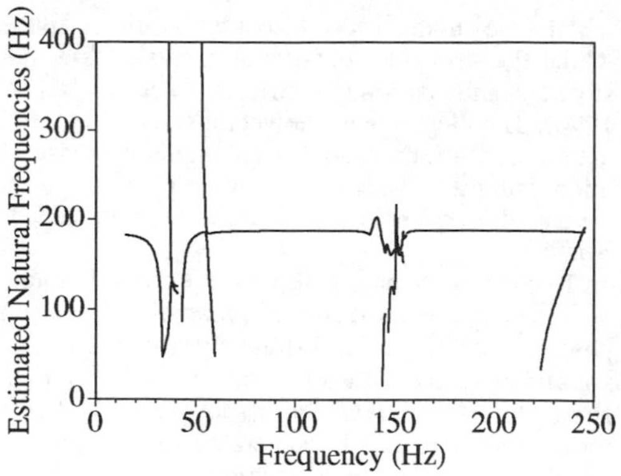

Figure 9. Case 4. Estimated Natural Frequencies of the Foundation Alone using a 9Hz Wide Sliding Frequency Window and Retaining 8 Singular Values.

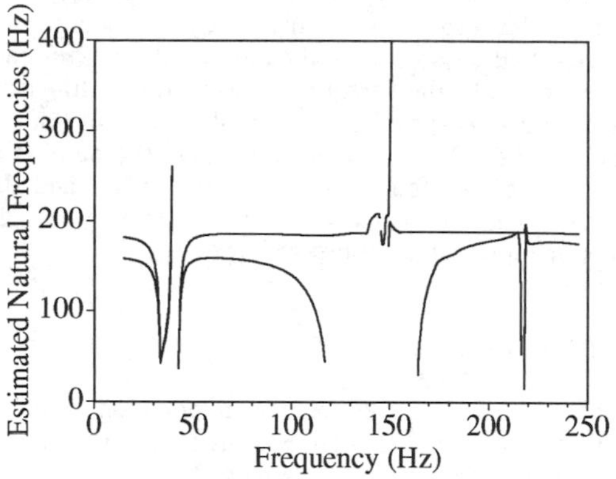

Figure 10. Case 4. Estimated Natural Frequencies of the Foundation Alone using a 9Hz Wide Sliding Frequency Window and Retaining 7 Singular Values.

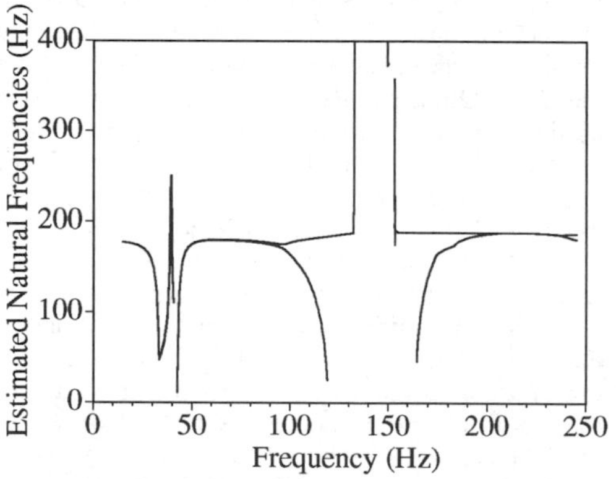

Figure 11. Case 4. Estimated Natural Frequencies of the Foundation Alone using a 9Hz Wide Sliding Frequency Window and Retaining 6 Singular Values.

5 A NUMERICAL EXAMPLE

The example considered is illustrated in Figure 2. A simple steel rotor 2.3 m long, of diameter 10 cm and mass 142 kg is mounted on two bearings of stiffness 175 and 350 MN/m respectively. The bearings have damping coefficients of 1 and 2 MNs/m respectively. These bearings are independently supported on a rigid symmetrical foundation of mass 250 kg and moment of inertia about its centre of gravity of 100 kgm^2. The foundation is mounted on springs directly below the bearings. These springs both have a stiffnesses of 175 MN/m. Damping in the rotor and foundation are assumed to be negligible. A model similar to this was previously analysed by Lees (1988) over a restricted frequency range. The response at the pedestals is computed over the frequency range 1 Hz to 250 Hz with a 0.5 Hz spacing. Figure 3 shows the response of the pedestal at the first bearings with a 0.1kgm unbalance at the centre of the rotor. Notice that 3 critical speeds are apparent within the frequency range. For this example the first three natural frequencies of the rotor and bearing on rigid pedestals are 38.1 Hz, 150 Hz, and 328 Hz. It is therefore likely that there will be one natural frequency of the foundation alone in the frequency range. Indeed from the model these natural frequencies are 188 and 342 Hz demonstrating that this conclusion is correct.

The forces at the bearing have been derived using the methods outlined above with an estimated value of the bearing stiffness and damping. Thus the bearing parameters used to generate the simulated response data are different to those used to estimate the force. Ideally the force estimation methods should be robust to these bearing modelling errors. In this example the stiffness and damping of bearing 2 have been assumed to be 50% too high. Figure 4 shows the estimated force into the foundation compared to the exact force for bearing 1. Errors occur close to 150 Hz, which is one of the natural frequencies of the rotor mounted on the bearings but with rigid pedestals.

Case 1. The foundation parameters are estimated using data over a 9 Hz frequency range (19 frequency points). The mass, damping and stiffness matrices for the foundation contain 9 independent parameters (assuming symmetry) and are estimated from the data at the 19 frequencies. The natural frequencies are then obtained from the estimated matrices. The 9 Hz range may be considered as a sliding window. Thus the natural frequencies of the foundation are estimated using the frequency range 10-19 Hz and the result assigned to the mid-frequency 14.5 Hz. The estimation is repeated using the range 10.5-19.5 Hz, with mid-frequency 15 Hz and so on. Figure 5 shows the natural frequencies estimated in this way against the mid-frequency. The natural frequency of the foundation at approximately 188 Hz is clearly identified over a large part of the frequency range. The quality of the foundation model reduces near to 150 Hz, because of the error in the force estimation. The second natural frequency varies considerably over the frequency range. In Figure 5, and also in the remaining figures, overdamped modes are not shown, gining rise to gaps in the responses.

Case 2. Figure 6 shows the effect of increasing the frequency window to 201 points or 100 Hz. A certain amount of smoothing occurs, but the only consistent natural frequency over a range of frequency bands is still the one at approximately 188 Hz.

Case 3. Thus far the examples have contained a systematic error, namely the error in the bearing model used to estimate the force from the pedestal response. Figure 7 shows the effect of adding random noise to the pedestal responses. Gaussian noise with zero mean and a standard deviation equal to 0.1% of the maximum absolute response has been added to both real and imaginary parts of the response. Figure 8 shows the estimated natural frequencies of the foundation using a 9 Hz frequency window. As expected the mode at approximately 188 Hz cannot be estimated using a frequency range remote from the natural frequency. However between 160 Hz and 200 Hz the mode is estimated well. There is a trend of estimating a once per revolution natural frequency. This is due estimating the foundation model over a narrow frequency range using data that essentially contains noise.

Case 4. Figures 9 to 11 show the estimated natural frequencies of the foundation using a similar procedure to case 1, but where fewer singular values are retained in the parameter estimation. Since 9 parameters are to be estimated, retaining 9 singular values gives the results shown in Figure 5. Before the singular value decomposition is performed the columns of the regression matrix are scaled using the maximum of the full frequency range (250 Hz) so that the mass, damping and stiffness terms are of approximately equal magnitude. Reducing the number of retained singular values changes the noise mode identified, but the true mode at approximately 188 Hz remains.

Case 5. Table 1 shows the natural frequencies identified using the whole frequency range of 10-250 Hz simultaneously to identify the foundation model. The number of retained singular values are changed. The real foundation mode is identified consistently. The second mode is unstable and is clearly not a physical mode.

Table 1. Case 5. Estimated Eigenvalues of the Foundation Alone, using all Frequencies.

No. Singular Values Retained	Eigenvalues 1 & 2	Natural Frequency	Eigenvalues 3 & 4	Natural Frequency
9	-0.0801±1185j	188.6 Hz	-948.9, 1078	real
8	-0.0804±1185j	188.6 Hz	-1138, 1300	real
7	-0.0786±1185j	188.6 Hz	-1216, 1218	real
6	-1.285±1196j	190.4 Hz	0.4352±1186j	unstable
5	-0.3829±1212j	192.9 Hz	0.2502±1187l	unstable

6 CONCLUSIONS

The preferred identification method uses a model of the rotor and bearings to estimate the forces exerted by the rotor on the foundations from the measured response of the pedestals. These forces may then be used to identify or update the model of the foundation. Errors in the force estimate will cause errors in the foundation parameter estimates. The frequencies where the force estimation is poor, namely near the frequencies of the system with the pedestals fixed, should be avoided.

The method to identify the foundation parameters seems robust with respect to the number of physical modes of the foundation in the measured frequency range. The noise modes may be successfully identified using strategies outlined in this paper.

ACKNOWLEDGEMENTS

The authors acknowledge the support and funding of Nuclear Electric Ltd and Magnox Electric Plc. Dr. Friswell gratefully acknowledges the support of the EPSRC through the award of an Advanced Fellowship.

REFERENCES

Feng, N.S. & Hahn, E.J. 1995. Including foundation effects on the vibration behaviour of rotating machinery. *Mechanical Systems and Signal Processing,* 9(3): 243-256.

Feng, N.S. & Hahn, E.J. 1996. Turbomachinery foundation parameters using foundation modal parameters. *21st Int. Seminar on Modal Analysis,* Leuven, Belgium: 1503-1513.

Friswell, M.I. & Mottershead, J.E. 1995. *Finite Element Model Updating in Structural Dynamics.* Kluwer Academic Publishers.

Genta, G. 1995. *DYNROT 7.0 A Finite Element Code for Rotordynamic Analysis.* Dipartimento di Meccania, Politecnico di Torino, Italy.

Lees, A.W. 1988. The least square method applied to investigating rotor/foundation interactions. *IMechE Conference, Vibrations in Rotating Machinery,* Edinburgh: Paper C366/065.

Lees, A.W. & Friswell, M.I. 1996. Estimation of forces exerted on machine foundations. *Int. Conference on Identification in Engineering Systems,* Swansea, March 1996: 793-803.

Lees, A.W. & Simpson, I.C. 1983. The dynamics of turbo-alternator Foundations. *IMechE Conference:* Paper C6/83.

Lees, A.W., Friswell, M.I., Smart, M.G. & Prells, U. 1996. Modelling the influence of the foundation to the mounted rotary machinery using ARMA methods. *Internal Report MECH-AM-96-05,* University of Wales Swansea.

Nuclear Electric 1994. *DYTSO5 User Guide.*

Prells, U., Lees, A.W., Friswell, M.I. & Smart, M.G. 1997. The Identification of the Influence of a Real Foundation on Machine Dynamics. *SIRM '97,* Kassel, Germany, March 1997.

Smart, M.G., Friswell, M.I., Lees, A.W. & Prells, U. 1996 Errors in Estimating Turbo-Generator Foundation Parameters. *21st Int. Seminar on Modal Analysis,* Leuven, Belgium: 1225-1235.

Smith, D.M. 1969. *Journal Bearings in Turbomachinery.* Chapman & Hall.

Vania, A. 1996. Identification of the modal parameters of rotating machine foundations. Politecnico di Milano, Dipartimento di Meccanica, Internal Report 9-96.

Zanetta, G.A. 1992. Identification methods in the dynamics of turbogenerator rotors. *IMechE Conference, Vibrations in Rotating Machinery,* Bath: Paper C432/092.

Modern Practice in Stress and Vibration Analysis, Gilchrist (ed.) © *1997 Balkema, Rotterdam, ISBN 90 5410 896 7*

Vibration testing: State of the art and challenges

P.Sas
Katholieke Universiteit Leuven, Division Production Engineering, Machine Design and Automation, Belgium

ABSTRACT : This paper is a synopsis of the keynote address delivered at the 3rd International Conference on modern Practice in Stress and Vibration Analysis'

Significant advances in the area of vibration testing have occurred over the past fifty years. The volume of this activity has increased over the last ten to twenty years in parallel with, and as a direct result of, the increase in numerical capability available to test and analysis. Over the years vibration testing has become an integrated part of the product engineering and design process as well a standard tool for condition monitoring. The field of vibration testing has certainly become a maturing technology where the growth occurs more in incremental fashion than in big quantum leaps. The paper will review this evolution and comment the current state of the art. Remaining challenges such as test and product variability will be stressed. A case study on vibration tests in microgravity condition will illustrate the potential of vibration testing as well as the relevance of the remaining challenges.

1. INTRODUCTION

A review of the domain of vibration testing leads to the conclusion that it covers a wide range of techniques and applications. Classifying them is difficult and will always be a controversial point. The classification put forward here is very general and divides vibration testing in two fields : Testing for Monitoring and Testing for Analysis and Modelling.

The monitoring field includes all applications and techniques where measured vibration responses are being used as input for some controlling or maintenance action. The analysis field includes all techniques where the vibration responses are being used as input for further analysis or modelling work. Some examples are given in table 1.

The analysis techniques have historically being used in engineering design and analysis in a traditional 'Test- Analyse and Fix' approach on real hardware (prototypes). In a modern design environment where extensive use is made of CAE tools such approach is no longer valid and integration of the testing and analysis tools into an overall CAE approach is necessary.

Challenges of the design of new products will evolve around the application of more complex materials, the reduction of development time, increased flexibility to market trends, increased functional and operational performance in less homogeneous, more hazardous environments, and realisation in smaller volumes at a reduced cost. CAE will play a major role in realising such goals. Yet those design challenges introduce new unknowns where the assumptions of the modelling process become precarious. Hence vibration testing and analysis will remain a vital

Table 1 : Classification of the vibration testing field.

Monitoring	Analysis - Modelling
Predictive maintenance Air bag sensing Vibration control Engine knock detection Active noise control Active vibration control Seismic monitoring	Modal analysis Operational (running) mode analysis Time-frequency analysis Signature analysis of rotating machinery Structural intensity analysis Statistical energy analysis

function in engineering for optimal structural dynamic performance.
Monitoring techniques on the contrary are being used in the operational life of a product and often are an integral part of the system to be monitored. Such boundary conditions induce requirements different from those in analysis work. As a result one can observe a clear difference in hard- and software developed in both domains although often the same basic techniques are being used.

As an example one could take a look at how modal analysis, operating mode analysis and signature analysis fit into the day-to-day work of engineers in NVH laboratories in the automotive industry (J.Leuridan). A principal activity of those laboratories today is the refinement of complete vehicles to achieve better vibration and noise quality based on prototype testing. A large amount of time is spent on applying routine procedures in order to fully understand the relation between operating conditions and the perceived problems. Operating conditions are assessed from road data and laboratory tests. The latter involve chassis dynamometer test of various nature (fast and incremental run-up / run-down) as well as road simulation procedures. Data are analysed in both the frequency and order domains. Performance is measured by vibration levels, acoustic pressure or intensity level, usually expressed in transmissibility or weighted for subjective qualification. The objectives of those tests are to assess whether problems are resonance related and to gain insight in the relation between the structural vibrations and the radiated noise. Recent developments in the field of vibration testing, such as the real-time visualisation of the operating mode shapes and the correlation analysis of data based on principal component analysis, have contributed to discovering and better qualifying vibration and acoustical problems. It is hardly a role that could be taken over by analytical simulations.

If problems seem to be related to particular resonance's and so to undesirable modal characteristics, an experimental modal analysis may well be in order. The objectives of the modal test will depend on the problem to be addressed, and range from

- the modal analysis of a body-in-white for global modes
- the analysis of a full body to understand structural behaviour of components (exhaust system, transmission, engine/gearbox mountings)
- the detailed analysis of local modes on selected areas
- acoustic (cavity) modal analysis to understand the dominant modes in the car interior.
- the qualification of non-linear behaviour and its possible quantification.

The requirement and approach to the analysis or tests depend on those objectives, but also on what the results will be used for. They may be used merely for animation of mode shapes to gain intuitive insights, to simulate the response for variations of operating conditions, to simulate the effect of structural modifications, to provide feedback to an analysis for correlating with predictions to a FEM analysis and possibly updating the latter, or to serve as input for a computational vibro-acoustical analysis.

From the above situation one can elaborate on the role of structural dynamic testing and analysis in a general engineering environment and not just for car design. Testing and analysis are in support of design simulation. Insufficient understanding of the various simulation procedures, the characterisation of new materials, the use of different construction methods for structures, etc... All generate unknowns and lead to an inefficient use of simulation, and therefore more iterations. A principal role for structural dynamic testing will be in providing the necessary data feedback from the laboratory and field systems to support the design and analysis process. Data feedback has to be understood in a broad sense. It refers to material testing results; to the development of simulation criteria and conditions that properly reflect the boundary conditions encountered during a product's life cycle; to system models identified for characterising components which are too difficult to model with numerical methods (FEM.), or for their verification. This role requires a sufficient level of integration of test and analysis in CAE.

2. A TECHNOLOGY DRIVEN FIELD

The field of vibration testing is by definition technology driven, sensors and analysing equipment are basic tools for any vibration test. When reviewing the technology evolution over the last decades it is striking to see that as well in sensors as in analysing equipment the major breakthrough was realised in the late seventies. This opinion might be biased by some sentimental memories of the author who witnessed this period as a student and PhD researcher, but reviewing the facts there is some truth in it. Certainly this period set the stage for the explosive growth that was to follow in the next decades.

The seventies saw the evolution and refinement of such basic sensing tools as the proximity probe, precision condenser and electret microphones, piezoelectric accelerometers and force gages, impedance heads and instrumented modal testing hammers. In this pre-IC era the 1 gram piezoelectric accelerometer did exist already and was a marvel of precision engineering. Also more 'advanced' measurement techniques such as laser vibrometry, acoustic intensity measurements or pulsed laser holography were already introduced in the seventies.

In the field of analysis equipment a revolution took place by the introduction of digital signal processing. Up to that time analysing vibration signals was a very tedious and time consuming procedure involving analogue filters, TFA analysers, tape recorder based delay correlation analysers and others. All this changed dramatically with the advent of the first commercial FFT analysers With instruments such as the Time-Data 1923 (based on a PDP11 mini-computer) and the HP5451A (based on a HP2100 mini-computer), both introduced in 1972, the time necessary to estimate frequency spectra and derived functions, such as FRF's, was dramatically reduced compared to traditional TFA analysers. In addition the systems proved to be more redundant than the traditional analogue equipment. The resulting functions were available in digital format, which, together with the availability of performant mini-computers, opened a completely new field of analysis possibilities. This opportunity didn't pass unnoticed, several companies, often university spin-offs, emerged, and started using those analysers in noise and vibration consulting work, in parallel they developed and commercialised application software for those analysers. The most successful of those companies were undoubtedly SDRC (USA) with roots in the University of Cincinnati and LMS International (Belgium) with roots in the K.U.Leuven. Together they dominate the world market in noise and vibration analysis software and this since more than a decade.

From this three distinct areas of applications of the FFT analyser emerged. The first one, modal analysis, captured the imagination of engineers. The ability to identify the modes of vibration of a structure and extract individual mode shapes from a set of measured frequency response functions was a remarkable achievement. The ability to display each mode in animation made the engineers ecstatic. It was a time of easy success, at least if you had one of those FFT analysers at your disposal and could operate them. The animated mode shapes were often a target on themselves, engineers were more than happy with the animated modes, since solving the vibration problem was considered to be straightforward once you had the mode shapes. We do know by now that this was a somewhat naive assumption.

Digital vibration control, though it did not have as much 'sex-appeal' emerged as the second significant application. A shaker could be controlled to reproduce a vibration environment in the laboratory with remarkable accuracy. Random vibration experienced in a vehicle on the road, shocks, or the periodic vibration produced by an engine speeding up - all could be simulated in the laboratory saving tremendous man hours of product qualification.

Signature analysis of rotating equipment was a, at least in the beginning, a distant third in size of application. The three-dimensional RPM spectral map is a remarkable method of separating vibration due to structural resonance's from that due to rotating machine components. The derived plots of order tracks provide a means of identifying vibration causing mechanisms. Engine running-up and coast down waterfall maps became available with the first Fourier analysers and have since then become a standard in the majority of analysers that followed. Especially for NVH work in the automotive field it is the most frequently used analysis technique.

The capabilities of the instrumentation available in the 70's can be best demonstrated by a list of the technical specifications of a dedicated measurement system that was developed in 1978-79 by researchers of the K.U.Leuven by order of the automotive industry (M.Mergeay et al.). The measurement system was based on two integrated HP5451C analysers each equipped with a HP1000 (0.5MIPS) mini-computer. The follow-up and further commercialisation of this system was the start of LMS International.

System specifications (1979):

- High productivity for data acquisition, analysis and automatic reporting within a multi-user system.
- up to 64 channel simultaneous data acquisition: 32 high frequency channels (0-800Hz) and 32 low frequency channels (0-50Hz).
- Acquisition of data under normal operating conditions or under artificial excitation.
- Automatic calibration of transducers and amplifiers
- Analysis capabilities : Modal analysis (SIMO), running mode analysis, 3D rpm waterfall mapping, order ratio diagrams, tracking of tacho references, user defined graphics, acoustic intensity, transmission path analysis
- Animated display of mode shapes and operating mode shapes, deformation vector or trace display
- User friendly man-machine interface
- Fast graphic output (> 500 plots/hour) on electrostatic graphic plotter
- Normalised graphics
- Modular software architecture

Those specifications are not so much different from those of the multi-channel analysis systems that are currently on the market. Of course progress has been made since then, the hardware became more powerful, more accurate and faster. The software capabilities have continuously being extended, new analysis methods and more powerful numerical algorithms have been added on a regular pace. But the basics haven't changed much. Major progress has certainly been realised in the field of user interfaces : color graphics, mouse manipulation, or window environments did not exist in the seventies. This had however some advantages: users were for example not misled by superfluous graphic cosmetics and a hard copy of the screen resulted always in a document that was easy to reproduce (black and white). In addition memory was scarce (64K RAM was a luxury, disc memory was typically 2.5Mbyte) such that software programmers had to use the memory in an optimal way. This made the software highly efficient such that the CPU times were acceptable even by current standards.

Another big improvement compared to the first generation systems is the size and weight of current systems. The first generation systems such as the HP5451B came in a 185cm high and 50cm wide instrument rack and weighted 150kg. Current portable systems are often PC based, and consist of a robust notebook PC with multiple-channel FFT board(s) in a docking station, bringing the total weight of a 16 channel system down to less than 5kg. Such multiple-channel FFT boards are based upon modern DSP (Digital Signal Processing) chips, which are optimised for vector operations and especially for the computation of the FFT.

Multiple-channel systems have also become much more affordable, the price of both transducers and analysers has decreased considerably over the years. As an example the price per channel for accelerometer measurements with a multi-channel system has fallen with a factor of more than 20 over the last two decades. This cost reduction is not only on the credit of the computer hardware, but also of the transducer hardware. Accelerometers for example are now being integrated with signal-conditioning circuitry on one IC (integrated circuit chip). Such an IC sensor, when produced in large series, is much lower in cost than classic electromechanical accelerometers. Sensors of this kind are being used in airbags at a cost of a few dollars/piece. Due to their limited sensitivity and bandwidth such sensors are however difficult to use for general vibration measurements. The more traditional electromechanical piezo transducers are more appropriate for such applications . But also for those traducers considerable cost reductions have been achieved.

Table 2 : Summary of modal parameter estimation algorithms

Algorithm	Domain	
Quadrature amplitude	frequency	SDOF
Kennedy-Pancu circle fit	frequency	SDOF
Frequency domain finite difference	frequency	SDOF
Frequency domain polynomial	frequency	MDOF
Complex exponential algorithm	time	MDOF
Least squares complex exponential	time	MDOF
Polyreference time domain	time	MDOF
Ibrahim time domain	time	MDOF
Eigensystem realisation	time	MDOF
Polyreference frequency domain	frequency	MDOF
Rational fraction polynomial	frequency	MDOF
Orthogonal polynomial	frequency	MDOF
Multiple ref. orthogonal polynomial	frequency	MDOF
Direct parameter identification	time	MDOF
Direct parameter identification	frequency	MDOF

The progress realised in analysis methods is substantial, numerous new methods and algorithms have been proposed over the years, to give a full overview is beyond the scope of this paper. Taking modal analysis as an example and more specifically the area of modal parameter estimation it is clear that the research effort over the last fifteen years has yielded many algorithms that are being used privately or being sold as part of a commercial software package. Table 2 represents a summary of the most widely used algorithms.

Modal parameter estimation techniques have certainly come to maturity and one could question whether there is a further need for research on parameter estimation methods, especially if one takes into account the inherent test and product variability. This point will be further discussed in the next paragraph.

Clearly, one technical need in the area of modal parameter estimation is a set of database benchmarks that can be used to identify particular performance characteristics of the various modal parameter estimation algorithms. Another need is the development of statistically based methods that can estimate error bounds on the modal parameters in terms of the error bounds on the data.

3. CHALLENGES

3.1 Instrumentation

As shown in previous paragraph considerable progress has been made in instrumentation especially in the field of miniaturisation. This trend will undoubtedly continue and the time when a single integrated circuit package will contain a three dimensional translational accelerometer, a three dimensional rotational accelerometer, a microphone and sensors for temperature and humidity is foreseeable. On a shorter term some of the open issues in the area of instrumentation are :

- Smart sensors with built-in-test, calibration and identification capabilities together with an interfacing methodology that will utilise such features
- Wireless transducers with data telemetry to avoid cabling problems
- Cable management systems that facilitate large number of channels (networking or multiplexing)
- Low mass rotational transducers
- Further development of non-contacting sensor technologies such as video and/or laser scanning technology to eliminate resolution/digitisation problems
- Smart monitoring sensors with local intelligence that can be integrated in the systems to be monitored
- Compact force actuators with sufficient amplitude at low frequencies.

3.2 Interfaces, data exchange

One of the biggest remaining challenges is to integrate the vibration testing field in a total engineering approach to solve structural dynamics problems. This requires more than a continued technology development. In most CAE environments, systems used for vibration testing will coexist with other systems such as those for design, numerical analysis, product optimisation and production engineering. Often the principles of concurrent engineering will have to be followed such that an easy exchange of data between systems and services is required. The ideal environment would be the virtual engineering office in which the engineer has on-line access to test-data of several labs world-wide, where he will have the necessary tools to analyse the data, where he can define and coordinate additional measurements, where he can run numerical simulations and combine the results into a global model which will allow design optimisation. In all this data-transmission via network (WWW) will be essential.

Today this ideal situation is not yet realised, parts of the puzzle are available, some still need to be developed, and further standardisation is required for the data exchange and data interfacing between software packages of different suppliers.

3.3 Product and test variability

One of the biggest challenges in vibration analysis is to take into account the influence of product and test variability on the dynamic characteristics of the system under test. Product variability refers to the difference in dynamic behaviour between structures that are assumed to be identical, for example identical cars being produced the same day on the same production line. Test variability covers as well the influence from changes in boundary conditions such as temperature, humidity as the influences from the test set-up.

A few examples to illustrate the dramatic influence of such variability:

A. Automotive structures : The production process of a car is such a complex procedure that, even with the current tight tolerance levels in automatic production lines, considerable differences exist in the structural dynamic behaviour between identical structures. Those differences make it impossible to determine optimal noise and vibration control measures. What is optimal for one car might be sub-optimal or even have an inverse effect on another one. Recently several survey tests have been performed to determine the range of this variability. The results of those tests have not been released for publication, but the existence of considerable variance is confirmed. One of the few published results is a study where two structure-borne and two airborne paths were measured on 99 identical automotive vehicles (M.S.Kompella et al.). The FRF's of those paths varied by as much as 5-10dB over the measured frequency range (0-1000Hz). The authors concluded that the statistical variations observed in this study suggested that a noise control treatment based on the analysis of a single vehicle may not have the desired noise control effect for other nominally identical vehicles.

In two more recent publications (J.A.Cafeo et al. and T.K.Hasselman et al) a survey is given where seven cars of the same make and model were each subjected to nine independent vibration tests with 0-50Hz as frequency bandwidth. The test were independent in the sense that all fixturing and instrumentation were removed and reinstalled between replicate tests in an effort to more realistically represent experimental variability. The sixty-three data sets were used in statistical analysis to measure the total variability and estimate the separate contributions of experimental and product variability to the total observed variability. Their measurements show a global variability of 5-10dB in acceleration response above 25Hz. Below that frequency the variations are significantly smaller. More variability was observed in the FRF's for the powertrain and exhaust response points than for points on the body. The test-to-test variability was smaller than the product variability.

B. Civil structures : A possible application of vibration testing is the monitoring of variations in the dynamic characteristics of a structure to detect and identify structural damage. Especially for large civil structures such as bridges or antennas which are difficult to visually inspect this looks like a promising application. Such a failure detection procedure will only be applicable when the variations in dynamic behaviour due to damage are larger than the variation due to changes in boundary conditions such as temperature or traffic. On the IMAC XV conference which was hold in February 97 several sessions were dealing with damage detection in civil structures by vibration responses. Several speakers concluded that the dynamic behaviour of civil structures such as bridges is more sensitive to temperature variations than to structural damage. Methods will have to be developed to filter out the environmental effects before using the data for damage detection. It is clear that the efficiency of such detection methods will be function of the accuracy of those filtering methods.

From those examples it is clear that test and product variability is a limiting factor for many vibration analysis methods and that methods to assess and model this variability are necessary in order to come to more robust design optimisations that are insensitive to this variability. In addition it is obvious that when performing measurements or using modal parameter estimation methods it is useless to search for the last digit in measurement accuracy or parameter estimation performance when the variability of the test structure might be orders of magnitude larger. One can therefore question the relevance of research effort spent on further improving parameter estimation methods for modal analysis.

One boundary condition which is especially relevant in space applications is gravity. The following paragraph reports on tests that have been conducted to quantify the influence of zero gravity on a series of specially designed test structures.

4. ZERO GRAVITY VIBRATION TESTS

The determination of structural frequency response functions are a vital element in predicting, modelling and controlling the vibration levels generated by various sources onboard spacecraft.
In spacecraft systems the structural transmissibility of vibrations from a disturbance source to a receiver point is traditionally modelled by FE methods. Those models are validated by extensive ground vibration testing on the full spacecraft or on components. Modal testing is thereby one of the standard test procedures. Such ground vibration tests are necessarily conducted in a 1g environment where, if possible, zero gravity is simulated by elaborate suspension systems. This is of course a crude approximation and there is a need for more detailed research and validation experiments on the influence of zero gravity on the structural dynamic behaviour. In future missions it is planned to instrument spacestructures systematically with a sufficient number of transducers to monitor their dynamic behaviour during flight. A first attempt to conduct such an experiment took place during the European Spacelab D2 mission in 1993, K.U.Leuven was involved in the data processing which took place in 94/95 (H.R.Stark et al). Structural frequency response functions (FRF's) have been determined Spacelab module on-ground and in-orbit. For the in-orbit experiment a payload specialist operated an impact hammer as excitation source. The prime use of those FRF's was to extend the database which is used for design and verification of spacecraft subsystems and equipment. A secondary use was an attempt to assess the influence of microgravity boundary conditions on the structural dynamic behaviour of the spacelab by analysing the difference in modal parameters between on-ground and in-orbit conditions. This task turned out to be impossible due to the complexity of the system (high modal density, non-linearities, high background noise, limited number of response points).

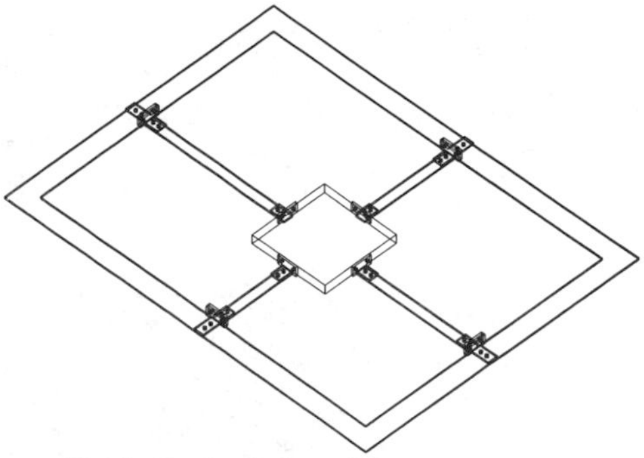

Fig.1 : Basic design of the test structure

Therefore, a more straightforward and reliable experiment was defined, aiming specifically at identifying these variations. This research took place in the framework of an ESA/PRODEX project. Simple test structures, representative for real spacecraft components, have been designed. Experimental modal analysis was performed in 1g gravity and in microgravity conditions. Microgravity conditions were realised in consecutive parabolic flights during the 23rd ESA Parabolic Flight Campaign (December 96). Guidelines for appropriate finite element modelling, have been derived from the comparison between 1g and zero gravity conditions.
Several pre-tests and FE simulations resulted in the following basic design (fig.1): a rectangular frame with a central mass connected to the outer frame by four slender bars (spokes). This test configuration is flat and asymmetric such that only out of plane degrees of freedom have to be

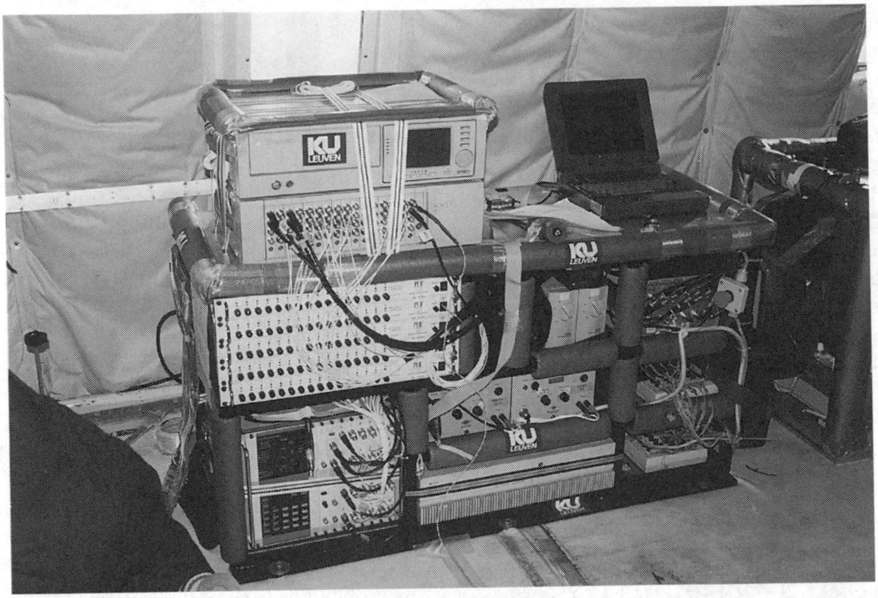

Fig.2 Measurement set-up

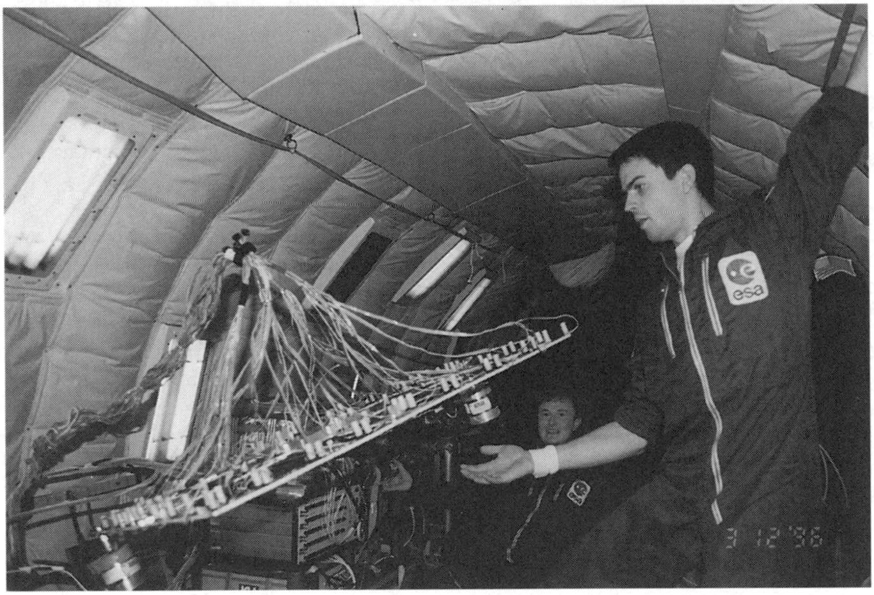

Fig.3 Free floating structure during microgravity

considered. Based on this concept nine test structures have been realised using various materials for the frame (aluminium, reinforced composite and highly damped sandwich material) and using bolted, glued or pinned joints for the connections between the frame and the spokes.

As data acquisition time is limited during parabolic flight, special care has to be taken

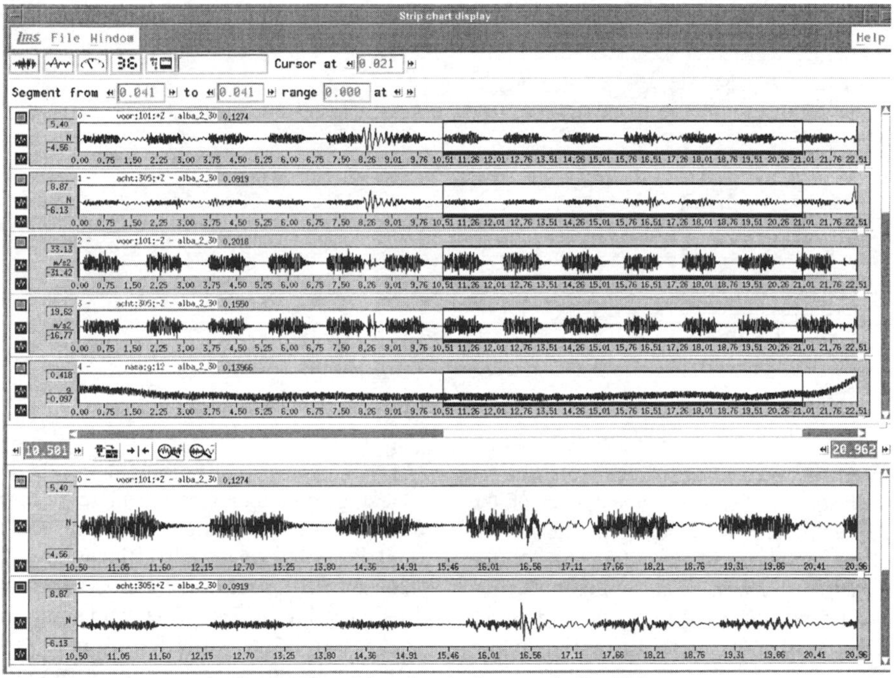

Fig.4 : Time histories during parabolic flight

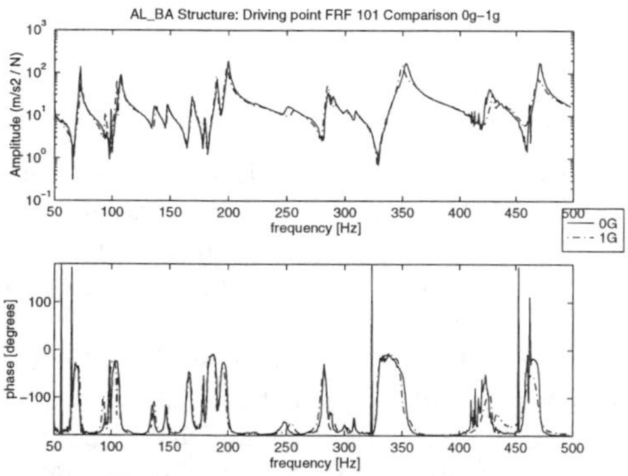

Fig. 5 : FRF comparison 0g - 1g

when defining measurement parameters and acquisition algorithms. One parabola yields indeed only a maximum of 20 seconds of micro gravity. To conduct a modal test in that time interval simultaneous acquisition of all inputs and responses is required during the 0g phase. Based on a preliminary FE analysis, 2 inputs and 80 response points were considered a minimum to estimate the modal characteristics in a frequency band of 0-500Hz for the selected test structure.

Two inertia shakers were used to generate the excitation forces, the structure was instrumented with 80 lightweight accelerometers to measure the vibration responses. Burst random was used as excitation signal (0-500Hz). During the flight the frequency responses (FRF's) were estimated on-line, simultaneously the time histories of all measured signals were digitally recorded on tape for off-line processing. Fig.2 gives an overview of the in- flight measurement set-up and of the instrumentation

During data acquisition the operators tried to stabilise the tested structure in the middle of the cabin to keep it free-floating (fig.3). However collisions or correcting actions are almost impossible to avoid in the 0g phase. Those additional inputs generate transient responses which bias the original responses. This affects the quality and the reliability of the on line measured FRF's. To minimise this bias error it was decided to edit the time responses and remove all affected time sequences. Figure 4 gives an example of this editing procedure. The first two traces represent the force cell signals, the third and fourth traces are accelerometer signals and the fifth trace is the gravity signal.

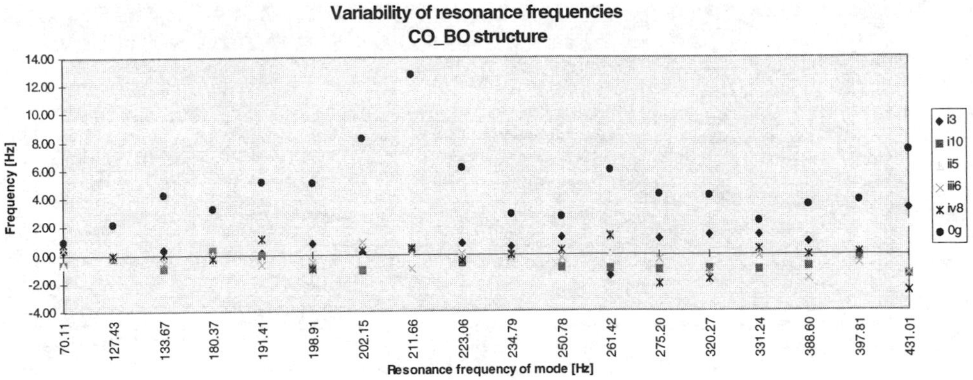

Fig.6 Variability of resonance frequencies

The lower two plots show a detail of the two force cell signals. The first three bursts of the detailed section are clean; during the next burst the structure collided with the cabin wall and the next bursts are affected by manipulation of the tests structure. Only the unaffected time sequences are used for the off-line estimation of the FRF's.

At the time of composing this manuscript the data processing and the detailed comparison of 0g and 1g results was still in progress, such that only some preliminary results are given, without further interpretation. Figure 5 shows a FRF comparison for the aluminium frame with pinned joint connections. This test structure has low damping and intermediate stiffness. The changes observed between 0g and 1g in the FRF are comparable to the variance measured in the repeatability tests. Highly damped structures show more significant differences.

As an illustration figure 6 shows the variance that was observed in the resonance frequencies of the composite frame with bolted joints. In this figure the notations i,ii,iii and iv refer to four different orientations of the test structure for the 1g tests. Those results seem to indicate that there is a consistent difference in resonance frequency between the 1g and 0g condition. As mentioned before those are only preliminary results and further analysis is required to confirm and further explain this.

5. CONCLUSIONS

Over the years the field of vibration testing has become a maturing technology. Measurement and analysis tools became more performant and affordable. Integration of those tools into a global CAE approach is in progress and will eventually lead to the virtual vibration engineering office. One remaining challenge however is the influence of product and test variability on the analysis result and how to take this variability into account in the modelling phase.

REFERENCES

J.A.Cafeo, R.V.Lust, S.Dodgett, D.J.Nefske, D.A.Feldmaier, S.H.Sung, 'A design-of-experiments approach to quantifying test-to-test variability for a modal test', Proc. IMAC-15, pp.599-604, 1997

T.K.Hasselman, J.D.Chrostowski, 'Effects of product and experimental variability on model verification of automobile structures', Proceedings IMAC-15, pp.599-604, 1997

M.S.Kompella, R.J.Bernhard, 'Measurements for the statistical variation of structural-acoustic characterstics of automotive vehicles', SAE paper931272, pp.65-81, 1993

J.Leuridan, 'Modal analysis: a perspective on integration', Proc. IMAC-10 (keynote address), San Diego, 1992

M.Mergeay, J.Simons, P.Sas, 'Multichannel Integrated Fourier System, Proc. ISMA-4, K.U.Leuven, 1979

H.R.Stark, C.Stavrinidis, D.H.Eilers, W.Heylen, L.Bregant, 'In orbit determinaton of satellite strcutural transfer functions', Proc. ISMA 19, K.U.Leuven, pp.1273-1281, 1994

Modern Practice in Stress and Vibration Analysis, Gilchrist (ed.) © *1997 Balkema, Rotterdam, ISBN 90 5410 896 7*

The Weibull distribution as a means of monitoring gear damage

D.C.D.Oguamanam, H.R.Martin & J.P.Huissoon
Department of Mechanical Engineering, University of Waterloo, Ont., Canada

ABSTRACT: The Weibull distribution is introduced as a viable statistical distribution for detecting defects in gears. The fourth moment of the distribution is used as the discriminating feature, and its characterizing parameters are estimated using both the Method of Moments and Maximum Likelihood Estimation techniques. It is shown that while fourth moment does identify defects, it does not distinguish them. Further, the computation times with the method of moments are consistently higher than corresponding values for the maximum likelihood estimation techniques.

1 INTRODUCTION

A single catastrophic event or an accumulation of initially undetected and rather innocuous events in a gear system can lead to the failure of the system. The effect of gear failure, which is highly dependent upon the role of the gear, ranges from benign to catastrophic.

Research into early detection of gear failure is still an ongoing process and is approached from either one or a combination of domains---time, quefrency, frequency, and time-frequency. Presently, there exists no one known method to effectively diagnose the various failures nor to provide prognosis.

In the conventional application of statistical distribution techniques to monitoring the condition of gearing systems, the normal distribution is fitted to the data recorded over a revolution of the gear. The kurtosis value, which is an indicative of the "peakiness" or "spikiness" of a distribution, is calculated and used to predict the state of the system.

Martin et al. (1990) have successfully used the Beta distribution, a distribution which has effectively provided information on surface finish (Whitehouse 1978), to monitor gear failures. The shape parameters of the distributions are estimated using the method of moments (MM). In this technique, the expected mean of the distribution is equated to the sample mean and the expected variance of the sample variance.

In the study by Oguamanam et al (1995), the shape parameters of the beta distribution are estimated using both MM and maximum likelihood estimation techniques (MLE). In the MLE technique, a set of parameters which can describe the given distribution is searched to determine that which has the maximum probability of generating the actual data. They observed that while the latter is more sensitive to defects, it is more computational intensive. However, the computation times are still admissible in on-line monitoring systems. This observation on computation times explains why the use of the MM is encouraged where it performs as well as the MLE technique (Milton & Arnold 1990).

In this paper, the use of the kurtosis of the Weibull distribution, a distribution that is widely used in reliability studies, as a discriminating feature for detecting damage in gears is investigated. The characterizing parameters of the distributions are estimated using both the MM and MLE techniques. The failures investigated, cracks and pits, are simulated in a ten-teeth gear pump.

In line with (Milton & Arnold 1990, Oguamanam et al 1995), the vibration signal obtained over one revolution is segmented among the ten teeth and each set is independently analysed. The advantage of this technique over the conventional method of analysing the total data is the increased probability of detecting damage before it has substantially progressed. Further, a mathematical justification for this approach is presented.

2 THEORETICAL BACKGROUND

The angular motion transmitted by a pair of properly designed and manufactured (rigid, perfect, and uniformly spaced involute teeth) gears in mesh is exactly uniform (Bartelmus 1984). However, design decisions and/or manufacturing defects may result in the presence of unsteady components in the relative angular motion of the gear pair. This is usually manifested in the system by noise and vibration.

The noise and vibration in the gear system, as in most rotating machinery, is treated as a random process (Bartelmus). Thus the process is described using probability statements and statistical averages (Bendat & Piersol 1985, Milton & Arnold 1990). The probability density function *pdf* of the Weibull distribution is defined as

$$p(x) = \begin{cases} \alpha\beta^{-\alpha}x^{\alpha-1}e^{-\left(\frac{x}{\beta}\right)^{\alpha}} & \text{if } x > 0 \\ 0 & \text{otherwise} \end{cases} \quad \textbf{(1)}$$

where α and β are the shape and scale parameters, respectively. Figure 1 depicts the Weibull distribution for various sets of parameters.

The parameters of the distribution can be estimated amongst others by the MM or MLE. The estimation of the parameters using the MM requires the determination of the mean and variance of the given data which are respectively calculated as

$$\mu = \frac{1}{n}\sum_{I=1}^{N} x_i \quad \textbf{(2)}$$

$$\sigma^2 = \frac{1}{N}\sum_{i=1}^{N}\left(x_i - \mu\right)^2 \quad \textbf{(3)}$$

The expected value and variance of the distribution are then calculated and equated to equations 2 and 3, respectively. The resultant equations are solved simultaneously to determine the values of the parameters. The expected value of the distribution is defined as

$$E[X] = \int_{-\infty}^{\infty} x p(x)dx \quad \textbf{(4)}$$

while the variance is defined as

$$E[(X - E[X])^2] = \int_{-\infty}^{\infty} (x - E[X])^2\, p(x)dx \quad \textbf{(5)}$$

The MLE, on the other hand, searches a set of parameters which can describe the given distribution and selects that which has the greatest probability of repeating or generating the actual data. Given a set of observations $(x_1, x_2, \ldots, x_N)$ that are to be described by a distribution of probability density function $p(x)$ with parameters $(\theta_1, \theta_2, \ldots, \theta_N)$, the likelihood function is expressed as

$$L(\hat{\theta}_1, \hat{\theta}_2, \ldots, \hat{\theta}_n) = \prod_{i=1}^{N} p(x_i) \quad \textbf{(6)}$$

where the $\hat{\theta}_i$s are estimates of the parameters.

The parameters of the distribution are calculated by taking the natural logarithm of the above equation. Then the partial derivatives with respect to the parameters are determined, and the resultant set of equations is solved for the maximizing parameter values. The derivation of the estimates of the parameters is presented in the following section.

2.1 Justification for the investigation of the data from individual gear tooth meshes

The vibration signal from an individual mesh is a random process with an arbitrary distribution. Using the central limit theorem, it can be shown that the polled (or aggregated) data will assume a normal distribution provided the number of samples is sufficiently large. The question of what constitutes an appropriate data size is a wide area of research in statistics and probability. However, it suffices to mention that the size of the sample is dependent upon the distribution of the individual samples.

If it is assumed that the distribution of the random variable describing the vibration signal from each mesh is Weibull, it can be shown that polling these random variables together does not necessarily result in a random variable with Weibull distribution. To prove this statement, assume that X_1 and X_2 are independent, identically distributed (iid) random variables with *pdfs* as defined in equation 1. The problem is therefore to determine whether the random variable X, a summation of X_1 and X_2, has a Weibull distribution. It can be shown by contradiction that this is not true, at least in the strict sense. In order to reduce the complexity of the proof, it is further assumed that the random variables X_1 and X_2 each have unit α. Thus the *pdf* reduces to

$$p(x) = \begin{cases} \beta^{-1} e^{-\left(\frac{x}{\beta}\right)} & \text{if } x > 0 \\ 0 & \text{otherwise} \end{cases} \quad \textbf{(7)}$$

The distribution of interest is $P(X_1 + X_2 = X)$, which is deduced as follows:

$$\begin{aligned} P(X_1 + X_2 = x) &= P(X_1 = \xi, X_2 = x - \xi) \\ &= \int_0^x \beta^{-1} e^{-\left(\frac{x}{\beta}\right)} \beta^{-1} e^{-\left(\frac{x-\xi}{\beta}\right)} d\xi \\ &= \beta^{-2} x e^{-\left(\frac{x}{\beta}\right)} \end{aligned} \quad \textbf{(8)}$$

This indicates that the random variable X has a gamma *pdf*. Further, (8) is not a special case of (1) because there is no possible combinations of α and β such that they are equal. Thus the random variable X cannot have a Weibull distribution.

3 DETERMINATION OF CHARACTERISTIC PARAMETERS AND KURTOSIS

The derivation parameters of the Weibull distribution using MLE results in the following two equations.

$$\frac{1}{\hat{\alpha}} + \frac{\sum_{i=1}^{N} \ln x_i}{N} - \frac{\sum_{i=1}^{N} x_i^{\hat{\alpha}} \ln x_i}{\sum_{i=1}^{N} x_i^{\hat{\alpha}}} = 0 \quad \textbf{(9)}$$

$$\hat{\beta} = \left(\frac{\sum_{i=1}^{n} x_i^{\hat{\alpha}}}{n} \right)^{\frac{1}{\hat{\alpha}}} \quad \textbf{(10)}$$

The roots of the above equation (i.e., function parameters) can be obtained using the root finding techniques of numerical analysis (Chapra & Canale 1988). Thomen et al (1969) have shown that an average of 3.5 iterations with the Newton method is needed to determine the parameters if the starting point suggested by Menon (1963) is used. This starting point is

$$\hat{\alpha}_0 = \left\{ \frac{\sigma \left[\sum_{i=1}^{N} (\ln x_i)^2 - \frac{\left(\sum_{i=1}^{N} \ln x_i \right)^2}{N} \right]}{\pi^2 (N-1)} \right\}^{\frac{1}{2}} \quad \textbf{(11)}$$

The estimation of the distribution parameters using MM involves equating the expected value (a discrete version of equation 4) to equation 2, and the expected variance to (3). The resulting expressions are

$$\mu = \frac{\hat{\beta}}{\hat{\alpha}} \Gamma\left(\frac{1}{\hat{\alpha}}\right) \quad \textbf{(12)}$$

and

$$\sigma^2 = 2 \frac{\hat{\beta}^2}{\hat{\alpha}} \Gamma\left(\frac{2}{\hat{\alpha}}\right) - \frac{\hat{\beta}^2}{\hat{\alpha}^2} \Gamma^2\left(\frac{2}{\hat{\alpha}}\right) \quad \textbf{(13)}$$

Solving equations 12 and 13 simultaneously results in

$$\Phi \Gamma^2\left(\frac{1}{\hat{\alpha}}\right) - \Gamma\left(\frac{2}{\hat{\alpha}}\right) = 0 \quad \textbf{(14)}$$

where the variable Φ is equal to

$$\frac{1}{2}\left(\frac{\sigma^2}{\mu^2} - 1\right)$$

Equation 14 is solved numerically for $\hat{\alpha}$ for any given Φ and the result is substituted into (12) to solve for $\hat{\beta}$.

The *kth* statistical moment, M_k, and central moment, m_k, are deduced from

$$\begin{aligned} M_k &= E[X^k] = \sum_{i=1}^{N} x_i^k \, p(x_i) \\ m_k &= E[(X - \hat{\mu})^k] = \sum_{i=1}^{n} (x_i - \hat{\mu})^k \, p(x_i) \end{aligned} \quad (15)$$

The following equations are obtained after some manipulation:

$$M_k = \hat{\beta}^k \frac{k}{\hat{\alpha}} \Gamma\left(\frac{k}{\hat{\alpha}}\right) \quad (16)$$

$$m_k = (-1)^{k-1} \hat{\mu}^k (k-1) + \sum_{r=0}^{k-2} \frac{k!}{(k-r)!r!} (-1)^r M_{k-r} \hat{\mu}^r \quad (17)$$

The kurtosis is a ratio of the fourth central moment of the Weibull distribution to the square of the variance. Thus

$$\kappa = \frac{m_4}{m_2^2} \tag{18}$$

4 EXPERIMENTAL SETUP AND PROCEDURE

The experimental setup, comprised of a drive motor, hydraulic gear pump, Metrabyte Das-16F I/O board, optical switch, Krohn-Hite model 3750 analog filter, torque and speed sensor, Kluite Semiconductor pressure transducer (model XTM-190-200 SG), Brüel and Kjaer accelerometer (model 4368), oscilloscope, variable orifice, and an IBM compatible PC.

The hydraulic gear pump was a Vickers single pump model (F3)-G30-7D30D-2A-1D-31. It had SAE rate capacity of 30 USgpm at 1200 rpm and 100 psi. There were two ten-teeth spur gears, each gear having a diametrical pitch of 5 (25 mm pitch circle radius), 20° pressure angle and 47 mm face width.

The gear pump was driven by a hydraulic servo controlled motor. A torque/speed sensor connected the motor to the gear pump. The outlet (or pressure side) of the gear pump was connected via a variable orifice to the oil reservoir, to provide loading. An optical switch was attached to the input shaft of the pump for time synchronous averaging. The optical switch triggered the A/D converter in the Metrabyte Das-16F I/O board to read the input from the accelerometer, that was mounted on the gear pump casing. These signals were digitised by the A/D converter, and prefiltered by an analog filter in the bandpass mode. After data collection, the collated data was analysed using the Weibull distribution software, written in C, that was resident in the microcomputer.

Prior to beginning the experiment to determine the condition of the gear teeth in the pump, the accelerometer was moved to various positions on the test-rig in order to determine the position that best detected vibration from gear meshing. This position was found to be at the top of the gear casing. This surface was then milled, drilled and threaded for effective mounting of the accelerometer.

Having attached the accelerometer, four experiments were designed to simulate various types of gear damage. A cracked tooth was simulated in the first experiment. This was achieved by cutting a 2 mm deep slot into the root of a tooth on the driven gear using an electric discharge machine (EDM). Preliminary runs of the pump revealed no defect because the load on the pump could not provide the envisaged cantilever beam action of the tooth. Thus about 10 mm was milled off the face width from either side of the tooth. The variable orifice was opened to 58.33% of its capacity while the pump speed was maintained at 850 rpm (about 70% of the rated speed) using the controller box. Data collection was then initiated with the sampling rate at 57800 samples/sec and the number of samples at 4096. The sampling rate is selected to permit the collection of 4080 samples in one revolution of the gear, thus providing 408 samples per tooth mesh.

In the second experiment, the driven gear had two defective teeth. One tooth had a crack (from the first experiment) while the other had newly introduced "gross pitting". The pits, which were spread across the face width along the pitch line, were introduced by arc welding. Five pits were used each 4 mm in diameter. The process as outlined in the first experiment was then repeated.

The third experiment was to investigate the

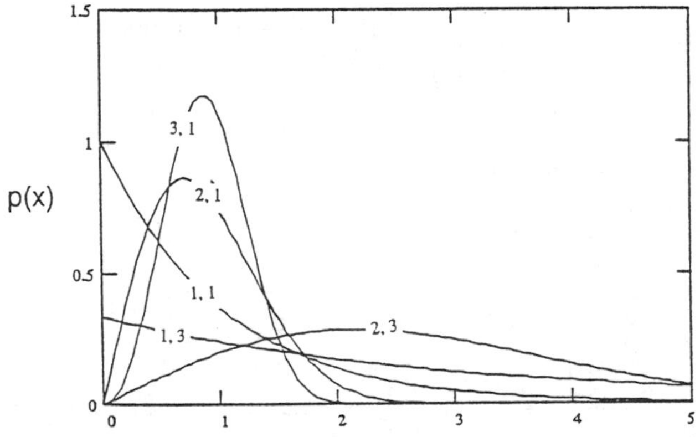

Figure 1. PDF of Weibull(α,β)

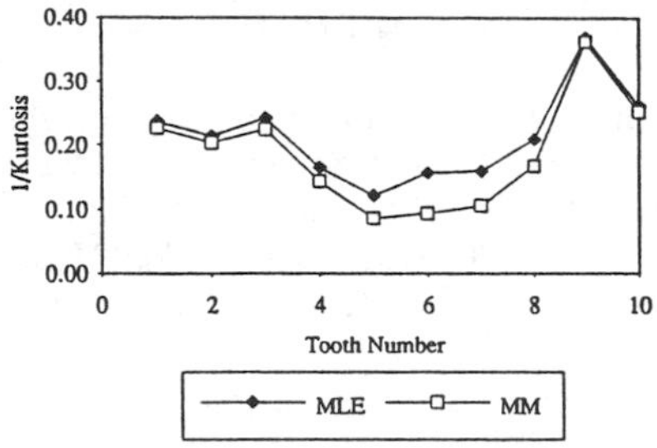

Figure 2. Cracked tooth

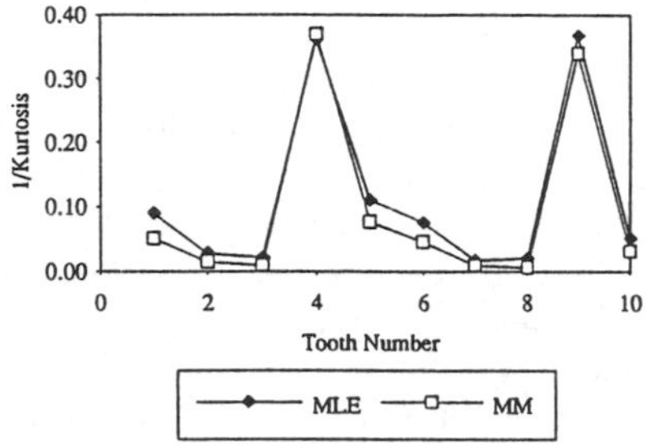

Figure 3. Cracked tooth and one grossly pitted tooth

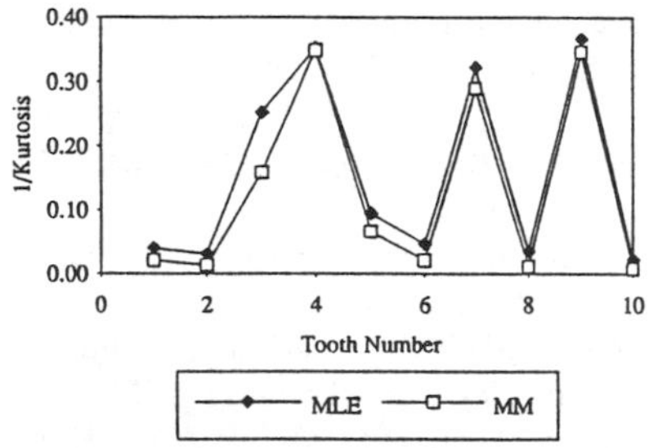

Figure 4. Cracked tooth, one grossly pitted tooth, and one mildly pitted tooth

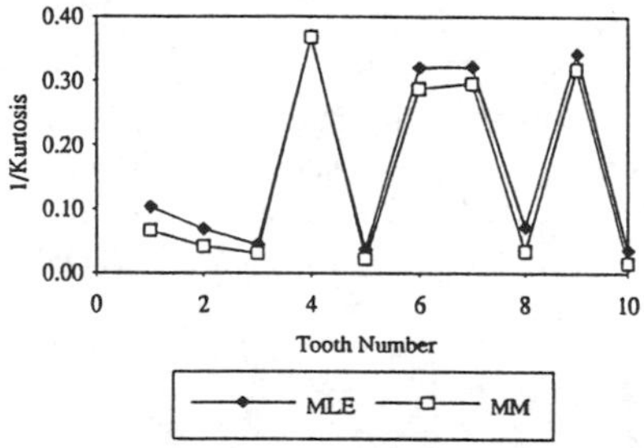

Figure 5. Cracked tooth, one grossly pitted tooth, and two mildly pitted teeth

detectable pitting threshold. To this end, a 4 mm diameter pit was introduced on the pitch circle at about the midpoint of the tooth face length and the process described in the first experiment was repeated.

The fourth experiment was aimed at investigating the possibility of detecting pitting in two adjacent teeth. Thus a 4 mm diameter pit was introduced on a tooth adjacent to the tooth used in the third experiment and the process described in the first experiment was repeated.

5 RESULTS AND ANALYSIS

In the first experiment, crack was simulated by the introduction of a crack on the ninth tooth. The plot of the inverse kurtosis against the tooth numbers is depicted in Figure 2. The shape parameters have been estimated using both MM and MLE techniques.

A peak is observed on the ninth tooth. The peak obtained with the MLE technique is slightly higher than that with the MM. The teeth without simulated defect are observed to display varying kurtosis values, and none is equal to that at the ninth tooth.

The same gear as that used in the first set of experiments was modified by pitting the fourth tooth in the second set of experiments. Figure 3 depicts plots of the inverse kurtosis against tooth number.

The plots indicate defects in tooth numbers four and nine. It is observed that, for the ninth tooth, the peak obtained when the characteristic parameters are estimated using MLE is higher than the corresponding peak obtained using MM. However, the peak on the fourth tooth weakly indicates the converse. Further, it is worth mentioning that the value of the peaks on the ninth tooth are approximately equal to those from the first experiment. This could be interpreted as no change in the damage severity.

In the third experiment, a single pit was introduced on the tooth surface of the seventh tooth. All the defective teeth are correctly identified (Figure 4). The kurtosis of the signals from all but one tooth (i.e. tooth number 3) without simulated defect reflect same. Tooth number 3, however, can be wrongly considered defective based on the MLE plot.

Comparing Figure 4 and Figure 3, it is observed that while the kurtosis value of the ninth tooth remained virtually unchanged, a slight drop is recorded in the fourth. However, it is perhaps premature to conclude that it is indicative of a smoothing of the pits.

The objective of the fourth experiment was to investigate the detectability of damage on adjacent teeth. Thus a pit was introduced on the sixth tooth of

Table 1. Computation Times

Experiment	Computation Time (mins)	
	MLE	MM
1	7.74	39.83
2	7.63	41.60
3	7.66	40.88
4	7.68	39.81

the same gear used in the third set of experiments. Figure 5 depicts the plots of the inverse kurtosis against time for both MLE and MM shape parameters estimation techniques. This figure shows that all four defective teeth, namely, four, six, seven, and nine are correctly detected.

The computation time for the results are tabulated in Table 1. The average time required with the MLE based method is about 7.68 mins while the MM based method averaged around 40.53 mins. The difference in these times is explained by the amount of iterations (or calculations) required by the MM based method. It suffices to say that a look-up library of scale parameters, α_1, for various values of Φ was developed to expedite the analysis. To our knowledge, there is no known algorithm to calculate the shape parameters of a Weibull distribution using MM.

6 CONCLUSION

The use of the Weibull distribution as a statistical distribution for gear damage detection has been successfully demonstrated. It is observed that the ability to detect defects is independent of the characteristic parameters estimation technique (MM and MLE). However, the MLE is suggested over the MM because of its lower process time, a contradiction of the observation on the Beta distribution (Oguamanam et al 19xx).

While the distribution provides kurtosis values that are indicative of gear damage, a comparison has to be made relative to other gear teeth. In other words, the distribution, unlike the normal distribution, does not provide a threshold value below or above which a defect could be concluded. Thus there is no avenue to discriminate the various defects, a major disadvantage of statistical distribution based gear monitoring systems.

7 ACKNOWLEDGEMENT

The authors would like to acknowledge the financial support of the Natural Sciences and Engineering Research Council of Canada (NSERC), Grant Numbers OGP0006468 and OGP0007729, and Frontier Hydraulics Ltd. for providing the gear pump.

REFERENCES

Bartelmus, W., 1984. Application of some statistical estimators of the vibration signal as criteria of assessment of mesh state. *Condition Monitoring '84, Swansea, 10-13 April 1984:* 256-265.

Beckman, R.J. & G.L. Tietjen 1978. Maximum likelihood estimate for the beta distribution. *Statistical Computation and Simulation J.* 7:253-258.

Bendat, J.S. & A.G. Piersol 1985. *Random data: analysis and measurement procedures.* New York: Wiley.

Chapra, S.C. & R.P. Canale 1988. *Numerical methods for engineers.* 2nd edition. New York: McGraw-Hill.

Martin, H.R., F. Ismail & S. Sakuta 1990. New statistical approach for gear damage detection. *2nd IMMDC Conf.* 329-334. Los Angeles.

Menon, M.V. 1963. Estimation of the shape and scale parameters of the Weibull distribution. *Technometrics.* 5(2):175-182.

Milton, J.S. & J.C. Arnold 1990. *Introduction to probability and statistics: principles and applications for engineering and the computing sciences.* 2nd edition. New York:McGraw-Hill.

Oguamanam, D.C.D., H.R. Martin & J.P. Huissoon 1995. On the application of the beta distribution in gear damage analysis, *Applied Acoustics.* 45(3): 247-261.

Shipley, E.E. 1967. Gear failures. *Machine Design.* 39(28):152-162.

Thoman, D.R., L.J. Bain & C.E. Antle 1969. Inferences on the parameters of the Weibull distribution. *Technometrics.* 11(3):445-460.

Whitehouse, D.J. 1978. Beta functions for surface typologies? *Annals of the CIRP.* 27(1):491-497.

Modern Practice in Stress and Vibration Analysis, Gilchrist (ed.) © *1997 Balkema, Rotterdam, ISBN 90 5410 896 7*

The estimation of rotor unbalance from shaft and pedestal vibration measurements

A.W. Lees & M.I. Friswell
University of Wales, Swansea, UK

ABSTRACT: A method is presented to determine the state of unbalance of a rotating machine using the measured pedestal and shaft vibration. The only requirements of the procedure are a good numerical model for the rotor. No assumptions are made concerning the operational mode shape of the rotor and the influence of the supporting structure is included in a consistent manner. For simplicity the analysis is presented in a single plane orthogonal to the rotor axis, but no difficulty is foreseen in extending the method to two directions. Examples are given for a two bearing system.

1 INTRODUCTION

Methods of balancing can be categorised into two groups, the influence coefficient method which only requires the assumption of linearity of both the machine and measuring system, and modal balancing which in addition requires a knowledge of the modal properties of the machine. The former of these approaches has the attraction of requiring less a priori knowledge of the system and techniques have been well developed to make optimum use of redundant information (Drechsler, 1980). The approach does however suffer from the significant disadvantage of requiring a number of test runs on site. For machinery with a high commercial output, this is a significant disadvantage.

Modal approaches require fewer test runs but, as mentioned, require prior knowledge of the machine. Two methods have been proposed in recent years (Gnielka, 1983 and Krodkiewski *et al.*, 1994) which offer the prospect of balancing without test runs. Gnielka (1983) uses prior knowledge of mode shapes and modal masses and compares measured results to a numerical model of the machine. The work of Krodkiewski *et al.* (1994) has similar requirements and seeks to detect changes in unbalance from running data. Both these approaches place reliance on the model of the machine. In a real machine, however, the modal properties of the machine may be significantly influenced by the properties of the structure on which it is supported, and this has proved extremely difficult to model (Lees and Simpson, 1983). Numerical models of rotating machinery have been used to great effect over a number of years (McCloskey and Adams, 1992 and Frigeri *et al.*, 1988), and their accuracy and range of effectiveness have been steadily developing. The principle limitation is now thought to be the unknown effects of the foundation and recently several authors have addressed this issue (Lees, 1988, Zanetta, 1992, Feng and Hahn, 1995).

In principle, if a numerical model of a machine were sufficiently reliable, rotor unbalance could be derived directly from the measured vibration levels at the bearing. In this paper it is shown how rotor unbalance may be derived from measured data using an accurate model of the rotor. Lees and Friswell (1997) used an approximate model of the bearings and measurements of the pedestal vibration to estimate the unbalance. In this paper measurements of shaft vibration are used so that no model of the bearing is required. No knowledge is required of the supporting structure: this is represented as stiffness and mass matrices whose coefficients are determined as a part of the calculation. It is shown that good estimates for the unbalance at each of the pre-determined balancing planes may be derived .

2 BEARING FORCES

The rotor is considered to be adequately modelled

and so the free-free modal properties of the complete rotor train may be calculated from the model. The free-free modes of the component rotors may be measured by suspending the rotor in slings and can thus verify the model. Knowledge of these free-free modes is simply one way of representing the dynamics of the rotor itself; mode shapes and frequencies are readily transformed into stiffness and mass matrices of the rotor. There is additional advantage in this approach since rotors are often identical amongst a class of machines whereas supporting structures are extremely variable (Lees and Simpson, 1983).

Given the rotor model, the motion of the rotor is the result of two sets of forces. Firstly, there is a set of unbalance forces which can arise anywhere along the rotor, but over the running speed range of a machine, the resultant forces can be represented as a limited number of modal contributions, which in turn can be represented as a series of discrete unbalances at a fixed number of balance planes along the rotor. Secondly, forces arise in the bearings and the time dependent part of these forces arise as a consequence of the unbalance forces. At any point x along the rotor, the displacement y_r is given by

$$y_r(\omega, x) = \int_0^L G(\omega, x, x') F(\omega, x') \, dx' \tag{1}$$

where $F(\omega, x')$ represents the force per unit length along the rotor, and $G(\omega, x, x')$ is the Green's function of the rotor, representing the response at point x arising from a unit force at x'. This function depends on frequency, and using the standard result

$$G(\omega, x, x') = \sum_{j=1}^{\infty} \frac{\psi_j^*(x)\psi_j(x')}{\omega_{nj}^2 - \omega^2} \tag{2}$$

where the mode shapes have been normalised to the rotor mass, so that

$$\int_0^L \psi_j^*(x)\rho(x)\psi_j(x)\, dx = 1 \tag{3}$$

and $\rho(x)$ is the mass per unit length of the rotor at x, ω_{nj} is the jth natural frequency of the rotor (free-free), ψ_j represents the corresponding mode shape and * denotes the complex conjugate transpose. The frequencies and mode shapes may be estimated using a discrete approximation via a finite element model of the rotor. In the above representation, no allowance has been made for damping within the rotor. There would be little difficulty in including such a term but in practice damping from the bearings and supporting structure will be dominant in turbo machinery. In the present calculation the damping is neglected for the sake of clarity.

The forces acting on the rotor are of two types, the unbalance at various locations which are acting at m locations $x_{e1}, \ldots, x_{em}$ which are taken as a set of pre-defined locations. Note that for a finite range of running speeds, the effective unbalance distribution may be represented as a summation over the balance planes (Kellenburger, 1972). The unknown bearing reaction forces act at the n bearing locations $x_{b1}, \ldots, x_{bn}$. In many cases $n = m$, but separate symbols are retained for generality and clarity. The bearing forces $F_{b1}, \ldots, F_{bn}$ have yet to be determined. The total force acting on the rotor becomes

$$F(\omega, x) = \sum_{i=1}^{m} \omega^2 e_i \, \delta(x - x_{ei}) + \sum_{i=1}^{n} F_{bi}(\omega)\delta(x - x_{bi}) \tag{4}$$

where δ is the Dirac delta function, and e_i is the unbalance (the product of unbalance mass and radius) at location x_{ei}. Inserting this expression into equation (1) yields the displacement of the shaft at any point x as

$$y_r(\omega, x) = \sum_{i=1}^{m} G(\omega, x, x_{ei})\omega^2 e_i + \sum_{i=1}^{n} G(\omega, x, x_{bi}) F_{bi}(\omega) \tag{5}$$

Given n (or $2n$ if two directions are considered) measured values of the rotor displacements $y_{rj} = y_r(x_{bj})$ then a set of simultaneous equations may be formed linking the bearing force with the shaft displacements at the bearing locations by setting $x = x_{b1}, \ldots, x_{bn}$ in turn. Hence

$$y_{rj}(\omega) = \sum_{i=1}^{m} G_{je_i}(\omega)\, \omega^2 e_i + \sum_{i=1}^{n} G_{ji}(\omega) F_{bi}(\omega) \tag{6}$$

where $G_{pq}(\omega) = G(\omega, x_{bp}, x_{bq})$ and $G_{pe_q}(\omega) = G(\omega, x_{bp}, x_{eq})$. In previous studies (Lees and Friswell, 1996) the unbalance was considered to be known. In the present work, however, the unbalance in each of the pre-set balancing planes may be represented as an unknown vector $\{e\}=\left[e_1,\ldots,e_n\right]^T$. Then for frequencies away from the free-free resonances of the rotor, the forces may be written as

$$\{F(\omega)\}=\left[G_{bb}(\omega)\right]^{-1}\left\{\{y_r(\omega)\}-\left[G_{be}(\omega)\right]\omega^2\{e\}\right\} \quad (7)$$

where

$$\left[G_{bb}(\omega)\right]=\begin{bmatrix} G(\omega,x_{b1},x_{b1}) & \cdots & G(\omega,x_{b1},x_{bn}) \\ \vdots & \ddots & \vdots \\ G(\omega,x_{bn},x_{b1}) & \cdots & G(\omega,x_{bn},x_{bn}) \end{bmatrix} \quad (8)$$

and

$$\left[G_{be}(\omega)\right]=\begin{bmatrix} G(\omega,x_{b1},x_{e_1}) & \cdots & G(\omega,x_{b1},x_{e_n}) \\ \vdots & \ddots & \vdots \\ G(\omega,x_{bn},x_{e_1}) & \cdots & G(\omega,x_{bn},x_{e_n}) \end{bmatrix} \quad (9)$$

Thus

$$\{F(\omega)\}=[p(\omega)]\{y_r(\omega)\}-[q(\omega)]\{e\} \quad (10)$$

where

$$[q(\omega)]=\omega^2\left[G_{bb}(\omega)\right]^{-1}\left[G_{be}(\omega)\right] \quad (11)$$

and

$$[p(\omega)]=\left[G_{bb}(\omega)\right]^{-1} \quad (12)$$

Note that the expression for the forces exerted on the rotor and the reaction on the bearing pedestal are independent of the foundation dynamics. The foundation will, of course, strongly influence, the rotor vibration y_r, but this is measured. A full discussion of this approach and the conditioning of the matrices has been given by the authors (Lees and Friswell, 1996, Friswell *et al.*, 1996).

3 INCLUDING THE FOUNDATION PARAMETERS

The dimension of the vector $\{F\}$ is the number of bearings (times two if both perpendicular directions are considered). The vector $\{F\}$ represents the force acting on the rotor from the bearing reaction, hence the force acting on the supporting structure is just $-\{F\}$. The matrix $\left[G_{be}\right]$ represents the relationship between balance planes and bearings and it is just a subset of the frequency dependent free-free rotor description, as described by equation (2).

Let the dynamic behaviour of the supporting structure be represented by the undetermined stiffness and mass matrices $[K]$ and $[M]$. A damping matrix may also be introduced, but since the damping of rotating machinery is usually dominated by the behaviour of the bearing oil film, the damping matrix may be neglected. Using these matrices, a further vector equation may be written to express the force vector as

$$-\{F(\omega)\}=[K]\{y_p(\omega)\}-\omega^2[M]\{y_p(\omega)\} \quad (13)$$

and an equation is formed at each frequency measured. Note that equation (13) requires the number of modes in the foundation to be equal to the number of bearings. In the case of a foundation with many modes within the running frequency range, the structural matrices should vary. This may be treated by subdivision of the frequency range, resulting in a larger least squares problem, but no additional difficulty in principle. Since the elements of the matrices are unknown, it is convenient to re-write this equation in the form

$$-\{F(\omega)\}=[w(\omega)]\{v\} \quad (14)$$

where $[w]$ contains all references to $\{y_p\}$, and the vector $\{v\}$ contains the elements of matrices $[K]$ and $[M]$. The ordering of this vector is arbitrary, but for the two bearing, single direction case we make the choice $\{v\}=\left[k_{11}\quad k_{22}\quad k_{12}\quad m_{11}\quad m_{22}\quad m_{12}\right]^T$, where k_{ij} represents the (i,j)th element of $[K]$. It has been assumed that both $[K]$ and $[M]$ are symmetric so that $k_{ij}=k_{ji}$ and $m_{ij}=m_{ji}$. In this case $[w(\omega)]$ is a 2x6 matrix taking the form

$$[w(\omega)]=\begin{bmatrix} y_{p1}(\omega) & 0 \\ 0 & y_{p2}(\omega) \\ y_{p2}(\omega) & y_{p1}(\omega) \\ -\omega^2 y_{p1}(\omega) & 0 \\ 0 & -\omega^2 y_{p2}(\omega) \\ -\omega^2 y_{p2}(\omega) & -\omega^2 y_{p1}(\omega) \end{bmatrix}^T \quad (15)$$

Equating the two force expressions (10) and (14) at each frequency yields the equation

$$-[w(\omega)]\{v\} = [p(\omega)]\{y_r(\omega)\} - [q(\omega)]\{e\} \tag{16}$$

Assembling these equations at each measured frequency monitored, gives

$$[W]\{v\} + [Q]\{e\} = \{P\} \tag{17}$$

where the matrices $[Q]$ and $[W]$ and the vector $\{P\}$ are formed as

$$[Q] = \begin{bmatrix} [q(\omega_1)] \\ [q(\omega_2)] \\ \vdots \\ [q(\omega_N)] \end{bmatrix}, \quad [W] = \begin{bmatrix} [w(\omega_1)] \\ [w(\omega_2)] \\ \vdots \\ [w(\omega_N)] \end{bmatrix}$$

and $$\{P\} = \begin{Bmatrix} [p(\omega_1)]\{y_r(\omega_1)\} \\ [p(\omega_2)]\{y_r(\omega_2)\} \\ \vdots \\ [p(\omega_N)]\{y_r(\omega_N)\} \end{Bmatrix}$$

where ω_i is the ith of N frequency points at which displacement is measured.

Equation (17) may be written in the form

$$\begin{bmatrix} W & Q \end{bmatrix} \begin{Bmatrix} v \\ e \end{Bmatrix} = \{P\} \tag{18}$$

This represents an over-specified set of equations for the unknown unbalance and foundation parameters, provided sufficient frequencies are measured. The least squares solution may be obtained using the Moore-Penrose pseudo-inverse as

$$\begin{Bmatrix} v \\ e \end{Bmatrix} = \begin{bmatrix} W^T W & W^T Q \\ Q^T W & Q^T Q \end{bmatrix}^{-1} \begin{bmatrix} W^T \\ Q^T \end{bmatrix} \{P\} \tag{19}$$

In many practical situations there may not be sufficient information to determine all parameters, but this may be overcome by the introduction of other physical information, such as restricting the foundation matrices to be banded, or by regularisation of the problem by other means. A convenient approach to this is by the use of Singular Value Decomposition (SVD), and limiting the number of singular values used for the matrix inversion. Use of this technique was not necessary for the examples in this paper but the interested reader is referred to the excellent book by Golub and Van Loan (1989) for further details of the method.

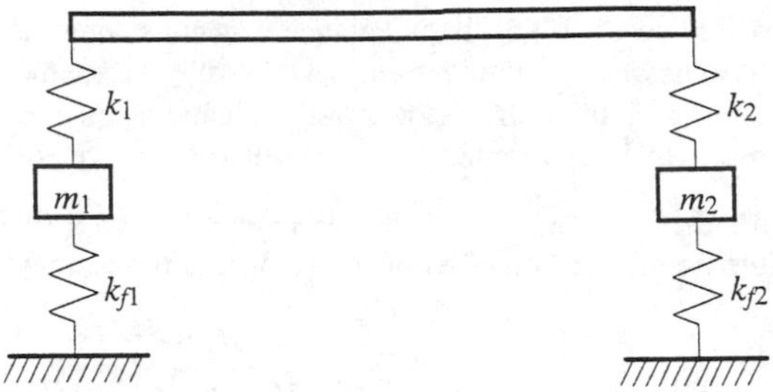

Figure 1. The two bearing example

Table 1. Summary of the cases considered for the two bearing example

Case	Frequency Range (Hz)	No of Freq. Steps	Noise Level (%)
1	120	120	0
2	120	120	2
3	120	120	5
4	120	50	5
5	80	50	5
6	50	50	5
7	80	30	5
8	120	120	20

The direct solution method has been used in the present calculations in which $[W]$ and $[Q]$ are both real. In practice these matrices will usually be complex, and in that case equation (19) must be divided into real and imaginary parts. The unknown parameters are real and are given by the solution of the equation

$$\begin{bmatrix} \mathrm{Re}[W] & \mathrm{Re}[Q] \\ \mathrm{Im}[W] & \mathrm{Im}[Q] \end{bmatrix} \begin{Bmatrix} v \\ e \end{Bmatrix} = \begin{Bmatrix} \mathrm{Re}[P] \\ \mathrm{Im}[P] \end{Bmatrix} \tag{20}$$

Equation (17) remains valid for the study of two orthogonal directions. If both orthogonal directions are considered, and there is damping present in the bearing model, $[W]$, $[Q]$ and $\{P\}$ all become complex. The parameters to be identified remain real. The foundation parameter vector $\{v\}$ will now contain direct and cross stiffness terms relating to the horizontal and vertical directions. The unbalance vector will now have 2 components at each balance plane. The resulting equation is solved as before.

Table 2. The mean estimated parameters for the two bearing example from a Monte-Carlo simulation. The standard deviations of the parameters are given in brackets

Case	k_{11} (MN/m)	k_{22} (MN/m)	k_{12} (MN/m)	m_{11} (kg)	m_{22} (kg)	m_{12} (kg)	e_1 (kg m)	e_2 (kg m)
1	177	177	0	95	135	0	2	3
2	178 (5)	176 (5)	0 (5)	101 (17)	134 (21)	1 (17)	1.94(0.01)	2.94 (0.02)
3	181 (13)	178 (15)	-3 (13)	114 (47)	142 (58)	-5 (47)	1.86 (0.04)	2.87 (0.04)
4	178 (13)	175 (19)	-1 (10)	104 (45)	131 (76)	2 (34)	1.86 (0.04)	2.87 (0.04)
5	81 (5)	82 (4)	111 (5)	-24 (20)	4 (21)	613 (22)	2.60 (0.04)	2.63 (0.05)
6	69 (5)	94 (3)	102 (3)	-616 (122)	222 (69)	538 (52)	2.23 (0.07)	2.79 (0.07)
7	82 (6)	83 (6)	111 (6)	-11 (30)	12 (31)	620 (30)	2.56 (0.05)	2.61 (0.05)
8	172 (52)	161 (63)	-8 (55)	117 (192)	113 (239)	7 (197)	1.46 (0.18)	2.47 (0.16)

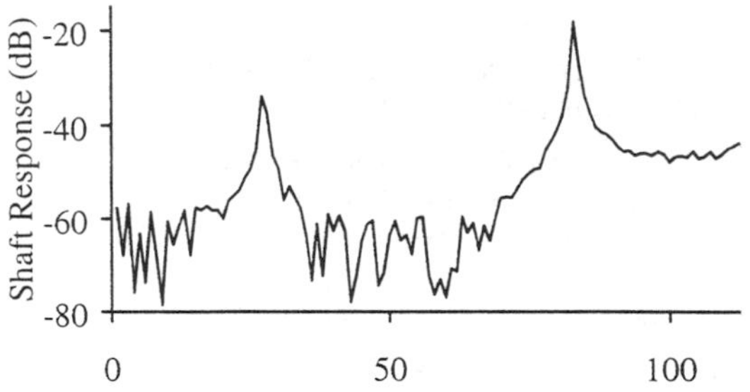

Figure 2. Absolute shaft response at the first bearing, with 5% noise

4 A SIMPLE EXAMPLE

The example considered is illustrated in Figure 1. A simple rotor 4 m long of mass 1450 kg is mounted on two bearings of stiffness 177 and 354 MN/m respectively. These bearings are independently supported on foundations which are each represented by a single mass and stiffness. These two support stiffnesses are both set at 177 MN/m whilst the two pedestal masses are 90 and 135 kg respectively. With these parameters the first two natural frequencies of the free-free rotor alone are 67 and 183 Hz. The rotor also has 2 rigid body modes. The bearing stiffnesses were held constant over the frequency range. It is recognised that this is unrealistic, but the simple model suffices to illustrate the important features of the method. Table 1 shows the range of tests considered. The first test uses data from 120 frequency points over the range of 0-120 Hz, which contains the first two modes. In the remaining cases, the effects of restricting the frequency range and the number of data points is examined in the presence of noise. Gaussian random noise is added to both the shaft and pedestal vibration as a factor times the high frequency pedestal vibration response. Figure 2 shows the shaft response at the first bearings for test case 3.

4.1 Results

Table 2 shows the results for the cases given in Table 1. Each calculation using Gaussian random noise was the average of fifty independent calculations. In case 1, the identification calculation is carried out with no noise present. No difficulty is experienced in the correct identification of both the elastic coefficients of the model and the unbalance at the pre-determined balance planes. Cases 2 and 3 cover the same frequency range, covering two natural frequencies, and both cases show excellent treatment of the noisy signals.

Case 4 utilises fewer frequency steps but retains the same frequency range. The effect of the noise produces slightly higher standard deviations, although the results are still acceptable. Case 8 considers the highest level of noise, but the mean unbalance predictions remain within 27 % of the exact result, with a standard deviation of only 9 %.

As the frequency range is reduced to 50 Hz, case 6, the accuracy of the model parameters becomes unacceptable. Note that this reduced frequency range covers only one of the natural frequencies of the system. It is worthy of note, however, that the predictions of the unbalance are only 12 % in error. Cases 5 and 7 cover a wider frequency range but with many fewer frequencies. It is not surprising that the noise degrades the estimate of unbalance.

4.2 Discussion

No damping has been included in the studies in this paper. The inclusion of damping presents no real difficulties, but for an analysis in a single plane, the prediction of phase will be examined in a subsequent paper. For a fully self consistent analysis, it is believed that consideration and measurements are necessary in both directions perpendicular to the rotor axis. This analysis follows logically from the methods presented above, but a full study of the sensitivity of this case is beyond the scope of the present paper.

A notable feature of the method is that in the presence of noise the unbalance estimates are more accurate than the mass and stiffness parameters. The mass parameters are the terms most prone to errors in the estimation procedure. Clearly this behaviour is related to the form of the equation errors which are minimised. To be estimated accurately the rotor has to excite a sufficient number of the resonances of the foundation alone. Even when the frequency range is insufficient to produce good estimates of the foundation parameters, good estimates for the unbalance are still produced. Since unbalance estimation is an important practical problem, the accuracy demonstrated above indicates that a useful technique has been outlined.

5 CONCLUSIONS

A method has been presented to derive unbalance components using measured data and a model of the rotor only. Measurements of the shaft vibration and the pedestal vibration are required at each bearing. The approach is easily generalised to two directions. Although stiffness and mass terms show moderate sensitivity to uncertainties, the method has been shown to produce unbalance estimates which are insensitive to measurement noise. Some statistical bias is introduced into the parameter estimates, although for practical levels of measurement noise this would be acceptable. Data is required covering as many modes as there are balancing planes.

ACKNOWLEDGMENTS

This work forms part of a project funded by Nuclear Electric Ltd. and Magnox Electric plc. to derive methods for inferring the influence of flexible turbo-generator foundations. The authors thank them for funding this work and for permission to publish this paper. Dr. Friswell gratefully acknowledges the support of the EPSRC through the award of an Advanced Fellowship.

REFERENCES

Drechsler, J. 1980. Processing surplus information in computer aided balancing of large flexible rotors. *I.Mech.E Conference on Vibrations in Rotating Machinery, Cambridge, UK*: 65-70.

Feng, N.S. and Hahn, E.J. 1995. Including foundation effects on the vibration behaviour of rotating machinery. *Mechanical Systems and Signal Processing*, 9(3): 243-256.

Frigeri, C., Zanetta, G.A. and Vallini, A. 1988. Some in-field experiences of non-repeatable behaviour in the dynamics of rotating machinery. *I.Mech.E. Conference on Vibrations in Rotating Machinery, Edinburgh, UK*: Paper C302/88, 395-404.

Friswell, M.I., Lees, A.W. and Smart, M.G. 1996. Model updating techniques applied to turbo-generators mounted on flexible foundations. *NAFEMS conference on Structural Dynamics Modelling: Test Analysis and Correlation, Cumbria, UK*: 461-472.

Gnielka, P. 1983. Modal balancing of flexible rotors without test runs: an experimental investigation. *Journal of Sound and Vibration,* 90(2): 157-172.

Golub, G. and Van Loan, C. 1989. *Matrix Computation*. John Hopkins University Press.

Kellenburger, W. 1972. Should a flexible rotor be balanced in N or N+2 planes? *Trans. ASME, Journal of Engineering for Industry*, 94: 548-560.

Krodkiewski, J.M., Ding, J. and Zhang, N. 1994. Identification of unbalance change using a non-linear mathematical model for rotor bearing systems. *Journal of Sound and Vibration*, 169(5): 685-698.

Lees, A.W. 1988. The least squares method applied to investigating rotor/foundation interactions. *I.Mech.E. Conference on Vibrations in Rotating Machinery, Edinburgh, UK*: Paper C306/065, 209-215.

Lees, A.W. and Friswell, M.I. 1996. Estimation of forces exerted on machine foundations. *International Conference on Identification in Engineering Systems, Swansea, UK*: 793-803.

Lees, A.W. and Friswell, M.I. 1997. The evaluation of rotor unbalance in flexibly mounted machines. *Journal of Sound and Vibration*, submitted.

Lees, A.W. and Simpson, I.C. 1983. The dynamics of turbo-alternator foundations. *I.Mech.E. Conference, London, UK*: Paper C6/83.

McCloskey, T.H. and Adams, M.L. 1992. Troubleshooting power plant rotating machinery vibration using computational techniques: case histories. *I.Mech.E. Conference on Vibrations in Rotating Machinery, Bath, UK*: Paper C432/137, 239-250.

Zanetta, G.A. 1992. Identification methods in the dynamics of turbo-generator rotors. *I.Mech.E. Conference on Vibrations in Rotating Machinery, Bath, U.K*: Paper C432/092, 173-182.

Modern Practice in Stress and Vibration Analysis, Gilchrist (ed.) © 1997 Balkema, Rotterdam, ISBN 90 5410 896 7

Some aspects of novelty detection methods

C. Surace
Department of Structural Engineering, Politecnico di Torino, Italy
K. Worden
Department of Mechanical Engineering, University of Sheffield, UK

ABSTRACT: In this article, an implementation of the so-called "novelty detection" method to monitor damage in structures is proposed. As opposed to the conventional damage identification procedures, the novelty detection approach is more general in that can be applied to systems of arbitrary complexity that present more than one normal operating condition. This would be particularly advantageous for applications in which the structure to be monitored is known to undergo alterations in dynamic behaviour due to other causes such as a change in mass and not just to damage.
The method described has been applied to a numerical simulation of a mechanical system with three discrete degrees of freedom. Using the results obtained it has been possible to analyse the effects of measurement noise on the performance of the method and assess the sensitivity of the method to structural alterations.

1 INTRODUCTION

One of the fundamental problems currently facing the aerospace industry is concerned with the health monitoring of structures. Maintenance of aircraft is expensive, relying on the regular inspection of craft for structural damage, usually after a fixed number of flight hours. Unlike the situation for hydraulics and avionics, structural damage is not continuously monitored; damage detection is accomplished by taking the craft out of service and applying Non-Destructive Test (NDT) procedures on the ground. There are clear advantages in developing a continuous monitoring system for structural damage.

The method proposed here for health monitoring is via *novelty detection*. This is a technique which has found recent applications in medicine/ physiology for detecting the early development of pathologies by Tarassenko, Roberts (1994). Unlike model-based damage detection methods which rely on the existence of an accurate virgin-state model of the system, novelty detection relies on the existence of a *description of normality* usually defined in terms of measurements or features established during a fault-free period of service. There are similarities with the 'traditional' method of condition monitoring which looks for the development of trends in data which should be constant. The main advantage of the method is that it relies on very few assumptions about the nature of faults, this is vital if the procedure is to diagnose damage of which it has no previous 'experience'.

The main problem in applying this new technology is that of determining the appropriate measurements or features. In the case of in-flight monitoring, these should be capable of signalling condition despite low signal-to-noise and in more or less ignorance of the excitation of the structure. One possibility is to use transmissibility functions. A major advantage of the method is that detailed modelling of the structure is not required, so the frequency band can be taken at the structural range or acoustic range with no real modification to the approach. This obviates the need for detailed FE analysis, a considerable expense and potential source of errors.

The method used here is based on an auto-associative neural network which learns a template or ideal pattern for normal operating conditions. This is later compared to patterns measured during the monitoring period. Simpler methods based on averaged patterns are possible. However, they are subject to restrictions on the form of the normal operating condition set in pattern space. It is shown in this paper how the neural network measures compare to simple averaging measures for various normal operating conditions.

2 NOVELTY INDEX

Novelty detection aims to establish simply whether or not a new pattern is significantly different from a previous pattern, while ignoring any insignificant differences such as random fluctuations due to noise. Correspondingly the distinction between a significant and negligible difference in patterns is made via the so-called *novelty index.*

For the application to structural damage detection presented here, a neural network approach has been adopted, employing an auto-associative multi-layer perceptron network required to reproduce at the output layer those patterns which are presented at the input. This would be a trivial exercise except that the network structure has a 'bottleneck' i.e. the patterns are passed through hidden layers which have fewers nodes than the input layer: this forces the network to learn the significant features of the patterns.

The network can be trained by adding random-generated noise to large number of identical patterns associated with the undamaged structure. These different versions are presented at the input to the network. Thus, the novelty index $\nu(\mathbf{z})$ can be formulated in terms of the distance between the pattern input to the network $\mathbf{z}$ and that obtained at the output $\hat{\mathbf{z}}$. There are several ways to define this distance. In this study the Euclidean distance was considered: $\nu(\mathbf{z})^2 = (\mathbf{z} - \hat{\mathbf{z}})^T(\mathbf{z} - \hat{\mathbf{z}})$.

If learning has been successful, than $\nu(\mathbf{z}) \approx 0$ when $\mathbf{z}$ represents normal condition (it can be demonstrated that in this case ν is proportional to the level of noise on data). If $\mathbf{z}$ corresponds to damage $\nu(\mathbf{z}) > 0$. It is reasonable to assume that ν will increase monotonically with the level of damage. However the way in which ν varies cannot be determined using this method since the result gives only a qualitative indication of whether or not damage is present in the structure.

3 CASE STUDY

In previous studies Worden (1997) and Surace, Worden and Tomlinson (1997), have established the

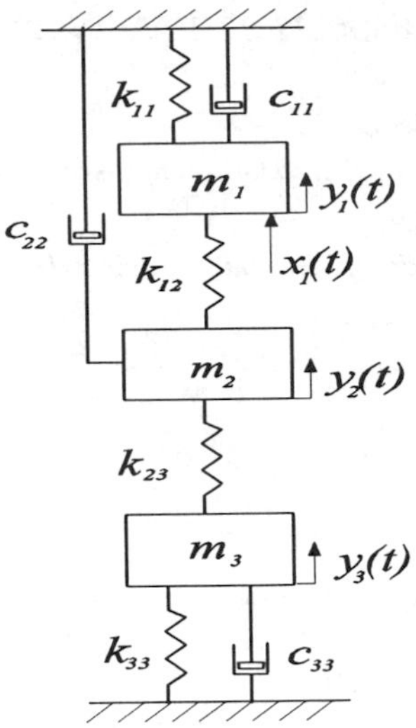

Figure 1: The three degree-of-freedom system simulated

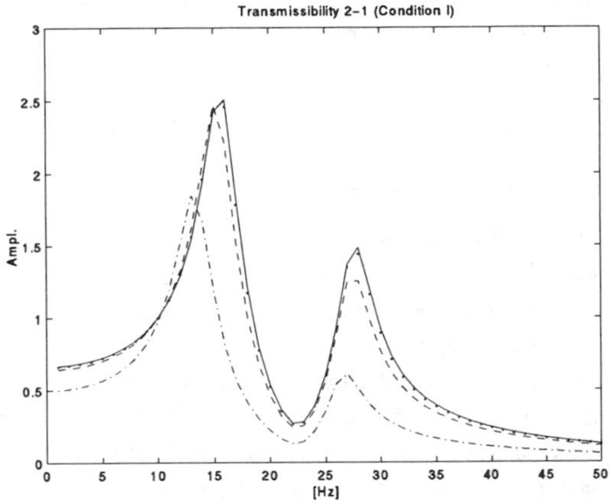

Figure 2: Transmissibility function corresponding to the structure in NCI (–) together with the functions relative to three damage conditions: 2% (. .) , 10% (- -) and 50% (-.-). reductions in k_{12} respectively.

validity of the novelty detection method described above to identify the presence of damage in simple structures for which a single normal operating condition exists. Instead the objective of this case study, conducted via numerical simulation of a lumped-mass structural system, was to evaluate the validity of this method to detect damage in structures which have at least two different normal operating conditions, for example a mechanical system such as an aircraft which may be re-

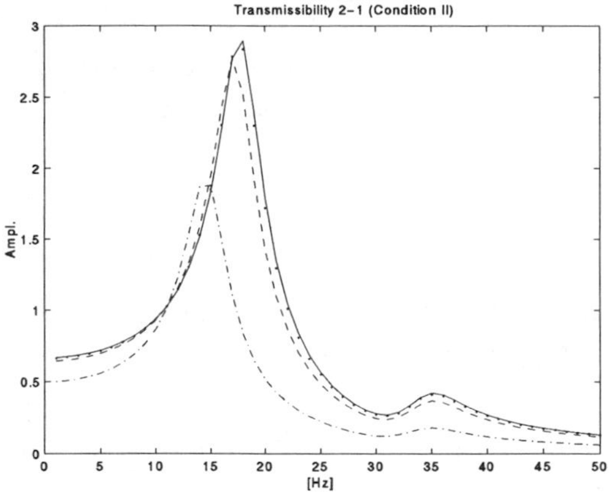

Figure 3: Transmissibility function corresponding to the structure in NCII (–) together with the functions relative to three damage conditions: 2% (. .) , 10% (- -) and 50% (-.-). reductions in k_{12} respectively.

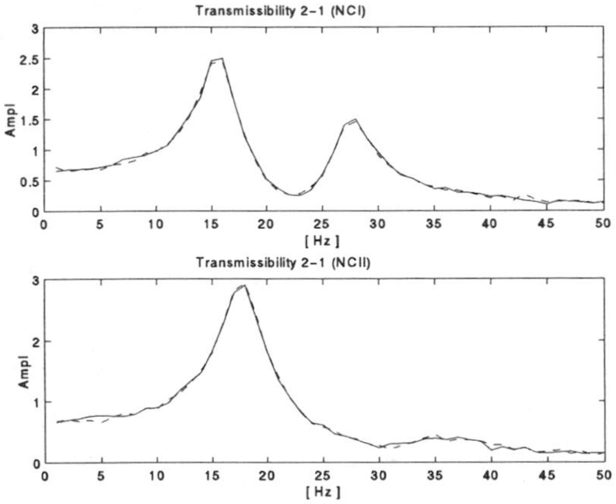

Figure 4: Sample patterns from the noise training set.

quired to operate with a set of different effective masses.

The three degree-of-freedom system with concentrated masses considered by Worden (1997), and shown in Figure 1 was simulated. In the first normal condition (NCI) the following values were used: $m_1 = m_2 = m_3 = 1$, $c_{11} = c_{22} = c_{33} = 1$, $k_{11} = k_{12} = k_{21} = k_{23} = k_{31} = k_{33} = 10^4$. In the second normal condition (NCII) the same parameters were used with the exception that mass m_3 was reduced by 50%. The fault in the system was simulated by decreasing the stiffness k_{12} by different degrees.

In both conditions, the response transmissibility

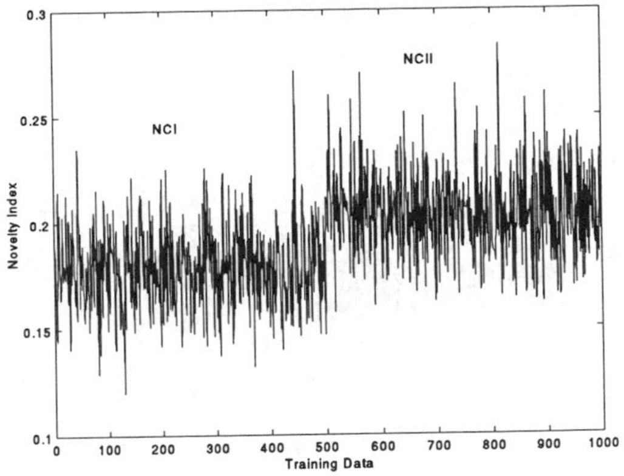

Figure 5: Novelty index calculated on the training set in NCI and in NCII

function between masses m_1 and m_2 was calculated in terms of the transverse displacement for a harmonic excitation applied at mass 1 (Figure 1).

In the simulations performed, the frequency of the harmonic exciting force was varied in the range 0 to 50 Hz in order to encompass the natural frequencies of the system.

Figure 2 illustrates the transmissibility function corresponding to the structure in NCI together with the functions relative to three damage conditions, denoted DCI, with 2%, 10% and 50% reductions in k_{12} respectively. Figure 3 shows equivalent functions corresponding to NCII together with damaged cases with the same reductions in stiffness denoted DCII.

The training vector was obtained by making 500 copies of the transmissibility functions corresponding to the undamaged structure for each of the two conditions, and then polluting each of these independently by adding Gaussian noise with an RMS value equal to 1% of the peak value of the transmissibility function, as shown in Figure 4.

For the task of pattern recognition, a neural network with 5 layers and node structure 50:40:30:40:50 was selected and trained for 100000 cycles, presenting the patterns in random order.

Firstly the novelty index formulated using the Euclidean distance was calculated on the training set. The mean values of ν in the two normal conditions considered were close but different: in NCI ν_1=0.178 and in NCII ν_2=0.204, as shown in Fig. 5. For this reason it was decided to normalise the novelty index as follows:

$$\nu'(\mathbf{z})^2 = \frac{(\mathbf{z} - \hat{\mathbf{z}})^T(\mathbf{z} - \hat{\mathbf{z}})}{\mathbf{z}^T\mathbf{z}} \tag{1}$$

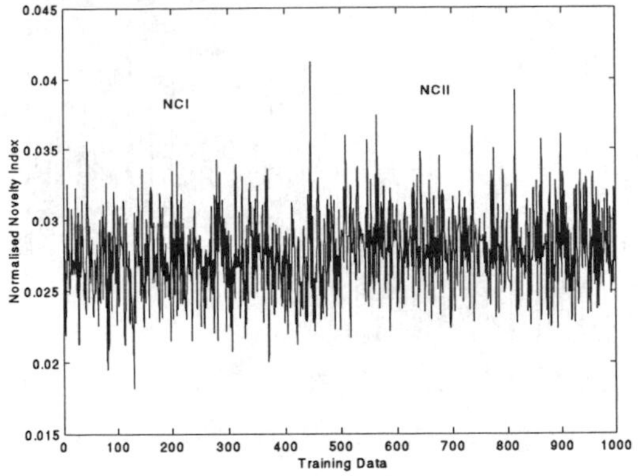

Figure 6: Normalised novelty index calculated on the training set in NCI and in NCII

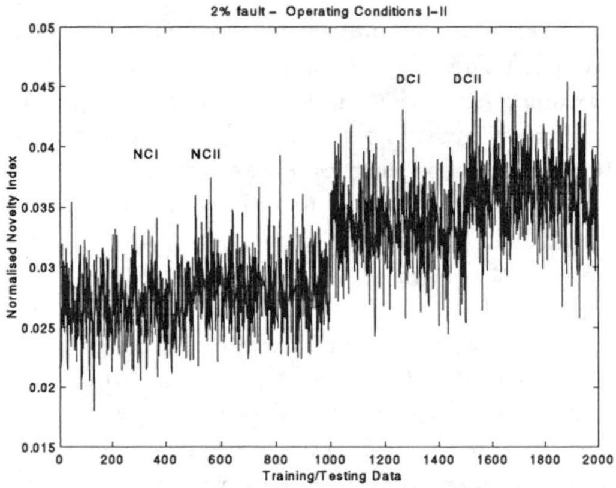

Figure 7: Normalised novelty index: training set NC and 2% fault data DC in the two operating conditions I and II.

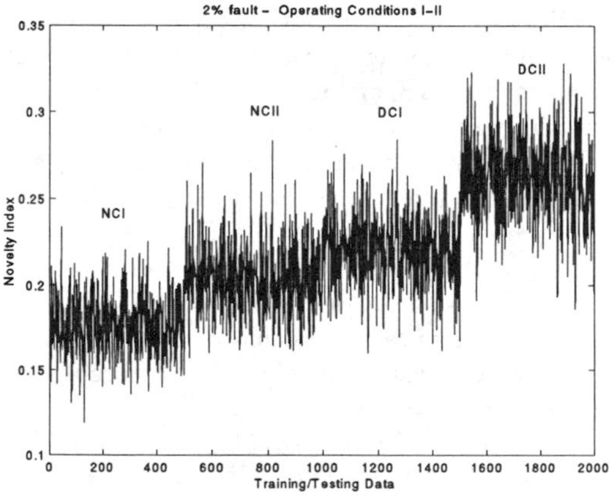

Figure 8: Novelty index: training set NC and 2% fault data DC in the two operating conditions I and II.

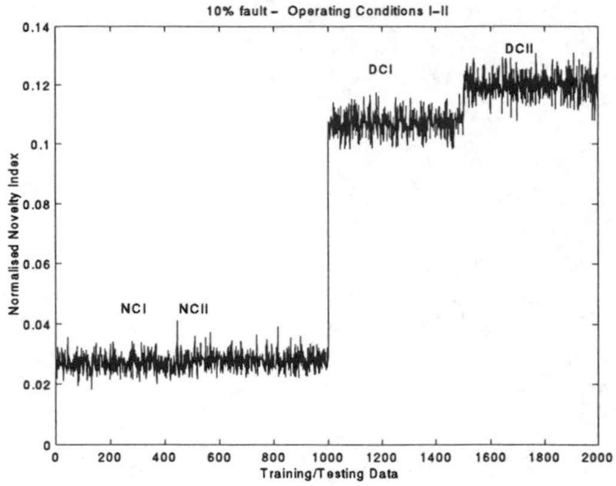

Figure 9: Normalised novelty index: training set NC and 10% fault data DC in the two operating conditions I and II.

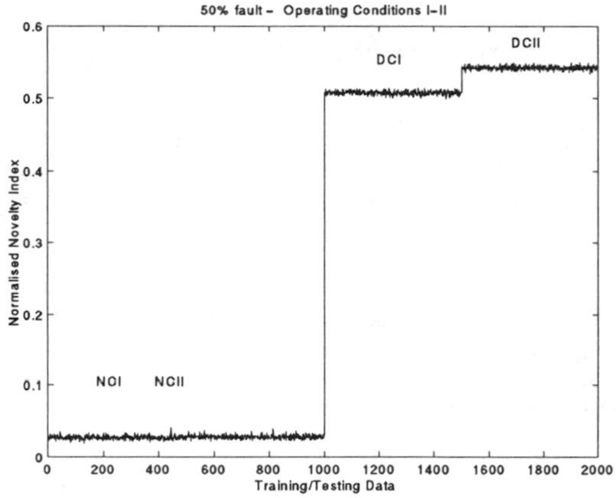

Figure 10: Normalised novelty index: training set NC and 50% fault data DC in the two operating conditions I and II.

The values of the normalised novelty index are presented in Figure 6. It can be observed that the mean values of ν'_1 and ν'_2 are almost coincident ($\overline{\nu'}_1$=0.0271, $\overline{\nu'}_2$=0.0282).

Then testing sets were constructed by concatenating 500 noise-corrupted copies of the transmissibility functions relative to the different damage scenarios considered for both operating conditions.

The results shown in Figure 7 correspond to a 2% stiffness reduction. As can be seen, the normalised novelty index permits identification of the presence of the fault whereas the conventional novelty index of eq. (1) is ambiguous in that it is difficult to distinguish between NCII and DCI (Fig. 8). In addition results related to 10% and 50% stiff-

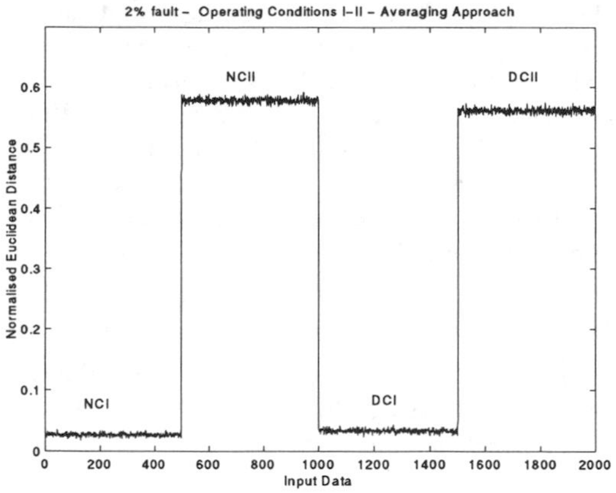

Figure 11: Normalised euclidean distance measured using the averaging approach: normal conditions and 2% fault data.

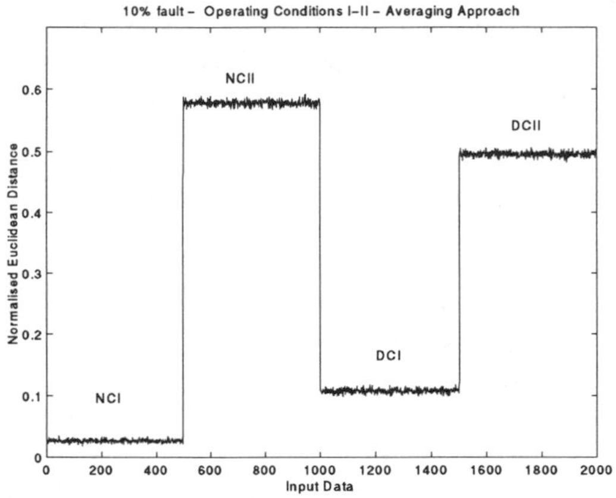

Figure 12: Normalised euclidean distance measured using the averaging approach: normal conditions and 10% fault data.

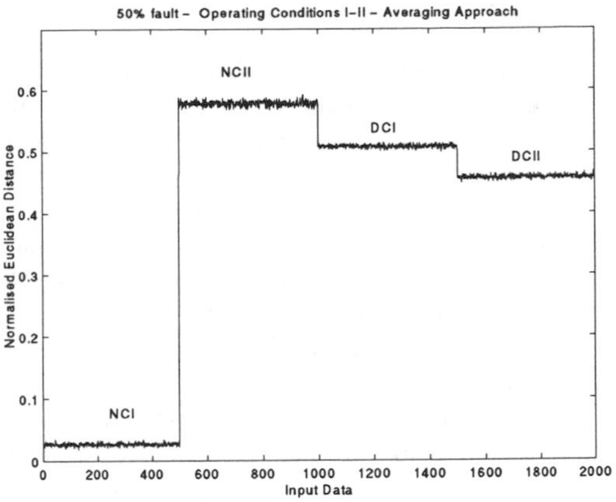

Figure 13: Normalised euclidean distance measured using the averaging approach: normal conditions and 50% fault data.

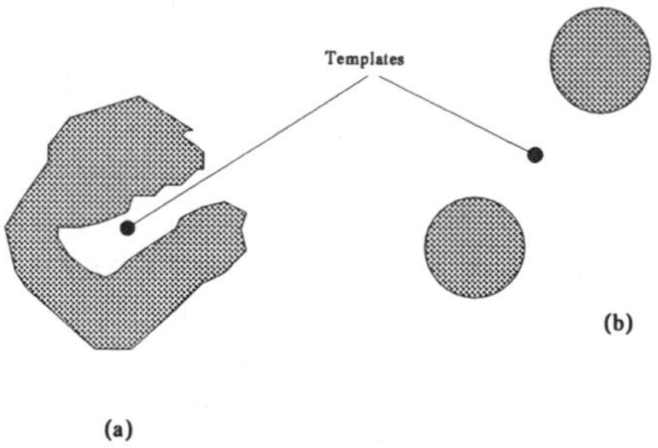

Figure 14: (a) Normal condition set non-convex; (b) normal condition set disconnected.

ness reductions are presented in Figures 9 and 10, demonstrating again that damage can be detected.

The functioning of the network can be interpreted with the observation that, since the patterns are passed through hidden layers which have fewer nodes than the input layer, the network is forced to learn just the significant prevalent features of the patterns. This means that the output of the training data is the transmissibility function of the undamaged structure without noise contamination.

This reasoning would suggest that a possible simplification of the procedure avoiding the use of Neural Networks could be achieved by using the Euclidean distance directly with transmissibility functions measured after performing as many averages as possible in order to reduce the influence of random noise.

However this alternative method proved to be equivalent with respect to the Neural Network approach only if a single normal condition was considered at a time. If the two normal conditions are considered simultaneously, this simple technique has the significant disadvantage in that it cannot discern between two different causes for alterations in the modal properties, as demonstrated in Figures 11 to 13. From these results, it is evident that the neural network method is superior to the the simpler averaging approach when different normal operating conditions of the system occur.

4 DISCUSSION

It remains to explain the results of this study. Fortunately, this appears to be straightforward; in fact, simple geometry suffices. Consider, the method based on average transmissibilities. A template for normal condition is constructed from an average of the normal condition patterns, i.e. in pattern space, the template is at the centroid of the normal condition set. Figure 14 shows two situations where the method will break down. In the first case (Figure 14a), the normal condition set is non-convex and therefore within any given radius there can be anomalous conditions as well as possibly normal conditions, this means that the method will suffer from false negatives. The second case is really an extreme case of the first (Figure 14b), the normal condition set is disconnected. This case corresponds to the situation explored in this paper. The method based on the neural network does not suffer from this problem. In geometrical terms, the auto-associative network defines a vector field on the pattern space $\mathbf{\Psi} = \mathbf{z} - \hat{\mathbf{z}}$. The object of the training exercise is to produce a network for which $\mathbf{\Psi} = \mathbf{0}$ on the normal condition set. Although the standard network approximation theorems are not set up with this situation in mind, it seems eminently plausible that a neural network can be found with the desired condition met on any given normal condition set. In fact, the AAN used to generate the novelty index can be straightforwardly embedded in a larger network which computes the novelty index directly at the expense of adding connections between layers which are not adjacent to each other.

5 CONCLUSIONS

An innovative technique using a novelty detection method for identifying the presence of damage in structures has been presented and illustrated with an application to a simulated 3 DOF system.

The method is based on discerning between changes in the modal properties resulting from the damage which should be detected and variations in the data resulting from statistical fluctuations and noise in the measurements which must be ignored.

Whereas conventional damage identification procedures have the limitation of requiring a sufficiently precise mathematical model of the structure, the novelty detection approach requires no model 'a priori', making the method more applicable to systems of arbitrary complexity.

Furthermore the technique has the advantage over conventional damage detection procedures in that in theory it can be applied to systems that present more than one normal operating condition. However to explore the potential for practical application of such a technique to detect damage in real structures further research with more advanced case studies is required.

REFERENCES

Roberts S., Tarassenko L. 1994. A Probabilistic Resource Allocating Network for Novelty Detection. *Neural Computation* 6:270-284.

Surace C., Worden K. 1997. A Novelty Detection Approach to Diagnose Damage in a Cracked Beam. *Proceedings of 15th IMAC, Orlando, U.S.A.*

Tarassenko L., Hayton P., Cerneaz N., Brady M. Novelty Detection for the Identification of Masses in Mammograms. *Preprint, Department of Engineering Sciences, University of Oxford.*

Worden K. 1997. Damage Detection using a Novelty Index. *Proceedings of 15th IMAC, Orlando, U.S.A.*

Worden K. 1997. Structural Fault Detection Using a Novelty Measure. *Journal of Sound and Vibration* 201(1):85-101.

Modern Practice in Stress and Vibration Analysis, Gilchrist (ed.) © *1997 Balkema, Rotterdam, ISBN 90 5410 896 7*

Towards a new method for identifying systems

S.D.Garvey & J.E.Penny
Department of Mechanical and Electrical Engineering, Aston University, Birmingham, UK
M.I.Friswell
Department of Mechanical Engineering, University of Wales, Swansea, UK

ABSTRACT: The identification problem addressed is one where a set of frequency response functions exists (from measurement) and system matrices of a minimal order are sought which accurately reproduce all of the dynamics of the FRFs. The method outlined comprises three stages : (a) create a model having a very large number of degrees of freedom and a particular structure which matches the measured FRFs to arbitrary accuracy over the range of frequencies measured, (b) keeping the same large number of degrees of freedom, allow the parameters in the system to relax to further reduce residual, (c) successively reduce the number of degrees of freedom in the model retaining as much fidelity to the measured response functions as possible and quantifying the loss in accuracy at every step.

1. INTRODUCTION

The identification problem addressed is one where a set of frequency response functions has been measured for a selected set of degrees of freedom and system matrices of a minimal order are sought which accurately reproduce all of the dynamics.

Modern methods for Modal Analysis often use time domain data derived from the frequency domain data by the Fourier Transform. The most popular among these include the Eigensystem Realisation Algorithm (ERA) [Juang *et al*, 1985] and the Polyreference Least Squares Complex Exponential method [H. Vanderauweraer *et al*, 1987]. Methods such as these perform well but are very computationally expensive and obtaining a uniform weighting on errors in the frequency domain causes additional cost. This might be an important feature if measured data were to be used to represent a component in a larger dynamic system. Moreover, they can be applied to high order systems but the accuracy of roots subsequently obtained can frequently be questionable. Methods also exist based directly on gradients of errors in the FRF data which offer the prospect of directly minimising errors in FRF data but these suffer badly from the tendency to become locked into local minima. Modern optimisation methods such as genetic algorithms [Penny *et al,* 1995] and synthetic annealing can be applied which do not suffer the propensity to converge to local minima. Generally, these are highly computationally intensive as a result of the fact that the methods used to derive revised sets of parameters are not based on a logic which is related to the nature of the problem.

This paper presents a possible new method of system identification comprising three-stages :

1. Create an initial model having a large number of degrees of freedom which accurately reproduces measured FRFs. This model has a very particular form which permits compact storage.

2. Carry out a relaxation stage on the system matrices simultaneously adjusting all unknowns in the model.

3. Reduce the number of degrees of freedom in the model one at a time, modifying the model each time to minimise the loss of fidelity to the original FRFs.

These steps are explained in turn below. The identification really takes place in the final stage when the order of the model is being reduced.

There is a flavour of "genetic-algorithm" about the process suggested here. However, for the purposes of elucidating the intuition behind this, the analogy to biological systems is very poor compared with the analogy to a large organisation which expands rapidly employing new people (degrees of freedom) in a purely incremental way to achieve the desired function (dynamic stiffness or receptance) and which subsequently "down-sizes" (reduces the order) by discovering that combining the duties of employees more sensibly, the labour force can be reduced to a minimum order.

2. SYSTEM MATRICES TO REPRODUCE FRFS.

Consider that a set of FRFs has been measured in the form of displacement / force. Then let $\mathbf{R}(\omega)$ be the $(N \times N)$ frequency-dependent matrix of receptances. It is recognised that, ordinarily, only one column of $\mathbf{R}(\omega)$ would be measured but we suppose here that the complete matrix is available and subsequently discuss the implications of only having a part of it.

In this first stage of the proposed identification process, we seek to find $(M \times M)$ matrices $\mathbf{K}_0$, $\mathbf{C}_0$ and $\mathbf{M}_0$ satisfying the following equations. Note that we expect $M \gg N$ and $P = M\text{-}N$.

$$\mathbf{K}_0 = \begin{bmatrix} \mathbf{K}_{NN} & \mathbf{K}_{NP} \\ \mathbf{K}_{PN} & \mathbf{K}_{PP} \end{bmatrix} \qquad \mathbf{C}_0 = \begin{bmatrix} \mathbf{C}_{NN} & \mathbf{C}_{NP} \\ \mathbf{C}_{PN} & \mathbf{C}_{PP} \end{bmatrix}$$

$$\mathbf{M}_0 = \begin{bmatrix} \mathbf{M}_{NN} & \mathbf{M}_{NP} \\ \mathbf{M}_{PN} & \mathbf{M}_{PP} \end{bmatrix} \tag{1}$$

$$\mathbf{D}_0(\omega) = \left[\mathbf{K}_0 + j\omega\mathbf{C}_0 - \omega^2\mathbf{M}_0\right] = \begin{bmatrix} \mathbf{D}_{NN}(\omega) & \mathbf{D}_{NP}(\omega) \\ \mathbf{D}_{PN}(\omega) & \mathbf{D}_{PP}(\omega) \end{bmatrix} \tag{2}$$

$$\mathbf{D}_\mathrm{T}(\omega) = \mathbf{D}_{NN}(\omega) - \mathbf{D}_{NP}(\omega)\left(\mathbf{D}_{PP}(\omega)\right)^{-1}\mathbf{D}_{PN}(\omega) \approx \left(\mathbf{R}(\omega)\right)^{-1} \tag{3}$$

The coordinates in the model have effectively been partitioned into N coordinates which might be termed *terminal coordinates* and P coordinates which might be termed *internal coordinates* using terms from dynamic substructuring. Forces may be applied to any or all of the first N coordinates but the forces on the P internal coordinates are implicitly zero. The assertion that no force can be applied to the internal coordinates is used to obtain the expression for $\mathbf{D}_\mathrm{T}(\omega)$ in (3).

The task of synthesising system matrices which reproduce the desired dynamics can be addressed in many different ways. One approach would be to fix $\mathbf{K}_{NN}$, $\mathbf{C}_{NN}$ and $\mathbf{M}_{NN}$ at the outset and to account for the rest of the residual in dynamic stiffness through the contributions from the internal degrees of freedom. To do this, three reference frequencies, ω_1, ω_2 and ω_3, would be selected and corresponding dynamic stiffness matrices $\mathbf{D}_{\mathrm{NN}1}$, $\mathbf{D}_{\mathrm{NN}2}$ and $\mathbf{D}_{\mathrm{NN}3}$ would be extracted (perhaps as local weighted averages of $\mathbf{D}_{\mathrm{NN}}(\omega)$ around ω_1, ω_2 and ω_3 respectively). Then solution of (4) would supply an initial $\mathbf{K}_{NN}$, $\mathbf{C}_{NN}$ and $\mathbf{M}_{NN}$.

$$\begin{bmatrix} \mathbf{I} & j\omega_1\mathbf{I} & -\omega_1^2\mathbf{I} \\ \mathbf{I} & j\omega_2\mathbf{I} & -\omega_2^2\mathbf{I} \\ \mathbf{I} & j\omega_3\mathbf{I} & -\omega_3^2\mathbf{I} \end{bmatrix} . \begin{bmatrix} \mathbf{K}_{NN} \\ \mathbf{C}_{NN} \\ \mathbf{M}_{NN} \end{bmatrix} = \begin{bmatrix} \mathbf{D}_{NN1} \\ \mathbf{D}_{NN2} \\ \mathbf{D}_{NN3} \end{bmatrix} \tag{4}$$

There is a strong likelihood that the $\mathbf{K}_{NN}$, $\mathbf{C}_{NN}$ and $\mathbf{M}_{NN}$ matrices thus obtained would neither be purely real nor positive (semi-)definite. This is not a concern at the present.

Alternatively, it is not necessary to compute $\mathbf{K}_{NN}$, $\mathbf{C}_{NN}$ and $\mathbf{M}_{NN}$ at the outset. It is clear that by retaining these matrices as unknowns, they span a 3-dimensional space of functions for each of the N^2 entries in $\mathbf{D}_{NN}$. Three functions spanning this space are denoted $f_1(\omega)$, $f_2(\omega)$ and $f_3(\omega)$. By restricting attention to functions of frequency which are "normal" to $f_1(\omega)$, $f_2(\omega)$ and $f_3(\omega)$, we can avoid using the internal coordinates to compensate for any components of error in $\mathbf{D}_\mathrm{T}(\omega)$ which can be dealt with directly by $\mathbf{K}_{NN}$, $\mathbf{C}_{NN}$ and $\mathbf{M}_{NN}$. A short appendix defines an inner-product $\langle\langle a(\omega), b(\omega)\rangle\rangle$ between any two frequency-dependent functions $a(\omega)$, $b(\omega)$ and correspondingly defines a norm $|a(\omega)| = \langle\langle a(\omega), a(\omega)\rangle\rangle^{0.5}$. Subject to the defined inner-product and norm, $f_1(\omega)$, $f_2(\omega)$ and $f_3(\omega)$ must obey (5).

$$\begin{aligned} &\langle\langle f_1(\omega), f_1(\omega)\rangle\rangle = 1 \quad , \quad \langle\langle f_1(\omega), f_2(\omega)\rangle\rangle = 0 \\ &\langle\langle f_2(\omega), f_2(\omega)\rangle\rangle = 1 \quad , \quad \langle\langle f_2(\omega), f_3(\omega)\rangle\rangle = 0 \\ &\langle\langle f_3(\omega), f_3(\omega)\rangle\rangle = 1 \quad , \quad \langle\langle f_3(\omega), f_1(\omega)\rangle\rangle = 0 \end{aligned} \tag{5}$$

It is straightforward to obtain $f_1(\omega)$, $f_2(\omega)$ and $f_3(\omega)$ using Gram-Schmidt orthonormalisation. Retaining $\mathbf{K}_{NN}$, $\mathbf{C}_{NN}$ and $\mathbf{M}_{NN}$ as unknowns is clearly attractive when it comes to fitting internal coordinates and we assume from here on that this is done. Now, we can compute a residual dynamic stiffness which we shall call $\mathbf{E}(\omega)$. A notation is required for the component of any frequency-dependent matrix, $\mathbf{X}(\omega)$, which is orthogonal to $f_1(\omega)$, $f_2(\omega)$ and $f_3(\omega)$ and the following is used ..

$$\mathbf{X}(\omega)^{\perp} \equiv \mathbf{X}(\omega) - \begin{bmatrix} \ll \mathbf{X}(\omega).f_1(\omega) \gg .f_1(\omega) + \\ \ll \mathbf{X}(\omega).f_2(\omega) \gg .f_2(\omega) + \\ \ll \mathbf{X}(\omega).f_3(\omega) \gg .f_3(\omega) \end{bmatrix} \quad (6)$$

Now, since $\mathbf{K}_{NN}$, $\mathbf{C}_{NN}$ and $\mathbf{M}_{NN}$ (right until the end of the process), the full detail of the residual dynamic stiffness, $\mathbf{E}(\omega)$, is not known but $\mathbf{E}(\omega)^{\perp}$ is known and is given by (7).

$$\mathbf{E}(\omega)^{\perp} = \left[(\mathbf{R}(\omega))^{-1}\right]^{\perp} - \mathbf{D}_T(\omega)^{\perp} \quad (7)$$

The first task in the method put forward in this paper comprises finding matrices $\mathbf{D}_{NP}(\omega)$, $\mathbf{D}_{PN}(\omega)$ and $\mathbf{D}_{PP}(\omega)$ such that for all ω ...

$$\mathbf{E}(\omega) = \mathbf{R}(\omega)^{\perp} + \mathbf{D}_{NP}(\omega)(\mathbf{D}_{PP}(\omega))^{-1}\mathbf{D}_{PN}(\omega) \approx 0 \quad (8)$$

The possibilities for determining $\mathbf{D}_{NP}(\omega)$, $\mathbf{D}_{PN}(\omega)$ and $\mathbf{D}_{PP}(\omega)$ are infinite and it is necessary to impose certain constraints on the problem in order to develop a practicable algorithm.

Firstly, for simplicity in this paper, we assume that all matrices are symmetrical so that $\mathbf{D}_{NP}(\omega) \equiv \mathbf{D}_{PN}^{T}(\omega)$ and $\mathbf{D}_{PP}(\omega) \equiv \mathbf{D}_{PP}^{T}(\omega)$ for all ω. The method is not constrained to such cases but the tedium of maintaining the full generality is not warranted here for an extension which is so trivial.

Secondly, we will insist that $\mathbf{D}_{PP}(\omega)$ is diagonal for all ω. There is only a slight loss of generality associated with this. Suppose that we have found some matrices $\mathbf{D}_{NP}(\omega)$, $\mathbf{D}_{PN}(\omega)$ and $\mathbf{D}_{PP}(\omega)$ which produce the relationships required and recall that $\mathbf{D}_{PP}(\omega)$ is formed as $(\mathbf{K}_{PP} + j\omega\mathbf{C}_{PP} - \omega^2\mathbf{M}_{PP})$. If we insist that the system of internal coordinates is proportionally damped, then there is some modal matrix, $\mathbf{U}_P$, for which ($\mathbf{U}_P^{T}\,\mathbf{K}_{PP}\,\mathbf{U}_P$, $\mathbf{U}_P^{T}\,\mathbf{C}_{PP}\,\mathbf{U}_P$ and $\mathbf{U}_P^{T}\,\mathbf{M}_{PP}\,\mathbf{U}_P$) are all diagonal. Then swapping the "original" $\mathbf{D}_{PP}(\omega)$ for ($\mathbf{U}_P^{T}\,\mathbf{D}_{PP}(\omega)\,\mathbf{U}_P$) and changing $\mathbf{D}_{NP}(\omega)$ to ($\mathbf{U}_P^{T}\,\mathbf{D}_{NP}(\omega)$) reproduces the same product $\mathbf{D}_{NP}(\omega)\,(\mathbf{D}_{PP}(\omega))^{-1}\,\mathbf{D}_{PN}(\omega)$ as before. The loss of generality mentioned above is connected with the fact that the (sub-)system represented by ($\mathbf{K}_{PP}$, $\mathbf{C}_{PP}$, $\mathbf{M}_{PP}$) might not be "proportionally" damped. a relatively minor variation on the method described here based on a state-space representation of the internal coordinates overcomes even this limitation but as with the assumption of symmetry, the extension to the fully general case does not justify the additional notation and text which would be necessary.

Thirdly, since we can arbitrarily scale the diagonal of $\mathbf{D}_{PP}(\omega)$ provided that the square root of this scaling is applied to $\mathbf{D}_{NP}(\omega)$ and $\mathbf{D}_{PN}(\omega)$, we shall choose to scale $\mathbf{D}_{PP}(\omega)$ such that all of the mass terms are unity i.e. $\mathbf{M}_{PP} = \mathbf{I}$.

The process of creating $\mathbf{D}_{NP}(\omega)$, $\mathbf{D}_{PN}(\omega)$ and $\mathbf{D}_{PP}(\omega)$ is then reduced to a stepwise task in which the norm of a residual error in dynamic stiffness is successively reduced by introducing two new internal coordinates at a time. Clearly there will be a series of $\mathbf{D}_{PP}(\omega)$ matrices ... $\mathbf{D}_{PP\_k}(\omega)$ for k = 1,2,3 ... and there will be corresponding series of $\mathbf{D}_{NP}(\omega)$ (and $\mathbf{D}_{PN}(\omega)$) matrices. Similarly, there will be a series of (components of) residuals $\mathbf{E}_k(\omega)^{\perp}$ and there will also be a series of $\mathbf{D}_T(\omega)$ matrices ... $\mathbf{D}_{T\_k}(\omega)$ for k = 1,2,3 ... beginning with $\mathbf{D}_{T\_0}(\omega) = \mathbf{D}_{NN}(\omega)$.

The matter of how to select parameters appropriately for a single step is addressed in the following section with the simple case of N=1 being addressed in the first instance to establish the intuitive foundation.

3. A SINGLE STEP IN SYNTHESIS OF FRFs.

The kth step of stage 1 of the synthesis process adds an incremental matrix of dynamic stiffness, $\Delta\mathbf{D}_{T\_k}(\omega)$, to $\mathbf{D}_{T\_k}(\omega)$ producing $\mathbf{D}_{T\_k+1}(\omega)$. The increment in dynamic stiffness is in the form of (9).

$$\Delta\mathbf{D}_{T\_k}(\omega) = -\left[\frac{\mathbf{d}_{ak}(\omega)(\mathbf{d}_{ak}(\omega))^{T}}{d_{bk}(\omega)} + \frac{\mathbf{d}_{ck}(\omega)(\mathbf{d}_{ck}(\omega))^{T}}{d_{dk}(\omega)}\right]$$

... where ...

$$\begin{aligned} \mathbf{d}_{ak}(\omega) &= (\mathbf{k}_{ak} + j\omega\mathbf{c}_{ak} - \omega^2\mathbf{m}_{ak}) \\ d_{bk}(\omega) &= (k_{bk} + j\omega c_{bk} - \omega^2) \\ \mathbf{d}_{ck}(\omega) &= (\mathbf{k}_{ck} + j\omega\mathbf{c}_{ck} - \omega^2\mathbf{m}_{ck}) \\ d_{dk}(\omega) &= (k_{dk} + j\omega c_{dk} - \omega^2) \end{aligned} \quad (9)$$

In (9), $\mathbf{k}_{ak}$, $\mathbf{c}_{ak}$ and $\mathbf{m}_{ak}$ are vectors having N entries each and they couple the first internal degrees of freedom in the "k"th pair of to the existing (N x N) dynamic stiffness matrix. Similarly for $\mathbf{k}_{ck}$, $\mathbf{c}_{ck}$ and $\mathbf{m}_{ck}$ and the second internal coordinate of the "k"th pair. Vectors $\mathbf{k}_{ak}$ and $\mathbf{k}_{ck}$ become columns (2k-1) and (2k) of $\mathbf{K}_{NP}$ respectively. The scalars k_{bk}, c_{bk}, k_{dk} and c_{dk} characterise the dynamic stiffness of the internal coordinates themselves.

We confine attention, initially, to the case where N=1. Then, for each $\Delta\mathbf{D}_{T\_k}(\omega)$, a total of ten new parameters must be defined. There is obviously some logic in determining $\Delta\mathbf{D}_{T\_k}(\omega)$ purely such that it has a peak to coincide with a peak in the residual dynamic stiffness $\mathbf{E}_k(\omega)^{\perp}$. Then individual peaks could be removed from $\mathbf{E}(\omega)^{\perp}$. After some limited experience with the FRF synthesis process described, the present authors advocate generating a number of candidate sets of parameters during any one step - each different set of parameters being initialised such that a different peak in $\mathbf{E}(\omega)^{\perp}$ is cancelled - and subsequently allowing each set of parameters to vary in order to find the nearest local maximum in $|\mathbf{E}(\omega)^{\perp}.\ \Delta\mathbf{D}_{T\_k}(\omega)|$ over a band of frequencies which may span the entire frequency range of interest or which may be confined to a region of frequencies around the peak. Once each individual set of parameters has been optimised, one set of parameters is chosen on the basis that it offers the maximum $|\mathbf{E}(\omega)^{\perp}.\ \Delta\mathbf{D}_{T\_k}(\omega)|$.

It is possible to produce a $\Delta\mathbf{D}_{T\_k}(\omega)$ which possess significant values only in the region of a given frequency by setting the following constraints (again, note that these constraints are only put in place for the purpose of developing a set of initial values).

$$k_{ak} = m_{ak} = 0 = k_{ck} = m_{ck}$$
$$c_{ck} = j\, c_{ak} \quad , \quad k_{bk} = k_{dk} \quad , \quad c_{dk} = 5.c_{bk} \tag{10}$$

From the original space of ten variables, seven have now been eliminated and a set of three equations is required to initialise the remaining variables. From (9), it is evident that with the above constraints in place, $\Delta\mathbf{D}_{T\_k}(\omega)$ can be simplified to (11).

$$\Delta\mathbf{D}_{T\_k}(\omega) = \frac{-(j\omega)^3 (c_{ak})^2 (4c_{bk})}{\left[(k_{bk}-\omega^2)+j\omega c_{bk}\right]\left[(k_{bk}-\omega^2)+5j\omega c_{bk}\right]} \tag{11}$$

Now, the peak value of $|\Delta\mathbf{D}_{T\_k}(\omega)|$ occurs in the region of $\omega^2 = k_{bk} = \omega_k^2$ and this provides a method whereby k_{bk} can be chosen. The actual value of $\Delta\mathbf{D}_{T\_k}(\omega)$ at the magnitude peak determines c_{ak} if c_{bk} is known. A value is selected for c_{bk} based on the rate of change of phase of $\Delta\mathbf{D}_{T\_k}(\omega)$ at $\omega = \omega_k$. The phase of $\Delta\mathbf{D}_{T\_k}(\omega)$, denoted here as $\angle\Delta\mathbf{D}_{T\_k}(\omega)$, is influenced by both denominator and numerator but the rate of change of phase with respect to ω, is controlled only by the denominator as given in (11) and this provides a means of determining c_{bk} directly.

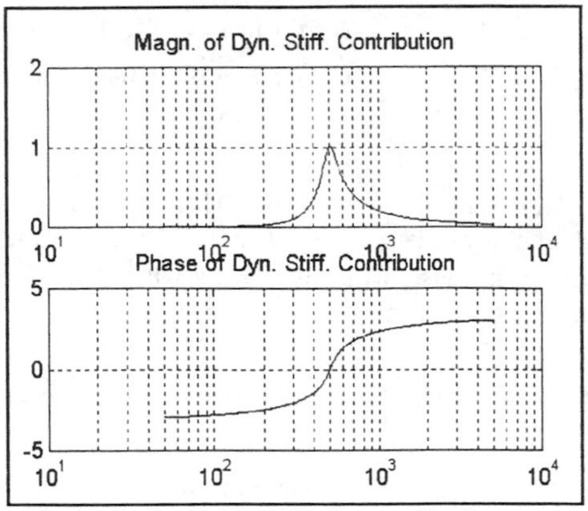

Fig 1. A dynamic stiffness contribution.

$$\frac{d}{d\omega}(\angle\Delta\mathbf{D}_{T\_k}) = \frac{12}{5c_{bk}} \quad at \ \omega = \omega_k. \tag{12}$$

The above strategy for initialising the ten parameters for a given $\Delta\mathbf{D}_{T\_k}(\omega)$ generally results in two complex parameters (c_{ak} and c_{ck}) with all others being purely real.

It is now evident why the internal coordinates are introduced in pairs. The first coordinate is chosen to have low damping and so produces a significant "peak" in the added dynamic stiffness. The second coordinate cancels out the contributions to dynamic stiffness made by the first coordinate at frequencies which are not close to the resonance frequency of the internal coordinates. Fig. 1 below shows the magnitude and phase of a scalar lump of dynamic stiffness contributed by 2 additional internal coordinates using the constraints of (10) and equations (11) and (12) to compute the characterising parameters from the following starting information : peak frequency = 500 (rad/s), rate of change of phase with frequency = 0.02 (rad/Hz), stiffness at peak = 1.0 (N/m).

4. AN EXAMPLE OF SYNTHESIS OF FRFs.

Suppose that a single point FRF has been measured and that modal data have been determined for that single point. The relevant data are summarised in table #1. From these, it is possible to reconstruct a frequency response curve and this is done for 301 frequencies between 50 rad/s and 3000 rad/s (including the end points). Equal logarithmic spacing was applied so that the first 5 frequencies were ... 50.0, 50.6871, 51.3836, 52.0897, 52.8054 rad/s.

Table 1. Modal Data for an Example FRF.

ω_n	ζ	u_k
10	0.1	1.0
15	0.2	1.0
90	0.12	2.0
150	0.08	3.0
250	0.08	1.0
350	0.08	3.0
500	0.16	5.0
520	0.03	12.0
540	0.11	16.0
560	0.11	25.0
1000	0.06	19.0
2000	0.06	160.0
5000	0.15	30.0
10000	0.15	140.0
15000	0.08	80.0

Fig. 2 shows the target dynamic stiffness, $(\mathbf{R}(\omega))^{-1}$ (**R** is a scalar function of ω in this case) generated from the above data.

Fig. 3 shows $\mathbf{D}_{T\_3}(\omega)^{\perp}$, which represents the residual dynamic stiffness (orthogonal to $f_1(\omega)$, $f_2(\omega)$ and $f_3(\omega)$), after 3 pairs of internal degrees of freedom have been added and Fig. 4 shows $\mathbf{E}_{10}(\omega)^{\perp}$. Fig. 5 shows a plot of the square of the norm of $\mathbf{E}_k(\omega)^{\perp}$ as a function of k and the rate of convergence is evident from this.

5. DEALING WITH MULTIPLE FRFS.

There are two aspects to the extension of the above synthesis process to multiple FRFs.

1. Estimating $(\mathbf{R}(\omega))^{-1}$ from a single column of $\mathbf{R}(\omega)$ poses some difficulty.

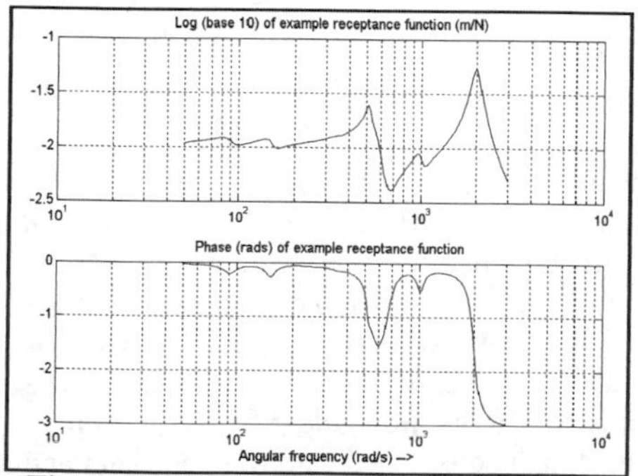

Fig. 2. Magn. and phase of $[(\mathbf{R}(\omega))^{-1}]^{\perp}$.

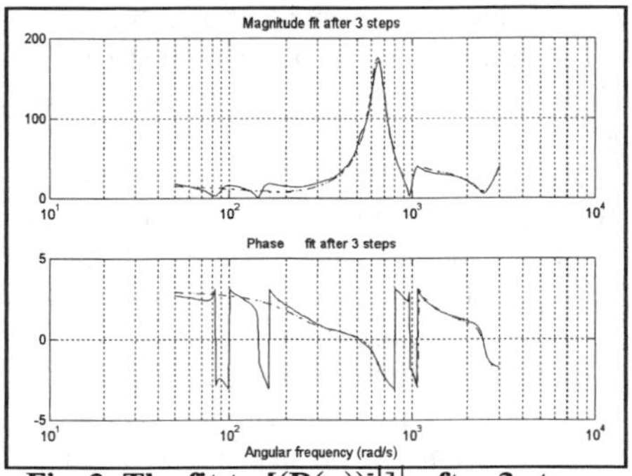

Fig. 3. The fit to $[(\mathbf{R}(\omega))^{-1}]^{\perp}$ after 3 steps.

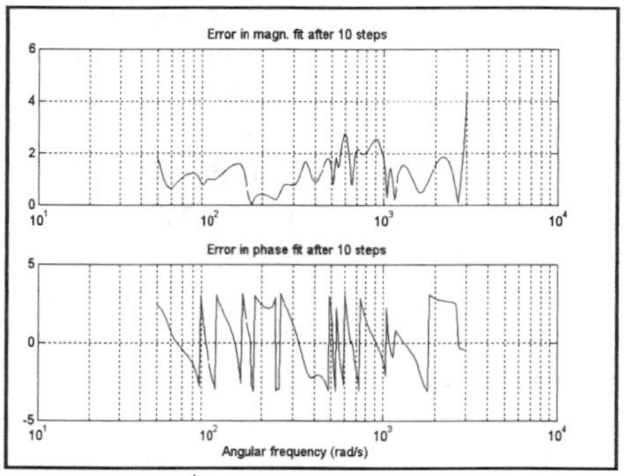

Fig. 4. $\mathbf{E}_{10}(\omega)^{\perp}$ (residual after 10 steps).

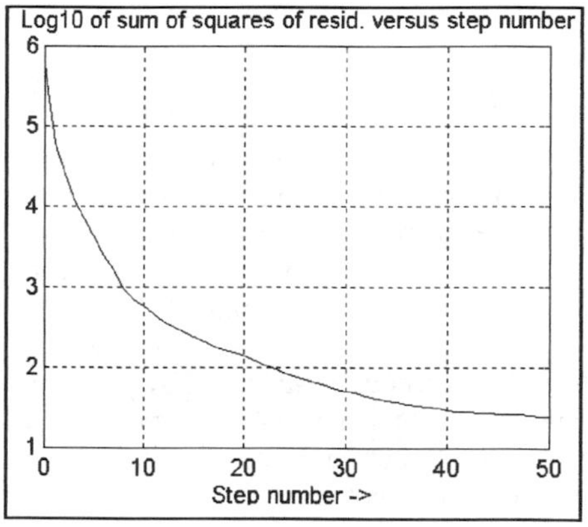

Fig. 5. Convergence of fit to $\mathbf{E}_0(\omega)^{\perp}$.

2. When an approximation for $(\mathbf{R}(\omega))^{-1}$ has been obtained, how should each pair of additional internal coordinates be obtained.

Of these, the second poses the least problem. If we know $(\mathbf{R}(\omega))^{-1}$ over a range of frequencies, then $\mathbf{E}(\omega)^{\perp}$ can be determined and we can conduct a singular value decomposition at each frequency. The "peaks" in magnitude of $\mathbf{E}(\omega)^{-1}$ would be identified as those frequencies where the largest singular value reached a local maximum. The method described above for initialising the ten parameters used for providing a contribution to a scalar dynamic stiffness residual could then be applied with relatively little modification. Vectors $\mathbf{c}_{ak}$ and $\mathbf{c}_{ck}$ would be constrained (initially) to be simple multiples of the vector corresponding to the largest singular value and through this constraint the problem of initialising parameters would be reduced to a scalar problem similar to that dealt with above. It is recognised that the unit-vector corresponding to the largest singular value would itself have a rate of change with respect to ω. This issue can be dealt with by allowing vectors $\mathbf{c}_{ak}$ and $\mathbf{c}_{ck}$ to be non-zero initially also.

The problem of determining $(\mathbf{R}(\omega))^{-1}$ from a single column of $\mathbf{R}(\omega)$ is very significant but it can possibly be addressed in the following ways

1. At every stage in the synthesis process (right from the start), there is an estimate for $\mathbf{R}(\omega)$ and $(\mathbf{R}(\omega))^{-1}$. Rather than having a pre-determined target $(\mathbf{R}(\omega))^{-1}$ for all ω at the outset of the synthesis, it is possible that the process could work with a target which was being revised continuously.
2. The entire synthesis process might take place piece-meal. Initially, the auto-response might be used to construct a scalar dynamic stiffness function of ω. Then the second degree of freedom might be incorporated utilising the minimum deviation from data built up for the first and so on.
3. It is possible that ficticious data could be generated for the missing FRF.

While the questions relating to how the proposed identification method might be applied to sets of FRFs are far from having been resolved, it is evident that solutions are possible.

6. STAGE 2 OF THE PROPOSED PROCESS.

In this stage, all of the parameters comprising the matrices are allowed to vary simultaneously in order to achieve two goals :

1. Impose certain constraints (positive definiteness, real numbers in matrices etc.)
2. Reduce (further) the norm of the residuals

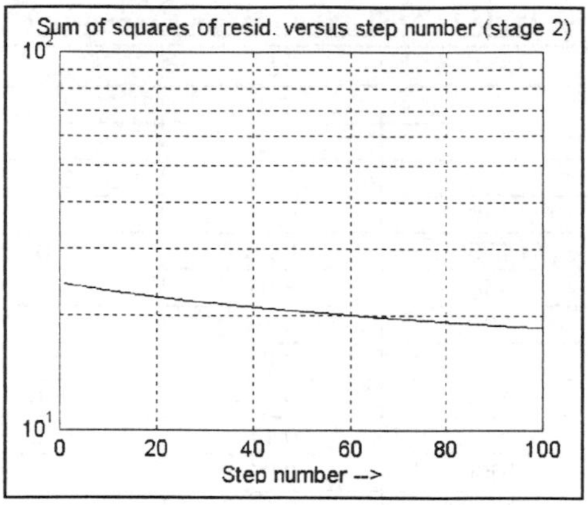

Fig. 6. Convgnce. in stage 2 (100 int DoFs)

In order to build in the constraints, a single cost-function which has components associated with the magnitude of the residual in dynamic stiffness as well as components associated with the degree of violation of the constraints. By steadily increasing the weighting on the component associated with the constraints, the constraints can be enforced to arbitrary precision.

A conjugate gradient method is advocated for the relaxation taking place in stage 2. It is thought that since there will be such a large degree of redundancy in the parameters (in that virtually the same responses could be achieved for any one set of parameters in a vast space), the problem of the optimal solution becoming trapped in local minima is largely averted.

For the example problem introduced in section 4, no constraints were considered. Fig. 6 shows the reduction in error as a function of step number in stage 2. Evidently, the error does not reduce dramatically with steps and the steps themselves are computationally lengthy. Evidently it is not sensible to run this stage for more than a few steps.

7. STAGE 3.

Having created a dynamic system which exhibits the same behaviour as the measured system but which has unnecessarily high order, it now remains to reduce this order. This section outlines a proposed method. Traditional methods such as static reduction are entirely inappropriate here since they are based on discarding degrees of freedom on the basis of high constrained natural frequencies and in the present case, the parameters of the internal degrees of freedom were all set initially such that they had dynamics within the range of frequencies of interest. Moreover, it would be important (in cases of large N) that during the reduction process, the very specific form of the matrices should be maintained.

The reduction of order would take place stepwise with each step removing a single internal degree of freedom (not a pair!) and then subsequently relaxing the model again to minimise the norm of the residual if the magnitude of the residual exceeded some nominal threshold value. Once again, conjugate gradient methods coupled with line-searches would be used for each minimisation. The method suggested for selection of which internal degree of freedom to remove is less obvious. As before, attention is focused on those components of dynamic stiffness which are orthogonal to to $f_1(\omega)$, $f_2(\omega)$ and $f_3(\omega)$) since the other components are easily corrected through an adjustment of $\mathbf{K}_{NN}$, $\mathbf{C}_{NN}$ and $\mathbf{M}_{NN}$.

Let $\mathbf{G}_i(\omega)$ be the dynamic stiffness function subtracted due to the removal of internal coordinate i and let $\mathbf{H}_{ij}(\omega)$ be the derivatives of $\mathbf{D}(\omega)^{\perp}$ with respect to each parameter, j, associated with the remaining internal degrees of freedom. The essence of the selection procedure is exposed treating $\mathbf{G}_i(\omega)$ and $\mathbf{H}_{ij}(\omega)$ as scalar functions of ω (i.e. N=1). Define x_{ij} and y_i according to (13) below.

$$x_{ij} = \left[\frac{\left| \ll \mathbf{G}_i, \mathbf{H}_{ij} \gg \right|^2}{\ll \mathbf{H}_{ij}, \mathbf{H}_{ij} \gg \ll \mathbf{G}_i, \mathbf{G}_i \gg} \right] \qquad (13)$$

$$y_i = \sum_{all\ j} x_{ij}$$

If all of the $\mathbf{H}_{ij}(\omega)$ were mutually orthogonal for a given i according to the inner product defined, and if they spanned a space which contained $\mathbf{G}_i(\omega)$ completely, then we would have $y_i = 1$ indicating that the increase in the residual caused by eliminating internal coordinate i could be offset completely by some linear combination of the derivatives.

In general the $\mathbf{H}_{ij}(\omega)$ are not orthogonal and neither do they span a space which contains $\mathbf{G}_i(\omega)$ completely. If they did (irrespective of mutual orthogonality) then internal degree of freedom i could be removed and the additional residual so introduced could be compensated completely by adjustments in the other parameters (assuming that the change in residual was sufficiently small that the derivatives remained effectively unchanged after the adjustments).

The quantity y_i simultaneously provides an account of the extent of the lack of orthogonality of the $\mathbf{H}_{ij}(\omega)$ and the proportion of $\mathbf{G}_i(\omega)$ which does lie within the subspace of possible adjustments. If the $\mathbf{H}_{ij}(\omega)$ tend to be "aligned" well with $\mathbf{G}_i(\omega)$, then y_i will be high and this indicates in favour of removing internal degree of freedom i since the sum of squares of norms of the corrective adjustments will be small which is ultimately the objective when removing any one of the internal coordinates.

We do not use y_i alone to select which internal coordinate to remove since this criterion does not take account of the size of « $\mathbf{G}_i(\omega)$ $\mathbf{G}_i(\omega)$ ». Instead, the internal coordinate having the largest value of (y_i^2 / « $\mathbf{G}_i(\omega)$ $\mathbf{G}_i(\omega)$) ») is selected.

Having selected internal coordinate i to be removed, we now provide an initial estimate for the corrections to each parameter associated with internal coordinates other than i. Let $\mathbf{p}$ be the full vector of parameters associated with remaining internal coordinates and let $\Delta\mathbf{p}$ be the initial correction which will be made on $\mathbf{p}$. Then $\Delta\mathbf{p}$ is given as ...

$$\Delta\mathbf{p} = \frac{\mathrm{conj}\left(\ll \mathbf{G}_i, \mathbf{H}_{ij} \gg \right)}{y_i} \cdot \frac{\mathbf{H}_{ij}}{\ll \mathbf{H}_{ij}, \mathbf{H}_{ij} \gg} \qquad (14)$$

Having made these corrections following the removal of one coordinate, a small number of conjugate gradient steps might be undertaken in order to account for the combination of non-linear effects due to finite corrections having been made and the fact that the $\mathbf{H}_{ij}(\omega)$ were not mutually orthogonal and hence using the above expression for $\Delta\mathbf{p}$ will inevitable cause other components of dynamic stiffness to be drawn in.

The proposed reduction method has been applied very crudely - beginning after a synthesis which had added 20 pairs of internal coordinates. The criterion discussed above was applied to determine which internal degree of freedom to remove at each step. Initial corrections following the removal of the coordinate were made on the basis of the expression for $\Delta\mathbf{p}$ above coupled with a line-search algorithm to ensure that the total error was reduced. Then, ten steps of conjugate gradient

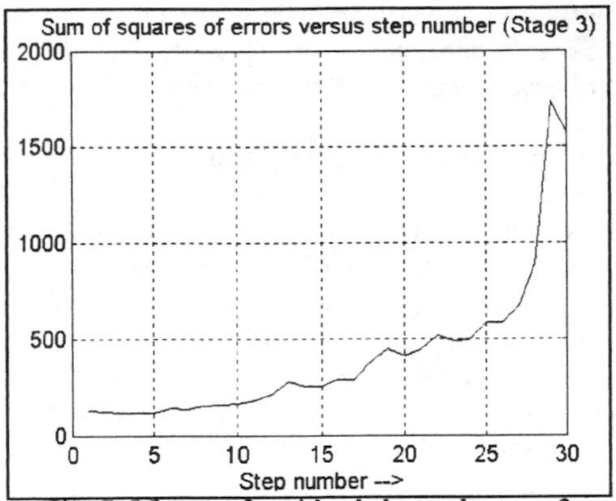

Fig. 7. Magn. of residual through stage 3.

minimisation were carried out following the removal of each individual coordinate. Fig. 7 shows the results obtained in terms of the sum of squares of residuals in dynamic stiffness at each step. Fig. 7 is directly comparable with Fig. 5 and Fig. 6.

It is difficult to judge what has been gained by having introduced 40 internal degrees of freedom and subsequently removed 30 of them as compared with having simply stopped at that point in stage 1 where only 10 internal degrees of freedom had been added. Comparison of Fig. 5 and Fig. 7 implies that the achievement was negative and yet the proportional improvements due to single steps of conjugate gradient relaxation were substantially better for the system which had had 40 internal coordinates subsequently reduced to 10. Insufficient investigation has been invested at this time to conclude either way.

CONCLUSIONS.

A novel method has been proposed whereby a set of system matrices of minimal order which reproduces a measured set of FRFs might be determined. The method has been demonstrated with for one particular case of a single auto-response with limited success (good results for the synthesis of dynamic stiffness characteristics but poor ones for the subsequent reduction of the models to low order).

Although it is acknowledged that the method proposed is highly computationally intensive, it has the attraction that at every stage, the process is dealing with a model of the correct form and there is a clear logical path between the starting point and the final model. If there can ever be a universally applicable identification algorithm needing no human judgement or intervention and having accuracy which is limited only by the computational time devoted to the task, it may be that a method such as that outlined here forms the basis.

REFERENCES

J.-N. Juang and R.S. Pappa, An Eigensystem Realization Algorithm for Modal Parameter Identification and Model Reduction, *AIAA Journal of Guidance, Control, and Dynamics,* 8(5), pp620-627, 1985.

H. Vanderauweraer and J. Leuridan, Multiple Input Orthogonal Polynomial Parameter-Estimation, *Mech. Systems and Signal Processing,* 1(3), pp.259-272, 1987.

J E T Penny, M I Friswell & G Linfield, 'The Location of Damage from Vibration Data using Genetic Algorithms' Proc 13th International Modal Analysis Conference, Nashville, Tenn. (1995) Feb.

APPENDIX I. Inner products.

The inner-product of two scalar functions $a(\omega)$ and $b(\omega)$ is denoted «$a(\omega)$, $b(\omega)$» and is defined for the present purposes as ...

$$\langle\langle a(\omega), b(\omega)\rangle\rangle = \int_{\omega_{min}}^{\omega_{max}} w(\omega)a(\omega).b(\omega).\mathrm{d}\omega \qquad (15)$$

In the above, $w(\omega)$ is a weighting function which might simply be a flat line independent of ω or - in the case of this paper - the sum of a number of delta-functions defined at discrete (logarithmically spaced) values of ω.

The extension to the case where one or both operands are matrices is trivial.

3 Linear and non-linear stress and vibration analysis

Modern Practice in Stress and Vibration Analysis, Gilchrist (ed.) © 1997 Balkema, Rotterdam, ISBN 90 5410 896 7

Analysis of the vibrations of a rotating plate subjected to aerodynamic loading

W. M. Ostachowicz
Institute of Fluid Flow Machinery, Polish Academy of Sciences, Gdansk, Poland

M. P. Cartmell
Department of Mechanical Engineering, University of Edinburgh, UK

ABSTRACT: This paper takes the general problem of a rotating plate as a generic representation of a beam or blade and introduces the novel effect of aerodynamic loading on the plate such that a flow induced pressure acts normal to the plate during its rotation. The results of the work have shown that certain key phenomena occur for specific conditions, namely that in general the setting angle (i.e. the orientation of the lateral axis of the plate, or beam, relative to the axis of rotation) does not radically affect the natural frequencies of the component, except in the case of high shaft speeds where sizeable shifts can occur. The other general conclusion is that natural frequencies do increase noticeably with Mach number and that at mid-range shaft speeds and mid-range Mach numbers the setting angle does have a noticeable tendency to shift the system natural frequencies.

1. INTRODUCTION

The vibratory response of a rotating plate subjected to aerodynamic fluid-flow interactions is of considerable interest to turbo-machinery designers, and a problem which has persistently eluded definitive analysis. Several studies have been made of rotating uniform cantilever beams, with substantial bearing on the programme of work currently under way due to Argento and Scott (1992), Bazoune and Khulief (1992), Gau and Shabana (1990), Kane et al (1987), and Simo and Vu-Quoc (1986). Flexibility encountered at the root of rotating beams was researched by Wright et al (1982), Abbas (1985), and Afolabi (1986), and in addition to this relevant research on aerodynamically forced blades has also been reported by Hoyniak and Fleeter (1986), Basu and Griffin (1986), Leissa (1981 & 1982), and Srinavasan and Fabunmi (1983). Recently Ostachowicz and Cartmell (1996) proposed extensions to the work of Vyas and Rao (1992) by considering the effects of variable angular velocity on the vibrations of a rotating cantilever beam.

In the work reported here the authors have expanded on some effects of rotating plate modelling as well as combined effects of rotation and fluid-flow / plate interaction. The plate is considered in bending and torsion, but coriolis forces are neglected. A Finite Element formulation is used with four nodes and three degrees of freedom per node. Fifteen elements were found to be sufficient for a compromise between good results and computational speed. The inertia forces, centrifugal terms, and aerodynamic excitation are represented by means of virtual work expressions. This paper attempts to describe and clarify some of the matters relating to the mechanics of the aerodynamically forced rotating beam (or plate) system in the form of a simple case study, based on a practical system built in the laboratory. It is hoped that the results will contribute to further studies by the authors and others alike.

2. DESCRIPTION OF THE MODEL

By considering the system shown in Figure 1 it can be seen that a plate of constant cross-section is mounted with setting angle θ on a hub of radius R_o. The normal forces loading the plate are assumed to be developed by the effect of flow induced pressure, where aerodynamic pressure is assumed according to the formulation of Dowell (1975),

$$\Delta p^{M}(y,t) = \rho_p U_p^2 \left\{ \frac{1}{M}\left[\frac{\partial w}{\partial y} + \frac{1}{U_p}\frac{\partial w}{\partial t} \right] \right\} \tag{1}$$

where ρ_o is the density of the fluid, U_p is the fluid velocity, M is the Mach number, and A a function which depends parametrically upon the Mach number. In the majority of work equation (1) is used in reduced form as shown above, which only takes account of conventional 'piston theory'. This approach is maintained here. The mass, centroidal, and elastic axes are all considered as coincident. The equation of motion for the system under investigation is as follows

$$D\left(\frac{\partial^4 w}{\partial x^4}+2\frac{\partial^4 w}{\partial x^2 \partial y^2}+\frac{\partial^4 w}{\partial y^4}\right)-N_x\frac{\partial^2 w}{\partial x^2}-$$

$$N_y\frac{\partial^2 w}{\partial y^2}-2N_{xy}\frac{\partial^2 w}{\partial x \partial y}+X\frac{\partial w}{\partial x}+Y\frac{\partial w}{\partial y}+h\rho\frac{\partial^2 w}{\partial t^2}-$$

$$\rho h\Omega^2\left(w\sin^2\theta-y\sin\theta\cos\theta\right)+$$

$$\rho_p\frac{U_p}{M}\left(U_p\frac{\partial w}{\partial y}+\frac{\partial w}{\partial t}\right) = 0 \qquad (2)$$

where $D = Eh^3/(12[1 - v^2])$, E denotes the modulus of elasticity, N_x, N_y, N_z define the in-plane stress resultants, X & Y the mass forces, and Ω the angular velocity of the rotor. Equation (2) is supplemented by two equations which determine the in-plane stress resultants

$$\frac{\partial N_x}{\partial x}+\frac{\partial N_{xy}}{\partial y}+X = 0 \qquad (3)$$

$$\frac{\partial N_{xy}}{\partial x}+\frac{\partial N_y}{\partial y}+Y = 0 \qquad (4)$$

and the mass forces are obtained from the centrifugal terms

$$X = \rho\,\Omega^2\,(x + R_o) \qquad (5)$$

$$Y = \rho\,\Omega^2\,y\cos^2\theta \qquad (6)$$

and the in-plane stress resultants are derived from standard plate theory, thus

$$N_x=\frac{Eh}{\left(1-\upsilon^2\right)}\left[\frac{\partial u}{\partial x}+\upsilon\frac{\partial v}{\partial y}\right] \qquad (7)$$

$$N_y=\frac{Eh}{\left(1-\upsilon^2\right)}\left[\frac{\partial v}{\partial y}+\upsilon\frac{\partial u}{\partial x}\right] \qquad (8)$$

$$N_{xy}=\frac{Eh}{\left(1-\upsilon^2\right)}\frac{(1-\upsilon)}{2}\left[\frac{\partial u}{\partial y}+\frac{\partial v}{\partial x}\right] \qquad (9)$$

From this it is possible to start to construct the Finite Element models, starting with a kineto-static analysis using a disc element (four nodes and two degrees of freedom per node) in conjunction with shape functions in u and v so that the stresses N_x, N_y, and N_{xy}, can be calculated. The plate is subjected to the distributed in-plane forces X and Y (equations (5) and (6)) and is divided into 15 equal finite elements (Figure 2), from which it can be seen that the in-plane displacements u and v, along the x and y axes respectively, may be calculated for each element using

$$\begin{Bmatrix} u \\ v \end{Bmatrix}=\left[N_e\right]\left\{q_e\right\} \qquad (10)$$

where $[N_e]$ is a matrix of appropriate shape functions relating to plane stress-strain analysis of a four node serendipity element, and the column vector $\{q_e\}$ contains the elemental degrees of freedom. The kineto-static problem is set up from

$$\left[K_s\right]\left\{q_s\right\}=\left\{f_s\right\} \qquad (11)$$

where $[K_s]$ is a global stiffness matrix for this analysis and $\{f_s\}$ is a column vector containing the distributed mid-plane forces X and Y. The column vector $\{q_s\}$ is calculated from equation (11) and from which it is then possible to calculate all the column vectors $\{q_e\}$ for each element. The stresses N_x, N_y, and N_{xy}, can then be calculated via equations (7)-(9).

At this point an MZC plate bending element is defined (Melosh, Zienkiewicz, and Chang element - see Zienkiewicz & Taylor, (1989)). This element has four nodes and three degrees of freedom per node, and uses a suitable shape function via standard analysis. The equation of motion for the system may be stated in matrix form, as is conventional, in the following manner. First of all the potential energy function is established so that the principle of virtual work can be utilised. The potential energy of a plate element may be stated in this form

$$U_1 = \frac{1}{2}\iint_{(A)}\left[N_x\left(\frac{\partial w}{\partial x}\right)^2 + N_y\left(\frac{\partial w}{\partial y}\right)^2\right]dxdy$$

$$+\frac{1}{2}\iint_{(A)} 2N_{xy}\frac{\partial w}{\partial x}\frac{\partial w}{\partial y}dxdy$$

$$+\frac{1}{2}D\iint_{(A)}\left(\frac{\partial^2 w}{\partial x^2}+\frac{\partial^2 w}{\partial y^2}\right)dxdy$$

$$-\frac{1}{2}D\iint_{(A)} 2(1-\upsilon)\frac{\partial^2 w}{\partial x^2}\frac{\partial^2 w}{\partial y^2}dxdy$$

$$+\frac{1}{2}D\iint_{(A)} 2(1-\upsilon)\left(\frac{\partial^2 w}{\partial x \partial y}\right)^2 dxdy \qquad (12)$$

(where D is as previously defined). The first term in equation (12) can be expanded as follows

$$U_1^{(1)} = \frac{1}{2}\iint_{(A)}\begin{Bmatrix}\frac{\partial w}{\partial x}\\ \frac{\partial w}{\partial y}\end{Bmatrix}^T\begin{bmatrix}N_x & N_{xy}\\ N_{xy} & N_y\end{bmatrix}\begin{Bmatrix}\frac{\partial w}{\partial x}\\ \frac{\partial w}{\partial y}\end{Bmatrix}dxdy \qquad (13)$$

and N_x, N_y, N_{xy}, $\frac{\partial w}{\partial x} and \frac{\partial w}{\partial y}$ can be substituted to obtain

$$U_1^{(1)} = \frac{1}{2}\{q_e\}^T\left[K_e^{(1)}\right]\{q_e\} \qquad (14)$$

where $\left[K_e^{(1)}\right]$ denotes the first term of the stiffness matrix of the element. The second term of equation (12) is of this form

$$U_1^{(2)} = \frac{1}{2}D\iint_{(A)}\left(\frac{\partial^2 w}{\partial x^2}+\frac{\partial^2 w}{\partial y^2}\right)dxdy$$

$$-\frac{1}{2}D\iint_{(A)} 2(1-\upsilon)\frac{\partial^2 w}{\partial x^2}\frac{\partial\ w}{\partial y^2}dxdy$$

$$+\frac{1}{2}D\iint_{(A)} 2(1-\upsilon)\left(\frac{\partial^2 w}{\partial x \partial y}\right)^2 dxdy \qquad (15)$$

and can also be written in matrix form via the procedure of Weaver & Johnston (1984) for example. On this basis the authors used existing formulas to construct the matrix $\left[K_e^{(2)}\right]$. The stiffness matrix for the plate element is the sum of $\left[K_e^{(1)}\right] and \left[K_e^{(2)}\right]$, therefore equation (12) can be written in the matrix / vector form

$$U_1 = \frac{1}{2}\{q_e\}^T[K_e]\{q_e\} \qquad (16)$$

Virtual work formulas are used to calculate the mass forces, centrifugal terms, and the aerodynamic excitation. A virtual displacement of an arbitrarily chosen point in a finite element may be stated as

$$\delta w = \{\delta q_e\}^T[N_e]^T \qquad (17)$$

By using the following virtual work equation

$$\delta L_e = \int_{(A)} \delta w\left(h\rho\frac{\partial^2 w}{\partial t^2}\right)dxdy$$

$$-\int_{(A)} \delta w\left(\rho h\Omega^2\left(w\sin^2\theta - y\sin\theta\cos\theta\right)\right)dxdy$$

$$+\int_{(A)} \delta w\left(\rho_p\frac{U_p}{M}\left(U_p\frac{\partial w}{\partial y}+\frac{\partial w}{\partial t}\right)\right)dxdy \qquad (18)$$

together with equation (17) it is possible to write the virtual work expression in compact form, thus,

$$\delta L_e = \{q_e\}^T\left(\{f_e\}+\left[K_e^{(3)}\right]\{q_e\}+[M_e]\{\ddot{q}_e\}\right)$$

$$+\{q_e\}^T\left(\left[K_e^{(4)}\right]\{q_e\}+[C_e]\{\dot{q}_e\}\right) \qquad (19)$$

where the following definitions are required for the various vector / matrix terms

$$\{f_e\} = h\rho\Omega^2 b\sin\theta\cos\theta\int_{-1}^{1}\int_{-1}^{1}(-\eta)[N_e]abd\xi d\eta$$

$$\left[K_e^{(3)}\right] = h\rho\Omega^2\sin^2\theta\int_{-1}^{1}\int_{-1}^{1}[N_e]^T[N_e]abd\xi d\eta$$

$$[M_e] = -h\rho\int_{-1}^{1}\int_{-1}^{1}[N_e]^T[N_e]abd\xi d\eta$$

$$[K_e^{(4)}] = -\rho_p U_p^2 \frac{1}{M} \int_{-1}^{1} \int_{-1}^{1} [N_e]^T [N_e] abd\xi d\eta$$

$$[C_e] = -\rho_p \frac{U_p}{M} \int_{-1}^{1} \int_{-1}^{1} [N_e]^T [N_e] abd\xi d\eta$$

Finally, using the well known formulation for global matrices (using equations (16 and (19) in this process) it is now possible to construct the equation of motion for the system in traditional compact form

$$[M]\{\ddot{q}\} + [C]\{\dot{q}\} + [K]\{q\} = \{f\} \tag{20}$$

In the case where {f} = 0 then the eigenvalue problem emerges. Both the free and forced cases have been investigated.

3. NUMERICAL RESULTS AND DISCUSSION

In this section a number of different cases are presented based on calculations using the preceding analysis. The 15 element plate is shown in Figure 2. The following data was taken for the calculations: $E = 2.1*10^{10}$ N/m^2, $\nu = 0.3$, $\rho = 7850$ kg/m^3 (material), $\rho_p = 1.293$ kg/m^3 (gas), and plate thickness h = 0.005m. All the results are graphed in nondimensionalised form according to the definitions, $\beta = \omega L^3 \sqrt{(\rho h/D)}$, $\bar{\Omega} = \Omega / \omega_{01}$, $\bar{R} = R_o/L$, D as already stated, $M = U_p/a_k$, and aspect ratio L/b. Fundamental definitions apply: ω being the general natural frequency variable, ω_{01} the natural frequency of fundamental bending of the nonrotating plate, L the length of the plate, h the thickness, b the width, ρ the plate material density, ρ_p the gas density, R_o the hub radius, Ω the shaft rotational velocity, U_p the gas velocity, and a_k the acoustic velocity. All the zero aero-excitation results presented here are deliberately intended to relate to the (zero aero-excitation) conditions investigated by Dokainish and Rawtani (1971). Figure 3 shows the variation in fundamental bending frequency with nondimensionalised angular velocity and zero aero-excitation. Curves are given for setting angles of 0° (solid in all Figures) and 45° (dotted in all Figures) and an aspect ration of L/b = 1. It can be seen that the nondimensionalised natural frequency β rises in a strictly increasing progression with angular velocity, showing a steeper rise for $\theta = 0°$ than for $\theta = 45°$. Higher modes (up to six investigated) all show similar qualitative behaviour. In Figure 4 the shaft speed is kept constant at unity and the Mach number quantity varied over the range $0 \le M \le 0.5$ for the fundamental bending mode. Little sensitivity is shown, however the fundamental torsion mode as shown in Figure 5 depicts a much stronger relationship, and the sixth bending frequency shown in Figure 6 shows an almost linear relationship between β and M. Figure 7 (M = 0, L/b = 1) illustrates the condition when the hub radius quantity is varied. Figure 8 shows the effect of shaft speed for the data of Fig 7 except that L/b = 3. Figure 9 is somewhat different as it deals with the case where rotational velocity = 1, hub radius parameter = 1, L/b = 3, and $0 \le M \le 0.5$ for $\theta = 0°$ and 45°. It appears that the fundamental bending frequency, β, is virtually independent of M. Figure 10 is for the case L/b = 3, M = 0, and variable hub radius. The results in Figure 11 depict the situation in which M = 0.2, angular velocity parameter = 2, hub radius quantity = 0.5, and L/b = 1 for $0 \le \theta \le 90°$, and shows an interesting decreasing relationship between β and θ. In all cases presented (and many others for which there is insufficient space) results for M = 0 show a very high degrees of qualitative and quantitative similarity with those of Dokainish and Rawtani (1971). This provides a reasonable degree of important cross-validation between the two sets of results. Results for $M \neq 0$ are original to this paper.

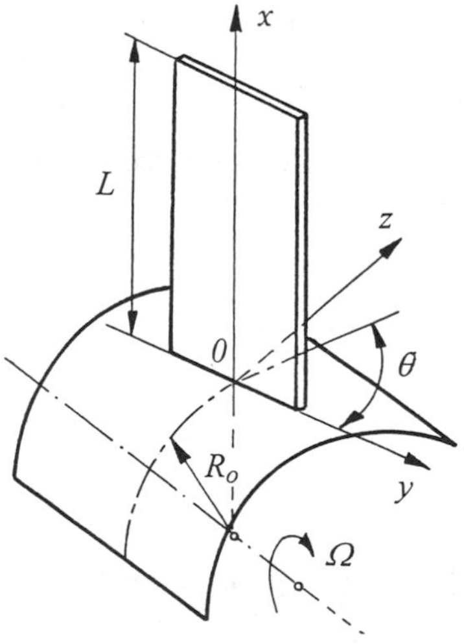

Figure 1 Rotating Plate of Constant Cross-Section

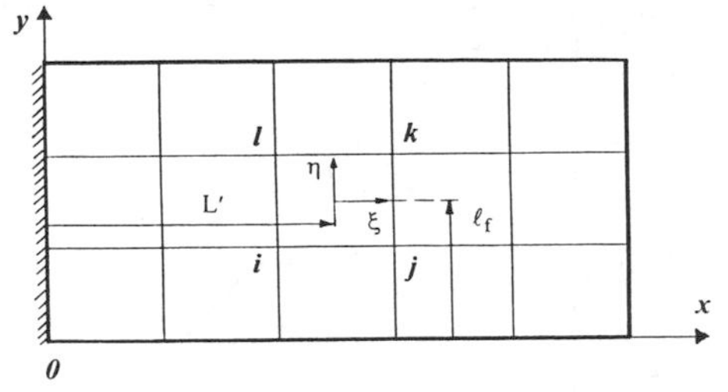

Figure 2 Cantilever Plate subdivided into a 5x3 mesh of four-node finite elements

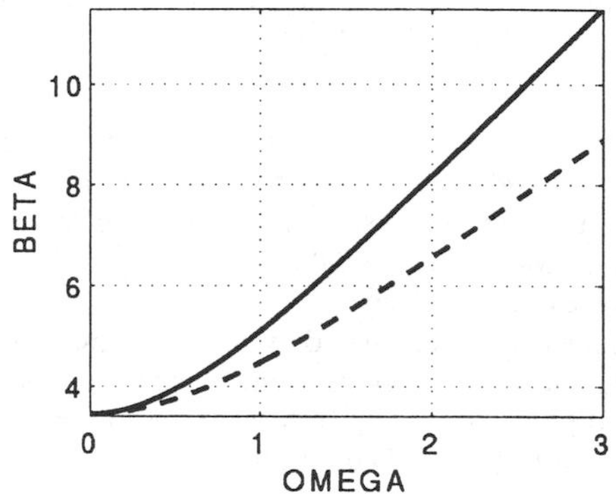

Figure 3 Influence of rotational velocity on the fundamental natural frequency

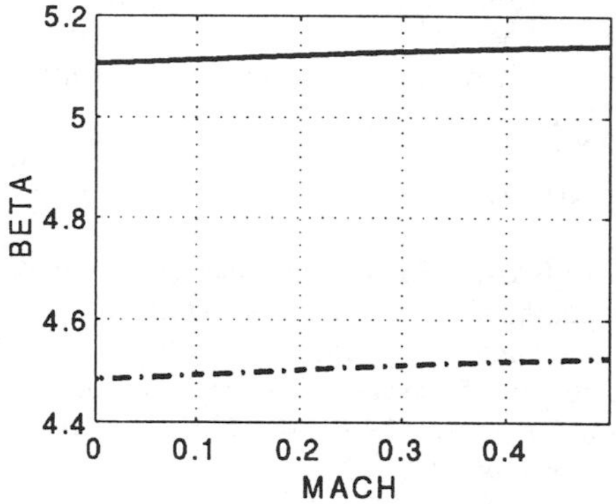

Figure 4 Influence of gas velocity on the fundamental natural frequency in bending

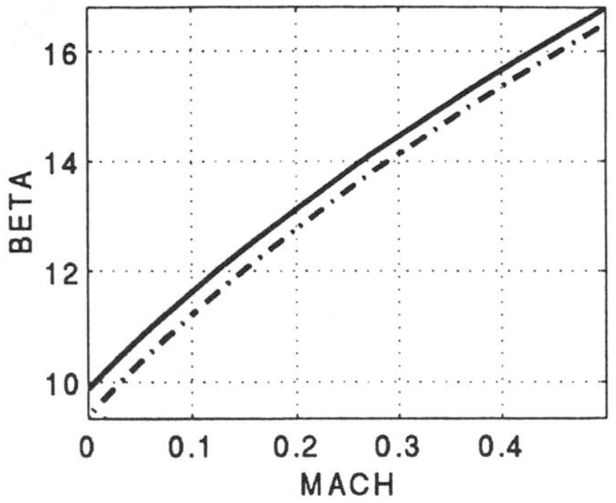

Figure 5 Influence of gas velocity on the fundamental natural frequency in torsion

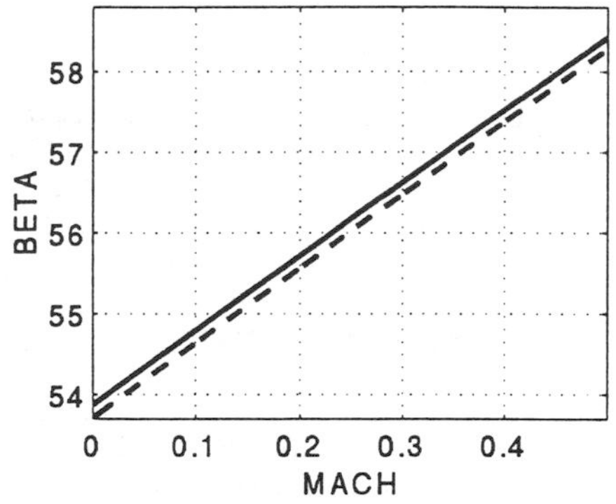

Figure 6 Influence of gas velocity on the sixth natural frequency in bending

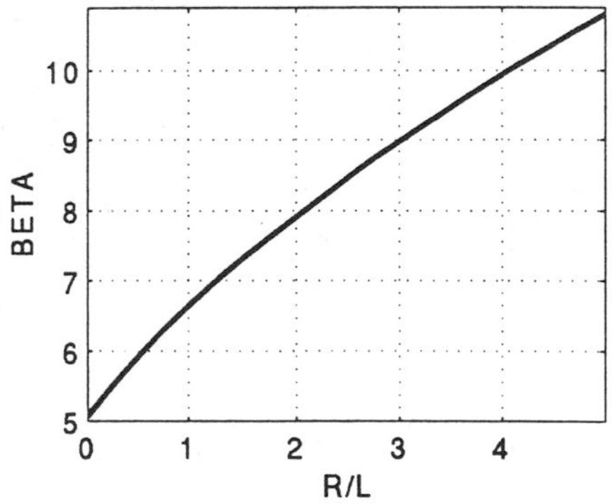

Figure 7 Influence of the nondimensionalised hub ratio on the fundamental natural frequency in bending

4. CONCLUSIONS

The research presented in this paper has taken the problem of a rotating plate subjected to aerodynamic loading, and has shown that certain key phenomena occur for given conditions. In most cases the setting angle is of minimal consequence on the natural frequencies, except for high shaft speeds where there is a sizeable shift for both

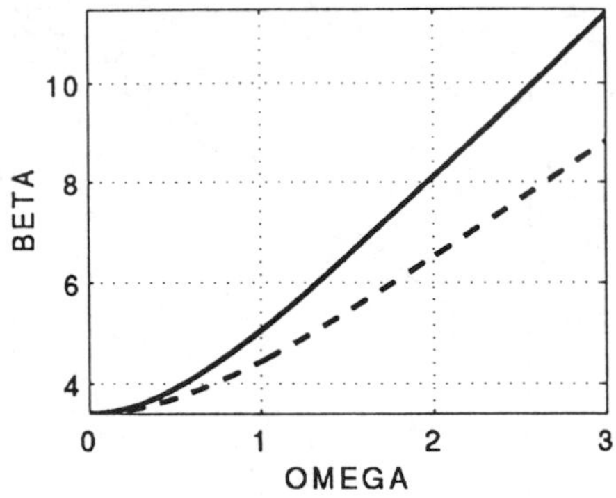

Figure 8 Influence of nondimensionalsed rotational velocity on the fundamental natural frequency in bending

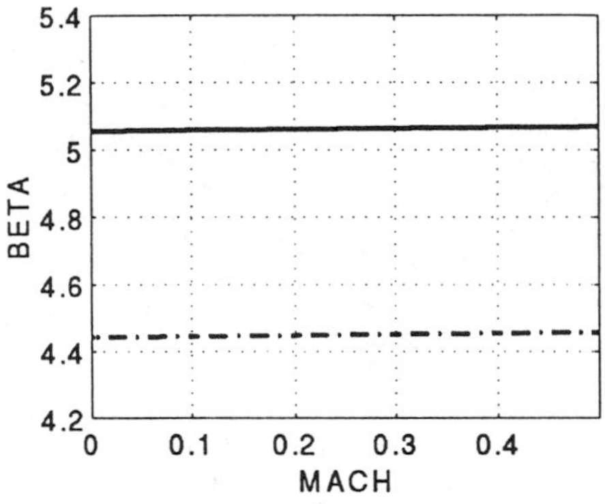

Figure 9 Influence of gas velocity on the fundamental natural frequency in bending

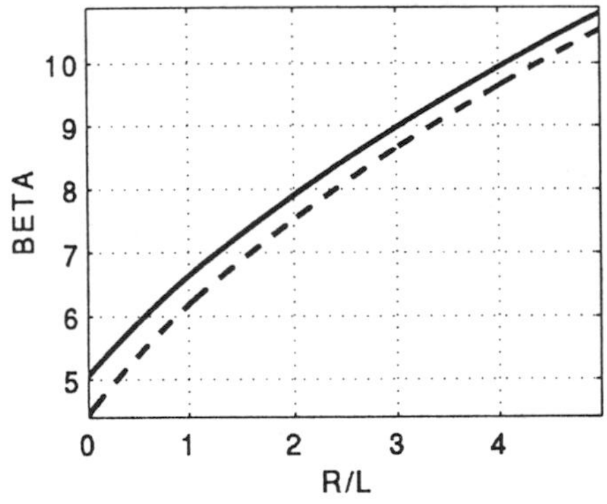

Figure 10 Influence of the hub radius on the fundamental natural frequency in bending

aspect ratios considered. For the case presented where aerodynamic loading is non-zero (i.e. $M \neq 0$) it can be seen that the natural frequencies of the

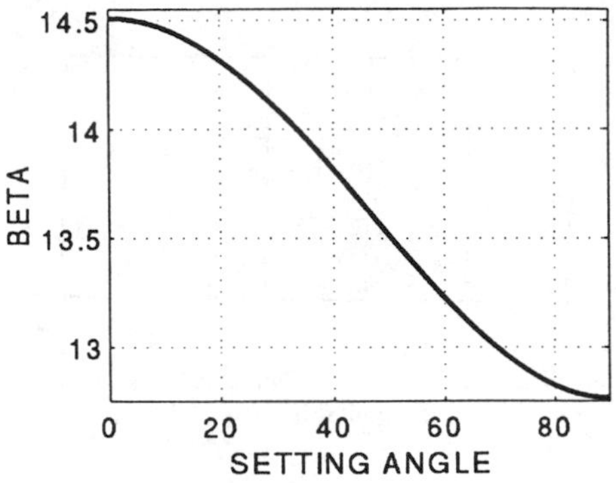

Figure 11 Influence of setting angle on fundamental natural frequency in bending

plate tend to increase with M (this being more accentuated in some cases than others). In the case of mid-range shaft speeds and simultaneous mid-range Mach numbers the setting angle shows a noticeable tendency to reduce the modal frequencies (i.e. the β values). Currently the authors are working on analytical models incorporating nonlinearities, with a view to modelling practical compressor blades.

ACKNOWLEDEGEMENT

The authors wish to acknowledge the financial support made available to them by the British Council and the Polish State Committee for Scientific Research (KBN) in order to carry out this study.

REFERENCES

Abbas, BAH, 1985, Dynamic analysis of a thick rotating blade with flexible roots. *The Aeronautical Journal*, 89: 10-16.

Afolabi D, 1986, Natural frequencies of cantilever blades with resilient roots. *J. of Sound and Vibration*, 110: 429-441.

Argento A, & Scott RA, 1992, Dynamic response of a rotating beam subjected to an accelerating distributed surface force. *J. of Sound and Vibration*, 157: 221-231.

Basu P & Griffin JH, 1986, The effect of limiting aerodynamic and structural coupling in models of mistuned bladed disk vibration. *ASME J. of Vibrations, Acoustic, Stress, and Reliability in Design*, 108: 132-139.

Bazoune A & Khulief YA, 1992, A finite beam element for vibration analysis of rotating

tapered Timoshenko beams. *J of Sound and Vibration*, 156: 141-164.
Dokainish MA & Rawtani S, 1971, Vibration analysis of rotating cantilever plates. *Int. J. of Numerical Methods in Engineering*, 3: 233-248.
Dowell EH, 1975, *Aeroelasticity of plates and shells*. Noordhoff International Publishing, Leyden.
Gau WH & Shabana AA, 1990, Effects of shear deformation and rotary inertia on the nonlinear dynamics of rotating curved beams, *ASME J. of Vibrations and Acoustics*,112: 183-193.
Hoyniak D & Fleeter S, 1986, Forced response analysis of an aerodynamically detuned supersonic turbomachine rotor. *ASME J. of Vibration, Acoustic, Stress and Reliability in Design*, 108: 117-124.
Kane TR, Ryan RR, & Banerjee AK, 1987, Dynamics of a cantilever beam attached to a moving base. *J. of Guidance and Control*, 10,(2): 139-151.
Lee SY & Kuo YH, 1991, Bending frequency of a rotating beam with an elastically restrained root. *ASME J. of Applied Mechanics*, 58: 209-214.
Leissa A, 1981, Vibrational aspects of rotating turbomachinery blades. *ASME Applied Mechanics Reviews*, 34: 629-635.
Leissa A, Lee JK, & Wang AJ, 1982, Rotating blade vibration analysis using shells. *Trans. ASME J. of Eng. for Power*, 104: 296-302.
Ostachowicz WM & Cartmell MP, 1996, Vibration analysis of a rotating beam with variable angular velocity. *Machine Vibration*, 5(4): 189-196.
Simo JC & Vu-Quoc L, 1986, On the dynamics of flexible beams under overall motions - the plane case: Part II. *J. of Applied Mechanics*, 53: 855-863.
Srinavasan AV & Fabunmi JA, 1983, Cascade flutter analysis of cantilevered blades, *ASME J. of Engineering for Power*, 105: 1-10.
Vyas NS & Rao JS, 1992, Equations of motion of a blade rotating with variable angular velocity, *J. of Sound and Vibration*, 156: 327-336.
Weaver W Jr & Johnston PR, 1984, *Finite Elements for Structural Analysis*, Prentice Hall Inc, New Jersey.
Wright AD, Smith CE, Thresher RW, & Wang JLC, 1982, Vibration modes of centrifugally stiffened beams, *ASME J. of Applied Mechanics,* 49: 197-202.
Zienkiewicz OC & Taylor RL, 1989, *Finite Element Method*, McGraw-Hill Book Co. London.

Modern Practice in Stress and Vibration Analysis, Gilchrist (ed.) © 1997 Balkema, Rotterdam, ISBN 90 5410 896 7

Modal clustering in the vibration of partially embedded beams

R.P.West
Department of Civil, Structural and Environmental Engineering, Trinity College Dublin, Ireland
M.N.Pavloić
Department of Civil Engineering, Imperial College of Science, Technology and Medicine, London, UK

ABSTRACT: The phenomenon of modal clustering is investigated for any beam which is partially embedded in an elastic Winkler foundation. An exact closed-form solution to the governing equations enables an understanding of the phenomenon to be developed. A specially written computer algorithm allows a parametric study to be undertaken from which a series of charts are developed which facilitate the prediction of the existence of clustered modes for a wide range of problems.

1 INTRODUCTION

Since Hetényi (1946) published his classical text on the vibration of beams which are fully embedded in an elastic homogeneous Winkler foundation, much research has been done on expanding this work in recent times (see West 1991). Indeed, many different foundation models have been proposed (Kneifati 1985) but even for the simple and often-used elastic Winkler foundation (which is composed of discrete linear uncoupled springs) the difficulties of establishing realistic values for the one soil parameter used (k, the coefficient of horizontal subgrade reaction) are well-known (West 1991). Nonetheless, analyses have been undertaken using mostly either exact solutions or finite elements in an attempt to understand the vibrational behaviour of beams or piles fully or partially embedded in elastic foundations (Eisenberger et al 1985).

This paper sets out the main findings of a comprehensive parametric study into partially embedded beams in a homogeneous elastic Winkler foundation where the existence of very closely spaced modes plays a large part in the phenomenological observations on the model's behaviour. While the homogeneity of the model restricts the current study to essentially cohesive soils (Terzaghi 1955), similar though less pronounced observations have been shown to apply to cohesionless soils (Pavlović and Wylie 1983).

2 GOVERNING EQUATIONS AND THEIR SOLUTION

If a partially embedded beam is modelled as a Bernoulli-Euler beam of stiffness EI, span l and mass per unit length of m, then the governing equation for the dynamic equilibrium of the non-embedded portion (region 1) at time t is well-known as (Warburton 1976)

$$\frac{\partial^4 y}{\partial x^4}+\frac{m}{EI}\frac{\partial^2 y}{\partial t^2}=0 \tag{1}$$

using the co-ordinate system in Figure 1. Similarly, the corresponding equation for the embedded portion (region 2) with an elastic foundation of uniform modulus of subgrade reaction k is

$$\frac{\partial^4 y}{\partial x^4}+\frac{m}{EI}\frac{\partial^2 y}{\partial t^2}+\frac{k}{EI}y=0 \tag{2}$$

The solution to (1) is easily found by separation of variables, yielding an equation of the form

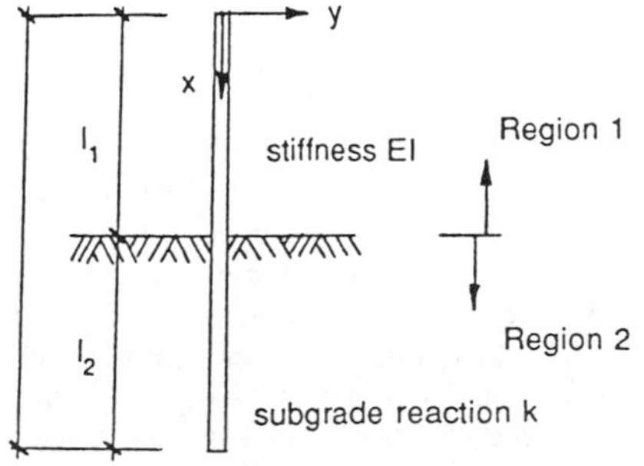

Figure 1: Co-ordinate system for a partially embedded beam.

$$y(x,t) = Y(x) \cdot T(t) \tag{3}$$

The temporal part of the solution, $T(t)$, represents simple harmonic motion and need not concern us here. The shape function is of prime interest and is given by

$$Y(x) = C_1 \cosh\frac{cx}{l} + C_2 \sinh\frac{cx}{l} + C_3 \cos\frac{cx}{l} + C_4 \sin\frac{cx}{l} \tag{4}$$

for which $\omega^2 = c^4 EI/ml^4$, where c is a constant (to be found) and ω is the natural circular frequency of the structural system.

However, the solution to (2) is slightly more complex where, if $\lambda = kl^4/EI$ is introduced and letting $\omega^2 = (d^4 + \lambda)EI/ml^4$, then, considering again only the shape function, three possible solutions exist depending on the sign of d^4:

$[d^4 > 0]$:

$$Y(x) = C_1' \cosh\frac{dx}{l} + C_2' \sinh\frac{dx}{l} + C_3' \cos\frac{dx}{l} + C_4' \sin\frac{dx}{l} \tag{5}$$

$[d^4 = 0]$:

$$Y(x) = C_1''\frac{x^3}{l^3} + C_2''\frac{x^2}{l^2} + C_3''\frac{x}{l} + C_4'' \tag{6}$$

$[d^4 < 0]$:

$$Y(x) = C_1'''\cosh\frac{d'x}{l}\cdot\cos\frac{d'x}{l} + C_2'''\sinh\frac{d'x}{l}\cdot\sin\frac{d'x}{l} + C_3'''\cosh\frac{d'x}{l}\cdot\sin\frac{d'x}{l} + C_4'''\sinh\frac{d'x}{l}\cdot\cos\frac{d'x}{l} \tag{7}$$

where $d' = d/\sqrt{2}$

Only one of these three equations applies, depending on the value of c (where $c^4 = (d^4 + \lambda)$).

In the solution to the combined problem, there are eight constants of integration to be found, four each from the two regions of the beam. This can be reduced to just four equations by applying the extremity boundary conditions (free, pinned or built-in end-fixity) to both sets of equations in turn and by then imposing continuity of deflection, slope, shear and moment at the junction of the two regions. When this is done what remains is a set of four homogeneous equations of the form

$$[\mathbf{M}]\underset{\sim}{\mathbf{A}} = \underset{\sim}{0} \tag{8}$$

It emerges that $[\mathbf{M}]$ is a matrix of hyperbolic and trigonometric functions in terms of c and $\underset{\sim}{\mathbf{A}}$ is the vector of constants of integration. This represents an eigenvalue problem which is solved by finding the eigenvalues, c, for which the determinant of $[\mathbf{M}]$ is zero. A typical set of equations defining $[\mathbf{M}]$, for the built-in free (henceforth B-F) case, are presented in Appendix A. A full set of matrices are available in West (1991).

The method used to find the eigenvalues has been specially derived based on the well-known Williams and Wittrick method (1970). As the problem is non-symmetric, this new method (West and Pavlović 1997a) involved the use of a dynamic increment which searched for changes in the sign-count of the upper triangular form of matrix $[\mathbf{M}]$. This was done near-infallibly in an extensive parametric study to be described here.

3 MODAL CLUSTERING

3.1 *General Trends*

If one introduces an embedment ratio parameter, β, defined by

$$\beta = \frac{l_2}{l} \tag{9}$$

then any problem geometry is defined completely by specifying the non-dimensional parameters β and λ and the boundary conditions at the extremities of the beam. When the algorithm is used to study the variation in the eigenvalues, c (and, hence, the natural circular frequencies, ω) as the stiffness parameter, λ, is increased, a typical result may be seen in Figure 2 in which the example of a B-F beam with $\beta = 0.5$ (50% embedded) is chosen. The ranges for which the different senses of d^4 apply are shown and, in particular, it is observed that sequential modes seem to cluster as they approach the $d^4 = 0$ curve which defines the boundary between the two main regions for d^4. If, for example, the mode shapes for the first two modes in the $d^4 > 0$ region are considered (Figure 3), that is at low values of λ, it can be seen that these shapes are associated with the well-known solutions for the first and second modes of an unembedded cantilever. On the other hand, these mode shapes change (for high λ) in the $d^4 < 0$ region where it is clear (Figure 3) that the shapes are associated with a B-B beam with a reduced span of l/2 (as the beam is 50% embedded). It may be concluded, therefore, that the plateau of eigenvalues in Figure 2 with values of λ less than the $d^4 = 0$ curve

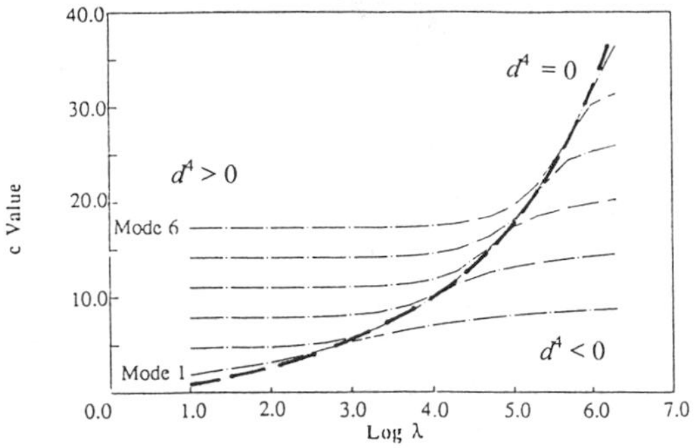

Figure 2: Variation in eigenvalue c with log λ for the first six modes of a built-in free (B-F) beam with 50% embedment (β = 0.5).

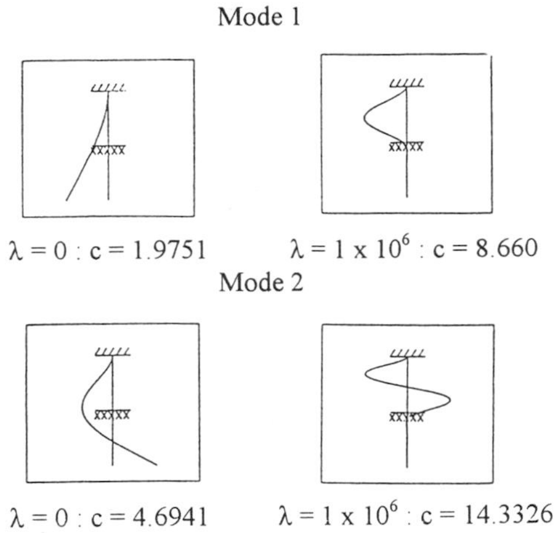

Figure 3: Mode shapes for modes 1 and 2 for a B-F beam with β = 0.5 and λ = 0 and 1×10^6

represent solutions which have behaviour predominantly influenced by unembedded beams whereas the plateau for λ values greater than the d^4 = 0 curve represent solutions which are asymptotic to the case where the foundation is sufficiently stiff in comparison to the beam stiffness to model a fully built-in condition at the junction between the two regions. This pattern of behaviour is consistently observed for all values of β except 0 and 1 exactly where the classical non-embedded and fully embedded solutions apply.

One implication of this behaviour of modes may be seen in Figure 4 where the first four eigenvalues are presented for three different embedded end-fixity conditions, given that the non-embedded end is built-in (β = 0.5 still). It may be observed that the embedded end-fixity condition becomes irrelevant for λ values not much higher than the d^4 = 0 threshold, that is, after the modes cluster. This confirms the observations above regarding the nature of the mode shapes in the $d^4 < 0$ region. The values of c to which the upper plateau is asymptotic is easily predicted as they are similar to the B-B cases on the lower plateau with shorter spans (depending on β).

3.2 *Finite-Element Verification*

These results, while of interest in themselves, become especially pertinent because in verifying the exact solutions above using a conventional finite-element package (ANSYS), it emerged that unless a relatively large number of elements are used, the finite-element results are inaccurate when modal clustering exists. If, for example, less than eight elements are used in the finite-element model the error in the clustered modes is small (<1%), but, surprisingly, some of the remaining modes can be as much as 15% in error, including the fundamental mode (West and Pavlović 1997b). It has been found that only if over 30 elements are used are all of the first six modes predicted within 1% accuracy when any two of the modes cluster. This result is unexpected for such a simple structural system. It is, therefore, desirable to be able to predict when the modes will cluster so that a more refined finite-element model than normal can be included in an analysis of the soil-structure interaction only where necessary. To this end, and to further investigate the phenomenological behaviour of partially embedded beams, a parametric study has been undertaken.

4 PARAMETRIC STUDY

4.1 β - λ - c *Surfaces*

If one expands the range of geometries of interest to include a wide variety of β values from 0 to 1 and a range of λ values from 0 to 5×10^6 in reasonable increments, the trends seen in Figure 2 can be reproduced in a three-dimensional plot of a surface of eigenvalues c against λ and β, for example, in Figure 5 for a B-F beam for just modes 1 and 2. A section through these surfaces at β = 0.5 will reproduce the mode 1 and 2 curves seen in Figure 2. It can be seen that at low λ there exists a so-called lower plateau for which the solution is broadly similar to a non-embedded beam (because λ is so low). The higher the mode the greater the extent of this plateau in terms of λ increasing (see Figure 2 where the $d^4 > 0$ region extends over greater λ as the mode number is increased). At high λ and relatively low β, it can be seen that the general trend is sloping upwards in the β direction on an inclined plane which is relatively λ insensitive, as explained previously.
This indicates behaviour akin to a built-in condition

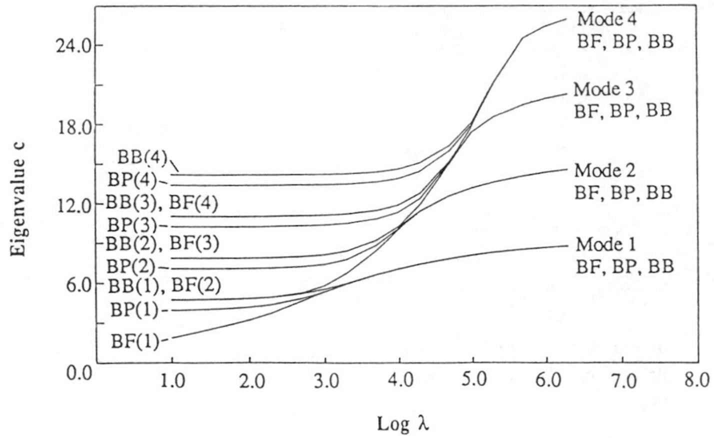

Figure 4: The first four eigenvalues, c, for any partially embeded beam which is built-in at its non-embedded end and for which $\beta = 0.5$.

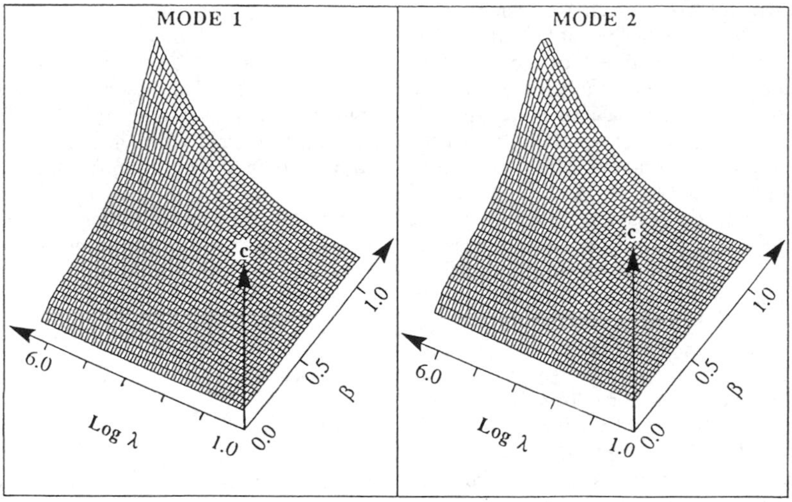

Figure 5: The β - λ - c surfaces for modes 1 and 2 for a B-F beam.

at the junction of the two regions of the beams. At the $\beta = 1$ boundary, the eigenvalues c continue to rise as λ increases, as expected and never reach a point where it becomes λ insensitive (that is, as λ increases for fixed β, c does not vary by much) because no suitable mode shape exists (for a reduced span with built-in at the junction). In the area between the lower plateau and this λ insensitive inclined plane, the area in which the clustering of modes occurs is confined to the β insensitive inclined surface (where β is increased, the c values do not vary much for fixed λ). This will be more evident when a comparison is made between the vertical ordinates (eigenvalue c) of sequential β - λ - c surfaces.

4.2 *Proximity Indices*

In order to predict for what ranges of λ and β any two modes will cluster a new parameter, a proximity index (PI), will be introduced defined by (for sequential modes i and j

$$\mathrm{PI} = \left(\frac{c_j(\lambda) - c_i(\lambda)}{c_j(\lambda = 0) - c_i(\lambda = 0)} \right) \times 100\% \tag{10}$$

A PI of 100% represents distinct sequential modes (normalised at $\lambda = 0$) and a PI of 0% represents identical eigenvalues.

The proximity index surface can now be

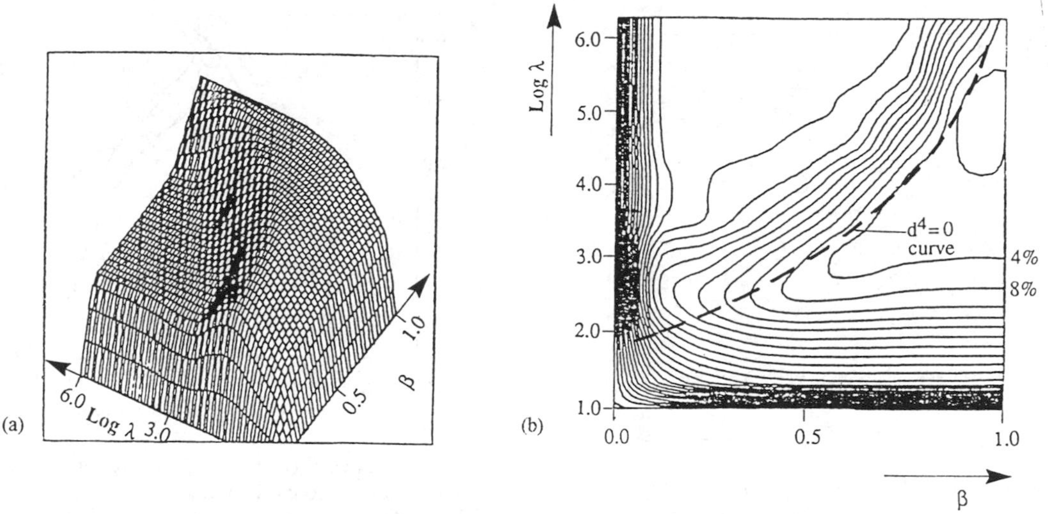

Figure 6: (a) Proximity index surfaces and (b) proximity index surface contours for modes 1 and 2 for a B-F beam. The contour interval is 4.16%.

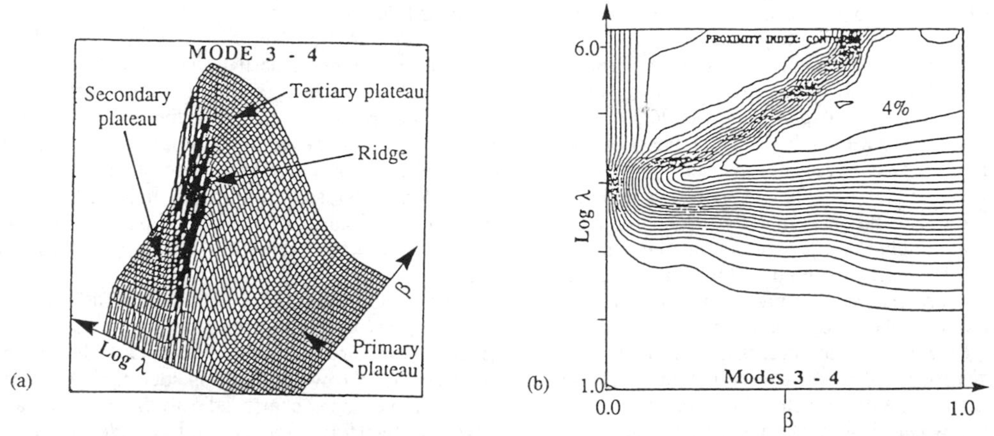

Figure 7: (a) Proximity index surfaces and (b) proximity index surface contours for modes 3 and 4 for a B-F beam.

generated, for example, for modes 1 and 2 for a B-F beam in Figure 6a. λ and β lie in the horizontal plane and the vertical ordinate represents the proximity index but plotted with a PI of 100% at the bottom for convenience. A contour plot of this surface (Figure 6b) shows that the modes have a PI of less than 4% at very high β and high λ. The position of the $d^4 = 0$ curve is also shown for it illustrates how, for any given value of β, modes begin to cluster less rapidly for increasing λ (in the $d^4 < 0$ region) but very quickly move apart after the $d^4 = 0$ threshold has been surpassed (evidence of this is also apparent in Figure 2).

When modes 3 and 4 are compared in Figure 7a, three distinct, almost horizontal, plateaux exist (Figure 7a). The 'primary' plateau (with a PI of approximately 100%) represents the fact that modes 3 and 4 are a fixed frequency apart for low λ (see, also, Figure 2 at low λ). The 'secondary' plateau represents the fact that modes 3 and 4 are also a fixed frequency apart at high λ (again, see Figure 2), but by a different amount to the primary plateau (hence the different PI value for this plane). The 'tertiary' plateau, which is close to a PI of 0%, represents that area over which the modes cluster very closely and this area increases considerably as

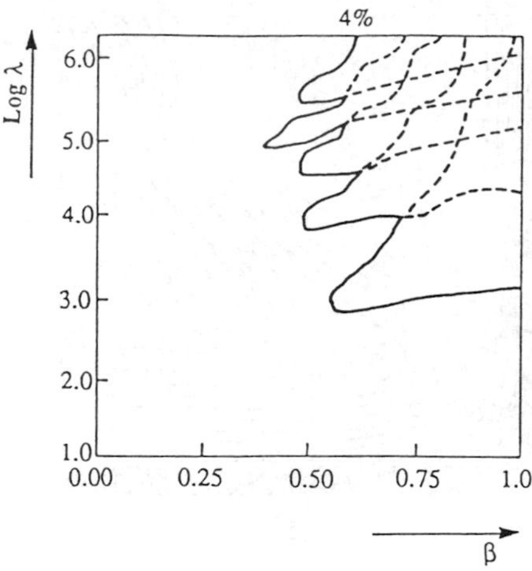

Figure 8: Envelope of highly clustered proximity index contours (at the 4% level) for modes 1-6 for a B-F beam.

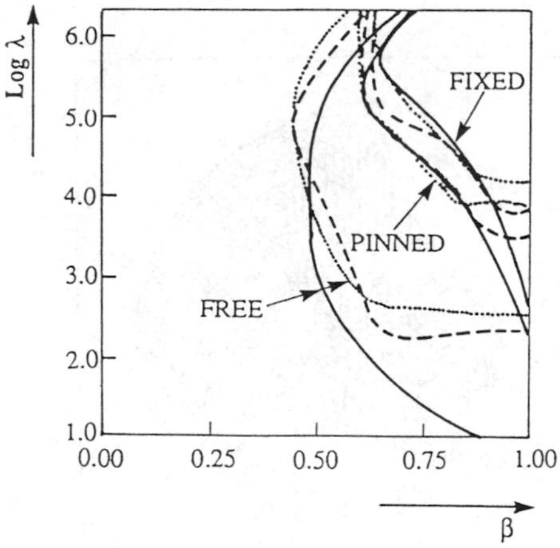

Figure 9: Summary of smeared highly clustered proximity index contours for the first six modes of a partially embedded beam with any end-fixity conditions (Non-Embedded End: free [———]; pinned [– – –] and fixed [· · · · · ·]).

β increases. As λ increases, the PI always drops off the tertiary onto the secondary plateau for all values of β (except the limiting cases of β = 0 and 1 exactly). That is, for high enough λ, a partially embedded beam always behaves as a non-embedded beam with a span somewhat less than *l*, depending on β, with a built-in condition at the junction of the two regions of the beam.

The contour plot for this case is shown in Figure 7b where it can be seen that the 4% contour has moved to a higher λ range as compared with the modes 1-2 case in Figure 6b, but it covers approximately the same total area. If one now designates the 4% contour as highly clustered then the five PI contour plots corresponding to the first six modes for a B-F beam can be compared (Figure 8). This envelope of very close contours can be used to predict when close clustering of any modes will occur for any value of λ and β. It is worth noting in this figure that in some cases more than two modes may cluster for a particular value of λ and β. This is particularly so at high β values where the individual PI contours for two sequential modes overlap with the next PI contour (shown by dashed lines in the figure). This is not the case in Figure 2 because, as seen from Figure 8, the PI contours for any pair of modes do not overlap for values of β as low as 0.5.

Finally, when the parametric study is completed and a smooth curve is smeared over each envelope to provide an indication of the likelihood of close clustering for any end-fixity conditions, the chart in Figure 9 is developed. This may be used to predict the existence of modal clusters, thereby warning the user that a refined finite-element mesh is required in the model if the modes of vibration are to be accurately represented.

5 EXAMPLE

Practical examples of partially embedded beams are piles which are used for a jetty in a deep fast flowing river. The piles are most vulnerable to vortex excitation (from the eddies that are formed in the wake of the flow) during construction when they stand alone, unbraced at the top. Indeed, there are many reported cases of failures of piles under such circumstances (for example, O'Connor and Greene 1986).

Consider the case where a 25m long BP3 Larssen box pile is used in a stiff clay with k = 2.0 MN/m^2 (under these circumstances λ = 8900). It will be assumed that the water depth is such that β = 0.55. The proximity index charts indicate that modes 3 and 4 will cluster (for a free-free fully flooded pile) and the first four natural frequencies are 0.011, 0.060, 0.103 and 0.112 Hz. A 32 element finite-element analysis confirms this but, the finite element analysis can allow for the added mass due to the entrained water in region 1 only, which revise these to 0.008, 0.044, 0.102 and 0.105 Hz. Using a method provided by Hallam et al (1978) these can be converted into flow speeds (which are less than 2 m/s) for which the pile is vulnerable to vortex excitation (see West 1991).

6 CONCLUSIONS

An exact closed-form solution to the problem of finding the natural frequencies of a beam partially embedded in an elastic Winkler foundation has been presented, using the non-dimensional parameters λ (related to the foundation and beam stiffnesses) and β (the degree of embedment). A summary chart has been produced from a comprehensive parametric study which enables modal clustering to be predicted for a variety of beam end-fixity conditions. This will assist designers in determining the fineness of the finite-element model required to determine reasonably accurate results in a dynamic analysis. If only natural frequencies are required, the complete solution is provided herein.

ACKNOWLEDGEMENTS

The assistance of Dr. Michael Heelis in producing the equations and diagrams for this paper is gratefully acknowledged.

REFERENCES

Eisenberger, M., Yankelevsky, D.Z. and Adin, M.A. (1985), 'Vibration of beams fully or partially supported on elastic foundations', *Earthq. Engng. Struct. Dyn.*, 13, 651-660.

Hallam, M.G., Neaf, N.J. and Wootton, L.R. (1978), *Dynamics of marine structures: Methods of calculating the dynamic response of fixed structures subject to wave and current action*, CIRIA Underwater Engng., Rpt. UR8.

Hetényi, M. (1946), *Beams on elastic foundations*, The University of Michigan Press, Ann Arbor, Michigan.

Kneifati, M.C. (1985), 'Analysis of plates on a Kerr foundation', *J. Engng. Mech.*, ASCE, 111 (11), 1325-1342.

Pavlović, M.N. and Wylie, G.B. (1983), Vibration of beams on non-homogeneous elastic foundations', *Earthq. Engng. Struct. Dyn.*, 11, 797-808.

O'Connor, M. and Greene, M. (1986), 'The new jetty at Moneypoint generating station', *IEI Trans.*, 110, 21-32.

Terzaghi, K. (1955), 'Evaluation of coefficient of subgrade reaction', *Geotechnique*, 5, 297-326.

Warburton, G.B. (1976), *The dynamical behaviour of structures*, 2nd Ed., Pergamon Press, Oxford.

West, R.P. (1991), *Modal clustering in the vibration of beams partially embedded in a Winkler foundation*, PhD thesis, Trinity College Dublin.

West, R.P. and Pavlović, M.N. (1997a), 'A fast iterative algorithm for eigenvalue determination', *Comp. and Struct.*, 63, 4, 749-758.

West R. P. and Pavlović, M.N. (1997b), 'Finite-element model sensitivity in the vibration of partially embedded beams', to be published.

Williams, F.W. and Wittrick, W.H. (1970), 'An automatic computational procedure for calculating natural frequencies of skeletal structures', *Int. J. Mech. Sci.*, 12, 781,791.

APPENDIX A: Half matrices for a built-in free beam

The 4x4 matrix [**M**] in equation [8] is composed of two parts. The first two columns depend on the non-embedded end fixity. For example, for a built-in condition they are

$$\begin{bmatrix} \cosh c\tilde{\beta} - \cos c\tilde{\beta} & \sinh c\tilde{\beta} - \sin c\tilde{\beta} \\ \sinh c\tilde{\beta} + \sin c\tilde{\beta} & \cosh c\tilde{\beta} - \cos c\tilde{\beta} \\ \cosh c\tilde{\beta} + \cos c\tilde{\beta} & \sinh c\tilde{\beta} + \sin c\tilde{\beta} \\ \sinh c\tilde{\beta} - \sin c\tilde{\beta} & \cosh c\tilde{\beta} + \cos c\tilde{\beta} \end{bmatrix} \quad \text{(A.1)}$$

where $\tilde{\beta} = 1 - \beta$ and $c = \sqrt[4]{\dfrac{\omega^2 m l^4}{EI}}$

The third and fourth columns depend on the fixity at the embedded end. For a free condition, for example, depending on the sense of d^4, they are

$[d^4 > 0]$:

$$\begin{bmatrix} -(\cosh d\beta + \cos d\beta) & -(\sinh d\beta + \sin d\beta) \\ \zeta(\sinh d\beta - \sin d\beta) & \zeta(\cosh d\beta + \cos d\beta) \\ -\zeta^2(\cosh d\beta - \cos d\beta) & -\zeta^2(\sinh d\beta - \sin d\beta) \\ \zeta^3(\sinh d\beta + \sin d\beta) & \zeta^3(\cosh d\beta - \cos d\beta) \end{bmatrix} \quad \text{(A.2)}$$

where $\beta = \dfrac{l_2}{l}$, $\zeta = \dfrac{d}{c}$ and $d^4 = c^4 - \lambda$

$[d^4 = 0]$:

$$\begin{bmatrix} -\beta & -l \\ \dfrac{l}{c} & 0 \\ 0 & 0 \\ 0 & 0 \end{bmatrix} \quad \text{(A.3)}$$

$[d^4 > 0]$:

$$\begin{bmatrix} -(\Lambda\cdot\lambda) & -(\Lambda\cdot\gamma + \Gamma\cdot\lambda) \\ \zeta'(-\Lambda\cdot\gamma + \Gamma\cdot\lambda) & 2\zeta'(\Lambda\cdot\lambda) \\ 2\zeta'^2(\Gamma\cdot\gamma) & 2\zeta'^2(\Lambda\cdot\gamma - \Gamma\cdot\lambda) \\ -2\zeta'^3(\Lambda\cdot\gamma + \Gamma\cdot\lambda) & -4\zeta'^3(\Gamma\cdot\gamma) \end{bmatrix} \quad \text{(A.4)}$$

where $\Lambda = \cosh d'\beta$, $\lambda = \cos d'\beta$, $\Gamma = \sinh d'\beta$, $\gamma = \sin d'\beta$, $d' = d/\sqrt{2}$ and $\zeta' = d'/c$

Modern Practice in Stress and Vibration Analysis, Gilchrist (ed.) © 1997 Balkema, Rotterdam, ISBN 90 5410 896 7

Finite element and photoelastic analysis of overlapped crankshafts

A.P.Sime, T.H.Hyde & N.A.Warrior
CDI Group, Department of Mechanical Engineering, University of Nottingham, UK

ABSTRACT: The paper outlines the application of the finite element technique to the analysis of operating stresses in overlapped crankshafts for marine diesel engine applications. Modelling aspects such as boundary conditions (to exploit the geometrical symmetry of the crank, and to represent the complex load and restraint conditions seen in service), mesh refinement, and the use of 2D approximations are considered. Stress distributions under the principal loading modes of radial bending, tangential bending and torsion are determined. As a means of validation, the stress levels under the bending loads are compared with results from photoelastic analyses of the same crankshaft geometry.
The results show that stress concentrations at the fillets are around 5, normalised with respect to the maximum bending or shear stress (depending on the loadcase) in the crankpin, and that the magnitude of the SCFs are similar for the loading cases considered.

1 INTRODUCTION

Overlapped crankshafts, where the term overlap refers to the extent of radial interference of the crankpin and journal, are used in heavy duty engine applications. A typical crank for a multi-cylinder V12 or V16 application may be 10 metres in length and is produced as one forging. The component is given a working life on the basis of wear and fatigue crack growth and in order to maximise life, the designer strives to minimise the stress concentration factors which exist in the vicinities of the journal/web and crankpin/web connections.

The finite element (FE) technique can be used to study the operating stresses in crankshafts (Heath & McNamara 1990, Guagliano, Terranova & Vergani 1993), but the user must define the modes of loading and must consider the effects of modelling approximations such as:

- accuracy of representation of bearing oil film load as loads and/or restraints
- validity of choice of planes of symmetry (as it is impractical to model a full multi-cylinder crankshaft)
- the validity of neglecting features such as oil galleries (it is generally considered impractical to model small bore holes due to the very complex meshes required to do this in 3D).

The modelled crank was produced by Mirrlees Blackstone [3] for a V-12 configuration engine. In this engine, two connecting rods are axially spaced on each crankpin. Therefore, the force transmitted through the big end bearing from a firing piston does not act along the whole crankpin, but rather along one half of the crankpin (see Fig. 1).

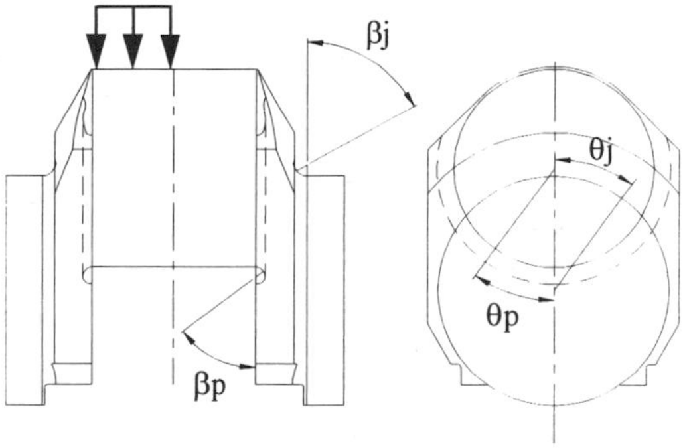

Figure 1. A typical vee-engine crankthrow.

The designer studies loadcases of radial bending, tangential bending and pure torsion (see Fig. 2) as the operating stresses will be a superposition of these cases. Stress concentration factors at various positions around the crankpin and journal fillets are found, normalised with respect to the maximum bending stress in the crankpin (radial and tangential bending) or maximum shear stress in the crankpin (pure torsion), and are compared with allowable levels in design codes, eg (CIMAC 1986).

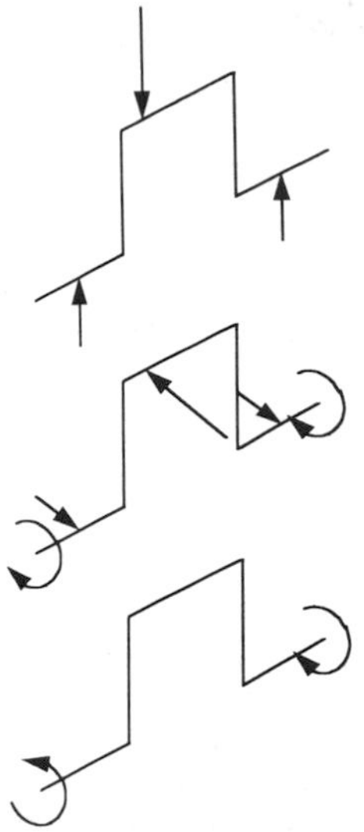

Figure 2. Loading modes.

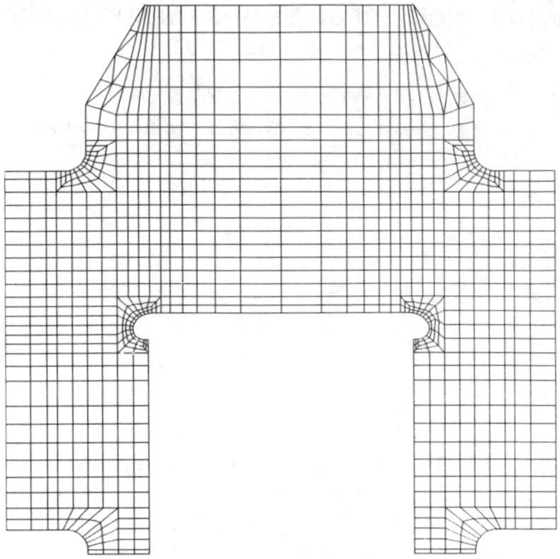

Figure 3. 2D mesh.

2 FE ANALYSIS

Principal stress distributions in the vicinity of the crankpin and journal fillet radii from radial and tangential bending FE analyses are compared with results from photoelastic stress analyses performed on 1/3 scale models of the same geometry. The Young's modulus and the Poisson's ratio for the photoelastic resin at the stress freezing temperature are 11.5 MPa and 0.5 respectively.

2.1 *Radial bending*

Maximum principal stresses are normalised by the maximum bending stress in the crankpin. Figs 5a and 5b show the comparison of fillet stress distributions from the radial bending analyses.

2.1.1 *2-dimensional (2D) analysis*

A 2D mesh (see Fig. 3) of 8-noded quadratic plane strain elements representing the centreline of the crankthrow was created and refined to a point where the minimum principal stresses, normal to the surface, were virtually zero on the free surfaces around the fillets and the elemental stresses were almost continuous and did not change with successive refinement.

In order to study the effects of different loading and boundary conditions in the FE approach the following models were created.

(i) A full crankthrow was loaded with a vertical point load at the quarter-point of the crankpin and was simply supported at the bearing centre-line positions. To represent the steel crank, a Poisson's ratio (υ) value of 0.3 was used. Stress concentration factors of 3.8 and 5 exist at the crankpin and journal fillets respectively.
(ii) A half model was built-in at the centreline of the crankpin, the original point load, was applied at the quarter-point of the crankpin and the reaction at the main bearing was represented by a point load at the centre of the journal. The fillet radii results are the same as in (i).
(iii) A pseudo-symmetrical half model was created and restrained in the axial direction at the centreline of the crankpin, the original quarter-point load, was applied at the mid point of the crankpin and the reaction at the main bearing was represented by a restraint at the centre of the journal. The fillet radii results are within 2% of those in (i).
(iv) In order to represent the bearing oil film loading a distributed load was applied to half of the crankpin. The results agree very closely with the results for the point loaded model, differing by no more than 1% in the fillet radii.
(v) 'Hybrid Elements' using a mixture of stress and displacement variables, were used to calculate stresses at Poisson's ratio of 0.4999. The results agree closely with (i), the peak crankpin fillet stress in the hybrid models is lower by 0.6%. The journal fillet peak stress is 2% higher.
(vi) The results under plane stress conditions are almost identical to (i).

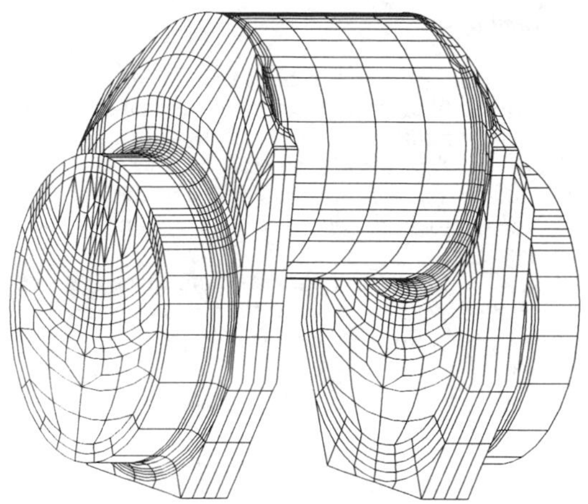

Figure 4. 3D mesh.

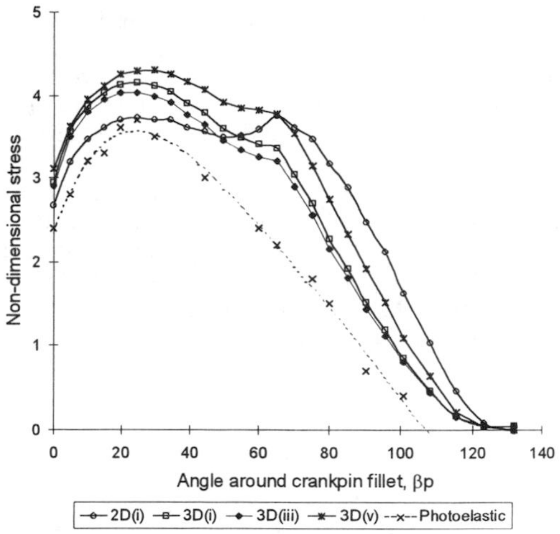

Figure 5a. Radial loading: 2D & 3D crankpin fillet stress distribution.

2.1.2 *3-dimensional (3D) analysis*

3D models were created using 20 noded quadratic elements, using the mesh density of the 2D models as a guide (see Fig. 4).

The results are shown as curves in Fig. 5a and 5b, for comparison with the 2D and photoelastic results.

(i) The full 3D crankthrow (υ=0.3) was loaded with a vertical point load at the quarter-point of the crankpin and was simply supported at the main bearing centres. Stress concentration factors of 4.1 and 4.5 exist at the crankpin and journal fillets respectively.

(ii) The results from the quarter crank analysis (where the model is built-in at the centreline and the connecting rod load and the bearing reaction are represented by point loads) are similar to (i), differing by less than 0.5% at the peak of the fillets.

(iii) A pseudo-symmetrical quarter model was created and restrained in the axial direction at the centreline of the crankpin. A load giving the same bending moment at the web centre as the original quarter-point load was applied at the mid point of the crankpin and the reaction at the main bearing was represented by a restraint at the centre of the journal. The fillet radii results are within 9% of those in (i).

(iv) The results for distributed load models differ from (i) at the crankpin fillet by +1.9% and at the journal fillet by +1.2% at the points of maximum stress.

(v) For υ = 0.499 using hybrid elements, it will be seen that for the crankpin fillet, the stresses increase by 3.5%, but at the journal fillet, the stresses decrease by 2.6%.

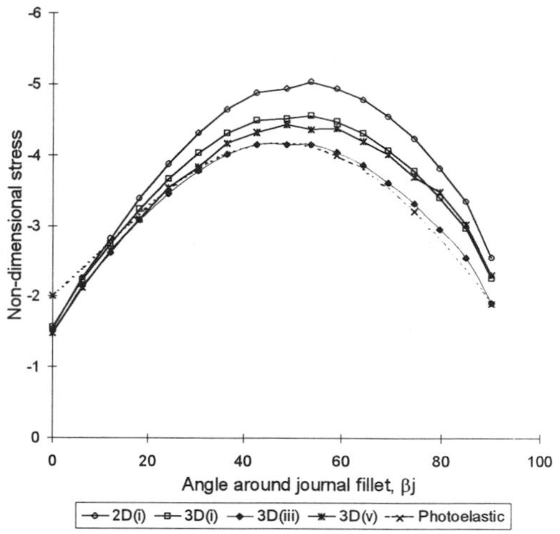

Figure 5b. Radial loading: 2D & 3D journal fillet stress distribution.

2.2 *Tangential bending*

The results from the photoelastic analysis are presented as distributions of the peak in-plane stresses around the crankpin and journal fillets, and are normalised by the maximum bending stress in the crankpin. As a comparison, stresses from the FE analysis are also presented as peak in-plane stresses. The full crankthrow is loaded with a point force acting radially into one side of the crankpin, perpendicular to the plane of the crankthrow (Fig. 2). Boundary conditions are defined as two point restraints on the journal at one extreme end to resist torque only and one simple support at the opposing end to provide vertical support. These restraint conditions are similar to those used in the photoelastic work. The results are presented in Figs 6a and 6b.

In the crankpin fillet the tensile stress concentrations are higher than the compressive stress concentrations (3.7 and 2.7 respectively). In the journal fillet much lower SCFs are calculated. It is believed that the stresses in the journal fillet are strongly influenced by the restraint conditions.

2.3 *Torsion*

The maximum principal stresses seen in the crankpin and journal fillet radii have been normalised with respect to the maximum shear stress seen in a plain cylinder of the same diameter as the crankpin, subjected to the torque applied to the journal free end. The photoelastic work did not cover any torsion analyses.

One face of the journal was built-in and at the other end of the crank a set of four loads was applied to create a torque about the centreline of the journals.

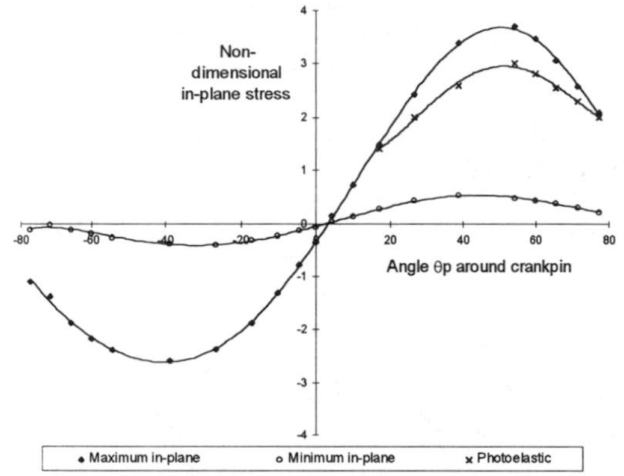

Figure 6a. Tangential loading: 3D crankpin in-plane stress distribution.

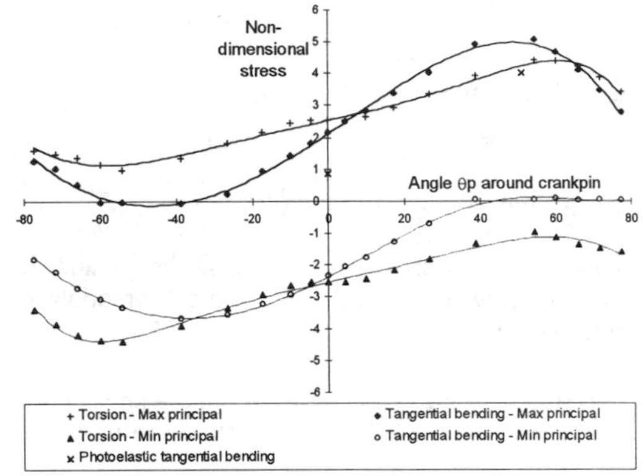

Figure 7a. Torsional loading: 3D crankpin maximum principal stress distribution.

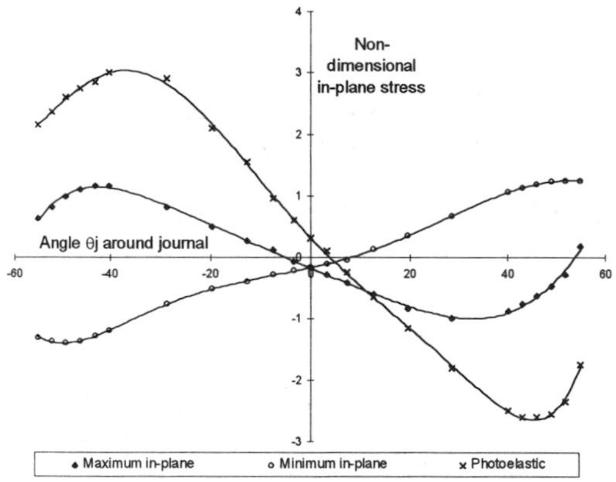

Figure 6b. Tangential loading: 3D journal in-plane stress distribution.

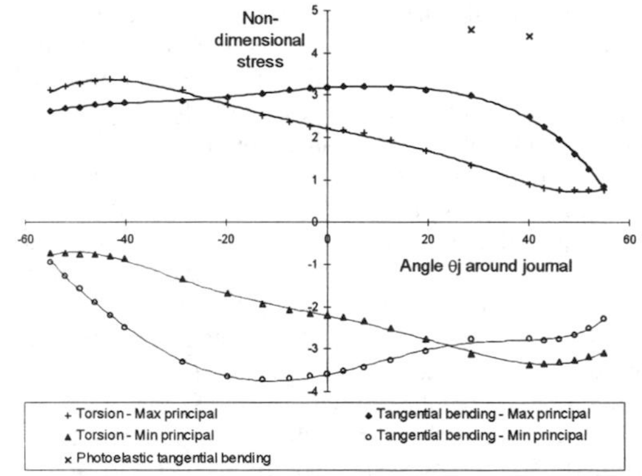

Figure 7b. Torsional loading: 3D journal maximum principal stress distribution.

The maximum principal stresses around the crankpin and journal from the torsion analysis are presented in Figs 7a and 7b and are also compared with the principal stresses from the tangential bending analysis.

For this geometry, in the crankpin fillet, the stress concentration factors for the tangential bending and torsion are 5.2 and 4.6 respectively. For the torsion loading a skew-symmetric stress distribution is seen with a compressive SCF, also of 4.6 at the opposite angle, minus θp, whereas the SCF for the tangential bending is significantly lower at 3.7. In the journal fillet, the stress concentration factors for tangential bending and torsion are 3.7 (in the compressive sense) and 3.4 respectively. As in the crankpin fillet, the torsion load results in a skew-symmetric stress distribution with a compressive SCF of 3.7. Section 2.2 suggests that the journal stress distribution under tangential bending is strongly influenced by the restraint conditions. The principal stress distributions around the journal reinforce this belief.

3 DISCUSSION

The radial bending results show that a distributed load can be approximated as a point load with very little effect on the results at the fillets.

The approximation in the quarter models, of building in the centre of the crankpin, only has a negligible effect on results at the fillets, when compared to the half and full models.

The 2D results agreed with the photoelastic work to within around 20%, but the peaks occur at almost exactly the same angles. It has been reported, (Guagliano, Terranova & Vergani 1993) that a 2D FE analysis of the central plane of the crank can reliably determine the stress distributions in non-overlapped crankshafts for the case of radial bending. In the present work, however, using overlapped geometries the 2D approximation displays a significantly different stress distribution in the crankpin fillet and cannot be reliably exploited.

In 3D the FE results are generally greater than the photoelastic results by 10%. The error at the SCF at the journal fillet increased from zero in the 2D work to 10% in the 3D, and the error at the SCF in the journal fillet decreased from 20% (2D) to 10% (3D).

When using a Poisson ratio of ~0.5, with hybrid elements, the SCFs increase slightly in the crankpin fillet (notably at the second peak) and decrease slightly in the journal fillet.

The results from tangential bending are seen to be strongly dependent on boundary conditions. Although the FE analysis was unable to exactly recreate the photoelastic constraints, the shapes of the SCF distributions around the crankpin are similar to the photoelastic results - the tensile stress concentrations are higher than the compressive stress concentrations and both SCFs are higher than those obtained in the photoelastic work. However, the shape of the curve for the peak in-plane distribution around the journal fillet agrees very poorly with the photoelastic work.

The differences between the pure torsion results and those for tangential bending arise from the extra local bending and shear in the tangential results which causes tension and compression at the back and front of the crankpin.

There are no photoelastic results to compare with the pure torsion results, but it is felt that the boundary conditions and loading under pure torsion are as accurate as those for radial bending and better than those for tangential bending. Because of the reasonable agreement shown in these other loadcases it is anticipated that the pure torsion results will be of similar accuracy.

4 CONCLUSIONS

- The stress concentration factor in radial bending is similar to that seen in tangential bending or torsional loading.
- A 2D representation does not give a reliable approximation to 3D stress concentration factors.
- The oil film load can be reliably modelled as a point load for analysis of stresses in the journal/web and crankpin/web fillet radii.
- The axial symmetry can be exploited to represent full cranks under radial bending with appropriate boundary conditions.
- The effect of Poisson's ratio is significant.

ACKNOWLEDGEMENTS

The authors are grateful to Prof Henry Fessler for his helpful contributions, to the EPSRC and Mirrlees Blackstone for funding the work under the CASE agreement and to HKS for the ABAQUS FE software and FEMSYS for the mesh generation software.

REFERENCES

1 A.R. Heath & P.M. McNamara. Crankshaft stress analysis - combination of finite element and classical analysis techniques, Journal of Engineering for Gas Turbines and Power, Trans. ASME, July 1990, Vol. 112, pp 268-275.

2 M.Guagliano, A.Terranova, L.Vergani, Theoretical and Experimental study of the stress concentration factor in diesel engine crankshafts, Journal of Mechanical Design, Trans. ASME, March 1993, Vol 115, pp 47-52.

3 Mirrlees Blackstone Ltd. Stockport,Cheshire,UK.
4 CIMAC M53, 1986.

APPENDIX 1

The crankshaft geometry is defined as follows:

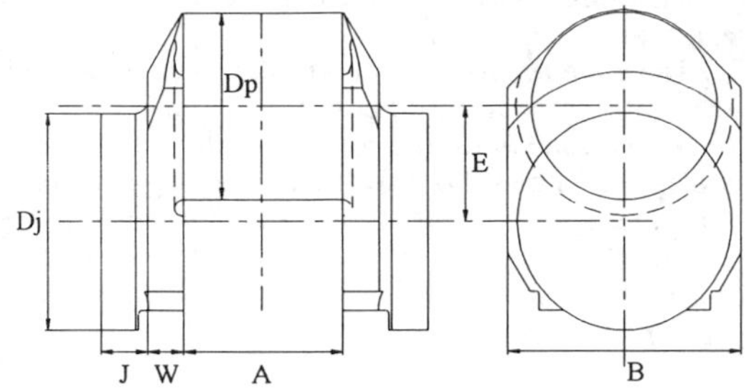

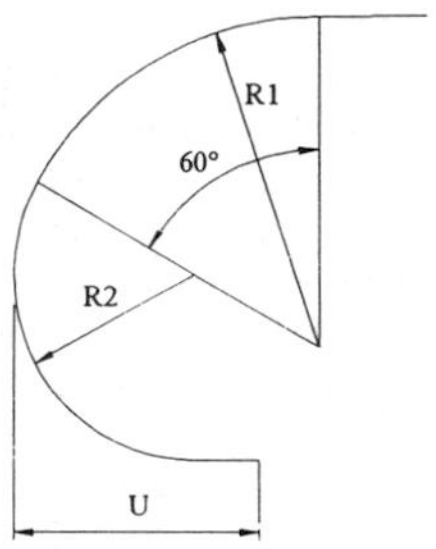

Crankpin fillet

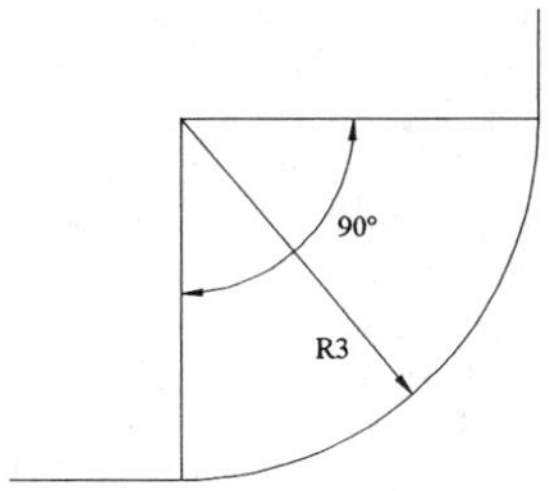

Journal fillet

Dj / Dp	1.154
E / Dp	0.615
J / Dp	0.256
W / Dp	0.192
A / Dp	0.820
B / Dp	1.257
U / Dp	0.049
R1 / Dp	0.064
R2 / Dp	0.036
R3 / Dp	0.064

Modern Practice in Stress and Vibration Analysis, Gilchrist (ed.) © 1997 Balkema, Rotterdam, ISBN 90 5410 896 7

Wave propagation in rotating thermo-piezoelectric solids*

J.Wauer
Institut für Technische Mechanik, Universität Karlsruhe, Germany

ABSTRACT: The propagation of plane waves in a pyroelectric solid is studied for the case when the entire medium is rotating with a uniform angular velocity. The usual electrically quasistatic Maxwell equations and the conventional theory of thermo–piezoelectricity are taken into account. The governing dispersion relation is obtained to determine the effects of speed rate and thermal conductivity on the finite phase velocity of the waves. The analysis is carried out for an infinite medium and also for a plate layer of finite thickness with the rotational axis in the midplane of the slab. Several limiting cases of interest are discussed. The computations are specified for a hexagonal solid of (6mm) class.

1 INTRODUCTION

The wave propagation in unbounded elastic media and the vibrations of finite elastic bodies are well–understood. In recent years considerable attention has been given to study similar problems for piezoelectric and pyroelectric material. Generalizing the formulation is straightforward (see Eringen & Maugin, 1990, for instance). The approach usually addopted in treating piezoelectric solids is simplification of Maxwell's equations by neglecting magnetic effects, conduction, displacement currents and free charges (see Toupin, 1963 and Tiersten, 1969, for instance). In the context of this conventional electrically quasistatic theory, Mindlin (1961) developed the conventional theory of thermo–piezoelectricity.

While the formulation of the basic equations for pyroelectric solids is not difficult, even for the planar case, the evaluation contains a lot of open questions; only some layer problems were treated by Paul & Renganathan (1985) and Yang & Batra (1995), for example.

In the present contribution, such two–dimensional problems are taken up again, but attention is focused on another aspect, namely the influence of a constant moderate speed rate. For magnetoelastic and magneto–thermoelastic bodies, such considerations exist (see Chauduri & Debnath, 1983a and 1983b, respectively), for piezoelectric solids, corresponding results will be presented in this study.

After formulating the general boundary value problem, the calculations are concretized to a material of hexagonal (6mm) class. The analysis is carried out for an unbounded medium and also for a planar layer of finite thickness with the rotational axis in its midplane. Of special interest are the results for the limiting case of a non–rotating layer to be compared with those obtained by Paul & Renganathan (1985).

2 GENERAL FORMULATION

Consider a rotating thermo–piezoelectric body of volume V possessing a smooth regular boundary S. The unit outward normal to S is n_i, and S is partitioned as

$$S_u \cup S_T = S_\phi \cup S_D = S_\theta \cup S_q = S. \tag{1}$$

The structural member is uniformly rotating with a moderate angular velocity $\mathbf{\Omega} = \Omega \mathbf{e}_n$ where $\mathbf{e}_n$ is the unit vector representing the direction of the axis of rotation. The displacement equation of motion in a rotating frame of reference has two additional terms: the centripetal accele-

* Dedicated to Professor Dr. Franz Ziegler, TU Wien upon the occasion of his 60th birthday anniversary.

ration, $\mathbf{\Omega} \times (\mathbf{\Omega} \times \mathbf{u})$ due to the time–varying motion only, and the Coriolis acceleration, $2\mathbf{\Omega} \times \mathbf{u}_{,t}$ where $\mathbf{u}$ is the dynamic displacement vector. Pre–deformations due to the centrifugal force are neglected and corresponding time–independent excitations are omitted. Only free pyroelectric oscillations are treated here.

In body–fixed rectangular Cartesian coordinates, the fundamental linear equations are the strain–displacement relations

$$2S_{ij} = u_{i,j} + u_{j,i}, \tag{2}$$

the equations of motion

$$\begin{aligned} \rho\left[\mathbf{u}_{,tt} + \mathbf{\Omega} \times (\mathbf{\Omega} \times \mathbf{u}) + 2\mathbf{\Omega} \times \mathbf{u}_{,t}\right]_j - T_{ij,i} &= 0, \\ T_{ij} &= T_{ji}, \end{aligned} \tag{3}$$

the law of conservation of internal energy for piezoelectric materials

$$\theta_0 s_{,t} + q_{i,i} = 0, \tag{4}$$

and Maxwell's equations (within a quasistatic theory) for piezoelectric material

$$\operatorname{div} \mathbf{D} = 0, \quad \mathbf{E} = -\operatorname{grad} \phi \tag{5}$$

representing Gauss' law and simplified Faraday's law. In these equations, ρ is the mass density, S_{ij} is the infinitesimal Green strain tensor, T_{ij} is the Cauchy stress tensor, s is the entropy density, $\mathbf{E}$ is the electric field intensity, ϕ is the electric potential and $\mathbf{D}$ is the electric displacement. Further, q_i is the heat flux vector within the material and θ_0 is the reference uniform temperature of the body. $(.)_{,t}$ and $(.)_{,i}$ denote partial differentiation with respect to time t and the spatial coordinates x_i, respectively.

In addition, the following constitutive equations are considered:

$$\begin{aligned} T_{ij} &= c_{ijkl} S_{kl} - \beta_{ij}\theta - e_{kij} E_k, \\ s &= d\theta + \beta_{ij} S_{ij} + p_k E_k, \\ D_i &= p_i\theta + e_{ijk} S_{jk} + \epsilon_{ij} E_j, \\ q_i &= -k_{ij}\theta_{,j} \end{aligned} \tag{6}$$

where the conventional restrictions on the material constants are valid (see Chandrasekharaiah, 1988). Herein, θ is the change in the temperature of a material particle, c_{ijkl} is the isothermal elastic tensor, e_{ijk} are the piezoelectric moduli, β_{ij} are the thermal stress moduli, p_k are the pyroelectric moduli, and ϵ_{ij} is the electric permittivity tensor. k_{ij} the thermal conductivity tensor and d is related to the specific heat: $d = (\rho c_v)/\theta_0$. Furthermore, a repeated index implies summation over the range of the index.

Finally, the prescribed (homogeneous) boundary conditions are

$$\begin{aligned} u_i &= 0 \text{ on } S_u, & T_{ij} n_j &= 0 \text{ on } S_T, \\ \phi &= 0 \text{ on } S_\phi, & D_i n_i &= 0 \text{ on } S_D, \\ \theta &= 0 \text{ on } S_\theta, & q_i n_i &= 0 \text{ on } S_q. \end{aligned} \tag{7}$$

The part S_u of the boundary of the body is clamped, S_T is traction–free, S_q is thermally insulated, S_θ is kept at the initial uniform temperature of the body, the normal component of the electric displacement vanishes on S_D, and the electric potential is zero on S_ϕ.

Eliminating T_{ij}, D_i, s and q_i and using Eqs (2) and $(5)_2$, one obtains

$$\begin{aligned} \rho\left[\mathbf{u}_{,tt} + \mathbf{\Omega} \times (\mathbf{\Omega} \times \mathbf{u}) + 2\mathbf{\Omega} \times \mathbf{u}_{,t}\right]_j & \\ -c_{ijkl} u_{k,li} + \beta_{ij}\theta_{,i} - e_{kij}\phi_{,ki} &= 0, \\ p_i\theta_{,i} + e_{kij} u_{i,jk} - \epsilon_{ij}\phi_{,ij} &= 0, \\ -k_{ij}\theta_{,ij} + \theta_0\left(d\theta_{,t} + \beta_{ij} u_{i,jt} - p_i\phi_{,it}\right) &= 0 \end{aligned} \tag{8}$$

where in the boundary conditions (7), the stresses T_{ij}, the electrical displacements D_i and the heat fluxes q_i also have to be expressed in terms of displacements u_i, electric potential ϕ and incremental temperature θ (and derivatives of them).

If a non–rotating body is assumed, the boundary value problem reduces to that formulated by Yang & Batra (1995).

3 CONCRETE APPLICATION

Consider a pyroelectric solid for a material of hexagonal (6mm) class. Two geometries are discussed as shown in Figure 1. In both cases, there is an inertial X, Y, Z reference frame with the speed rate of the solid about the Y axis:

$$\mathbf{\Omega} = \Omega \mathbf{e}_Y. \tag{9}$$

The material with its crystal axes x, y, z is arranged in such a way that the y axis of the body–fixed frame and the rotational Y axis coincide.

The first problem concerns an infinity medium and plane waves in the x–direction are considered so that

$$(\,.\,)_{,y} = (\,.\,)_{,z} = 0. \tag{10}$$

In addition, a plate layer of finite thickness with surfaces parallel to the x, y midplane is dealt

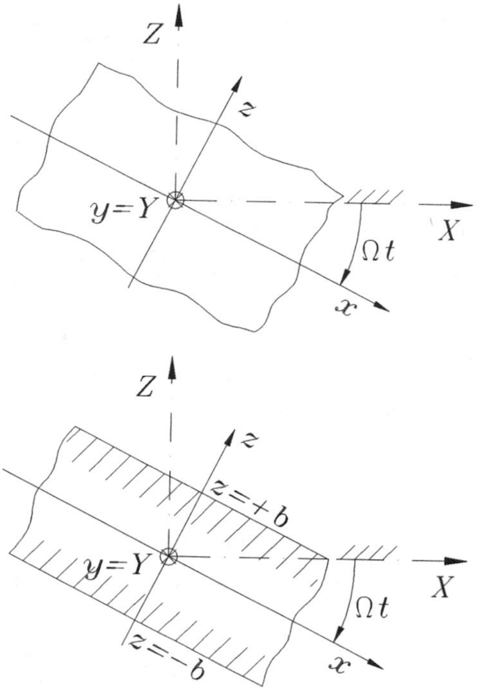

Figure 1. Geometry and reference frames.

with. Here, the fully two–dimensional motion in the cross–sectional xz plane ($-\infty < x < +\infty$ and $-b < z < +b$) is analyzed which is a combination of travelling waves in the x–direction and standing waves (oscillations) in the z–direction (due to reflections at the two bounding surfaces).

For simplification, the displacement coordinate v parallel to the y axis is neglected in both specifications.

Using Voigt's notation, the constitutive equations (6) reduces for the considered two–dimensional motion to

$$
\begin{aligned}
T_{xx} &= c_{11}u_{,x} + c_{13}w_{,z} + e_{31}\phi_{,z} - \beta_1\theta,\\
T_{yy} &= c_{12}u_{,x} + c_{13}w_{,z} + e_{31}\phi_{,z} - \beta_1\theta,\\
T_{zz} &= c_{13}u_{,x} + c_{33}w_{,z} + e_{33}\phi_{,z} - \beta_3\theta,\\
T_{yz} &= 0,\\
T_{xy} &= 0,\\
T_{zx} &= c_{44}(w_{,x} + w_{,z}) + e_{15}\phi_{,x},\\
D_x &= e_{15}(w_{,x} + u_{,z}) - \epsilon_{11}\phi_{,x},\\
D_y &= 0,\\
D_z &= e_{31}u_{,x} + e_{33}w_{,z} - \epsilon_{33}\phi_{,z} + p_3\theta,\\
s &= \beta_1 u_{,x} + \beta_3 w_{,z} - p_3\phi_{,z} + d\theta.\\
q_x &= -k_{11}\theta_{,x}, \quad q_z = -k_{33}\theta_{,z}.
\end{aligned}
\tag{11}
$$

The equations of motion, Gauss' charge equation, and the equation for the flow of heat energy reduce to

$$
\begin{aligned}
&\rho(u_{,tt} + 2\Omega w_{,t} - \Omega^2 u) - c_{11}u_{,xx} - (c_{13} + c_{44})w_{,xz}\\
&\quad - c_{44}u_{,zz} - (e_{31} + e_{15})\phi_{,xz} + \beta_1\theta_{,x} = 0,\\
&\rho(w_{,tt} - 2\Omega u_{,t} - \Omega^2 w) - c_{44}w_{,xx} - (c_{13} + c_{44})u_{,xz}\\
&\quad - c_{33}w_{,zz} - e_{15}\phi_{,xx} - e_{33}\phi_{,zz} + \beta_3\theta_{,z} = 0,\\
&e_{15}w_{,xx} + (e_{15} + e_{31})u_{,xz} - \epsilon_{11}\phi_{,xx} + e_{33}w_{,zz}\\
&\quad - \epsilon_{33}\phi_{,zz} + p_3\theta_{,z} = 0,\\
&k_{11}\theta_{,xx} + k_{33}\theta_{,zz}\\
&\quad - \theta_0(\beta_1 u_{,xt} + \beta_3 w_{,zt} - p_3\phi_{,zt} + d\theta_{,t}) = 0.
\end{aligned}
\tag{12}
$$

For convenience, the following simplest boundary conditions are assumed for the layer problem:

$$u = w = \phi = \theta_{,z} = 0 \text{ at } z = \pm b, \tag{13}$$

The choice of the considered boundary conditions is not restrictive. The results for other boundary conditions differ only quantitatively.

For a non–rotating plate layer, the boundary value problem basically reduces to that analyzed by Paul & Renganathan (1985); only more complicated boundary conditions were assumed. A non–rotating piezoelectric laminate (neglecting the thermal influences) was examined by Heyliger et al. (1995).

For the concrete analysis, a dimensionless formulation is appropriate. For this purpose, introduce the scaling

$$\bar{t} = \Omega_0 t, \quad \bar{x} = \frac{x}{b}, \quad \bar{z} = \frac{z}{b} \tag{14}$$

where

$$\Omega_0^2 = \frac{c_{33}}{\rho b^2}, \tag{15}$$

the non–dimensional variables

$$\bar{u} = \frac{u}{b}, \quad \bar{w} = \frac{w}{b}, \quad \bar{\phi} = \frac{\epsilon_{33}}{b\,e_{33}}\phi, \quad \bar{\theta} = \frac{p_3}{e_{33}}\theta \tag{16}$$

and parameters

$$
\begin{aligned}
T_\theta &= \frac{\theta_0\Omega_0 d b^2}{k_{33}}, \quad \alpha_{\theta w} = \frac{\theta_0\Omega_0 b^2\beta_3 p_3}{k_{33}e_{33}},\\
\varepsilon_{\theta\phi} &= \frac{\theta_0\Omega_0 p_3^2 b^2}{k_{33}\epsilon_{33}}, \quad \varepsilon_{w\phi} = \frac{e_{33}^2}{\epsilon_{33}c_{33}}, \quad \varepsilon_{w\theta} = \frac{\beta_3 e_{33}}{p_3 c_{33}},\\
\eta &= \frac{a}{b}, \quad \bar{\Omega} = \frac{\Omega}{\Omega_0}, \quad \gamma_{ij} = \frac{c_{ij}}{c_{33}}, \quad \delta_{ij} = \frac{e_{ij}}{e_{33}},\\
\beta &= \frac{\beta_1}{\beta_3}, \quad \epsilon = \frac{\epsilon_{11}}{\epsilon_{33}}, \quad \kappa = \frac{k_{11}}{k_{33}}.
\end{aligned}
\tag{17}
$$

If now derivatives with respect to $\bar{x}$, $\bar{z}$ and $\bar{t}$ are denoted by stars, dashes and overdots, respectively, and all overbars are dropped for convenience, one obtains

$$
\begin{aligned}
&-\gamma_{11}u^{**}-(\gamma_{13}+\gamma_{44})w^{**}-(\delta_{31}+\delta_{15})\varepsilon_{w\phi}\phi^{*\prime}\\
&\quad-\gamma_{44}u''+\beta\varepsilon_{w\theta}\theta^{*}+\ddot{u}+2\Omega\dot{w}-\Omega^2u=0,\\
&-\gamma_{44}w^{**}-w''-(\gamma_{13}+\gamma_{44})u^{*\prime}-\delta_{15}\varepsilon_{w\phi}\phi^{**}\\
&\quad-\varepsilon_{w\phi}\phi''+\varepsilon_{w\theta}\theta'+\ddot{w}-2\Omega\dot{u}-\Omega^2w=0,\\
&\delta_{15}w^{**}+(\delta_{15}+\delta_{31})u^{*\prime}-\epsilon\phi^{**}+w''-\phi''+\theta'=0,\\
&-\kappa\theta^{**}-\theta''+T_\theta\dot{\theta}+\beta\alpha_{\theta w}\dot{u}^{*}\\
&\quad+\alpha_{\theta w}\dot{w}'-\varepsilon_{\theta\phi}\dot{\phi}'=0
\end{aligned}
\tag{18}
$$

and

$$u=w=\phi=\theta'=0 \text{ at } z=\pm 1. \tag{19}$$

Obviously, this is a linear boundary value problem of 8th order in each space coordinate and fifth order in time.

4 EVALUATION AND RESULTS

Classical methods are applied. Assuming appropriate solutions in product form, the dispersion relation characterizing the wave propagation in an infinite rotating medium and the (approximate) eigenvalue equation associated with the planar dynamics of the plate layer are derived.

4.1 *Unbounded solid*

Considering plane wave conditions, the governing field equations (18) simplify to

$$
\begin{aligned}
&-\gamma_{44}u''+\ddot{u}+2\Omega\dot{w}-\Omega^2u=0,\\
&-w''-\varepsilon_{w\phi}\phi''+\varepsilon_{w\theta}\theta'+\ddot{w}-2\Omega\dot{u}-\Omega^2w=0,\\
&w''-\phi''+\theta'=0,\\
&-\theta''+T_\theta\dot{\theta}+\alpha_{\theta w}\dot{w}'-\varepsilon_{\theta\phi}\dot{\phi}'=0.
\end{aligned}
\tag{20}
$$

Assuming the appropriate time and space dependence

$$q(x,t)=C_q\,e^{i(kx-\omega t)},\quad q\in\{u,w,\phi,\theta\} \tag{21}$$

where k is the characteristic wave number and ω represents the circular frequency, the simplified field equations (20) reduce to a homogeneous system of algebraic equations for the C_q $(q\in u,w,\phi,\theta)$. The vanishing determinant is the dispersion relation

$$
\begin{vmatrix}
a_{11} & a_{12} & a_{13} & a_{14}\\
a_{21} & a_{22} & a_{23} & a_{24}\\
a_{31} & a_{32} & a_{33} & a_{34}\\
a_{41} & a_{42} & a_{43} & a_{44}
\end{vmatrix}=0 \tag{22}
$$

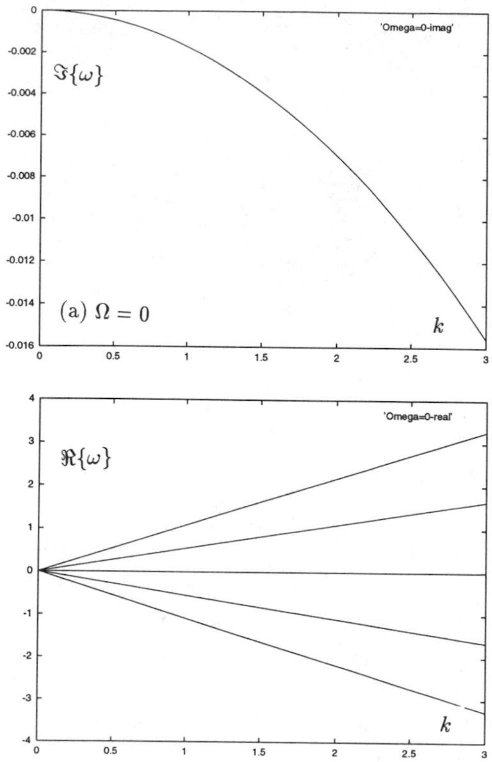

Figure 2. Circular frequeny versus wave number.

to be found where

$$
\begin{aligned}
a_{11}&=\gamma_{44}-\frac{\omega^2+\Omega^2}{k^2},\quad a_{12}=2i\frac{\Omega\omega}{k^2},\\
a_{13}&=a_{14}=0,\\
a_{21}&=-2i\frac{\Omega\omega}{k^2},\quad a_{22}=1-\frac{\omega^2+\Omega^2}{k^2},\\
a_{23}&=\varepsilon_{w\phi},\quad a_{24}=-i\frac{\varepsilon_{w\theta}}{k},\\
a_{31}&=0,\quad a_{32}=-a_{33}=-k,\quad a_{34}=-i,\\
a_{41}&=0,\quad a_{42}=\frac{\alpha_{\theta w}\omega}{k},\\
a_{43}&=-\frac{\varepsilon_{\theta\phi}\omega}{k},\quad a_{44}=1+i\frac{T_\theta\omega}{k^2}.
\end{aligned}
\tag{23}
$$

For the quantitative evaluation, the piezoeceramics $BaTiO_3$ is considered. The data can be found in the literature (Berlincourt et al., 1964 and Condon & Odishaw, 1972). One obtains the following non–dimensional parameters:

$$
\begin{aligned}
\varepsilon_{w\phi}&=0.1881,\quad \varepsilon_{w\theta}=0.5718,\quad T_\theta=576.96,\\
\alpha_{\theta w}&=0.3234,\quad \varepsilon_{\theta\phi}=0.2120,\\
\gamma_{11}&=1.0274,\quad \gamma_{13}=0.6027,\quad \gamma_{44}=0.3014,\\
\delta_{31}&=-0.2486,\quad \delta_{15}=0.6571,\quad \beta=1.6981,
\end{aligned}
$$

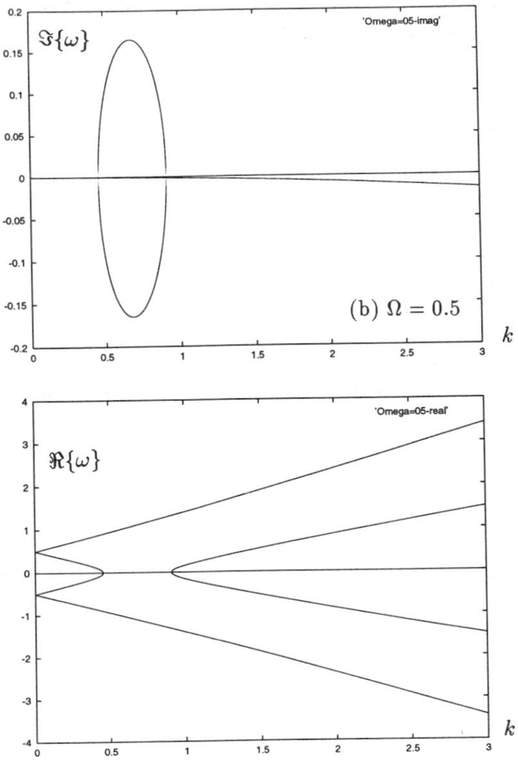

Figure 2. (Continued).

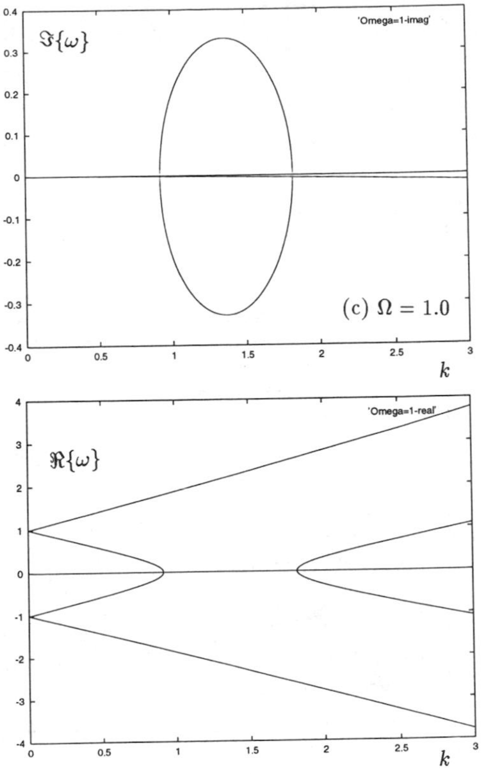

Figure 2. (Continued).

$$\kappa \;=\; 1.0, \quad \epsilon = 0.8849. \tag{24}$$

The wave number k and speed rate Ω are varied.

Figure 2 shows the real and the imaginary part of the frequency parameter ω versus the wave number k (which is assumed to be real-valued here) for three different Ω–values. Even for the non–rotating case, the waves become dispersive due to the accompanying thermal conduction process. With increasing rotational speed, this effect strengthens. In general, there are five circular frequencies ω for a fixed parameter set and given values of k and Ω (recalling the fact that also the reduced field equations (20) are of 5th order in time). One of them which is purely negative corresponds to the thermal creep motion while the other ones are two pairs of conjugate complex values associated with the piezoelectric waves which, in general, are damped. But within the used scaling (for $\Re\{\omega\}$), it can not be visualized. An interesting new feature is the fact that with increasing speed, instabilities may occur (characterized by positive imaginary parts $\Im\{\omega\}$). For rotating elastic disks, this phenomenon is well–known if the disk is clamped at the outer surface (see Seemann & Wauer, 1990). But it is also known that for a disk with traction–free outside boundaries and clamped at the center, there is a stiffening effect and such a structural member is stable in the whole speed range. Therefore, the results obtained here for an infinite pyroelectric medium have to be checked by a computation embedded in a geometrically nonlinear theory of piezoelectricity. It will be done in a future paper.

4.2 *Plate layer*

Based on the classical eigenvalue theory for elastic waveguides, the procedure to find the governing eigenvalue equation is described by the author in two recent papers (Wauer, 1996 and 1997) for similar problems as considered here. It is also applicable for the present layer problem but even for the simple boundary conditions (19), it is expensive. Therefore, a wave solution

$$q(x,z,t) \;=\; \cos kx e^{\lambda t} f_q(z), \quad q \in \{w, \phi, \theta\},$$

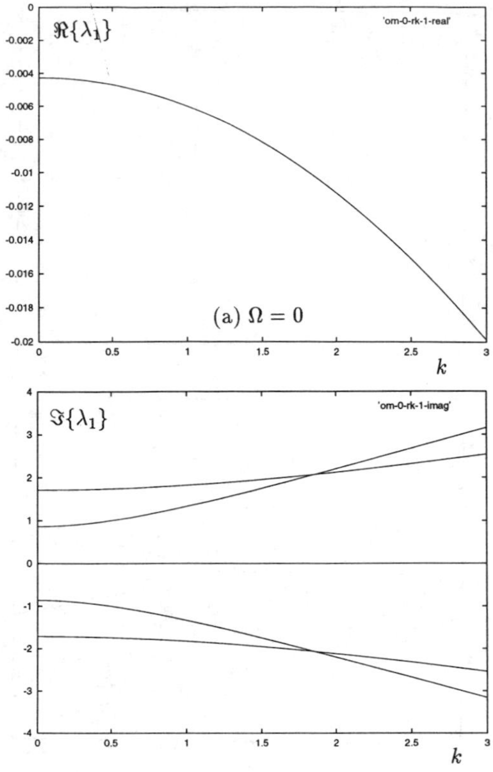

Figure 3. Eigenvalue versus wave number.

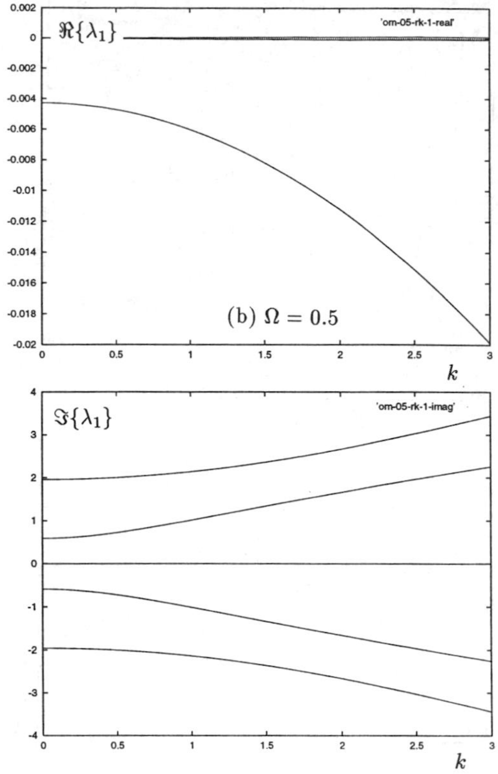

Figure 3. (Continued).

$$u(x,z,t) \;=\; \sin kx e^{\lambda t} f_u(z) \tag{25}$$

combined with a Ritz series expansion

$$\begin{aligned} f_q(z) &= \sum_{j=1}^{N\to\infty} C_{qj} \sin \frac{j\pi}{2}(z+1), \quad q \in \{u, w, \phi\}, \\ f_\theta(z) &= \sum_{j=1}^{N\to\infty} C_{\theta j} \cos \frac{j\pi}{2}(z+1) \end{aligned} \tag{26}$$

fulfilling all boundary conditions (19) is assumed (possibly appearing zero–eigenvalues are not considered here). The vanishing determinant of the resulting system of $4 \times N$ algebraic equations (applying the classical Galerkin projection) yields approximations for the eigenvalues λ_j to be found.

In the present contribution, only the results of a special one–term approximation

$$\begin{aligned} f_q(z) &= C_{qj} \sin \frac{j\pi}{2}(z+1), \quad q \in \{u, w, \phi\}, \\ f_\theta(z) &= C_{\theta j} \cos \frac{j\pi}{2}(z+1) \end{aligned} \tag{27}$$

are presented which leads to an eigenvalue equation in form of a vanishing 4×4 determinant of the type (22) with

$$\begin{aligned} a_{11} &= \gamma_{11}k^2 + \gamma_{44}r_j^2 + \lambda_j^2 - \Omega^2, \\ a_{12} &= 2\Omega\lambda_j, \quad a_{13} = 0, \quad a_{14} = -\beta\varepsilon_{w\theta}k, \\ a_{21} &= -2\Omega\lambda_j, \quad a_{22} = \gamma_{44}k^2 + r_j^2 + \lambda_j^2 - \Omega^2, \\ a_{23} &= (\delta_{15}k^2 + r_j^2)\varepsilon_{w\phi}, \quad a_{24} = -\varepsilon_{w\theta}r_j, \\ a_{31} &= 0, \quad a_{32} = -(\delta_{15}k^2 + r_j^2), \\ a_{33} &= \epsilon k^2 + r_j^2, \quad a_{34} = -r_j, \\ a_{41} &= 0, \quad a_{42} = \alpha_{\theta w}r_j\lambda_j, \quad a_{43} = -\varepsilon_{\theta\phi}\lambda_j, \\ a_{44} &= \kappa k^2 + r_j^2 + T_\theta\lambda_j \end{aligned} \tag{28}$$

where $r_j = j\pi/2$. Again, five complex–valued eigenvalues λ_{j1} to λ_{j5} result (recalling the fact that also the general field equations (18) are of 5th order in time) for every ordinal number $j = 1, 2, \ldots, \infty$.

Figure 3 shows (for $j = 1$) the real and the imaginary part of the eigenvalue λ_j again versus the wavenumber k for different values of speed parameter Ω. Similar as for an infinite medium, also for the layer two pairs of conjugate complex eigenvalues and one additional (negative) real eigenvalue appear for each ordinal number j.

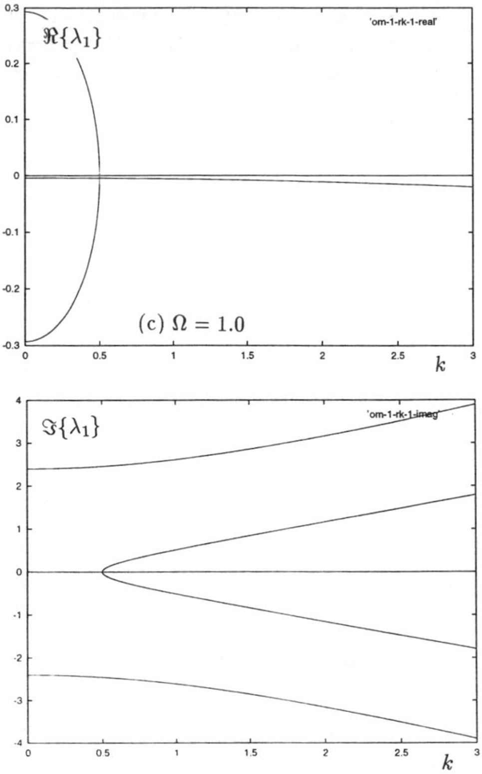

Figure 3. (Continued).

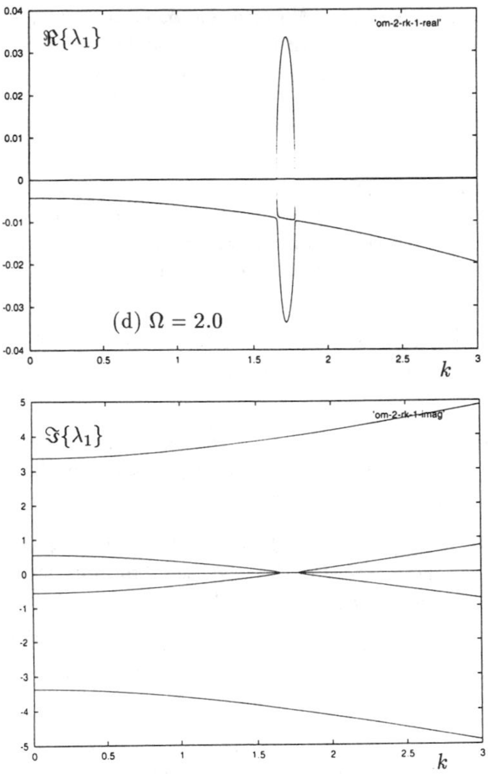

Figure 3. (Continued).

For $\Omega = 0$, the results differ from those obtained (for other boundary conditions) by Paul and Renganathan (1985) because they formulated an eigenvalue problem containing only the eigenvalue λ^2 and not λ simultaneously but an additional parameter.

For a rotating plate layer, the results are similar as for the unbounded solid. Since the layer is clamped at the outer surfaces, it can be supposed that here no stabilizing stiffining effect occurs and the dynamic buckling phenomenon found is plausible.

5 CONCLUSIONS

The propagation of plane waves in rotating pyroelectric bodies has been examined. It has been shown that a linear boundary value of 5th order in time and 8th order in the space coordinates results, which under the given (not very restrictive) assumptions possesses space–independent coefficients. It follows that basically, an analytical or semi–analytical calculation of the wave characteristics is possible.

The obtained results verify that eventhough no mechanical damping influences are taken into account, the circular frequencies (for the waves in unbounded solids) or the eigenvalues (characterizing the dynamics of a layer) are complex in general, i. e., there is a thermal damping. But for real problems, this damping is very small.

The rotational speed influences the wave characteristics significantly. The most interesting phenomenon is a destabilizing effect which for a layer with clamped edges seems to plausible while for an infinite medium, the result must be checked by a calculation which takes into consideration centrifugal stiffening effects.

ACKNOWLEDGEMENTS

Most of the work originates from a stay at Virginia Polytechnic Institute and State University. The author thanks Professors Daniel J. Inman and Raymond H. Plaut for their invitation. Financial support was provided by the Volkswagen foundation. Furthermore, the assistance of Dr.-Ing. V. Mehl carrying out the computations is greatfully acknowledged.

REFERENCES

Berlincourt, D. A., D. R. Curran & H. Jaffe 1964. Piezoelectric and piezomagnetic materials and their function as transducers. In W. P. Mason (ed.), *Physical Acoustics*: 169–270. New York: Academic Press.

Chandrasekharaiah, D. S. 1988. A generalized linear thermoelasticity theory for piezoelectric media. *Acta Mech.* 71: 39–49.

Chauduri, S. K. R. & L. Debnath 1983a. Magnetoelastic plane waves in infinite rotating media. *J. Appl. Mech.*50: 283–288.

Chauduri, S. K. R. & L. Debnath 1983b. Magneto-thermo-elastic plane waves in rotating media. *Int. J. Engng. Sci.* 21: 155–163.

Condon, E. V. & H. Odishaw 1972. *American Institute of Physics*, chpt. 9–118. New York: McGraw-Hill.

Eringen, A. C. & G. A. Maugin 1990. *Electrodynamics of continua I.* New York: Springer.

Heyliger, P., S. Brooks & D. Saravanos 1995. In S. Sture (ed.), *Engineering Mechanics, Vol. 2*: 722–725. New York: ASCE.

Mindlin R. D. 1961. On the equations of motion of piezoelectric crystals. In *Problems of Continuum Mechanics*: 282–290. Philadelphia: SIAM.

Paul, H. S. & K. Renganathan 1985. Free vibrations of a pyroelectric layer of hexagonal (6mm) class. *J. Acoust. Soc. Am.* 78:395–397.

Seemann, W. & J. Wauer 1990. In J. M. Kim & W.-J. Yang (eds.), *Dynamics of rotating machinery*: 35–50. New York: Hemisphere.

Toupin, R. A. 1963. A dynamical theory of dielectrics. *Int. J. Engng. Sci.* 1: 101–126.

Tiersten, H. F. 1969. *Linear piezoelectric plate vibrations.* New York: Plenum Press.

Wauer. J. 1996. Free and forced magneto-thermo-elastic vibrations in a conductive plate layer. *J. Therm. Stresses* 19: 671–691.

Wauer J. & S. Suherman 1997. Thickness vibrations of a piezo-semiconducting plate layer. To be published.

Yang, J. S. & R. C. Batra 1995. Free vibrations of a linear thermopiezoelectric body. *J. Therm. Stresses* 18:247–262.

Modern Practice in Stress and Vibration Analysis, Gilchrist (ed.) © *1997 Balkema, Rotterdam, ISBN 90 5410 896 7*

Theoretical scheme for a vibration based detection of a crack in stepped beams

B.P.Nandwana & S.K.Maiti
Department of Mechanical Engineering, Indian Institute of Technology, Bombay, India

ABSTRACT: A theoretical scheme with potential for detection of location and size of a crack in slender stepped beams based on measurement of transverse natural frequencies is presented. The crack is modelled by a rotational spring. The characteristic equation obtained with this modelling is manipulated to give a variation of the stiffness with crack location for the three lowest natural frequencies. The three curves intersect at a common point giving the crack location and the spring stiffness. The crack size is then obtained using the standard relationship between the stiffness and the crack size. A numerical study to demonstrate the utility and accuracy is presented considering a two step beam. The accuracy of prediction of location is good for crack size greater than 10% of section depth for both the cantilever and simply supported beams. The maximum error is about 1.5%. The error in prediction of crack size is also satisfactory.

1 INTRODUCTION

The development of a crack in a component changes its vibration parameters which may be structural (i.e. mass, stiffness and flexibility) or modal (i.e. natural frequencies, modal damping values and mode shapes). The study of changes in one or more of these parameters can form the basis for detection of location and size of crack. Such vibration based methods of detecting a crack can offer some advantages. They can help to determine location and size of a crack from the global vibration data, which can be collected from a single point on the component thereby avoiding the necessity of scanning the whole length.

In a method using vibration measurements for the detection of cracks appropriate modelling has an important role. One technique suggested is to reduce the section modulus at the crack section (Petroski 1981). This has been adopted to study cracked rotors (Grabowski 1979, Mayes & Davies 1976, Christides & Barr 1984). Another technique proposed in case of the transverse vibrations of slender beams is to employ a rotational spring at the crack location (Rizos et al. 1990, Liang et al. 1991). An axial spring can be adopted if longitudinal vibrations are being studied (Adams et al. 1978). The central concept in the case of rotational spring is that a crack gives rise to an additional flexibility, and hence, there is a jump in the slope of a slender beam at the crack section. The stiffness of the spring is so selected that the rotation due to a moment at the crack section is equal to the jump in slope due to the extra flexibility. The stiffness is available in a close form for an edge crack oriented normal to the length in a beam of uniform cross-section (Ostachowicz & Krawkczuk 1991). This modelling is found convenient for solving inverse problems without any iteration.

The methods based on the structural parameters use modal test data (Kam & Lee 1992), flexibility (Pandey & Biswas 1994), etc. Attempts have also been made to employ changes in curvature mode shapes for the detection (Pandey et al. 1991). These methods suffer from some disadvantages in that they require data collection from a large number of points on the component.

The methods based on natural frequency has received a considerable attention in the literature. This is perhaps because the natural frequency can be measured easily and monitoring is possible from any location on the component. A method has been proposed for detection of damage in an one dimensional component (Adams et al. 1978). This is based on the receptance of the system. Another method suitable for two-dimensional components using sensitivity analysis has been proposed (Cawley & Adams 1979).

The technique of rotational spring has been applied for detection of the crack location through measurement of amplitudes at two points on the component (Rizos et al. 1990). The same technique has been applied for detection of crack location and size in uniform beams employing natural frequencies (Liang et al. 1991). The later investigators indicate that for a given natural frequency and crack location, the characteristic equation can be solved to obtain the stiffness.

The method based on the rotational spring has always been applied to beams of uniform cross-section. There is a need to examine if it can be applied to more realistic configurations, e.g. continuously varying cross-section, or as a first step to stepped beams. In this paper the method is examined for a stepped beam.

2 FORMULATION

Figure 1 shows a stepped beam with crack in the first step. The crack section, located at a distance ℓ from the fixed end, is represented by a rotational spring of stiffness K_t. This divides the step containing the crack into 2 segments. The remaining uncracked step(s) are treated as one segment each. Therefore, a beam with n steps will have $n + 1$ segments.

The governing equation of flexural vibration is given by

$$\frac{d^2}{dx^2}\left(EI\frac{d^2U}{dx^2}\right) + \omega^2 \rho A U = 0 \tag{1}$$

where U is displacement, ω is natural frequency of the vibration of the beam, E is the Young's modulus of elasticity, I is the second moment of full area of cross-section, ρ is the mass density, A is the cross-sectional area and x-axis is aligned with the axis of the beam. Introducing separate displacement variable for each segment and dimensionless length parameter $\xi = x/L$ where L = total length of the beam, the equations for the four segments are:

$$\frac{d^4U_i}{d\xi^4} + \lambda_i^4 U_i = 0, \qquad i = 1, 2, \ldots 4 \tag{2}$$

where $\lambda_1^4 = \lambda_2^4 = \omega^2 \rho A_1 L_1^4 / E I_1$, $\lambda_3^4 = \omega^2 \rho A_2 L_2^4 / E I_2$ and $\lambda_4^4 = \omega^2 \rho A_3 L_3^4 / E I_3$. U_i stands for the displacement of the segment i. These equations are valid for the regions $0 \le \xi \le \beta$, $\beta \le \xi \le \beta_1$, $\beta_1 \le \xi \le \beta_2$ and $\beta_2 \le \xi \le 1$ for i=1, 2, 3 and 4 respectively. Here $\beta = \ell/L$ is non-dimensional crack location and $\beta_1 = L_1/L$ and $\beta_2 = L_1/L$ are nondimensionalised coordinates for the steps. With an additional step there will be one more equation.

The solutions of the four segments can be written in the following form.

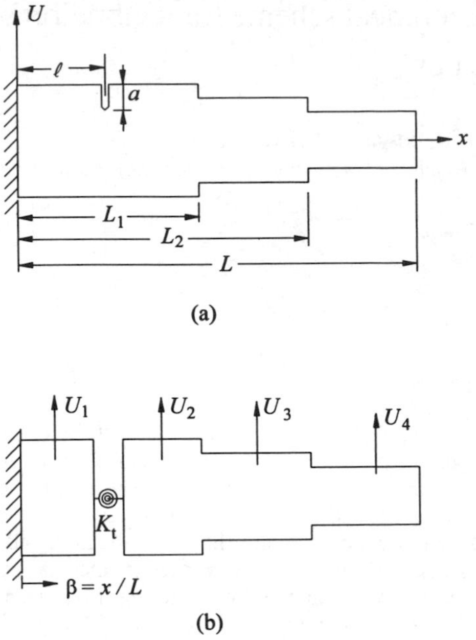

Figure 1. Stepped cantilever beam. (a) Geometry of the beam. (b) Rotational spring representation of the crack.

$$U_i = A_{i1} \cos \lambda_i \xi + A_{i2} \cosh \lambda_i \xi + A_{i3} \sin \lambda_i \xi + A_{i4} \sinh \lambda_i \xi \tag{3}$$

where A_{ij}, i, j = 1 to 4 are arbitrary constants. With every additional step another four constants will appear.

The boundary conditions at the beam ends for cantilever configuration are

$$U_1 = 0, \qquad U_1' = 0, \qquad \text{at } \xi = 0 \tag{4}$$

$$U_4'' = 0, \qquad U_4''' = 0, \qquad \text{at } \xi = 1 \tag{5}$$

The compatibility conditions of displacement, slope, moment and shear force at the junction of the two steps, $\xi = \beta_1$ and $\xi = \beta_2$ are

$$U_2 = U_3, \quad U_2' = U_3', \quad (EI_1)U_2'' = (EI_2)U_3'', \quad (EI_1)U_2''' = (EI_2)U_3''' \tag{6}$$

$$U_3 = U_4, \quad U_3' = U_4', \quad (EI_2)U_3'' = (EI_3)U_4'', \quad (EI_2)U_3''' = (EI_3)U_4''' \tag{7}$$

The continuity of displacement, moment and shear forces at the crack location ($\xi=\beta$) can be written in the following form:

$$U_1 = U_2, \quad U_1'' = U_2'', \quad U_1''' = U_2''' \tag{8}$$

The crack is supposed to give rise to a jump in slope (Liang et al. 1991). The transition can be written in the following form.

$$\frac{dU_1}{dx} + \frac{d^2}{dx^2}(EI_1U_1)\frac{1}{K_t} = \frac{dU_2}{dx}$$

Writing in terms of ξ

$$U_1' + \frac{\lambda_1}{K}U_1'' - U_2' = 0 \tag{9}$$

where $K = \dfrac{K_t L}{EI_1}$ is non-dimensional stiffness of the rotational spring representing the crack.

From the conditions (4) to (9) the following system of equations is obtained.

$$[\mathbf{D}]\ \{\mathbf{A}\} = \{\mathbf{0}\} \tag{10}$$

where $\{\mathbf{A}\}$ is the vector of the arbitrary constants A_{ij} and the matrix $[\mathbf{D}]$ is as follows.

$$[\mathbf{D}] = \begin{bmatrix} [\mathbf{B_L}] & [\mathbf{0}] & [\mathbf{0}] & [\mathbf{B_R}] \\ [\mathbf{0}] & [\mathbf{S_L^1}] & [\mathbf{S_R^1}] & [\mathbf{0}] \\ [\mathbf{0}] & [\mathbf{0}] & [\mathbf{S_L^2}] & [\mathbf{S_R^2}] \\ [\mathbf{C_L}] & [\mathbf{C_R}] & [\mathbf{0}] & [\mathbf{0}] \end{bmatrix} \tag{11}$$

The non-zero submatrices of $[\mathbf{D}]$ are given by

$$[\mathbf{B_L}] = \begin{bmatrix} 1 & 1 & 0 & 0 \\ 0 & 0 & 1 & 1 \\ 0 & 0 & 0 & 0 \\ 0 & 0 & 0 & 0 \end{bmatrix}$$

$$[\mathbf{B_R}] = \begin{bmatrix} 0 & 0 & 0 & 0 \\ 0 & 0 & 0 & 0 \\ -\cos\lambda_4 & \cosh\lambda_4 & -\sin\lambda_4 & \sinh\lambda_4 \\ \sin\lambda_4 & \sinh\lambda_4 & -\cos\lambda_4 & \cosh\lambda_4 \end{bmatrix}$$

$$[\mathbf{S_L^i}] = \begin{bmatrix} \cos\alpha_i^l & \cosh\alpha_i^l & \sin\alpha_i^l & \sinh\alpha_i^l \\ -\sin\alpha_i^l & \sinh\alpha_i^l & \cos\alpha_i^l & \cosh\alpha_i^l \\ -\cos\alpha_i^l & \cosh\alpha_i^l & -\sin\alpha_i^l & \sinh\alpha_i^l \\ \sin\alpha_i^l & \sinh\alpha_i^l & -\cos\alpha_i^l & \cosh\alpha_i^l \end{bmatrix}$$

$$[\mathbf{S_R^i}] = \begin{bmatrix} -\cos\alpha_i^r & -\cosh\alpha_i^r & -\sin\alpha_i^r & -\sinh\alpha_i^r \\ F_i\sin\alpha_i^r & -F_i\sinh\alpha_i^r & -F_i\cos\alpha_i^r & -F_i\cosh\alpha_i^r \\ G_i\cos\alpha_i^r & -G_i\cosh\alpha_i^r & G_i\sin\alpha_i^r & -G_i\sinh\alpha_i^r \\ -H_i\sin\alpha_i^r & -H_i\sinh\alpha_i^r & H_i\cos\alpha_i^r & -H_i\cosh\alpha_i^r \end{bmatrix},$$

$$i = 1, 2$$

$$[\mathbf{C_L}] = \begin{bmatrix} \cos\alpha & \cosh\alpha & \sin\alpha & \sinh\alpha \\ -\cos\alpha & \cosh\alpha & -\sin\alpha & \sinh\alpha \\ \sin\alpha & \sinh\alpha & -\cos\alpha & \cosh\alpha \\ -\dfrac{K}{\lambda_1}\sin\alpha - \cos\alpha & \dfrac{K}{\lambda_1}\sinh\alpha + \cosh\alpha & \dfrac{K}{\lambda_1}\cos\alpha - \sin\alpha & \dfrac{K}{\lambda_1}\cosh\alpha + \sinh\alpha \end{bmatrix}$$

$$[\mathbf{C_R}] = \begin{bmatrix} -\cos\alpha & -\cosh\alpha & -\sin\alpha & -\sinh\alpha \\ \cos\alpha & -\cosh\alpha & \sin\alpha & -\sinh\alpha \\ -\sin\alpha & -\sinh\alpha & \cos\alpha & -\cosh\alpha \\ \dfrac{K}{\lambda_1}\sin\alpha & -\dfrac{K}{\lambda_1}\sinh\alpha & -\dfrac{K}{\lambda_1}\cos\alpha & -\dfrac{K}{\lambda_1}\cosh\alpha \end{bmatrix}$$

where $\alpha_1^l = \lambda_2\beta_1$, $\alpha_1^r = \lambda_3\beta_1$, $\alpha_2^l = \lambda_3\beta_2$, $\alpha_2^r = \lambda_4\beta_2$,
$F_1 = \lambda_3/\lambda_2$, $G_1 = (\lambda_3/\lambda_2)^2\,(I_2/I_1)$, $H_1 = (\lambda_3/\lambda_2)^3\,(I_2/I_1)$,
$F_2 = \lambda_4/\lambda_3$, $G_2 = (\lambda_4/\lambda_3)^2\,(I_3/I_2)$, $H_2 = (\lambda_4/\lambda_3)^3\,(I_3/I_2)$,
and $\alpha = \lambda_1\beta$

For a crack in any other span the Eqns. (10) and (11) have the same form, only the relative positions of the various submatrices in $[\mathbf{D}]$ change. The characteristics equation obtained from Eqn. (10) can be written as:

$$|\Delta| = 0 \tag{12a}$$

where $|\Delta| = \det[\mathbf{D}]$. Alternatively, it can be rewritten in the form

$$\frac{K}{\lambda_1}|\Delta_1| + |\Delta_2| = 0 \quad \text{or,} \quad K = -\lambda_1\frac{|\Delta_2|}{|\Delta_1|} \tag{12b}$$

where $|\Delta_1|$ and $|\Delta_2|$ have the same form as $|\Delta|$ except for the differences in the last row.

The above method can be easily adopted for a simply supported beam. Only $[\mathbf{B_L}]$ and $[\mathbf{B_R}]$ matrices are then modified as follows:

$$[\mathbf{B_L}] = \begin{bmatrix} 1 & 1 & 0 & 0 \\ -1 & 1 & 0 & 0 \\ 0 & 0 & 0 & 0 \\ 0 & 0 & 0 & 0 \end{bmatrix}$$

$$[\mathbf{B_R}] = \begin{bmatrix} 0 & 0 & 0 & 0 \\ 0 & 0 & 0 & 0 \\ \cos\lambda_4 & \cosh\lambda_4 & \sin\lambda_4 & \sinh\lambda_4 \\ -\cos\lambda_4 & \cosh\lambda_4 & -\sin\lambda_4 & \sinh\lambda_4 \end{bmatrix}$$

3 METHODOLOGY FOR CRACK DETECTION

For a cracked beam the first three natural frequencies are measured. Using one of these frequencies and assuming a particular value for β and the non-dimensionalised stiffness K is computed from Eqn. (12). Thereby a variation of stiffness with crack location is obtained. Similar curves can be plotted for another two natural frequencies. Since physically there is only one crack, the position where the three curves intersect, gives the crack location (Liang et al. 1991) and the spring stiffness representing the crack. For simply supported beam two such locations are predicted because of symmetry. The crack size is then obtained using the relationship between stiffness K and crack size a.

4 NUMERICAL STUDY

For a case study a beam with two steps (Figure 2) only is considered. Both cantilever and simply supported end conditions have been examined. The material data used are as follows: modulus of elasticity E=2.1×10^{11} N/m^2, density ρ = 7860 kg/m^3 and poisson's ratio ν = 0.3. A number of crack locations and sizes are considered. The natural frequencies for both the uncracked and cracked geometries are computed by finite element method. For this purpose, the beam is discretised mostly by 8-noded isoparametric elements (Figure 2). Around the crack tip 12 quarter point singularity elements are used. The natural frequencies thus obtained are shown in Tables 1 and 2 for cantilever and simply supported beams respectively.

While applying the method it is found that the three curves do not intersect at a common point in a number of cases, e.g. Figure 3. In order to get a common intersection point for the three K vs. β plots, it is important that the modulus of elasticity E, which goes as input to relation (12) for each mode must be calculated using the procedure which is henceforth termed as *zero setting*.

In this procedure two sets of frequency parameters are involved, say (λ^{t*}), the theoretical frequency parameter for the uncracked beam for a particular mode and λ*, the input frequency parameter for the same mode. (λ^{t*}) is obtainable in either close form for simple geometries from any standard textbook on vibration (e.g., Timoshenko 1955) or by solving numerically the corresponding characteristic equation for uncracked case, e.g., in the present case. λ*, on the other hand, is determined from the input uncracked natural frequency (ω*) through $\lambda^{*4} = \omega^{*2}\rho AL^4/EI$. These two parameters ($\lambda^{t*}$ and λ^*) must be the same, but they are generally found to differ. This calls for a minor adjustment or *zero setting*. The modulus of elasticity (E) is so adjusted that $\lambda^{t*4} = \omega^{*2}\rho AL^4/EI$. This gives a value for E, which is given as input to Eqn. (12). These results expressed equivalently indicate that for a cracked beam, in a particular mode, the frequency parameter λ is obtain-

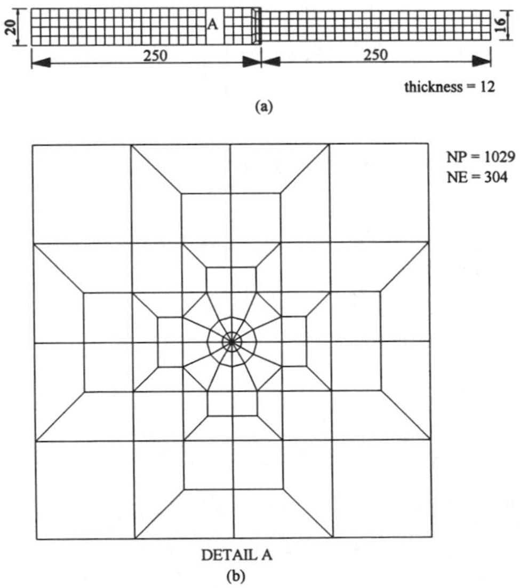

Figure 2. Typical discretisation for stepped beam.

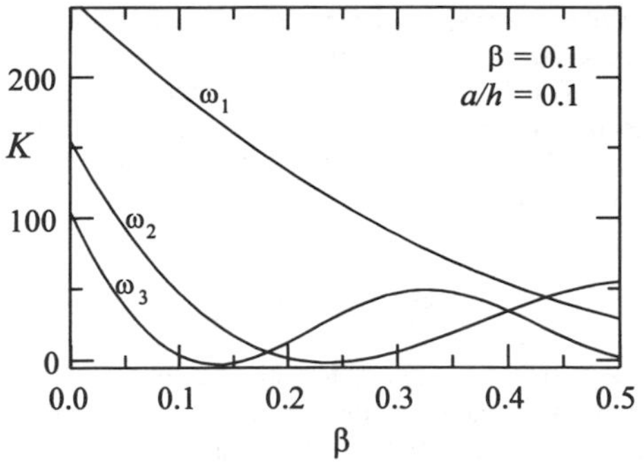

Figure 3. Variation of rotational stiffness with crack location without zero setting.

able from $\lambda^* = \lambda^{t^*} (\omega/\omega^*)^{1/2}$ where ω is input natural frequency for the cracked beam. This λ can be alternatively supplied to Eqn. (12). Thus for the method to work, natural frequencies of both cracked and uncracked beam are needed. The input data can be based on experiments in actual practice or it can be based on , for example, finite element calculations for numerical studies.

With this sort of 'zero setting' the curves are plotted for all the cases. Typical plots are shown in Figure 4. Figure 3 may be misread to give a prediction for crack location at $\beta= 0.45$ as against the actual $\beta= 0.1$. There is a substantial improvement in the prediction only after the zero setting (Figure 4). For all the cases the three plots for three natural frequencies intersect and give the location of the crack. In some cases, the intersection point can not be easily read out from the graph because of the scale. With a magnification this difficulty may be eliminated.

To eliminate any subjective error involved in the graphical procedure to determine the crack location an alternative numerical method is employed. The intersection point (K_1, β_1) for the pair of curves corresponding to ω_1 and ω_2 is obtained. The similar intersections (K_2, β_2) and (K_3, β_3) are obtained for the other two pairs (ω_1 and ω_3; ω_2 and ω_3). The average of the three intersection points is taken as the prediction.

Table 1. Comparison of predicted and actual crack location and size for stepped cantilever beams.

Actual		Natural frequencies (rad/s)			Predicted				
Location	Size	ω_1	ω_2	ω_3	Location		Stiffness	Size	
β	(a/h)				β	Error (%)	(K)	(a/h)	Error (%)
uncracked		454.90	2346.50	6511.61					
Crack in first step									
0.0	0.1	453.02	2337.21	6483.42	-0.0029	-0.29	463.84	0.0701	-2.99
0.1	0.1	452.18	2340.89	6508.30	0.1017	0.17	233.10	0.1001	0.01
0.2	0.1	452.98	2346.12	6504.87	0.2006	0.06	236.79	0.0993	-0.07
0.3	0.1	453.63	2345.72	6488.26	0.2992	-0.08	237.27	0.0992	-0.08
0.4	0.1	454.13	2342.56	6494.52	0.4025	0.25	235.18	0.0997	-0.03
0.4	0.2	451.91	2331.40	6446.92	0.4025	0.25	60.28	0.2002	0.02
0.4	0.3	447.98	2312.03	6366.98	0.4022	0.22	25.73	0.3011	0.11
0.4	0.4	441.60	2281.77	6248.37	0.4017	0.17	13.15	0.4036	0.36
0.4	0.5	431.25	2235.36	6079.77	0.4008	0.08	7.17	0.5152	1.52
Crack in second step									
0.5	0.1	454.03	2333.65	6508.31	0.4999	-0.01	222.66	0.1151	1.51
0.6	0.1	454.60	2337.26	6505.31	0.6000	-0.00	298.29	0.0989	-0.11
0.7	0.1	454.80	2340.87	6488.15	0.7011	0.11	297.92	0.0990	-0.10
0.8	0.1	454.88	2344.67	6493.88	0.8064	0.64	282.96	0.1017	0.17
0.9	0.1	454.90	2346.42	6509.52					
0.9	0.2	454.90	2345.89	6501.39					
0.9	0.3	454.89	2344.91	6486.40	0.9146	1.46	19.69	0.3741	7.41
0.9	0.4	454.87	2343.25	6460.72	0.9062	0.62	13.51	0.4376	3.76
0.9	0.5	454.85	2340.38	6415.22	0.9026	0.26	8.25	0.5318	3.18

That is, K = ⅓ΣK_i and ⅓$\Sigma\beta_i$. The intersection points (K_1, β_1), (K_2, β_2) and (K_3, β_3) must be selected judiciously. The crack size is obtained using the relationship between K and crack size a/h (Ostachowicz & Krawkczuk 1991).

$$K = \frac{b h^2 L}{72 \pi I (a/h)^2 f(a/h)} \tag{13}$$

where

$$f(a/h) = 0.6384 - 1.035(a/h) + 3.720(a/h)^2 - 5.177(a/h)^3 + 7.553(a/h)^4 - 7.332(a/h)^5 + 2.490(a/h)^6 \tag{14}$$

and b and h are width and height respectively of the beam.

Tables 1 and 2 give a comparison of the computed crack location and size with the actual values. The method is able to predict the location of crack of size more than 10% of section depth. In the case of cantilever beam it fails when the crack is located very near the free end but in such cases also the prediction is possible for larger crack size (30% or more). The error in prediction of location is always less than about 1.5%. The crack size is also predicted satisfactorily. The maximum error in crack size prediction is less than 6.25% except for larger crack sizes (more than 30%) where it is around 12%.

5 CONCLUSIONS

A theoretical scheme for detection of location and size of a crack in slender stepped beam has been presented. The details of the method are given. Accuracy obtainable is illustrated by a case study involving a two-step beam. The method can predict the location of crack of size more than 10% of section depth. The error in prediction of location is always less than about 1.5%. The crack size is also predicted satisfactorily. The procedure can be easily adapted for more steps and crack located in any of the segments. Though experimental results are not presented here sample experiments carried out with other configurations show encouraging trends (Oza 1997).

Table 2. Comparison of predicted and actual crack location and size for stepped simply supported beams.

Actual		Natural frequencies (rad/s)			Predicted				
Location	Size	ω_1	ω_2	ω_3	Location		Stiffness	Size	
β	(a/h)				β	Error (%)	K	(a/h)	Error (%)
uncracked		1021.57	4220.43	9161.39					
Crack in first step									
0.1	0.1	1021.29	4214.84	9140.27	0.0953	-0.47	220.73	0.072	-2.81
0.2	0.1	1020.54	4204.89	9127.30	0.1991	-0.09	236.03	0.069	-3.06
0.3	0.1	1019.62	4203.01	9153.42	0.3009	0.09	237.30	0.069	-3.08
0.4	0.1	1018.89	4211.25	9158.10	0.3984	-0.16	239.65	0.069	-3.11
0.4	0.2	1011.23	4185.38	9147.53	0.3995	-0.05	61.16	0.14	-6.02
0.4	0.3	997.78	4141.33	9129.07	0.3999	-0.01	26.04	0.215	-8.47
0.4	0.4	976.36	4074.67	9099.97	0.4001	0.01	13.26	0.297	-10.31
0.4	0.5	942.46	3977.23	9054.69	0.4005	0.05	7.18	0.389	-11.11
Crack in second step									
0.5	0.1	1015.54	4218.36	9114.47	0.4977	-0.23	225.13	0.08	-2.01
0.6	0.1	1017.52	4218.37	9142.99	0.6005	0.05	300.18	0.069	-3.12
0.7	0.1	1018.65	4209.46	9161.19	0.7009	0.09	296.28	0.069	-3.07
0.8	0.1	1020.03	4207.27	9133.63	0.8007	0.07	295.14	0.069	-3.06
0.9	0.1	1021.15	4215.10	9138.26	0.9019	0.19	291.18	0.07	-3.01
0.9	0.2	1019.91	4199.55	9072.99	0.8984	-0.16	78.84	0.138	-6.24
0.9	0.3	1017.67	4171.44	8957.63	0.8979	-0.21	33.75	0.212	-8.85
0.9	0.4	1013.94	4125.02	8775.18	0.8979	-0.21	17.18	0.292	-10.79
0.9	0.5	1007.62	4047.43	8493.42	0.8983	-0.17	9.27	0.384	-11.62

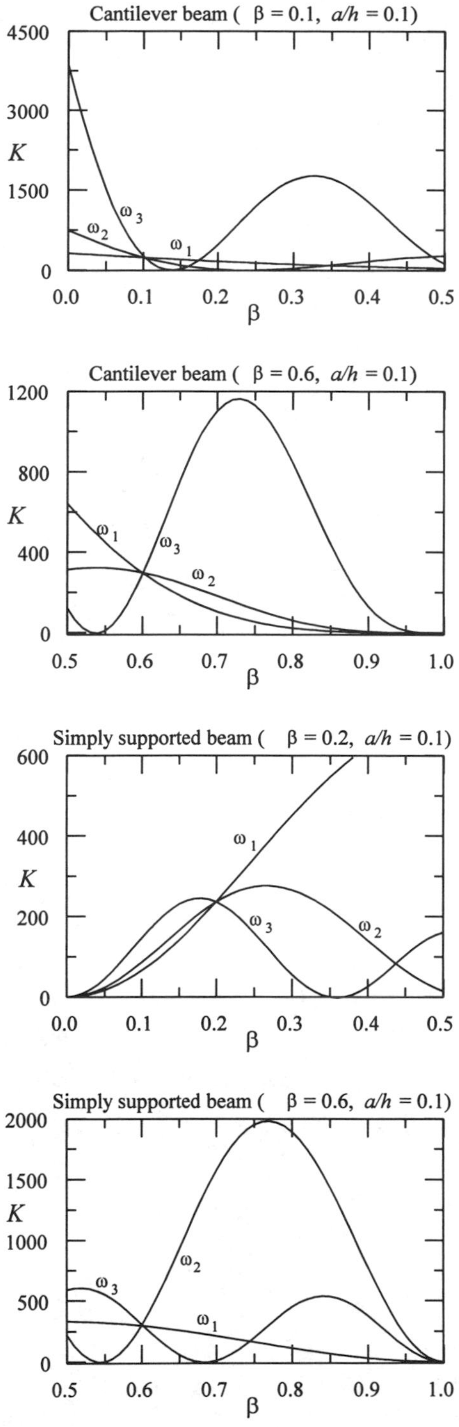

Figure 4. Variation of rotational spring stiffness with crack location for typical cases.

REFERENCES

Adams, R. D., P. Cawley, C. J. Pye & B. J. Stone 1978. A vibration technique for non-destructively assessing the integrity of structures. *J. Mech. Engng Sci.* 20: 93-100.

Cawley, P. & R. D. Adams 1979. The location of defects in structure from measurements of natural frequencies. *J. Strain Anal.* 14: 49-57.

Christides, S. & A. D. S. Barr 1984. One-dimensional theory of cracked Bernoulli-Euler beams. *Int. J. Mech. Sci.* 26: 639-648.

Grabowski, B. 1979. The vibration behaviour of a turbine rotor containing a transverse crack. *J. Mech. Des.* 102: 140-146.

Kam, T. Y & T. Y. Lee 1992. Detection of cracks in structures using modal test data. *Engrg. Fract. Mech.* 42: 381-387.

Liang, R. Y., F. K. Choy & J. Hu 1991. Detection of cracks in beam structures using measurements of natural frequencies. *J. Franklin Inst.* 328: 505-518.

Mayes, I. W. & W. G. R. Davies 1976. The vibration behaviour of a rotating shaft system containing a transverse crack. *Institution of mechanical engineers conference publication, Vibrations in rotating machinery*: Paper C168-76.

Ostachowicz, W. M. & M. Krawkczuk 1991. Analysis of the effect of cracks on the natural frequencies of a cantilever beam. *J. Sound Vibr.* 150: 191-201.

Oza, D. M. 1997. Study of vibration of uniform beam with inclined crack to evolve a method of crack detection. *M. Tech. dissertation, Mech. Engg. Dept., IIT Bombay.*

Pandey, A. K., M. Biswas & M. M. Samman 1991. Damage detection from changes in curvature mode shapes. *J. Sound Vibr.* 145: 321-332.

Pandey, A. K. & M. Biswas 1994. Damage detection in structures using changes in flexibility. *J. Sound Vibr.* 169: 3-17.

Petroski, H. J. 1981. Simple static and dynamic models for the cracked elastic beam. *Int. J. Fract.* 17: R71-R76.

Rizos, P. F., N. Aspragathos & A. D. Dimarogonas 1990. Identification of crack location and magnitude in a cantilever beam from the vibration modes. *J. Sound Vibr.* 138: 381-388.

Timoshenko, S. 1955. *Vibration problems in engineering*. New Delhi: Affiliated East-West Press.

Modern Practice in Stress and Vibration Analysis, Gilchrist (ed.) © 1997 Balkema, Rotterdam, ISBN 90 5410 896 7

Hertzian contact vibrations due to random excitation and surface roughness

M. Pärssinen
The Marcus Wallenberg Laboratory for Sound and Vibration Research, KTH, Stockholm, Sweden

ABSTRACT: The problem of a hemispherical rider sliding with constant velocity over a rough surface is considered. The Fokker-Planck equation is applied to the Hertzian contact problem in order to numerically generate statistical moments of resulting stationary vibration level under external random white noise load and internal surface roughness excitation. For the derivation it is assumed that these two random excitation mechanisms are statistically uncorrelated. Under this assumption statistical moments are computed by numerical integration of the derived probability density function. The results, which are generated using the exact Hertzian contact stiffness, are compared to a similar previous published Fokker-Planck solution, which utilises an approximate expression for the Hertzian contact stiffness. The previous published results showed good qualitative agreement with the results presented in this paper. The results presented herein though should give more accurate numerical values for resulting statistical moments of the mean vibratory separation.

1 INTRODUCTION

Nayak (1972) applied the Fokker-Planck equation (Caughey 1963) to the Hertzian contact problem by considering the excitation to be a white-noise stationary signal. Approximate closed form expressions for the resulting displacement and velocity standard deviations was derived. This procedure was later refined. By approximating the Hertzian nonlinearity with a third order polynomial, and utilising the stationary solution to the Fokker-Planck equation, a probability density function for the displacement was derived (Hess et al. 1992). It was numerically shown that an excitation consisting of either an external or an internal (surface roughness) random signal with zero mean value level resulted in an increase of the mean value separation of the interacting bodies. By assuming that the adhesion theory of friction is valid also the mean value of friction force was found to decrease. The computed results were shown to be in good agreement with measured friction forces at a sliding contact. In the perspective of the above mentioned works it is of interest to consider the Fokker-Planck solution, without approximating the Hertzian stiffness term.

It should though be noted that the Hertzian contact model originally applies to elastic smooth contacts. Its validity on rough surface contact problems has previously been examined (Greenwood and Tripp 1966). The contact was considered to be built up by a multitude of small spherically shaped asperities with equal contact radius, but with their heights being statistically distributed. It was shown that, at sufficiently high loads, the rough surface contact problem could be represented as a smooth Hertzian contact.

For a general text on the Hertzian contact problem, with a historical perspective, the reader may consult Johnson (1982).

2 THE MODEL

The problem of a hemispherical rider moving with velocity V over a rough surface is considered. The elastic Hertzian contact model will be applied in order to account for the mutual approach of points distant to the deformation region in the two interacting bodies. The rough contact problem will be modelled by considering the static problem of bringing two non-conforming bodies (with smooth surfaces) in contact. Using Hertzian contact theory a relationship between the applied static load P_0 and the resulting deflection δ can be derived (Johnson 1985) as seen below.

$$P_0 = k \cdot \delta^{3/2} \quad (1)$$

$$k = \frac{4}{3}\sqrt{RE'^2} \quad (2)$$

In equation (2) R represents the hemispherical contact radius of curvature. For a sphere in contact with a plane this would be the radius of the sphere. For two general non-conforming bodies one has to consider their equivalent radius of curvature (Johnson 1985) . E' is the equivalent modulus of elasticity, which can be derived under the condition that the pressure acting on the first body is equal to the pressure acting on the second. Under the assumption that the materials in the two interacting bodies are elastic and isotropic homogenous the equivalent elastic modulus of elasticity is given by equation (3).

$$\frac{1}{E'} = \frac{1-\nu_1^2}{E_1} + \frac{1-\nu_2^2}{E_2} \quad (3)$$

In the above equation E_1, E_2 and ν_1, ν_2 are the elastic modulus and Poisson's ratio of the respective bodies. Equation (1) can be applied to dynamic problems provided that the frequency contents of the dynamic excitation and of the resulting deflections are small compared to the first internal resonances of the interacting bodies. It is thus assumed that the deformation is taking place in the contact region with the rest of the body moving as a rigid mass. Under this assumption a model of the resulting vibration of the rider can be set up as seen in Figure 1.

Applying Hertzian contact theory to the dynamic case the deformation is thus assumed to be localised to the contact area, with the rest of the rider moving as a rigid body. The surface roughness is represented by the co-ordinate y_i, giving a vertical excitation as the rider moves over the rough surface. In accordance with Hertzian contact theory the resulting contact deformation is modelled by the non-linear spring. The deformation δ in equation (1) is thus in Figure 1 represented by the difference $y_i - y$. This difference of co-ordinates is defined so that loss of contact occurs when $y > y_i$. Furthermore

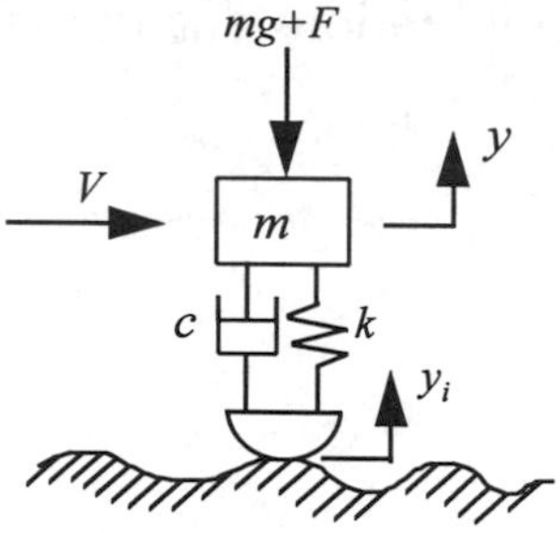

Figure 1. Model of contact between hemispherical rider and rough surface.

the non-linear spring stiffness is zero under loss of contact conditions. In order to implement the Fokker-Planck equation as previously been done (Hess et al. 1992) the damping term though is assumed to be not affected by loss of contact conditions, which is an approximation. The damping c is thus assumed to be constant at all times.

2.1 Governing equations

The procedure for deriving the governing equation of motion will basically be the same as in a previous publication (Hess et al. 1992). The final equation of motion though differs as the Hertzian non-linear stiffness herein will not be approximated. Also the two cases of external random excitation and internal surface roughness excitation will be handled simultaneously. The statistical properties of the resulting displacement due to a linear combination of external and internal excitation will be discussed in Section 2.2.

Using the force-displacement relationship equation (1) and the model as seen in Figure 1 the dynamics of the rider is governed by equations (4a,b).

$$m\ddot{y} = c(\dot{y}_i - \dot{y}) + k(y_i - y)^{3/2} - mg - F \qquad y \le y_i \quad (4a)$$

$$m\ddot{y} = c(\dot{y}_i - \dot{y}) - mg - F \qquad y > y_i \quad (4b)$$

The dynamics during contact is given by the first of equation (4a), and during loss of contact by equation (4b). This ordering of equations will also be the case in subsequent equations (6a,b) and (9a,b). As previously mentioned it should be noted that loss of contact is modelled as only affecting the stiffness term, without having any effect on the damping.

The displacement y will now be referred relative to the static compression y_0, which is the magnitude of the static contact compression in absence of external dynamic excitation or internal surface roughness excitation.

$$y \to y - y_0 \qquad y_0 = \left(\frac{P_0 + mg}{k}\right)^{2/3} \tag{5}$$

Note that the static compression y_0 is equivalent to the static deflection δ in equation (1).

By also considering the external force term F to consist of a static and a time-varying term $F = P_0 + P(t)$ the equations of motion can be rewritten as seen below.

$$m\ddot{y} + c(\dot{y} - \dot{y}_i) - k(y_0 + y_i - y)^{3/2} = -P_0 - mg - P(t) \qquad y \le y_i \tag{6a}$$

$$m\ddot{y} + c(\dot{y} - \dot{y}_i) = -P_0 - mg - P(t) \qquad y > y_i \tag{6b}$$

A change of variables to the dimensionless variables q, q_i and u is now introduced.

$$q = \frac{u}{y_0}, \quad q_i = \frac{y_i}{y_o} \tag{7}$$

$$u = \frac{y - y_i}{y_0} = q - q_i \tag{8}$$

It is seen that the dimensionless displacement q is equal to u for the special case of no surface roughness excitation ($q_i \equiv 0$). The subsequent derivation though will be made using the dimensionless deflection u, which is a measure of the separation of the interacting bodies.

By substituting variables u and q_i in equations (6a,b) and dividing the resulting equations with the factor my_0 the following is achieved:

$$\ddot{u} + \frac{c}{m}\dot{u} - \frac{k}{m}y_0^{1/2}(1-u)^{3/2} + \frac{1}{my_0}(P_0 + mg) = -\frac{P(t)}{my_0} - \ddot{q}_i \qquad u \le 1 \tag{9a}$$

$$\ddot{u} + \frac{c}{m}\dot{u} + \frac{1}{my_0}(P_0 + mg) = -\frac{P(t)}{my_0} - \ddot{q}_i \qquad u \le 1 \tag{9b}$$

The loss of contact condition, which in equations (9a,b) is determined by the sign of $1-u$, may equivalently be indicated using the Heaviside step function H, which is defined by equation (10).

$$H(u<0) = 0$$
$$H(u \ge 0) = 1 \tag{10}$$

Equations (9a,b) may thus be combined to give a single equation of motion (11), with excitation defined by equation (12).

$$\ddot{u} + \frac{c}{m}\dot{u} + \left\{-\frac{k[1 - H(u-1)]}{m} y_0^{1/2}(1-u)^{3/2} + \frac{1}{my_0}(P_0 + mg)\right\} = F_{1+2}(t) \tag{11}$$

$$F_{1+2}(t) = F_1(t) + F_2(t),$$
$$F_1(t) = -\frac{P(t)}{my_0}, \qquad F_2(t) = -\ddot{q}_i. \tag{12}$$

It should be noted that a zero deflection u represents zero deviation from the static deflection. A negative u-value indicates compression with respect to the static deflection. The special case $u \ge 1$ represents loss of contact. Also it is seen in equation (11) that the static terms are now considered to be part of the non-linear stiffness term. This is crucial as the equation of motion now is in a form suitable for application of the stationary solution to the Fokker-Planck equation (Caughey 1963), without having to introduce approximations for the non-linear stiffness term.

2.2 *A linear combination of uncorrelated random excitation and surface roughness excitation*

The statistical properties of the two excitation mechanisms will be the same as specified by Hess et al. (1992). By further assuming that these mechanisms are statistically uncorrelated the statistical properties of the sum of these excitations will herein be considered.

The external random excitation $P(t)$ is considered to be a white noise process with a zero mean value and a constant one-sided spectral density W_0.

$$\langle P(t)\rangle = 0,\ \langle P(t)\cdot P(t-\tau)\rangle = \frac{1}{2}W_0\delta(\tau) \tag{13}$$

In equation (13) the brackets $\langle\ldots\rangle$ denote expectation value, and $\delta(\ldots)$ is the Dirac delta function. Using equation (12) the statistical properties of $F_1(t)$ can be derived. The result is seen in equation (14).

$$\langle F_1(t)\rangle = 0,\ \langle F_1(t)\cdot F_1(t-\tau)\rangle = \frac{1}{2}Z_1\delta(\tau), \quad Z_1 = \frac{W_0}{m^2 y_0^2} \tag{14}$$

The surface roughness spectral density function $S_{y_i y_i}$ is assumed to be a smooth function of the spatial wavenumber k (Hess et al. 1992).

$$S_{y_i y_i} = \frac{L}{\pi k^4} \tag{15}$$

In the above equation L is a constant which is to be fitted to measurements of the surface roughness spectrum. Using the relation equation (7) between q_i and y_i it was shown (Hess et al. 1992) that the excitation $F_2(t) = -\ddot{q}_i$ in equation (12) is a white noise process with a zero mean value level.

$$\langle F_2(t)\rangle = 0,\ \langle F_2(t)\cdot F_2(t-\tau)\rangle = \frac{1}{2}Z_2\delta(\tau), \quad Z_1 = \frac{4LV^3}{y_0^2} \tag{16}$$

V is the velocity of the rider moving over the rough surface, and y_0 is the static compression defined by equation (5).

By assuming that the external excitation and the internal surface roughness excitation - represented by $F_1(t)$ and $F_2(t)$ - are statistically uncorrelated the statistical properties of the sum $F_{1+2}(t) = F_1(t)+F_2(t)$ can be derived trivially.

$$\langle F_{1+2}(t)\rangle = \langle F_1(t) + F_2(t)\rangle = \langle F_1(t)\rangle + \langle F_2(t)\rangle \tag{17}$$

$$\langle F_{1+2}(t)\cdot F_{1+2}(t-\tau)\rangle = = \langle F_1(t)\cdot F_1(t-\tau)\rangle + \langle F_2(t)\cdot F_2(t-\tau)\rangle \tag{18}$$

As the statistical properties of $F_1(t)$ and $F_2(t)$ are given by equations (14) and (16) the statistical properties of $F_{1+2}(t)$ are given by equation (19).

$$\langle F_{1+2}(t)\rangle = 0,\ \langle F_{1+2}(t)\cdot F_{1+2}(t-\tau)\rangle = \frac{1}{2}Z_3\delta(\tau), \quad Z_3 = \frac{1}{y_0^2}\left(\frac{W_0}{m^2} + 4LV^3\right) \tag{19}$$

3 STATIONARY SOLUTION USING THE FOKKER-PLANCK EQUATION

The Fokker-Planck equation (Caughey 1963) is an expression for the evolvement in time of the second order probability function for Markoff processes. If the conditional probability tends to a limiting stationary probability density function, the stationary form of the Fokker-Planck equation is achieved. Applying the stationary form of the Fokker-Planck equation to the single-degree-of-freedom system equation (20) with the non-linear stiffness $g(u)$, it is possible to show (Caughey 1963) that the stationary probability density function $p_s(x_1)$ for the displacement is achieved as equation (22). This is for the case of the excitation being a white noise random process with properties specified by equation (21).

$$\ddot{u} + \beta\dot{u} + g(u) = F(t) \tag{20}$$

$$\langle F(t)\rangle = 0,\ \langle F(t)\cdot F(t-\tau)\rangle = \frac{1}{2}Z_3\delta(\tau) \tag{21}$$

$$p_s(u) = C\cdot\exp\left\{-\frac{4\beta}{Z_3}\int_0^u g(u_1)du_1\right\} \tag{22}$$

Furthermore it can be shown that if a stationary solution exists it is unique. The constant C in equation (22) is given by the normalising condition (23).

$$\int_{-\infty}^{\infty} p_s(u)du = 1 \tag{23}$$

No results concerning the resulting probability density function for the velocity is needed here as it can be shown that the stationary velocity and displacement are statistically independent. Having the expression for the stationary probability density function, the mean value $\langle u \rangle$, the mean square value $\langle u^2 \rangle$ and standard deviation σ_u can be computed as shown below.

$$\langle u \rangle = \int_{-\infty}^{\infty} u \cdot p_s(u) du \tag{24}$$

$$\langle u^2 \rangle = \int_{-\infty}^{\infty} u^2 \cdot p_s(u) du \tag{25}$$

$$\sigma_u^2 = \langle u^2 \rangle - \langle u \rangle^2 \tag{26}$$

The stationary solution (22) to the Fokker Planck equation was applied to the Hertzian contact problem by Nayak (1972). An approximate closed form expression for the standard deviation of the displacement level was derived by a Taylor series expansion of the resulting term in the exponential of the probability density function (22).

The differential equation (20), will now be identified with the derived differential equation of motion (11). This is accomplished by identifying the statistical properties (21) of the excitation term with equation (19). By comparing equation (11) with equation (20) equations (27) and (28) are achieved.

$$\beta = \frac{c}{m} \tag{27}$$

$$g(u) = -\frac{k[1 - H(1-u)]}{m} y_0^{1/2} (1-u)^{3/2} + \frac{1}{m y_0}(P_0 + mg) \tag{28}$$

By performing the integration involved in equation (22) the stationary probability density function for the dimensionless displacement u is achieved. The result of this is shown in equations (29-30).

$$p_s(u) = C \cdot \exp\left\{ -\frac{4\beta y_0^2}{\left(W_0 / m + 4mLV^3\right)} \cdot G(u) \right\} \tag{29}$$

$$G(u) = \frac{2k y_0^{1/2}}{5} [(1 - H(u-1))(1-u)^{5/2} - 1] + (P_0 + mg)\frac{u}{y_0} \tag{30}$$

The cases of sole external excitation is achieved by letting the product LV^3 be zero, and the case of sole internal surface roughness excitation is achieved by letting the spectral density W_0 be zero. Note that the corresponding probability density function $p_s(q)$ is equal to $p_s(u)$ for the special case of no internal surface roughness excitation, as his case implies that $q \equiv u$. The following relationship between the mean values of u, y and q should also be noted.

$$\langle u \rangle = \left\langle \frac{y - y_i}{y_0} \right\rangle = \langle q \rangle - \langle q_i \rangle = \langle q \rangle - 0 = \frac{\langle y \rangle}{y_0} \tag{31}$$

The mean value of displacement y can thus be calculated by multiplying the mean value of u or q with the static displacement y_0. Note that this y-value still is measured relative to the static displacement. A positive mean value level $\langle y \rangle$ thus indicate that the mean value compression of the Hertzian spring is less than the static compression. Note also that the mean value of u is equal to the mean value of q.

The adhesion theory of friction is assumed to be valid. The instantaneous friction force is thus assumed to be proportional to the area of contact. Using Hertzian contact theory the following relation was shown (Hess et al. 1992).

$$\left\langle \frac{F}{F_0} \right\rangle = 1 - \langle u \rangle \tag{32}$$

F_0 is the friction force in absence of normal vibration and F is the resulting friction force under non-zero normal vibration conditions. As expected it is seen that a positive mean value deflection $\langle u \rangle$ (indicating an increase of mean separation of the interacting bodies) results in a mean decrease of friction force. Expression (32) will be used for relating the change of friction force to the computed mean value deflection due to normal vibrations.

4 NUMERICAL PARAMETER STUDY

The resulting statistical moments will be computed utilising numerical integration, by combining the integrand (29) with integrals (23-25). The i:th statistical moment $\langle u^i \rangle$ of the dimensionless deflection u is achieved as seen in equation (33).

$$\langle u^i \rangle = \int_{-\infty}^{\infty} u^i \cdot p_s(u) du . \tag{33}$$

It should be noted that integration over the loss of contact interval $+1 \rightarrow \infty$ can be performed analytically. It therefore remains to find a numerical approximation to the integration over the interval $-\infty \rightarrow +1$.

$$\langle u^i \rangle = \underbrace{\int_{-\infty}^{1} u^i \cdot p_s(u) du}_{\textit{Perform numerically}} + \underbrace{\int_{1}^{\infty} u^i \cdot p_s(u) du}_{\textit{Perform analytically}} \tag{34}$$

The numerical integration involved in equation (34) is performed using an adaptive recursive Newton Cotes 8 panel rule (Forsythe 1977). For cases without surface roughness excitation the non-dimensional displacement u is equal to q. The convention will be to use the q-notation in such cases in order to point out that the displacement is not taken relative to the surface roughness co-ordinate.

Using the results derived herein the computed statistical moments will be used to evaluate the previous published solution (Hess et al. 1992), in which the Hertzian non-linear contact stiffness was approximated by a third order polynomial. From here on that solution will be referred to as the approximate Fokker-Planck solution.

4.1 *Parameters*

The special cases of sole external excitation or sole surface roughness excitation were investigated by changing a number of parameters (Hess et al. 1992). These are given in Table 1. It should be noted that the damping is given by the non-dimensional damping ratio ζ, which is defined below.

Table 1. Base set of parameters.

m=0.5 kg	ζ = 0.01 m	R=0.01 m	P_0=5 N
(External excitation)		(Surface roughness)	
W_0=0.0005 N^{2}s		L=0.5 m^{-1}	V=0.1 m/s

$$2\omega_0\zeta = \frac{c}{m} \tag{35}$$

The factor ω_0 is the small amplitude natural frequency, which is derived by linearising equation (9a) under the assumption that $u << 1$. By a Taylor's series expansion the non-linear stiffness term is approximated by the zeroth and first order terms.

$$\begin{aligned} &\frac{k}{m} y_0^{1/2}(1-u)^{3/2} = \\ &= \frac{k}{m} y_0^{1/2} - \frac{3k}{2m} y_0^{1/2} u + O(u^2) \approx \frac{k}{m} y_0^{1/2} - \omega_0^2 u \end{aligned} \tag{36}$$

By combining equations (35) and (36) a relation between the viscous damping c and the non-dimensional damping ratio ζ is achieved.

$$c = \zeta\sqrt{6mky_0^{1/2}} \tag{37}$$

4.2 *Results*

In Table 2 computed statistical moments and related results are presented for the case of no internal surface roughness excitation. The base parameter values are shown in Table 1. The same parameter values as used by Hess et al. (1992) have been chosen. In Table 3 corresponding results are shown for the case of no external random excitation. As the mean value deflection gives the resulting decrease of friction (32) it is of interest to compare the mean value deflections presented in Tables 2 and 3 with corresponding previous published results. In Tables 4 and 5 the relative deviations between the Fokker-Planck solution derived herein and the previous published approximate Fokker-Planck solution (Hess et al. 1992). From the results shown in Tables 4 and 5 it is seen that the approximate Fokker-Planck solution consistently gives an overestimation of the resulting mean value deflection compared to the Fokker-Planck solution which utilises the exact Hertzian stiffness. Among the calculated cases the maximum magnitude of the relative deviation deviation is 5.4 %.

Table 2. Mean value levels and decrease of friction under external random excitation.

Case		$\langle q \rangle$	$\langle F / F_0 \rangle$
1	Base*	0.0573	0.9427
	m (kg)		
2	0.4	0.0872	0.9128
3	0.8	0.0238	0.9762
	ζ		
4	0.0075	0.0853	0.9147
5	0.0200	0.0243	0.9757
	R (m)		
6	0.0500	0.0830	0.9170
7	0.0001	0.0223	0.9777
	P_0 (N)		
8	3.5	0.0870	0.9130
9	10.0	0.0228	0.9772
	W_0 ($N^2 s$)		
10	0.00065	0.0823	0.9177
11	0.00025	0.0243	0.9757

*Base refers to Table 1.

Table 3. Mean value levels and decrease of friction under surface roughness excitation.

Case		$\langle q \rangle$	$\langle F / F_0 \rangle$
1	Base*	0.0573	0.9427
	m (kg)		
2	0.4	0.0871	0.9129
3	0.8	0.0233	0.9767
4	0.0075	0.0853	0.9147
5	0.0200	0.0243	0.9757
	R (m)		
6	0.0500	0.0830	0.9170
7	0.0001	0.0223	0.9777
	P_0 (N)		
8	3.5	0.0870	0.9130
9	10.0	0.0228	0.9772
	L (m^{-1})		
10	0.65	0.0823	0.9177
11	0.25	0.0243	0.9757
	V (m/s)		
12	0.11	0.0851	0.9149
13	0.08	0.0249	0.9751

*Base refers to Table 1

5 SUMMARY AND CONCLUSIONS

By applying the stationary solution of the Fokker-Planck equation to the Hertzian contact problem the resulting stationary probability density function for the relative displacement has been derived. This

Table 4. Relative deviation between mean values computed herein $\langle q \rangle_1$ and previous computations $\langle q \rangle_2$ (Hess et. al 1992) for external random excitation.

Case		$\langle q \rangle_1$	$\langle q \rangle_2$	$\frac{\langle q \rangle_2 - \langle q \rangle_1}{\langle q \rangle_1}$
1	Base*	0.0573	0.0584	1.9 %
	m (kg)			
2	0.4	0.0872	0.0890	2.1 %
3	0.8	0.0238	0.0250	5.0 %
	ζ			
4	0.0075	0.0853	0.0871	2.1 %
5	0.0200	0.0243	0.0255	4.9 %
	R (m)			
6	0.0500	0.0830	0.0847	2.1 %
7	0.0001	0.0223	0.0235	5.4 %
	P_0 (N)			
8	3.5	0.0870	0.0888	2.1 %
9	10.0	0.0228	0.0240	5.3 %
	W_0 ($N^2 s$)			
10	0.00065	0.0823	0.0840	2.1 %
11	0.00025	0.0243	0.0255	4.9 %

*Base refers to Table 1

Table 5. Relative deviation between mean values computed herein $\langle u \rangle_1$ and previous computations $\langle u \rangle_2$ (Hess et. al 1992) for external random excitation.

Case		$\langle u \rangle_1$	$\langle u \rangle_2$	$\frac{\langle u \rangle_2 - \langle u \rangle_1}{\langle u \rangle_1}$
1	Base*	0.0573	0.0584	1.9 %
	m (kg)			
2	0.4	0.0871	0.0889	2.1 %
3	0.8	0.0233	0.0245	5.2 %
	ζ			
4	0.0075	0.0853	0.0871	2.1 %
5	0.0200	0.0243	0.0255	4.9 %
	R (m)			
6	0.0500	0.0830	0.0847	2.1 %
7	0.0001	0.0223	0.0235	5.4 %
	P_0 (N)			
8	3.5	0.0870	0.0888	2.1 %
9	10.0	0.0228	0.0240	5.3 %
	L (m^{-1})			
10	0.65	0.0823	0.0840	2.1 %
11	0.25	0.0243	0.0255	4.9 %
	V (m/s)			
12	0.11	0.0851	0.0869	2.1 %
13	0.08	0.0249	0.0261	4.8 %

*Base refers to Table 1

derivation is similar to a derivation in a previous published paper (Hess et al. 1992), in which the Hertzian contact stiffness was approximated. The derivation herein though utilises the exact Hertzian stiffness, including loss of contact. For the derivation presented here it was furthermore assumed that the external random excitation and the internal surface roughness excitation are statistically uncorrelated. Under this assumption a single probability density function sufficiently included the combined effects of the two excitation mechanisms. Computed statistical moments, using the exact Hertzian non-linear contact stiffness, have been compared to the previous published approximate Fokker-Planck solution (Hess et al. 1992). The comparison included the displacement mean values. The results shows that this previous published solution introduces an overestimation of the relative mean value level. Among the computed cases this relative error varied up to 5.4 %.

The results presented herein, as well as the previous published result (Hess et al. 1992), successfully predicts the effect of the Hertzian non-linear contact stiffness on the mean value deflection. The mean value distance between the two interacting bodies is increased under the influence of either an external or internal zero mean random excitation mechanism. This effect was furthermore experimentally verified with reasonable agreement between theory and experiment (Hess et al. 1992). For future reference though the results presented herein should give more accurate numerical values for resulting statistical moments of the mean vibratory separation.

ACKNOWLEDGMENT

The co-operation and funding provided by Scania AB is gratefully acknowledged.

REFERENCES

Caughey, T.K. 1963. Derivation and application of the Fokker-Planck equation to discrete non-linear dynamical systems subjected to white random excitation. *Journal of the Acoustical Society of America*. 35: 1683-1692.

Greenwood, J.A. & J.H. Tripp 1967. The elastic contact of rough spheres. *Journal of Applied Mechanics*. 34: 153-159.

Forsythe, G.E., M.A. Malcolm & C.B. Moler 1977. *Computer methods for mathematical computations*. Englewood Cliffs, New Jersey : Prentice-Hall, Inc.

Hess, P.D., A. Soom & C.H. Kim 1992. Normal vibration and friction at a Hertzian contact under random excitation: Theory and experiment. *Journal of Sound and Vibration*. v153(3): 491-508.

Johnson, K.L. 1982. One hundred years of Hertzian contact. *Proc. Instn. Mech. Engnrs*. v196: 363-378.

Johnson, K.L. 1985. *Contact mechanics*. Cambridge: Cambridge University Press.

Nayak, P.R. 1972. Contact vibrations. *Journal of Sound and Vibration*. v22(3): 297-321.

4 Geometric and material non-linearity

Modern Practice in Stress and Vibration Analysis, Gilchrist (ed.) © 1997 Balkema, Rotterdam, ISBN 90 5410 896 7

Fast modelling of ring stiffened cylinders for nonlinear buckling analysis

P.C. Chatterjee
Department of Naval Architecture and Ocean Engineering, University of Glasgow, UK

ABSTRACT: The objective of this paper is to introduce fast modelling techniques for those with interest in extracting useful nonlinear finite element results in a limited time scale. Parametric language based command files are found very useful for changing mesh discretisation and setting up slightly different models. Symmetrical structures such as cylinders can be analysed quickly and efficiently by using these techniques and as an example, a series of ring stiffened cylinders are considered. The cylinder models have constant weight, length, radius and shell thickness but varying ring width, spacing and ring thickness. Geometric imperfections are discussed in detail but not included in FE analyses to demonstrate fast modelling. Nonlinear buckling analysis results are explained with the help of the current stiffness parameter and negative pivots. Finally, eigenvalue buckling and nonlinear bifurcation predictions are compared to the experimental results.

1 INTRODUCTION

The recent advances in the Finite Element (FE) world are most impressive. Just over twenty years ago, it was necessary to draw the structure geometry on a drawing board before meshing it. The next step was to copy the details of the node co-ordinates and element connectivities on coding sheets. Input had to be transferred to punched cards and output - if the user was fortunate - was endless pages of lineflow covered with columns of numbers.

The advent of powerful and affordable computers has brought a revolution in attitudes to FE computing in recent years. However, the task involved in creating data files for finite element solvers is still time consuming. Nowadays, under heavy work load, many engineers would prefer not to spend time in tweaking data files. Importing CAD geo-metries into FE pre-processors is an effective and popular option, but that may not help to analyse slightly different structures. Various models with different geometry are often required to finalise one design. Modifications and manipulations of an existing model in FE pre-processors can be complex and not always convenient.

This paper presents a parametric language based approach for fast modelling of symmetrical structures, such as stiffened cylinders. The advantage of using the symmetry for fast modelling is lost when initial imperfections are included in FE models. In the following sections, "How to capture the buckling characteristics of ring stiffened cylinders even without modelling the imperfections" is explained in detail. Fast modelling techniques are described later.

2 BUCKLING OF STIFFENED CYLINDERS UNDER EXTERNAL PRESSURE

Thin-walled stiffened cylindrical shells subjected to external pressure hold a prime position in the field of structural engineering. Submarines and submersibles, columns and pontoons in offshore oil production platforms are typical examples of their applications. These structures may be characterised by a collapse mechanism involving the plasticity of material or instability in the elastic field or the combination of both nonlinear effects. Elastic buckling has peculiar aspects related to the sudden occurrence of the phenomenon without any fore-warning signals, such as large displacements and strains. Bushnell (1985) believes that most of the failures did not occur because of a lack of analysis capability. There were computer programs that, given appropriate input, would yield accurate predictions. But the structures failed because very few had sufficient familiarity with buckling phenomena to identify proper numerical tests which would warn of impending disaster.

The membrane stiffness of a thin shell is, in general, several orders of magnitude greater than the bending stiffness. In fact, it can absorb a great deal of membrane strain energy without deforming too much. It must deform much more in order to absorb an equivalent amount of bending strain energy. If the shell is loaded in such a way that most of its strain energy is in the form of membrane compression and if there is a way that this stored-up, membrane energy can be converted into bending energy, the shell may fail rather dramatically in a process called 'buckling', when it exchanges its membrane energy for bending energy.

The buckling of thin stiffened cylinders under external pressure can be studied by three different types of investigation: an analytical approach based on theoretical formulations, the simulation of the phenomenon by numerical methods, and experimental tests on scaled models (Kendrick 1985). The analytical investigation offers some advantages in terms of time and equipment investments, but the results are heavily influenced by the simplified hypothesis assumed. They can nevertheless be employed for a preliminary scantling of the structure.

The experimental analysis, at least, is a powerful tool to verify the numerical results and to collect helpful information about the behaviour of the actual structure. However, this requires much effort and very expensive facilities.

A numerical investigation supports the opportunity of following the history of the structure through deformed configurations before its collapse. It can take into account different phenomena.

3 GEOMETRIC IMPERFECTIONS

The problem of buckling of thin cylindrical shells under axial compression has received far more attention than many other problems in structural mechanics because of the extraordinary discrepancy between test and theory which remained unexplained for so many years. It was discovered later (Donnell & Wan 1950) that the discrepancy arises from the extreme sensitivity of the critical load to initial imperfections.

A reasonable measure of geometrical quality is the ratio of initial deviation from the perfect cylindrical shape to shell thickness. Geometric imperfections arising from fabrication processes are unavoidable and cannot be disregarded in many cases. They can be included in finite element modelling after calculating the Fourier coefficients:

$$R = R_m + \sum \left(A_n \cos n\theta + B_n \sin n\theta\right) \sin\left(\frac{\pi x}{L_c}\right) \quad (1)$$

where imperfections are considered at the middle of the cylinder and a half-sine shape is assumed along the length. Lengthwise imperfections can be taken into account with more complex formulae. If these imperfections are included, the structure loses its symmetry and the modelling is no more straightforward. Therefore, they should be avoided for fast modelling if possible. Modelling of thin cylindrical shells under axial compression without initial imperfections cannot be justified. But, in general, ring stiffened cylinders are not imperfection sensitive under external pressure loading and it is possible to achieve good results after applying fast modelling techniques, as we shall see later.

4 AVAILABLE OPTIONS FOR ANALYSING RING STIFFENED CYLINDERS WITHOUT GEOMETRIC IMPERFECTIONS

4.1 *Eigenvalue Buckling*

Eigenvalue buckling analysis is a technique that can be applied to relatively "stiff" structures to estimate the maximum load that can be supported prior to structural instability or collapse. The basic assumptions are that the linear stiffness matrix does not change prior to buckling and that the stress stiffness matrix is simply a multiple of its initial value. Accordingly, the technique can only be used to predict the load level at which a structure becomes unstable, provided the pre-buckling displacements have negligible influence on the structural response.

The eigenvalue buckling load can be an overestimate of the actual capacity of a structure when the pre-buckling response is not linear. Boote et al. (1996) found that the experimental collapse load was 3.5 times less than their eigenvalue buckling result. However, the lowest buckling mode shape is important because it predicts the most dominant mode shape in the actual structure. The number of circumferential and longitudinal waves found in the buckling mode shape can help in building a 'sector' model for nonlinear analysis. Sector models work with a section of the whole cylinder for reducing the mesh size. A π/n sector model with symmetrical boundary conditions on both sides is widely used (Morandi 1995) for nonlinear analysis where n is the number of circumferential waves.

4.2 *Nonlinear Analysis*

When geometric imperfections are included, a 'traditional' nonlinear analysis strategy can be useful in locating the limit point on the load-deflection curve, as shown in Figure 1. The Current Stiffness Parameter (CSP), which is a scalar quantity designed to characterise the overall structural stiffness at various stages of a nonlinear solution, becomes negative if the analysis is continued after reaching the limit point. Including imperfections in the model may not be easy but the analysis is more or less straightforward. The limit point calculated can be compared with the experimental collapse load if results are available.

The aforementioned analysis strategy cannot be adopted for initially perfect cylinders because their limit points cannot be correlated to collapse loads of slightly imperfect cylinders. In most cases, the lowest bifurcation load is of more engineering significance.

There is no such thing as true bifurcation buckling in the case of real structures which contain unavoidable imperfections. Although true bifurcation buckling is fictitious, it is convenient and often leads to a good approximation of the actual failure load of slightly imperfect structures and their mode shapes.

The lowest bifurcation load of an initially perfect cylinder predicts the actual collapse load more accurately in comparison to the linear eigenvalue analysis for various reasons. First of all, the bifurcation load calculation is a geometrically nonlinear analysis (with or without nonlinear material model) where the changing effect of structural deformation on the structural stiffness and on the position of applied loads is considered. In other words, pre-buckling deformations need not be linear. Secondly, the stress stiffness matrix changes as the analysis proceeds and it is not assumed as a multiple of its initial value.

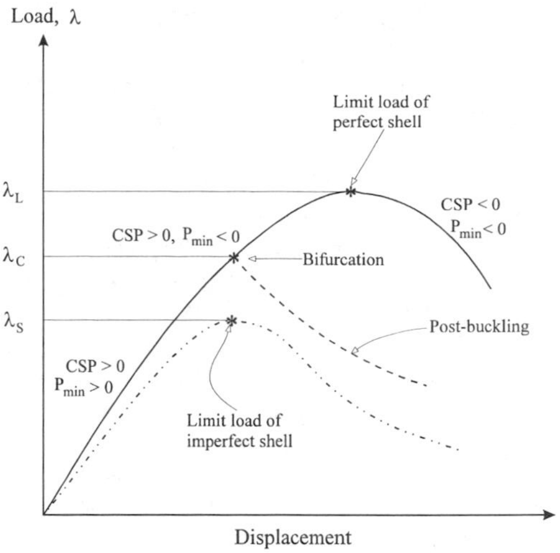

Figure 1. Limit and bifurcation loads of perfect and imperfect shells.

5 LOWEST BIFURCATION BUCKLING LOAD CALCULATION

'Negative pivots' are very important in locating the bifurcation points. A pivot refers to the diagonal element of the upper triangular matrix which is formed after Gaussian elimination. The frontal solution technique adopted by almost all FE packages is based on Gaussian elimination. A well conditioned stiffness matrix can produce a negative pivot if the solution system passes through a bifurcation or limit point. A negative CSP value, together with a negative pivot, corresponds to a limit point, but a positive CSP and a negative pivot usually indicate a bifurcation point (Figure 1). However, the scalar product of the lowest eigenmode extracted from the tangent stiffness matrix and the applied load vector is necessary to confirm a limit or a bifurcation point.

Once the lowest bifurcation load is identified, the collapse loads of slightly imperfect cylinders can be estimated from Koiter's (1963) general elastic post-buckling theory:

$$\lambda_s / \lambda_c \approx 1 - 2\left(-\rho\,\alpha\,w_{imp}\right)^2 \qquad (2)$$

where w_{imp} is the normalised imperfection amplitude, α and ρ are constants that depend on the imperfection shape. For the plastic range the theory of Hutchinson & Koiter (1970) can be used.

6 FAST MODELLING TECHNIQUES

It has been mentioned before that modifications and manipulations of an existing model in FE pre-processors can be complex and not always convenient. In the author's opinion, one of the best ways to model stiffened cylinders or any other symmetrical structure is through command or script based computer files. A number of FE pre-processors 'record' (i.e. save in a text file) the actions taken by a user while a session is in progress. If necessary, these session files can be 'replayed' to repeat previous actions. But this type of file does not help to change the mesh or build models for slightly different structures. It is necessary to replace some numerical values by variables for using a session or log file as a program. In MS-DOS environment, "copy c:\project\data1.txt a:" can be saved as a batch file and can be used to copy data1.txt to a floppy. But it cannot be used to copy any other file. If data1.txt is replaced by an MS-DOS variable (e.g. copy c:\project\%1 a:), the batch file can be used to backup any file in c:\project directory. The same concept can be used to convert session or log files of FE pre-processors for increasing the scope of their applications.

7 FACILITIES AVAILABLE IN MYSTRO

MYSTRO is the powerful pre- and post-processor for a general purpose, well established FE software called LUSAS. A parametric language facility in MYSTRO is available which is based on the syntax of the C language for use in command files. The parametric language helps to introduce variables and control structures (e.g. if, elseif and else statements, for loops, etc.) to increase the scope of a command file so that it can be used to build various models. A complete set of data files for the FE solver and model files for visual presentation can be generated within a few minutes from scratch using these command files.

It is possible to assign values to variables interactively, even whilst a command file is running. A dialog box allows the parametric programmer to prompt the user for variable input each time the command file is run. This is useful for creating similar FE models with different dimensions. The *inquire* command is used to display the dialog box. For example, *inquire "Cylinder Geometry" radius "Cylinder Radius" shell_thk "Shell Thickness"* will

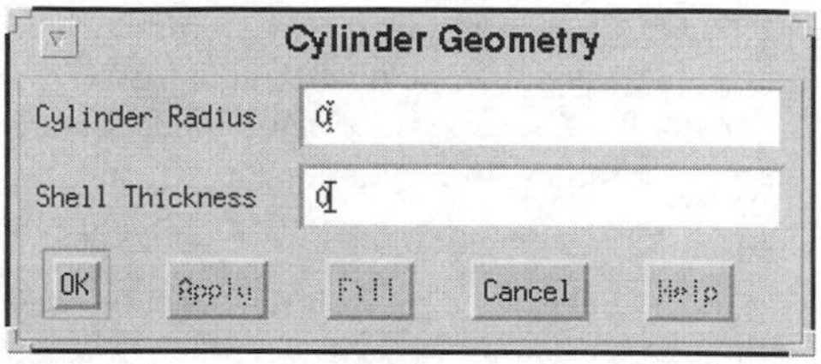

Figure 2. A sample dialog box requesting input.

produce the dialog box in Figure 2 requesting the user input of two variables: radius and shell_thk.

8 RING STIFFENED CYLINDER MODELS

To demonstrate the fast modelling techniques, a series of ring stiffened cylinders have been selected from the experimental studies of Seleim & Roorda (1986) at the University of Waterloo, Canada. They did buckling experiments with ten ring stiffened cylinders, machined from aluminium alloy. Their objective was to observe the change in buckling behaviour as the stiffening pattern, on an otherwise constant shell, progresses from many equally spaced weak rings to a few heavy rings. In other words, the size, spacing and number of rings were varied so that the weight of all ten models remained constant at 6.57 kg. All the models had the same internal diameter of 254 mm, overall length of 920 mm and wall thickness of 2 mm. The test specimens were machined from 25.4 mm thick aluminium tubes to reduce the initial geometric imperfections. The models in the work of Seleim & Roorda (1986) had exterior rings for convenience and ease of machining. Out of ten models, three pairs (i.e. Model 3 and 10; 2 and 9; 7 and 8) were identical. They were built with a special interest to see whether the experimental results could be closely reproduced.

The experimental results obtained by Seleim & Roorda (1986) are presented in Table 1 where Z is a parameter which can be used to separate short cylinders from moderately long cylinders:

$$Z = \left(\frac{L}{R}\right)^2 \left(\frac{R}{t}\right) \sqrt{\left(1 - \mu^2\right)} \qquad (3)$$

where L = length of cylinder between consecutive rings, R = cylinder radius, t = shell thickness and μ = Poisson's ratio.

Table 1. Experimental collapse loads.

Expt. model no.	No. of rings	Collapse load (MPa)	Z value
2, 9	3	1.533, 1.358	181.1
4	5	2.238	80.4
6	7	3.266	45.4
7, 8	9	2.946, 3.524	29.0
5	11	3.194	20.2
3, 10	13	2.837, 3.124	14.8
1	17	3.036	8.9

9 NUMERICAL ANALYSIS

The eigenvalue buckling analysis is important for finding the most dominant mode shape and associated circumferential wave pattern. Table 2 presents typical MYSTRO commands used in a command file to automatically build seven FE models. Initially, one model was created using the graphical user interface of MYSTRO. The session file was converted into a command file by adding variables and control structures. The command file was then used for the automatic generation of other six FE models. The command file in Table 2 is incomplete because the details are software specific and fall outside the scope of this paper. Nevertheless, it is not difficult to generate similar files for analysing symmetrical structures from the information given in Table 2.

For eigenvalue buckling analysis, semiloof curved thin shell elements for the cylinder skin and compatible semiloof thin beam elements for ring stiffeners have been used. The idealisation of ring stiffeners as beam elements significantly reduce the problem size which is important for full cylinder modelling. For nonlinear bifurcation analysis rings are also modelled with shell elements but sector models are used instead of full cylinders.

The results from the eigenvalue buckling analysis (Table 3) are in good agreement with the experimental collapse loads for Models 1, 2, 4, 6 and 9. Except Model 1, the Z values for these models are high, indicating that they are moderately long cylinders. Model 1 is an exception because there are too many weak rings and they are not effective in defining separate shell bays and consequently the Z value calculation may not be appropriate.

On the other hand, the eigenvalue buckling loads for Models 3, 5, 7, 8 and 10 are clearly overestimates and not in agreement with the experimental collapse loads. It is worth noting that these models have relatively small Z values within 14 and 30.

In Figure 3, the rings are not distinctly visible in the lowest eigenmode of Models 2 and 9 because they are modelled as beam elements. Figure 4 presents a corresponding circumferential wave pattern. A similar wave pattern is depicted in Figure 5 for models with 11 or more rings.

Table 2. An example of a command file.

```
{
! Exclamation marks denote comment lines
! Declare the variables, for example:
int      nRing
real     radius
! Prompt the user for values
inquire "Input Geometry" angle "Arc Angle" nRing "No of ...
Rings" radius "Internal Radius"
! Transformation datasets
DEFINE TRANSFORMATION TRANSLATION ITSET=1 ...
X=0 Y=(length/(nRing+1)) Z=0
! Define points, lines and surfaces (i.e. features)
DEFINE POINT PN=1 X=(radius+shellThk/2) Y=0 Z=0
DEFINE LINE ARC_MINOR BY_SWEEPING LN=1 PN=1...
ITSET=3
COPY LINE LN=1 LNINC=* NTIMES=1 ITSET=1
DEFINE SURFACE CYLINDER BY_JOINING SN=1 ...
FN1=L1 FN2=L2
! Define mesh, material, geometry and assign them to features
DEFINE MESH BY_NAME IMSH=* FEATYP=Surface ...
LNAME=qsl8 MSHTYP=3 NDIVX=nBayLength ...
NDIVY=nArc
ASSIGN MESH FEATYP=Surface SN=1 IMSH=1
DEFINE MATERIAL IMAT=* MATTYP=1 LPTPF=1 ...
E=76000 NU=0.33
ASSIGN MATERIAL FEATYP=Surface SN=1 IMAT=1
DEFINE GEOMETRY IGMP=1 LGTPF=7 E=0 T=shellThk
ASSIGN GEOMETRY FEATYP=Surface SN=1 IGMP=1
! Define loading and supports and assign them to features
DEFINE LOADING ILDG=1 LTPF="Uniformly Dist. Load"...
WZ=pressure
ASSIGN LOADING FEATYP=Surface SN=1 ILDG=1 ...
LCID=1 FACTOR=1.0
DEFINE SUPPORTS ISUP=1 U=1 V=1 W=1 THL1=1 ...
THL2=1
ASSIGN SUPPORT FEATYP=Line LN=1 ISUP=1 LCID=1
! Copy features with transformation datasets if necessary
COPY SURFACE SN=1 SNINC=* NTIMES=nRing ITSET=1
COPY SURFACE SN=ALL SNINC=* ...
NTIMES=360/angle-1 ITSET=3
! Define parameters for eigenvalue buckling analysis
DEFINE CONTROL ICTRL=1 ITYPE1=5 NROOT=nRoots ...
NIVC=nRoots*2 SHIFT=0.0 NORM=0 ISTURM=1 ...
IEIGSL=0 MAXMIN=0 IBUCKL=0 ...
RTOL=tolerance NITEM=50
! Assign the analysis type to the loadcase ID
ASSIGN CONTROL ICTRL=1 LCID=1
}
```

Table 3. Eigenvalue buckling loads.

Expt. model no.	No. of rings	EB load (MPa)	Circumf. wave no.
2, 9	3	1.584	6
4	5	2.489	7
6	7	3.456	8
7, 8	9	4.606	9
5	11	4.508	3
3, 10	13	3.830	3
1	17	3.173	3

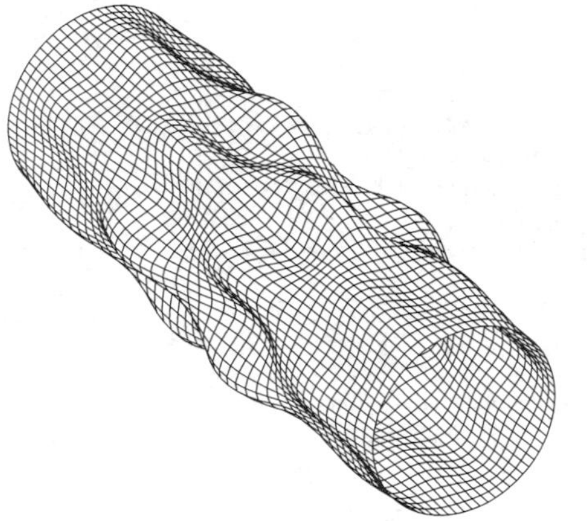

Figure 3. The lowest eigenmode of cylinders with 3 rings.

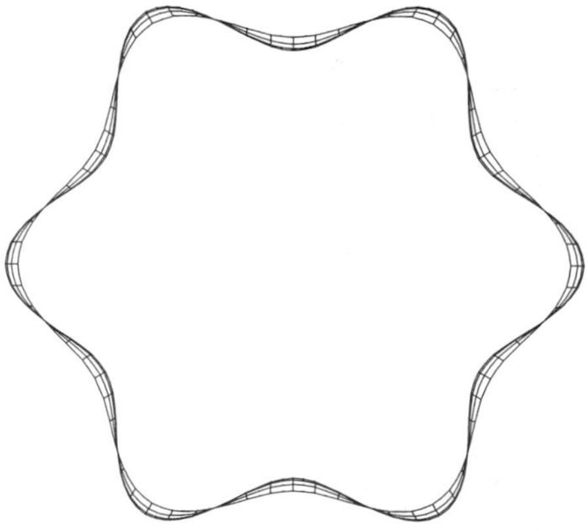

Figure 4. Another view of the cylinder in Figure 3 along the axis of symmetry.

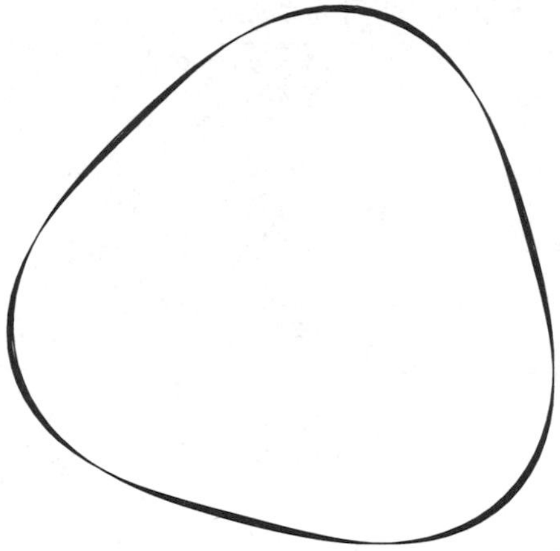

Figure 5. Three circumferential waves in cylinders with 11 or more rings.

9.1 *Bracketing the Lowest Bifurcation Point*

The eigenmodes found are very important to establish the sector models for nonlinear buckling analysis. Figure 6 presents a sector of Model 3 with 13 rings (Model 10 is identical) subjected to symmetric boundary conditions. It should be noted that the mesh discretisation has been varied most easily with the help of command files. The identification of a bifurcation or limit point is called 'bracketing' in LUSAS (FEA 1996). The Total Lagrangian approach for geometric nonlinearity is adopted here to account for deformations in load increments. A nonlinear material model for the aluminium alloy 6061 is also included in the bracketing analysis.

Figure 7 presents the nonlinear load-displacement curve of models with 13 rings (i.e. Models 3 and 10) and at a node where the maximum resultant displacement occurs. The results are obtained from a 'traditional' nonlinear analysis guided by the current stiffness parameter. The collapse load of a perfect

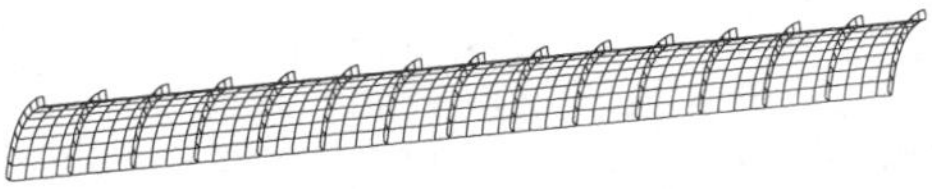

Figure 6. A sector model for cylinders with 13 rings.

ring stiffened shell is found much higher than the experimental results. As explained before, this limit point load is of no engineering significance. The same FE model is run with bracketing options and the bifurcation load found is very close to the experimental results. The bifurcation point is marked on Figure 7. Figure 8 presents the load-displacement curve where the bi-section method is used to locate the lowest bifurcation load.

The sector angles of other models and the results from bracketing analyses are presented in Table 4. In some cases, $2\pi/n$ sector angle is considered instead of π/n for avoiding narrow width FE models with unforeseen boundary effects. The bifurcation loads are found to predict more accurately the experimental collapse loads compared to eigenvalue buckling results. Numerical values are given in Table 5 and presented graphically in Figure 9.

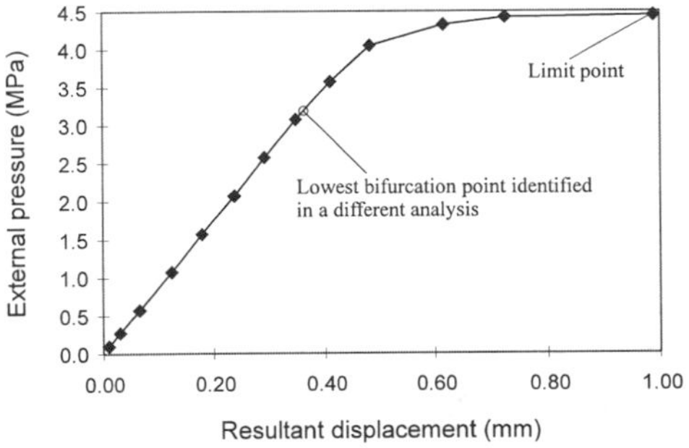

Figure 7. Limit and bifurcation points for cylinders with 13 rings.

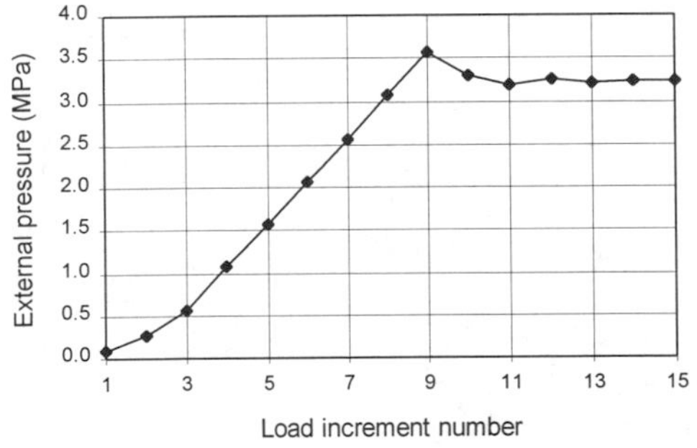

Figure 8. Bifurcation point identification.

In general, the load carrying capacity drops down with an increase in geometric imperfections. For identical cylinders, the minimum experimental collapse loads have not been considered in Table 5 to avoid results from cylinders with more imperfections.

Table 4. Nonlinear bifurcation buckling loads.

Expt. model no.	No. of rings	Bifurcation load (MPa)	Sector angle (deg.)
2, 9	3	1.579	60 (= $2\pi/6$)
4	5	2.373	51.4 (= $2\pi/7$)
6	7	3.274	45 (= $2\pi/8$)
7, 8	9	4.049	40 (= $2\pi/9$)
5	11	3.349	60 (=$\pi/3$)
3, 10	13	3.243	60 (=$\pi/3$)
1	17	3.080	60 (=$\pi/3$)

Table 5. A comparison between numerical predictions and experimental results in MPa.

Expt. model	No. of rings	EB load	Bifurc. load	Expt. result
2, 9	3	1.584	1.579	1.533
4	5	2.489	2.373	2.238
6	7	3.456	3.274	3.266
7, 8	9	4.606	4.049	3.524
5	11	4.508	3.349	3.194
3, 10	13	3.830	3.243	3.124
1	17	3.173	3.080	3.036

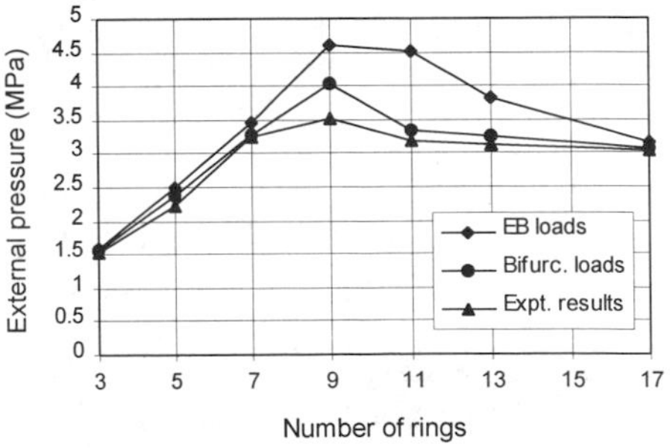

Figure 9. Eigenvalue buckling and bifurcation loads provide an upper bound to the load-carrying capability of ring stiffened imperfect shells.

10 CLOSURE

This paper does not cover some other aspects of buckling of stiffened cylinders such as material imperfections and residual stresses. The paper has concentrated on the out-of-circularity which is the most important imperfection.

The objective is to introduce fast modelling techniques for those with interest in extracting useful nonlinear FE results in a limited time scale. Script based command files are very useful in this context for changing mesh discretisation and setting up slightly different models.

The lowest bifurcation point is found very effective in predicting the collapse load of slightly imperfect cylinders. The exceptions are those with 9 rings. In fact, fine mesh models with 80° sector angle do not improve the numerical results. However, it should be noted that the experimental models in the studies of Seleim & Roorda (1986) had local deformations and they may be responsible for the discrepancy shown in Figure 9.

According to Brush & Almroth (1975), the bifurcation load for the perfect structure may be close to the buckling load of the imperfect one, depending on the shape of the secondary equilibrium path for the perfect structure. If the secondary path drops downward from the bifurcation point, the agreement may not be close, unless the imperfection is quite small. The secondary unstable equilibrium paths will be investigated in the next phase of this study using 'branching' algorithms (Shi & Crisfield 1992).

REFERENCES

Boote, D., F. Massimo, D. Mascia & R. Iaccarino 1996. Submarine pressure hull collapse: a correlation between numerical and experimental analysis. *Proceedings of Offshore Mechanics and Arctic Engineering*. IA:307-317.

Brush, D.O. & B.O. Almroth 1975. *Buckling of bars, plates, and shells*. New York: McGraw-Hill.

Bushnell, D. 1985. *Computerized buckling analysis of shells*. Dordrecht: Martinus Nijhoff.

Donnell, L.H. & C.C. Wan 1950. Effects of imperfections on buckling of thin cylinders and columns under axial compression. *Journal of Applied Mechanics*. 17:73-83.

Finite Element Analysis (FEA) Limited 1996. *LUSAS User Guide*.

Hutchinson, J.W. & W.T. Koiter 1970. Postbuckling theory. *Applied Mechanics Review*. 23:1353-1356.

Kendrick, S. 1985. Ring-stiffened cylinders under external pressure. In R. Narayanan (ed.), *Shell Structures Stability and Strength: 57-95*. Elsevier.

Koiter, W.T. 1963. Elastic stability and postbuckling behaviour. In R.E. Langer (ed.), *Proc. Symp. Nonlinear Problems: 257-275*. Madison: Wisconsin.

Morandi, A.C., P.K. Das & D. Faulkner 1995. Ring frame design in orthogonally stiffened cylindrical structures. *Proceedings of Offshore Technology Conference, Houston, 1-4 May 1995*. OTC 7801:959-968.

Seleim, S.S. & J. Roorda 1986. Buckling behaviour of ring-stiffened cylinders; experimental study. *Thin-Walled Structures*. 4:203-222.

Shi, J. & M.A. Crisfield 1992. A simple indicator and branch switching technique for hidden unstable equilibrium paths. *Finite Elements in Analysis and Design*. 12:203-213.

Modern Practice in Stress and Vibration Analysis, Gilchrist (ed.) © *1997 Balkema, Rotterdam, ISBN 90 5410 896 7*

Modelling transformation behaviour and its effects on residual stress in a finite element system

M. Sedighi & C. McMahon
Department of Mechanical Engineering, University of Bristol, UK

ABSTRACT: Heat treatment in the manufacture of steel parts involves coupled thermal, mechanical and phase transformation phenomena that affect the phase distribution and material properties within the parts, and also the distribution of residual stresses after manufacture. A number of analytical approaches to the modelling of heat treatment have been developed over the years, but most have used dedicated software or special purpose finite element codes. This paper describes an alternative approach, using the user subroutine and coupled analysis capabilities of a commercial code. The multiple interactions between stress, phase transformation and thermal fields are first described, and then the techniques used to model these interactions are presented. The results of two case studies are given. In the first, the approach is compared with published results for a eutectoid steel. In the second, the quenching of a medium carbon low-alloy steel is modelled.

1 INTRODUCTION

Heat treatment of steels is a routine process for production of components to introduce properties which can increase the life of parts and improve their performance. The purpose of modelling the heat treatment procedure is to allow distortion, microstructure and hardness distribution to be estimated, and, most important, to predict the distribution of residual stress inside the body.

A number of analytical approaches to the modelling of heat treatment have been developed over the years, but most have involved the development of dedicated software or special purpose finite element codes. Nowadays, the capability of commercial codes is developing rapidly, and residual stress modelling can benefit from this. In this work, the Ansys suite has been used (Ansys Inc., 1996a), but in principle any commercial software with a similar capability can be used. The approach that has been taken is to use a coupled thermal and structural analysis, and to model the phase transformation kinetics by means of user subroutines that allow the transformation behaviour of the steel to be modelled. These subroutines also allow material properties to be modified according to the phase composition of the material at any stage of the analysis, by using a mixture law. The effects of these property changes are incorporated into the analysis by using a number of features of the analysis package. To verify the approach, results have been compared with published data for a eutectoid steel achieved by a dedicated program, and are presented here together with results for a two-dimensional axisymmetric analysis of a medium carbon low alloy steel.

2 OVERALL APPROACH TO MODELLING OF THERMAL TRANSIENTS

During the quenching process, three interacting physical process take place in the material. These are temperature change, phase transformation and stress-strain change. These processes interact in the following six ways, as shown in Figure 1 (Kamamoto, 1985):

1. Thermal stress;
2. Heat release due to the work done in deformation;
3. Dependency of phase transformation on temperature;
4. Heat release with phase transformation;
5. Phase transformation effects on stress state:
   - Volumetric dilatation with phase transformation;
   - Transformation plasticity;
6. Stress dependency of phase transformation.

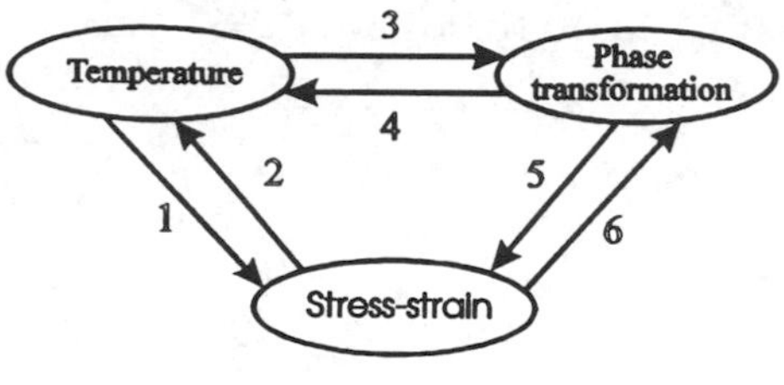

Figure 1. Couplings in quenching

Each of these interactions will now be considered.

2.1 *Thermal stress*

Simulation of a time-temperature history during quenching involves the solution of the heat conduction equation with the heat transfer coefficient as a boundary condition. A general form of the temperature distribution through a body in Cartesian coordinates can be expressed by:

$$\rho C\frac{\partial T}{\partial t}=q+\frac{\partial}{\partial x}\left(k_x\frac{\partial T}{\partial x}\right)+\frac{\partial}{\partial y}\left(k_y\frac{\partial T}{\partial y}\right)+\frac{\partial}{\partial z}\left(k_z\frac{\partial T}{\partial z}\right) \quad (1)$$

where T is temperature, k thermal conductivity, ρ density, C specific heat, $\dot{q}$ the heat generation rate, and t time.

At the boundary, the heat flux $\dot{q}$ given by (Thuvander, 1992):

$$-k\frac{\partial T}{\partial n}=\dot{q}=h(T_s-T_e) \quad (2)$$

where $\partial T/\partial n$ is the temperature gradient perpendicular to the surface, T_s is the surface temperature, T_e is the quenchant temperature and h is the heat transfer coefficient.

Because of the thermal gradient through the specimen, interaction between the temperature and stress-strain fields is linked by $\varepsilon^{th} = \alpha\,\Delta T$, where ε^{th} is the thermal strain, α is the coefficient of thermal expansion, and ΔT is temperature change.

2.2 *Heat release due to deformation*

Deformation heat release arises from the work done by the plastic strain, owing to the thermal stresses and dilatation, that leads to a heat input into the material which should be taken into account in coupled thermal-stress analysis. However, it is suggested by Denis (1987) that the plastic strain is sufficiently low for this heat input to be small, and therefore the term can be neglected.

2.3 *Dependency of phase transformation on temperature*

When a steel specimen is cooled down, austenite transforms to ferrite, pearlite, bainite, or martensite. in the process of phase transformation. The kinetics of transformation are complex, and are generally modelled by means of a set of experimental curves called the time-temperature-transformation (TTT) diagram, which presents isothermal transformation behaviour. Figure 2 shows a TTT diagram for a eutectoid steel.

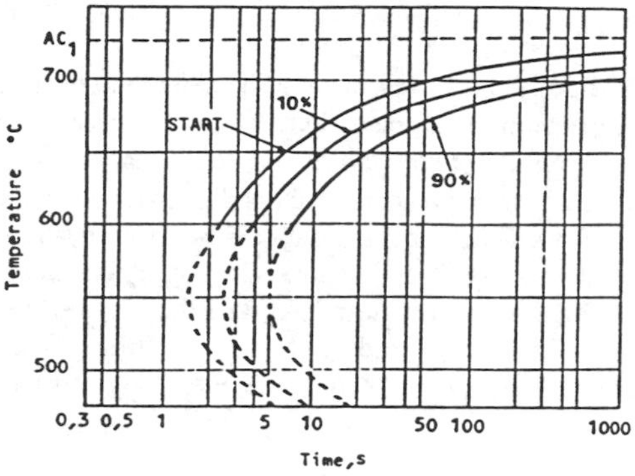

Figure 2. TTT diagram of a eutectoid steel - XC80 (Denis 1987)

The diffusional transformation from austenite into ferrite, pearlite and bainite has two stages - incubation and growth. In anisothermal transformation, the incubation period is supposed to be over when the Scheil sum reaches unity (Bourdouxhe, 1992):

$$S=\int_0^t\frac{dt}{\pi(t)}=\sum_{i=1}^{N}\frac{\Delta t_i}{\pi(T_i)}=1 \quad (3)$$

where $\pi(T)$ is the isothermal incubation time depending on temperature and Δt_i is the time increment at step i.

For the growth period, Johnson and Mehl have studied the relationship between the transformed volume fraction and time. Their proposed equation is as follows (Fernandes, 1984):

$$y = 1-exp(-bt^n) \quad (4)$$

where y is the amount of volume fraction, t is the time, and b and n temperature dependent constants which can be obtained from the TTT diagram.

For martensitic transformation, the growth is modelled by the Koistinen-Marburger equation (Koistinen, 1959):

$$y = 1-exp[-\alpha(M_s-T)] \quad (5)$$

where M_s is the martensite start temperature and coefficient $\alpha=1.1\times10^{-2}\ K^{-1}$ for most steels.

2.4 *Heat release with phase transformation*

The heat generation during phase transformation is related to the rate of transformation $\Delta v/\Delta t$ and to the temperature-dependent enthalpy of transformation, ΔH (Fernandes, 1984):

$$\dot{q} = \Delta H \frac{\Delta \upsilon}{\Delta t} \quad (6)$$

2.5 *Phase transformation effects on the stress state*

The effects of phase transformation on stress are due to two different phenomena, volumetric dilatation and transformation plasticity, which will be discussed separately.

In transformation, owing to the different density of austenite and the transformed phases, there is an increase in volume called *volumetric dilatation* and its strain increment can be written as equation 7 (HEARTS, 1994):

$$d\varepsilon^{tr} = \alpha^{tr}\ dy_k \quad (7)$$

where α^{tr} is the coefficient of transformation and dy_k is the fraction of the transformed phase.

The second mechanical interaction between stress and phase transformation is known as transformation plasticity (TRIP). TRIP is one of the sources for micro-residual-stresses (i.e. at a grain level). Fischer (1992) defined TRIP as the irreversible deformation behaviour of a transforming specimen even under a load state with an equivalent stress significantly lower than the yield stress. HEARTS (1994) has used equation 8 for calculation of transformation plasticity strain increment for all phases:

$$d\varepsilon^{tp} = 3K(1-y)\ dy\ \sigma_e \quad (8)$$

where y is the volume fraction of the transformed phase, σ_e is the effective stress and K is a coefficient of transformation plasticity obtained from experimental stress-strain data.

2.6 Stress dependency of phase transformation

The effect of stress on transformation has been studied by many authors. This paper is not going to discuss this interaction in detail. Only the formulation of the two types of transformation will be overviewed. For diffusion dependent transformation, a model based on shifting of the curves of a TTT diagram as a function of the stress state has been developed (Denis, 1987). This shifting can be applied by modification of the temperature dependent parameters b and n in equation 4. The relation between the stress state and the shifting function can be expressed by $D = g(\sigma_e)$

$$\begin{cases} t_{d\sigma} = (1-D)\, t_d \\ n_\sigma = n \\ b_\sigma = \dfrac{b}{(1-D)^{n_\sigma}} \end{cases} \quad (9)$$

where t_d is the start time of transformation.

For martensitic transformation, HEARTS (1994) and Denis (1985) have introduced equation 10 relating the changes in M_s (martensite start temperature) to the mean stress and the effective stress:

$$\Delta M_s = A\sigma_m + B\sigma_e \quad (10)$$

where $\sigma_m = (\sigma_x+\sigma_y+\sigma_z)/3$, σ_e is the effective stress, and the coefficients A and B are dependent on the material.

3 ISSUES IN INCORPORATING RESIDUAL STRESS ANALYSIS IN A COMMERCIAL CODE

As noted, dedicated software or special purpose finite element codes have been developed for the modelling of the thermo-mechanical transients of heat treatment in recent years. Two such programs are HEARTS (1994) and QUEST7 (Sjöstrom, 1982). The present work uses ANSYS(Ansys, Inc.1996a) as the primary computational tool. Different mathematical expressions for the physical behaviours noted above are coded as user subroutines. These subroutines are related to volume fraction calculation, changing material properties, volumetric dilatation, heat release due to transformation, transformation plasticity and the effect of stress state on transformation kinetics. In this section, the approach to the coupling of the thermal and stress analysis will first be discussed, and then the approaches to the modelling of the interaction will be outlined. These modelling approaches are as follows:

3.1 *Coupling of analyses*

In general, there are two approaches to calculation for thermal-stress analysis, the direct and the indirect method. In the direct method, the analysis performs both thermal and stress analyses in parallel. This method is more accurate in the case of bilateral coupling. But in the indirect method, a thermal analysis is first conducted and then the nodal temperature distribution is stored in a result file, and this data is then used as a "body force" load for structural analysis, the link being the coefficient of thermal expansion.

In the case of phase transformation analysis, assuming no effect of the stress state on the phase transformation kinetics, the indirect method is more convenient and less time consuming. In the indirect method, the elements' phase transformation histories are saved as an additional result file and are used when stress analysis is performed.

3.2 *Phase transformation and the prediction of microstructure*

Tracing phase changes is the most important and time consuming part of the analysis. This job is done between time steps using the element temperature to calculate the amount of phase fraction transformed in the step by using a subroutine. This subroutine can also be used to extract the CCT (continuous cooling transformation) diagrams for different steels from the TTT diagram data. CCT diagrams allow cooling laws to be taken into account in the transformation process, and are obtained from TTT diagrams by using the isothermal transformation temperatures, and the start time and times to 10% and 90% transformation for those temperatures. In this model, the approach used by Fernandes (1984) has been adopted.

For each element, first of all, the Scheil nucleation sum is calculated by using equation 3. Different Scheil values are used for pearlite and bainite transformation as suggested by Manning (1946). If the Scheil sum reaches unity the parameters n and b from equation 4 are calculated by means of the experimental TTT diagram data (Thuvander, 1992).

$$\begin{cases} n_i = \dfrac{\ln \dfrac{\ln 0.1}{\ln 0.9}}{\ln \dfrac{t_{90\%}}{t_{10\%}}} \\ b_i = -\, t_{10\%}^{-n_i} \ln 0.9 \end{cases} \tag{11}$$

By using the previous transformed phase fraction y_{i-1} an imaginary time t^* is calculated. This time is incremented by the length of time step to give the amount of phase fraction. This value can be modified by knowledge of the available austenite $y_{\gamma i-1}$ and the temperature dependent maximum amount of the phase y_{max}(Fernandes, 1984):

$$\begin{cases} t_i^* = \left[-\dfrac{\ln(1-y_{i-1})}{b_i} \right]^{1/n_i} \\ y_i^* = 1-\exp\left[-b_i(t_i^*+\Delta t_i)^{n_i}\right] \\ y_i = y_i^*(y_{\gamma_{i-1}}+y_{i-1})y_{\max} \end{cases} \tag{12}$$

3.3 *Change in material properties*

The analytical approach is made complicated by the phase transformation behaviour of steel. As a manufactured component cools, different parts of the component cool at different rates, and thus may have different combinations of phases at any one time. The mechanical and physical properties of the material for a given temperature are therefore not constant, but dependent on the time-temperature history of the material - it is not possible to use a single set of properties since the combination of phases in the material will vary depending on the cooling rate. The analytical work in this study is being approached using a linear mixture law which has been used by other authors who have studied thermal stress behaviour. This states that a material which is undergoing structural change due to phase transformation may be assumed to consist of a mixture of N constituents (Inoue, 1984). When denoting the volume fraction of the i-th constituent as ξ_i, mechanical and physical properties x of the material may be expressed as a linear combination of the properties x_i of the constituents:

$$x = \sum_{i=1}^{N} \xi_i\, x_i \qquad \sum \xi_i = 1 \tag{13}$$

So, in order to obtain the physical and mechanical properties for a steel for any time-temperature history, the properties of the constituent phases are required over the range of temperatures under study. In our case the properties of different phases are used at the beginning of analysis as a series of temperature dependent arrays. Between time steps, the value of properties of each element are updated by running a subroutine using a linear mixture law.

3.4 *Volumetric dilatation*

There are two approaches which can be used for taking into account the effect of volumetric dilatation in a commercial finite element code. The first approach is to update the thermal expansion coefficient value for each element between steps to combine the thermal effect and transformation volume changes for one unit of temperature:

$$\alpha_{total} = \alpha_{thermal} + \alpha_{transformation} \qquad (14)$$

In general:

$$\begin{cases} \Delta\varepsilon = \alpha\,\Delta T \\ \Delta\varepsilon = \dfrac{\Delta v}{3v} \end{cases} \qquad (15)$$

so

$$\alpha_{tr} = \frac{\frac{\Delta v}{v}}{3\Delta T} \quad , \quad \frac{\Delta v}{v} = \left(\frac{\rho_A}{\rho_k} - 1\right)\Delta y_k \qquad (16)$$

where T is temperature, v is volume, $\Delta\varepsilon$ is strain increment, α_{tr} is transformation expansion coefficient, Δy_k is amount of new phase transformed in this step, ρ_A is austenite density and ρ_k is density of transformed phase. This approach is simple but due to step changes in the overall expansion coefficient, there is some noise in the stress-time results.

The second approach is to use alternative facilities which are normally provided in commercial codes for swelling, like piezoelectric enlargement or a neutron bombardment swelling effect. In this case the latter - material enlargement due to neutron bombardment or any assumed similar effect - was used. The swelling rate may be a function of temperature, time, neutron flux level, and stress. The fluence (flux × time) is input by using a body force load command. A linear stepping function is used to calculate the change in swelling strain within a load step (Ansys, Inc.1996a)

$$\Delta\varepsilon_{sw} = \frac{d\varepsilon_{sw}}{d(\phi t)}(\Delta(\phi t)) \qquad (17)$$

where ϕt is the fluence and the swelling strain rate equation is as defined in a subroutine. In this subroutine the value of fluence is updated between steps for each element by using equation 14.

$$\phi t_i = \phi t_{i-1} + \sum_{k=1}^{6} dy_k\, \varepsilon_k^{tr}(T) \qquad (18)$$

where T is temperature, and dy and ε^{tr} are the transformed fraction and transformation strain of constituent k in step i respectively.

3.5 *Heat release*

Between the time steps, the increased amount of phase fraction multiplied by the value of the temperature dependent enthalpy of that phase (equation 6) results in a heat generation rate which can be input to the code by using body force command.

3.6 *Transformation plasticity*

Two approaches were examined to incorporate TRIP within the commercial code. The first approach was used by Rammerstorfer (1981) and by Thuvander (1992). Thuvander says that on a macro level transformation plasticity can be interpreted as a drop in yield stress during transformation. In other words it consists of reducing the yield strength for the step intervals where transformation takes place. In practice, the subroutine which changes the material properties between steps, reduces the overall element yield strength by taking the yield strength of the transformed fraction to be equal to zero. This reduction causes a plastic strain which can be called transformation plasticity.

The second approach for incorporation of TRIP is to use a creep facility which is normally provided as a structural capability of commercial codes. In general, creep stain rate may be a function of stress, strain, temperature or neutron flux level etc. These relations are provided by some commercial codes as pre-defined expressions. In Ansys, other expressions may be incorporated into the program by the user. In our analysis, creep strain rate is only a function of stress state and transformed phase fraction. By a linear dependence on the volume fraction of the new phase (Denis, 1987) equation 19 can be used instead of equation 8.

$$\Delta\varepsilon_{cr} = \Delta\varepsilon_{tr} = K\,\Delta y\,\sigma_e \qquad (19)$$

Since the swelling due to neutron bombardment was already used in the model for incorporating volumetric dilatation, equation 20 is used for taking into account the effect of stress state and volume fraction (Ansys, Inc. 1996b).

$$\Delta\varepsilon_{cr} = K\phi\sigma_e \qquad (20)$$

3.7 *Effect of stress on transformation kinetics*

If no account is taken of the effect of stress on transformation kinetics, all of the coupling can be incorporated by using indirect thermal-structural analysis. But to introduce this effect, the analysis should be run using the direct method with an element suitable for coupled analysis. The problem which arises is that, at the end of each step, post-processing should be run to obtain the mean stress and equivalent stress for all of the elements. This causes a time consuming procedure for restarting the data files. The mean stress and equivalent stress are used to shift the TTT diagram in equation 8 to modify the n and b parameters. Then these two parameters can be used in same manner achieved in equations 11 and 12 for calculation of volume fraction with the effect of stress on transformation.

4 APPLICATION EXAMPLE

4.1 *Verification, pearlitic steel*

For investigation of the validity of the model, a published work (Denis, 1987) was used for verification. The material is a eutectoid carbon steel; XC80 (0.82C, 0.27Si, 0.73Mn, 0.02P, 0.007S, 0.02Cr, 0.04Ni, 0.05Al, 0.03Cu, 0.008N) and the specimen is a cylinder 13 mm in diameter in which only pearlite transformation occurs. All of the physical and mechanical properties, phase transformation kinetics data, heat transfer coefficient data, and coefficients of transformation expansion and transformation plasticity for austenite-pearlite are taken from the original work (Denis, 1987). Geometry containing 13 axisymetric elements, as shown in Figure 3, was selected for thermal and stress analysis. In the thermal analysis, a temperature dependent heat transfer coefficient was applied as material heat flux properties for element no. 13. In the stress analysis, the *y*-axis is an axis of symmetry and all of the elements' top nodes are coupled in the y direction to create a generalised plain-strain condition similar to a long bar.

For comparison between Denis's work and the model, three sets of results are presented. Figure 4 shows the stress history at the surface and centre of the bar without taking account of the effect of TRIP and stress effects on transformation kinetics. Figure 5 has taken into account TRIP, but not stress effects. Figure 6 presents the axial stresses for an analysis containing all of the interactions including the effect of stress on transformation.

4.2 *Cooling of a medium carbon, low alloy steel*

The second steel studied is a low alloy medium carbon steel EN15 (Woolman, 1964). The steel can be heat treated to produce a tensile strength of 620 to 775 MPa and in small sizes up to 930 MPa. The

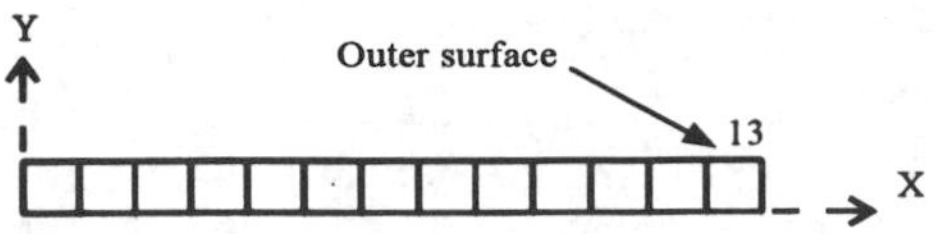

Figure 3: The geometry of the finite element model

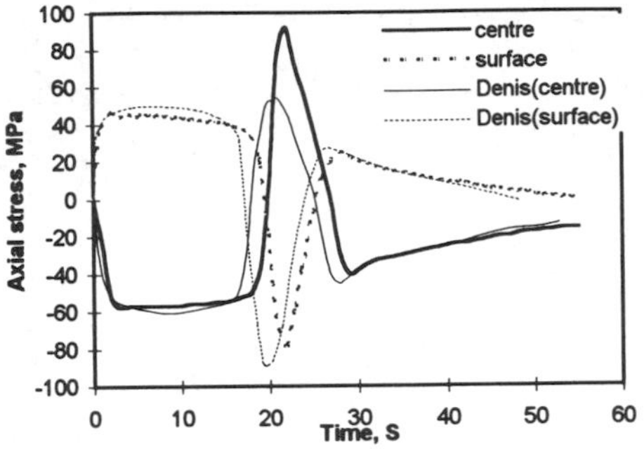

Figure 4: Axial stress vs. time with effect of volumetric dilatation due to transformation

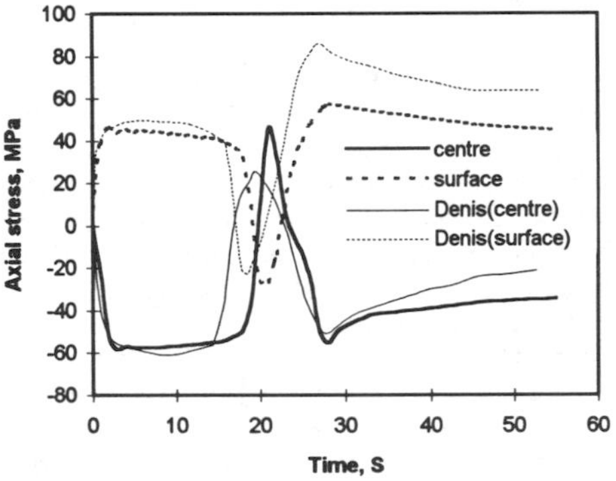

Figure 5: Axial stress vs. time with effect of transformation plasticity

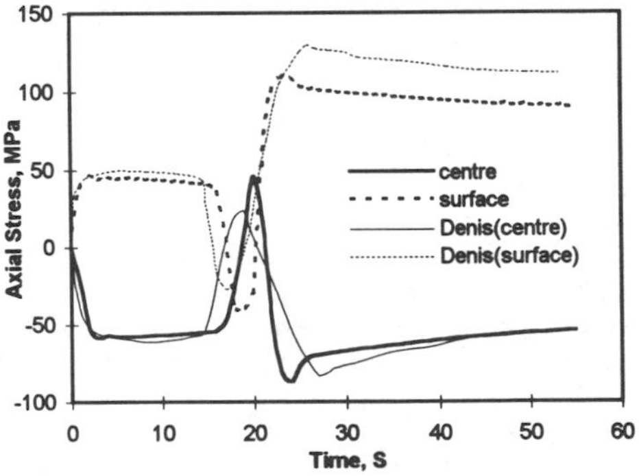

Figure 6: Axial stress vs. time with effect of stress on transformation kinetics

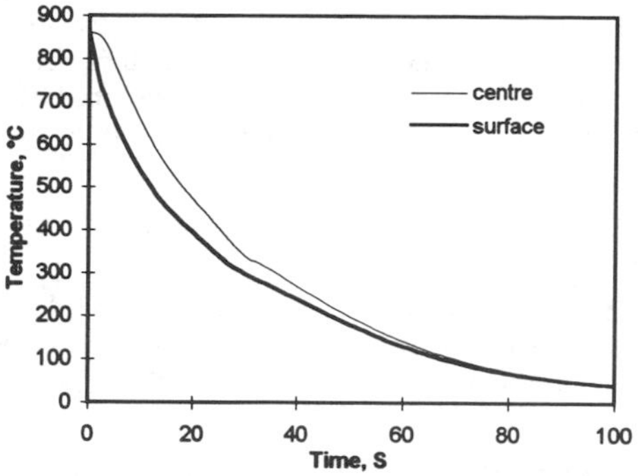

Figure 7: Cooling curves for the EN15 bar, with effect of transformation

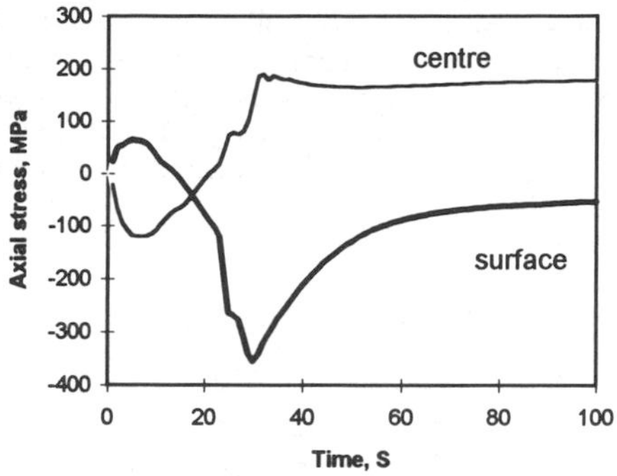

Figure 8: Axial stress vs. time without effect of transformation plasticity

material chemical composition is (0.37C, 0.2Si, 1.44Mn, 0.018P, 0.015S, 0.16Cr, 0.16Ni, 0.11Mo). In this case the model is a cylinder 20mm in diameter. This example is also treated, as explained in the previous example, with axisymetric elements (Figure 3), but in this case 10 are used. The physical and mechanical properties of the phases involved during quenching are taken from other literature (HEARTS (1994), Sjöstrom (1982), Thuvander (1992), and Woolman (1964)) by taking into account the effect of carbon content and alloying elements. Since the main purpose of this paper is to explain the analysis approach with a commercial finite element code, so a simplified heat transfer condition is applied for the residual stress data presented. The value of heat transfer coefficient used was 1000 W/m^2K (and bulk quenchant temperature 25 °C) which is used as convection condition on the bar surface, element no. 10. In the remainder of the research, variation in heat transfer coefficient around the bar and with temperature, for different quenchant conditions, is being considered. The enthalpy release for transformation from austenite to the phases is taken from Sjöstrom (1982). The temperature-dependent coefficients of volumetric dilatation are calculated from the density of phase regarding equation 16. The coefficient of transformation plasticity, K, equation 20, is taken from the similar steel, S45C (Hearts, 1994). Regarding the effect of stress on transformation kinetics, the coefficients D (equation 8) and A & B (equation 9) are taken from Denis (1987) and Denis (1985).

Figure 7 shows the cooling law at the centre and surface of the bar. Quenching starts at 860°C and, owing to the transformation heat release, the slope of the cooling curves changes to a slower cooling rate. A more considerable heat release is seen when martensite transformation takes place. Figure 8 shows the axial stresses at the surface and centre of the bar without the effect of transformation plasticity.

When cooling starts, the stresses rise due to the temperature gradient between the surface and centre, but since the thermal expansion coefficient and volumetric dilatation are two opposing driving forces the effect is reversed once transformation commences.

In Figure 8, since TRIP is not incorporated, the resultant residual stresses at room temperature are at a higher level. When the transformation plasticity effect is taken into account (Figure 9), the eventual stress level is lower, and the maximum stress state occurs at the beginning of cooling because of the thermal expansion effect. Figure 9 also shows for comparison the residual stress measured by X-ray diffraction method at the surface of an EN15 bar (2x20mm) quenched in oil.

CONCLUSION

This paper has described the incorporation of steel phase transformation behaviour into a coupled

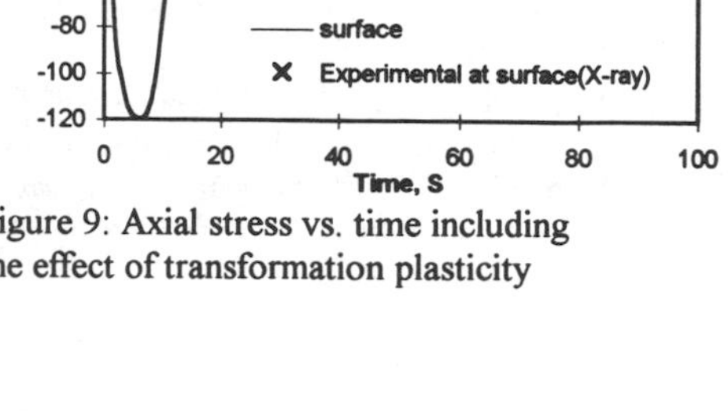

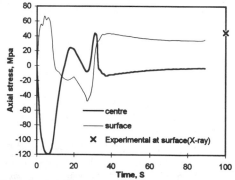

Figure 9: Axial stress vs. time including the effect of transformation plasticity

thermo-mechanical analysis of quenching. It has shown that the physical interaction that takes place in quenching may be modelled well with a commercial finite element code. This suggests that routine analysis of quenching behaviour of steel parts will be feasible in the future.

ACKNOWLEDGEMENT

The work presented in this paper is as a part of a PhD project and has been carried out at the Department of Mechanical Engineering, Bristol University, UK. It was sponsored by the Iranian Ministry of Culture and Higher Education.

REFERENCES

Ansys Inc. 1996a. *Ansys theory manual.* Urbana, USA

Ansys Inc. 1996b. *Ansys element manual.* Urbana, USA

Bourdouxhe M., Denis S., and Simon A. 1992. Computation of phase changes and deformations in long products undergoing thermal treatments. *Proc. Third Int. Conf. on Residual Stresses (ICRS 3).* Tokushima, Japan, 202-207

Denis S. Guitier E. Simon A. & Beck G. 1985. Stress phase transformation interactions, basic principles, modelization and their role in the calculation of internal stresses, *J. Material Science & Technology,* 1:805-813

Denis S. Sjöstrom S. & Simon A. 1987. Coupled temperature, stress, phase transformation calculation model: numerical illustration of internal stresses evolution during cooling of a eutectoid carbon steel cylinder. *J. Metallurgical Transactions A* 18A:1203-1212

Fernandes F.M.B. Denis S. & Simon A. 1984. Mathematical model coupling phase transformation and temperature evolution during quenching of steels. *Int Symp.Calculation of Internal Stresses in Heat Treatment,* 275-297

Fischer F.D. 1992. Transformation plasticity (TRIP) in steels, an overview. *Proc. Third Int. Conf. on Residual Stresses (ICRS 3).* Tokushima, Japan. 169-176.

HEARTS 1994. *Theoretical Manual, Heat Treatment Simulation Program "HEARTS" (version 2.1).* CRC Research Institute Inc., Japan

Inoue T. & Wang Z. 1984. Coupling phenomena between stress, temperature and metallic structures in the process with phase transformation. *Int. Symp. Calculation of Internal Stresses in Heat Treatment.* 298-310

Kamamoto S. Nishimori T. & Kinoshita S. 1985. Analysis of residual stress and distortion due to quenching in large-scale low-alloy steel shaft. *Int. Symp. Calculation of Internal Stresses in Heat Treatment.* 1:134-154

Koistinen D.P. & Marburger. 1959. *Acta Metallurgica.* 7:59-67

Manning G.K. & Lorig C.H. 1946. The relation between transformation at constant temperature and transformation during cooling. *Trans. AIME;* 167:442-466

Rammerstorfer F.G. Fischer D.F. & Mitter W. 1981. On thermo-elastic-plastic analysis of heat treatment processes including creep and phase changes. *J. Computers and Structures,* 13:771-779

Sjöstrom S. 1982. *The Calculation of Quench Stresses in Steel,* PhD thesis, Linkoping University, Sweden

Thuvander A. & Melander A. 1992. Calculation of distortion during quenching of a low alloy steel. *Material Science Forum.* 102-104:767-782

Woolman J. & Mottram A.I.M. 1964. *The mechanical and physical properties of the British EN steels,* London: Pergamon Press

Modern Practice in Stress and Vibration Analysis, Gilchrist (ed.) © *1997 Balkema, Rotterdam, ISBN 90 5410 896 7*

Nonlinear vibration of plates by the hierarchical finite element and continuation methods

P. Ribeiro & M. Petyt
Institute of Sound and Vibration Research, University of Southampton, UK

ABSTRACT: In this paper the hierarchical finite element (HFEM) and harmonic balance methods are applied to derive the equations of motion of thin, isotropic plates, in steady-state forced vibration with large amplitude displacements. Symbolic computation is used in the derivation of the model. The equations of motion are solved by the Newton and continuation methods. The stability of the obtained solutions is investigated by studying the evolution of perturbations of the solutions. The convergence properties of the HFEM, the influence of the number of degrees of freedom and of in-plane displacements are discussed. The HFEM results are compared with published experimental and numerical results and good agreement is found.

1 INTRODUCTION

As the excitations to which modern structures are subjected are increasing and their weight decreasing, large vibration amplitudes are becoming more frequent. Particularly in the aircraft industry, the increase of performance capabilities resulted in greater acoustic excitation levels. Moreover, designs in which the exhaust gas directly impinges on the structure cause increased amplitudes of excitation. As a result, large amplitude, geometrically non-linear vibration of the aircraft skin-panels, and consequently reduced fatigue life, may occur.

The effect of geometric nonlinearity on plates with fixed ends is of the hardening spring type, i.e., the "resonance frequency" increases with the amplitude of vibration. The "mode shape" changes during the period of vibration and is amplitude dependent (Benamar, 1990; Han and Petyt, 1996).

Especially in nonlinear analysis, the number of degrees of freedom (d.o.f.) has a substantial influence on the time needed to derive and solve the model. In an effort to reduce the number of d.o.f., Han and Petyt (1997) applied the hierarchical finite element method (HFEM) to study the free vibration of plates with geometrical non-linearity.

In the HFEM, convergence tends to be achieved with fewer d.o.f. than in the *h*-version of the finite element method. Better approximations are accomplished by adding higher order shape functions to the existing model, without redefining the mesh. The linear matrices possess the embedding property, i.e., the associated element matrices for a number of shape functions $p=p_1$ are always submatrices for $p=p_2$, $p_2 \geq p_1$. The existing nonlinear matrices of an approximation of lower order, p_1, can be used in the derivation of the nonlinear matrices of the improved approximation, p_2.

In nonlinear vibrations, the frequency response curve (FRF) can have multi-valued regions, turning and bifurcation points. Lewandowski (1991) successfully applied a continuation method to describe the FRF curve in the case of non-linear vibration of beams.

In this paper the HFEM method is used to construct the model of thin, rectangular, isotropic, fully clamped plates. The derived equations of motion are solved by a continuation method. Free and forced vibration are analysed.

2 MATHEMATICAL MODEL

The mathematical model used is the one developed by Han and Petyt (1997). For each element, the middle plane displacements (Figure 1) are expressed in the form:

$$\begin{Bmatrix} u_0 \\ v_0 \\ w_0 \end{Bmatrix} = [N] \begin{Bmatrix} q_p \\ q_w \end{Bmatrix} \tag{1}$$

$$[N] = \begin{bmatrix} \lfloor N^u \rfloor & 0 & 0 \\ 0 & \lfloor N^u \rfloor & 0 \\ 0 & 0 & \lfloor N^w \rfloor \end{bmatrix} \tag{2}$$

$$\lfloor N^u \rfloor = \lfloor g_1(\xi) g_1(\eta) \; g_1(\xi) g_2(\eta) \cdots g_{p_i}(\xi) g_{p_i}(\eta) \rfloor \tag{3}$$

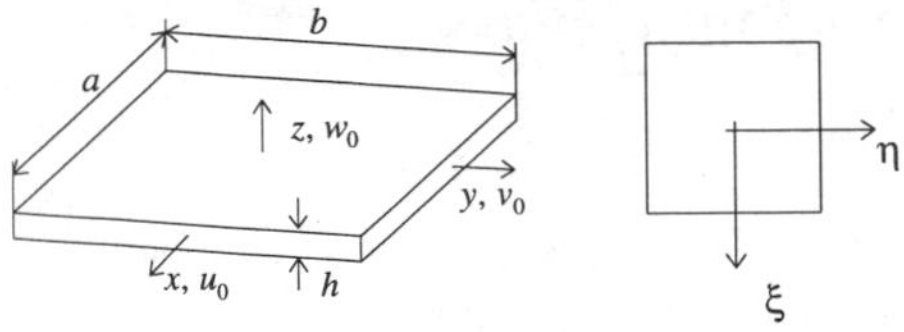

a b

Figure 1 - a) Rectangular plate: x, y and z - global coordinate system; u_0, v_0 and w_0 - middle-plane displacements; a, b and h - plate dimensions. b) ξ, η - local coordinate system.

$$\lfloor N^w \rfloor = \lfloor f_1(\xi)f_1(\eta)\ f_1(\xi)f_2(\eta) \cdots f_{p_i}(\xi)f_{p_i}(\eta) \rfloor \quad (4)$$

Where p_o and p_i are the number of out-of-plane and of in-plane shape functions used in the model; $\{g\}$ and $\{f\}$ are the vectors of in- and out-of-plane shape functions; q_p and q_w are the generalised in- and out-of-plane displacements and $[N]$ the matrix of shape functions. The set of shape functions used is the Rodrigues' form of Legendre polynomials (Han, 1993).

Since only one element is going to be used to model the whole plate,

$$\xi = 2x/a, \quad \eta = 2y/b \quad (5)$$

The equations of motion, without damping, are derived by equating the sum of the virtual work of the inertia forces and of the elastic restoring forces to the virtual work of the external forces. Considering only transverse applied external forces, using von Kármán nonlinear strain displacement relationships and neglecting shear deformation and rotatory inertia, one obtains

$$\int_\Omega \left(\{\delta\varepsilon_o^p\}^T + \{\delta\varepsilon_L^p\}^T\right)[A]\left(\{\varepsilon_o^p\} + \{\varepsilon_L^p\}\right) d\Omega$$

$$\int_\Omega \{\delta\varepsilon_o^b\}^T [D]\{\varepsilon_o^b\} d\Omega + \int_\Omega \rho h(\delta u_0 \ddot{u}_0 + \delta v_0 \ddot{v}_0 + \delta w_0 \ddot{w}_0)\, d\Omega$$

$$= \lfloor \delta u_0 \quad \delta v_0 \quad \delta w_0 \rfloor \iint_\Omega [N]^T \begin{Bmatrix} 0 \\ 0 \\ \overline{P}_d(x,y,t) \end{Bmatrix} d\Omega, \quad (6)$$

$$[A] = \frac{Eh}{(1-\nu^2)} \begin{bmatrix} 1 & \nu & 0 \\ \nu & 1 & 0 \\ 0 & 0 & \frac{1}{2}(1-\nu) \end{bmatrix}, \quad (7)$$

$$[D] = h^2[A] \quad (8)$$

Where $\{\varepsilon_o^p\}$ and $\{\varepsilon_o^b\}$ are the linear membrane and bending strains; $\{\varepsilon_L^p\}$ is the geometrically nonlinear membrane strain; E, ν, ρ denote Young's modulus, Poisson's ratio and density; $\overline{P}_d(x,y,t)$ is the distributed applied force (N/m^2).

The strain-displacement relationships are

$$\{\varepsilon_o^p\} = \begin{bmatrix} u_{0,x} \\ v_{0,y} \\ u_{0,y} + v_{0,x} \end{bmatrix}, \{\varepsilon_L^p\} = \begin{bmatrix} (w_{0,x})^2/2 \\ (w_{0,y})^2/2 \\ w_{0,x}w_{0,y} \end{bmatrix}, \{\varepsilon_o^b\} = \begin{bmatrix} -w_{0,xx} \\ -w_{0,yy} \\ -2w_{0,xy} \end{bmatrix} \quad (9)$$

Where $,x$ denotes differentiation with respect to x.

Substituting equations (9) into equation (6) and allowing the virtual generalised displacements to be arbitrary gives:

$$\begin{bmatrix} M_p & 0 \\ 0 & M_b \end{bmatrix} \begin{Bmatrix} \ddot{q}_p \\ \ddot{q}_w \end{Bmatrix} + \left(\begin{bmatrix} K1_p & 0 \\ 0 & K1_b \end{bmatrix} + \begin{bmatrix} 0 & K2 \\ 0 & 0 \end{bmatrix} + \begin{bmatrix} 0 & 0 \\ K3 & 0 \end{bmatrix} + \begin{bmatrix} 0 & 0 \\ 0 & K4 \end{bmatrix} \right) \begin{Bmatrix} q_p \\ q_w \end{Bmatrix} = \begin{Bmatrix} 0 \\ \overline{P} \end{Bmatrix} \quad (10)$$

$[M_p]$ and $[M_b]$ are the in-plane and bending inertia matrices; $[K1_p]$ and $[K1_b]$ the in-plane and bending linear stiffness matrices; $[K2]$, $[K3]$ and $[K4]$ the nonlinear stiffness matrices) and $\{\overline{P}\}$ is the vector of generalised external forces.

With the introduction of mass proportional hysteretic damping - which depends on the damping factors β_p, β - these equations become

$$\begin{bmatrix} M_p & 0 \\ 0 & M_b \end{bmatrix} \begin{Bmatrix} \ddot{q}_p \\ \ddot{q}_w \end{Bmatrix} + \begin{bmatrix} \frac{\beta_p}{\omega} M_p & 0 \\ 0 & \frac{\beta}{\omega} M_b \end{bmatrix} \begin{Bmatrix} \dot{q}_p \\ \dot{q}_w \end{Bmatrix} + \left(\begin{bmatrix} K1_p & 0 \\ 0 & K1_b \end{bmatrix} + \begin{bmatrix} 0 & K2 \\ K3 & K4 \end{bmatrix} \right) \begin{Bmatrix} q_p \\ q_w \end{Bmatrix} = \begin{Bmatrix} 0 \\ \overline{F} \end{Bmatrix} \quad (11)$$

All integrals involved in calculating the inertia and stiffness matrices in equation (11) were evaluated using symbolic computation (Redfern, 1994). All submatrices are symmetric except $[K2]$ and $[K3]$, which are related by $[K3] = 2[K2]^T$.

Neglecting in-plane inertia and damping the following equations result from equation (11)

$$[M_b]\{\ddot{q}_w\} + \frac{\beta}{\omega}[M_b]\{\dot{q}_w\} + [K1_b]\{q_w\} + [Knl]\{q_w\} = \{\overline{P}\}, \quad (12)$$

where

$$[Knl] = [K4] - 2[K2]^T[K1_p]^{-1}[K2].$$

If the external excitation is harmonic, $\{\overline{P}\}=\{P\}$ cos(ωt), the steady state response $\{q_w(t)\}$ may be expressed in a first approximation, as:

$$\{q_w(t)\}=\{w_c\}\cos(\omega t)+\{w_s\}\sin(\omega t) \tag{13}$$

This equation is inserted into the equations of motion (12) and the harmonic balance method (HBM) is applied. The equations of motion obtained are of the following form:

$$\{F\}=\left(-\omega^2\begin{bmatrix}M_b & 0\\ 0 & M_b\end{bmatrix}+\begin{bmatrix}0 & \beta M_b\\ -\beta M_b & 0\end{bmatrix}\right.$$
$$\left.+\begin{bmatrix}K1_b & 0\\ 0 & K1_b\end{bmatrix}\right)\begin{Bmatrix}w_c\\ w_s\end{Bmatrix}+\begin{Bmatrix}F_1\\ F_2\end{Bmatrix}-\{P\}=\{0\}, \tag{14}$$

where

$$\{F_1\}=\frac{2}{T}\int_0^T[Knl]\{q_w\}\cos(\omega t)\mathrm{d}t$$
$$=\left(\frac{3}{4}[KNL1]+\frac{1}{4}[KNL3]\right)\{w_c\}+\frac{1}{4}[KNL2]\{w_s\}, \tag{15}$$

$$\{F_2\}=\frac{2}{T}\int_0^T[Knl]\{q_w\}\sin(\omega t)\mathrm{d}t$$
$$=\frac{1}{4}[KNL2]\{w_c\}+\left(\frac{1}{4}[KNL1]+\frac{3}{4}[KNL3]\right)\{w_s\}, \tag{16}$$

$[KNL1]$ is a function of $\{w_c\}$ only, $[KNL2]$ is a function of both $\{w_c\}$ and $\{w_s\}$ and $[KNL3]$ is a function of $\{w_s\}$ only[1]. These three matrices are, as well as $[M_b]$ and $[K1_b]$, symmetric. The vector of generalised displacements is defined by

$$\{w\}=\begin{Bmatrix}w_c\\ w_s\end{Bmatrix} \tag{17}$$

The total number of degrees of freedom of the model is $n=2p_o^2$, for a damped model, and $n=p_o^2$, for an undamped model.

3 THE CONTINUATION METHOD

To solve the system of equations (14), Newton's method is used in the nonresonant region. For each frequency, the first approximation of $\{w\}$ is the $\{w\}$ from last point of the FRF curve. By solving the system of equations

$$[J]\{\delta w\}=-\{F\}, \tag{18}$$

$\{\delta w\}$ is obtained and $\{w\}$ is corrected.

$[J]$ is the Jacobian of $\{F\}$ defined by:

$$[J]=\partial\{F\}/\partial\{w\}. \tag{19}$$

The process is repeated until convergence is achieved. The frequency is then changed to another fixed value and the same method applied.

In the vicinity of resonance frequencies there are multiple solutions, which are difficult to obtain with the Newton method alone. In these regions a continuation method is utilised. This continuation method was first employed to study the vibration of beams by Lewandowski (1991).

The continuation method is composed of two main loops. In the external loop a predictor to the solution is defined. For that, the two last determined points of the backbone curve - ($\{w\}_i$, ω_i^2) and ($\{w\}_{i-1}$, ω_{i-1}^2) - are used. The prediction of $\{w\}_{i+1}$ is thus obtained in the following way:

$$\{w\}_{i+1}=\{w\}_i+\Delta\{w\}_{i+1}$$
$$\Delta\{w\}_{i+1}=\left(\{w\}_i-\{w\}_{i-1}\right)\frac{dwaux}{wm} \tag{20}$$

dwaux is the amplitude of the first increment vector, $\Delta\{w\}_{i+1}$ and *wm* is the amplitude of the vector $\left(\{w\}_i-\{w\}_{i-1}\right)$. A prediction for ω_{i+1}^2 must also be calculated. This results from the equation

$$\omega_{i+1}^2=\omega_{i+1}^2+\Delta\omega_0^2 \tag{21}$$

$$\Delta\omega_0^2=\pm s/\left(\{\delta w\}_1^T\{\delta w\}_1\right)^{1/2} \tag{22}$$

s and $\{\delta w\}_1$ will be defined afterwards. The sign in equation (22) is chosen following that of the previous increment, unless the determinant of $[J]$ has changed sign. In the last case a sign reversal is applied. To calculate $\{\delta w\}_1$ in (22), the last known frequency of the FRF curve is used.

Now, the approximated solution must be corrected. This correction is carried out in an internal loop. Applying Newton method to equation (14):

$$[J]\{\delta w\}-[M]\{w\}_{i+1}\delta\omega^2=-\{F\} \tag{23}$$

Note that, unlike equation (18), this time variations in the frequency are also considered. The fact that both the generalised displacements and the frequency are unknowns, allows one to pass the turning points of the backbone curve. However,

[1] With this formulation, $[KNL2]$ must be calculated using $2\lfloor N_{,x}^w\rfloor\{w_c\}\lfloor N_{,x}^w\rfloor\{w_s\}$.

there is one extra unknown: the frequency of vibration. Consequently, another equation is needed. This is obtained by constraining the distance between the two successive points of the FRF curve, the arc-length s, to a fixed value, by the following constraint equation

$$s^2 = \left\|\Delta\{w\}_{i+1}\right\|^2 \tag{24}$$

From equation (23), one has

$$\{\delta w\}=\delta\omega^2\{\delta w\}_1+\{\delta w\}_2 \tag{25}$$

$\{\delta w\}_1$ and $\{\delta w\}_2$ result from the equations

$$[J]\{\delta w\}_1 = [M]\{w\}_{i+1} \tag{26}$$

$$[J]\{\delta w\}_2 = -\{F\} \tag{27}$$

Then the corrected value of $\{w\}$ will be

$$\{w\}_{i+1} = \left(\{w\}_{i+1}\right)_{previous} + \Delta\{w\}_{i+1} \tag{28}$$

with

$$\Delta\{w\}_{i+1} = \left(\Delta\{w\}_{i+1}\right)_{previous} + \{\delta w\} \tag{29}$$

Substituting $\Delta\{w\}_{i+1}$ from equation (29) into the constraint equation (24) gives the relation for $\delta\omega^2$

$$a_1\left(\delta\omega^2\right)^2 + a_2\,\delta\omega^2 + a_3 = 0, \tag{30}$$

where

$$a_1 = \{\delta w\}_1^T\{\delta w\}_1,\ a_2 = 2\left(\Delta\{w\}_i + \{\delta w\}_2\right)^T\{\delta w\}_1,$$

$$a_3 = \left(\Delta\{w\}_i + \{\delta w\}_2\right)^T\left(\Delta\{w\}_i + \{\delta w\}_2\right) - s^2 \tag{31}$$

Equation (30) has two solutions. To avoid a return to the known part of the curve, the angle between the incremental amplitude vector of the previous iteration and the one of the present iteration should be positive. If both angles are positive the appropriate root is the one that is closer to the linear solution of equation (30).

The corrected value of the natural frequency is given by:

$$\Delta\omega^2_{i+1} = \Delta\omega^2_i + \delta\omega^2 \tag{32}$$

$$\omega^2_{i+1} = \omega^2_m + \Delta\omega^2_{i+1} \tag{33}$$

The iterations are repeated until the inequalities

$$\left|\left(\omega^2_{i+1} - \omega^2_i\right)/\omega^2_{i+1}\right| \langle error1 \tag{34}$$

$$\left\|\{w\}_{i+1} - \left(\{w\}_{i+1}\right)_{previous}\right\| / \left\|\{w\}_{i+1}\right\| \langle error2 \tag{35}$$

$$\left\|\{F\}\right\| < error3 \tag{36}$$

are satisfied. If the roots of equation (30) are complex, if too many iterations are necessary in order to achieve convergence or if $\left(\omega_{i+1} - \omega_i\right)$ is greater than the value desired by the user, then the arc-length is reduced and the process restarted from the previous know point of the backbone curve.

4 STABILITY OF THE SOLUTIONS

To investigate the local stability of the harmonic solution a small disturbance is added to the steady state solution

$$\{\tilde{q}\} = \{q_w\} + \{\delta q_w\} \tag{37}$$

and its evolution is studied. If $\{\delta q_w\}$ dies out with time then $\{q_w\}$ is stable, if it grows then $\{q_w\}$ is unstable.

Inserting the disturbed solution (37) into equation (12), expanding the nonlinear terms by means of Taylor series around $\{q_w\}$ and ignoring terms of order higher than $\{\delta q_w\}$, the variational equations (38) are obtained:

$$\begin{aligned}&[M_b]\{\delta\ddot{q}_w\} + \frac{\beta}{\omega}[M_b]\{\delta\dot{q}_w\} + [K1_b]\{\delta q_w\}\\ &+\frac{\partial\left([Knl]\{q_w\}\right)}{\partial\{q_w\}}\{\delta q_w\} = \{0\}.\end{aligned} \tag{38}$$

The coefficients $\partial\left([Knl]\{q_w\}\right)/\partial\{q_w\}$ are periodic functions of time. With symbolic manipulation, they can easily be expanded in a Fourier series. If $\{q_w\}$ is of the form (13), then:

$$\frac{\partial\left([Knl]\{q_w\}\right)}{\partial\{q_w\}} = [[p_1] + [p_2]\cos(2\omega t) + [p_3]\sin(2\omega t)], \tag{39}$$

$$[p_1] = \frac{1}{T}\int_0^T \frac{\partial}{\partial\{q_w\}}\left([Knl]\{q_w\}\right)\,\mathrm{d}t, \tag{40}$$

$$[p_2] = \frac{2}{T}\int_0^T \frac{\partial}{\partial\{q_w\}}\left([Knl]\{q_w\}\right)\cos(2\omega t)\mathrm{d}t, \tag{41}$$

$$[p_3] = \frac{2}{T}\int_0^T \frac{\partial}{\partial\{q_w\}}([Knl]\{q_w\})\ \sin(2\omega t)dt. \qquad (42)$$

Multiplying equations (38) by the transpose of the modal matrix $[B]$ and using modal coordinates $\{\xi\}$, one arrives at:

$$\{\delta\ddot{\xi}\} + \frac{\beta}{\omega}[I]\{\delta\dot{\xi}\} + [\omega_j^2]\{\delta\xi\} + [B]^T([p_1] + [p_2]$$
$$\cos(2\omega t) + [p_3]\sin(2\omega t))\,[B]\{\delta\xi\} = \{0\} \qquad (43)$$

where $[\omega_j^2]$ is the diagonal matrix of linear natural frequencies

This is a system of extended coupled Hill's equations. The first derivative term can be eliminated by introducing a new vector of variables

$$\{\delta\xi\} = e^{-\frac{1}{2}\frac{\beta}{\omega}[I]t}\{\delta\bar{\xi}\}, \qquad (44)$$

obtaining, because matrix $e^{-\frac{1}{2}\frac{\beta}{\omega}[I]t}$ commutes with any other matrix and is non-singular,

$$\{\delta\ddot{\bar{\xi}}\} + \left([\omega_j^2] - \frac{1}{4}\left(\frac{\beta}{\omega}\right)^2[I] + [B]^T\ ([p_1] + [p_2]\right.$$
$$\left.\cos(2\omega t) + [p_3]\sin(2\omega t))[B]\right)\{\delta\bar{\xi}\} = \{0\} \qquad (45)$$

where $[I]$ is the identity matrix.

Now, following Hayashi (1964), the solution of (45) will be expressed in the form:

$$\{\delta\bar{\xi}\} = e^{\lambda t}\ (\{b_1\}\cos(\omega t) + \{a_1\}\sin(\omega t)) \qquad (46)$$

which should allow one to determine, in a first approximation, the first order simple unstable region.

Inserting (46) into (45) and applying the HBM results in

$$(\lambda^2[I] + \lambda[M_1] + [M_0])\begin{Bmatrix} b_1 \\ a_1 \end{Bmatrix} = \begin{Bmatrix} 0 \\ 0 \end{Bmatrix} \qquad (47)$$

where

$$[M_1] = \begin{bmatrix} 0 & 2\omega[I] \\ -2\omega[I] & 0 \end{bmatrix} \qquad (48)$$

$$[M_0] = \begin{bmatrix} [B]^T[J_{11}][B] - \left(\omega^2 + \left(\frac{1}{2}\frac{\beta}{\omega}\right)^2\right)[I] + [\omega_{0j}{}^2] & [B]^T[J_{12}][B] \\ [B]^T[J_{21}][B] & [B]^T[J_{22}][B] - \left(\omega^2 + \left(\frac{1}{2}\frac{\beta}{\omega}\right)^2\right)[I] + [\omega_{0j}{}^2] \end{bmatrix} \qquad (49)$$

$$[J_{11}] = \frac{\partial\{F_1\}}{\partial\{w_c\}}, \quad [J_{12}] = \frac{\partial\{F_1\}}{\partial\{w_s\}},$$
$$[J_{21}] = \frac{\partial\{F_2\}}{\partial\{w_c\}}, \quad [J_{22}] = \frac{\partial\{F_2\}}{\partial\{w_s\}}. \qquad (50)$$

To determine the characteristic exponents, λ, equations (47) are transformed into (Takahashi, 1979)

$$\begin{bmatrix} 0 & [I] \\ -[M_0] & -[M_1] \end{bmatrix}\begin{Bmatrix} X \\ \Gamma \end{Bmatrix} = \lambda\begin{Bmatrix} X \\ \Gamma \end{Bmatrix}, \qquad (51)$$

where $\{X\}$ is a vector formed by $\{b_1\}$, $\{a_1\}$. The values of λ are the eigenvalues of the double size matrix in the previous equation. If the real part of $\lambda_r - \beta/(2\omega)$ is positive for any λ_r, then the solution is unstable, otherwise it is stable.

For undamped systems, Lewandowski (1991) demonstrated that the sign of the determinant of the Jacobian of $\{F\}$ provides important conclusions about the stability of the solution. This demonstration is easily extended to systems with mass proportional damping as follows.

Matrices $[I]$ and $[M_0]$ are symmetric and matrix $[M_1]$ is skew-symmetric. Consequently, the eigenvalues of equation (47) are either purely imaginary or purely real (Lewandowski, 1991). If λ is imaginary the solution is always stable; if λ is real the stability limit is defined by

$$\lambda = \beta/(2\omega). \qquad (52)$$

Inserting (52) in (47) one arrives at

$$[B]^T[J][B]\begin{Bmatrix} b_1 \\ a_1 \end{Bmatrix} = \begin{Bmatrix} 0 \\ 0 \end{Bmatrix}, \qquad (53)$$

A non-trivial solution of (53) exists if

$$\det([B]^T[J][B]) = 0 \Leftrightarrow |B|^2|J| = 0 \Leftrightarrow |J| = 0. \qquad (54)$$

Thus, in the stability limit, the determinant of the Jacobian of $\{F\}$, $|J|$, is zero.

$|J|$ is a polynomial in $\{w_c\}$, $\{w_s\}$ and ω; therefore, it is a continuous function in those coefficients. All the experimental and numerical analysis of nonlinear vibration of plates, indicate that the shape of vibration, defined in this model by $\{w_c\}$ and $\{w_s\}$, is a continuous function of the amplitude and the frequency of vibration. Thus, $|J|$ varies in a continuous way through the FRF curve. If there is a change in its sign between two consecutive points of the FRF curve, then $|J|=0$ for a particular point between these two. In that particular point, the stability limit might have been crossed.

So, a complete study of the first order solution's stability is carried out in the following way:

1 - determination of the stability of the first solution by finding the characteristic exponents. For low amplitudes of vibration, in the nonresonant area, this is not necessary. In fact, in these conditions the solution is always stable, as can be demonstrated by a perturbation method (Szemplińska, 1990).

2 - calculation of $|J|$, which is needed in the continuation method and, when the Newton method is applied, can be easily calculated from $[J]$. If $|J|$ changes sign or if $|J|$ is approximatelly zero then calculate the characteristic exponents to verify if the stability of the solution changed.

5 APPLICATIONS

5.1 *Plates analysed*

The described methods were applied to two different plates, with all edges immovable and clamped. Both plates were made on steel, with the properties: $E = 21.0\times10^{10}$ N/ m^2, $\nu = 0.3$, $\rho = 7800$ Kg/ m^3.Their geometric properties are defined in Table 1.

Table 1 - Geometric properties of plates 1 and 2

Plate	a (mm)	b (mm)	h (mm)
1	486	322.9	1.2
2	500	500	2.0833

The external applied force is a harmonic plane wave at normal incidence.

5.2 *Convergence properties of HFEM*

Convergence studies for free and forced vibration were carried out using Plate 1. Since the plate, the external excitation and the boundary conditions are symmetric with respect to both axes x and y, only modes for which the transverse displacement is symmetric with respect to x and y are excited. Thus only symmetric out-of-plane shape functions need to be included in the model. However, both symmetric and antisymmetric in-plane shape functions must be used. This is so, because the in-plane displacements are anti-symmetric with respect to one axis but they are symmetric with respect to the other axis: $u(x,y)=u(x,-y)=-u(-x,y)$ and $v(x,y)=v(-x,y)=-v(x,-y)$. The amplitude of the external applied force was 10 N/m^2. In all the figures in this paper the amplitudes of vibration were calculated for $(x, y) = (0, 0)$.

In Figures 2, 3 and 4 the backbone curves and the frequency response function (FRF) curves are displayed for different values of p_o and for $p_i = 6$.

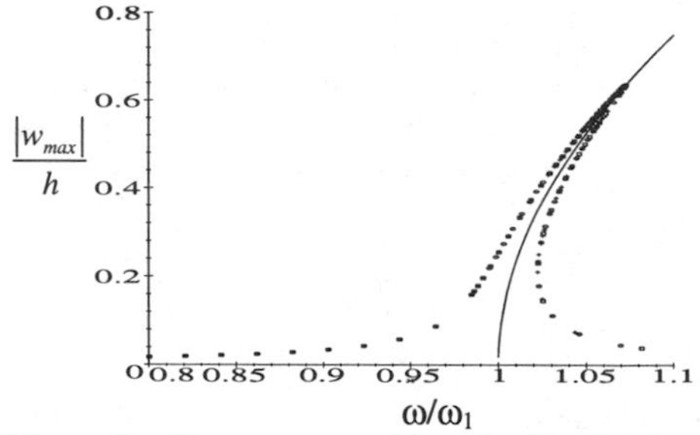

Figure 2 - Convergence with p_o in the vicinity of the first mode. FRFs:□ $p_o = 2$, o $p_o = 3$, + $p_o = 4$; Backbone curves: —. ω_1 - first linear frequency.

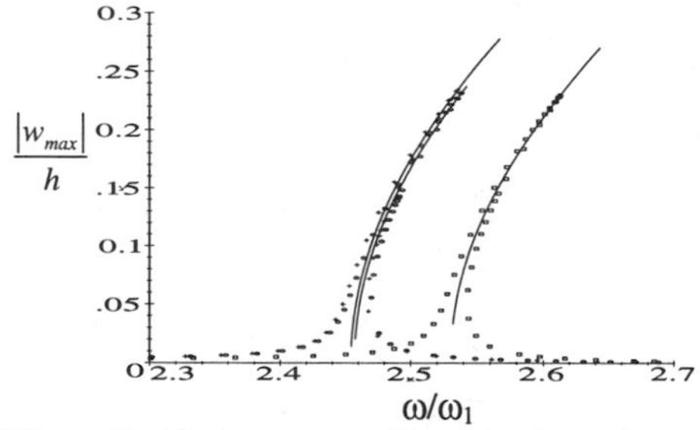

Figure 3 - Convergence with p_o in the vicinity of the second mode. Legend as in Figure 2.

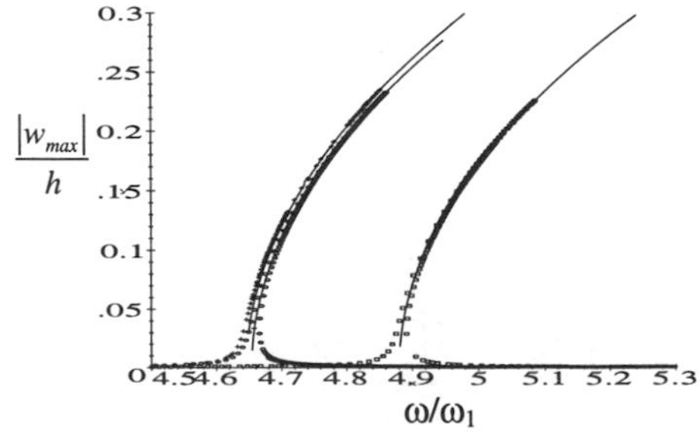

Figure 4 - Convergence with p_o in the vicinity of the third mode. Legend as in Figure 2.

The results obtained with p_o = 2 are accurate around the first mode, for the second and third modes three out-of-plane shape functions provide a quite reasonable approximation to the solution.

In Figures 5, 6 and 7 the backbone curves and the FRF curves for forced vibration are displayed for different values of p_i, in all the cases $p_o = 3$.

From figures 5, 6 and 7 it is concluded that, for the amplitudes considered, the results obtained with p_i = 4 are quite accurate. It is also shown that the exclusion of the in-plane displacements (p_i = 0) increases the stiffness of the model. The influence of the in-plane displacements is particularly visible around the first mode, probably due to the larger amplitudes of vibration attained at the point (0, 0) with this mode.

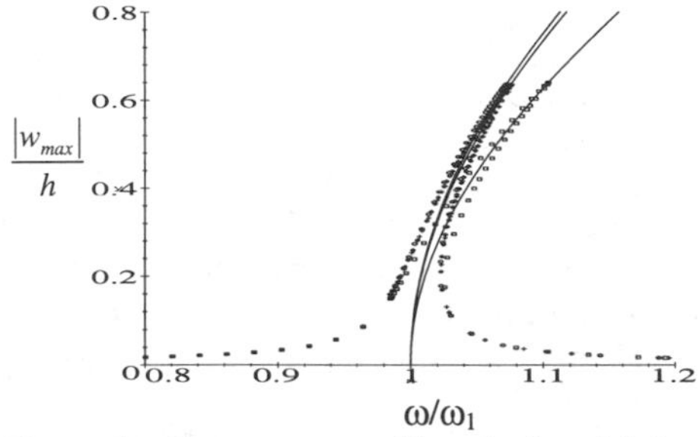

Figure 5 - Convergence with p_i in the vicinity of the first mode. FRFs:□ $p_i = 0$, + $p_i = 4$, ◊ $p_i = 5$, o p_i=6; Backbone curves: —.

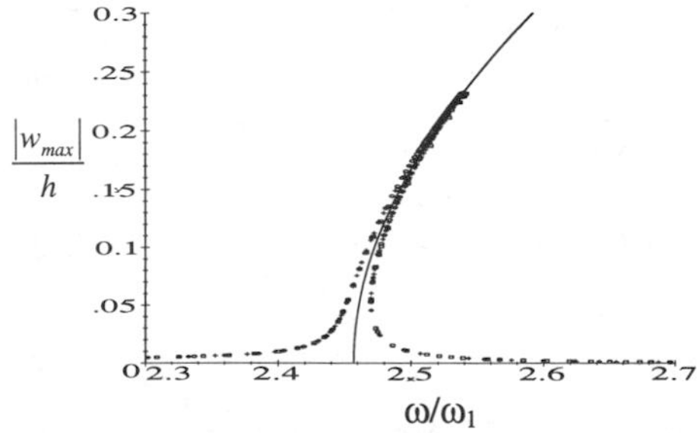

Figure 6 - Convergence with p_i in the vicinity of the second mode. FRFs:□ p_i = 0, + p_i = 3, ◊ p_i = 4, o p_i=5; Backbone curves: —.

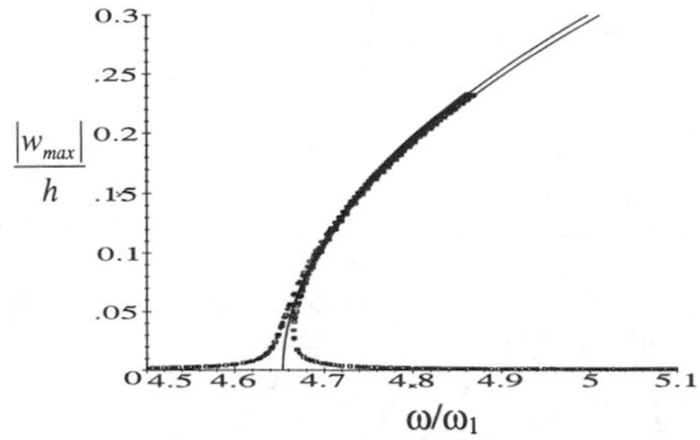

Figure 7 - Convergence with p_i in the vicinity of the third mode. Legend as in Figure 6.

5.3 *Comparison with experimental and theoretical results*

In order to validate the model, the results using p_o = 3 and p_i = 6 were compared with experimental and other theoretical results.

For plate 2, comparison is made in Tables 2 and 3 between the HFEM free and forced vibration results and results from the literature. In figure 8 the first resonance frequencies obtained with the HFEM, for plate 1, are compared with the experimental ones.

Table 2 - Comparison of frequency ratios ω/ω_1 of immovable fully clamped square isotropic plates.

$\frac{w_{max}}{h}$	Han & Petyt (1997) p_o=p_i=7	Rao et al (1993)	HFEM $\frac{w_{max}}{h}$	HFEM ω/ω_1
0.2	1.0068	1.0095	0.2099	1.0079
0.6	1.0600	1.0825	0.6007	1.0632
1	1.1599	1.2149	1.0011	1.1670

Table 3 - Frequency ratio ω/ω_1 of immovable fully clamped isotropic square plates under uniform harmonic distributed force P_0=0.2*.

$\frac{w_{max}}{h}$	Elliptic function solution*	Perturbation solution*	Finite element*	HFEM $\frac{w_{max}}{h}$	HFEM ω/ω_1
±0.6	0.8951	0.8956	0.8905	+.5992	0.8962
	1.2117	1.2119	1.2083	-.5997	1.2112
±1	1.0822	1.0845	1.0700	+1.000	1.0800
	1.2540	1.255	1.2429	-1.001	1.2490

* From Mei & Decha-Umphai, 1985. $P_0 = cF_0/\rho h^2\omega_1^2$ $c=\iint \phi dxdy / \iint \phi^2 dxdy$, ϕ - normalised mode shape. F_0 - amplitude of external applied force (N/m²).

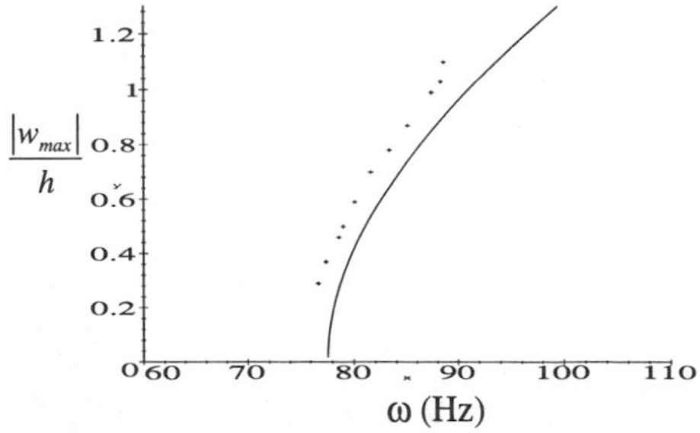

Figure 8 - Comparison between the first resonance frequency predicted by the HFEM, —, and the measured one, + (Benamar, 1990).

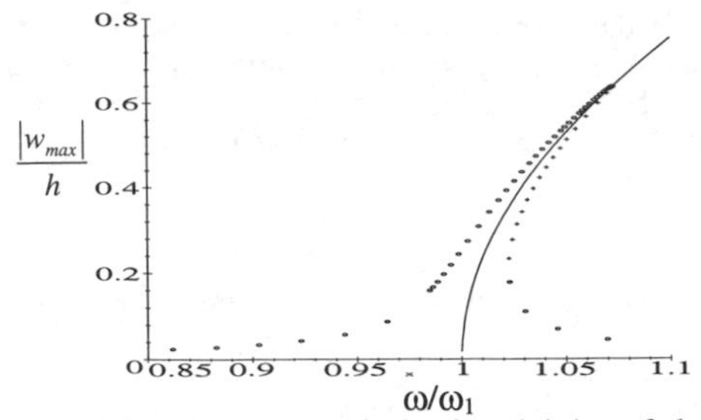

Figure 9 - Stability study in the vicinity of the first mode. o stable solutions, + unstable solutions.

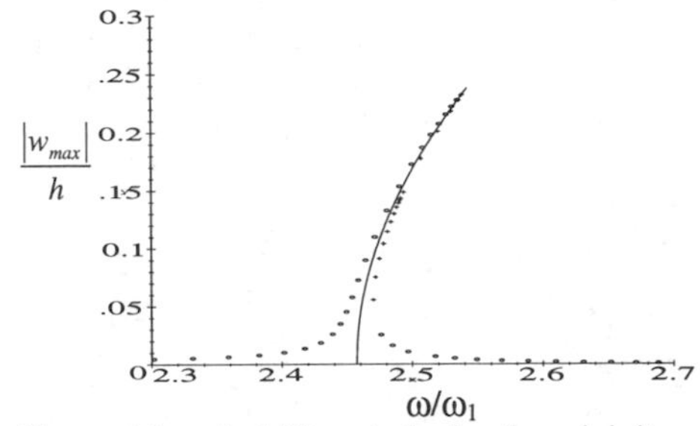

Figure 10 - Stability study in the vicinity of the second mode. Legend as in Figure 9.

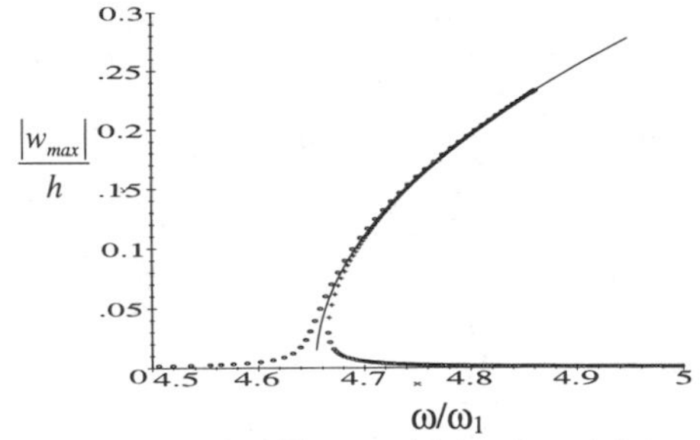

Figure 11 - Stability study in the vicinity of the second mode. Legend as in Figure 9.

5.4 *Study of Stability*

The method described in section 4 was applied to study the stability of the solutions. The results obtained are shown in the figures 9-11.

6 CONCLUSIONS

Using the HFEM, with a small number of d.o.f. the FRF curves were determined up to the 3rd order mode. By comparison with experimental and numerical results, the model was validated.

With the continuation method the FRF curves were completely and automatically described, inclusive of multi-valued regions.

For mass proportionally damped systems it was proved that the determinant of the Jacobian of $\{F\}$, $|J|$, is zero in the stability limit. Thus, one only has to determine the characteristic exponents of the first solution and after an indication that $|J|$ was zero. The eigenvalue problem which defines the characteristic exponents, was quickly solved due to the reduced number of degrees of freedom of the HFEM model.

REFERENCES

Benamar, R 1990. *Nonlinear dynamic behaviour of fully clamped beams and rectangular isotropic and laminated plates*. Ph.D. Thesis. University of Southampton. UK.

Han, W. 1993. *The Analysis of isotropic and laminated rectangular plates including geometrical non-linearity using the p-version finite element method*. Ph.D. Thesis. University of Southampton. UK.

Han, W. & Petyt, M. 1997. Geometrically nonlinear vibration analysis of thin, rectangular plates using the hierarchical finite element method. *Computers and Struct.* 63 (2): 295-308.

Hayashi, C 1964. *Nonlinear Oscillations in Physical Systems*. New York: McGraw-Hill.

Lewandowski, R. 1991. Non-linear, steady-state analysis of multispan beams by the finite element method. *Computers and Struct.* 39(1,2): 83-93.

Mei, C & Decha-Umphai, K 1985. A finite element method for non-linear forced vibrations of rectangular plates. *AIAA Journal.* 23(7): 1104-1110.

Rao, Sheikh & Mukhopadhyay 1993. Large-amplitude finite element flexural vibration of plates/stiffened plates. *J. Acoust. Soc. Am.* 93(6): 3250-3257.

Redfern, D. 1994. *The Maple Handbook*. New York: Springer-Verlag.

Szemplińska-Stupnicka, W. 1990. *The Behaviour of Non-linear Vibrating Systems*. Dordretch: Kluwer Academic Publishers.

Takahashi, K. 1979. A method of stability analysis for non- linear vibration of beams. *J. of Sound and Vibr.* 67(1): 43-54.

Modern Practice in Stress and Vibration Analysis, Gilchrist (ed.) © 1997 Balkema, Rotterdam, ISBN 90 5410 896 7

Contact problems using Overhauser splines

S.Ulaga, M.Ulbin & J.Flašker
University of Maribor, Faculty of Mechanical Engineering, Smetanova, Slovenia

Abstract: A contact detection algorithm using Overhauser spline is presented, solving contact problems with finite element method. Contact area defined by objects boundaries could be approximated in different ways. It can be modelled using straight lines between nodes, finite element shape functions or interpolation functions over boundary nodes respectively. As line description gives a poor approximation of the boundary, shape functions are widely used to describe the analysed geometry. C^0 inter-element continuity of the geometry description is provided by the isoparametric elements. As the 'smooth' and accurate geometry description is crucial to contact stress analysis, decoupling of the geometry from the polynomial shape functions and implementation of the Overhauser splines is suggested in the present work. Single parametric curve is used to model the contacting surface offering the C^1 continuos description of the boundary enabling exact boundary condition imposition. On the other hand geometry description using splines provides the data required for actual contact area determination and exact contact size calculation. Implementation of suggested approach as it is used in developed finite element code is presented. Benefits and drawbacks of proposed approach are discussed. The developed code is then used for solving contact problems in gears.

1 INTRODUCTION

Contact detection e.g. identification of the region of contact is usually the first task in contact problem analysis. Contact detection algorithm depends on object boundary approximation. It could be approximated using straight lines between nodes, with finite element shape functions, some other approximation functions over boundary nodes or some other special technique i.e. (Belytschko 1991). Straight lines approximation is too basic so finite element shape functions are commonly used (English 1993). Shape functions geometry definition is unambiguous only inside the element and not at corner nodes between elements.

Therefore boundary is defined as discrete segments where each segment is defined with element shape function. Instead of the piecewise continuos and non-smooth boundary presentation provided by the element shape function, boundary of the object can be described by unique, smooth and C^1 continuos parametric spline. For presented contact determination algorithm boundaries of contacting surfaces were approximated using Overhauser splines (Brewer 1977) instead of finite element shape functions. Introducing redundant boundary definition eliminates problem of inter-element discontinuities of shape functions.

2 OVERHAUSER SPLINE

The implementation of the Overhauser spline is simple, if applied to the discretised boundary. While most of the other spline functions require definition of the control points and tangent vectors, Overhauser spline requires only the positions of the control points and spline is drawn exactly through all but first and last of the control points. General equation of the Overhauser spline is shown in Equation 1.

$$\bar{p}(u) = \sum_{s=1}^{m-2} \bar{p}_s(u) \qquad u=[0,1] \qquad (1)$$

Overhauser spline segment is shown in Figure 1. Four points define each segment of the spline. Spline is defined only between p_i and p_{i+1}, which means that additional point, must be supplied at the start and at the end of the spline. Introducing the Overhauser spline is only possible when contact surface can be predicted and spare points must be available before and after contact surface. Nodes on surface are control points used for Overhauser spline definition.

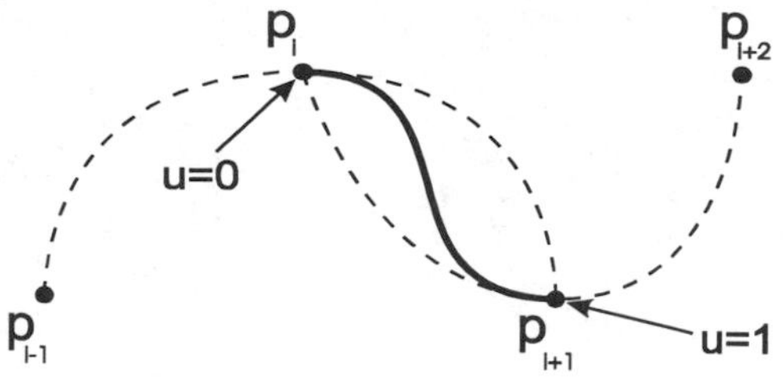

Figure 1: Overhauser spline

Overhauser spline is defined with four parametric functions and it is constrained by four control points:

$$\bar{p}_{i+1}(u)=\left[F_1(u),\,F_2(u),\,F_3(u),\,F_4(u)\right]\begin{bmatrix}\bar{p}_{i-1}\\ \bar{p}_{i}\\ \bar{p}_{i+1}\\ \bar{p}_{i+2}\end{bmatrix} \qquad (2)$$

$$\begin{aligned} F_1(u) &= -\frac{1}{2}u^3 + u^2 - \frac{1}{2}u \\ F_2(u) &= \frac{3}{2}u^3 - \frac{5}{2}u^2 + 1 \\ F_3(u) &= -\frac{3}{2}u^3 + 2u^2 + \frac{1}{2}u \\ F_4(u) &= \frac{1}{2}u^3 - \frac{1}{2}u^2 \end{aligned} \qquad (3)$$

In Equation 2 and Equation 3 the parameter u has the value between 0 and 1 because the equations are valid only for one segment of the spline.

3 CONTACT DETECTION ALGORITHM

For contact determination a suitable algorithm that uses spline representation of the contacting surfaces is required. In Figure 2 target boundary p is shown. As a consequence of the exact load the contactor point was deformed in that way that its relative position moves from point S to point Q. Line SQ is therefore the penetration line. First of all it must be decided whether the contactor point Q is inside the target body or not.

When using finite element shape functions first task is identifying the segment of the target body where the contact occurs. This is done by a rough check of the co-ordinates of a potentially contacting point against the maximum and minimum co-ordinates of an imaginary envelope constructed from the co-ordinate range of a potentially contacting target elements surface nodes. If potentially contacting point is inside of the envelope the point is tested with respect to identified segment.

When potentially contacting point is outside the envelope the point is no longer the contact candidate. The test by an imaginary envelop fails if the penetration in iteration step is too big so that the point penetrates beyond envelope. The point is identified as lying outside the body.

If the contact surface is defined with the Overhauser splines, the test is simple and exact. First the shortest distance between a point Q and the spline p must be determined.

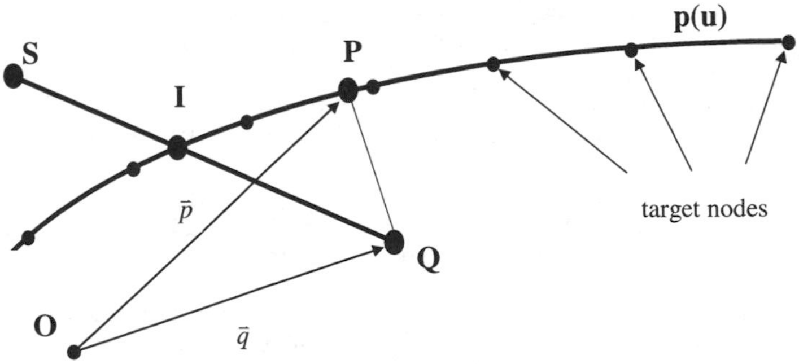

Figure 2: Contact detection

The orientation of the normal vector to the spline that goes through the test point shows whether the test point has penetrated the target body or not. Therefore it is necessary that the sequence of control points of the spline is unique and such that the normal of the spline always points inside the target body. Shortest distance between test point Q and the spline p can be found by solving

$$(\vec{p}(U) - \vec{q})\,\vec{p}^{\,u}(U) = 0 \tag{4}$$

In Equation 4 U has the values between 0 and the number of segments of the curve. U can be divided into integer part, which is represented by i in the Equation 2 and the remainder, which is represented by u in the Equation 3. Equation 4 can be solved using Newton-Raphson method (Nakamura 1993):

$$U_{i+1} = U_i + \frac{f(U_i)}{f^{u}(U_i)} \tag{5}$$

where

$$\begin{aligned} f(U) &= (\vec{p}(U) - \vec{q})\,\vec{p}^{\,u}(U) \\ f^{u}(U) &= \vec{p}^{\,u}(U)\,\vec{p}^{\,u}(U) + (\vec{p}(U) - \vec{q})\,\vec{p}^{\,uu}(u) \end{aligned} \tag{6}$$

In Equation 6 first and second derivative of the curve $\vec{p}$ are:

$$\vec{p}^{\,u}_{i+1}(u) = \left[F^u_1(u),\ F^u_2(u),\ F^u_3(u),\ F^u_4(u)\right] \begin{bmatrix} \vec{p}_{i-1} \\ \vec{p}_i \\ \vec{p}_{i+1} \\ \vec{p}_{i+2} \end{bmatrix} \tag{7}$$

$$\begin{aligned} F^u_1(u) &= -\frac{3}{2}u^2 + 2u - \frac{1}{2} \\ F^u_2(u) &= \frac{9}{2}u^2 - 5u \\ F^u_3(u) &= -\frac{9}{2}u^2 + 4u + \frac{1}{2} \\ F^u_4(u) &= \frac{3}{2}u^2 - u \end{aligned} \tag{8}$$

and

$$\vec{p}^{\,uu}_{i+1}(u) = \left[F^{uu}_1(u),\ F^{uu}_2(u),\ F^{uu}_3(u),\ F^{uu}_4(u)\right] \begin{bmatrix} \vec{p}_{i-1} \\ \vec{p}_i \\ \vec{p}_{i+1} \\ \vec{p}_{i+2} \end{bmatrix} \tag{9}$$

$$\begin{aligned} F^{uu}_1(u) &= -3u + 2 \\ F^{uu}_2(u) &= 9u - 5 \\ F^{uu}_3(u) &= -9u + 4 \\ F^{uu}_4(u) &= 3u - 1 \end{aligned} \tag{10}$$

The next problem is finding the intersection between penetration line SQ and surface p as seen in Figure 2. When usual approximations with finite element shape functions are used, the segment on which penetration line penetrates the boundary must be identified first. After that Newton-Raphson method is used to evaluate intersection. Intersection between Overhauser spline and any line can be found by inserting the spline function into the line Equation 11:

$$\mathrm{A}\,x(U)+\mathrm{B}\,y(U)+C=0 \tag{11}$$

where $x(U)$ and $y(U)$ are components of the vector $\bar{p}$:

$$\bar{p}(U)=\begin{bmatrix} x(U) \\ y(U) \end{bmatrix} \tag{12}$$

which can be again solved using the Newton-Raphson method where functions $f(U)$ and $f^{u}(U)$ are now:

$$\begin{aligned} f(U) &= A\,x(U)+B\,y(U)+C \\ f^{u}(U) &= A\,x^{u}(U)+B\,y^{u}(U) \end{aligned} \tag{13}$$

By performing the above calculations it is possible to solve all boundary contact calculations. The determined penetration value, normals and tangents are then included into augmented matrix. Overhauser spline is extendible into Overhauser surface and can be used in similar fashion using Equation 14 instead of Equation 1.

$$\bar{p}(u,v)=\sum_{s=1}^{m-2}\sum_{t=1}^{n-2}\bar{p}_{s,t}(u,v) \tag{14}$$

Implementation of the Overhauser class makes contact calculation independent of the rest of the finite element code and require no changes to the main finite element code. Finite element program stays fully operational and implementing the contact problem does not change any other function. The same program can now be used for any analysis and in addition offers the possibility of solving contact problems.

4 CONTACT SIZE EVALUATION

Accurate geometry description of contacting surfaces is extremely important in contact problem analysis, but due to its particular formulation it is often in conflict with the conventional isoparametric finite element approach.

In some technical applications (such as gears, bearings, orthopaedic prostheses) the knowledge of the exact size of the contacting area is an important issue. Using isoparametric finite element codes the element size restricts the determination accuracy of the actual size of the contact area in equilibrium. While the element shape functions give only a piecewise continuity along the boundary, parametric splines provide C^1 continuity and therefore accurate boundary description, which leads to exact determination of the contact size and position of the contact area.

The proposed algorithm is capable of the contact size determination for different shapes of the contact surfaces and for single as well as multiple contact.

In the first step of the algorithm the intervals between the control points of the Overhauser's spline are detected, where the sign of the contact changes (*no contact* $\Rightarrow$ *contact* or *contact* $\Rightarrow$ *no contact*). If the contact is detected, the exact points of intersection between contactor and target splines are determined. Next the length of single contacting intervals is evaluated by integration along the boundary between the calculated intersection points Nakamura [4]. The procedure is repeated for every load increment and one can follow the contact development as the load increases.

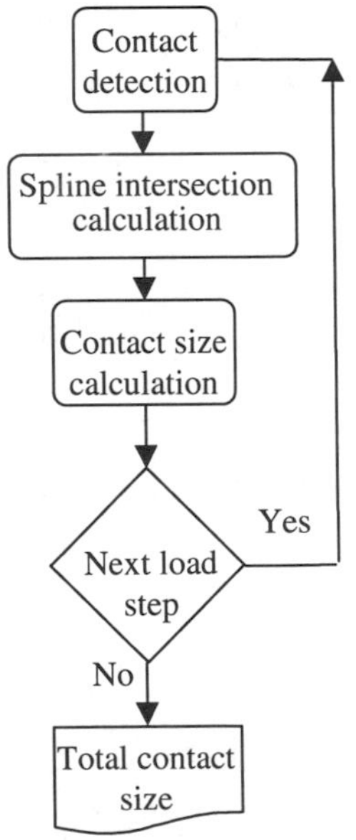

Figure 3: Contact size determination

In Figure 3 a simple float chart is presented. Contact area is calculated in every load step giving the total of the contact area at the end.

5 NUMERICAL RESULTS

Contact pressure distribution on gear teeth was obtained using CONTACT'97 code. Finite element meshes as shown in Figure 4 were used. Shape of the gear teeth was obtained from the data in Table 1 (Ulbin 1996). The particular position of the meshing gear teeth was chosen so to get the normal forces acting parallel to the co-ordinate axes. This idealisation is required to enable the comparability with the results obtained by standard DIN 3990 procedure. Surface pressures and the normal stresses were compared.

Figure 4: Contacting teeth

The driven gear was fixed, while the nodal force was applied to the driving gear. A commercial code I-DEAS and a free-meshing technique were used to create the mesh (Figure 4).

The normal stresses are shown in Figure 5. Results are in good agreement with the results obtained by the Hertz theory.

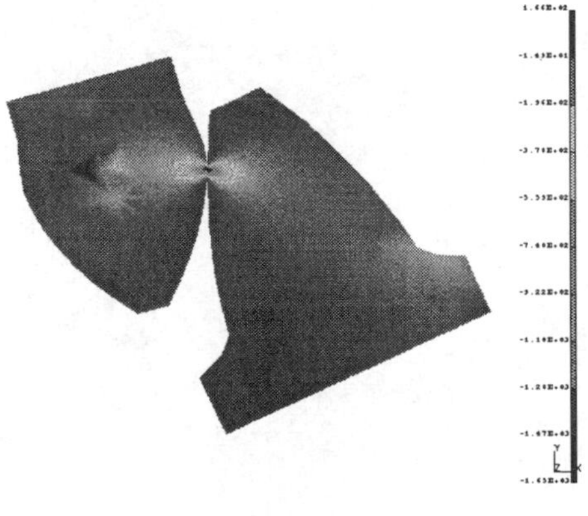

Figure 5: σ_x stresses

As for the point C, similar calculations were performed for the other characteristic points. Comparison of the numerical and Hertz theory results is shown in the Table 1:

Table 1: Results comparison

Characteristic point	Hertz theory	FE
A	1576	1597
B	1945	1957
C	1655	1650
D	1431	1455
E	999	1062

Comparison of the analytically and numerically obtained values of the surface pressure is presented in Figure 6.

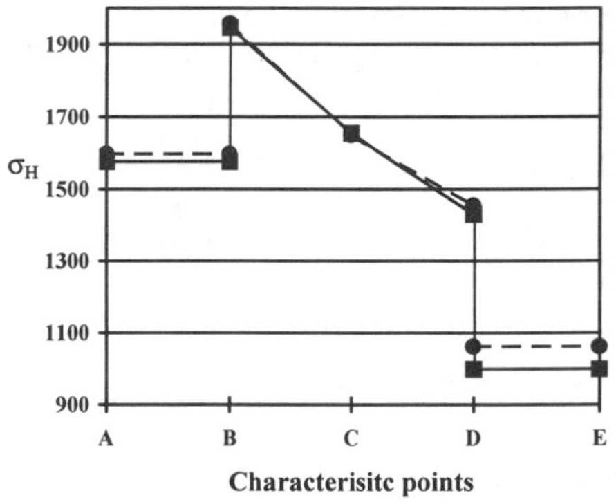

Figure 6: Comparison of the analytical and numerical results

6 CONCLUSIONS

Using spline functions for the approximation of the contact area certainly results in better and robust algorithms for contact detection. But it introduces redundant data, which requires more careful data handling. Apart form that spline approximation is applicable only to the contact problems where the contact surfaces are known in advance or can be predicted.

Developed computer program for contact problem analysis was used for contact problems in gears, where the approximated contact surfaces are obvious. Exact contact area is determined using the above algorithms, after the contact problem is solved using Lagrange multiplier approach and finite element method.

REFERENCES

[1] Belytscho, T., Neal, M. O. 1991. Contact-Impact by the Pinball Algorithm with Penalty and Lagrangian Methods, *International Journal for numerical methods in Engineering*, Vol. 31, 547-572

[2] Brewer, J., Anderson, D. C. 1977. Visual interaction with Overhauser curves and surfaces. *Computer Graphics*, Vol. 11, No. 2, pp. 132-137.

[3] English, G. R. Lagrange Multiplier Method for Contact and Friction: Implementation and Theory, *PhD thesis, University of Liverpool*, Department of Mechanical Engineering, 1993.

[4] Nakamura, S. 1993. Applied Numerical Methods in C, *Prentice-Hall International*, Inc.

[5] Ulbin M. 1996 – Contribution to the research of the contact problems with gears using finite element method (In Slovene, abstract in English), *PhD thesis, University of Maribor*, Slovenia.

5 Vibroacoustics

Modern Practice in Stress and Vibration Analysis, Gilchrist (ed.) © *1997 Balkema, Rotterdam, ISBN 90 5410 896 7*

Modelling of damping in porous media at low frequencies

H.J. Rice
Department of Mechanical and Manufacturing Engineering, Trinity College, Dublin, Ireland

P. Göransson
The Aeronautical Research Institute of Sweden (FFA) & The Royal Institute of Technology (KTH), Marcus Wallenberg Laboratory for Sound and Vibration, Stockholm, Sweden

ABSTRACT: Damping induced by fluid entrainment in porous media is the major mechanism of vibroacoustic energy dissipation within propellor aircraft cabins. Although such systems are often satisfactorily modelled using an equivalent fluid formulation with a mass enhancement to represent the porous frame, at propellor blade pass frequencies, the reactive forces generated by the frame stiffness and damping also become signifcant and the structure must be analysed as a full two phase system. In this paper, two materials commonly used as fuselage thermal insulation are considered and a simple seismic experimental test configuration is modelled to verify the analysis method. The structural frame properties are determined through vacuum testing and the specific fluid induced effects are developed from flow resistivity measurements. A finite element model of the two phase system is formulated and the response of the model is shown to compare excellently with experimental test results carried out under athmospheric conditions.

1 INTRODUCTION

A persistent problem which has presented itself to the modern Euoprean Aerspace industry is the development of reliable finite element models for aircraft cabin vibro-acoustic behaviour as economic and legislative restraints increasingly impose themselves on cabin design. A schematic layout of a typical fuselage is shown in Figure 1 where it is seen that the transmission of noise into the cabin must pass through the insulation material imbedded in the double wall fuselage structure where most of the dissipative effects occur. As studies show that dynamic behaviour at low frequencies of undamped double wall structures are complex it is clear that modelling of the damped system will require careful consideration of the insulation layer dynamics.

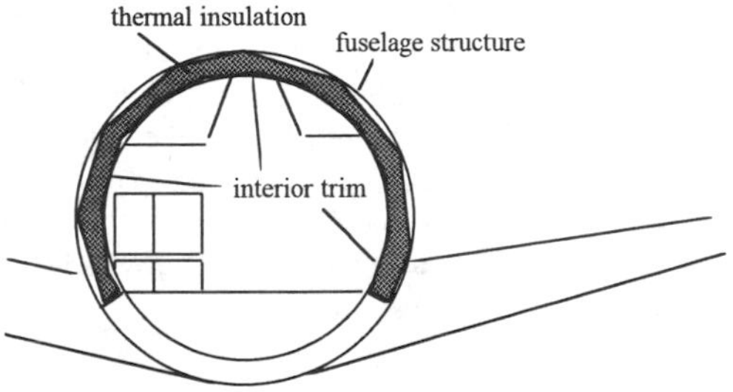

Figure 1: Schematic of Aircraft Cabin Cross-section

The thermal layer consists of a generally anisotropic layup of glass fibres which constitutes the frame or solid phase and the entrapped air which is the fluid phase.

Modelling of porous materials using a parallel, rigid fibre model was first considered by Lord Rayleigh (Rayleigh 1883) and a sound theoretical foundation was presented in the extensive works on a general three dimensional continuum theory by Biot (Biot 1956, Biot and Willis 1957). Biot showed that in addition to two types of coupled dilatational fluid-frame waves in the medium, there also exists a third distortional, shear wave for a three dimensional continuum model of the porous medium. Further recent work in this field can be found in (Pride et al. 1992, Allard 1996).

It is clear that the prediction of elasto-acoustic waves through porous media require consideration of both the fluid and structure states and the coupling between these. Examples of such formulations, based on Biot's theory, include those Ghaboussi and Wilson

(Ghaboussi and Wilson 1972) for a fluid-saturated porous elastic solid with a compressible fluid and by Zienkiewicz and Shiomi (Zienciewicz and Shiomi 1984) for the analysis of soil-incompressible fluid interaction. Recently, alternative formulations have been discussed by Gajo, Saetta and Vitaliani (Gajo et al 1994), Johanssen, Allard and Brouard (Johanssen et al 1995) and Göransson (Göransson 1997).
An effective method for dealing with the frequency dependence of material damping based on thermodynamic relaxation functions was proposed by Lesieutre , see for example (Lesieutre 1992). This was further developed by Dovstam (Dovstam 1995, 1997) who formulated an isotropic Augmented Hooke's Law (AHL) in the frequency domain. Examples of research on non-thermodynamic damping models include works by Bagley and Torvik (Bagley and Torvik 1983) and Enelund and Olsson (Enelund and Olsson 1995). Recently Dalenbring and Dovstam have identified the AHL-parameters for a plexi-glass plate from experimental data (Dalenbring and Dovstam 1997).

In this paper an application of such a prediction method is performed in order to model the frequency dependancy of the solid phase. In-vacuo testing of a material sample similar to that performed by Pritz (Pritz 1986) is first done to determine frame properties and subsequent analyses (with discretisation schemes) are then conducted to predict the additional dissipative effects of the entrained fluid. The validity of the predictive code can then be assessed by directly comparing the numerical predictions to experimental results taken under athmospheric testing conditions.

2 ANALYSIS

2.1. *Governing Equations*

The time harmonic elasto-acoustic wave propogation behaviour of light weight porous materials may following (Biot 1956,1957) be posed as

Pore fluid wave equation:

$$\frac{h^2}{\alpha_p}p_{,ii}+\frac{\omega^2h^2}{R}p+\omega^2\left(\frac{hQ}{R}-\frac{h\beta_c}{\alpha_p}\right)u_{f_{i,i}}=0 \qquad (1)$$

Frame structure wave equation:

$$\sigma_{f_{ij,j}}-\frac{Q^2}{R}u_{f_{j,ji}}+\omega^2\left(\alpha_f-\frac{\beta_c^2}{\alpha_p}\right)u_{f_i}+\left(\frac{h\beta_c}{\alpha_p}-\frac{hQ}{R}\right)p_{,i}=0$$

(2)

where

$$\alpha_p=\left(\rho_{22}+\frac{ib}{\omega}\right);\beta_c=\left(\rho_{12}-\frac{ib}{\omega}\right);\alpha_f=\left(\rho_{11}+\frac{ib}{\omega}\right)$$

$$\rho_{22}=h\rho_0+\rho_a;\rho_{11}=\rho_f+\rho_a \qquad (3)$$

and

p is the acoustic pressure disturbance,
ω is the circular frequency
$\sigma_{f_{ij}}$ is the Cauchy linear, elastic stress tensor,
u_f is the elastic displacement of the solid frame of the porous medium,
Q is a coupling factor related to the dilatational deformation of the material,
R is the compressional bulk modulus of the fluid,
b is the viscous drag coefficient between the fluid movement and the elastic frame
h is the porosity, i.e volume of fluid/total volume of material,
ρ_a is an inertial coupling factor between fluid and frame deformation,
ρ_0 is the ambient fluid density,
ρ_f is the bulk density of the porous material,

Note that the viscous drag coefficient, b, is related to the static flow resistance by

$$b=\Phi_{static}F(\lambda_p) \qquad (4)$$

For the frequency ranges considered in the present study

$$F(\lambda_p)\approx 1 \qquad (5)$$

Further detailed discussion on the coupling conditions along the interface between the mixed medium (fluid in pores and solid frame) and other homogeneous, non-porous media (elastic solid, exterior fluid, elastic thin shell, etc.) are contained in (Göransson 1996).

Dissipative mechanisms present in the transmission of

vibroacoustic energy in these systems are also attributable to the material damping occurring in the solid part or frame component of the porous medium. Recent research results (Dovstam 1995) on constitutive modelling of material damping based on stress-strain relaxation formulations derived from thermodynamic principles may be used in conjunction with the porous model given above by introducing additional loading on the frame as

$$\sigma_f = \hat{\mathbf{H}}\mathbf{E}_f \tag{6}$$

where

$$\mathbf{E}_f^T = \left[\varepsilon_{11} \mid \varepsilon_{22} \mid \varepsilon_{33} \mid 2\varepsilon_{12} \mid 2\varepsilon_{13} \mid 2\varepsilon_{23}\right] \tag{7}$$

$$\varepsilon_{ij} = u_{f_{i,j}} + u_{f_{j,i}} \tag{8}$$

and the material constitutive matrix is given as

$$\hat{\mathbf{H}} = \lambda\left(1+d_\lambda(s)\right)\mathbf{H}_\lambda + G\left(1+d_G(s)\right)\mathbf{H}_G \tag{9}$$

$$s = -i\omega \tag{10}$$

$$d_\lambda(s) = \sum_{i=1}^{N_\alpha} \frac{(3\varphi_i^2 + 4\varphi_i\mu_i)}{\lambda\alpha_i}\frac{s}{s+\beta_i} \tag{11}$$

$$d_G(s) = \sum_{i=1}^{N_\alpha} \frac{2\mu_i^2}{G\alpha_i}\frac{s}{s+\beta_i} \tag{12}$$

$$\mathbf{H}_\lambda = \begin{bmatrix} 1 & 1 & 1 & \\ 1 & 1 & 1 & \mathbf{0} \\ 1 & 1 & 1 & \\ & \mathbf{0} & & \mathbf{0} \end{bmatrix} \tag{13}$$

$$\mathbf{H}_G = \begin{bmatrix} 2 & 0 & 0 & 0 & 0 & 0 \\ & 2 & 0 & 0 & 0 & 0 \\ & & 2 & 0 & 0 & 0 \\ & symm & & 1 & 0 & 0 \\ & & & & 1 & 0 \\ & & & & & 1 \end{bmatrix} \tag{14}$$

completes the linear, frequency dependant, isotropic material damping model of an elastic material. In the above equations $\mathbf{H}_\lambda$ represents dilitational processes whilst $\mathbf{H}_G$ represents shearing mechanisms. Further details of this model are discussed in (Dovstam 1996, Dalenbring and Dovstam 1997).

2.2 *Numerical Discretisation*

In order to model the experimental test vehicle and other geometries, the above dynamical model may be discretised using a weighted residual finite element formulation described in (Göransson 1996). This may be summarised as follows

The dynamical state within the medium may be characterised by the pore fluid pressure p, the pore fluid displacement potential ψ and the frame displacement field $\mathbf{u}_f$. The elemental discretisation may then be performed according to

$$p(\mathbf{x},t) = \sum_{n=1}^{N_p} N^n(\mathbf{x})P^n(t) \tag{15}$$

$$\psi(\mathbf{x},t) = \sum_{n=1}^{N_p} N^n(\mathbf{x})\Psi^n(t) \tag{16}$$

$$u_{f_i}(\mathbf{x},t) = \sum_{n=1}^{N_p} N^n(\mathbf{x})U_{f_i}^n(t) \tag{17}$$

where N_p is the number of nodes in each element and $N^n(\mathbf{x}); \mathbf{x} \in \Omega_p$ are the associated shape functions. The weight functions are identically given by

$$\delta p(\mathbf{x}), \delta\psi(\mathbf{x}), \delta u_{f_i}(\mathbf{x}) = \sum_{n=1}^{N_p} N^n(\mathbf{x}) \tag{18}$$

The unknown elemental quantities may now be partitioned according to

$$\mathbf{U}_p^{(e)^T} = \left\{\mathbf{U}_f^{(e)^T}, \Psi^{(e)^T}, \mathbf{P}^{(e)^T}\right\} \tag{19}$$

and symmetric elemental mass, stiffness and damping matrices defined as

$$\mathbf{M}_p^{(e)} = \begin{pmatrix} \rho_{11}\mathbf{M}_f^{(e)} & \rho_{12}\mathbf{C}_{pf}^{(e)T} & 0 \\ \rho_{12}\mathbf{C}_{pf}^{(e)} & \rho_{22}\mathbf{B}^{(e)} & 0 \\ 0 & 0 & 0 \end{pmatrix}$$

$$\mathbf{K}_p^{(e)} = \begin{pmatrix} \mathbf{K}_f^{(e)} & 0 & -(hQ/R)\mathbf{E}_{pf}^{(e)T} \\ 0 & 0 & h\mathbf{B}^{(e)T} \\ -(hQ/R)\mathbf{E}_{pf}^{(e)} & h\mathbf{B}^{(e)} & -(h^2/R)\mathbf{M}^{(e)} \end{pmatrix}$$

$$\mathbf{D}_p^{(e)} = \begin{pmatrix} b\mathbf{M}_f^{(e)} & -b\mathbf{C}_{pf}^{(e)T} & 0 \\ -b\mathbf{C}_{pf}^{(e)} & b\mathbf{B}^{(e)} & 0 \\ 0 & 0 & 0 \end{pmatrix} \tag{20}$$

Note that $\mathbf{K}_f^{(e)}$ is computed from the AHL constitutive model, see equation (9). Details of the calculation sub-matrices are described in (Göransson 1996).

With these matrix definitions, the elasto-acoustic state of the porous medium, driven at a certain frequency can be expressed as the solution of

$$\left\{\mathbf{K}_p^{(e)} - \omega^2\mathbf{M}_p^{(e)} - i\omega\mathbf{D}_p^{(e)}\right\}\mathbf{U}_p^{(e)} = \mathbf{F}_p^{(e)}(\omega) \tag{21}$$

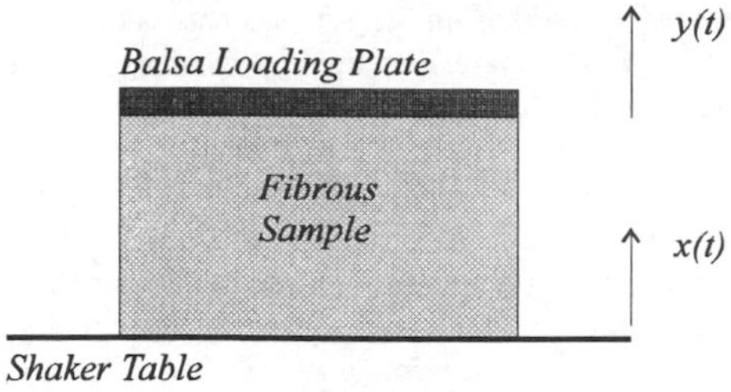

Figure 2: Seismic Mass Test Setup

3 EXPERIMENT

3.1 *Setup*

In order to validate the above theory, a simple seismic test shown schematically in figure 2 was conducted using two grades of fibrous thermal insulation materials.

The dynamics of the fibrous material was assessed by measuring the frequency response of the plate with respect to the shaker table motion. The response of the plate was sensed by a non-contacting capacitative displacement transducer for the first material sample and a laser vibrometer for the second sample. The laser was preferred in the second experiment as the resonant frequency of the system was higher. The motion of the base was measured using an accelerometer. In both cases the sample was mounted in vacuum chamber capable of maintaining an absolute pressure of 120 mTorr (760 Torr = 1atm) and hence it was possible to measure the dynamic response with and without the effect of the entrained air. A HP 35665A analyser was used to provide the excitation signal (bandlimited random) and the log and average the response spectra. Further details of the test rig are described in (Grogan 1997).

3.2 *Procedure and Modelling*

Many of the dynamic properties required to model the fibrous material such as density and (static) stiffness based entities may be directly measured as can flow related properties such such as porosity and resistivity. However, dynamic frame properties such as frequency dependent stiffness and damping must be observed through a dynamic test. The testing of the samples in vacuo were therefore used to identify these frame properties by fitting AHL models and, as the additional properties relating to fluid flow were independently observable, a model of the complete coupled fluid structure system could then be formulated and compared with subsequent athmospheric test results. It should be noted, in addition , that the frame properties are also liable to exhibit some degree of non-linearity. Discussion of underlying non-linear models is given elsewhere (Pritz 1990, Rice et al 1997). For the experiments conducted here, low amplitude testing levels only were used to minimise possible error associated with these non-linear effects.

Once the material properties were determined, a finite element model of the complete system using two layers of sixteen (4×4) noded isoparametric elements with incomplete bi-quadratic
interpolation based on equations (15) to (20) was formulated to simulate the response of the system under harmonic excitation. To represent the acoustic field outside the porous sample, an anechoic termination was assumed at the outer boundary of the finite element mesh modelling the acoustic waves originating from the motion of the porous sample and propagating outwards.

3.3 *Vacuum Testing/System Identification*

As stated earlier the primary use of the the vacuum test was to identify the dynamic stiffness and damping parameters associated with the absorbent frame. All of the other properties could be measured or assessed independently

3.3.1 *Elastic and Flow Resistance Parameters*

The static stiffness and flow resistance of of material A were taken from work reported by Cummings (Cummings 1996) whilst in the case of material B static tests on the material were conducted by the authors. Their respective static properties are shown in table 1 below.

Table 1: Static Material Properties

Material	A	B
Young's Mod *Pa*	225	135
Bulk Density kg/m^3	13.58	10.88
Flow Res. *Rayls/m*	33268	18000
Porosity	.994	.995
Tortousity	1.12	1.12

For material A, the static flow resistance was assumed to be higher than the nominally measured 23400 *Rayls/m*, because of the relatively heavier seismic mass used for this test as the gravitational loading was observed to increase its bulk density from an uncompressed value of 10.2 kg/m^3 which caused an increase in the static flow resistance. For material B, the seismic mass used was considerably lighter and hence the nominal (uncompressed) values as determined from material tests were used.

3.3.2 *Sample A*

In the tests conducted on sample A a rectangular block of dimension $55mm \times 55mm \times 18mm$ was used supporting a seismic mass of value 17.29 *g*. As the response transducer was mounted to measure the relative motion of the loading plate with respect to the base motion, as frequency response function defined by

$$H_A(\omega) = \frac{G_{x(y-x)}(\omega)}{G_{xx}(\omega)} \qquad (22)$$

was measured using spectra generated from 100 ensemble averages of band limited random data. The numerical model was then run without the fluid effects and the AHL parameters adjusted to fit the experimental data thus providing a model for the structural dynamic stiffness and damping. The parameters returned were

Table 2: AHL Parameters for Material A

β_1 / Hz	μ_1 / Pa	α_1 / Pa	φ_1 / Pa
6.65	15.4	1	0

The simulated FRF is compared to the measured curve in figure 3

It is interesting to compare the resonance frequency which would be estimated based on the static elastic

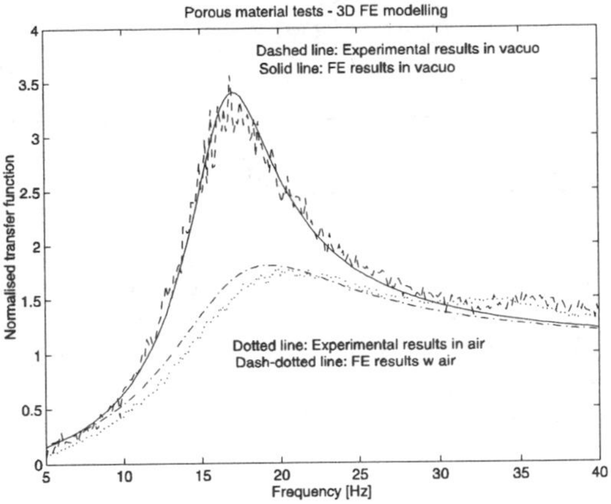

Figure 3: Frequency Response Curves for Test A

Youngs modulus and the effective dynamic modulus, obtained when the AHL terms are taken into account. This incorporates a low frequency relaxation process and has the effect of raising the dynamic stiffness and thus increases the effective resonance frequency from 7.5 *Hz* to about 17 *Hz*.

The loss factor for this damping model, i.e. the sum of the imaginary contributions of d_λ and d_G, are shown in figure 4.

3.3.3 *Sample B*

The process was then repeated on sample B using a sample with dimensions $55mm \times 55mm \times 19mm$ and a loading mass of 0.95 *g* was used, giving a considerably higher resonance frequency than the

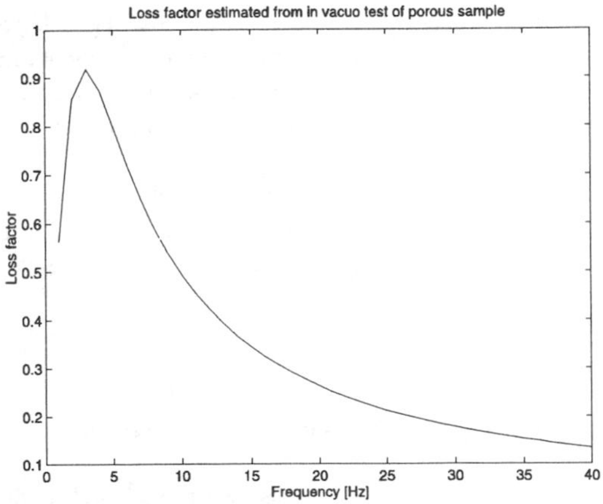

Figure 4: Structural Loss Factor Associated with Test A.

Table 3: AHL Parameters for Material B

β_1 / Hz	μ_1 / Pa	α_1 / Pa	φ_1 / Pa
1.0	14.2	1.0	0.0
β_2 / Hz	μ_2 / Pa	α_2 / Pa	φ_2 / Pa
1400	14.15	1.0	0.0

material A set up. As the resonant frequency was higher it proved more satisfactory to use a laser vibrometer to measure the loading mass response and a direct transmissibility frequency response measure was used instead defined by

$$H_B(\omega) = \frac{G_{xy}(\omega)}{G_{xx}(\omega)} \tag{23}$$

The in-vacuo experimental results for material B is shown in figure 5 together with the numerical simulation of the same conditions. The fit shown was found for the following AHL parameters according to Once again, the low frequency relaxation process causes the resonant frequency to be raised from a statically estimated value of 21 *Hz* to about 48 *Hz*. Similar results have also been shown in the tests conducted on plexi-glass plating (Dalengring 1997). In the present case a second relaxation process was also introduced so as to correct the value of the loss factor in the region of 50 *Hz*. The loss factor incorporating these two processes is shown in figure 6.

It is interesting to note that the two materials have similar low relaxation frequencies, accounting for the dynamic resonance frequencies, as well as similar damping parameters μ_1. It should be remembered though that the choice of the material parameters for damping is arbitrary. It is only the damping functions d_λ and d_G which are unique in the sense that their real part should provide the necessary increase in resonance frequency and their imaginary contributions should result in the correct loss factor levels. However, using a "traditional" real loss factor and the static value for elastic modulus would in this case give erroneous results in the resonance frequency region.

3.4 *Athmospheric Testing / Model Validation*

For both samples the tests were repeated under athmospheric pressure so that the influence of the volume fluid-structure interaction within the porous

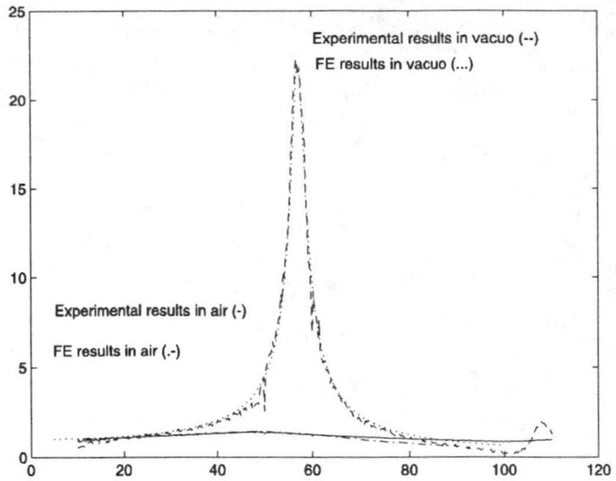

Figure 5: Frequency Response Curves for Test B

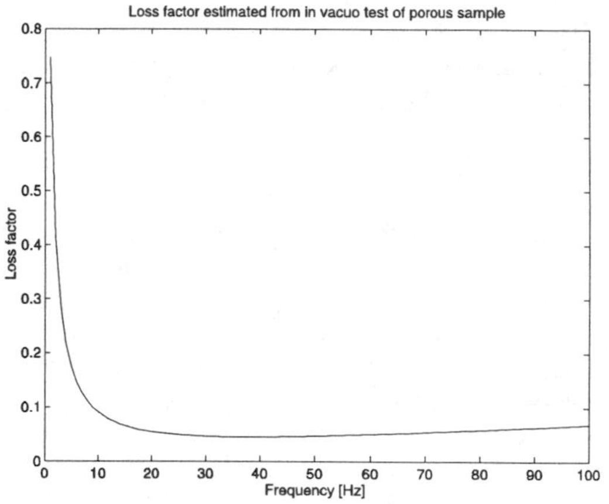

Figure 6: Loss Factor for Material B

medium could be observed. This is manifested as a dramatic increase in the apparent damping, see figures 3 and 5. This could be compared directly to the results from the numerical model with the fluid/structure effects enabled which is seen to give excellent agreement with the experimental results for both test cases. Note that in both tests the increase in damping as well as the absolute level of the response is well predicted by the enhanced numerical model.

Although these agreements have been verified with two specific and relatively simple vibro/acoustic systems and given that there is scope for potential experimental error due mainly to the difficulty in physically handling this relatievly limp material, the prospect of using this type of Biot/AHL finite element analysis on more complex structures is extremly encouraging on the present results.

4 CONCLUSIONS

Using static flow resistance values, static elastic Youngs moduli, and frequency dependent frame dynamical properties established by vacuum tests the vibrational response of two different porous materials have been measured and successfully simulated using a finite element formulation.

The use of careful complementary experimental testing (in order to identify dynamical structural parameters) has shown itself to be a critical tool in the development of modelling strategies for porous material dynamics.

REFERENCES

Allard,J.F. 1996, *Propagation of sound in porous media: modelling of sound absorbing materials.* Chapman&Hall, London

Bagley, R.L. Torvik, P.J. 1983, 'Fractional calculus- A different approach to the analysis of viscoelastically damped structures.' *AIAA J* **21**, 741-748.

Biot, M.A. 1956, Theory of propagation of elastic waves in a fluid-saturated porous solid. I. Low frequency range. *Journal of the Acoustical Society of America* **28**, 168-178.

Biot, M.A. and Willis, D.G. 1957, 'The elastic coefficients of the theory of consolidation.' *J Appl. Mech.* **24,** 594-601.

Cummings, A. 1996, University of Hull, *Personal communication.*

Dalenbring, M. and Dovstam, K. 1997, 'Damping function estimation based on modal receptance models and neural nets.' *To be submitted*

Dovstam, K. 1995, 'Augmented Hooke's Law in frequency domain. A three dimensional, material damping formulation.' *Int. J Solids Structures* **32**, 2835-2852.

Dovstam, K. 1997, 'Receptance model based isotropic damping functions and elastic displacement modes.' *To appear, Int. J Solids Structures*

Enelund, M. and Olsson, P. 1995, 'Damping described by fading memory models.' *AIAA Paper* #95-1181.

Gajo, A., Saetta, A. and Vitaliani, R. 1994, 'Evaluation of three- and two-field finite element methods for the dynamic response of saturated soil.' *Int J Numerical Methods in Engineering* **37**, 1231-1247.

Ghaboussi, J. and Wilson, E.L. 1972, 'Variational formulation of dynamics of fluid saturated porous elastic solids.' *Proc. ASCE* **98**, EM4, 947-963.

Göransson, P. 1997, 'A 3D, symmetric, finite element formulation of the Biot equations for a fluid saturated, linear, elastic porous medium.' *To Appear, Int J Numerical Methods in Engineering*

Grogan, K.W. 1997, "Identification of the Non-linear Dynamics of Fibrous Materials", MSc Thesis, University of Dublin.

Johanssen, T.F., Allard, J.F. and Brouard, B. 1995, 'Finite element method for predicting the acoustic properties of porous samples.' *Acta Acoustica* **3**, 487-491.

Lesieutre, G.A. 1992, 'Finite elements for dynamic modelling of uniaxial rods with frequency dependent material properties.' *Int J Solids Structures* **29**, 1567-1579.

Pride, S.R., Gangi, A.F. and Morgan, F.D. 1992, 'Deriving the equations of motion for porous isotropic media.' *J Acoust. Soc. Am* **92**(6), 3278-3290.

Pritz,T. 1986, 'Frequency dependance of frame dynamic characteistics of mineral and glass wool materials', *Journal of Sound and Vibration* **106**, 161-169.

Pritz,T. 1990, 'Non-linear of frame dynamic characteristics of mineral and glass wool materials', *Journal of Sound and Vibration* **136**, 263-274.

Lord Rayleigh 1883, 'On porous bodies in relation to sound.' *Philosophical Magazine* **16**, 181-186.

Rice, H.J., Torrance, A. Eikelman, G. 1997, "A Model of Stiffness Non-Linearity in Fibrous Damping Materials", *To be submitted, Journal of Sound and Vibration.*

Zienkiewicz, O.C. and Shiomi,T. 1984, 'Dynamic behaviour of saturated porous media; the generalised Biot formulation its numerical solution.' *International Journal for Numerical and Analytical Methods in Geomechanics* **8,** 71-96.

Modern Practice in Stress and Vibration Analysis, Gilchrist (ed.) © 1997 Balkema, Rotterdam, ISBN 90 5410 896 7

State space methods in an eulerian symmetrical formulation for vibroacoustics

F.Cura', G.Curti & F.Scarpa
Department of Mechanics, Technical University of Turin, Italy

ABSTRACT: In this paper the eulerian symmetrical formulation is applied for the analysis of interior coupled structural - acoustics problems. In this method the uncoupled structural and acoustic modal basis are used for the determination of the generalised state space matrices of the coupled system; the obtained matrices have the well known properties of linear undamped gyroscopic systems, leading to an eigenproblem with complex solutions. A QZ algorithm is used for eigenvalue and eigenvector computation; following a formulation developed by Meirovitch for linear undamped gyroscopic systems, the same eigenvectors are used for the computation of the state vector and the frequency response functions of structural inertance and sound pressure level. A formulation for the generalised structural excitation is also developed. The experimental and numerical comparisons supports the reliability of this method for vibroacoustic interior analysis.

1. INTRODUCTION

The dynamic interaction between an acoustic cavity and a coupled elastic structure has been widely investigated in the last thirty years; the increasing demand of acoustic comfort in the automotive field, and the need of avoiding "boom noise" problems in interior cockpits has led many researchers to develop analytical and numerical predictive tools to investigate this kind of problems. The unsymmetrical eulerian formulation is normally used in Finite Element Methods for the description of coupled fluid - structure interaction problems; this formulation consists on the use of structural displacements and acoustic pressures as generalised co-ordinates for the coupled structural - acoustic system. In this way the generalised mass and stiffness matrices are unsymmetrical, and the related eigenvalue problem can be readily solved using appropriate solution algorithms, as the Lanczos method. The symmetrical eulerian formulation employs the acoustic velocity potential instead of fluid pressure, leading in this way to generalised full diagonal mass and stiffness matrices and to a skew - symmetric coupling matrix; the obtained state matrices have the same typology of those related to linear undamped gyroscopic systems, with complex eigensolutions (complex conjugate eigenvectors and eigenvalues with null real part). Bokil and Shirahatti (1994), following an idea proposed by Gorman, Dowell and Smith (1977), applied the Meirovitch algorithm to the analysis of acousto-structural coupled systems in an eulerian symmetrical formulation. In this algorithm (Meirovitch, 1975, 1976) the state space matrices related to linear undamped gyroscopic systems are transformed to obtain a symmetrical real eigenvalue problem; the eigensolution are computed following a technique proposed by Martin and Wilkinson (1968). In this paper we applied a QZ algorithm (Bowdler, Martin and Reinsch 1968) devoted to the analysis of generalised system matrices; the related routine is implemented in MATLAB® package. With this algorithm it is possible to obtain reliable complex eigensolutions; the extracted eigenvectors are complex conjugate, and the related eigenfrequencies have null real part. The frequency response functions can be readily computed with a modal expansion using the real and imaginary part of the eigenvectors. Due to the fact that the equations of motion of the coupled system are written in terms of the uncoupled structural and acoustic modal basis, the final state matrices can have small dimensions (twice the total number of acoustic and structural modes considered in the range of frequency analysis). In this way, it is convenient also to compute the frequency response functions of the acoustic pressure and inertance via a direct response, using the usual methods applied in state space analysis (Brogan 1991).
Hence, the eulerian symmetrical formulation can be readily used to perform acousto - structural coupled analysis using the known uncoupled modal basis of the structure and the cavity. The object of this paper is to present the results obtained by application of state space methods to the eulerian symmetrical

formulation; the performed computations shows that the used formulation is reliable for coupled acousto - structural interactions problems. A detailed account of the theoretical approach itself is beyond the scope of this paper; however, a brief presentation of the formulation will be presented here for sick of completeness. For more detail the reader can refer to (Gorman, Dowell & Smith 1977, Bokil & Shirahatti 1994).

2. THEORETICAL BACKGROUND

In the eulerian symmetrical formulation the set of partial differential equations describing the motion of the coupled system is transformed in a system of ordinary differential equations of the second order using the following modal expansions for the structure and the acoustic cavity:

$$w = \sum_{m=1}^{M} q_m \psi_m \tag{1}$$

$$p = -\rho_0 \sum_{n=0}^{N} a_n F_n \tag{2}$$

where w and p are the structural displacement and the acoustic pressure, ρ_0 is the density of the fluid at rest, q_m and a_n are the modal participation factors describing the modal expansions related to the elastic structure and acoustic cavity. ψ_m and F_n are the eigenmodes of the *in vacuo* flexible structure and the acoustic cavity with *hard wall* conditions. The set of ordinary differential equations leads to the following second order dynamic system:

$$[M]\{x\} + [C]\{x\} + [K]\{x\} = \{Q\} \tag{3}$$

where:

$$\{x\} = \begin{Bmatrix} a_n \\ q_m \end{Bmatrix} \tag{4}$$

The modal mass and stiffness matrices are full diagonal; the components of the modal stiffness matrix are affected by the presence of the mode 0 of the cavity, which represents the Helmholtz stiffening effect of the fluid on the structure (Bokil & Shirahatti 1994). The coupling matrix is skew - symmetric; their elements are computed by a cross integration between the structural and acoustic modes on the surface of the structure. As can be seen, these matrices are similar to the ones related to linear undamped gyroscopic systems; it can be demonstrated that the related eigenfrequencies are imaginary, with null real part and the eigenvectors are complex conjugate pairs (Wilkinson 1965). The presence of structural and acoustic modal damping factors affects the coupling matrix, filling in the terms in the main diagonal (Gorman, Dowell & Smith 1977); in this case the eigenfrequencies become complex conjugate with real part, and the related eigenmodes are in pairs of left and right eigenvectors (Meirovitch 1990). In this formulation the generalised forcing function is assumed to act on the flexible structure; the generalised force can be represented by an exterior acoustic pressure with uniform distribution (Bokil & Shirahatti 1994) or a structural concentrated force. In this paper we present the application of both the two forcing functions: the formulation of the exterior acoustic pressure can be found in literature (Gorman, Dowell & Smith 1977, Bokil and Shirahatti 1994), while the one related to the structural force has been developed following (Soedel 1981).

The system in equation (3) can be represented in the following State Space form:

$$\{y(t)\} = [A]\{y(t)\} + [B]\{T(t)\} \tag{5}$$

where:

$$[A] = \begin{bmatrix} 0 & I \\ -M^{-1}K & -M^{-1}C \end{bmatrix} \qquad [B] = \begin{bmatrix} 0 \\ M^{-1} \end{bmatrix}$$

$$\{T(t)\} = \begin{bmatrix} 0 \\ Q(t) \end{bmatrix} \tag{6}$$

The state vector is represented by:

$$\{y(t)\} = \begin{Bmatrix} a_n \\ q_m \\ a_n \\ q_m \end{Bmatrix} \tag{7}$$

The eigenproblem related to equation (5) admits the following eigenvalues and eigenvectors pairs:

$$\lambda_r = \pm i\omega_r \tag{8}$$

$$\begin{cases} \{u_r\} = \{y_r\} + i\{z_r\} \\ \{\bar{u}_r\} = \{y_r\} - i\{z_r\} \end{cases} \tag{9}$$

As suggested by Meirovitch (1975), the eigenproblem (5) can be rewritten if the following form:

$$[M^*]\{y(t)\} + [G^*]\{y(t)\} = \{0\} \tag{10}$$

where:

$$[M^*]=\begin{bmatrix} K & 0 \\ 0 & M \end{bmatrix} \qquad [G^*]=\begin{bmatrix} 0 & -K \\ K & C \end{bmatrix} \qquad (11)$$

The eigenform presented above allows the following modes normalisation:

$$\begin{cases} \{y_r\}^T [M^*] \{y_s\} = \delta_{rs} \\ \{z_r\}^T [M^*] \{z_s\} = \delta_{rs} \end{cases} \qquad (12)$$

The eigenvectors calculated from (11) have also the following interesting property:

$$\{z_s\}^T [G^*] \{y_r\} = -\{y_s\}^T [G^*] \{z_r\} = \omega_r \delta_{rs} \qquad (13)$$

The knowledge of complex conjugate eigenvalues and eigenvectors allows the computation of the state vector via a modal expansion (Meirovitch 1976). Another way of extracting the state vector is to perform a direct response of the systems (5) or (10) (with the generalised state vector forcing function on the right term). Due to the fact that generally the modal mass, stiffness and coupling matrices have not large dimensions, it is reasonable to use a direct response to compute the frequency response functions of the physical variables of interest (structure acceleration, sound pressure level, etc.); in this case the CPU times are very similar, and the accuracy of the results is actually the same.

3. NUMERICAL CONSIDERATIONS

Bokil and Shirahatti (1994) applied a technique suggested by Meirovitch (1975) to transform the unsymmetrical eigenvalue problem (10) to the following symmetric one:

$$\lambda [M^*] \{y\} = [V] \{y\} \qquad (14)$$

where:

$$[V] = [G^*]^T [M^*]^{-1} [G^*] \qquad (15)$$

The system (14) is represented by two real and symmetric matrices and can be solved using and appropriate numerical algorithm (Martin & Wilkinson 1968). This technique is efficient, and allows the computation of real eigenfrequencies and eigenvectors; following (12) and (13) it is possible to normalise and choose the real and imaginary parts of the complex conjugate eigenvectors of the system (10). Using a QZ algorithm (Bowdler, Martin and Reinsch 1968) we can extract directly the complex conjugate modes, and normalise the real and imaginary parts following (12); the obtained eigenvectors follow the property (13). The state vector can be represented by a linear combination of real and imaginary parts of system eigenvectors (Meirovitch 1976).

The computation of the state vector via direct response can be performed starting from system (5). In this case some balancing of the state matrices can be useful to avoid numerical problems during inversion process; in our experience, an efficient numerical way for the direct response computation is to transform the system (5) in a canonical controllable form (Brogan 1991) and to use the routine *frsp* of MATLAB® package, which demonstrates a really good numerical reliability. This routine implements a series of balancing and Hessemberg transformations of the starting State Space matrices in order to reduce round - off and numerical propagation errors.

4. TEST CASES

In this paper is considered the case of a simply supported plate bounding at one end a rectangular acoustic cavity. In the first example a cubic cavity is coupled with a square brass flexible panel with an acoustic excitation; in the second the same panel is interfaced to a rectangular cavity with 1 meter of length and the plate is excited by a structural forcing function.

4.1 Simply supported plate coupled with a cubic cavity

This case was experimentally represented in a famous paper (Guy & Bhattacharya 1973). A cubic closed acoustic cavity ($0.2 \times 0.2 \times 0.2$ m) is iterfaced at one end by a square brass simply supported plate (0.2×0.2 m), with a thickness of 0.0009144 m. The plate is excited by an exterior speaker (i.e., the forcing function is represented by a uniform pressure distribution on the plate). To make a more complete comparison, the computed and experimental natural

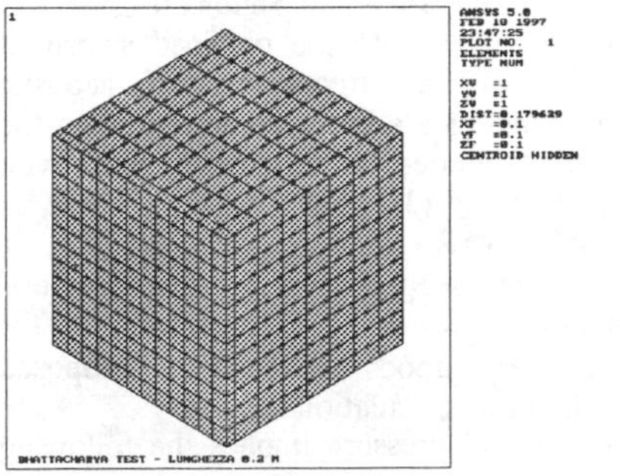

Figure 1. A FEM model of the test case 1.

Tabel 1. Uncoupled and coupled natural frequencies of test case 1.

Uncoupled analytical frequencies (Hertz)	Uncoupled FEM frequencies (Hertz)	Measured uncoupled frequencies (Hertz)	Coupled computed frequencies (Hertz)	Coupled FEM frequencies (Hertz)	Measured coupled frequencies (Hertz)
78.05	77.9	78	86.9	86.4	91
195.1	194.4	-	194.6	193.8	-
312.2	309.8	-	312.2	309.1	-
390.2	388.8	390.3	390.3	387.7	397
507.4	501.9	-	507.3	501.3	-
663.5	659.6	-	663.3	658.5	-
702.5	691.1	702.5	702.5	690.2	730
780.6	770.5	-	780.6	769.7	-
850	*852.6*	-	851.3	853.5	-
-	-	-	852.3	857.9	864

Figure 2. Plate mode corresponding to the second coupled acoustic frequency (852 Hertz).

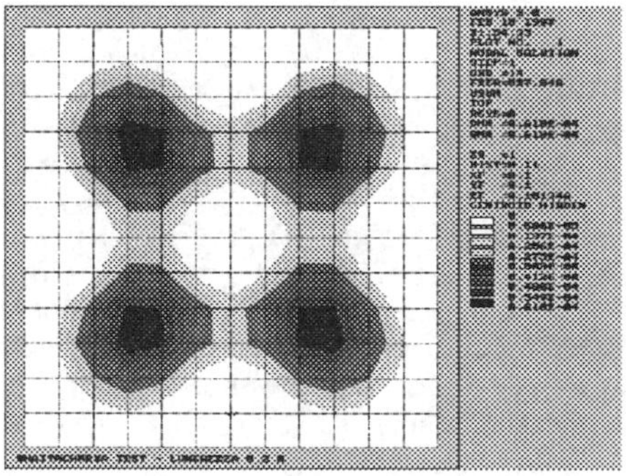

Figure 3. The structural mode computed by a Finite Element analysis.

frequencies of the coupled system are compared with those obtained by a FEM calculation, using the model shown in Figure 1.

In Tabel 1 are illustrated the coupled and uncoupled frequencies of the example system; in italic are shown the uncoupled acoustic frequencies of the cavity. We can remark a good agreement between the computed and experimental results; in particular, this kind of geometry layout shows very well the stiffening effect of the fluid on the flexible plate for the first coupled frequency (an increase of the first structural frequency from 77 to 87 Hertz). This result was also achevied by Bokil and Shirahatti (1994). It is interesting to investigate the obtained structural normal modes of the frequencies of acoustic derivation; this modes are linear combinations of the existing structural modes, in particular of those with (1,1), (1,3), (3,1) and (3,3) half - wave numbers, as can be seen in Figure 2.

In Figure 3 we can see the corresponding structural mode computed by a Finite Element analysis; the agreement is very good between the proposed method and the FEM calculation is good.

An acoustic external pressure implies the following generalised forcing function (Bokil & Shirahatti 1994):

$$Q_m = -p^e \frac{ab}{\pi^2 mn}\left(1-(-1)^m\right)\left(1-(-1)^n\right) \qquad (16)$$

where a and b are the dimensions of the simply supported plate, p^e is the absolute value of the external pressure and m, n are the half - wave numbers of the structural modes. From (16) it can be seen that the acoustic forcing function admits non zero values only for odd - odd half - wave numbers and decreases for higher modes. Equation (16) is used for Transmission Loss computation, as reported in Figure 4.

As it can be seen, there is a good agreement between experimental and numerical results for the Transmission Loss diagram. At 87 Hertz we can observe a negative transmission of the acoustic wave pressure near the centre in the cavity face opposite to the plate; this fact is due to the high velocity of the panel in correspondance of the first coupled mode

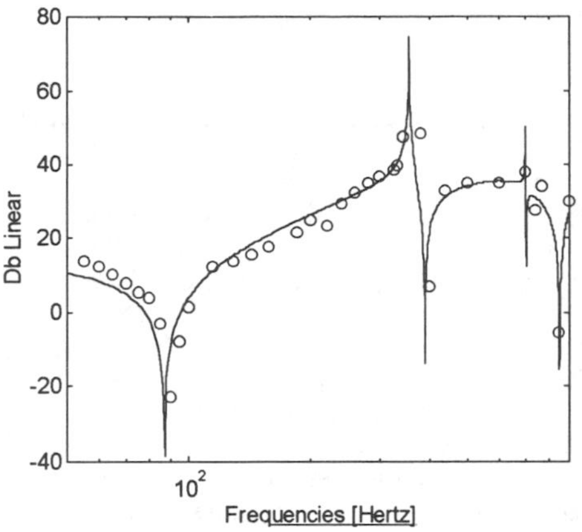

Figure 4. Transmission Loss vs. frequencies. - proposed method; o experimental results.

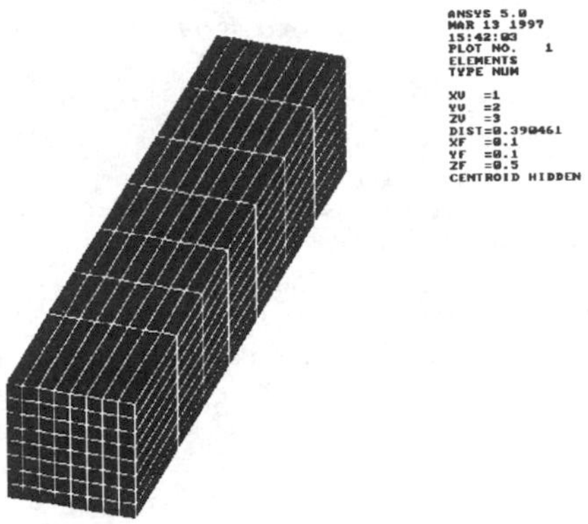

Figure 5. Coupled system of 1 meter length of cavity

(Bokil & Shirahatti 1994). It must be remarked that the same good match between experimental data and numerical calculation has been recorded by Bokil and Shirahatti (1994) using the Martin and Wilkinson algorithm (1968).

4.2 *Simply supported plate coupled with a rectangular cavity of 1 meter length*

For this test case we prepared a Finite Element model of a simply supported plate with the same dimensions and material properties of example 1, coupled with a rectangular cavity of 1 meter length, as shown in Figure 5.

In this case the behaviour of the acoustic cavity follows the monodimensional acoustic cavity theory (Morse & Ingard 1968) for low frequencies analysis; in fact, in tha frequency analysis range of interest (from 0 to 400 Hertz) the acoustic pressure modes

Tabel 2. Comparison between Finite Element and proposed method results.

Coupled FEM frequencies	Coupled computed frequencies
78.02	78.7
173.9	172.12
193.19	195.14
193.19	195.14
306.77	312.23
353.66	340.3
386.21	391.2

have a cosinusoidal variation along the longitudinal co-ordinate. In Tabel 2 a comparison between FEM and calculated coupled frequencies is shown.

The agreement between the results given by the different methods is good; a slight difference can be recorded for the coupled frequency derived from the second acoustic one (353.66 vs. 340 Hertz); this difference shifts only the location of the peaks in the frequency response functions of structural inertance and sound pressure level.

The system is assumed to be excited by a structural generalised forcing function; the formulation of this force is the following:

$$Q_m{}^e = -p^e sin\left(\frac{m\pi x_{app}}{a}\right)\cdot sin\left(\frac{n\pi y_{app}}{b}\right) \qquad (17)$$

The forcing function (17) is different from the acoustic one (16); in fact it has all non zero components, except the case the excitation point is located in a nodal line. In this way the frequency response functions will presents more peaks than the corresponding acoustic excited ones.

In Figure 6 it is shown the frequency response function of the structural acceleration normalised to the excitation force.

The agreement between Finite Element and calculated results is very good. A frequency shift is present in correspondance of the coupled frequency derived from the second acoustic one; as it can be seen, the level are almost the same, except for a higher peak for the second coupled frequency corresponding to the first acoustic one.

In Figure 7 the sound pressure levels in the centre of the bottom of the acoustic cavity show good agreement; a antiresonance peak is registered in the frequency response function calculated by the proposed method, and the same frequency shift shown for the inertance diagram of Figure 6 is present also in this frequency response function of the acoustic pressure.

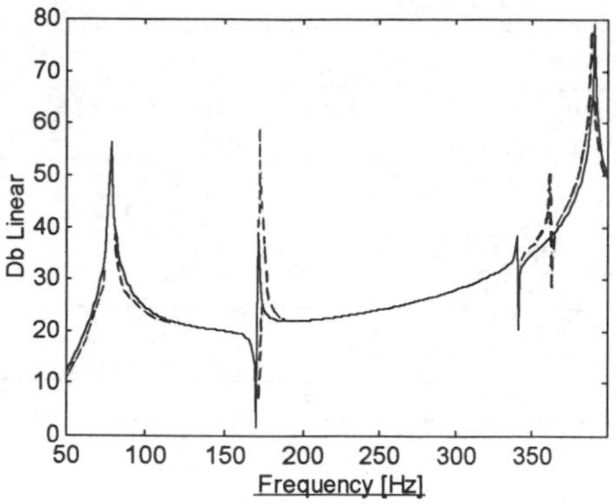

Figure 6. Comparison between Finite Element and calculated frequency response function of structural accelaration. Continuous line - proposed method; dotted line - FEM results.

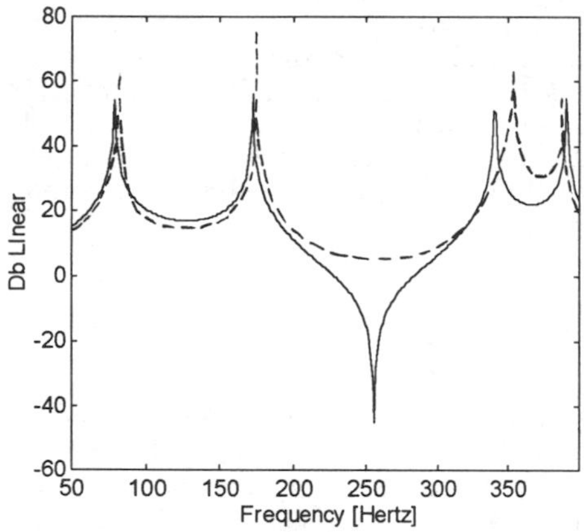

Figure 7. Frequency response function of sound pressure level. Continuous line - proposed method; dotted line - FEM results.

5. CONCLUSIONS

The application of eulerian symmetrical formulation in State Space form shows a good results reliability with the experimental and numerical comparison. The use of QZ algorithm routines widely diffused in commercial linear algebra packages allows the computation of complex eigenvalues and eigenvectors ready to be used in modal expansions for the calculation of state vectors. Due to the State Space form of this formulation, application of algorithms of modern control theory can be applied for further analysis and investigations on the dynamic interaction between elastic structures and interior acoustic cavities.

REFERENCES

Bokil, V.B. & Shirahatti, U.S. 1994. A Technique for the Modal Analysis of Sound Interaction Problems. *J. of Sound and Vibration*. 173(1): 23-41.

Bowdler, H., Martin, R.S., Reinsch, C. & Wilkinson, J.H. 1968. The QR and QL algorithms for symmetric matrices. *Numerische Mathematik* 11: 293-306.

Brogan, W.L. 1991. *Modern Control Theory.* Prentice Hall, 3rd edition.

Dowell, E.H., Gorman, G.G. III & Smith, D.A. 1977. Acoustoelasticity: General Theory, acoustic natural modes and forced response to sinusoidal excitation, including comparaisons with experiments. *J. of Sound and Vibration.* 52(4): 519-542.

Guy, R.W. & Bhattacharya, M.C. 1973. The transmission of sound through a cavity - backed plate. . *J. of Sound and Vibration.* 27(2): 207-223.

Meirovitch, L. 1974. A New Method of Solution of the Eigenvalue Problem for Gyroscopic Systems. *AIAA J.* 12(10): 1337-1342.

Meirovitch, L. 1976. A Modal Analysis for the Response of Linear Gyroscopic Systems. *J. of App. Mech.* 42(2): 446-450.

Meirovitch, L. 1990. *Dynamics and Control of Structures.* John Wiley & Sons.

Modern Practice in Stress and Vibration Analysis, Gilchrist (ed.) © *1997 Balkema, Rotterdam, ISBN 90 5410 896 7*

Axial dynamic stiffness of cylindrical vibration isolators – The 'exact' linear solution

L. Kari
The Marcus Wallenberg Laboratory for Sound and Vibration Research, KTH, Stockholm, Sweden

ABSTRACT: A linear model of the axial dynamic stiffness for cylindrical vibration isolators in the audible frequency range is presented, where influences of material damping, higher order modes and structure borne sound dispersion are investigated. The problems of simultaneously satisfying the boundary conditions at the lateral and radial surfaces of the cylinder are removed, by adopting the mode matching technique, using the dispersion relation for an infinite cylinder and approximately satisfying the boundary conditions at the lateral surfaces by a circle-wise fulfillment or a subregion method. The rubber material is assumed to be nearly incompressible with deviatoric viscoelasticity based on an extended fractional order derivative model, its main advantage being the small number of material parameters. The results are verified by experiments on a rubber cylinder, equipped with bonded circular steel plates, in the frequency range 50 - 5 000 Hz. The model and the measurements are shown to agree strikingly well within the whole frequency range.

1 INTRODUCTION

Structural vibrations at audible frequency range, that is structure borne sound, often radiate sound causing a major environmental problem. To diminish the transmitted structure-borne sound energy, thereby reducing noise pollution, the receiving structures are disconnected from the source by vibration isolators. The increased interest in noise abatement requires suitable models predicting transmitted sound through vibration isolators. The longitudinal modes in cylinders up to a few kHz are of particular interest. The influences of material damping, higher order modes and structure-borne sound dispersion are features that must be modeled to treat the problem properly.

The dispersion relations for linear elastic waves in an infinite isotropic homogeneous solid cylinder have been known for over a century, Pochhammer (1876) and Chree (1889). However, due to the complexity of Pochhammer-Chree dispersion relation, a number of approximate theories have been developed. Though the approximations work for finite cylinders, their dispersion and higher order modes are valid only for an upper frequency limit, which may be as low as a few hundred Hz for a typical rubber vibration isolator.

The aim is to model dynamic properties over a broad frequency range, up to at least 5 000 Hz, expressed in terms of the axial driving point stiffness and the axial transfer stiffness, analytically derived by mode matching. Although strictly valid only for infinitesimal prestrains, it generates an improved understanding of the influences of losses, higher order modes and structure-borne sound dispersion.

2 THE METHOD

2.1 *Notations*

Tensors are denoted by boldface and their components by lightface letters, repetition of indices is avoided as tensors rather than their components are employed. Whenever representation for a particular co-ordinate system is desired, conversion is obtained from the tensor components, as described in Fung (1965). Tr, T, dev and div abbreviate trace, transpose, deviation and divergence. Any function denoted ${}^{(\tau)}f$ is related to f as ${}^{(\tau)}f\,(t) = f\,(t\text{-}\tau)$. The operators $\circ$, $*$, ∇, $\Re$, $\Im$ and $\overline{\{\cdot\}}$ denote composition, complex conjugate, covariant derivative, real part, imaginary part and closure of a set. Moreover, the operators $\otimes$, $\cdot$ and : denote tensor product, single and double contraction. Finally, $\boldsymbol{i}$, N and Z_+ denote metric tensor, set of natural numbers and of positive integers.

2.2 *Constitutive assumptions*

The cylinder material is assumed to be isotropic, homogeneous, nearly incompressible and non-aging, while obeying the principle of fading memory, Cristensen (1982). Since the effects of material damping, higher order modes and transmitted structure-borne sound dispersion is the study focus, analysis is confined to isothermal conditions, infinitesimal strains and prestrains. Crystallization and non-linear friction are not considered. A convolution integral, expressed as a constitutive relaxation relation, is additively decomposed into a spherical part

$$\operatorname{tr}\sigma = 3\kappa_\infty \operatorname{div}\boldsymbol{u} + \int_{-\infty}^{t} 3\,{}^{(\tau)}\mathring{\kappa}\,\frac{\partial \operatorname{div}\boldsymbol{u}(\tau)}{\partial\tau}\,d\tau \tag{1}$$

and a deviatorical part

$$\operatorname{dev}\sigma = 2\mu_\infty \operatorname{dev}\circ\nabla\boldsymbol{u} + \int_{-\infty}^{t} 2\,{}^{(\tau)}\mathring{\mu}\,\frac{\partial \operatorname{dev}\circ\nabla\boldsymbol{u}(\tau)}{\partial\tau}\,d\tau \tag{2}$$

where σ is the stress tensor and $\boldsymbol{u}$ the infinitesimal displacement. Compression and shear relaxation functions are additively decomposed as

$$\kappa = \kappa_\infty h + \mathring{\kappa}, \qquad \mu = \mu_\infty h + \mathring{\mu}, \tag{3}$$

where $\lim_{s\to\infty}\mathring{\kappa}(s) = \lim_{s\to\infty}\mathring{\mu}(s) = 0$, h is step function, $\kappa_\infty = \lim_{s\to\infty}\kappa(s)$ and $\mu_\infty = \lim_{s\to\infty}\mu(s)$ are equilibrium elastic moduli. Temporal Fourier transformations, $(\tilde{\cdot}) = \int_{-\infty}^{\infty}(\cdot)e^{-i\omega t}dt$, of the constitutive relaxation relations yield

$$\operatorname{tr}\tilde{\sigma} = 3\hat{\kappa}\operatorname{div}\tilde{\boldsymbol{u}} \tag{4}$$

and

$$\operatorname{dev}\tilde{\sigma} = 2\hat{\mu}\operatorname{dev}\circ\nabla\tilde{\boldsymbol{u}}, \tag{5}$$

where $\hat{\kappa} = \kappa_\infty + i\omega\,\tilde{\mathring{\kappa}}$ and $\hat{\mu} = \mu_\infty + i\omega\,\tilde{\mathring{\mu}}$ are the complex bulk and shear moduli.

In general, compression and shear relaxation functions are independent. However, a simple and suitable model for rubber material assumes that they are dependent as

$$\kappa = b\,\mu_\infty h, \tag{6}$$

where the positive real valued constant $b \gg 1$, typically $\sim 10^{2\text{-}5}$. An extended fractional-order visco-elastic model applied in this nearly incompressible example gives longitudinal and transversal wave numbers as

$$k_L = \omega\sqrt{\frac{\rho}{b\mu_\infty}\,\frac{1}{1+\frac{4\hat{\mu}}{3b\mu_\infty}}} \tag{7}$$

and

$$k_T = \omega\sqrt{2\rho\lim_{\varpi\to\omega}\frac{\int_{0^-}^{\infty}{}^{\text{dev}}p(\alpha)\,(i\varpi)^\alpha\,d\alpha}{\int_{0^-}^{\infty}{}^{\text{dev}}q(\alpha)\,(i\varpi)^\alpha\,d\alpha}}, \tag{8}$$

where ${}^{dev}p$ and ${}^{dev}q$ are deviatoric material functions. In contrast to the synchronous material model, the loss factor for the longitudinal wave number is not normally overestimated in the rubber region.

2.3 *Formulation of the general problem*

A practical field representation at the junctions of the plates and mounting structures is by variables acting at the junction centers. In Figure 1, the stress tensor field is represented by the force and moment tensors, with the displacement tensor field represented by the displacement and rotation tensors. This simplification is appropriate when $2a\,|\,\boldsymbol{k} - (\boldsymbol{k}\cdot\boldsymbol{n})\boldsymbol{n}\,| \ll \pi$, where $\boldsymbol{k}$ is junction wave number, $\boldsymbol{n}$ the unit outward normal to the junction and a is cylinder radius.

A suitable description of the vibration isolators structure-borne sound properties provides the dynamic axial driving point and transfer stiffness, defined as

$$\tilde{k}_{11} = \tilde{\boldsymbol{f}}\cdot\boldsymbol{n}|_1 \,/\, \tilde{\boldsymbol{d}}\cdot\boldsymbol{n}|_1 \tag{9}$$

and

$$\tilde{k}_{12} = \tilde{\boldsymbol{f}}\cdot\boldsymbol{n}|_2 \,/\, \tilde{\boldsymbol{d}}\cdot\boldsymbol{n}|_1, \tag{10}$$

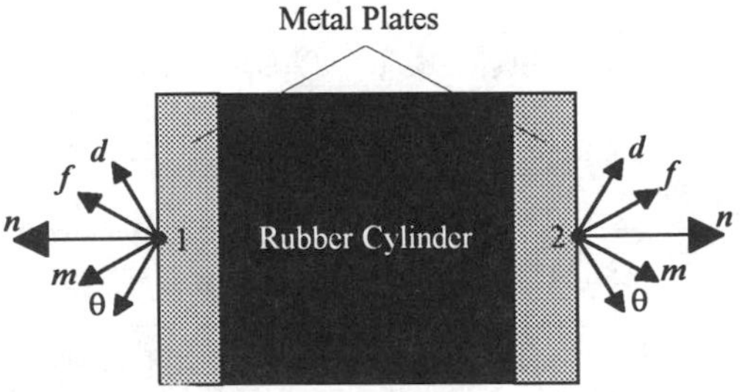

Figure 1. Representation of fields.

provided that $\tilde{\boldsymbol{d}}-(\tilde{\boldsymbol{d}}\cdot\boldsymbol{n})\boldsymbol{n}|_1=\tilde{\theta}|_1=\tilde{\boldsymbol{d}}|_2=\tilde{\theta}|_2=\boldsymbol{0}$ and that $\tilde{\boldsymbol{d}}|_1\neq\boldsymbol{0}$. In addition,

$$\tilde{k}_{22}=\tilde{\boldsymbol{f}}\cdot\boldsymbol{n}|_2/\tilde{\boldsymbol{d}}\cdot\boldsymbol{n}|_2 \tag{11}$$

and

$$\tilde{k}_{21}=\tilde{\boldsymbol{f}}\cdot\boldsymbol{n}|_1/\tilde{\boldsymbol{d}}\cdot\boldsymbol{n}|_2, \tag{12}$$

provided that $\tilde{\boldsymbol{d}}-(\tilde{\boldsymbol{d}}\cdot\boldsymbol{n})\boldsymbol{n}|_2=\tilde{\theta}|_2=\tilde{\boldsymbol{d}}|_1=\tilde{\theta}|_1=\boldsymbol{0}$ and that $\tilde{\boldsymbol{d}}|_2\neq\boldsymbol{0}$. Reciprocity implies $\tilde{k}_{21}=\tilde{k}_{12}$ and the particular vibration isolator symmetry $\tilde{k}_{22}=\tilde{k}_{11}$, which are subsequently used. Through similar procedures other components of dynamic stiffness are defined.

As this is the vibration isolator dynamic stiffness description only rigid body motions of the junctions at 1 and 2 are allowed. Although higher order modes are admissible in the cylinder, the plates may be rigid. This is plausible as longitudinal and transversal wave numbers in metal are small.

Consider the vibration isolator cylinder, a simple body consisting of continuously distributed rubber material and occupying a fixed open set $\mathcal{B}\subset\mathrm{R}^3$, where fixed $\mathcal{B}$ defines the cylinder reference configuration in its natural state; stress-free and undeformed.

The mixed boundary conditions, in the frequency domain, are the displacements

$$\tilde{\boldsymbol{u}}=\tilde{\boldsymbol{d}}, \qquad \text{given on } \partial_d^1\mathcal{B}, \tag{13}$$

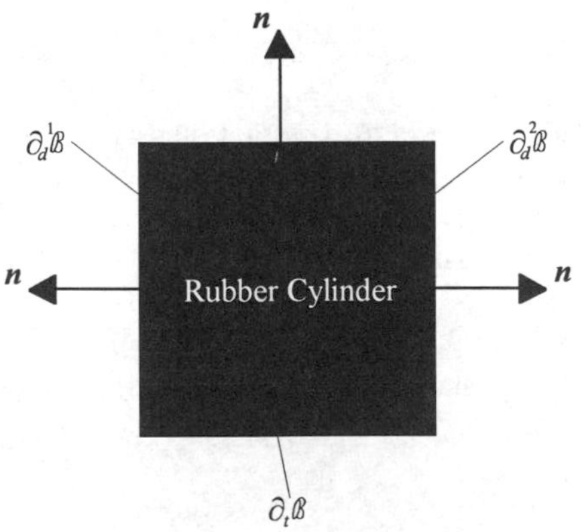

Figure 2. The rubber cylinder. Mixed boundary condition.

$$\tilde{\boldsymbol{u}}=\boldsymbol{0}, \qquad \text{given on } \partial_d^2\mathcal{B}, \tag{14}$$

and, the traction

$$[b\,\mu_\infty(\operatorname{div}\tilde{\boldsymbol{u}})\,\boldsymbol{i}+2\hat{\mu}\operatorname{dev}\circ\nabla\tilde{\boldsymbol{u}}]\cdot\boldsymbol{n}=\boldsymbol{0}, \qquad \text{given on } \partial_t\mathcal{B}, \tag{15}$$

where $\tilde{\boldsymbol{d}}-(\tilde{\boldsymbol{d}}\cdot\boldsymbol{n})\boldsymbol{n}|_{\partial_d^1\mathcal{B}}=\boldsymbol{0}$, $\overline{\partial_d^1\mathcal{B}\cup\partial_d^2\mathcal{B}\cup\partial_t\mathcal{B}}=\partial\mathcal{B}$, $\partial\mathcal{B}$ is the boundary of the cylinder and $\boldsymbol{n}$ is the unit outward normal to $\partial\mathcal{B}$. In addition, all the involved fields must be single-valued. It should be noted that there are geometrical singularities at $\overline{\partial_d^1\mathcal{B}}\cap\overline{\partial_t\mathcal{B}}$ and $\overline{\partial_d^2\mathcal{B}}\cap\overline{\partial_t\mathcal{B}}$. To ensure a unique solution, the additional physical condition needed is supplied by the requirement that the acoustical energy, contained in any finite neighborhood enclosing the singularity, must be finite. This requirement is known as an edge condition and has been discussed by Mittra and Lee (1971) in reference to electric and magnetic wave guides. A fruitful manipulation, having a sound basis in practice, is to soften the sharp edges with non-zero, infinitesimal radii of curvature.

The dynamic stiffness becomes

$$\tilde{k}_{11}=[\int_{\partial_d^1\mathcal{B}}[b\,\mu_\infty(\operatorname{div}\tilde{\boldsymbol{u}})\,\boldsymbol{i}+2\hat{\mu}\operatorname{dev}\circ\nabla\tilde{\boldsymbol{u}}]:(\boldsymbol{n}\otimes\boldsymbol{n})dS$$
$$-\omega^2\pi a^2 l_{mp}\rho_{mp}\tilde{\boldsymbol{d}}\cdot\boldsymbol{n}]/\tilde{\boldsymbol{d}}\cdot\boldsymbol{n} \tag{16}$$

and

$$\tilde{k}_{12}=\int_{\partial_d^2\mathcal{B}}[b\,\mu_\infty(\operatorname{div}\tilde{\boldsymbol{u}})\,\boldsymbol{i}+2\hat{\mu}\operatorname{dev}\circ\nabla\tilde{\boldsymbol{u}}]:(\boldsymbol{n}\otimes\boldsymbol{n})dS/\tilde{\boldsymbol{d}}\cdot\boldsymbol{n} \tag{17}$$

provided rigid body motions of the metal plate, $\tilde{\boldsymbol{d}}-(\tilde{\boldsymbol{d}}\cdot\boldsymbol{n})\boldsymbol{n}|_{\partial_d^1\mathcal{B}}=\tilde{\boldsymbol{u}}|_{\partial_d^2\mathcal{B}}=\boldsymbol{0}$ and that $\tilde{\boldsymbol{d}}\neq\boldsymbol{0}$, where ρ_{mp} and l_{mp} are plate density and thickness.

Helmholtz decomposition gives $\boldsymbol{u}=\operatorname{grad}\phi$ $+\operatorname{curl}\psi$ where ϕ and ψ are the scalar and tensor potentials. Through the gauge transformation $\psi'=\psi-\operatorname{grad}\phi'$, where $\operatorname{div}\psi'=0$, the Helmholtz equations read

$$\nabla^2\tilde{\phi}+k_L^2\tilde{\phi}=0 \tag{18}$$

and

$$\nabla^2\tilde{\psi}'+k_T^2\tilde{\psi}'=\boldsymbol{0}. \tag{19}$$

Formal derivation of the closed form solution to this problem is laborious, arising from possible geometrical singularities, the constitutive equation form and the imposed boundary conditions. In addition, boundary conditions on $\partial_d^1\mathcal{B}$ and $\partial_d^2\mathcal{B}$ are locally non-mixed, which refers to the problem of a non-separable nature.

The widespread technique of mode-matching, Mittra and Lee (1971), is probably the most forthright method of solving this problem, where the axial dependence must be separated and the remaining problem solved, resulting in the eigenmodes of the cross-section. Finally, provided these functions constitute a complete set, the total field is obtained by superposition of the eigenmodes then matching them to the boundary conditions on $\partial_d^1\mathcal{B}$ and $\partial_d^2\mathcal{B}$.

2.4 *Derivation of eigenvalues and eigenmodes*

Consider an infinite cylinder where a convenient representation of the geometry is in a cylindrical co-ordinate system with the *z*-axis directed along the main axis.

The axial dependence is readily separated as $(\tilde{\cdot}) = \frac{1}{2\pi}\int_{-\infty}^{\infty}(\tilde{\tilde{\cdot}})e^{-ik_z z}dk_z$, where $(\tilde{\tilde{\cdot}})$ is the spatial Fourier transformation of $(\tilde{\cdot})$, yielding the eigenmodes of the infinite cylinder to be used in the finite problem. The Helmholtz equations become

$$\underline{\nabla}^2\tilde{\tilde{\phi}} + k_L^2\,\tilde{\tilde{\phi}} = 0 \tag{20}$$

and

$$\underline{\nabla}^2\tilde{\tilde{\psi}}' + k_T^2\,\tilde{\tilde{\psi}}' = \mathbf{0}, \tag{21}$$

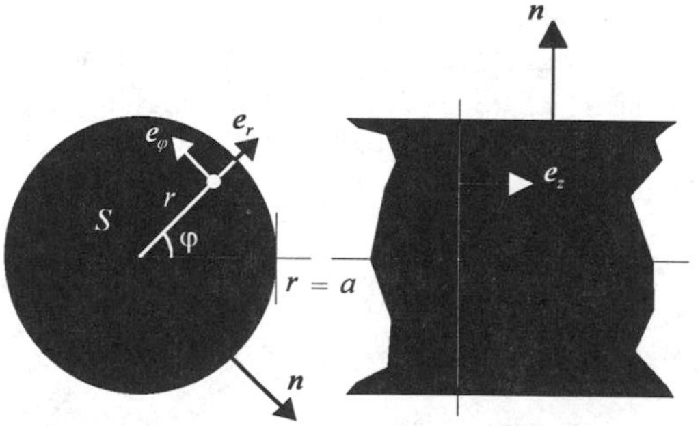

Figure 3. The geometry of infinite rubber cylinder.

where $\underline{\nabla}$ denotes a covariant derivative in the cylindrical co-ordinate system, with the formal replacement of $\frac{\partial}{\partial z} \leftarrow -ik_z$, and $\underline{\nabla}^2$ as the corresponding Laplace operator, the Helmholtz decomposition reads

$$\tilde{\tilde{\boldsymbol{u}}} = \underline{\mathrm{grad}}\,\tilde{\tilde{\phi}} + \underline{\mathrm{curl}}\,\tilde{\tilde{\psi}}, \tag{22}$$

where $\tilde{\tilde{\psi}}' = \tilde{\tilde{\psi}} - \underline{\mathrm{grad}}\,\tilde{\tilde{\phi}}'$, $\underline{\mathrm{div}}\,\tilde{\tilde{\psi}}' = 0$ and $(\underline{\cdot})$ means the substitution $\nabla \leftarrow \underline{\nabla}$ that is inherent in the operator $(\cdot)$. The boundary condition is

$$[b\,\mu_\infty(\underline{\mathrm{div}}\,\tilde{\tilde{\boldsymbol{u}}})\,\boldsymbol{i} + 2\hat{\mu}\,\mathrm{dev}\circ\underline{\nabla}\tilde{\tilde{\boldsymbol{u}}}]\cdot\boldsymbol{n}\Big|_{r=a} = \mathbf{0} \tag{23}$$

The relations (20) - (23), the gauge transformation together with the requirements of single valuedness and non-singularity, result in a general transcendental equation. With respect to the original problem, only the axially symmetric and non-torsional part is important, due to boundary conditions (13) and (14). The particular transcendental equation reads

$$[k_T^2 - 2k_{\perp T}^2]^2\vartheta(k_{\perp L}a) + 4k_{\perp L}^2[k_T^2 - k_{\perp T}^2]\vartheta(k_{\perp T}a) = 2k_{\perp L}^2 k_T^2, \tag{24}$$

where $\vartheta = x\,J_0(x)\,/\,J_1(x)$, also known as the Onoe function of first kind and first order, and J_n is the Bessel function of first kind and order n. The axial wave number is given by

$$k_z^2 = k_L^2 - k_{\perp L}^2 \tag{25}$$

or

$$k_z^2 = k_T^2 - k_{\perp T}^2. \tag{26}$$

The relations (24) - (26) constitute the dispersion relation $k_{z,n} = k_{z,n}(\omega)$ for the eigenmodes of the infinite cylinder, where $n \in \mathbb{Z}_+$ and labels different solutions; the eigenvalues $k_{z,n}$, $k_{\perp L,n}$ and $k_{\perp T,n}$. In terms of the potentials physical components, the sufficient and somewhere non-vanishing eigenmodes are $\tilde{\tilde{\phi}}_n \propto J_0(k_{\perp L,n}r)$ and $\tilde{\tilde{\psi}}_{\varphi,n} \propto J_1(k_{\perp T,n}r)$, $r \in [\,0, a\,[$, with the corresponding eigenvalues $k_{\perp L,n}$ and $k_{\perp T,n}$, respectively.

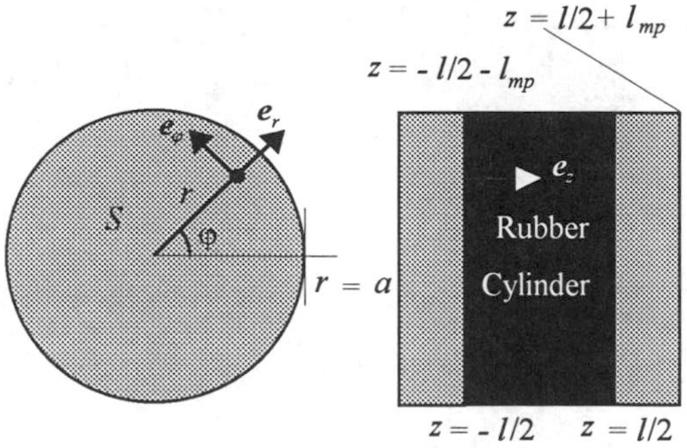

Figure 4. The geometry of the cylindrical vibration isolator.

2.5 *Derivation of dynamic stiffness*

Consider again the finite vibration isolator, where the geometry is represented in a cylindrical co-ordinate system with the z-axis directed along the main axis. Arbitrary axially symmetric and non-torsional stress and strain fields are obtained by superposition of the eigenmodes derived above, provided that these functions constitute a complete set. In the case of elastic materials, Love (1927) points out that the real eigenvalues are finite in number at a given frequency, thus the corresponding eigenfunctions cannot themselves form a complete set. Likewise, the purely imaginary eigenvalues are finite in number at a given frequency. However, Adem (1954) shows that there are also an infinite number of complex eigenvalues at a given frequency, rendering a complete set possible. The completeness extension to viscoelastic materials is provided by analytical continuation. Thus, the following expressions can be formulated for the potential fields within the finite cylinder

$$\tilde{\phi} = \sum_{n=1}^{\infty}[A_n^+ e^{-ik_{z,n}z} + A_n^- e^{ik_{z,n}z}]J_0(k_{\perp L,n}r) \tag{27}$$

and

$$\tilde{\psi}_\varphi = \sum_{n=1}^{\infty}[B_n^+ e^{-ik_{z,n}z} + B_n^- e^{ik_{z,n}z}]J_1(k_{\perp T,n}r), \tag{28}$$

where $r \in [0,a[$ and $z \in]-l/2, l/2[$. The coefficients are interrelated as $A_n^+ = \text{P}_n B_n^+$ and $A_n^- = -\text{P}_n B_n^-$, where

$$\text{P}_n = \frac{k_{\perp T,n}^2 - k_{z,n}^2}{2ik_{\perp L,n}k_{z,n}} \frac{J_1(k_{\perp T,n}a)}{J_1(k_{\perp L,n}a)}, \tag{29}$$

because of the boundary condition (15). The boundary condition (13) reads

$$\sum_{n=1}^{\infty}[C_n^+ e^{ik_{z,n}l/2} + C_n^- e^{-ik_{z,n}l/2}]U_n^r = 0 \tag{30}$$

and

$$\sum_{n=1}^{\infty}[C_n^+ e^{ik_{z,n}l/2} - C_n^- e^{-ik_{z,n}l/2}]U_n^z = -\tilde{d}^z, \tag{31}$$

where

$$U_n^r = -k_{\perp L,n}[k_{\perp T,n}^2 - k_{z,n}^2]J_1(k_{\perp T,n}a)J_1(k_{\perp L,n}r) - 2k_{\perp L,n}k_{z,n}^2 J_1(k_{\perp L,n}a)J_1(k_{\perp T,n}r), \tag{32}$$

$$U_n^z = -ik_{z,n}[k_{\perp T,n}^2 - k_{z,n}^2]J_1(k_{\perp T,n}a)J_0(k_{\perp L,n}r) + 2ik_{\perp L,n}k_{\perp T,n}k_{z,n}J_1(k_{\perp L,n}a)J_0(k_{\perp T,n}r), \tag{33}$$

$C_n^+ \text{X}_n = B_n^+$, $C_n^- \text{X}_n = -B_n^-$, $\text{X}_n = 2ik_{\perp L,n}k_{z,n}J_1(k_{\perp L,n}a)$ and $r \in [0,a[$. The boundary condition (14) reads

$$\sum_{n=1}^{\infty}[C_n^+ e^{-ik_{z,n}l/2} + C_n^- e^{ik_{z,n}l/2}]U_n^r = 0 \tag{34}$$

and

$$\sum_{n=1}^{\infty}[C_n^+ e^{-ik_{z,n}l/2} - C_n^- e^{ik_{z,n}l/2}]U_n^z = 0, \tag{35}$$

where $r \in [0,a[$.

It is not possible to formulate a simple orthogonal relation between the eigenmodes on the cross-section. The most straightforward way to obtain the coefficients from the relations (30) - (35) is probably through the point-matching technique but in most cases, more accurate results are achieved by the subregion method, which is a generalization of point-matching.

2.5.1 *Dynamic stiffness*

Point matching and subregion methods result in $\boldsymbol{Ax} = \boldsymbol{b}$ when the infinite series are truncated after M terms, where $\boldsymbol{A}$ is a known system matrix, $\boldsymbol{x}$ is an unknown coefficient vector and $\boldsymbol{b}$ is a known vector. Both methods exactly satisfy the equations of motion and the traction free boundary condition. Regarding the displacement boundary conditions, point matching fulfills the conditions at circles and the subregion method fulfills conditions in the mean within two subsequent circles, provided the equation systems are exactly determined and the rank of $\boldsymbol{A}$ is full. The dynamic stiffness (16) and (17) become

$$\tilde{k}_{11} = \pi a \omega^2 [2\rho \sum_{n=1}^{M} [D_n^+ e^{ik_{z,n} l/2} + D_n^- e^{-ik_{z,n} l/2}] S_n^{zz} - a l_{mp} \rho_{mp}] \quad (36)$$

and

$$\tilde{k}_{12} = 2\pi a \rho \omega^2 \sum_{n=1}^{M} [D_n^+ e^{-ik_{z,n} l/2} + D_n^- e^{ik_{z,n} l/2}] S_n^{zz}, \quad (37)$$

where

$S_n^{zz} = [2k_{\perp L,n}^2 - k_{\perp T,n}^2 + k_{z,n}^2] J_1(k_{\perp L,n} a) J_1(k_{\perp T,n} a) / k_{\perp L,n}$, $D_n^+ \tilde{d} = C_n^+$ and $D_n^- \tilde{d} = C_n^-$. Inasmuch as the series (27) and (28) fully represent the potential fields within the cylinder, the expressions (36) and (37) model the dynamic stiffness to any desired accuracy for a sufficient number of eigenmodes.

3 RESULTS AND DISCUSSION

To examine the methods in practice, a real vibration isolator has been analyzed, presenting numerical as well as measurement results. The formulations given above are implemented on a PC - Pentium Pro®. The computer code is written in LAHEY FORTRAN 90® with all calculations performed in double precision. Graphically, the results are presented by means of MATLAB®.

3.1 *Measurement*

A compression moulded cylindrical vibration isolator $l = 50.0$ mm long and $a = 50.0$ mm radius, equipped with circular steel plates 2.6 mm thick and $a = 50.0$ mm radius is used as test object. In order to facilitate safe mounting, additional plates 19.0 mm thick and $a = 50.0$ mm radius are attached to the plates. Total thickness is $l_{mp} = 21.6$ mm. The rubber material is vulcanized NR filled with small amounts

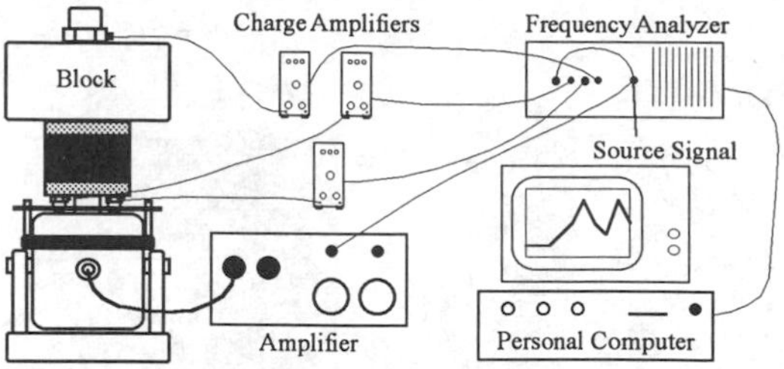

Figure 5. Measurement set up.

of non reinforcing carbon black. The nominal hardness is 40° IRH and the density 1 050 kg/m^3. The vibration isolator axial dynamic transfer stiffness is measured by an indirect method with the isolator mounted between a block and the moving table of an electro-dynamic vibration generator membrane. The moving table and block motions are measured by piezo-electric accelerometers. Data collection is performed by a 4-channel frequency analyzer, also supplying the signal to the generator via an amplifier with measurements verified by a personal computer.

3.2 *Calculated stiffness*

The fractional Kelvin-Voigt material model embodying both generalized and non-generalized material functions is applied to the nearly incompressible model. The least square estimated parameters are: ${}^{dev}p = \delta$, ${}^{dev}q = 2\mu_\infty[\delta + (\mu_\nu/\mu_\infty)^\alpha w]$, $\mu_\infty = 5.94\ 10^5$ N/m^2, $\mu_\nu = 13.0$ Ns/m^2, $w = [{}^{(\alpha_1)}h - {}^{(\alpha_2)}h]/\Delta\alpha$, $\Delta\alpha = \alpha_2 - \alpha_1$, $\alpha_1 = 0.080$, $\alpha_2 = 0.625$ and $b = 2.22\ 10^3$, together with $\rho = 1\ 050$ kg/m^3.The estimated material parameters are realistic: the fractional derivatives lie within the allowed limits, the equilibrium Poisson ratio is 0.4998, the density equals the stated value, while the equilibrium bulk and shear moduli slightly exceed the stated values. The shear modulus in the extended frequency range 1 to 10 000 Hz is in Figures 6 and 7 while the bulk modulus is 1.32 10^9 N/m^2.

The calculated stiffness is determined by point matching, using 400 equidistant collocation radii and $M = 100$; the equation system is two fold over-determined. The frequency points coincide with the measurement points.

The calculated stiffness is in Figures 8 - 13. Figure 13 shows the driving point stiffness while Figures 8 - 12 show the transfer stiffness. In addition and for comparison, the results of the axial transfer stiffness measurement are in Figures 8 - 12. The curves from

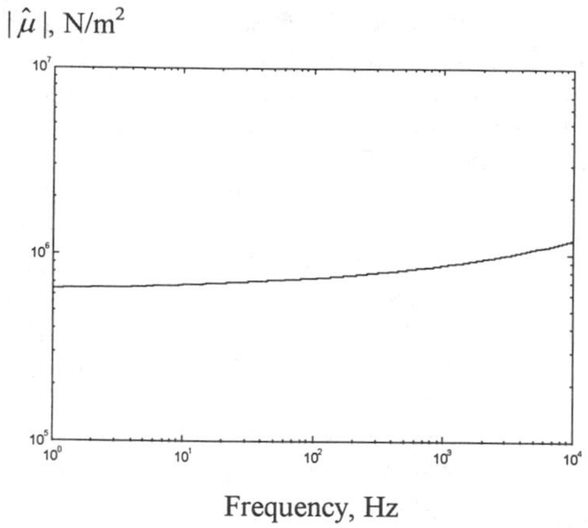

Figure 6. Magnitude of shear modulus. 1 - 10 000 Hz.

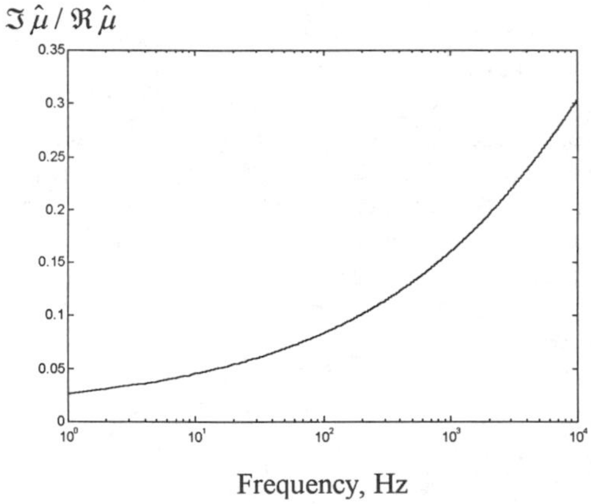

Figure 7. Loss factor of shear modulus. 1 - 10 000 Hz.

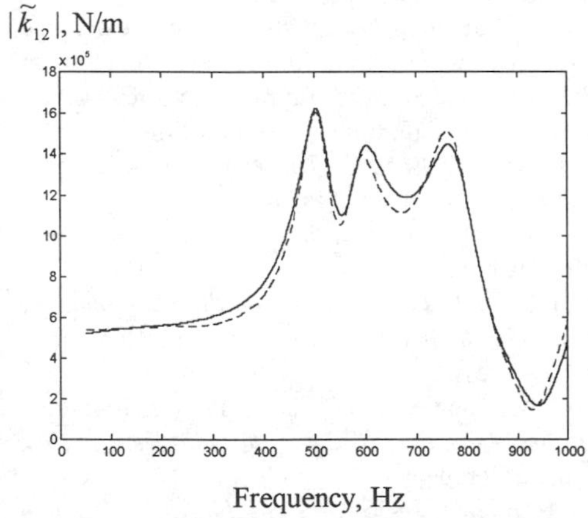

Figure 8. Calculated (solid) and measured (dashed) magnitude of transfer stiffness. 50 - 1 000 Hz.

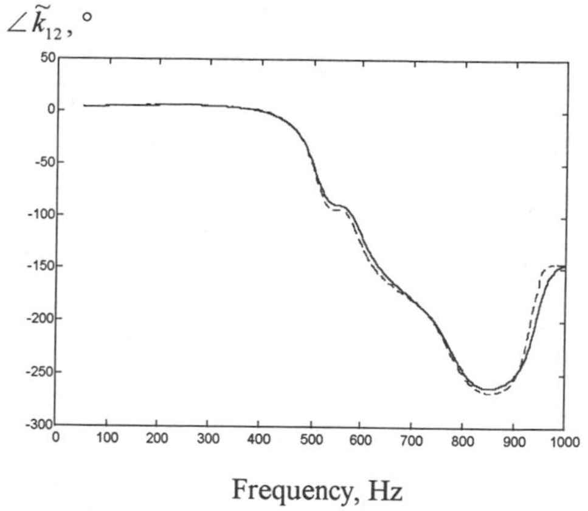

Figure 9. Calculated (solid) and measured (dashed) unwrapped phase of transfer stiffness. 50 - 1 000 Hz.

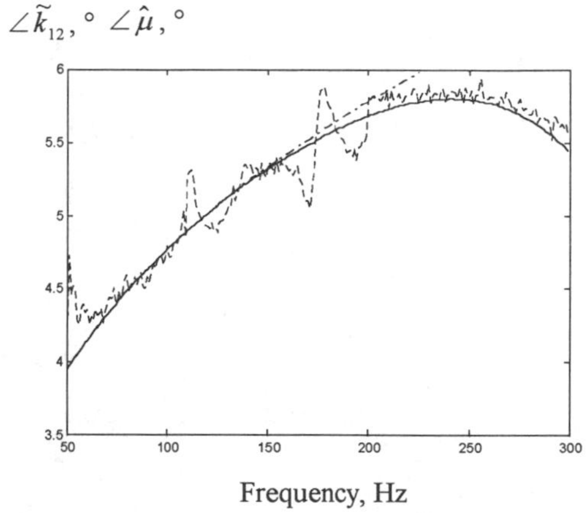

Figure 10. Phase of calculated (solid), measured (dashed) transfer stiffness and of shear modulus (dash-dotted). 50 - 300 Hz.

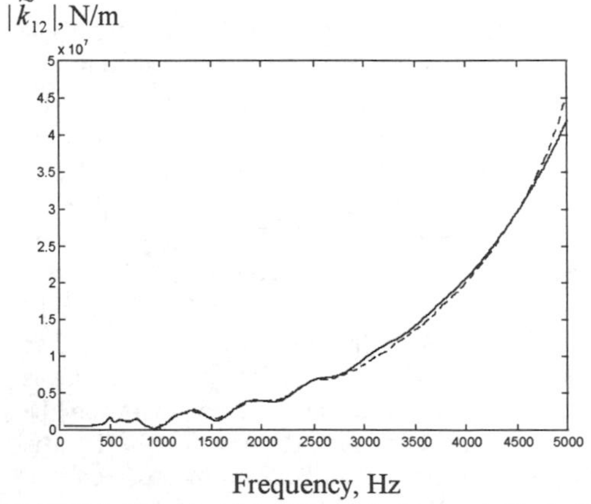

Figure 11. Calculated (solid) and measured (dashed) magnitude of transfer stiffness. 50 - 5 000 Hz.

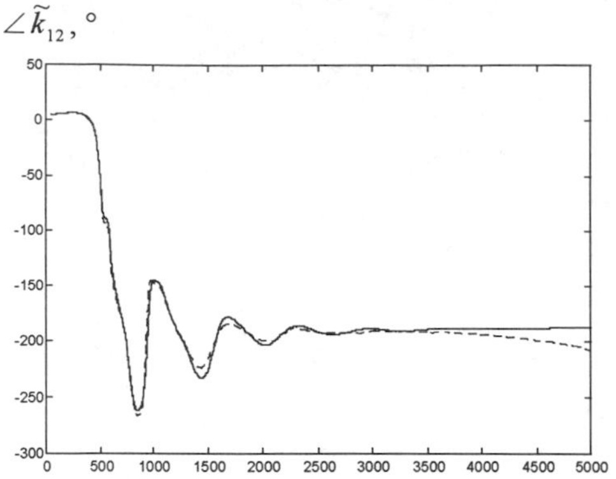

Figure 12. Calculated (solid) and measured (dashed) unwrapped phase of transfer stiffness. 50 - 5 000 Hz.

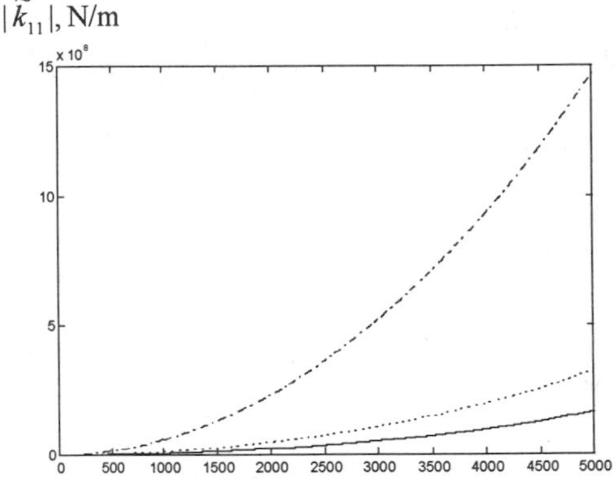

Figure 13. Calculated magnitude of driving point stiffness. Bonded steel plates not included (solid), 2.6 mm (dotted) and (2.6 + 19.0) mm bonded steel plates included (dash-dotted). 50 - 5 000 Hz.

the measurement are plotted in dashed lines while those from the calculations are plotted in solid lines.

The measured and calculated axial transfer stiffness agree very well. In particular, at the low frequency plateau, the peaks and the troughs in Figure 8 are almost exactly reproduced. In Figure 9, the calculated phase corresponds to the measured phase with only minor discrepancies. The low frequency fragment of the phase curve in Figure 10 results mainly from the material properties. For comparison, the phase curve of the shear modulus is plotted in a dash-dotted line closely following the phase curve of the dynamic stiffness to 150 Hz. The ensuing deviations are due to geometrical effects. Apparently, the material damping of the present material model is consistent with the measurement. The fluctuations of the measured phase for $f < 200$ Hz, however small, result mainly from uncanceled secondary motions of the vibration generator membrane. Magnitude and phase curves in the whole frequency region 50 - 5 000 Hz are shown in Figures 11 and 12, where measurements and calculations match very well. The small deviations close to 5 000 Hz, slightly more conspicuous for the phase, result from non rigid motions of the plates.

The driving point stiffness depends upon the dynamic properties of the plates. Three isolator plate configurations are provided in Figure 13; plates not included are plotted in solid lines, 2.6 mm plates included are plotted in dotted lines and (2.6 + 19.0) mm plates are plotted in dash-dotted lines.

4 CONCLUSIONS

In presenting a linear axial dynamic stiffness model for cylindrical vibration isolators in the audible frequency range of particular complexity and interest in noise abatement, with material damping, higher order modes and structure borne sound dispersion extensively investigated, the problems of simultaneously satisfying the boundary conditions at the lateral and radial surfaces of the cylinder are removed. The presented model is shown to agree strikingly well with measurements.

5 REFERENCES

J. ADEM 1954. *Quarterly of Applied Mathematics* **12**, 261-75. On the axially-symmetric steady wave propagation in elastic circular rods.

C. CHREE 1889. *Transactions of the Cambridge Philosophical Society* **14**, 250-369. The equations of an isotropic elastic solid in polar and cylindrical coordinates, their solutions and applications.

R. M. CHRISTENSEN 1982. *Theory of Viscoelasticity*, Second Edition. Academic Press.

Y. C. FUNG 1965. *Foundations of Solid Mechanics*. Prentice Hall.

A. E. H. LOVE 1927. *A Treatise on the Mathematical Theory of Elasticity*, Forth Edition. Cambridge University Press.

R. MITTRA and S. W. LEE 1971. *Analytical Techniques in the Theory of Guided Waves*. MacMillan Company.

L. POCHHAMMER 1876. *Journal für die reine und angewandte Mathematik* **81**, 324-36. Über die Fortpflanzungsgeschwindigkeiten kleiner Schwingungen in einem unbegrenzten isotropen Kreiszylinder.

Modern Practice in Stress and Vibration Analysis, Gilchrist (ed.) © *1997 Balkema, Rotterdam, ISBN 90 5410 896 7*

Optimisation of an active noise control system for reducing the sound transmission through a double-glazing window

P. De Fonseca
Fund for Scientific Research & Katholieke Universiteit van Leuven, Division of Production Engineering, Machine Design and Automation, Belgium

P. Sas & H. Van Brussel
Katholieke Universiteit van Leuven, Division of Production Engineering, Machine Design and Automation, Belgium

ABSTRACT: The implementation of an active noise control system in the cavity of a double-glazing window has been investigated. In a preliminary theoretical analysis the spatial distribution of both the control loudspeakers and the error microphones of the control system has been optimised with respect to the reduction of the acoustic potential energy in the cavity between the two glass plates using a simple genetic algorithm. The comparison of the reductions achieved by the optimised control system, with the reductions achievable with an idealised control system, which assumes a perfect knowledge of the dynamic characteristics of the double-glazing window, permits an evaluation of the performance of the optimisation process. It appears from the experimental analysis that the optimised control configuration yields far better increases in the sound transmission loss than a randomly chosen configuration. The obtained reductions in the transmitted sound power illustrate the potentials of active noise control for improving the sound insulation characteristics of double-glazing windows.

INTRODUCTION

Low-frequency noise causes much more inconveniences for man than indicated by the traditionally used dB(A)-curve, which is not representative for the real annoyance in the low-frequency band (Kuwano et al., 1989). In addition, nowadays people observe an ever increasing low-frequency component in the environmental noise spectrum (Berglund et al., 1996). This is mainly due to the growing economic activity during the last decades, although also other types of low-frequency noise sources emerge in the western society (e.g. house music). On the other hand, double-glazing windows have typically a very limited sound transmission loss in the low-frequency range, sometimes even smaller than the transmission loss of comparable single-glazing windows.

Therefore, this research project focuses on the possibilities of improving the low-frequency sound insulation characteristics of a double-glazing window by means of active noise control.

The efficient implementation of an active noise control system requires a detailed identification of the dynamic behaviour of the considered vibro-acoustic system. A modal analysis, analytical and experimental, forms the basis for the further study of the influence of active noise control on the sound transmission through the double-glazing window. A simplified model of the active noise controller has been developed and included in the vibro-acoustic model. This yields an integrated simulation model which can be used to determine an optimised set of sensor and actuator locations. The reduction of the acoustic potential energy in the cavity between both glass plates has been chosen as the objective function to maximise by means of a genetic algorithm. Finally, the optimised control configuration is implemented in the test set-up and the obtained sound transmission reductions are compared to those obtained with an arbitrarily chosen configuration, and with those simulated with the vibro-acoustic model.

I. VIBRO-ACOUSTIC ANALYSIS OF THE DOUBLE-GLAZING WINDOW

A. Description of the test set-up

The detailed experimental study of the active control of the sound transmission through a double-glazing window in the low-frequency band requires an appropriate test set-up. In order to justify a reliable comparison between the situations with and without active noise control, the sound power should be transmitted from the sending room to the receiving room only through the double-glazing window, and not through the surrounding framework. Therefore, the test set-up is incorporated in the partition of the transmission room in the Laboratory of Acoustics at the K.U. Leuven.

The double-glazing window under investigation consists of two 3 *mm* thick glass plates of 1,23 *m* by 1,48 *m*. The edges of the glass plates are attached to a concrete wall with a layer of putty. The distance between both plates is kept constant at 0,1 *m* by means of a stiff wooden frame. The inner dimensions of this frame (1,1 *m* x 1,35 *m*) determine the boundaries of the acoustic cavity between the glass plates. This wooden frame allows an easy integration of the control loudspeakers and microphones, serving respectively as transducers in the experimental identification phase and as error sensors in the active control system. An incident sound field is generated by a woofer in the sending room, the so-called primary sound source. The driving signal for this primary source is a bandlimited random noise signal generated by the LMS Cada-X measurement software. The transmitted sound power is measured by scanning a one-dimensional intensity probe along the glass plate at the reception side.

The present test set-up allows only to measure the sound intensity radiated by the glass plate in the receiving room, and not the incident sound power in the sending room. In order to eliminate the influence of disturbances, e.g. a small temperature change in the laboratory, it is necessary to measure the radiated sound intensity without the active noise control system again for each different control configuration. This procedure assures a reliable comparison between the situations with and without active noise control, and also between the performances of the control system in different configurations.

B. Sound transmission through a double-glazing window

The low-frequency sound transmission characteristics of double-wall partitions are well-known (Desmet and Sas, 1995). Also in the considered double-glazing window, several coupled structural-acoustical resonances appear already at low frequencies due to the flexibility of the glass plates. Together with a 'mass-air-mass'-like phenomenon these resonances result in a substantial

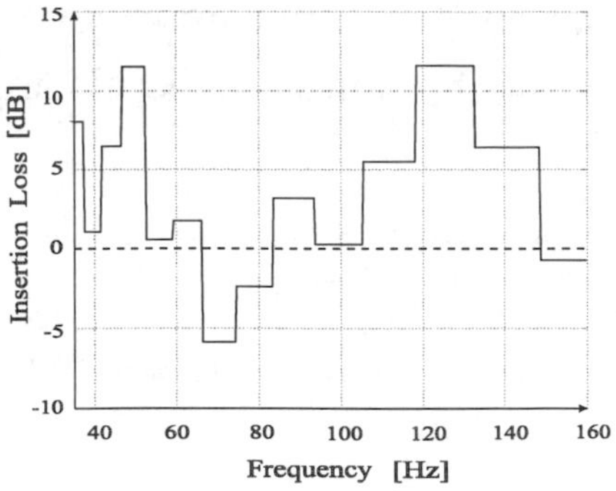

Figure 1 : Insertion loss of the double-glazing window

decrease of the low-frequency transmission loss. At this mass-air-mass resonance, the glass plates oscillate in anti-phase against the stiffness of the compressible acoustic medium in the cavity.

Figure 1 shows the measured insertion loss of the considered double-glazing window. The insertion loss (IL) is defined as the difference between the transmitted sound intensity of a double- and a single-glazing window, obtained by removing one glass plate:

$$IL[dB]=10\log\left(\frac{I_{single}}{I_{double}}\right) \qquad (1)$$

Obviously, the insertion loss at the 'mass-air-mass'-like resonance (around 70 Hz) and at the coupled resonances (e.g. around 100 and 160 Hz) decreases substantially. In order to improve the low-frequency transmission loss, the potential of an active noise control system, implemented in the cavity, has been investigated.

C. Simulation of the dynamic behaviour

The first step in the analysis of the dynamic behaviour of the double-glazing window is the construction of a finite element model. This model contains both the structural subsystem, i.e. the glass plates, and the acoustical subsystem, i.e. the acoustic medium in the cavity between the glass plates, and the fluid-structure interaction between both subsystems. This fluid-structure interaction takes into account the additional loading of the structural subsystem due to the presence of the acoustical medium, and the excitation of the acoustical medium by the vibrating structural subsystem. Neglecting any type of damping the vibro-acoustic system is described by :

$$(K - \omega^2 M)\{X\}$$

$$= \left(\begin{bmatrix} K_s & K_c \\ 0 & K_a \end{bmatrix} - \omega^2 \begin{bmatrix} M_s & 0 \\ M_c & M_a \end{bmatrix} \right) \begin{Bmatrix} u \\ p \end{Bmatrix} = \begin{Bmatrix} F_s \\ F_a \end{Bmatrix} \quad (2)$$

The column vector of unknowns {X} contains the displacements u in the structural nodes and the pressures p in the acoustical nodes. K_s, K_a and M_s, M_a are the structural and acoustical stiffness respectively mass matrices, while the fluid-structure interaction is modelled in the coupling matrices K_c and M_c. F_s and F_a are external excitation vectors in the structural respectively acoustical nodes.

The modal transformation is used to reduce the number of degrees-of-freedom of the original system:

$$\begin{Bmatrix} u \\ p \end{Bmatrix} = \begin{bmatrix} \phi_s \\ \phi_a \end{bmatrix} \{q\} = [\Phi]\{q\} \quad (3)$$

in which the columns of the mode shape matrix $[\Phi]$ are the undamped eigenvectors of the coupled vibro-acoustic system, orthonormalised with respect to the mass matrix. These eigenvectors consist of two parts, the first being related to the structural subsystem and the second to the acoustical subsystem.

By assuming proportional damping, the modal model becomes :

$$[a]\{q\} = \left([k] + j\omega[c] - \omega^2 [m]\right)\{q\} = \{f\} \quad (4)$$

in which the modal mass matrix [m] equals the *m*-dimensional unity matrix. The diagonal modal stiffness matrix contains the squares of the coupled natural frequencies ω_i :

$$[k] = \text{diag}\left(\omega_i^2\right), \quad (5)$$

and the diagonal modal damping matrix equals :

$$[c] = \text{diag}(2\varsigma_i\omega_i) \quad (6)$$

with ς_i the modal damping ratio of the ith coupled mode. Pre-multiplying the excitation force vector with the hermitian (superscript ^{H}) of the mode shape matrix yields the modal excitation vector :

$$\{f\} = [\Phi]^H \{F\} \quad (7)$$

D. Experimental validation of the vibro-acoustic model

The response vector $\{u^T \ p^T\}^T$ in (3) for the considered double-glazing window is determined by calculating the mode shape matrix $[\Phi]$ in a coupled finite element modal analysis. The natural frequencies ω_i also result from this finite element analysis. However, in order to improve the agreement in dynamic behaviour between the numerical model and the real double-glazing window, the natural frequencies ω_i and the modal damping ratios ς_i of the modal system matrix [a] in (4) result from an experimental modal analysis. Therefore the frequency response functions (FRF's) are measured in 418 points on the glass plate in the sending room by means of a scanning laser vibrometer and with excitation by the primary sound source driven with a burst random signal. With the same excitation signal also 36 FRF's of the acoustic pressure in the cavity between the two glass plates are measured. An experimental modal analysis is performed on both sets of measurements individually. The resulting mode shapes and eigenfrequencies of both analyses agree very well with each other. The deviation between the numerically calculated and the experimentally determined eigenfrequencies amounts to less than 10 % for most eigenmodes. Also the agreement between the numerical and the experimental mode shapes is reasonably good.

As the translation of the incident sound field in the sending room into nodal structural forces on the glass plate is not straightforward, the uncontrolled modal response vector $\{q_p\}$ is identified in a least squares form from a vector $\{X\}_{meas}$ of measured glass plate displacements and cavity pressures :

$$\min \left\| \{X\}_{meas} - [\Phi]\{q_p\} \right\|_2 \quad (8)$$

in which the subscript 2 refers to the usual 2-norm.

This procedure enables to construct a qualitatively reliable model as it is partially based on experimental results. Moreover, the model is easy to handle because the huge number of nodal degrees-of-freedom (more than 7000) in the finite element model has been reduced to only 85 modal coordinates, corresponding to the first 85 eigenmodes of the coupled vibro-acoustic system.

II. MODELING OF THE ACTIVE NOISE CONTROL SYSTEM

A. Development of a general feedforward controller model

From a control point of view, the most common approach in active noise control is the use of an adaptive feedforward algorithm. The control system drives a number of control actuators based on the information in a reference signal. In order to obtain a good controller performance over the frequency range of interest, this reference signal should be fully coherent with the primary or disturbing excitation. In the present simplifying model the reference signal is assumed to be exactly equal to the disturbing excitation.

Considering a linear system, the total dynamic response is the superposition of the response to the disturbing excitation and the response to the secondary excitation by the control actuators :

$$\begin{Bmatrix} u \\ p \end{Bmatrix}_{total} = \begin{Bmatrix} u \\ p \end{Bmatrix}_{p} + \begin{Bmatrix} u \\ p \end{Bmatrix}_{c} = \begin{bmatrix} \Phi_s \\ \Phi_a \end{bmatrix} [a]^{-1} \begin{bmatrix} \Phi_s^H & \Phi_a^H \end{bmatrix} \left(\begin{Bmatrix} F_s \\ F_a \end{Bmatrix}_p + \begin{bmatrix} P_s \\ P_a \end{bmatrix} \{L_c\} \right) \quad (9)$$

The subscript *p* refers to the primary disturbing excitation, while the subscript *c* refers to the secondary or control excitation. The vector $\{L_c\}$ *(cx1)* contains the source strength of each of the *c* control actuators. The matrix $[P]$ indicates the nodal position of each control actuator. It also consists of two parts because generally the control actuators can be either structural (e.g. shakers) or acoustical actuators (e.g. loudspeakers).

The controller determines the strengths $\{L_c\}$ of the control loudspeakers by minimising a cost function Z, which depends on the signals of the error sensors $\{E\}$. When the control configuration contains *e* error sensors, the vector of error signals $\{E\}$ *(ex1)* is:

$$\{E\} = \begin{bmatrix} S_s & S_a \end{bmatrix} \begin{Bmatrix} u \\ p \end{Bmatrix}_{total} \quad (10)$$

Similarly to $[P]$, the matrix $[S]$ indicates the nodal position of each of the *e* error sensors. The cost function Z is the sum of the squared error signals. Mathematically the following minimisation problem has then to be solved :

$$\min Z = \{E\}^H \{E\} \quad (11)$$

Substituting (9) and (10) into (11) yields :

$$\min Z = \left(\{F_p\} + [P]\{L_c\}\right)^H [\Phi]\left[[a]^{-1}\right]^H [\Phi]^H [S]^H [S][\Phi][a]^{-1}[\Phi]^H \left(\{F_p\} + [P]\{L_c\}\right) \quad (12)$$

The cost function Z reaches its minimum when the derivative to the vector containing the variable secondary source strengths equals zero :

$$\frac{\partial Z}{\partial \{L_c\}} = 0 \quad (13)$$

The resulting $\{L_c\}$ that minimises Z, is ($^+$ denotes the pseudo-inverse) :

$$\{L_c\} = -[h]^+ [S][\Phi][a]^{-1}[\Phi]^H \begin{Bmatrix} F_s \\ F_a \end{Bmatrix}_p \quad (14)$$

with $[h] = [S][\Phi][a]^{-1}[\Phi]^H [P]$ (15)

When a vibro-acoustic system is subject to a primary excitation and controlled by a given control configuration (i.e. with a certain amount of error sensors *e* and of control loudspeakers *c* at positions $[S]$ respectively $[P]$), the resulting response is obtained by combining (9) and (14).

B. Application to the double-glazing window

From the general controller model, developed in the previous paragraph, it is clear that both structural and acoustical sensors and actuators can be considered for the implementation on a general vibro-acoustic system. When the general model is used for the description of the active noise control system implemented on a double-glazing window, some restrictions have to be made on the type and the positions of the sensors and the actuators.

First of all the active noise control system should not influence the transparency of the window. This requirement implies that all actuators and sensors should be placed in the vicinity of the edges of the glass plates, in case of structural control, or along the border of the cavity between both glass plates, in case of acoustical control. However, in case of structural control, the amplitude of the displacement of the more or less clamped glass plates near their edges is rather limited. Hence the possibility of using shakers as actuators and accelerometers as sensors for the active control of the sound transmission through the double-glazing window can intuitively be discarded.

Secondly, the performance of the active control system, in terms of increase of the sound transmission loss of the double-glazing window, should be more or less independent of the acoustic characteristics of the interior of the buildings in which it is incorporated. No major modifications of the interior of the buildings can be allowed for the implementation of the actively controlled double-glazing window. Consequently, the use of error microphones and control loudspeakers in the radiated sound field is not possible.

Therefore, the choice of the error sensors and the control actuators for the active noise control system is limited to microphones and loudspeakers along the border of the cavity between the two glass plates. This yields an actively controlled double-glazing window which remains almost as compact as an ordinary double-glazing window, which does not require any modifications of the rooms in which it is incorporated, and whose performances are expected to be more or less independent of the acoustic properties of its environment.

III. OPTIMISATION OF LOUDSPEAKER AND MICROPHONE POSITIONS

The effectiveness of an active control system is strongly dependent on the number and positions of both the error sensors and control actuators. A major objective of this research work is therefore the optimisation of the positions of the error microphones and the control loudspeakers along the border of the cavity between the two glass plates for several control systems (i.e. with different numbers of microphones and different numbers of loudspeakers). Instead of using a very time consuming experimental trial-and-error procedure, the optimum control configuration can be determined from the vibro-acoustic model by optimising some objective function. Clearly, the sound energy, radiated into the receiving room is the most appropriate objective function to minimise. When the optimisation has to be performed over a large number of frequency lines, the computational burden of this objective function becomes an insurmountable disadvantage. In order to limit the computational effort, the average reduction of the acoustic potential energy is used instead, yielding the following objective function Π :

$$\Pi = \frac{1}{f_N - f_0} \sum_{f=f_0}^{f_N} \{p(f)\}^H \{p(f)\} \qquad (16)$$

The vector $\{p(f)\}$ contains the nodal pressure distribution in the cavity when the double-glazing window is subject to a primary excitation and is controlled by an active noise control system with a certain number of error microphones and control loudspeakers at certain positions along the border of the cavity.

The objective function (16) has been optimised by means of a genetic algorithm (Goldberg, 1989, and Hansen et al., 1996). Genetic algorithms belong to the so-called directed random search techniques. The form of guidance is based on Darwin's "survival of the fittest"-theories. For the present study, a genetic algorithm using a binary coding of the variables in the optimisation process has been used. The feasible positions for the control loudspeakers and the error microphones are restricted to 78 discrete positions in the wooden frame between the glass plates. Each position is coded in a 7-bit binary representation. One gene consists of the binary representations for the respective control loudspeakers and error microphones one after each other. The initial mating pool contains 26 randomly generated genes. They are all attributed a weight based on the value of their *fitness*-function F, defined as the percentage reduction of the objective function :

$$F = \frac{\Pi_o - \Pi_c}{\Pi_o} .100\% \qquad (17)$$

in which the subscript *o* refers to the situation without active noise control and the subscript *c* to the situation with active noise control.

Three genetic operators are included in the genetic algorithm : crossover, translation and mutation. Translation and mutation processes occur with a smaller probability than the crossover process.

The positions have been optimised for different types of control configurations. Due to the limited calculation speed of the controller hardware, the number of error microphones is restricted to four, and the number of control loudspeakers to two (a so-called 4i2o control system). This results in more than 180 billion possible configurations. Taking into account that the calculation of the objective function takes at least a few seconds on a UNIX workstation, exhaustive search is unfeasible. The genetic algorithm is therefore used to generate a near-optimum configuration within a reasonable time span.

IV. EVALUATION OF THE PERFORMANCE OF THE OPTIMISED CONFIGURATION

In this section, an acoustic energy control system is developed (referred to by the subscript "en"), which directly minimises the acoustic potential energy by assuming an huge (theoretically infinite) number of error sensors in each point of the cavity. It provides a good theoretical limit to evaluate the performance of a control system with a small number of conventional error sensors in an optimised configuration.

The cost function Z_{en}, which is minimised by this acoustic energy controller, is the total acoustic energy in the cavity. Using the discrete finite element description of the double-glazing window, this acoustic potential energy is directly related to the sum of the squared acoustic pressures in each node of the model. Mathematically the following minimisation problem has to be solved at each frequency line in order to determine the control source strengths $\{L_c\}_{en}$:

$$\min Z_{en} = \{p\}_{total}^H \{p\}_{total}, \tag{18}$$

where

$$\{p\}_{total} = \{p\}_p + \{p\}_c = [\Phi_a][a]^{-1}[\Phi]^H\left(\{F\}_p + [P]\{L_c\}_{en}\right) \tag{19}$$

Using a similar methodology as in the third paragraph, the resulting vector $\{L_c\}_{en}$ that minimises Z_{en} is given by :

$$\{L_c\}_{en} = -[h]_{en}^+[\Phi_a][a]^{-1}[\Phi]^H\{F\}_p \tag{20}$$

$$\text{with } [h]_{en} = [\Phi_a][a]^{-1}[\Phi]^H[P] \tag{21}$$

Figure 2 shows the simulation results of a comparison between the acoustic potential energy in the cavity of the double-glazing window for the situation without active control (dashed line), with an active control system with two control loudspeakers and four error microphones in locations which are optimised using the genetic algorithm (full line), and with an active acoustic energy control system (dotted line), as described above, with two control loudspeakers in the same locations as in the previous case.

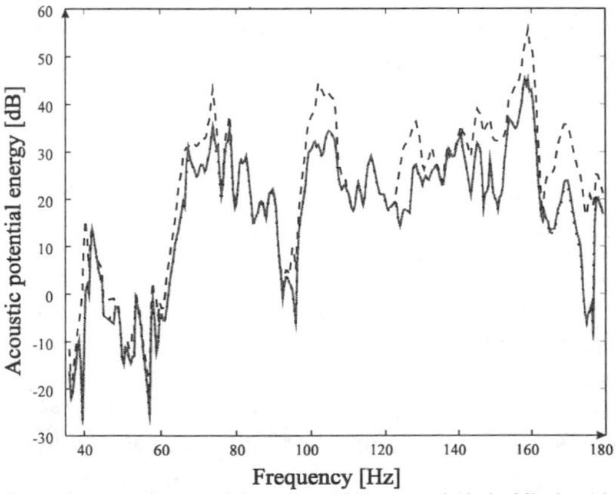

Figure 2 : acoustic potential energy without control (dashed line), with an optimised configuration (full line), and with acoustic energy control (dotted line)

The full line coincides over almost the entire frequency band with the dotted line, indicating that the control system with an optimised microphone and loudspeaker configuration nearly reaches the maximum possible reduction with the given loudspeaker positions. The percentage reduction of the acoustic potential energy averaged over the frequency band of interest is actually 85,68 % for the case with the acoustic energy control system, and 85,33 % for the case with the ordinary control system in an optimised configuration. Clearly the genetic algorithm performs extremely well as the microphone positions are concerned in the simultaneous optimisation of microphone and loudspeaker positions. It achieves almost 99 % of the attainable reduction by selecting the present microphone configuration for this loudspeaker configuration. When the microphones of the conventional control system are located in arbitrarily chosen positions along the border of the cavity and the loudspeakers are kept in the optimised positions, the performance of the control system is far from being optimal. It succeeds in reducing the acoustic potential energy with only 54 %, or 63 % of the maximum attainable reduction with the given loudspeaker positions. This result clearly illustrates the importance of optimising both sensor and actuator locations in an active (noise) control system.

From an optimisation point of view, the concept of the energy controller proves to be a very simple but efficient way to evaluate the quality of the best solution found by the genetic algorithm. Of course it can be used in combination with any other optimisation routine.

V. EXPERIMENTAL RESULTS

The experimental study of the performance of the optimised active noise control system on a double-glazing window is the final purpose of this research work. The main conclusion of the theoretical analysis is that significant global reductions in the acoustical potential energy can be achieved with an optimised control configuration. The experimental investigations on the double-glazing window confirm these theoretical results. All tests are performed with the active noise controller developed at the K.U.Leuven. The adaptive feedforward algorithm is implemented on a digital signal processing board (DSP, type TMS320C30) inserted in a standard PC. For sake of simplicity, the driving signal for the primary sound source is used as the reference signal by the feedforward controller. In a general case the reference signal is provided by a reference microphone which detects the incident sound field before it reaches the window.

Figure 3 compares the measured reduction of the sound intensity in the receiving room for an active noise control system with four error microphones and two control loudspeakers (4i2o) in an arbitrarily chosen configuration (left), and the calculated reduction of the acoustical potential energy in the cavity for the same configuration (right). It appears that the vibro-acoustic model yields qualitatively reliable predictions of the attainable reductions for a certain active control configuration. However some shifts of peaks in the spectra may be observed, for example between 35 and 40 Hz, between 65 and 70 Hz, and between 110 and 120 Hz. These shifts might be due to the absence of both the sending and the receiving room in the model of the double-glazing window. The assumption that these rooms are perfectly reverberant is not really fulfilled in the very low-frequency range.

A comparison between the figures 1 and 3 clearly indicates that the active noise control system achieves high reductions in the transmitted sound intensity, especially in those frequency bands where the insertion loss of the double-glazing window is negative or close to zero, which means that the double-glazing window transmits more or at least as much sound as the corresponding single-glazing window. As explained in section I these minima in the sound insulation characteristic are mainly due to the coupled resonances of the double-panel partition. When there is a strong interaction between the structural and the acoustical subsystem, the active noise control system in the cavity can efficiently influence the radiating glass plate, and hence substantially reduce the transmitted sound power by attenuating the sound field in the cavity.

Figure 4 shows the measured radiated sound intensities with (full line) and without active noise control (dashed line) for two different control configurations. The figure on the left gives the result for the optimised configuration of a control system with two control actuators and two loudspeakers (2i2o). Global reductions of 5 to 10

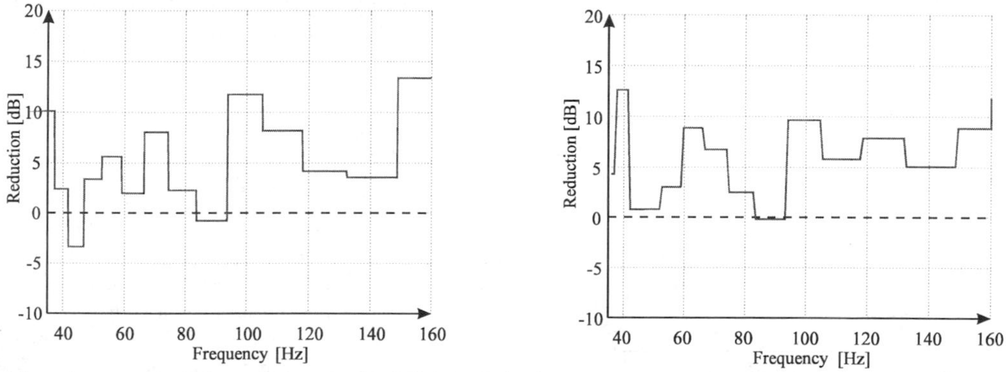

Figure 3 : measured sound transmission reduction (left), and calculated acoustic energy reduction (right) for a 4i2o control system in an arbitrarily chosen configuration.

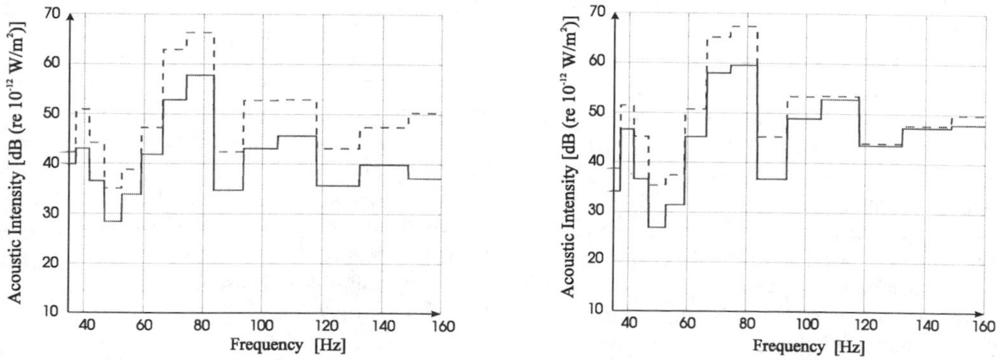

Figure 4 : measured radiated sound intensities without (dashed line) and with (solid line) a 2i2o active noise control system in an optimised configuration (left) and in an arbitrarily chosen configuration (right).

dB are obtained over the entire frequency range for a broadband disturbing excitation signal. The figure on the right shows the measured intensity spectra for an arbitrarily chosen configuration of the same control system. The system performs very poorly, especially in the upper half of the frequency range of interest where the dynamic behaviour of the double-glazing window is more often dominated by more than one eigenmode at a certain frequency line. The comparison of both parts in figure 4 clearly proves the importance of optimising the positions of the error microphones and the control loudspeakers in an active noise control system implemented in a double-glazing window. Similar measurements on the same window but with different numbers of error microphones and control loudspeakers (for example 1i1o or 2i1o) reveal identical conclusions. The performance of the control system improves with an increasing number of error microphones and control loudspeakers, but it reaches very quickly its maximum because of the limited calculation capacity of the DSP. Moreover, the comparison of the figures 3 and 4 shows that a 2i2o control system in an optimised configuration turns out to be more effective than a 4i2o control system in an arbitrarily chosen configuration.

CONCLUSION

Compared to the single-glazing windows, double-glazing windows have a high sound transmission loss, except at low frequencies. This paper illustrates that an active noise control system, implemented in the cavity between both glass plates, yields an improved low-frequency sound transmission loss, especially at those frequencies where the insertion loss of the double-glazing window is negative or close to zero. The effectiveness of the control system is strongly dependent on its configuration. A control system with a smaller number of error microphones and control loudspeakers but in an optimised configuration performs better than a more complex control system in an arbitrarily chosen configuration.

An integrated simulation model, incorporating the description of the dynamic behaviour of both the double-glazing window and the feedforward controller, has been developed. Using this model, the locations of both the control loudspeakers and the error microphones have been optimised simultaneously by means of a genetic algorithm. In order to evaluate the quality of the control configuration, obtained with the genetic algorithm, an ideal acoustic energy controller has been developed. The comparison of this theoretically maximum achievable reduction with the reductions obtained with a conventional controller in an optimised configuration allows to estimate the performance of the optimisation process. From this comparison it turns out that a properly tuned genetic algorithm is well suited for optimising the spatial distribution of sensors and actuators in an active control system. Moreover, the experimental results prove the necessity of this optimisation process for designing a control system which efficiently reduces the sound transmission through the double-glazing window over a broad frequency band.

ACKNOWLEDGEMENT

This text presents research results of the Belgian program on Interuniversity Poles of attraction by the Belgian State, Prime Minister's Office, Science Policy Programming. The scientific responsibility is assumed by its authors. The authors also wish to thank Professor A. Cops of the Physics Department of the K.U. Leuven for making available the transmission measurement room of the Laboratory for Acoustics.

REFERENCES

Berglund B.B., Hassmén P.H. and Soames Job R.F., Sources and effects of low-frequency noise, J. Acoust. Soc. Am. 99 (5), pp. 2985-3002, 1996.

Desmet W. and Sas P., Vibro-acoustic analysis of the low-frequency insertion loss of finite double-panel partitions, Proc. of the International Congress MV2 - Commett program 95, pp. 353-369, Lyon, 1995.

Goldberg D.E., Genetic Algorithms in Search, Optimisation and Machine Learning, Addison-Wesley, 1989.

Hansen C.H., Simpson M.T. and Wangler C.T., Application of genetic algorithms to active noise and vibration control, Proc. Fourth International Congress on Sound and Vibration, pp. 371-388, St.-Petersburg, 1996.

Kuwano S., Namba S. and Miura H., Advantages and disadvantages of A-weighted sound pressure level in relation to subjective impression of environmental noises, Noise Control Eng. J., 33, pp. 107-115, 1989.

6 Model updating and modal analysis

Modern Practice in Stress and Vibration Analysis, Gilchrist (ed.) © *1997 Balkema, Rotterdam, ISBN 90 5410 896 7*

Improvement of dynamic finite element models reduced by superelement techniques with updating

M.W. Zehn & O. Martin
Otto-von-Guericke-Universität Magdeburg, Institut für Mechanik, Germany

ABSTRACT: The present paper considers the effects of the superelement technique on the updating of results of reduced finite element (FE) models. Model reduction techniques are necessary for the application of the model structure in an automatic control circuit (e.g., for smart structures). Besides they are very useful to reduce the processing time for solving the eigenvalue problem for very large FE models. The disadvantage of model reduction techniques is that they can cause erroneous results. It is the aim of this paper to show to which extent updating procedures can correct FE models reduced by the superelement technique and how the choice of substructures will influence the results of the updating process. An example of an industrial part modelled with different numbers of superelements at the end of the paper make evident how the procedure works.

INTRODUCTION

In dynamic FE modeling it quite often happens that analytical predictions do not agree with test results. And in a simple combination of the results two wrongs do not make a right. Both results can be faulty for various objective and subjective reasons. Model updating seeks to correct the FE model so that the agreement between predictions and test results is improved. The mathematical backbone of all parameter correction methods is the solution of an inverse problem. Owing to incompleteness of the modal data (measured and calculated) the problem is not unique (ill-posed problem). The improvement of dynamic analysis models for real engineering structures represented by a large number of degrees of freedom FE models needs special consideration in procedures and algorithms employed. All steps towards model improvement depend upon richness and reliability of the information obtained by FE analyses and vibration testing (VBT). The whole process of model updating is beset with difficulties. The initial FE model and algorithms already employ different software products (CAD, FEA, EMA (experimental modal analysis)) and data must be interchanged. We cannot go into detail at this point but this underpins that model updating is not straightforward.

If in addition a substructure/superelement technique is used some additional problems arise. How many and which parts of the structure should be grouped together to substructures? Is there a need for additional external degrees of freedom (ex. DOF) and where should these be located? In the following we will elaborate these influences on model updating. It is common knowledge, that if the initial FE model is incapable to represent the modal space one works in, it is impossible to improve the FE model with model updating. The goal model is a validated FE model that is capable to perform dynamical analyses in a certain frequency range of interest. An adequate rigid body behaviour of the FE model should be ensured (mass, moments of inertia, principal axes and centre of gravity).

BASIC EQUATIONS

After FE discretisation the dynamic free vibration problem is represented by a system of differential equations (1) and a matrix eigenvalue problem, equation (2),

$$\mathbf{M}\,\ddot{\mathbf{x}}(t) + \mathbf{D}\,\dot{\mathbf{x}}(t) + \mathbf{K}\,\mathbf{x}(t) = \mathbf{0} \tag{1}$$

$$(\mathbf{K} - \lambda_{a_i}\mathbf{M})\Phi_{a_i} = \mathbf{0}$$
$$\Phi_a = [\Phi_{a_1}, \ldots, \Phi_{a_i}, \ldots, \Phi_{a_{m_a}}]$$
$$\Lambda = diag(\lambda_{a_i}) \quad {}^{(1} \tag{2}$$
$$\mathbf{M},\ \mathbf{D},\ \mathbf{K} \in \Re^{n_a \times n_a}$$
$$\Phi \in \Re^{n_a \times m_a} \quad \Lambda \in \Re^{n_a \times 1}$$

where **M**, **D**, **K**, **Λ**, and **Φ** are the mass, damping, stiffness, spectral, and modal matrix respectively. **Φ** is ortho-normalized with

[1] Subscript „a“ refers to calculated values whereas subscript „e“ indicates measured ones
m_a is the number of calculated eigenvalues and eigenmodes

$$\Phi^T \cdot \mathbf{M} \cdot \Phi = \mathbf{E} = \text{diag}(1)$$
$$\Phi^T \cdot \mathbf{K} \cdot \Phi = \Lambda = \text{diag}(\lambda a_i) \tag{3}$$

The work for the solution of the eigenvalue problem can be enormous when the complete system matrices are used. Therefore a wide variety of reduction methods exist. For a structure the relationship between the internal DOF x_i and the external DOF x_e of the complete system is written as

$$\mathbf{x} = \begin{bmatrix} \mathbf{x}_e \\ \mathbf{x}_i \end{bmatrix} = \begin{bmatrix} \mathbf{I} \\ \mathbf{T} \end{bmatrix} \cdot \mathbf{x}_e = \overline{\mathbf{T}} \cdot \mathbf{x}_e \tag{4}$$

With equation (4) the dimension of the system of differential equations represented by equation (1) can be reduced as follows:

$$\overline{\mathbf{T}}^T \cdot \mathbf{M} \cdot \overline{\mathbf{T}} + \overline{\mathbf{T}}^T \cdot \mathbf{D} \cdot \overline{\mathbf{T}} + \overline{\mathbf{T}}^T \cdot \mathbf{K} \cdot \overline{\mathbf{T}} = \mathbf{0} \tag{5}$$

$$\overline{\mathbf{M}} \cdot \ddot{\mathbf{x}}_e + \overline{\mathbf{D}} \cdot \dot{\mathbf{x}}_e + \overline{\mathbf{K}} \cdot \mathbf{x}_e = \mathbf{0} \tag{6}$$

The selection of external DOF relies mostly on substructuring of the model, experience, and experimental results. The degree of reduction which still produces sensible solutions dependents on the reduction method used. The calculation of the transformation matrix T and the reduction itself can be very time consuming if the reduction is applied on the whole system at once. In this case the condensation of parts of the structure (substructures) can be more advantageous. When a substructure technique is applied every structure j is reduced with its own transformation matrix T_j. The reduced matrices M_j, D_j and K_j are the so called macro- or superelements. The problem is to determine a transformation matrix. For static problems the transformation matrix can always be calculated exactly. The exact calculation of the transformation matrix for dynamic problems depends upon the eigenfrequencies of the structure. But it is our aim to solve the eigenvalue problem for the whole structure without using the complete system matrices. This aim places a demand upon an approximation for the transformation matrix T. In Zehn (1983) it was shown that the eigenvalue problem for the substructure j

$$\left(\begin{bmatrix} \mathbf{K}_{ii} & \mathbf{K}_{ie} \\ \mathbf{K}_{ei} & \mathbf{K}_{ee} \end{bmatrix} - \lambda \begin{bmatrix} \mathbf{M}_{ii} & \mathbf{M}_{ie} \\ \mathbf{M}_{ei} & \mathbf{M}_{ee} \end{bmatrix} \right) \begin{bmatrix} \mathbf{x}_i \\ \mathbf{x}_e \end{bmatrix} = \mathbf{0} \tag{7}$$

can be seperated from equation (1). λ is the eigenvalue of the complete system and x_i, x_e are the proportions of the eigenvectors of the complete system for structure j. This equation can of course not be used for the calculation of the eigenvalues and eigenvectors of the complete system since the influence of all the other substructures is missing. In Zehn (1983) it was shown, that all condensation methods which can be used for the complete system, can also be used for every substructure (equation (5)). The first row of equation (7) gives a relationship between the internal and external DOF for the j-th substructure.

$$\mathbf{x}_i = -(\mathbf{K}_{ii} - \lambda \mathbf{M}_{ii})^{-1} \cdot (\mathbf{K}_{ie} - \lambda \mathbf{M}_{ie}) \cdot \mathbf{x}_e$$
$$\mathbf{x}_i = \mathbf{T} \cdot \mathbf{x}_e \tag{8}$$

Using this relationship leads to the equation of motion with the reduced system matrices

$$\overline{\mathbf{M}} = \mathbf{M}_{ee} + \mathbf{M}_{ei} \cdot \mathbf{T}(\lambda) + \mathbf{T}^T(\lambda) \cdot \mathbf{M}_{ie} + $$
$$+ \mathbf{T}^T(\lambda) \cdot \mathbf{M}_{ii} \cdot \mathbf{T}(\lambda) \tag{9}$$

The reduced damping and stiffness matrix are calculated in the same manner. Different approximations of **T** lead to a number of procedures, see Zehn (1988). For a series expansion of **T** (equation (8)) we can write

$$\begin{aligned} \mathbf{T}(\lambda) &= -(\mathbf{K}_{ii} - \lambda \mathbf{M}_{ii})^{-1} \cdot (\mathbf{K}_{ie} - \lambda \mathbf{M}_{ie}) \\ &= (\mathbf{E} - \lambda \cdot \mathbf{K}_{ii}^{-1} \cdot \mathbf{M}_{ii})^{-1} \cdot \mathbf{K}_{ii}^{-1} \cdot (\lambda \cdot \mathbf{M}_{ie} - \mathbf{K}_{ie}) \\ &\approx \left(\mathbf{E} + \lambda \cdot \mathbf{K}_{ii}^{-1} \cdot \mathbf{M}_{ii} + \lambda^2 \cdot (\mathbf{K}_{ii}^{-1} \cdot \mathbf{M}_{ii}) + \ldots\right) \cdot \mathbf{K}_{ii}^{-1} \cdot (\lambda \mathbf{M}_{ie} - \mathbf{K}_{ie}) \\ &= -\mathbf{K}_{ii}^{-1} \cdot \mathbf{K}_{ie} + \sum_{k=1}^{\infty} \lambda^k \cdot (\mathbf{K}_{ii}^{-1} \cdot \mathbf{M}_{ii})^k \cdot \mathbf{K}_{ii}^{-1} \cdot (\lambda \cdot \mathbf{M}_{ie} - \mathbf{K}_{ie}) \\ &= \mathbf{T}_1 + \mathbf{B}(\lambda) \end{aligned} \tag{10}$$

If we cut off this series after the first term T_1 we get the transformation matrix for the static condensation. It is the simplest and for many cases a satisfactory approximation, which is independent of natural frequencies or eigenvectors. Consequently, it is most suitable in an updating procedure; for we can avoid repeated reassembling or new determination of the superelements within the algorithm. From the cut off in the series, equation (10), we obtain an error criterion for the superelement.

$$(\mathbf{K}_{ii} - \delta \cdot \mathbf{M}_{ii}) \cdot \Psi = 0 \qquad \frac{\lambda_i}{\delta_1} = \mu_{max} < 1 \tag{11}$$

where δ_1 is the cut-off frequency for each substructure. One has to select some additional internal points to satisfy inequality in equation (11). μ_{max} is therefore a quality measure for each substructure. With the static condensation method the relationship between the master DOF vector and all DOF vector can be expressed as

$$\mathbf{x} = \begin{bmatrix} \mathbf{I} \\ \mathbf{T}_1 \end{bmatrix} \mathbf{x}_e \tag{12}$$

If we include the mode shape vectors ψ_j from equation (11) into equation (12) a modal synthesis formulation may be written as

$$\mathbf{x} = \begin{bmatrix} \mathbf{I} & \mathbf{0} \\ \mathbf{T} & \Psi \end{bmatrix} \cdot \begin{bmatrix} \mathbf{x}_e \\ \mathbf{a} \end{bmatrix} \tag{13}$$

Before starting with updating one should spare no effort and use engineering experience to verify and improve the FE model in a conventional way (discretisation, addaptive refinement, dimensional reduction influence, mass properties, boundary conditions, material properties, simple analytical checks, comparison with similar parts or prototypes, etc...). Proposals for optimal measurement stations can be determined from the FE model.

CORRECTION OF THE FE MODEL BY MEASURED DATA

The objective of model updating is the correction of a finite element model that is able to perform correct dynamic analyses in a certain frequency range of interest.

Natke (1992) described model updating as indirect system identification, for it includes a discrete model.

model updating	
local updating	global updating
local improvement of physically relevant parameters global parameter adaption, estimation	FE model → representation model in the frequency range of interest

In the following we employ sensitivity formulations. The relations of particular response quantities expressed in a vector or matrix **R** with respect to changes in FE model properties (vector or matrix **P** of parameters) are call sensitivities.

$$S_{ij} = \frac{\partial R_i}{\partial P_j} \tag{14}$$

S - sensitivity matrix R - response quantities
P - FE model parameters

Since we focus on global model updating in this paper, candidates for response quantities **R** are:

- natural frequencies
- natural mode shape vectors
- MAC values
- modal displacements
- mass properties of rigid body
- ...

and for FE model parameter **P** , e.g.:

- mass distribution in FE model
- stiffness
- material properties
- boundary conditions
- ...

If we replace equation (14) by finite difference approximation, it can be expressed as

$$\mathbf{S}_{ij} \approx \frac{\Delta \mathbf{R}_i}{\Delta \mathbf{P}_j} = \frac{\mathbf{R}_i(\mathbf{P}_j + \Delta \mathbf{P}_j) - \mathbf{R}_i(\mathbf{P}_j)}{\Delta \mathbf{P}_j} \tag{15}$$

Equation (15) can be justified by a Taylor series of **R**, if we cut it off after the first term

$$\mathbf{R}(\mathbf{P} + \Delta\mathbf{P}) = \mathbf{R}(\mathbf{P}) + \frac{\partial \mathbf{R}(\mathbf{P})}{\partial \mathbf{P}} \frac{\Delta \mathbf{P}}{1!} + \cdots$$

$$\Delta\mathbf{R} = \mathbf{S}\Delta\mathbf{P} \tag{16}$$

$$\mathbf{RES} = \mathbf{S}\Delta\mathbf{P} - \Delta\mathbf{R} \rightarrow \mathrm{Min}$$

The finite difference approach leads together with the linearisation in equation (16), to an iterative procedure. If we apply different sensitivities, then equation (16) can be written as

$$\begin{bmatrix} \Delta\mathbf{R}_{\lambda} \\ \Delta\mathbf{R}_{\phi} \\ \Delta\mathbf{R}_{MAC} \\ \Delta\mathbf{R}_{CMDae} \\ \Delta\mathbf{R}_{Mass} \\ \vdots \end{bmatrix} = c_1 \begin{bmatrix} c_{S_\lambda} \cdot \mathbf{S}_{\lambda} \\ c_{S_\phi} \cdot \mathbf{S}_{\phi} \\ c_{S_{MAC}} \cdot \mathbf{S}_{MAC} \\ c_{S_{CMD}} \cdot \mathbf{S}_{CMD} \\ c_{S_{Mass}} \cdot \mathbf{S}_{Mass} \\ \vdots \end{bmatrix} \cdot \Delta\mathbf{P} \tag{17}$$

Scaling factors $c_{s\lambda}$, $c_{s\phi}$, c_{sMAC}, and c_{sMass} are introduced to overcome ill-conditioning of the total sensitivity matrix, while c_1 is a speed up or reduction parameter that can be used to increase (> 1.0) or decrease (< 1.0) the number of iteration steps.

The common results of vibration testing and FE modal analyses to characterise the linear dynamics of a structure are modal parameters (natural frequencies, damping, and mode shape vectors). Although the code, that we have implemented in a FE package, is prepared to handle all of equation (17), we will concentrate in the following on natural frequencies as response quantities **R,** and global mass and stiffness of the structure as parameters **P** for model correction. The measuremnt of natural frequencies is in general more accurate than the measurement of natural mode shapes. For the first derivatives of the eigenvalue problem in equations (1) with respect to FE model parameter $\mathbf{P}_j$ we obtain the expression

$$\frac{\partial \lambda_i}{\partial \mathbf{P}_j} = \phi_i^T \left(\frac{\partial \mathbf{K}}{\partial \mathbf{P}_j} - \lambda_i \frac{\partial \mathbf{M}}{\partial \mathbf{P}_j}\right)\phi_i = \mathbf{S}_{\lambda_{ij}} \tag{18}$$

which is one form of equation (14).

With the finite difference approximation, see equations (15) and (16), equation (17) yields

$$\frac{\Delta\lambda_i}{\Delta \mathbf{P}_j} = \phi_i^T\left(\frac{\Delta \mathbf{K}}{\Delta \mathbf{P}_j} - \lambda_i \frac{\Delta \mathbf{M}}{\Delta \mathbf{P}_j}\right)\phi_i \approx \mathbf{S}_{\lambda_{ij}} \tag{19}$$

$$\Delta\lambda_i = \phi_i^T \Delta\mathbf{K}\phi_i - \lambda_i \phi_i^T \Delta\mathbf{M}\phi_i$$

The use of equation (19) for updating (damping neglected) yields an iterative algorithm.

Now we look at the parameter estimation problem. If we choose elements in mass and/or stiffness matrices as FE model parameters **P** we obtain changes in the stiffness and mass matrix for each iteration step

$$M^* = M + \Delta M \qquad \text{and} \qquad K^* = K + \Delta K \tag{20}$$

that implies that analytical modal responces also change to

$$\Phi^* = \Phi + \Delta\Phi \qquad \text{and} \qquad \Lambda^* = \Lambda + \Delta\Lambda \tag{21}$$

Placing equation (20) into equation (1) we can write

$$(M + \Delta M) \cdot \ddot{x}(t) + (K + \Delta K) \cdot x(t) = 0$$

$$(E + \Phi^T \cdot \Delta M \cdot \Phi) \cdot \ddot{q}(t) + (\Lambda + \Phi^T \cdot \Delta K \cdot \Phi) \cdot q(t) = 0 \tag{22}$$

Both models described with equation (1) and (22) respectively are identical, if

$$\Phi^T \Delta M \Phi = \mathbf{0} \text{ and } \Phi^T \Delta K \Phi = 0 \tag{23}$$

Φ is not quadratic, because of incompleteness in modal information. Hence, conditions (23) are not only satisfied by $\Delta\mathbf{M}=\mathbf{0}$ and $\Delta\mathbf{K}=\mathbf{0}$, other solutions shown by Wahl et al. (1994) exist with

$$\begin{aligned} \Delta M &= \overline{M} - \Phi^{+^T} \cdot \Phi^T \cdot \overline{M}\Phi \cdot \Phi^+ \\ \Delta K &= \overline{K} - \Phi^{+^T} \cdot \Phi^T \cdot \overline{K} \cdot \Phi \cdot \Phi^+ \end{aligned} \tag{24}$$

Now we look for possibilities for changing very large FE models for a given (measured) modal information. Many different updating methods were described and tested by different authors. We refer to, e.g., Imregun et al. (1991), Mottershead et al. (1993), and Friswell and Mottershead (1995), etc. This paper focuses on procedures that can be implemented in an FE code to correct very large compact FE models. We assume that the basic FE model is of acceptable quality (validated by methods shown in Zehn et al., 1994), and therefore only small changes will be necessary. In order to avoid repeated reassembling, when dealing with large FE models, decomposition and storage of system matrices is compulsory for the sake of solution time and storage capacity.

For absolute mass update ($\Delta\mathbf{K}=\mathbf{0}$, and $\Delta\mathbf{M}$ is assumed diagonal $\Delta\mathbf{M}\rightarrow\Delta\mathbf{m}$) equation (19) can be rearranged to ($1 \le m \le \min\{m_a, m_e\}$)

$$\begin{bmatrix} \Delta\lambda_1/\lambda_{e1} \\ \Delta\lambda_2/\lambda_{e2} \\ \vdots \\ \Delta\lambda_M/\lambda_{eM} \end{bmatrix} = \begin{bmatrix} \phi_{11}^2 & \phi_{21}^2 & \cdots & \phi_{n_a1}^2 \\ \phi_{12}^2 & \ddots & & \\ \vdots & \vdots & \ddots & \vdots \\ \phi_{1m}^2 & \phi_{2m}^2 & \cdots & \phi_{n_am}^2 \end{bmatrix} \cdot \begin{bmatrix} \Delta m_1 \\ \Delta m_2 \\ \vdots \\ \Delta m_{n_a} \end{bmatrix}$$

$$\Delta\overline{\Lambda} = \overline{\Phi} \cdot \Delta m \tag{25}$$

For absolute stiffness update ($\Delta\mathbf{M} = \mathbf{0}$, and $\Delta\mathbf{K}$ is assumed diagonal $\Delta\mathbf{K}\rightarrow\Delta\mathbf{k}$) equation (19) can be rearranged to

$$\Delta\Lambda = \overline{\Phi} \cdot \Delta\mathbf{k} \tag{26}$$

For both absolute stiffness and mass updating (19) can be rearranged to

$$\Delta\Lambda = \left[\overline{\Phi}; \quad -\Lambda_e \cdot \overline{\Phi}\right] \cdot \begin{bmatrix} \Delta\mathbf{k} \\ \Delta\mathbf{m} \end{bmatrix} = \tilde{\overline{\Phi}} \cdot \Delta\mathbf{x} \tag{27}$$

For large FE models, n_a is much greater in size than m. Hence, equations (25), (26), or (27) are underdetermined systems of linear equations (which emphasises that the problem is not unique). Since these equations cannot be satisfied exactly, we may attempt to satisfy them as best as we can. That is to minimise the residual vector in equation (16) using the Euclidean (or least square) norm $\|\mathbf{RES}\|_2$. That leads to the following solution for equations (25), (26), or (27). With equation (18) we can write

$$\Delta\mathbf{P} = \mathbf{S}^T(\mathbf{S} \cdot \mathbf{S}^T)^{-1}\Delta\Lambda = \mathbf{S}^+\Delta\Lambda \tag{28}$$

where $\mathbf{S}^+$ is the pseudo inverse of **S**, which can be calculated by solving the linear equation system

$$\mathbf{S} \cdot \mathbf{S}^T \cdot \mathbf{S}^{+^T} = \mathbf{S} \tag{29}$$

Restrictions for certain nodes or areas of the model can be realised by setting columns in **S** (equation (25), (26), or (27)) to zero.

After each updating step we get changes in the FE model. These changes lead to changes in the modal properties of the FE model, which means that the modal properties must be calculated after each updating step. The eigenvalue problem is solved by subspace iteration, see Zehn (1981). In order to avoid the time consuming decomposition in every updating step the updating algorithm is included in the subspace iteration. A special technique in case of stiffness changes is necessary to keep the stiffness matrix unchanged within the updating algorithm. The algorithm is as follows:

1) Cholesky decomposition of stiffnes matrix

$$K = \mathbf{Z}^T \cdot \mathbf{Z} \tag{30}$$

2) Setting of start vectors $X^{(0)}$ and $M^* = M$

3) Multiplication

$$\overline{Y}^{(i)} = M^* \cdot X^{(i-1)} \tag{31}$$

4) Solution of linear equation by forward and back substitution and calculation of subspace matrix

$$\begin{aligned} &\mathbf{Z}^T \cdot \overline{\overline{\mathbf{X}}}^{(i)} = \overline{\mathbf{Y}}^{(i)} \\ &\overline{\mathbf{X}}^{(i)T} \cdot \mathbf{K} \cdot \overline{\mathbf{X}}^{(i)} = \overline{\overline{\mathbf{X}}}^{(i)T} \cdot \overline{\overline{\mathbf{X}}}^{(i)} = \overline{\mathbf{K}}^{(i)} \\ &\mathbf{Z} \cdot \overline{\mathbf{X}}^{(i)} = \overline{\overline{\mathbf{X}}}^{(i)} \end{aligned} \tag{32}$$

To avoid repeated Cholesky decomposition in case of stiffness changes the following algorithm can be applied.

$$\begin{aligned} &\mathbf{K} \cdot \Delta\overline{\mathbf{X}}^{(i)} = \Delta\overline{\mathbf{Y}}^{(i)} = -(\sum \Delta\mathbf{k}) \cdot \overline{\mathbf{X}}^{(i)} \\ &\mathbf{Z}^T \cdot \Delta\overline{\overline{\mathbf{X}}}^{(i)} = \Delta\overline{\mathbf{Y}}^{(i)} \\ &\mathbf{Z} \cdot \Delta\overline{\mathbf{X}}^{(i)} = \Delta\overline{\overline{\mathbf{X}}}^{(i)} \\ &\overline{\mathbf{X}}^{(i)} = \overline{\mathbf{X}}^{(i)} + \Delta\overline{\mathbf{X}}^{(i)} \end{aligned} \tag{33}$$

where $\sum \Delta\mathbf{k}$ is the accumulated stiffness change over i iteration steps.

The change in the subspace matrix $\overline{\mathbf{K}}^{(i)}$ as a result of the changed stiffness matrix can be approximated as follows:

$$\begin{aligned} \overline{\mathbf{K}}^{\bullet(i)} \approx \overline{\mathbf{K}}^{(i)} &- \overline{\mathbf{X}}^{(i)T} \cdot \sum \Delta\mathbf{k} \cdot \overline{\mathbf{X}}^{(i)} \\ &- \Delta\overline{\mathbf{X}}^{(i)T} \cdot \mathbf{K} \cdot \Delta\overline{\mathbf{X}}^{(i)} \end{aligned} \tag{34}$$

The product $\Delta\overline{\mathbf{X}}^{(i)T} \cdot \mathbf{K} \cdot \Delta\overline{\mathbf{X}}^{(i)}$ is easy to calculate between forward and backward substitution in equations (33).

$$\Delta\overline{\mathbf{X}}^{(i)T} \cdot \mathbf{K} \cdot \Delta\overline{\mathbf{X}}^{(i)} = \Delta\overline{\overline{\mathbf{X}}}^{(i)T} \cdot \Delta\overline{\overline{\mathbf{X}}}^{(i)} \tag{35}$$

5) Subspace operation

$$\begin{aligned} &\overline{\mathbf{K}}^{\bullet(i)} \cdot \mathbf{Q}^{(i)} = \mathbf{E} \cdot \mathbf{Q}^{(i)} \cdot \Xi^{(i)} \\ &\Xi^{(i)} = \mathrm{diag}(\xi_k^{(i)2}) \end{aligned} \tag{36}$$

$$\begin{aligned} &T^{(i)} = \mathbf{Q}^{(i)} \cdot \mathrm{diag}(\xi_k^{(i)-1}) \\ &X^{(i)} = \overline{X}^{(i)} \cdot T^{(i)}, \\ &X^{(i)} \to \Phi, \quad \xi_k^{(i)-1} = \overline{\lambda}_k \to \lambda_k \end{aligned} \tag{37}$$

Checking for convergence of eigenvalues, else go to 3)

6) Fitting in automatically selected and/or prescribed co-ordinates from measured mode shape vector into $\overline{\Phi}$ or $\mathbf{X}$

7) Solution of equation (28) for (25)

$$\Delta\mathbf{P} = \Delta\mathbf{m} \text{ or } \Delta\mathbf{k} \text{ or } [\Delta\mathbf{k}; \Delta\mathbf{m}] \tag{38}$$

8) Adding of mass changes

$$M^* = M + E \cdot \Delta m \tag{39}$$

9) Checking for convergence, stop update iteration, or else go to 2) and solve changed eigenvalue problem

$$X^{(0)} = \Phi \tag{40}$$

EXAMPLE

We will demonstrate the effects of superelement technique and updating on a base plate, which is the foundation for several electrical-mechanical parts to be mounted on. It is an aluminium alloy casting. The base plate, see figure 1, has stiffeners, out of plane areas, variable thickness, and holes. It is modelled with semiloof shell elements using the FE system COSAR[3] pre-processor COSMESH.

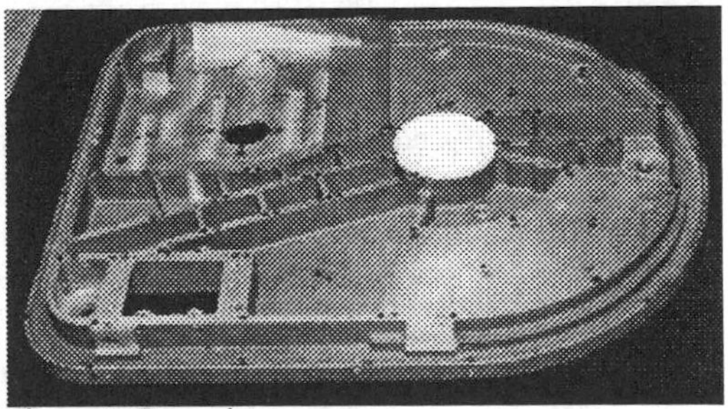

Figure 1: Base plate

Owing to the process of casting there are some uncertainties if shell elements are used. On the other hand it is a good compromise if we regard the number of elements that otherwise would be necessary if

[3] COSAR and COSMESH are trademarks of FEMCOS mbH, Germany

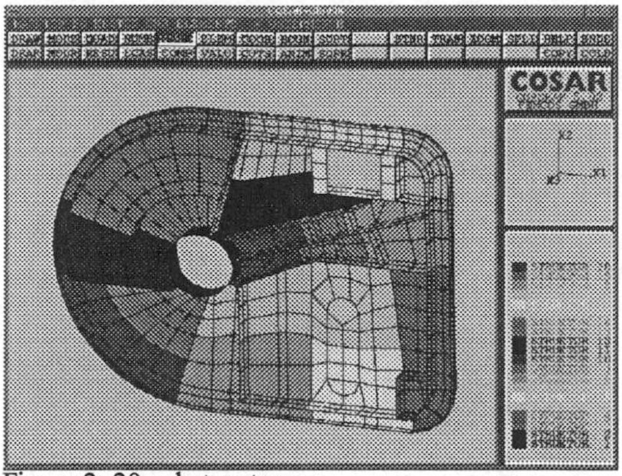

Figure 2: 20 substructures

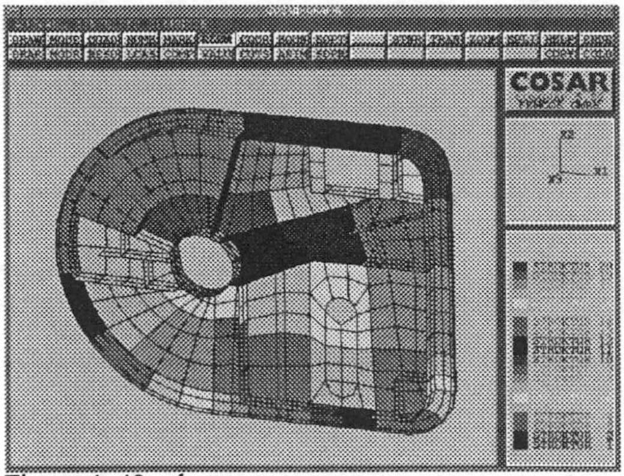

Figure 4: 49 substructures

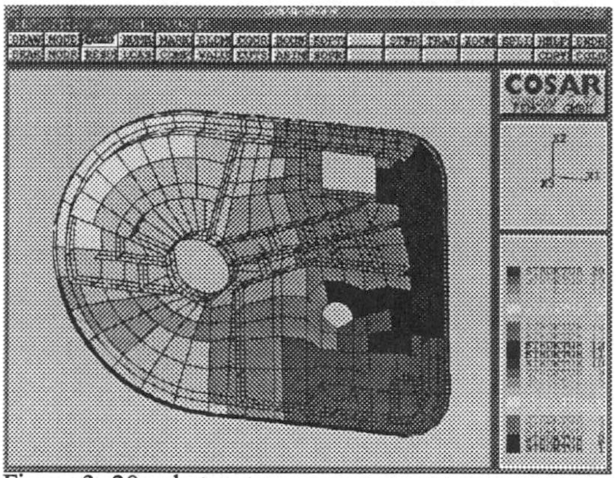

Figure 3: 20 substructures

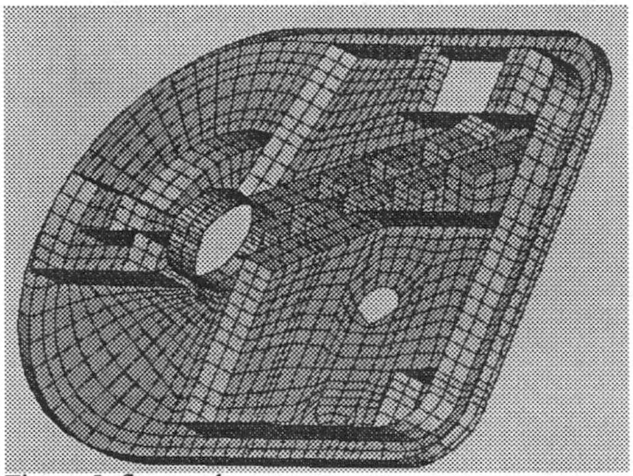

Figure 5: fine mesh

the base plate is modelled entirely with 3D elements. Three different cases are regarded. The base plate is divided once into 2 different sets of 20 substructures (figures 2 and 3) and once into 49 substructures (figure 4). For comparison the base plate is also modelled as one structure, once with a coarse mesh (9896 elements) and once with a fine mesh (26451 elements), figure 5. Mass, centre of gravity, and moments of inertia calculated in the FE system and obtained by experiment are nearly identical, see table 1. The base plate is calculated suspender free, consequently 6 rigid body modes appear.

Results for the single structure approximation are surprisingly good, see tables 4 and 5). A comparison with a fine mesh solution brought nearly the same results. So the coarser mesh is used for further investigations. Natural frequencies and mode shape vectors up to the 12th nonrigid body mode (1891.7 Hz) correspond to the test results. With rigid body modes 20 natural frequencies and mode shape vectors were calculated. For EMA we have used n_e=68 measurement points and LMS [4] equipment for measurement and experimental modal analysis.

The task here is to show if the good results for the single structure can also be achieved with the reduced FE models (figures 2, 3 and 4) and if the results can be improved with the presented model updating procedure.

Table 1: mass, volume and principle moments of inertia

	mass	volume	J1	J2	J3
measured	5.23 kg	1.984 dm^3	0.074 kgm^2	0.107 kgm^2	----- kgm^2
FE model	5.38 kg	2.038 dm^3	0.064 kgm^2	0.109 kgm^2	0.173 kgm^2

Table 2: Minimum value of a substructure, determined with equation (11)

	min. value of δ_1
substructring acc. figure 2	1.619E+09
substructring acc. figure 3	1.269E+09
substructring acc. figure 4	7.853E+08

Table 3: total mass changing in model updating

	total mass change in model updating
20 superel. updating (2)	0.0 kg
20 superel. updating (3)	-0.3226 kg
49 superel. updating (4)	0.0 kg

[4] LMS is trademark of LMS International, Belgium

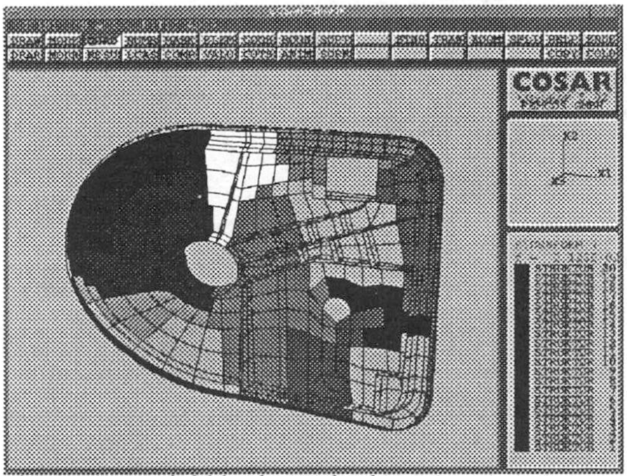

Figure 6: nonrigid body mode shape 1

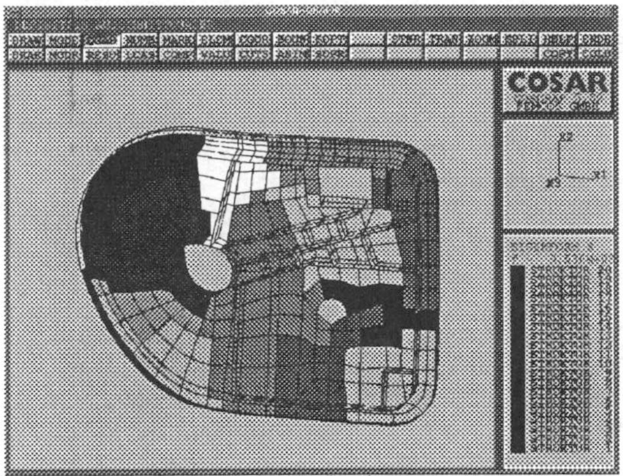

Figure 7: nonrigid body mode shape 2

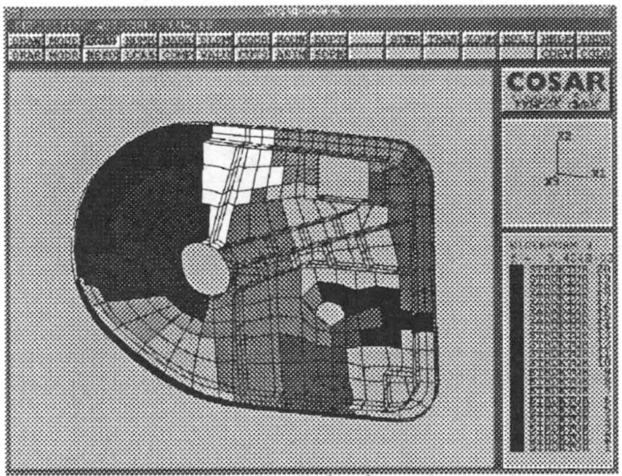

Figure 8: nonrigid body mode shape 3

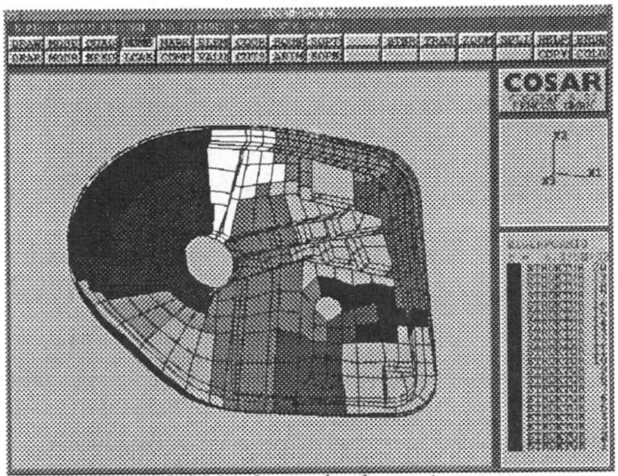

Figure 9: nonrigid body mode shape 4

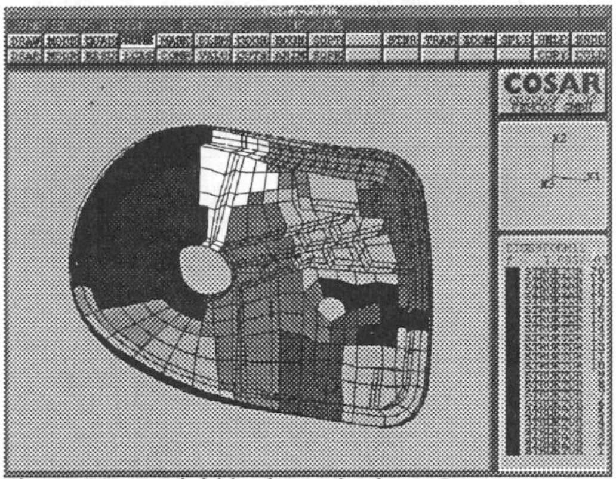

Figure 10: nonrigid body mode shape 5

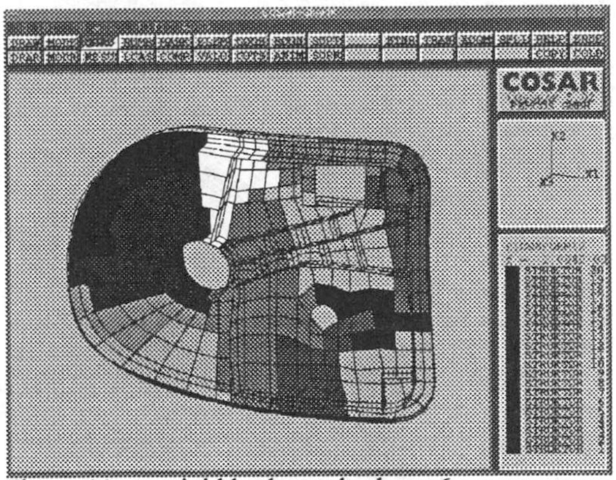

Figure 11: nonrigid body mode shape 6

Table 4 shows that the presented model updating procedure leeds to an improvement for most eigenfrequencies in all three cases. The costs for this improvement is a loss in mode shape pairing especially for the last case (figure 4). A possible explanation for this is the fact that for this case no additional external DOF were inserted into the model leading to a poorer quality of some substructures, see table 2.

Table 4: Natural frequencies of base plate in [Hz] (w - with update, w/o - without update)

	1.	2.	3.	4.	5.	6.
1 structure (fine)	207	360	531	630	722	1026
1structure (coarse)	210	365	536	640	730	1034
20 superel.(2) w/o	207	324	521	605	742	1022
20 superel.(2) w	223	367	395	659	751	846
20 superel.(3) w/o	213	354	540	634	709	1084
20 superel.(3) w	224	396	590	669	751	1093
49 superel.(4) w/o	228	360	571	693	765	1058
49 superel.(4) w	222	381	576	683	725	766
measurement	223	396	591	669	750	1094

Table 5: MAC values [%] for single structure w/o updating

measurement													
		1.	2.	3.	4.	5.	6.	7.	8.	9.	10.	11.	12.
	1.	**100**	0	0	1	0	1	0	0	0	2	0	1
F	2.	0	**100**	0	3	0	0	1	1	0	0	3	2
E	3.	0	0	**98**	0	0	1	0	0	3	0	11	1
	4.	2	1	1	**99**	0	0	1	1	1	2	7	6
m	5.	0	0	1	0	**98**	1	0	0	1	0	0	0
o	6.	1	0	0	0	0	**99**	2	2	0	3	0	4
d	7.	0	0	0	1	1	0	**83**	4	12	3	2	0
e	8.	1	0	0	1	1	2	8	**86**	7	0	0	0
l	9.	0	0	3	3	1	1	0	14	**85**	1	0	9
	10.	1	0	0	1	0	2	0	1	1	**93**	6	0
	11.	0	2	9	4	1	1	1	0	0	3	**94**	1
	12.	1	2	0	5	0	5	0	2	8	0	1	**98**

Table 6: MAC values [%] for 20 superelements structure (figure 2) w/o updating

measurement							
F		1	2.	3	4.	5.	6.
E	1.	97	0	0	2	1	5
m	2.	1	97	0	0	0	0
o	3.	0	0	98	0	0	1
d	4.	2	0	0	98	3	0
e	5.	3	0	0	2	92	2
l	6.	3	3	0	1	5	84

Table 7: MAC values [%] for 20 superelements structure (figure 3) w/o updating

measurement							
F		1	2.	3.	4	5.	6.
E	1.	100	1	1	4	29	2
m	2.	0	99	1	3	0	1
o	3.	0	0	97	1	0	1
d	4.	1	0	3	90	1	2
e	5.	29	0	1	20	95	0
l	6.	2	1	3	0	0	88

Table 8: MAC values [%] for 49 superelements structure (figure 4) w/o updating

measurement							
F		1.	2.	3.	4.	5.	6.
E	1.	98	0	2	5	1	0
m	2.	17	86	6	4	2	4
o	3.	0	8	84	0	2	4
d	4.	6	0	4	96	6	2
e	5.	1	0	2	5	91	0
l	6.	23	10	17	0	6	2

Table 9: MAC values [%] for 20 superelements structure (figure 2) with updating

measurement							
F		1.	2.	3.	4.	5.	6.
E	1.	81	1	1	1	0	4
m	2.	20	6	18	2	2	8
o	3.	1	87	6	1	0	0
d	4.	11	1	15	64	2	3
e	5.	0	15	16	5	58	3
l	6.	6	36	0	21	3	13

Table 10: MAC values [%] for 20 superelements structure (figure 3) with updating

measurement							
F		1.	2.	3.	4.	5.	6.
E	1.	99	0	1	3	25	2
m	2.	1	98	3	1	0	1
o	3.	1	1	92	1	1	2
d	4.	5	5	7	95	5	0
e	5.	3	0	9	5	84	1
l	6.	1	3	1	2	0	55

Table 11: MAC values [%] for 49 superelements structure (figure 4) with updating

measurement							
F		1.	2.	3.	4.	5.	6.
E	1.	92	0	0	6	3	1
m	2.	31	75	1	4	5	5
o	3.	16	1	69	32	0	0
d	4.	4	0	15	48	35	1
e	5.	30	19	4	2	5	4
l	6.	33	1	37	0	8	0

CONCLUSIONS

In this paper a procedure for model updating is presented, and employed for a structure that is modelled with a reduced DOF FE model by using the superelement technique. The embedding of the algorithm within the FE system (eigenvalue problem solver) has some major advantages. E.g., the amount of data transfer and storage capacity is reduced. It is very easy to apply the changed model to other dynamic simulations. It was shown that the presented procedure is able to update FE models reduced by the superelement technique.

REFERENCES

Friswell, M.L., Mottershead, J., 1995, *Finite Element Model Updating in Structural Dynamics*, KLUWER ACADEMIC PUBLISHERS, Dordrecht/Boston/London.

Gabbert, U., Zehn, M.W., 1992, „Universelles FEM-Programmsystem COSAR - ein zuverlässiges und effektives Berechnungswerkzeug für den Ingenieur“, In: *Beiträge zur II. COSAR-Konf.*, Universität Magdeburg.

Gabbert, U., Zehn, M.W., 1995, „Adaptaiv Remeshing Based on Error Estimations - Foundation, Implementation and Application", *Proc. NEFEMs 5th Int. Conf. „Reliability of Finite Element Methods for Enginering Applications*, Amsterdam.

Gabbert, U., Zehn, M.W., Wahl, F., „Improved Results in Structural Dynamic Calculations by Linking Finite Element Analysis (FEA) and Experimental Modal Analysis (EMA)", *ASME - Proc. of 1995 Design Eng. Techn. Conferences, Boston/USA*, Vol. 3, Part C, pp.1321 - 1328, 1995.

Gabbert, U., Zehn, M.W., Wahl, F., „A Modal Updating for Large Finite Element Problems.", *Second Int. Conference in Structural Dynamics Modelling Test, Analysis and Correlation*, Cumbria/UK, NAFEMS, 1996.

Imregun, M., Visser, W.J., 1991, „A review of modal updating techniques", *Shock and Vibration Digest*, Vol. 23, No. 1.

Mottershead, J., Friswell, M.L., 1993, „Model updating in structural dynamics: a survey", *Journal of Sound and Vibration*, Vol. 167, No 2.

Natke, H.G., 1992, *Einführung in Theorie und Praxis der Zeitreihen- und Modalanalyse*, Vieweg, 3. Auflage.

Wahl, F., Jungbluth, R., 1994, *Gezielte Modifikation mechanischer Strukturen auf der Grundlage gemessener modaler Größen*, Arbeitsbericht an die Deutsche Forschungs Gemeinschaft (DFG), Universität Magdeburg.

Zehn, M.W., 1981, *Eigenschwingungsberechnung dreidimensionaler Bauteile mit der Methode der Finiten Elemente*, Dissertation, TH Magdeburg.

Zehn, M.W., 1983, „Substruktur-/ Superelementtechnik für die Eigenschwingungsberechnung dreidimensionaller Modelle mit Hilfe der FEM", *Technische Mechanik*, Band 3, Heft 4.

Zehn, M.W., 1989, „Dynamische FEM-Strukturanalyse mit Substrukturtechnik", *Technische Mechanik*, Band 9, Heft 4.

Zehn, M.W., Schmidt, G., „Updating of large FE-models including superelement technique" , *Identification in Engineering Systems, Proc. Int. Conf.*, Swansea/UK, March 1996, Ed. By M.I.Friswell and J.E.Mottershead, pp 156 - 164.

Zehn, M.W., Wahl, F., „Zusammenspiel von experimenteller Modalanalyse und Finite Elemente Berechnung bei der Modellierung dynamisch beanspruchter Bauteile", 5.Tagung "Dynamische Probleme - Modellierung und Wirklichkeit", 10-11.10.1996 Hannover, *Mitteilungen des Curt-Risch-Instituts*, 1996, Herausgeber H.G.Natke, S.115-135.

Zehn, M.W., Wahl, F., Schmidt, G., 1994, „Möglichkeiten der Modellkontrolle und Modellreduzierung bei strukturdynamischen Untersuchungen mittels FEM unter Einbeziehung von Meßwerten", *VDI Berichte*, Nr. 1145.

Zehn, M.W., Wahl, F., Schmidt, G., 1995, „Verification and adjustment of dynamic structural finite element models by experimental modal analysis data", *7th International Conference CMEM 95*, Capri/Italy, Computational Mechanics Publications, Southampton Boston.

Modern Practice in Stress and Vibration Analysis, Gilchrist (ed.) © 1997 Balkema, Rotterdam, ISBN 90 5410 896 7

The modelling of a three-storey aluminium space frame and updating of finite element parameters at the joints

J.E. Mottershead & S. James
The University of Liverpool, Department of Mechanical Engineering, UK

ABSTRACT: An aluminium space frame structure is modelled at the joints by part-rigid beam elements and updated over the range of the first ten natural frequencies by correcting two joint parameters. It is demonstrated that the correction cannot be achieved solely on the basis of stiffness modifications, which means that there is no alternative to an equivalent model based on both stiffness and mass adjustments.

1 INTRODUCTION

An important aspect of the model updating problem (Mottershead and Friswell, 1993; Friswell and Mottershead, 1995) is parameterisation of the finite element model. Of course, it is necessary that the model predictions will be sensitive to the chosen parameters, but ideally we would wish to obtain an updated model which is physically representative of the test structure. In practice the complex geometry of many structures, especially at the joints, restricts the scope of the analyst in choosing parameters with physical meaning. Then an 'equivalent' model is sought which is representative of the test structure over a range of dynamic conditions.

Mottershead *et al.* (1996) used geometric (offset) parameters to update the model of a welded joint. Ahmadian, Mottershead and Friswell (1996) used parameters related to the eigenvalue decomposition of the joint substructure stiffness matrix. And Ahmadian, Mottershead and Friswell (1997) identified a constant-stiffness equivalent model of the rubber seal around a car window.

The aluminium space frame, which is the subject of this article, consists of uniform tubular spars joined at the ends by a screw into a spherical node. Whilst the tubes are well defined, the joints are too complicated to be modelled with geometric accuracy using finite elements, and for this reason it is necessary to carry out updating of the analytical model. In what follows, the modelling and parameterisation of the joints are considered. The physical complexity of the joints and a sensitivity analysis leads to the selection of two parameters, one of which is unrepresentative. But it is demonstrated how these two can be used to improve the performance of the model over the range of the first ten modes of vibration.

2 THE ALUMINIUM SPACE FRAME

The three-storey aluminium frame was built using the Meroform M12 Construction System. The structure consists of 22 mm aluminium tubes connected by standard Meroform aluminium nodes. The components are shown in Figure 1 and the complete structure is illustrated in Figure 2. The length of all the horizontal and vertical tube-members between the centres of the joints is 707 mm. Two opposing vertical sides of the frame are stiffened by diagonal members so that the bending frequencies in the x-z plane are separated from those in the y-z plane, and two families of twisting modes are characterised by out-of-plane bending of members in the x-z or y-z planes respectively.

A mathematical model was assembled using cubic beam elements to represent the aluminium tubes. There were 120 elements in total and 612 degrees-of-freedom. Four beam elements were used to model each tube over the complete length between the centres of the Meroform nodes. At the ends of the tubes there are stiff connectors which screw into the nodes, and these together with the nodes themselves were considered to be perfectly rigid. The stiffness matrix for a beam in bending in the x-y plane, having a flexible length ℓ separated by a rigid portion of length a from the first node, can be written as,

$$
\boldsymbol{k} = \frac{EI_x}{\ell^3}\begin{bmatrix} 12 & (12a+6\ell) & -12 & 6\ell \\ (12a+6\ell) & (12a^2+12a\ell+4\ell^2) & -(12a+6\ell) & (6a\ell+2\ell^2) \\ -12 & -(12a+6\ell) & 12 & -6\ell \\ 6\ell & (6a\ell+2\ell^2) & -6\ell & 4\ell^2 \end{bmatrix}, \tag{1}
$$

and the corresponding element mass matrix is,

$$
\boldsymbol{m} = \frac{\rho A\ell}{420}\begin{bmatrix} 156 & (156a+22\ell) & 54 & -13\ell \\ (156a+22\ell) & (156a^2+44a\ell+4\ell^2) & (54a+13\ell) & -(13a\ell+3\ell^2) \\ 54 & (54a+13\ell) & 156 & -22\ell \\ -13\ell & -(13a\ell+3\ell^2) & -22\ell & 4\ell^2 \end{bmatrix} \tag{2}
$$

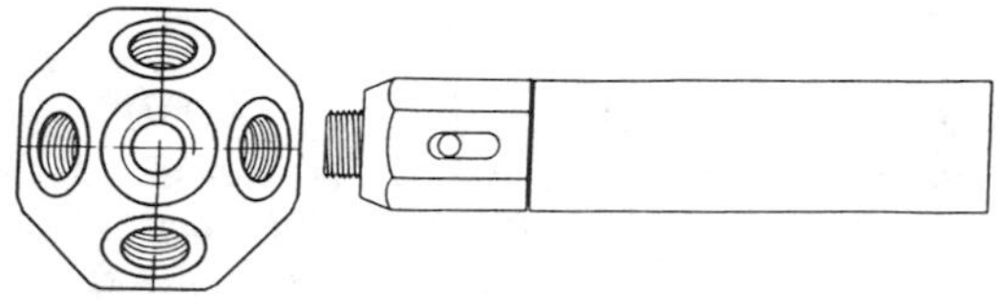

Figure 1. Meroform components

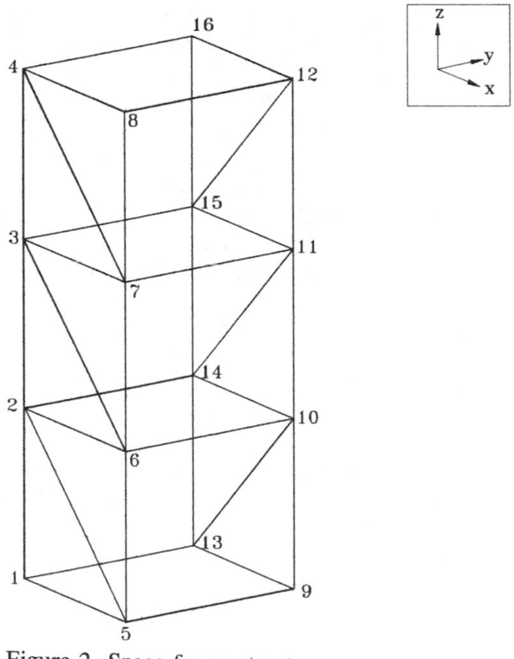

Figure 2. Space frame structure

Lumped masses and rotary inertias for the aluminium nodes and the beam-end connectors were included in the assembled finite element model.

The physical structure was excited by two electromagnetic exciters supplied with independent sequences of random noise at the first-floor level. The vibration response was measured using twelve data collection channels from accelerometers on the structure. Natural frequencies and mode shapes were extracted from the measured frequency response functions by using a multi-degree-of-freedom curve-fitting procedure from the LMS CADA-X system. The measured natural frequencies and finite element predictions are given in Table 1 where it is seen that the seventh predicted frequency exceeds its measured counterpart by 14.2% whilst the ninth prediction is 8.6% below the measured value. Figures 3 and 4 show the first and third measured modes respectively. The first mode is the first bending shape in the less-stiff y-z plane. The third measured mode is the first torsion mode with-out-of-plane bending of those members in the x-z plane. The diagonal MAC terms were all better than 96% with very small off-diagonal terms. Whilst it is to be expected that the discrepancies in the frequencies can be attributed to modelling errors at the joints, it is clear that the adjustment should not be in the nature of a simple reduction in the joint stiffness, because some of the predicted frequencies are too high but others are too low. The beam elements that meet at the joints are in need of a subtle redistribution of stiffness because the flexibility of the Meroform nodes has an effect which is not represented by the part-rigid beam elements. In the following section we consider the parameterisation of the finite element model to achieve the desired stiffness redistribution by applying model updating.

3 PARAMETERS FOR MODEL UPDATING

One approach to parameterising the finite element model is to decompose the element stiffness matrix into its own eigenvalues and eigenvectors,

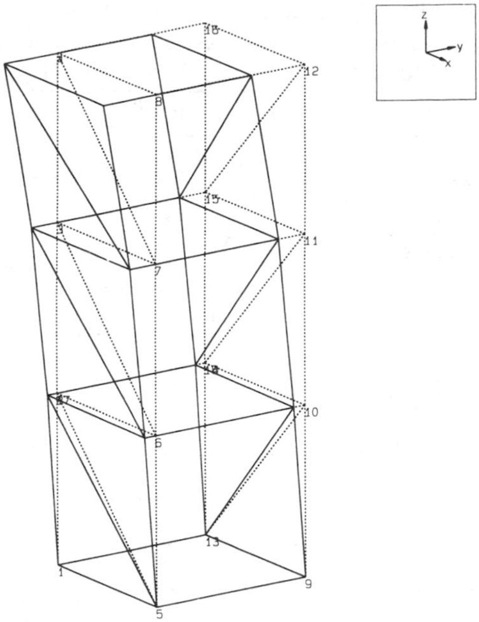

Figure 3. 1st Vibration mode

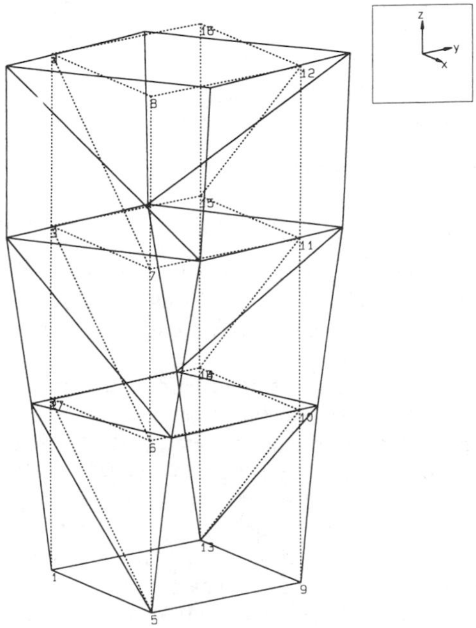

Figure 4. 3rd Vibration mode

$$k = \Psi^T \begin{bmatrix} \mathbf{0} & \\ & Q \end{bmatrix} \Psi, \tag{3}$$

where Q is the diagonal matrix of non-zero stiffness

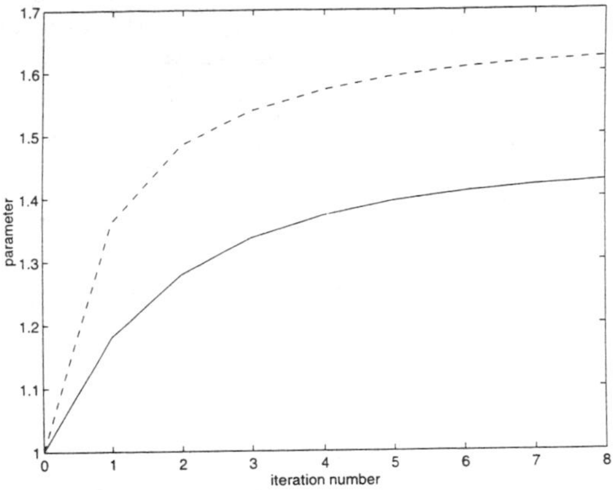

Figure 5. Updating parameters

eigenvalues and Ψ is the matrix of stiffness eigenvectors. Gladwell and Ahmadian (1996) and Ahmadian, Gladwell and Ismail (1997) developed a general method based on the above decomposition for updating in families of finite elements. A simpler approach would be to choose the terms in Q as the updating parameters. For a cubic beam element there would be two bending eigenvalues (in each of the two bending planes), a twisting eigenvalue and an axial extension/compression. We already observed that the lower bending and twisting mode-shapes of the aluminium frame are characterised by bending of the individual tube members. So it would be expected that the element stiffness eigenvectors in bending would participate strongly in the first several structural mode-shapes. This means that the natural frequencies of the frame are likely to be very sensitive to the bending terms in Q. By updating the part-rigid beams that meet at the joints through adjusting the first bending term in Q the bending stiffness of the element is modified without affecting the torsional and axial stiffnesses or the stiffness of the other bending modes. Of course the stiffness eigenvectors Ψ all remain unchanged so long as the adjustment is restricted to the terms in Q. A 'spin-off' advantage of updating the element stiffness eigenvalues is that the usual (structural) eigenvalue sensitivity,

$$\frac{\partial \lambda_i}{\partial q_j} = \Phi_i^T \left(\frac{\partial K}{\partial q_j} - \lambda_i \frac{\partial M}{\partial q_j}\right)\Phi_i , \tag{4}$$

is a constant because,

$$\frac{\partial K}{\partial q_j} = \Psi_j \, \Psi_j^T , \tag{5}$$

Table 1. Table of predicted natural frequencies (Hz)

Measured	6.46	20.58	23.13	32.95	35.83	43.45	52.0	61.51	94.82	100.44
Finite element	6.55	21.14	23.32	32.97	36.80	43.23	59.43	65.71	86.65	92.21
% error	1.4	2.7	0.8	0.06	2.7	-0.5	14.2	6.8	-8.6	-8.2

Table 2. Table of sensitivities

	PARAMETER				
natural frequency	1st bending eigenvalue	torsional eigenvalue	2nd bending eigenvalue	beam-end mass	node mass
6.55	1 367	3	43	-794	-230
21.14	14 104	38	487	-8 667	-2 405
23.32	15 371	197	539	-11 413	-3 184
32.97	33 168	1 104	1 135	-23 050	-6 533
36.80	43 311	113	1 446	-28 349	-7 118
43.23	59 076	1 285	2 006	-40 057	-10 184
59.43	2 778	45	396	-61 695	-18 367
65.71	23 781	321	1 088	-78 944	-23 483
86.65	76 390	6 132	6 262	-2 860	-888
92.21	91 230	10 733	8 099	-6 826	-1 714

and $\boldsymbol{\Psi}_j$ remains unchanged as the eigenvalue q_j is adjusted through the iterations of the updating process.

An alternative to updating the element stiffness eigenvalues would be to adjust the length of the rigid part of the beams at the joints (but keeping the overall length the same). Since this would result in a change to the element stiffness eigenvectors it can be seen that strictly the adjustment of the $\boldsymbol{Q}$ terms does not cover the case of increasing or reducing the extent of the rigid part. However, a small change to length of the rigid part of the beam will result in a dominant change to the terms in $\boldsymbol{Q}$ and a smaller change in the eigenvectors $\boldsymbol{\Psi}$. Thus the case of modifying the rigid and flexible lengths must be closely approximated by a linear combination of eigenvalue stiffness matrices,

$$\boldsymbol{k} = \sum_{j=1}^{6} \alpha_j \, \boldsymbol{k}_j \ , \qquad (6)$$

where, $\boldsymbol{k}_j = q_j \, \boldsymbol{\Psi}_j \, \boldsymbol{\Psi}_j^T$ (7)

In other words, a presently unknown modification to the stiffness eigenvalues will closely (but not exactly) represent an adjustment to the rigid and flexible lengths.

A further strategy for updating, though less attractive from a physical point of view, would be to make an adjustment to the mass at the joint. Although the joints are thought to be well modelled so far as mass is concerned, it is recognised that an increase in mass has precisely the same effect on the structural eigenvalues as a reduction in stiffness (and *vice-verca*). Thus a useful equivalent model can be produced by making a mass adjustment to correct the effect of a model which fails in its representation of stiffness.

The sensitivity of the first ten natural frequencies to changes in the stiffness eigenvalues, and to lumped masses from the beam-end connectors and the Meroform nodes is given in Table 2 (for the tube with circular cross section the bending modes in the two planes *x-z* and *y-z* are identical). It is clear that

Table 3. Table of updated natural frequencies (Hz)

Measured	6.46	20.58	23.13	32.95	35.83	43.45	52.0	61.51	94.82	100.44
Updated	6.62	21.21	22.80	32.59	36.66	42.87	52.74	59.38	90.18	96.05
% error	2.4	3.1	-1.4	-1.1	2.3	-1.3	1.4	-3.5	-4.9	-4.4

the ninth and tenth natural frequencies, which were under estimated by finite elements, are strongly sensitive to the first bending eigenvalue of the element stiffness. On the other hand the over-estimated seventh and eighth natural frequencies are most sensitive to the beam-end masses. The two sets of sensitivities are also fairly independent of each other. The sensitivity analysis indicates that these two parameters might be adjusted to reconcile the finite element predictions with the first ten natural frequencies.

It should be noted that neither the second bending eigenvalue nor the torsional eigenvalue will bring about the desired correction to the seventh and eighth natural frequencies. This means that the extent of the rigid part of the beam (at the joints) will not be a good parameter for updating either. Indeed, there is no combination of stiffness parameters from the finite element model that can produce the necessary correction to the seventh and eighth natural frequencies, and at the same time correct the ninth and tenth natural frequencies in the opposite direction. Thus the use of a mass parameter is unavoidable.

4 MODEL UPDATING RESULTS

The finite element model was updated by minimising an objective function having the form,

$$J = (\delta\boldsymbol{\lambda} - \boldsymbol{S}\delta\boldsymbol{\theta})^T \boldsymbol{W}_\lambda (\delta\boldsymbol{\lambda} - \boldsymbol{S}\delta\boldsymbol{\theta}) + \delta\boldsymbol{\theta}^T \boldsymbol{W}_\theta \delta\boldsymbol{\theta} \quad , (8)$$

where $\delta\lambda$ is the vector of small changes in structural eigenvalues, $\boldsymbol{S}$ is the matrix of eigenvalue sensitivities, and $\delta\boldsymbol{\theta}$ is the vector of parameter changes necessary to eliminate the discrepancy $\delta\lambda$. $\boldsymbol{W}_\lambda$ and $\boldsymbol{W}_\theta$ are positive-definite weighting matrices, and the terms in these matrices must be selected on the basis of engineering understanding to produce acceptable results $\boldsymbol{\theta}$.

The results presented in this article were produced by selecting the weighting matrices as follows,
$\boldsymbol{W}_\theta = 10^{-6} \times (500\ ,\ 5)^T$,
and,
$\boldsymbol{W}_\lambda = (500, 500, 20, 10, 8, 5, 4, 7, 1.2, 1)^T$.

Thus the joint stiffness parameter was weighted 100 times more strongly than the joint mass, and the lower eigenvalues were weighted more strongly than the higher ones (to counteract any unwanted weight caused by the large numerical values of the higher natural frequencies). The results are given in Table 3 where it is seen that the prediction error has been reduced to less than 5% of any of the measurements. The evolution of the parameters over eight updating iterations is shown in Figure 5 where the dashed line represents the stiffness parameter and the full line is for the updated mass at the joints. In this case an equivalent model of the joint stiffness has been obtained by adjusting two parameters: the first bending eigenvalue of the stiffness matrix (which changes by +43%) and the beam-end masses (which change by +62%). Thus the mis-modelling of stiffness in the original finite element model has been compensated to some extent by a mass adjustment in the updated model. This illustrates a common problem in updating, of finding physically meaningful parameters to adjust. Although a mass adjustment would not be the first choice to correct an error in stiffness, it is clear that an increase in mass has a similar effect to reducing a stiffness. In this example the mass adjustment is made precisely at the location of the aluminium nodes.

5 CONCLUSIONS

The finite element model of a three-storey aluminium space frame has been updated over the range of the first ten natural frequencies by adjusting two parameters. A sensitivity analysis has demonstrated that there is no combination of stiffness parameters available from the finite element model that alone can reconcile the numerical predictions with measured natural frequencies. Finally two parameters: the first bending eigenvalue of the stiffness matrix, and the beam end masses are used to bring about a reduction of the prediction error to within 5% on any of the measurements.

The space frame structure contains many updated beams in different spatial configurations which suggests that the equivalent model produced in this exercise might be useful for a different structure built from the same components: but this requires further investigation.

REFERENCES

Ahmadian, H., Gladwell, G.M.L. and Ismail, F. 1997. Parameter selection strategies in finite element model updating, *Trans. ASME. J. Vib. Acoust,* 119 (1), 37-45.

Ahmadian, H., Mottershead, J.E. and Friswell, M.I. 1996. Joint modelling for finite element model updating, *14th IMAC, Dearborn,* 591-596.

Ahmadian, H., Mottershead, J.E. and Friswell, M.I. 1997. Parameterisation and identification of a rubber seal, *15th IMAC, Florida*, 142-146.

Friswell, M.I. and Mottershead, J.E. 1995. *Finite Element Model Updating in Structural Dynamics*, Kluwer, Dordrecht.

Gladwell, G.M.L. and Ahmadian, H. 1996. Generic element matrices for finite element model updating, *Mech. Sys. Sig. Proc.*, 9, 601-614.

Mottershead, J.E. and Friswell, M.I. 1993. Model updating in structural dynamics: a survey, *J. Sound Vibration*, 162(2), 347-375.

Mottershead, J.E., Friswell, M.I., Ng, G.H.T. and Brandon, J.A. 1996. Geometric parameters for finite element model updating of joints and constraints, *Mech. Sys. Sig. Proc.*, 10(2), 171-182.

Modern Practice in Stress and Vibration Analysis, Gilchrist (ed.) © *1997 Balkema, Rotterdam, ISBN 90 5410 896 7*

Modal testing of a full-scale concrete floor: Test specifics and QA procedures

A. Pavic, P. Reynolds & P. Waldron
Centre for Cement and Concrete, University of Sheffield, UK

ABSTRACT: Quality assurance (QA) procedures adopted by the Centre for Cement and Concrete for the modal testing of full-scale civil engineering structures are outlined. The various phases of a typical test are described, with particular reference to the testing of a 600 tonne ribbed post-tensioned concrete floor. Particular attention has been paid to QA checks required specifically for "out-of-the-lab" site conditions, typical for civil engineering applications of modal testing. It is shown that using these procedures, the acquisition of good quality modal testing data was achieved for the said structure in a very limited amount of time on site. It is also shown that, contrary to common belief, modal testing of this structure using an electrodynamic shaker was quicker and more reliable than using hammer impact excitation.

1 INTRODUCTION

Modal testing is a well developed technique for the dynamic investigation of small and medium sized mechanical engineering structures and components. Although theoretically possible, the use of this technology for the dynamic investigation of full-scale civil engineering concrete structures has many practical problems. The sheer size of civil structures and other practicalities usually restrict the testing of such structures to the use of low energy hammer impact or shaker excitation. In addition, the time constraints imposed on the field testing of civil engineering structures are normally severe, with structures either being in use or in the process of construction. Finally, real concrete structures almost invariably have relatively low natural frequencies accompanied by heavily damped and closely spaced modes of vibration. These are difficult to estimate since the response signals are buried in measurement noise from the open environment in which the structures are usually tested.

This paper describes the modal testing procedures for a 600 tonne ribbed post-tensioned concrete floor in a new building development under construction in the centre of London. This real-life structure served as a test-bed for the application of suitably adapted modal testing procedures initially recommended by the UK Dynamics Testing Agency (DTA) for testing mechanical structures typically under laboratory conditions. However, since the ultimate aim of the modal testing was to provide experimental data for finite element model updating feasibility studies, it was important to obtain the highest quality data possible.

Firstly, the equipment and excitation methods used are presented. A brief comparison between hammer impact excitation and shaker excitation is given, in terms of both the testing procedures and the quality of the results.

Secondly, a QA system is described, which has been adopted by the Centre for Cement and Concrete (CCC) for the modal testing of full-scale civil engineering structures. It is designed to enable high quality modal testing data to be acquired in a very limited time on site, and addresses specific issues related to the modal testing of civil engineering structures.

2 DESCRIPTION OF TEST STRUCTURE

The structure tested was one of the floors in a high profile six-storey office building development in the centre of London. The configuration of the floor is shown in Figure 1. The 'ribbed zone' of the floor comprised a post-tensioned concrete ribbed slab of overall depth 350 mm spanning 14.5 m. The ribs were 650 mm wide, spaced at approximately 1 m centres acting integrally with a 110 mm thick concrete slab. Around the perimeter of the ribbed zone was a 'ring zone' of approximate width 1.25 m, consisting of a 350 mm deep solid slab. Finally, the 'core area' of the floor comprised a reinforced

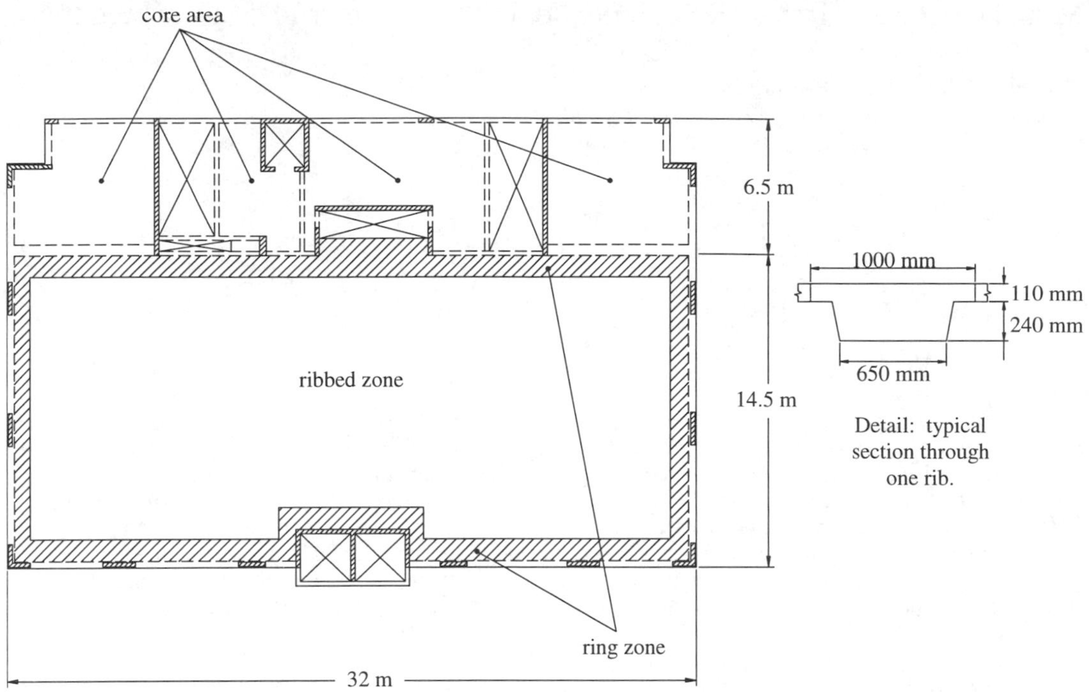

Figure 1. London test floor layout (all dimensions approximate).

concrete solid slab of depth 300 mm with various openings to permit the passage of stairs, lifts and services.

The floor was supported around its perimeter by a system of monolithically cast in-situ beams and columns.

3 TESTING EQUIPMENT

The equipment used by the CCC for modal testing of the floor was as follows.

Two measurable excitation sources were used. Firstly, a manually operated 5.4 kg instrumented hammer (Dytran model 5803A), and then an electrodynamic shaker (APS Dynamics model 113 with model 114-EP power amplifier). The shaker was operated in 'reaction mode', i.e. it was placed onto the structure and force was generated by applying electrodynamic force to reaction masses attached to the shaker armature. In order to determine the excitation force from the shaker in reaction mode, the acceleration of the shaker armature (and hence indirectly the force applied to the structure) was measured using an accelerometer.

Four high sensitivity (1 V/g), low frequency piezoelectric accelerometers were used for the testing (two Endevco model 7754-1000 with nominal frequency range 0.2 to 500 Hz and two Dytran model 3100B24 with frequency range 1.5 to 500 Hz). All accelerometers have built-in line drive micro-electronic pre-amplifiers in order to drive very long signal cables (50-500 metres), as appropriate for civil engineering applications. For shaker excitation, one of the accelerometers was attached to the shaker armature to measure the vertical excitation, and the other three were placed on carefully selected response points on the structure. For hammer testing, all four accelerometers measured response at various points on the structure.

Two data acquisition devices were used on site. A Diagnostic Instruments DI-2200 dual channel portable spectrum analyser was used to sample the excitation and response signals, and to calculate Frequency Response Functions (FRFs) in real-time. In addition, a 16 Channel Racal StorePlus VL analogue tape recorder was used to store the analogue transducer signals, allowing further digital re-sampling away from site.

Two notebook PCs were used, one for the generation of shaker excitation signals, and the other for the site processing of the modal testing data. This was required as a part of the established QA system.

A schematic diagram of the hammer and shaker testing equipment is shown in Figure 2.

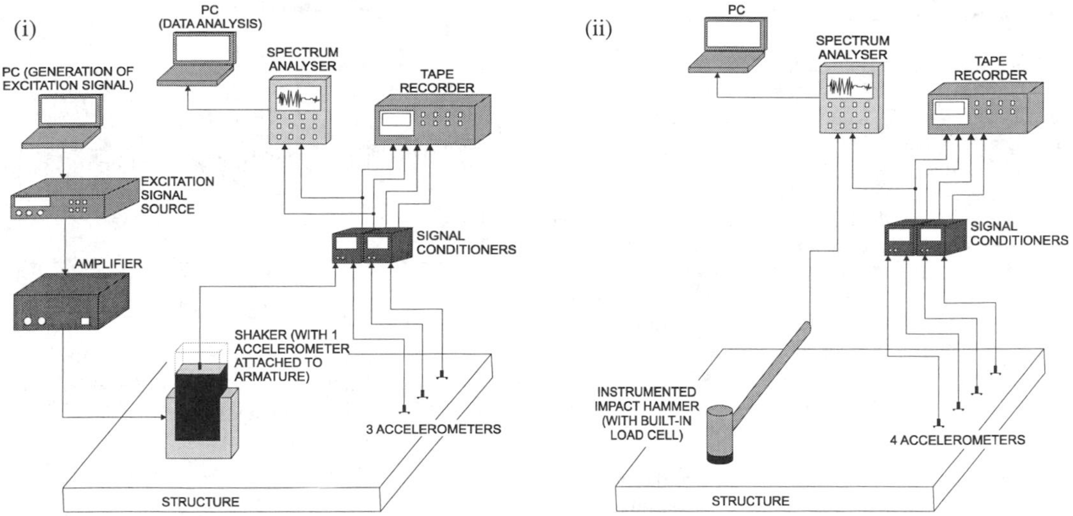

Figure 2. Schematic diagram of modal testing equipment with (i) shaker excitation, and (ii) hammer impact excitation.

4 EXCITATION

During the floor tests, various forms of excitation were used. Firstly, the instrumented hammer was used to provide a measurable impact excitation. This excitation technique had been successfully used previously by the testers. Secondly, following the recent acquisition of a more expensive electrodynamic shaker, two different forms of shaker excitation were applied - burst swept sine and band-limited burst random. This second form of excitation was used as fast transient broadband excitation methods were required in order to allow the minimum possible testing time on site. Only 48 hours were available over 4 consecutive days, and security arrangements dictated the assembly and removal of the testing equipment every day.

4.1 *Hammer impact excitation*

Prior to the field test described here, hammer impact excitation had been successfully used to test floor structures with a mass of up to 1200 tonnes.

The hammer testing technique was to apply repetitive impact excitation at every test point in turn (i.e. standard roving excitation) and to measure the response of the structure at several test points simultaneously, allowing several columns of the FRF matrix to be obtained. The typical procedure for measuring an FRF at one test point was as follows.

1. Set the data acquisition parameters (frequency range, frequency resolution, sensitivities, windows, etc.) to their optimum settings (Pavic & Waldron 1996).
2. Start the tape recorder with correct input attenuation, tape speed and frequency range selected.
3. Start the spectrum analyser acquisition routine with the required trigger level and pre-trigger time.
4. Instruct the hammer operator to strike the structure. An optimum strength of hammer hit should be established during QA checks and the hammer operator should be able to consistently reproduce similar hits. Following the hit, the hammer operator should remain as still as possible for the duration of the data acquisition. In order to minimise post-hit movement and to maximise hammer control, the hammer operator should be seated during hammer operation.
5. Wait for the full acquisition time to pass and inspect the FRF calculated by the spectrum analyser.
6. Repeat steps 4 and 5 for the required number of impacts, checking the stability of the FRF after each. Experience with testing large floors in open environments suggests that an average of not more than 10 impacts is required.
7. If the final FRF (after all impacts) is judged to be acceptable, store it. If the final FRF is judged to be unreliable then the measurement at that point should be repeated, and the poor FRF should be stored for further investigation.
8. Record the spectrum analyser file name and tape record number on a pre-prepared record sheet for future reference, and instruct the hammer operator to move to the next excitation location.

4.2 *Shaker excitation*

One of the main considerations for the modal testing using shaker excitation was the selection of excitation signal. Many published papers describe the benefits and drawbacks of various forms of excitation and the decision as to which to use was based partly upon the available literature (Brown et al. 1977, Olsen 1984). Due to the very limited development time available prior to the tests, the decision was also partly based upon which excitation signals could be generated using existing hardware and software. The two forms of excitation used in these tests were:

1. Burst Swept Sine Excitation (Chirp). A burst swept sine wave (either linear or logarithmic sweep as described by Olsen (1984)) was generated by an in-house computer program and downloaded to a digital function generator. A trapezoidal window was applied in the time domain to eliminate any transient response of the shaker (Figure 3). The burst swept sine signal could then be triggered manually and output directly to the shaker amplifier.

2. Band-Limited Burst Random Excitation. A random signal was generated using a digital data acquisition card, passed through a digital band-pass filter and sent as bursts of controlled duration through a digital to analogue converter to the shaker amplifier. The process was controlled by a computer program written in-house, and a typical resulting waveform is shown in Figure 4.

The actual testing procedure for both forms of excitation was very similar to the hammer impact testing procedure. Once again, roving excitation was used, and similar set-ups of data acquisition and time domain window parameters were used. The common perception that roving shaker excitation is time consuming did not apply since the shaker was operated in ‘reaction mode’. When testing large floors, moving the shaker from one test point to the next was as quick as moving a hammer operator from one test point to the next.

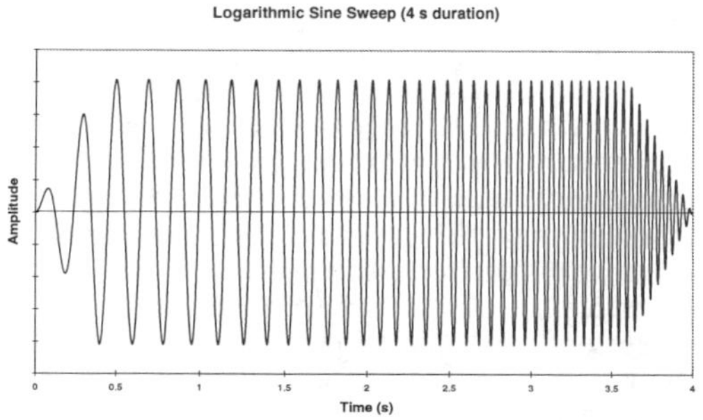

Figure 3. Typical logarithmic burst sine sweep excitation signal.

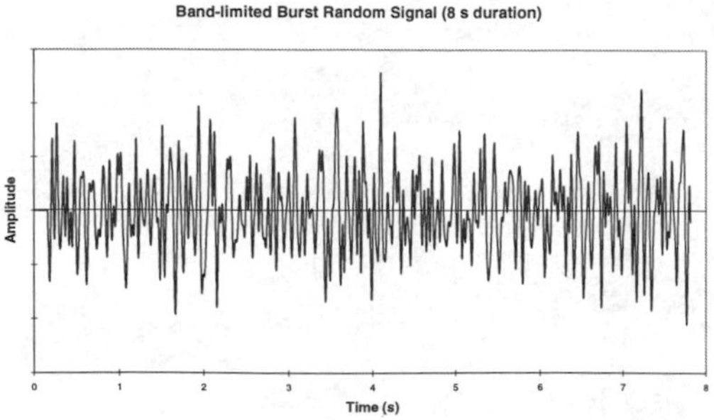

Figure 4. Typical band-limited burst random excitation signal.

4.3 *Comparison of hammer versus shaker excitation*

It was stated in the previous section that the actual testing procedure for using the shaker (with transient forms of excitation) was very similar to that using the hammer. It was found that, due to the increased control of the frequency content of the excitation, fewer averages were required to produce a stable FRF. Therefore, contrary to common perception, the total testing time when using shaker excitation was actually less than when using hammer excitation.

The processed results from the hammer and shaker tests clearly showed the estimated modal parameters from shaker testing were more consistent and of a higher quality than those obtained using hammer impact excitation. A visual comparison of the “smoothness” of one of the mode shapes is shown in Figure 5.

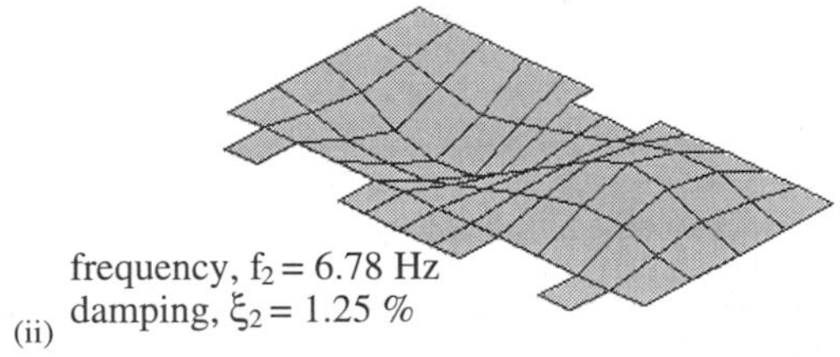

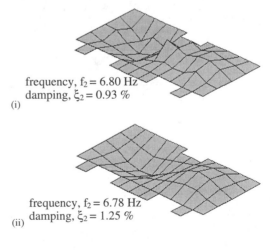

Figure 5. Second mode shapes measured using (i) hammer excitation and (ii) shaker excitation.

5 QUALITY ASSURANCE PROCEDURES

Observing recommendations made in the DTA Primer (DTA 1993b), the modal testing of the structure was prepared and conducted in 4 phases:

1. the preparatory phase,
2. the exploratory phase,
3. the measurement phase, and
4. the post-test data analysis and modal parameter estimation phase.

5.1 *The preparatory phase*

The preparatory phase for modal testing consists of defining the objectives of the test, and performing some limited analysis to determine the means by which the most comprehensive and highest quality data may be obtained. In the case of full scale modal testing in civil engineering, it also includes practical issues such as preparation of the test structure, checking calibration, packing and setting up of the equipment on site, and provision of transport and accommodation for test personnel. These latter issues are not normally relevant for laboratory testing of small mechanical structures, but careful planning of these seemingly trivial aspects can be crucial to the success of tests on civil structures.

The preparatory phase for the London tests is described below.

1. Definition of test objectives. The purpose of this test was to obtain experimental data for correlation with finite element models of the structure of varying complexity. In addition, feasibility studies regarding the use of the data with finite element model updating techniques were to be performed. Therefore the objectives of the test were simply to obtain the highest quality data possible in the limited time available, and to obtain the modal properties of all modes of vibration of the structure in the frequency range of interest. This corresponded to DTA Level 3 testing (DTA 1993b).

2. Preliminary numerical modelling. A finite element model of the structure was prepared using the most recent design information available. This was used to give an indication of the expected natural frequencies and mode shapes, and the test points were selected on the basis of this analysis. The adequacy of the selected test grid with regard to modal spatial aliasing was checked by constructing an auto-MAC plot (NAFEMS 1992) (Figure 6) using only finite element co-ordinates corresponding to test co-ordinates. In addition, files were produced to enable on-site animated display of analytical mode shapes, enabling re-selection of excitation or response points if any problems with the pre-selected points were encountered on site.

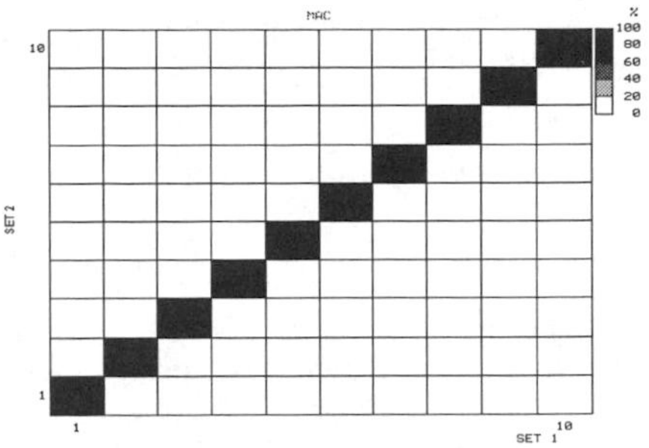

Figure 6. Auto-MAC for test grid co-ordinates shows no spatial aliasing for first ten FE calculated modes.

3. Preparation and packing of equipment. All equipment was checked prior to packing, and an adequate supply of consumables (batteries, tapes, floppy disks, etc.) was provided. Spare parts were provided for most critical items of equipment, the failure of which could possibly cause abandonment of the entire test. In order to ensure that no equipment was forgotten, pre-prepared checklists were used.

4. Accommodation and transport. As the test structure was located in central London, the accommodation and transport had to be carefully planned and booked well in advance. The complete absence of parking for a van near the test structure and hotel led to a combination of one-way hire vans and numerous taxis being used throughout the tests. This also emphasises the importance of all equipment being as portable as possible.

5. Preparation of test structure. Boundary conditions can rarely be prescribed for full-scale civil engineering structures and hence their "in-use" state must be tested. This is somewhat different to the DTA recommendations (DTA 1993a) which were, obviously, prepared with situations typical to the mechanical or aerospace engineering disciplines in mind. Also, non-structural elements are normally present which may or may not significantly affect modal testing results. These considerations must be borne in mind during modal testing and examined in detail during the post-test analysis phase. The floor under test was supported in part by non-structural facade walls, and it had suspended ceilings and building services attached. These were left in place. Since the building was under construction, the contractor was instructed to remove any non-permanent debris from the floor under test.

6. Setting up test equipment. As the equipment was set up on site, detailed notes were made on pre-prepared forms recording the usage of cables and transducers, and the channels on the various data acquisition devices to which they were connected. This was to enable the tracking of any rogue data to

faulty equipment, even after the equipment had been packed and transported back to base.

5.2 *The exploratory phase*

Following arrival on site and having set up the equipment, it was necessary to inspect the test structure and perform preliminary quality assurance tests to ensure that the highest quality data was obtained in the limited time on site.

Firstly an inspection of the test structure and surrounding environment was made. Detailed photographs, video footage and written notes were taken to provide a description of the actual condition of the structure. Pre-prepared forms were used to record information such as the existence of non-structural elements, the temperature and any background vibration sources (e.g. ground borne, acoustical, construction activity).

Next, subjective vibration response tests due to human heel-drops were performed by the test personnel at various points on the test structure. Although providing very little useful data, they gave the experienced test personnel a 'feeling' for the structure helping to understand better the way in which the structure behaved and the response levels which were likely to be measured.

To ensure that the highest quality data were recorded in the measurement phase, a number of QA checks were performed and recorded beforehand on pre-prepared test forms. These checks were performed as closely as possible in accordance with DTA and ISO recommendations (DTA 1993b, ISO 1994), although it was necessary to adapt some of them for use on civil engineering structures. By performing these checks, a higher degree of confidence in the measurements was ensured.

1. Excitation and Response Check. Time domain records of (hammer and shaker) excitation and response signals were acquired with fast sampling rates to determine transducer voltages. Equipment gain settings and sensitivities were set accordingly. The spectra of these signals were also examined to ensure that they looked reasonable. For shaker excitation, this process was performed for all excitation signals considered for use in the tests. For hammer excitation, a suitable strength of hits was agreed with the hammer operator so that over-ranging or clipping of the hammer signal would not occur. This can be difficult with an inexperienced hammer operator since (s)he will instinctively want to hit hard to excite 'properly' the structure, which in civil engineering is frequently massive.

2. Determination of Optimum Data Acquisition Parameters. Various settings of frequency range, frequency resolution and force exponential window were tried on the spectrum analyser in order to obtain the best quality FRF data. The required number of frequency domain averages was also investigated. The processability of the data was checked by immediately downloading the data to a PC and performing single-degree-of-freedom circle fit and line fit vibration parameter estimations. The quality of the circle and line fits gave an indication of the processability of the data, and the data acquisition parameters which gave the best quality fits were used for subsequent measurements (Pavic & Waldron 1996).

3. "Immediate Repeatability" Check. The digital data acquisition parameters were set as determined above and two FRFs were acquired using the selected number of averages, one immediately followed by another. In theory, the two such FRFs should be identical. However, for civil engineering structures, due to measurement noise and the very low level of excitation, differences in the two FRFs are always present, and in order to ensure that these differences are not excessive, it is prudent to perform this check. For the London test floor, engineering judgement was exercised to decide whether the effects of background noise were too great (Figure 7). The differences in the two FRFs in Figure 7 are typical for civil engineering structures and are considered to be acceptable since they lead to a fairly consistent set of estimated vibration parameters.

4. Homogeneity Check. In order to check whether the structure behaved linearly, excitation was applied at two different force levels and FRFs calculated. For a linear system, two such FRFs should be identical. For the floor in question, shaker excitation was applied at the maximum forcing level, and at half of that level. Due to the low level of excitation, the FRF measured at half of the full forcing level was much 'noisier' than the other, but the main characteristics of the FRFs were sufficiently close to assume that the check passed (Figure 8). For hammer impact excitation, hard and soft strength hammer blows were used, ensuring that the hard hits did not overload the hammer electronics. The ISO 7626-5 (ISO 1994) recommendation of using a hard hit to soft hit ratio of ten is normally impossible for civil engineering structures, since the soft hit would

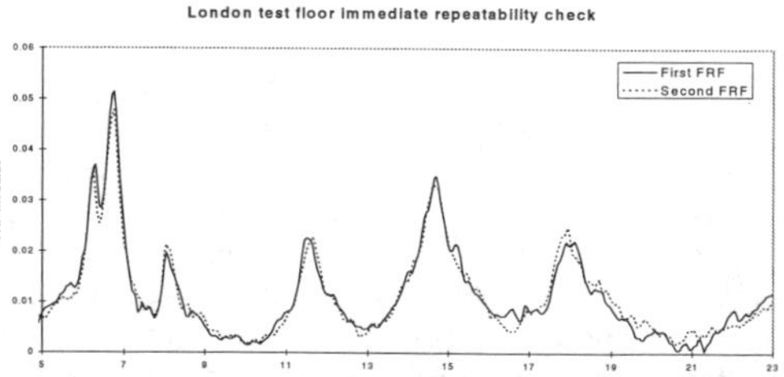

Figure 7. London test floor immediate repeatability check (burst random excitation).

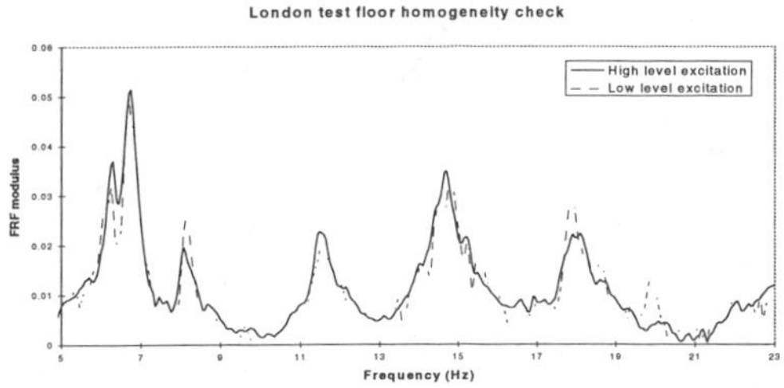

Figure 8. London test floor homogeneity check (burst random excitation).

be completely buried in background and measurement noise. Also, it is virtually impossible for the hammer operator to apply properly such a hit.

5. Reciprocity Check. This was another linearity check. Excitation was applied at one point on the floor and the response was measured at another. The excitation and response points were swapped and the measurement repeated. This was performed for three pairs of points. Maxwell's theorem states that the pairs of FRFs should be identical. Again, background noise affected the FRFs but they were sufficiently close to assume that the check passed.

6. Coherence Function Check. Coherence functions were measured for point mobility, and transfer mobilities between close and remote test points. At frequencies corresponding to peaks on the FRFs, the value of coherence was sufficiently high to assume that the effects of uncorrelated background noise were not too severe.

7. FRF Shape Check. Three point mobility FRFs were measured and the shapes of the FRF plots examined using log-log scales, for which several characteristics were visible indicating a good or bad FRF, as described by Ewins (1995). Experience is necessary for this check as it relies on the observer detecting any anomalies in the shape of the FRF which may indicate problems in the test structure, equipment set-up or the form of excitation.

8. "End of Test Repeatability" Check. At the end of the measurement phase, an FRF was measured using the exact set-up used for the immediate repeatability check. This FRF was then compared to one measured in the immediate repeatability check. The FRFs should be identical and significant differences would have indicated problems with noise, changing environmental conditions, or with the characteristics of the structure changing slowly through the test (e.g. due to temperature variations).

Finally, the shaker testing required determination of the optimum shaker excitation signal for the structure tested. The two excitation techniques were used in separate measurement 'swipes' during the measurement phase since they both provided good results and it was an interesting research exercise to compare them. The QA checks described above were performed for both forms of shaker excitation.

5.3 *The measurement phase*

The measurement phase is the main experimental data production stage of the modal testing process. Before embarking on these measurements, all necessary QA checks should have been performed and the testers should be satisfied with the quality of the data. If the data acquired during the checks are not satisfactory then remedial measures should be taken. A good example of this can be quoted from the London tests. The original plan was to perform the tests during the day. However, noise from construction activity on site was transmitted through the building to the floor under test, resulting in poor quality exploratory measurements and requiring an unusually large number of averages. Therefore, the decision was made to start testing in the evening once all construction activity had ceased. A noticeable improvement in the quality of the measurements was observed.

During the measurement phase, meticulous notes were made recording all measurement events, equipment set-up parameters, file names, tape numbers and any other information which may have been relevant to the data analysis phase. Pre-printed forms were used ensuring maximum possible speed of testing on site with no important information being omitted. Obviously, of equal importance was the careful labelling of computer files, disks and tapes to ensure that they could be properly matched with the written records.

As the testing was in progress, the development of the FRFs at every test point was carefully monitored. By observing that the FRFs were developing in a logical manner and that they were becoming more stable, filtering out measurement noise with an increased number of averages, a higher degree of confidence was established in the measured data. If, for any reason, the FRFs were not logical or there was suddenly a large discrepancy between two averages at a test point, further investigation was carried out or the measurement at that particular test point was repeated.

After each third of the test points had been measured, the FRF data was downloaded to a PC and multi-degree-of-freedom curve fitting algorithms were run. The software used was the ICATS suite of vibration parameter estimation software, developed at Imperial College, UK (ICATS 1997). By examining the development of measured mode shapes during testing, the quality of the measurements was assessed and if any problems had been encountered, remedial measures could have been taken. In addition, any anomalous test points could have been identified and the measurements at those points repeated. Since a return to site was not possible, as is frequently the case in the testing of civil engineering structures, it was important that this processing was performed on site during testing so that problems could be resolved immediately.

5.4 *The post-test data analysis and modal parameter estimation phase*

The test data was stored in two forms.

1. Digital FRF data which had been measured on site using a portable dual channel spectrum analyser.

2. Analogue transducer signals stored on a Racal StorePlus VL instrumentation tape recorder. These analogue records were replayed, sampled and processed after returning from site. The storage of analogue signals also allowed experimentation with different digitising parameters, without requiring an increase in time on site.

At all stages of processing, all computer files including FRF data, test grid data, estimated modal parameters and animation files were carefully labelled and recorded. Processed files were normally kept in the same directory on a PC as the unprocessed data so that the origin of any estimated modal parameters could be traced.

When replaying results from the analogue tape recorder, the same quality assurance procedures were used as for the actual site measurements. The development of the FRFs was carefully monitored, and any anomalies noted. Of course, at this stage re-measurement of any anomalous points was not possible, but any records which were spoiled by external noise, for example, should already have been re-measured and re-recorded on site.

Performing modal parameter estimation on modal testing data recorded from large-scale civil structures is frequently a problematic process. Due to the very low-level excitation used, the measured FRFs are normally noisy and parameter estimation algorithms can frequently give spurious results. Therefore, several multi-degree-of-freedom parameter estimation algorithms were used, and several runs of each algorithm were performed as recommended by the DTA (DTA 1993b). By examining trends in modal properties, a greater degree of confidence in the results was obtained.

6 CONCLUSIONS

Quality assurance procedures have been set up and adopted for the modal testing of full scale civil engineering structures. These are slightly different than in standard mechanical and aerospace applications due to the specifics of civil engineering structures such as their size and noisy open environments. Following the QA procedures, high quality modal testing data (three swipes of 33 test points) were acquired for a 600 tonne ribbed post-tensioned concrete floor in only four days on site.

Modal test data acquired using an electrodynamic shaker with a rated force capability of 133 N in 'reaction mode' was more consistent and of a higher quality than that acquired using hammer impact excitation with a 5.4 kg instrumented hammer.

Surprisingly, modal testing of the post-tensioned concrete floor using the portable roving shaker in reaction mode was approximately 20% faster than hammer impact excitation, due to ease of movement of the shaker across the structure and the increased control over the frequency content of the excitation.

ACKNOWLEDGEMENTS

The authors would like to thank Access Flooring Association, PSC Freyssinet (UK) and Taywood Engineering, the research partners collaborating with the CCC in a research project funded by the DoE Partners in Technology scheme.

REFERENCES

Brown, D., G. Carbon & K. Ramsey 1977. *Survey of Excitation Techniques Applicable to the Testing of Automotive Structures*. Society of Automotive Engineers, Paper No. 770 029.

DTA 1993a. *DTA Handbook - Volume 3 ~ Modal Testing*. London: Dynamic Testing Agency.

DTA 1993b. *Primer on best practice in dynamic testing*. London: Dynamic Testing Agency.

Ewins, D.J. 1995. *Modal Testing: Theory and Practice*. Somerset: Research Studies Press.

ICATS 1997. *MODENT, MODESH, MODACQ and MESHGEN Reference Manual*. London: ICATS.

ISO 1994. *Vibration and shock - Experimental determination of mechanical mobility - Part 5: Measurements using impact excitation with an exciter which is not attached to the structure*. ISO 7626-5:1994.

NAFEMS, 1992. *A Finite Element Dynamics Primer*. D. Hitchings (ed.). Glasgow: NAFEMS.

Olsen, N. 1984. Excitation Functions for Structural Frequency Response Measurements. *Proc. 2nd Int. Modal Analysis Conf., Orlando, February 1984:* 894-902.

Pavic, A. & P. Waldron 1996. Guidelines on modal testing of full-scale concrete floors using instrumented hammer impact excitation. *Joint Institution of Structural Engineers/City University Int. Seminar - Structural Assessment, the Role of Large and Full Scale Testing, London, July 1996.*

Modern Practice in Stress and Vibration Analysis, Gilchrist (ed.) © *1997 Balkema, Rotterdam, ISBN 90 5410 896 7*

Modelling of machine dynamics using digital signal processing techniques

K. Kelly, P. Young & G. Byrne
Department of Mechanical Engineering, University College Dublin, Ireland

ABSTRACT: In finish machining, even with high quality machines, the influence of the dynamic behaviour of the machine tool & workpiece on the surface quality can be significant. This is especially true when interrupted cutting, or machining of ultra-hard or non-homogenous materials is taking place. Traditional approaches to modelling this dynamic behaviour have tended to use combinations of springs, masses & dampers, usually with only one or two degrees of freedom. In the work reported here, digital signal processing techniques are used to incorporate the full dynamic characteristics (which are experimentally obtained) of the system, allowing the regenerative cutting process to be more accurately considered. Various experimental methods for obtaining the requisite data are evaluated and means of compensating for the inherent measurement difficulties are given. Finally, sample data from a turning centre is used to generate a simulated surface, thus illustrating the concept proposed here.

1 INTRODUCTION

The integrity of a surface produced during machining is influenced by many factors. In fundamental terms it may be viewed as deriving from the interaction of three systems:

a) The machine
b) The cutting insert(s)
c) The workpiece material

The interaction between these systems will be determined by the nature of the process itself and the choice of machining parameters. Many researchers to date, have attempted to model particular aspects of the surface integrity during machining, with some success. The work reported here forms part of a comprehensive model of surface generation in the turning of AlSi9Cu3 with PCD (poly-crystalline diamond) tooling (Kelly et al. 1996), being developed as part of a European research project (NEMPRO - BE8210). The emphasis in this modelling work, an outline of which is given below, is on producing a simulation which is capable of predicting the surface integrity, for a choice of machining parameters, in a timely manner.

This paper deals specifically with the influence of the dynamic characteristics of the machine tool on the surface integrity and the means to model this accurately.

1.1 Influence of machine vibration on surface integrity

The influence of the machine behaviour can be viewed as vibration arising from two separate sources. The first, and usually the more significant, is the vibratory motion between tool and workpiece caused by imbalance in the rotating components of the machine tool. This can be specified as a series of (rpm-dependent) frequencies with characteristic amplitudes and phases.

The second source of vibration comes from the response of the machine (its dynamic compliance) to a varying force input, i.e. a varying chip load. This response can be calculated if the dynamic stiffness function of the machine is known for the particular direction of interest, see figure 1.

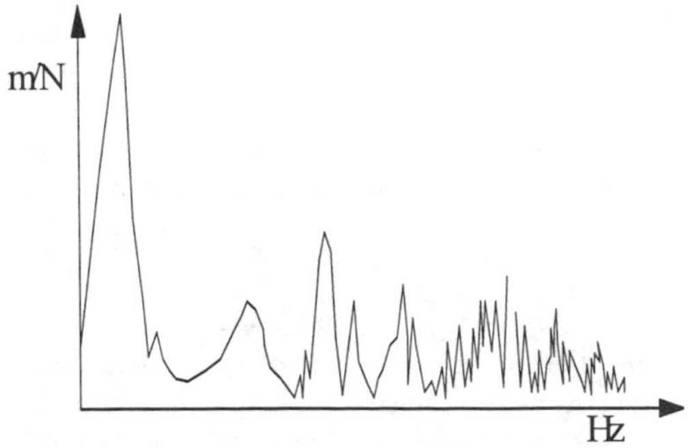

Figure 1. Typical dynamic compliance function

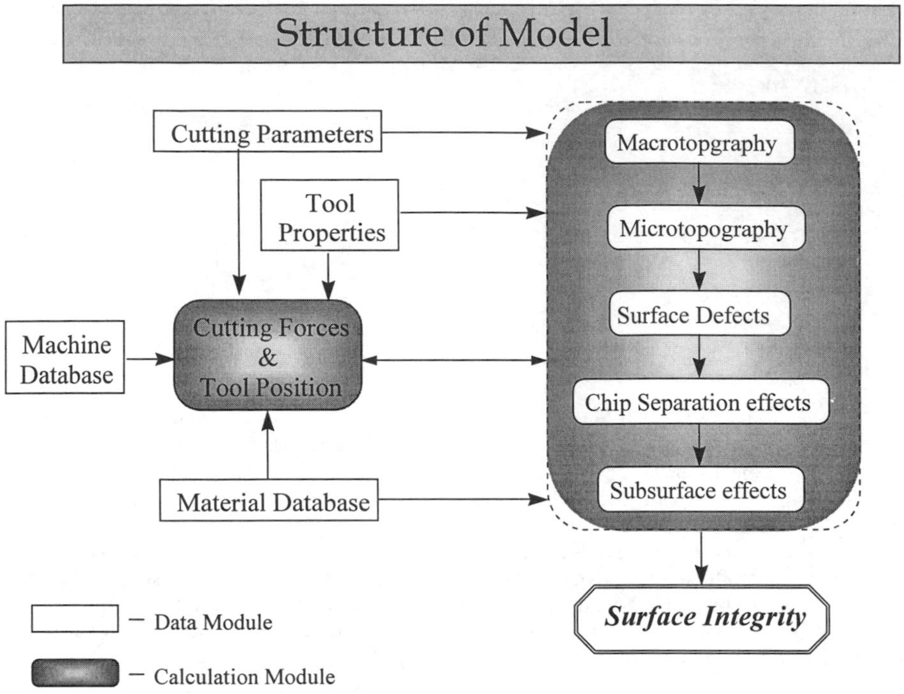

Figure 2. Model Structure

Many researchers have approximated these response functions using mass-spring-damper systems with one or two degrees of freedom (Jang & Seireg 1989, Tobias 1965). Depending on the height and distribution of these peaks, as well as the frequency range of interest, this representation will often produce sufficiently accurate results. In cases where there is no dominant peak, or where several peaks occur in close proximity, this method is not satisfactory. This study uses a time domain representation of the compliance function to specify the dynamic characteristics of the machine.

2 MODELLING STRATEGY

As explained above, the process of surface generation may be viewed as the interaction of three systems:

a) Machine,
b) Cutting Tool,
c) Workpiece Material,

where the degree and manner of interaction is controlled by the choice of cutting parameters. This interaction will determine the position of the cutting tool relative to the workpiece, which will itself, in conjunction with the behaviour of the material in the cutting zone, determine the shape and properties of the surface, i.e. the machined surface integrity.

To model the surface integrity generation in single point turning, the generic model structure shown in figure 2 has been adopted. In this model structure, the machine, workpiece & tool systems are characterised by a set of parameters in data modules.

The information in these data modules is then used to calculate the position of the cutting tool relative to the workpiece and the cutting forces as functions of time. The same databases are utilised again to calculate the integrity of the surface in a further calculation module.

It may be seen from figure 2 that, in this model structure, the complete surface integrity profile is approximated as a series of discrete aspects which can be superimposed to form the complete profile. The approach that has been adopted here assumes, in the first instance, that the tool profile (both macro- and micro-) is transferred to the workpiece surface. This surface is then modified to include surface defects, such as craters, tears, microchips, porosities etc. After this other material related effects such as squeezing, sideflow, spanzipfel, etc. are included, before finally considering sub-surface effects.

3 DYNAMIC CHARACTERISATION OF THE MACHINE TOOL

A discrete time system (in this case representing the dynamics of the machine tool) is, formally, a transformation that maps input signals to output signals. Such a system may be represented by the equation:

$$y = T(u) \tag{1}$$

and the *k*th value of the output is

$$y(k) = [T(u)](k) \tag{2}$$

The unit pulse response, or the impulse response function (IRF) is the output of the system when the input signal is the delta dirac function, i.e. u = δ:

$$h = T(\delta) \tag{3}$$

The output of the system for a general input function u, can be defined as:

$$y = h * u \tag{4}$$

The analogy in the frequency domain is $H(e^{j\theta})$, which may be obtained by transforming equation 4 into the frequency domain.

$$Y(e^{j\theta}) = H(e^{j\theta}).U(e^{j\theta}) \tag{5}$$

and is referred to as the frequency response function (FRF).

Thus we may characterise the dynamic behaviour of the machine using either a time-domain or a frequency-domain representation.

4 MEASUREMENT THEORY

4.1 Time Domain Measurement.

Time domain measurement uses a form of excitation which covers the frequency range of interest. This may be through random or burst-random inputs using an electromagnetic shaker, transforming the measurements into the time domain to calculate the IRF. Impact testing approximates the impulse response by applying a measured impulse, usually with a hammer, to the system and measuring the response until the system has returned to equilibrium. Again this may be used to calculate an FRF, or to provide a direct estimate of the impulse response. The recorded response differs from the theoretical impulse as the applied force is not applied instantaneously and has a finite application time. This limits the maximum frequency of the excitation and related noise in the recorded response is not easily identifiable.

4.2 Frequency domain measurement

Measurement in the frequency domain of the FRF involves applying an excitation of known frequency and amplitude to the system, and recording the amplitude and phase of the response. The principle of super-position, for a linear system, means that the value of the FRF at that particular frequency is given by the relative amplitudes of input and output signals and the phase lag between them. Recording these values over a frequency range will define the system within the limits of the applied inputs.

Experimentally, two main methods of applying this excitation are used, Swept and Stepped Sine. Normally the input signal is produced using a standard signal generator and converted into a force input to the system using an electromagnetic shaker.

4.2.1 Swept Sine

A 'Swept Sine' test uses a constant amplitude sine wave where the frequency increases, or decreases, linearly throughout the test. The response is sampled at regular intervals, giving a discrete set of points which define the FRF. Testing time is reasonably fast, but the input to the system is based on a constant amplitude electric signal to the exciter. As a result the input force to the system will vary depending on the dynamic properties of the system, exciter and connection medium. For more accurate results, the input force and response should both be measured.

4.2.2 Stepped Sine

"Stepped Sine" testing increments the frequency of the input by a discrete amount. The main advantage is that the force input to the system can be used as part of a feedback control system which ensures that the amplitude of the force into the system is constant across all frequencies. If a unit amplitude is used for the force input then only the response need be monitored to define the amplitude of the FRF. Readings may also be made by hand if analogue to digital conversion is not available.

4.2.3 Frequency Domain Averaging

The main advantage of operating in the frequency domain when measuring FRF's is that noise may be reduced by averaging measurements of a stationary system. Output noise appears as a random addition to the system response, and over a significant number of samples sums to zero. By averaging the frequency components over a number of measurements, the resulting FRF is a truer representation of the actual system.

The stepped sine measurement technique allows this averaging to be performed as an integral part of the measurement, and this method was used to produce the FRF's for this work.

4.3 Obtaining the impulse response function from frequency domain data

The finite length and resolution of sampled FRF's lead to inaccuracies in generating IRF signals. The nature of the Fourier transform, and its inverse mean that the resolution in one domain leads to limitations in length in the other domain and vice versa.

In terms of mapping from the time to frequency domain these problems manifest themselves as aliasing (measurement resolution) and leakage (measurement length). If care is not taken, responses at frequencies outside the measured range are mapped into the range of interest, distorting the FRF. The limited sampling length results in some of the contents of one frequency point to 'leak' into neighbouring points.

In similar fashion, the inverse transform alters the IRF due to the frequency spacing between points and the finite length of the sampled FRF. While the resolution of the FRF may be addressed in the experimental procedures, the finite length is often dictated by limitations of sensors and processing equipment.

Mathematical solutions to compensate for this limitation must be applied to improve the IRF estimation.

4.4 Compensation for time-domain leakage

It may be shown (Keqin 1993) that in the high frequency region of the spectrum, above some truncation frequency f_c, the transfer function takes the approximate form of a residual inertia:

$$H(\omega) \approx -\frac{1}{M\omega^2} \quad (\omega > \omega_c) \tag{6}$$

where M is the effective residue mass and $\omega_c = 2\pi f_c$ An improved estimate of the impulse response function, h*(t), may be obtained as follows:

$$h^*(t) = h(t) + \frac{1}{2\pi}\int_{-\infty}^{-\omega_c} -\frac{e^{j\omega t}}{M\omega^2}d\omega + \frac{1}{2\pi}\int_{\omega_c}^{\infty} -\frac{e^{j\omega t}}{M\omega^2}d\omega \tag{7}$$

$$= h(t) - \frac{1}{M\pi}\int_{\omega_c}^{\infty}\frac{1}{\omega^2}Cos\,\omega t\,d\omega$$

$$= h(t) + Cg(t)$$

where , C = -1/(πM) and

$$g(t) = \int_{\omega_c}^{\infty}\frac{1}{\omega^2}Cos\,\omega t\,d\omega \tag{8}$$

Because the impulse response function must represent that of a physically realisable system, i.e. it must be causal, the value of the impulse response function at time zero must be zero;

$$h^*(0) = h(0) + Cg(0) \tag{9}$$

The value of the constant C can be calculated by the formula:

$$C = -{}^{h(0)}\!/_{g(0)} \tag{10}$$

Integration by parts of the integral in equation 8 gives the following solution for g(t):

$$g(t) = \frac{Cos(\omega_c t)}{\omega_c} - t\left\{\frac{\pi}{2} - Si(\omega_c t)\right\} \tag{11}$$

where

$$Si(t) = \int_0^t \frac{Sin(x)}{x}dx \tag{12}$$

5 CALCULATION THEORY

5.1 Influence of material composition on cutting forces

Some work has been reported in the literature on the influence of workpiece material composition on cutting force variation (Zhang & Kapoor 1992). The central tenet of their work is that the variation in the cutting force during machining can be largely attributed to the local variation in the hardness of a 'sample' (the uncut chip area), due to the non-homogenous nature of the workpiece and the difference in hardnesses of the constituent phases.

Zhang & Kapoor applied their method to modelling the machining of steel. The approach is based on a statistical description of the local variation in material composition, generated from the results of a detailed metallographic analysis. By sampling randomly from this distribution, the local composition at a given point is defined and from this the instantaneous cutting force can be calculated.

Computation of the cutting force involves calculating a 'sample hardness' based on the constitution of the uncut chip area, dividing this by the bulk mean hardness to generate a factor which is used to scale the nominal value of specific cutting force. Because the variation in the local composition occurs in accordance with a gaussian distribution, the variation in hardness of the 'sample' is also gaussian. The excitation will thus occur across a broad range of frequencies. Broad-band excitation of a system with multiple degrees of freedom will produce a complicated response which may not be obviously related to the excitation or the response function. For illustrative purposes, the excitations demonstrated in this discussion are confined to a single frequency, making their effect on the surface much more visible, in the time/spatial domain.

5.2 Cutting force calculation

The instantaneous cutting forces are dependent on the instantaneous volumetric removal rate, the friction between the tool and workpiece and the local composition of the workpiece material. The relationship between the average cutting forces and the instantaneous volume of material being removed is often expressed in terms of the instantaneous depth of cut (Kienzle 1952), or in terms of the uncut chip area (Jang & Seireg 1989), i.e. the area on the rake face of the tool that is in contact with the workpiece at any given instant. Typically this relationship is expressed as:

$$F_c = b.h.k_c \tag{13}$$

or

$$F_c = A_i.k_c \tag{14}$$

where F_c is the instantaneous cutting force, b is the feed, h is the instantaneous depth of cut, A_i is the uncut chip area and k_c is referred to as the specific cutting force (i.e. the cutting force required for an uncut chip area of unity).

The value of the specific cutting force is itself a function of the uncut chip area as, at lower values of uncut chip area, the friction of the tool edge with the workpiece surface is proportionately more significant. The specific cutting force will also be dependent on the local composition of the material being machined. Obviously, this variation in specific cutting force is of much greater significance for highly inhomogenous materials. This relationship can be established from machining tests.

5.3 Calculation of tool path

Because of the finite stiffness of the machine tool, the process forces will cause a deflection/deformation of the structure. It is the variation in the process forces, usually initiated by geometric irregularity of the workpiece, kinematic vibration of the machine or material inhomogeneity which causes regenerative vibration.

The response of the machine tool to an excitation force may be calculated by convolving the impulse response of the machine tool with the excitation force signal:

$$r(t) = \int_0^t h(\tau) f(t-\tau) d\tau \tag{15}$$

or in discrete numerical form:

$$r(i) = \sum_{k=0}^{n} h_k . f(i-k) \tag{16}$$

This value, $r(i)$ will be the deviation from the nominal tool path in the particular direction under consideration. This deviation will lead to a change in the uncut chip area at the next instant in time which will in turn influence the cutting force.

The uncut chip area may be calculated from geometrical considerations.

Thus if the impulse response for the system is known, or can be calculated, the response of the system may be deduced by convolving it with a time history of the cutting force. This convolution is performed at each time increment and the result used to update the force history, enabling the next position in the tool path to be calculated.

5.4 Calculation of machined surface topography

In this paper, the influence of the dynamic excitation of the machine structure is the only factor under consideration. The surface topography produced by the turning operation may be predicted from a knowledge of the actual tool path relative to the workpiece surface. In turning, the nominal helical path is modified to account for deviations caused by the process forces exciting the machine tool structure.

The surface topography is therefore calculated purely from geometrical considerations, the parameters being supplied by:

i) the initial geometry of the workpiece (i.e. a cylinder)
ii) the geometry of the cutting insert (i.e. a circular arc)
iii) the prescribed tool path (i.e. a helix)
iv) dynamic excitation of the machine tool structure by the process forces

6 EXAMPLE CALCULATION

6.1 Sample material

Let us consider a notional material whose specific cutting force varies about the nominal value as shown in figure 3 below.

The specific cutting force shown above will vary from the nominal value once in every 'n' samples. We can therefore vary the frequency of excitation as a function of 'n', allowing us to assess the impact of various frequencies of excitation on the finished surface.

In the work reported here, the specific cutting force will be assumed to be independent of uncut chip area. In reality there will be a dependence on the uncut chip area. Other factors such as built-up edge, strain-hardening, porosity, coolant/lubricant and condition of the cutting insert will also affect the local cutting force.

6.2 Sample machine tool dynamic signature

Consider the following 'signature' of the dynamic compliance of a turning centre in the radial direction

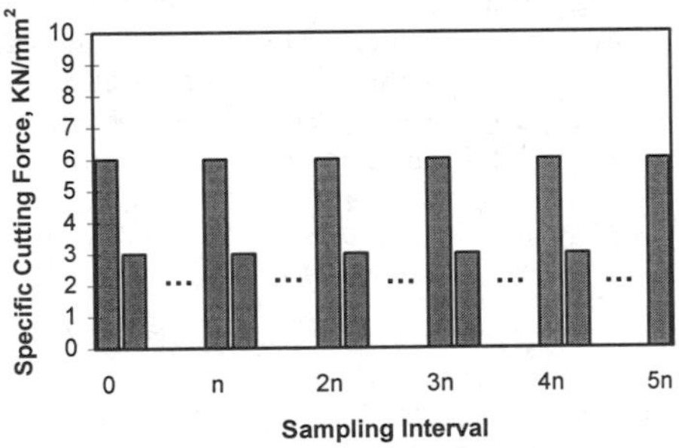

Figure 3. Specific cutting force variation for notional material

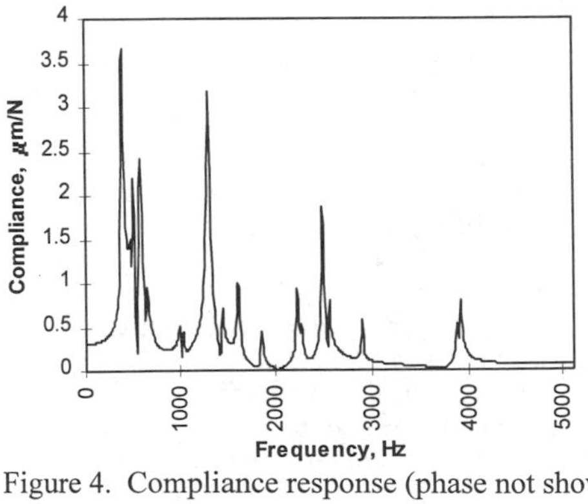

Figure 4. Compliance response (phase not shown) vs. frequency

(i.e. normal to the workpiece surface). Again, for clarity, we will restrict ourselves to motion in a single direction.

The time domain response, calculated using the discrete inverse fourier transform, and also the corrected impulse response calculated using the method outlined above are shown in figures 5 & 6 below.

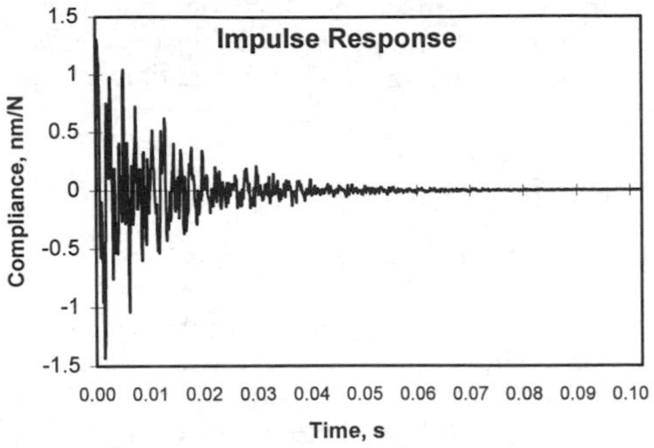

Figure 5. Impulse response function

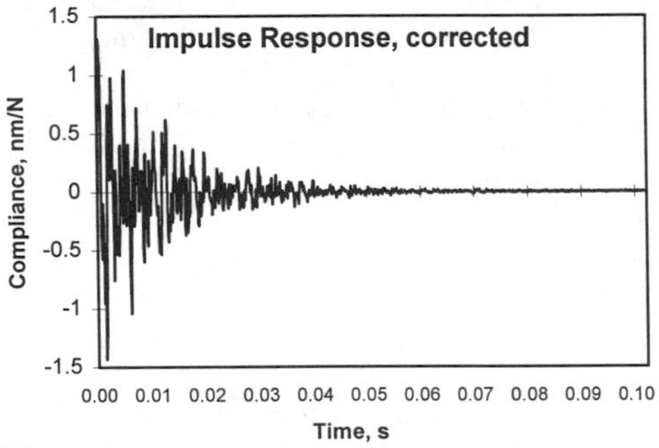

Figure 6. Corrected impulse response function

Table 1. Simulation parameters

Parameter	Value
Machine dynamics	As above
Specific cutting force	n = 8, 15
Workpiece diameter	100 mm
Tool nose radius	0.8 mm
Feed	0.12 mm
Depth of cut	0.1 mm
Spindle speed	260 rpm
Resolution, feed direction *	250 points per mm
cutting direction	311 points per rev
Displayed surface *	5 feedmarks x 10% revolution

* The simulation resolution is specified in terms of the number of points per (nominal) feedmark width in one direction, and the number of points per revolution in the other. The amount of surface displayed is 5 feedmark widths by a specified percentage of the circumference.

6.3 Machining parameters for simulation

The parameters for the simulation are given in table 1.

7 RESULTS & DISCUSSION

A graph of a simulated surface, with specific cutting force varying once in every 15 samples is shown in figure 7. Figure 8 shows a simulated surface, with specific cutting force varying once in every 8 samples. A surface produced with zero excitation is shown in figure 9 for comparison. The cutting direction is from right to left and the feed direction is from front to back in figures 7, 8 & 9.

In the case of the simulated surface shown in figure 7, the specific cutting force varies once in every 15 samples. The resolution in the cutting direction is 311 samples per revolution. One would expect therefore to find approximately 21 impulses supplied to the system per revolution. In the portion of the surface displayed (10%), one would expect approximately 2 impulses per feedmark. This is indeed what we observe. The relatively low

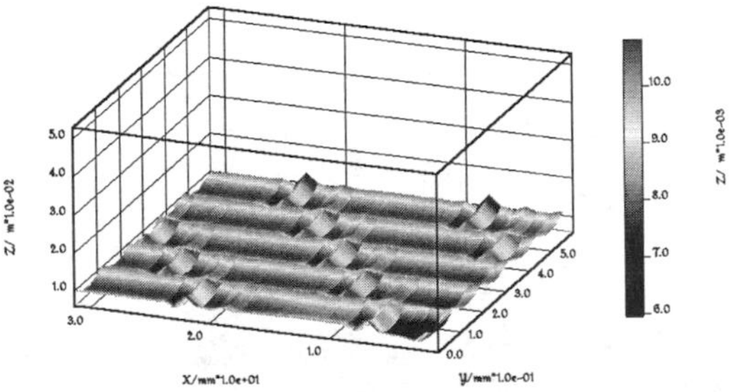

Figure 7. Simulated surface with specific cutting force variation, once in 15 samples.

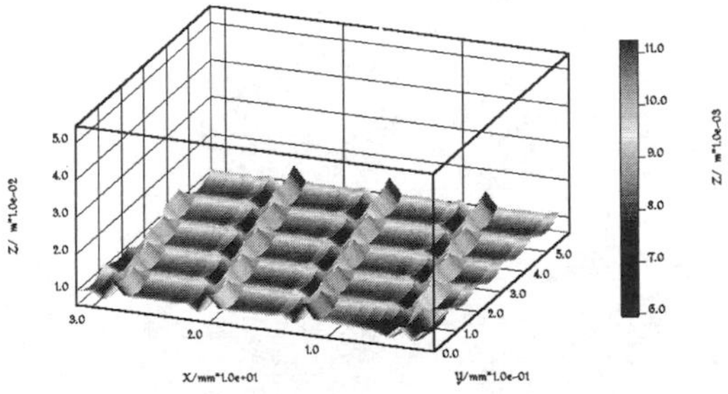

Figure 8. Simulated surface with specific cutting force variation, once in 8 samples.

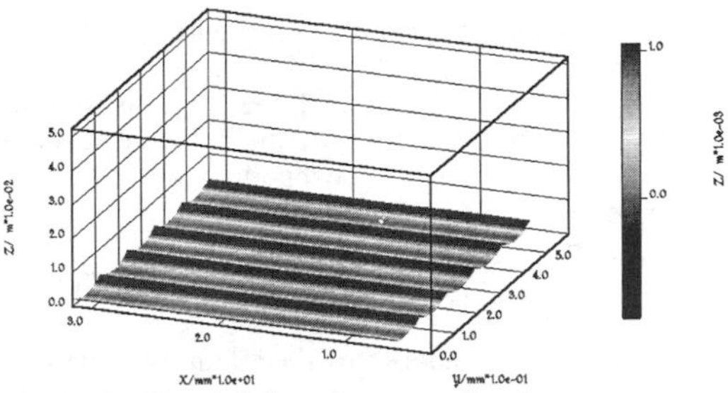

Figure 9. Simulated surface, no vibration

damping characteristic of machine tool structures may be observed in the ripple after each impulse. The regenerative nature of the cutting process, i.e. the influence of previous cuts on the system dynamics, may be observed also in the small ripples in the adjacent feedmarks, next to the impulse on the previous revolution. This is a direct consequence of the change in the chip load due to the influence of the vibration on the previous pass of the cutting insert.

In figure 8, a simulated surface is shown where the specific cutting force during simulation varies once in every 8 samples. Simulation conditions (cf. Table 1) are otherwise identical to those used in the simulation of the surface in figure 7. With a variation of one sample in 8, the expected number of impulses per feedmark shown, calculated as before, is approximately 4. This may be verified against the displayed surface in figure 8. As before the regenerative nature of the process may be seen. In this case because of the smaller phase lag between impulses in adjacent feedmarks, this contribution is manifested as a steadily increasing response throughout cutting, to each series of impulses.

Real materials, which will be inhomogenous to a greater or lesser degree, will provide excitation over a wide range of frequencies. This excitation during machining of the machine tools structure will lead to complex surface topography. While the surfaces shown above are due to single frequency excitation, the principles and methodology will also hold true for multiple frequency excitation, due to the principle of superposition. More complex excitation of the structure during machining of real materials may therefore be undertaken using the above methodology.

8 CONCLUSIONS

Various methods to obtain the impulse response function of a machine tool have been discussed. In particular, algorithms to compensate for inherent limitations of the testing procedures have been presented

A new application of the impulse response function to the simulation of the dynamic response of a machine tool to excitation by workpiece inhomogeneity has been developed. This forms one module of a larger model which simulates the surface integrity produced when turning AlSi9 with PCD tooling.

Sample results show the relationship between local workpiece composition and the generated surface profile, and also illustrate the regenerative nature of the cutting process.

The methods outlined in this paper provide the framework for accurate simulation of the resultant geometry in a way that conventional, one or two degree of freedom systems, cannot at present provide.

9 REFERENCES

Jang, D. & Seireg, A. 1989. Dynamic Simulation for Predicting Surface Roughness in Turning, *Machine Engineering Dynamics Application & Vibration*, ASME DE Vol 18-2, Montreal, Quebec, Canada

Kienzle, O. 1952. Die Bestimmung von Kräften und Leistungen an spanenden Werkzeugen und Werkzeugmaschinen. VDI-Z. 94(1952) 11/12, p. 299-305

Kelly, K. *et al*, Modelling of Surface Integrity Generation in Single Point Turning, *Proc. 13th Conference of the Irish Manufacturing Committee*, Sept 1996.

Keqin, X. 1993. Two Updated Methods for Impulse Response Function Estimation. *Mechanical Systems & Signal Processing*. 7(5), 451-460.

Tobias, A. 1965. *Machine Tool Vibration*. Blackie & Son Ltd.

Zhang, G. & Kapoor, S. 1991. Dynamic generation of Machined surfaces, Part 1: Description of a Random Excitation System, *Trans. ASME Journal of Engineering for Industry,* Vol 113, p. 144-148, May 1991

Zhang, G. & Kapoor, S. 1991. Dynamic generation of Machined surfaces, Part 2: Construction of surface topography, *Trans ASME, Journal of Engineering for Industry,* Vol 113, p. 149-153, May 1991

7 Dynamical response synthesis

Modern Practice in Stress and Vibration Analysis, Gilchrist (ed.) © *1997 Balkema, Rotterdam, ISBN 90 5410 896 7*

Towards a dynamic version of Saint Venant's principle

B. Karp & D. Durban
Technion, Israel Institute of Technology, Haifa, Israel

ABSTRACT: A new approach towards the formulation of a dynamic version of Saint Venant's principle is suggested. The main idea underlying this work is to look for different dynamic loadings that induce similar response far from the loaded area, rather than evaluating the energy decay rate (which is not necessarily decaying). The dynamic response of a waveguide serves as an example for the application of the proposed principle. It is demonstrated that under certain circumstances the spatial distribution of the load does not affect the deformation far from the loaded edge.

INTRODUCTION

Saint Venant's principle (SVP) has been proposed by Saint Venant in 1853 as a practical assumption, useful to derive approximate solutions for long elastic bars. The semi-inverse method for solving quasi-static problems of bars is considered to be satisfactory under the assumption of SVP. Love's definition of SVP (Love, 1944) is usually accepted as a clear definition enabling rigorous mathematical evaluation; '...a principle, first definitely enunciated by Saint-Venant, and known as the "principle of the elastic equivalence of statically equipollent systems of loads". According to this principle, the strains that are produced in a body by the application, to a small part of its surface, of a system of forces statically equivalent to zero force and zero couple, are of negligible magnitude at distances which are large compared with the linear dimensions of the part.' Love (1944, p. 131).

The usefulness of the SVP lies in that it is not necessary to consider the actual details of boundary conditions, which are in many problems complicated or unknown precisely. This simplification encouraged many scientists to evaluate the validity of SVP for media and response other than linear quasi-static elasticity like: viscoelasticity, non-linear elasticity, elastoplasticity, micropolar continuum, anisotropic elasticity, heat transfer and elastodynamics. Heat transfer and elastodynamic problems are governed by parabolic and hyperbolic equations, respectively, thus markedly differ from elastostatic problems. Recent reviews on SVP may be found in Horgan & Knowles (1983), Horgan (1989) and Horgan & Simmonds (1994).

The idea of application of SVP to dynamic problems was initially suggested by Boley (1955); 'Such a principle, of course, could be similarly employed in the solution of dynamical problems.' Since then, several studies dealt with the possible validity of SVP in elastodynamic problems, but not with its direct application in solving problems. Some of the basic papers are by Novozhilov & Slepian (1965), Kennedy & Jones (1969), Grandin & Little (1974). Energy decay estimates, employed to assess a dynamic version of SVP, may be found in Flavin & Knops (1987), Knops (1989) and Chirita & Quintanilla (1996). Most of the papers follow Love's interpretation, thus looking for decaying fields induced by dynamic self-equilibrating end loads.

The purpose of the present paper is to examine a new definition of a dynamic SVP and to evaluate its validity in dynamic response of structures. In the next paragraph we summarize the well-known results relating to propagation of elastic waves in linearly elastic waveguides. Next we discuss the importance of the exact spatial distribution of dynamic steady state edge loads in determining the dynamic field far from the loaded end. Finally, we suggest a new definition of SVP in elastodynamics in the light of the original Saint Venant statement, along with a quantitative evaluation of its applicability.

WAVES IN A WAVEGUIDES

The dynamical analogue of quasi-static problems of bars are the problems of waveguides: in both cases there is a characteristic length scale of the structure. We summarize here some known results needed for the forthcoming discussion.

The general solution for the displacement field $\mathbf{V}(x,z,t)$ for a strip composed of a linearly elastic material ($-h \leq x \leq h$, $z \geq 0$) under plane strain conditions has the form

$$\mathbf{V}(x,z,t) = \mathbf{U}(x)\, e^{i(\xi z - \omega t)} \tag{1}$$

where $\mathbf{U}(x)$ is the profile distribution (eigenfunction), ξ is the *wave number*, ω is the circular *frequency*, t and z stand for the time and the coordinate in the axial direction of the strip and h is the half width of the strip. The wave number ξ may be real, complex or imaginary. Real wave numbers represent *propagating waves*. Complex and imaginary wave numbers represent *attenuating waves*, where the characteristic attenuation length is inversely proportional to the imaginary part of the wave number - thus named *attenuation constant*. We consider $\mathbf{V}(x,z,t)$ to be a response of a semi-infinite strip to a steady state cyclic load applied at $z = 0$.

The dispersive nature of waves in waveguides is conveniently presented in *Attenuation maps*, after Mindlin (1960). Examples for such maps for strips in plane strain and cylindrical waveguides may be found in Graff (1975) as well as in other monographs (the maps usually named frequency spectrum). For convenience, a map for a strip with lubricated faces and non-dimensional frequency and wave number is shown in Figure 1. For lubricated faces the tangent tractions and normal displacement on the faces $x = \pm h$ are zero. The right direction of the horizontal axis indicates wave numbers with real value, while imaginary values are in the left direction. In that particular case there are no complex wave numbers. The ordinate is the frequency. For interpretation of this map the reader referred to Mindlin (1960) or to standard monographs on elastic waves.

Due to the dispersion phenomenon, the waveguides posses a property that only a finite number of propagating waves are possible at any exciting frequency, while there is an infinite number of attenuating (non-homogeneous) waves. The number of possible propagating waves grows with the increase of the frequency. The deformation in the near vicinity to the loaded edge, $z = 0$, (the *near field*) is a result of the propagating and attenuating waves. In the *far field*, the deformation results from the propagating waves alone. The extent of the near field is dictated by the attenuating wave with the smallest attenuation constant.

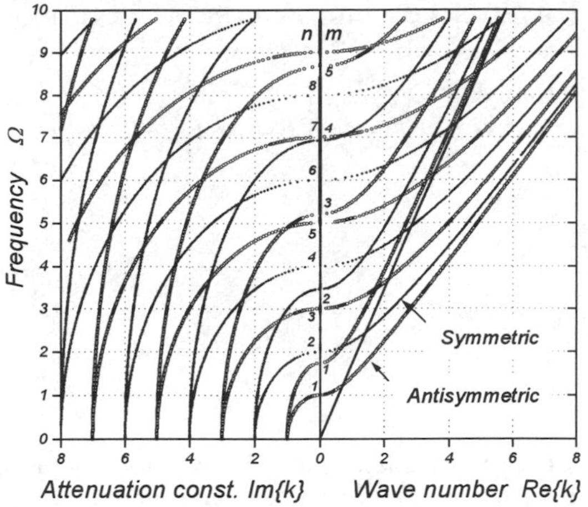

Figure 1 - Attenuation map for a strip with lubricated faces, for symmetric modes (thin lines composed of small dots) and antisymmetric modes (thick lines composed of hollow circles). The attenuation constant is the imaginary part of the wave number. $k = 2h\xi/\pi$ is the non-dimensional wave number. $\Omega = 2h\omega/\pi C_T$ is the non-dimensional frequency where C_T is the velocity of a transverse wave in an infinite medium. Here Poisson's ratio $\nu = 0.25$ and (m,n) are integers for transversal and longitudinal cut-off frequencies, respectively.

It is a common property of propagating waves to transfer energy in space. The resulting *energy flux* in a propagating wave serves as a measure of the strength of the wave (Achenbach, 1973). For a simple unidimensional longitudinal plane wave the expression for the energy flux per unit area P is

$$P = \frac{1}{2}\frac{E\mathcal{A}^2\,\omega^2}{C} \tag{2}$$

where E is Young's modulus, $\mathcal{A}$ is the amplitude of the wave, and C is the phase velocity which is given by $C = \omega/\xi$. From expression (2) it follows that for a given material, any set of three parameters among four relevant (energy flux P, amplitude $\mathcal{A}$, frequency ω and wave number ξ) determine uniquely the nature of the wave. Similar relation between the four parameters in waveguides obtained by Miklowitz (1978).

It should be noted, as shown by Novozhilov & Slepian (1965), that propagating waves are excited even when the applied load is self-equilibrated at any moment, thus transporting energy to infinity without any decay. This behaviour may be regarded as an illustration for the invalidity of the quasi-static definition of SVP to elastodynamic problems. It will be shown that extension of SVP to the elastodynamic response can be made according to its original usage, rather than by following the mathematical definition coined by Love.

SAINT VENANT'S PRINCIPLE IN ELASTODYNAMICS

In order to validate solutions obtained by a semi-inverse method, Saint Venant considered whether it is important to know the exact profile distribution of edge loads. As can be understood from Saint Venant's assumption he concluded that only static equivalents of the applied load are the *parameters that determine* the state of deformation far from the loaded surface.

We suggest here to follow the same line of thought. We would like to answer the question of importance of the exact spatial distribution of dynamic steady state edge loads in determining the dynamic deformation far from the loaded edge.

From waveguide dynamics it is understood that only propagating waves determine the deformation in the far field. Let us assume that the frequency of the applied end load is monochromatic and lies in the region where no more than one propagating wave is possible. That region is bounded by zero frequency and by the first or second cutoff frequencies, depending on the boundary conditions on the long faces $x = \pm h$. Let us further assume that several different edge loads acting on the waveguide, each at a time, have the same frequency ω. From Attenuation map it may be observed that all the loads will induce only one propagating wave with the same wave number, regardless of their spatial distribution. Having the same propagating wave number and frequency implies, via the eigenfunction expansion method, an identical eigenfunction $\mathbf{U}(x)$, but with different amplitude $\mathcal{A}$.

From equation (2) it can be deduced that imposing the requirement of equal energy flux of the waves, with identical frequency and wave number, exactly alike amplitude is obtained. It is obvious (from energy balance) that in order to maintain a propagating wave with energy flux P, the applied load should deliver an average power W equal to the energy flux P. Thus, end loads acting on a waveguide having identical frequency and power, will induce an identical propagating wave. Such loads may be named *dynamically equivalent loads*. An identical dynamic field will develop in the far field under action of dynamically equivalent loads, albeit their different spatial distribution on the surface.

A direct analogy between the observations made in the previous paragraph and SVP in elastostatics can be drawn. The meaning of Saint Venant's principle is that only the static equivalents determine the deformation far from the loaded area. This may explain the original phrase cited by Love (1944) "... elastic equivalence of statically equipollent systems of loads". Hence, the dynamic version of SVP may be stated as "*elastic equivalence of dynamically equipollent systems of loads*". Dynamically equipollent system of loads are those having same frequency and power.

Now we shall inquire whether the presumptions we have made represent any real problem. The linearity of the waveguide is quite common assumption and do not put a strong limitation on the applicability of the suggested principle. The strip is semi-infinite in the sense that there should exist a far field. The length of the strip is regarded infinite in comparison to its width when the length is much greater than its width and when the frequency is not too close to cut-off frequency. Additional reason for that assumption is that in our analysis the reflected waves were absent. This restriction may be fulfilled if the observation of the real strip made in time large enough in comparison to transient period and small enough before a reflected wave approaches the far field observed. While in a strip with free faces at least two propagating waves are possible, in strips with clamped faces and strips with mixed boundary conditions on the long faces there is a region on the frequency scale where not more than one propagating wave is possible (Graff, 1975).

The main conclusion is that if we have any solution to a problem for which the above assumptions are valid, we also have approximate solutions for the same structure on which different loads are applied, as long as all the loads have an identical frequency and average power. This was the original purpose of SVP in elastostatics to conclude from solution of one problem to another.

There is some possibility to extend the above result to a strip with free faces, where at least two propagating waves are possible; symmetric and antisymmetric. This extension is made possible by additional restriction on the applied load. Symmetrical loads with respect to the center line of the strip will induce only symmetrical waves. Thus, the space of determining parameters is expanded to include the frequency, the average power and being the load symmetric or antisymmetric.

The complicated nature of the dynamic version of SVP, as compared to the static version, is not surprising. In the dynamic case there is additional coordinate of time, thus deserving more consideration. The complexity may be appreciated by observing that the abscissa in Attenuation maps represent solutions for the static case, $\omega = 0$. There are additional ways to waive the assumptions on the waveguide (that are not presented here) by adding restrictions on the load.

Some quantitative measure for validity of a dynamic version of SVP may be added. How far is "far enough from the loaded surface"? The near field was defined by the attenuating waves that were induced as a result of a difference between the eigenfunction $\mathbf{U}(x)$ of the propagating wave and the spatial distribution of the applied load. The smallest attenuation rate is governed by the smallest attenuation constant, which yield the largest distance at which the attenuating waves have non-negligible effect. This measure for applicability is just the same as in the quasi-static case. Very small attenuation constant means SVP is not applicable. In other words - the spatial distribution of the load have a significant affect on most parts of the structure. In quasi-static problems there are several cases including orthotropic strips (Choi & Horgan, 1978) or shells (Goetchel & Hu, 1984). In the dynamic case we have to add the cases where the frequency is in the vicinity of cut-off frequency, as can be seen from Attenuation maps. For frequency close to any cut-off frequency, the imaginary part of one of the wave numbers is very small (approaching zero) leading to large attenuation distance.

CONCLUDING REMARKS

It has been demonstrated that under certain circumstances the exact spatial distribution of an edge load does not affect the dynamic field far from the loaded surface. This enables one to deduce from solution of one problem to approximate solution of another one. The conditions for that are identical frequency and power of the applied load which define dynamically equivalent system of loads. That conclusion is limited to some range of frequencies in to steady state problems. The extent of applicability is governed by the smallest attenuation constant.

The suggested approach overcomes the difficulties encountered in attempts to use the dynamic version of SVP as defined in the literature. However, there is still work to be done before the suggested principle may safely be applied to solution of dynamic problems.

REFERENCES

Achenbach J.D., 1973, *Wave propagation in elastic solids*, North-Holland Pub. Co.

Boley B.A., 1955, 'Application of Saint Venant's principle in dynamical problems', *J. Appl. Mech. (Trans. ASME)*, **22**, 204-206.

Chirita S., Quintanilla R., 1996, 'On Saint Venant's principle in linear elastodynamics', *J. Elasticity*, **42**, 201-15.

Choi I., Horgan C.O., 1978, 'Saint-Venant end effects for deformation of sandwich strips', *Int. J. Solids. Struct.*, **14**, 187-95.

Flavin J.N., Knops R.J., 1987, 'Some spatial decay estimates in continuum dynamics', *J. Elasticity*, **17**, 249-64.

Goetschel D.B., Hu T.H., 1985, 'Quantification of Saint-Venant's principle for a general prismatic member', *Comp. Struct.*, **21**, 869-74.

Graff K.F., 1975, *Wave motion in elastic solids*, Clarendon Pr., Oxford.

Grandin H.T., Little R.W., 1974, 'Dynamic Saint-Venant region in a semi-infinite elastic strip', *J. Elasticity*, **4**, 131-46.

Horgan C.O., 1989, 'Recent developments concerning Saint-Venant's principle: An update', *Appl. Mech. Rev.*, **42**, 295-303.

Horgan C.O., Knowles J.K., 1983, Recent developments concerning Saint-Venant's principle', *Adv. Appl. Mech.*, **23**, 179-269.

Horgan C.O., Simmonds J.G., 1994, 'Saint-Venant end effects in composite structures', *Comp. Eng.*, **4**, 279-86.

Kennedy L.W., Jones O.E., 1969, 'Longitudinal wave propagation in a circular bar loaded suddenly by a radially distributed end stress' *J. Appl. Mech. (Trans. ASME)*, **36**, 470-8.

Knops R.J., 1989, 'Spatial decay estimates in the vibrating anisotropic elastic beam', in *Waves and stability in continuous media*, Rionero S. ed., Series on Advances in Mathematics for Applied Sciences - Vol. 4, World Scientific.

Love A.E.H., 1944, *A Treatise on the mathematical theory of elasticity*, Dover Pub. New York.

Miklowitz J., 1978, *The theory of elastic waves and waveguides* , North-Holland Pub. Co.

Mindlin R.D., 1960, 'Waves and vibrations in isotropic elastic plates', in *Structural Mechanics*, ed. Goodier J.N. & Hoff N.J., Pergamon, New York.

Novozhilov V.V., Slepian L.I., 1965, 'On Saint-Venant's principle in the dynamics of beams', *PMM*, **29**, 261-81.

Modern Practice in Stress and Vibration Analysis, Gilchrist (ed.) © *1997 Balkema, Rotterdam, ISBN 90 5410 896 7*

Dynamic response of inelastic structures subjected to earthquake ground motion

B. M. Broderick
Department of Civil, Structural and Environmental Engineering, Trinity College, Dublin, Ireland

ABSTRACT: A number of advanced analysis features, suitable for application to finite-element based time-history analysis of the inelastic seismic response of structures subjected to strong ground motion, are described. These features are illustrated by case-studies of the response of moment-resisting frame buildings and freeway structures, which are used to assess the applicability and relevance of each of the analytical techniques described in terms of their ability to contribute towards a more accurate assessment of failure mode. An accurate prediction of the response of a structure requires accurate modelling at a number of different levels. This paper concentrates on just two of these: material response and ground motion. The applicability of both bilinear and multi-surface steel models and of a uniaxial cyclic model for concrete are described in the context of inelastic cyclic response. Because of the variability of earthquake ground motions, multiple analyses employing a range of accelerograms should be employed in any project. These accelerograms, however, must be selected and scaled in a judicious manner if an accurate comparison of the response due to each is to be made. Approaches for both of these tasks are described.

1 INTRODUCTION

Recent years have seen the occurance of severe damaging earthquakes in some of the most developed areas of the world. These include the Loma Prieta (1989) and Northridge (1994) earthquakes which produced much structural damage in the San Francisco and Los Angeles areas respectively, and the Kobe, Japan earthquake of early 1995. Each of these events caused a number of spectacular structural failures which indicated that common methods of design analysis, which do not rely upon time-history analyses, may be deficient in predicting the actual mode of failure of a structure during severe ground motions.

Design methods rely upon a simplification of the seismic forces which allows linear elastic static analysis to be performed, despite the explicit assumption in most seismic design codes that oscillations well into the inelastic range are to be expected under the design seismic event. For more complex and sensitive projects, however, both dynamic and inelastic analysis techniques are employed, allowing the complete response of a structure to recorded or artificial ground acceleration records to be obtained. This approach allows both the sensitivity of the seismic response to all components of the ground motion and the energy dissipation associated with inelastic deformations to be included in the response prediction.

2 INELASTIC MATERIAL RESPONSE

To accurately determine the response of a structural member to severe seismic loads, constitutive relationships which reflect the inelastic cyclic behaviour of the relevant structural materials must be employed. This implies that for steel, concrete or composite members, the strain history dependence of the stress-strain behaviour should be reflected in accurate but efficient material models.

2.1 Inelastic Steel Models

When subjected to constant strain amplitude cycling, mild steel exhibits a response which converges to a stabilized saturation loop dependent only on the amplitude of cycling (Elnashai and Izzuddin, 1993). The cyclic stress-strain curve curve joining the tips of the stabilized cycles is shown in Figure 1. Due to the difference in shape between the cyclic and virgin curves, the transient response under constant strain amplitude cycling is characterized by softening for small amplitudes and hardening for large amplitudes. Relaxation to zero mean-stress accompanies the process of cyclic softening or hardening if, during the transient response, the mean-stress attains a non-zero value. Therefore, a reasonably accurate cyclic model for steel must represent the virgin response, the steady state cyclic response and the transient behaviour involving softening, hardening, and mean-stress relaxation.

Elnashai and Izzuddin (1993) present a review of several models developed to represent the uniaxial stress-strain response of steel under cyclic loading. Each of these models attempts to account for the change in yield point due to previous excursions into the inelastic range. One of the simplest models available is the bilinear model shown in Figure 2,

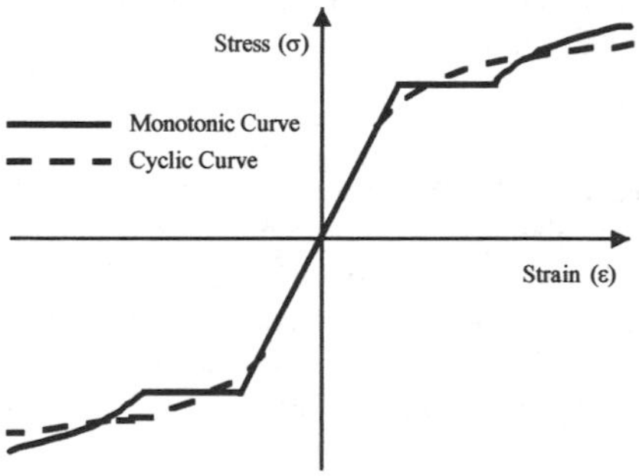

Figure 1: Inelastic steel response

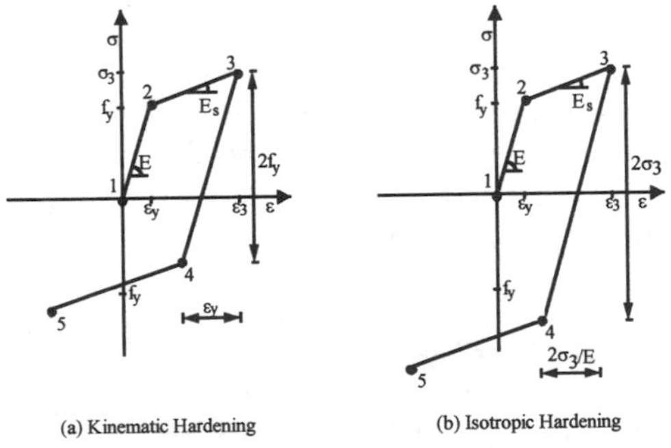

Figure 2: Bilinear steel models

which may be employed with either a kinematic or isotropic hardening rule. Kinematic hardening implies a translation in the yield surface, as shown in Figure 2(a), while isotropic hardening assumes a growth in the yield surface, as shown in Figure 2(b). While the Bauschinger effect is more accurately modelled by a kinematic hardening rule, this may only be achieved at the expense of inaccuracies in the virgin response.

This deficiency may be overcome by the use of a multi-surface plasticity model employing a number of intermediate stress-strain relationships between the yield and bounding surfaces. (Popov and Petersson, 1978). To achieve this, a weighting function is employed in addition to the monotonic and the cyclic curves (Figure 3). Although capable of representing the detailed behaviour of a class of steel, by virtue of the presence of a yield point, plateau and nonlinear hardening region, as well as cyclic response characteristics, the multi-surface plasticity model requires a relatively elaborate calibration procedure.

2.2 *Case Study: Composite Beam-Columns*

A number of partially-encased composite beam-column test specimens, detailed to provide the higher rotation ductility capacities desired in seismic design, were tested in both pseudo-dynamic and cyclic tests (Elnashai and Broderick, 1994). Here, the results of two tests are described: Test ICA2, in which cyclic displacements were imposed, and Test ICA3 in which the specimen was subjected to seismic loads in the form of the NS component of the El Centro (1940) earthquake. Correlative analyses were performed using both bilinear kinematic-hardening and multi-surface steel material models.

As the multi-surface plasticity model defines the stress-strain response of steel in terms of a series of cubic polynomials, both the ordinate and slope at six locations in each of the three curves illustrated in Figure 3 must be specified. Values for the virgin curve were obtained from steel coupon tests. The values for the cyclic curve and weighting function were obtained through observation and calibration with the force-displacement and strain results of Test ICA2. Figure 4 compares the experimental and analytical results for Test ICA2; it is clear that a closer calibration was possible using the multi-surface model. The shape of the hysteretic curve in Figure 4(c) retains the spindle shape of the experimental result, comparing favourably with the pinched appearance of the bilinear model in Figure 4(b).

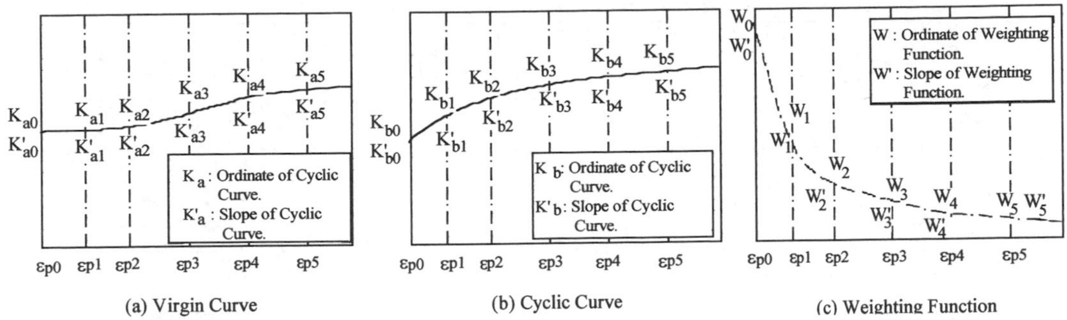

Figure 3: Multi-Surface Model Curves

Table 1: Multi-Surface Model Parameters

	Plastic Strain Ordinate ε_p					
	ε_y	0.015	0.030	0.060	1.50	3.00
K_a	280	288	300	315	405	440
K'_a	1000	1000	2000	4500	4500	1250
K_b	270	330	360	400	450	510
K'_b	160000	25000	17300	10200	1500	1400
W	1.00	0.80	0.65	0.50	0.35	0.25
W'	-20	-12	-10	-1.5	-1.4	-0.5

Table 2: Analytical and Experimental Results

	Plastic Hinge Length		Rotational Ductility		Energy Dissipation	
Test Title	Bi-linear	Multi-surface	Bi-linear	Multi - surface	Bi-linear	Multi - surface
ICA1	0.78	0.96	1.09	1.00	0.90	0.95
ICA2	0.57	1.05	1.07	1.00	0.88	1.00
ICA3	0.75	1.21	1.57	1.35	0.96	1.05
ICA4	0.51	0.83	1.36	1.17	1.03	1.08
ICA5	0.86	1.14	1.23	1.18	1.04	1.04
ICA6	0.94	1.25	1.13	0.95	0.95	0.94
ICA7	0.70	0.93	1.19	1.20	0.83	0.92

For both cyclic tests, the capacity of the member is well predicted by either analytical model. The yield point however is more accurately predicted, both in terms of load and displacement, by the multi-surface model. It is in this area of the cyclic response, immediately after yield has occurred, that the bilinear steel model is most deficient, hence significant discrepancies between the actual response and analytical response which it predicts are observable.

To simulate the conditions of the pseudo-dynamic Test ICA3, the accelerogram of the earthquake was applied as a ground motion at the base of the analytical model. As with the cyclic tests, the ultimate capacity of the member is well predicted by the analytical models, being slightly higher than the experimental result for the case of the multi-surface model and slightly lower in the case of the bilinear model. While the displacement response was well predicted by both analytical models, in comparisons with other tests in the ICA series, the multi-surface model was seen to be superior in predicting the displacement response after a number of large amplitude cycles have occurred (Broderick and Elnashai, 1994).

For all seven tests in the ICA series, Table 2 presents the results obtained using both material models, normalized by the equivalent experimental results. The parameters employed in this regard, namely the plastic hinge length, the achieved rotation ductility and the cumulative energy dissipation are especially relevant in assessing the performance of a seismically-resistant structure. Overall, these results show that the multi-surface model is more capable of duplicating the experimental results, both quantitatively and qualitatively. The use of the bilinear model consistently underestimates the length of the plastic hinge, whereas in comparison, the multi-surface model results are scattered more evenly around and closer to the experimental results. The bilinear model also over-estimates the rotational ductility demands on the test specimens. However, both models accurately predict the hysteretic energy dissipation of the specimens; showing that the use of a bilinear model is justified when high local accuracy is not required and only the overall performance of a structural system requires assessment.

In general, whereas modelling of steel plasticity using bilinear kinematic-hardening models is adequate for many applications in static analysis, the bilinear model has three main shortcomings in its inability to model (i) the presence of a horizontal yield plateau, (ii) the reduction in strain-hardening

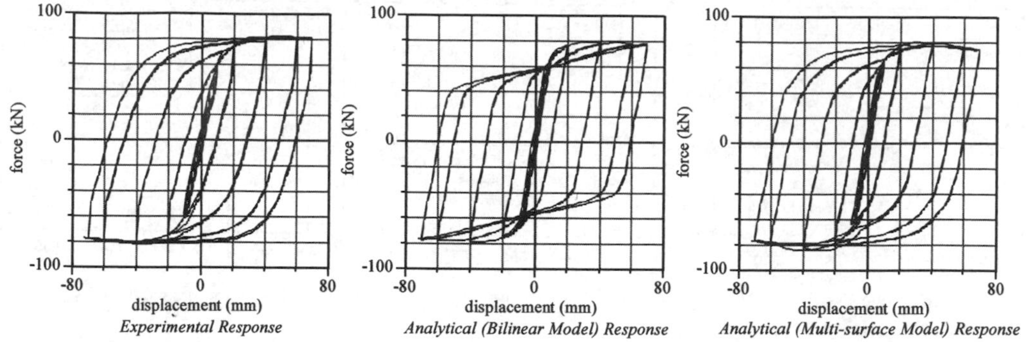

Figure 4: Cyclic test experimental and analytical results

slope with the increase in strain amplitude and (iii) experimentally-observed cyclic degradation. Under such conditions, the multi-surface plasticity formulation provides an appropriate modelling tool, accounting for the effect of small and large amplitude cyclic loading as well as mean stress relaxation. However, its accuracy is dependent on the rigour of the process of calibration to the measured virgin and cyclic curves, as well as the accuracy of the weighting function. In the context of capacity design under earthquake loading, it is important to pin-point the onset of plastic flow, plastic redistribution and reduction in load-bearing capacity in order to evaluate the section, member and system ductilities. It is therefore important to utilize stress-strain models that exhibit the important behavioural patterns observed in monotonic, variable amplitude cyclic and dynamic testing; phenomena which are well-catered for by the multi-surface model. For 'design basis' earthquake analysis, the bilinear model may be adequate, since the spread of plasticity will be, at most, moderate.

2.3 *Inelastic Concrete Response*

For the inelastic analysis of concrete structures under seismic loading, an accurate material model should reflect (i) the dependence of the behaviour of concrete on its multi-axial stress state, (ii) the confinement effects of non-concrete parts of reinforced and composite members and (iii) the material's principal cyclic response characteristics.

Any concrete element under general loading will experience triaxial stresses which give rise to strengths in excess of those associated with the uniaxial state. In reinforced concrete beam and column members, this may be attributed to the passive confinement provided by transverse steel reinforcement. This suggests that an accurate response analysis should require the use of a triaxial concrete model and three-dimensional analysis. In the context of the analysis of frame structures however, the linear elements employed allow uniaxial material models to be employed without significant loss of accuracy. This is achieved if the principal effects due to the multi-axial stress state are reflected in the uniaxial model. In the case of concrete, these are increased maximum compression stress and strain or curvature ductility (Figure 5).

An accurate uniaxial cyclic concrete material model should therefore capture the significant three-dimensional effects affecting the behaviour in the principal direction, namely the effect of confinement on both the peak stresses and strains and the post-peak σ-ε curve. One such model is that developed by Mander et al. (1988), in which an envelope curve defines the longitudinal compressive stress, f_c, as:

$$f_c = \frac{f'_{cc}\, x\, r}{r - 1 + x^r} \qquad (1)$$

in which,

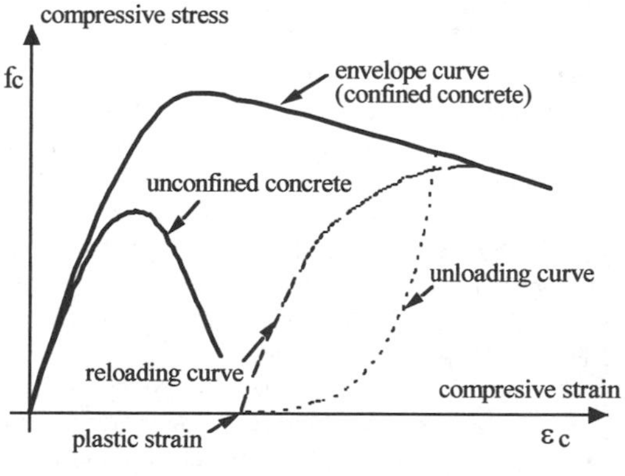

Figure 5: Uniaxial cyclic model for concrete

$$x = \frac{\varepsilon_c}{\varepsilon_{cc}} \text{ and } r = \frac{E_c}{E_c - E_{sec}} \qquad (2)$$

where

$$E_{sec} = \frac{f'_{cc}}{\varepsilon_{cc}} \text{ and } E_c = 5000\sqrt{f_{co}} \qquad (3)$$

and f'_{cc} is the compressive strength of the confined concrete, ε_c is the longitudinal concrete strain, ε_{co} is the critical strain of the unconfined concrete, f_{co} is the compressive strength of the unconfined concrete, E_c is the modulus of elasticity of the concrete and ε_{cc} is the critical strain of the confined concrete. The compressive strength of the confined concrete, f'_{cc} is obtained from that of the unconfined concrete, f_{co} by specifying a confinement factor, k_c.

$$f'_{cc} = k_c f_{co} \qquad (4)$$

As concrete confinement is achieved due to the passive confining pressures (f_l) exerted by the lateral reinforcement, separate confinement factors should be specified for each member, or part of member, where alternative reinforcement details are employed. Within each beam or column member, confinement will be most effective at those sections where reinforcing links are located. Between these, the area of effectively confined concrete is smaller. To take account of this, a confinement effectiveness coefficient, α_c, is used to convert the confining pressure, f_l, into an effective confining pressure, f'_l. This value of f'_l is then used to determine the confined concrete compressive strength and critical strain from their unconfined values according to experimentally-observed relationships. The confining pressure itself, f_l, is obtained by distributing the yield resistance of each stirrup over the member length.

Compared with the large body of information available on the behaviour of concrete under monotonic loading, considerably less work has been completed on its cyclic response. For seismic response analysis, however, this behaviour is of primary interest, not least on account of the permanent plastic strains which remain after unloading from large amplitude deformations. Most cyclic material models for concrete specify three separate curves to describe the stress-strain response of the material (Madas, 1993). As illustrated in Figure 5, these are (i) an 'envelope curve', equivalent to the confined monotonic curve; (ii) an 'unloading curve', in which the predicted irrecoverable plastic strain evaluated at the point where load reversal occurs is employed as a target point and (iii) a 'reloading curve' which traces the return of the stress-strain response to the envelope curve.

A succinct review of a number of cyclic models which meet these requirements is given by Madas (1993), from which that of Mander (1988) is selected as the most appropriate for the seismic response analysis of beam-column members.

2.4 *Case Study: Circular Bridge Piers*

The uniaxial concrete model has been employed in a study of the response of a freeway structure damaged during the 1994 Northridge earthquake (Broderick and Elnashai, 1995). In the portion of the structure analysed, all bents were composed of 4ft diameter reinforced concrete piers. To model the behaviour of these piers, a circular reinforced concrete section was employed. Figure 6 illustrates the subdivision of this section into a number of monitoring areas using polar co-ordinates and a radial linear variation in the thickness of the monitoring areas. The concrete component of the cross-section is divided into both confined and unconfined (cover) portions; the confinement factor, $k_c = f'_{cc}/f_{co}$, being determined from Eurocode 8, as

$$k_c = 1.0 + 5.0\alpha_c f_l/f_{co} \qquad (f_l/f_{co} < 0.05)$$

$$k_c = 1.125 + 2.5\alpha_c f_l/f_{co} \qquad (f_l/f_{co} > 0.05) \qquad (5)$$

in which, the confining pressure f_l is obtained from

$$f_l = \frac{\rho_s f_{sy}}{2}\left(1 - \sqrt{\frac{S_{sp}}{1.25 d_{cc}}}\right) \qquad (6)$$

where, $\rho_s = 0.00125$ is the volumetric ratio of the confining reinforcement, $f_{sy} = 454$ N/mm is its yield strength, $S_{sp} = 304.8$mm is its longitudinal spacing and $d_{cc} = 1117.6$mm is the diameter of the confined core. Evaluation of (6) gives $f_l = 0.151$ N/mm. By assuming a confinement effectiveness coefficient, α_c, of 0.90, equation (5) gives $k_c = 1.02$.

Gravity loads were applied in addition to all three components of the accelerograms recorded at Santa Monica, approximately 10km from the structure. A relatively low rotation ductility demand was experienced in the piers, while greater shear forces were experienced in the transverse as opposed to longitudinal direction. The results of the analyses showed that the structural failure which occurred could be attributable to an excessive shear force demand, coupled with a reduction in shear resistance capacity due to transient variations in axial load (Broderick and Elnashai, 1995). This interaction was only apparent when all three components of the ground motion were applied; in contrast, application of either or both horizontal components alone led to an underestimation of one or more aspects of the response. By applying a sophisticated analytical model, a full characterisation of the behaviour of the structure during the earthquake was possible and the causes of failure - which were attributable to the influence of a change in boundary conditions on the response hierarchy - properly identified.

2.5 *Case Study: Composite Beam-Columns*

Figure 7 illustrates the subdivision of the partially-encased composite section employed in the analyses described previously into a number of monitoring areas. Each of these areas possesses a monitoring point at which the direct stresses are determined. For each step of an analysis, equilibrium is imposed by determining these stresses at two Gauss sections in each beam-column element. The concrete component of the partially-encased section is divided into unconfined, partially-confined and fully-confined portions. Beyond the transverse links the concrete is assumed to be unconfined, while a maximum level of confinement is displayed close to the web. The confinement factors, k_c, are defined for each of these portions of the section, as are the depths of the unconfined area and the confinement parabola.

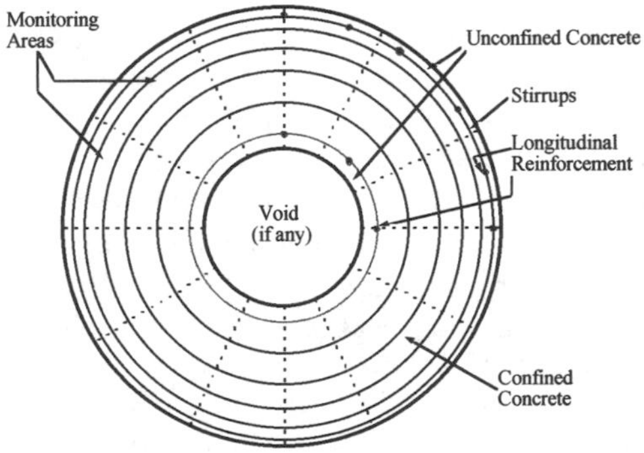

Figure 6: Circular RC bridge pier section

Elghazouli (1992) compared the values of k_c given by a number of models and concluded that the draft Eurocode 8 expressions (5) were applicable to these members also. For the test specimens, equilibrium considerations give the confining pressure as

$$f_l = \frac{t_w S_s f_{sy} + 2A_r f_{ry}}{B S_s} \quad (7)$$

where, t_w = 6.1 mm is the thickness of the steel web, S_s = 40 mm is the longitudinal spacing of the transverse stirrups, f_{sy} = 282 N/mm is the yield strength of the steel web, f_{ry} = 347 N/mm is the yield strength of the transverse stirrups, A_r = 28.3 mm^2 is the cross-sectional area of the transverse stirrups and B = 152.4 mm is the overall breadth of the steel section core. Evaluation of (7) gives f_l = 14.5 N/mm^2. By assuming a confinement effectiveness coefficient, α_c, of 0.60 to allow for the partial confinement experienced by much of the concrete encasement, and substituting the value f_{co} = 28 N/mm^2 from concrete cube tests, equation (5) gives

$$k_c = 1.125 + 2.5(0.6)(14.5/28.0) = 1.9$$

which is the value applied in the case study of Section 2.2.

3 ACCELEROGRAMS FOR SEISMIC ANALYSIS

When the site of a structure is known, suitable earthquake ground motions may be selected from records reflecting the ambient seismological and geotechnical conditions. Commonly, these are obtained from databases of historical records or from carefully generated artificial records. Otherwise, where the response to general earthquake loading is

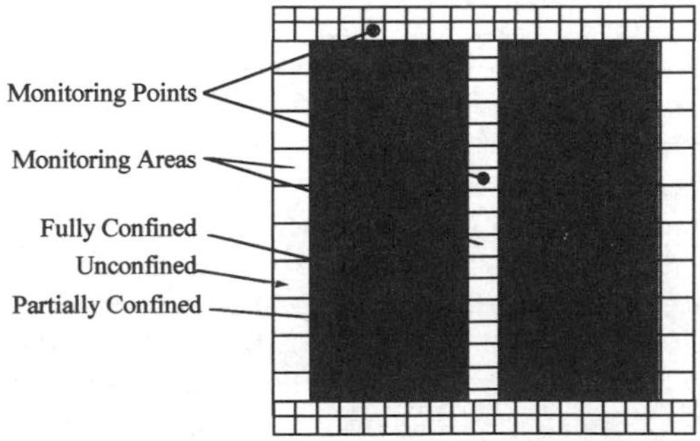

Figure 7: Partially-encased composite section

being investigated, it is necessary to consider a wider range of possible earthquake loads.

3.1 *Selection of Accelerograms*

Zhu et al. (1988) categorise earthquake ground motions as (a) 'normal' ground motions exhibiting significant energy content over a broad range of frequencies, (b) ground motions producing accelerograms possessing many large-amplitude, high-frequency oscillations, and (c) records in which the significant response is contained in a few long duration acceleration pulses. These characteristics can be attributed to variations in local soil conditions, epicentral distance and the magnitude and duration of the event.

The peak ground acceleration to peak ground velocity ratio (a/v) offers a simple yet meaningful means of identifying the characteristics of individual accelerograms. Because the velocity characteristic is obtained from an integration of the accelerograms, while the peak acceleration may be associated with high frequency waves, the peak ground velocity is instead associated with the moderate to low frequency waves of the accelerogram. Hence ground motions of type (b) above will possess high a/v ratios, while those of type (c) motions will lead to low a/v ratios.

It can therefore be expected that records possessing high a/v ratios will be critical for stiffer structures, whereas low a/v records will place greater demands on more flexible structures. Design codes implicitly reflect this by prescribing different design spectra for structures located on various soil types. These spectra, which define the strength requirements for a structure, are usually defined in terms of peak ground acceleration. The National Building Code of Canada, however, expresses its design spectra in terms of peak ground velocity, with a correction being applied in the acceleration-dependent low period range (< 0.5 seconds). To this end ground motions are classified as either low a/v (a/v < 0.8 g/ms^{-1}), normal a/v (0.8 $g/ms^{-1} \leq a/v \leq 1.2$ g/ms^{-1}), or high a/v (1.2 $g/ms^{-1} < a/v$).
The difference between the strength supply ensured by design codes and the strength demands imposed by a particular earthquake determines the level of ductility demand experienced. It has been shown (Zhu et al., 1988) that the design spectra provisions of the NBCC lead to more uniform displacement ductility demands. The implied variety of ductility demands experienced by structures designed to different codes places increased emphasis on the need to include a meaningful range of accelerograms in any analysis procedure.

3.2 *Accelerogram Scaling Techniques*

The important features of seismic response such as strength and ductility demand are highly dependent on ground motion intensity. Therefore, when evaluating the performance of a structure designed to resist code-prescribed seismic loads, the seismic energy imparted to the structure by the imposed base accelerations should be equal to that implied in the code design spectrum.

Earthquake ground motions as recorded, however, display wide variations in intensity. To ensure that each of the selected accelerograms impose similar levels of demand, it is necessary to scale the ground motions to a common level. As the seismic loads experienced during an earthquake are proportional to the instantaneous acceleration, recorded ground motions are normally scaled to a common peak acceleration. This method of scaling is simple to apply and agrees with how design codes normally define seismic loads. For structures with a fundamental period greater than 0.5 seconds, however, the peak ground velocity is more relevant; it being important therefore to scale earthquake records in a manner which reflects the period of the structure under consideration. This has been confirmed by a number of studies on large selections of earthquake records and is especially relevant for the important parameter of ductility demand.

In an investigation of the behaviour of reinforced concrete buildings under earthquake loads, Kappos (1991) scaled earthquake records to possess equal spectrum intensities, SI, defined as the area under a pseudo-velocity spectrum curve between the periods of 0.1 and 2.5 seconds:

$$SI(\beta) = \int_{0.1}^{2.5} S_v(T,\beta)dT \tag{8}$$

$$S_v(T_i,\beta) = \frac{S_a(T_i,\beta).T_i}{2\pi} \tag{9}$$

in which β represents the fraction of critical damping, S_V is the pseudo-spectral velocity, T is the response period and S_a is the spectral acceleration.

Scaling to equal spectrum intensities ensures that earthquake records possess equal energy contents

Table 3: Ground Motion Records and Properties

Record Label	Epicentral Distance	Soil Type	M_L	a_{gmax} (g)	v_{gmax} (m/s)	a/v Ratio $gm^{-1}s^{-1}$	0.25g / a_{gmax} (A)	SI_{EQ} (m)	SI_{EC8} /SI_{EQ} (B)	Total Scale (AxB)	$a_{gScaled}$ (g)
Fruili	52 km	Rock	6.4	0.159	0.080	1.99	1.56	143.7	1.20	1.89	0.301
Gazli	14 km	Med Stiff	7.3	0.724	0.606	1.20	0.35	71.3	2.42	0.84	0.608
L. Prieta EW	97 km	Soft	7.1	0.213	0.216	0.99	1.17	116.3	1.48	1.74	0.371
El Centro	8 km	Stiff	6.6	0.344	0.365	0.94	0.73	100.5	1.72	1.25	0.430
Spitak	27 km	Med Stiff	6.8	0.182	0.237	0.77	1.37	95.6	1.80	2.47	0.500
L. Prieta NS	97 km	Soft	7.1	0.250	0.433	0.58	1.00	192.3	0.90	0.90	0.225

between the periods 0.1 and 2.5 seconds and significantly reduces response spectral dispersion in the range 0.5 - 3.0 seconds, producing a more consistent level of displacement ductility demand.

3.3 *Case Study: Accelerograms for the Analysis of Composite Frames*

For the analysis of a range of composite frames, six earthquake records were selected on the basis of their a/v ratios; the selection consisting of two records in each of the a/v ranges above (Broderick, 1994). Details of the events are given in Table 3.

Given that the frames investigated possessed natural periods between 0.7 and 1.0 seconds, scaling to a peak ground velocity would have been most appropriate. However, the object of the study was to obtain a comparison with the provisions of Eurocode 8, which defines the seismic actions in terms of peak acceleration. As all the chosen records possess a unique a/v ratio, velocity scaling of the ground motions would destroy the equivalence between the records and the code design spectrum. However, inspection of (8) and (9) indicates the dependence of spectrum intensity on spectral acceleration, hence it agrees with the provisions of the code. Further, scaling to a common spectrum intensity has similar a effect to velocity-scaling in that it ensures less response variation in the period range of interest.

The scaling procedure employed consists of two stages (Table 4). The accelerograms are first scaled so that their maximum accelerations each equal the design value of 0.25g. Secondly, the spectrum intensities of the scaled accelerograms are evaluated and compared with that of the code spectrum; the accelerograms being then rescaled by the ratio between these values. Comparison of the mean and standard deviation values for the full set of earthquakes shows that the overall effect of the scaling operation is to increase the maximum ground accelerations, but to reduce their degree of variation.

REFERENCES

Broderick, B.M., 1994. Seismic testing, analysis and design of composite frames. *PhD Thesis, Imperial College,* University of London.

Broderick, B.M. and Elnashai, A.S., 1994. Seismic resistance of composite beam-columns in multistorey structures. Part 2: Analytical model and discussion of results. *J. Const. Steel Res.* 30: 231-258.

Broderick, B.M. and Elnashai, A.S., 1995. Analysis of the failure of Interstate 10 freeway ramp during the Northridge Earthquake. *Earthq. Engng. & Struct. Dyn.*22: 189-208.

Elghazouli, A.Y., 1992. Earthquake resistance of composite beam-columns. *PhD Thesis, Imperial Colege,* University of London.

Elnashai, A.S. and Broderick, B.M., 1994. Seismic resistance of composite beam-columns in multi-storey structures. Part 1: Experimental studies. *J. Const. Steel Res.* 30: 201-230.

Elnashai, A.S. and Izzuddin, B.A., 1993. Modelling of material nonlinearities in steel structures subjected to transient loading. *Earthq. Engng. & Struct. Dyn.* 22: 509-532.

Kappos, A.J., 1991. Analytical prediction of the collapse earthquake for RC buildings: suggested methodology. *Earthq. Engng. & Struct. Dyn.* 20, 167-176.

Madas, P.J., 1993. Advanced modelling of composite frames subjected to earthquake loading. *PhD Thesis, Imperial College,* University of London.

Mander, J.B., Priestly, M. and Park, R., 1988. Theoretical stress-strain model for confined concrete. *J. Struct. Engng. ASCE.* 114: 1804-1826.

Popov E.P. and Petersson H., 1978. Cyclic metal plasticity: experiments and theory. *J. Mech. Engng. ASCE.* 1371-1388.

Zhu, T.J., Heidebrecht, A.C. and Tso, W.K., 1988. Effect of a/v ratio on ductility demand of inelastic systems. *Earthq. Engng. & Struct. Dyn.*16: 63-79.

Modern Practice in Stress and Vibration Analysis, Gilchrist (ed.) © 1997 Balkema, Rotterdam, ISBN 90 5410 896 7

Prediction of vibration propagation in built-up plate structures

R. Haettel
The Marcus Wallenberg Laboratory for Sound and Vibration Research, KTH, Stockholm, Sweden

ABSTRACT: Calculation procedures are developed for prediction and control of vibration propagation throughout structures composed of coupled plates. A mathematical description of the flexural wave propagation in those structures is proposed according to a waveguide model. The calculation procedures, based on a transfer matrix formulation, facilitate a study of how various parameters affect the vibration transmission throughout the structures. Measurements are performed on a scale model to validate the theoretical model. Influence of local damping as well as exciting force locations are readily examined for the scale model.

1. INTRODUCTION

Structure-borne sound is the most important path for acoustical transmission in ships, airplanes and railway cars which are typical examples of built-up plate assemblies. Vibrational energy due to sources such as working machines can be transmitted far away from its origin and finally reach areas , such as passenger compartments, where that energy is radiated as noise. Therefore, accurate and efficient prediction tools are required for application in vibration control at the first-stage design of marine and aircraft structures or engine foundations. Some methods, such as FEM - Finite Element Method- or SEA -Statistical Energy Analysis- have been used on plate structures more or less successfully. They both present limitations, especially on the frequency range of application. Obtained from an analytical method, a waveguide model is proposed here to predict the propagation of structure-borne sound in built-up plate structures.

The waveguide model, based on the concept that vibrational power propagates mainly as flexural waves throughout the structures, compares well with experimental results as shown in [1], [2] and [3].

Mathematical expressions obtained from the flexural wave equation, solved for single plate elements, provide a calculation technique using transfer matrices relating exciting moments and angular displacements at the plate junctions. These parameters enable the computation of power dissipated in the plate elements. The calculation procedures are implemented in computationally efficient programs demonstrating how various parameters affect power transmission throughout the structures. By means of computer simulations and experiments, local damping influence and exciting force locations are readily analysed for a defined structure.

2. MODEL FORMULATION

2.1 *Analysis of a plate system*

Consider the structure, in Fig. 1, consisting of a set of four thin rectangular plates 1, 2, 3 and 4, rigidly coupled at junctions 2, 3 and 4. The extreme sides of the system are assumed to be simply supported in the width (parallel to the y-axis) and the length (parallel to the x-axis). All the plates have the same width L_y, are made of an isotropic material, and may be defined with different lengths $L_{x,p}$ and thicknesses t_p (where plate index p = 1, 2, 3 and 4). A damping η_p is taken into account by using a complex Young's modulus $E_p = E_0 \cdot (1+i\eta_p)$. At each junction, the mounting, which is assumed to be lossfree, is such that free rotation is allowed while displacement is zero. The force action on the structure produces flexural waves whose differential equation for the pth plate is expressed by

$$\nabla^2(\nabla^2 W_p) - \kappa^4 W_p = F/D_{Pp} \quad (1)$$

where W_p is the displacement of the pth plate in the z-direction, F an input force, if plate p is submitted to an external excitation, and D_{Pp} the bending stiffness of plate p, defined by

$$D_{Pp} = \frac{E_p \cdot t_p^3}{12(1-\nu^2)} \quad (2)$$

where ν is Poisson's ratio.

Other wave types such as longitudinal or rotational are not taken into account. The structure is considered only in its steady state , ie long enough after the excitation starts to excite the system to neglect all transitory effects.

Since the boundaries along the x-axis are assumed to be simply supported, the displacement W_p is separable according to

$$W_p(x,y) = \sum_n w_{p,n}(x) \cdot \varphi_{p,n}(y) \quad (3)$$

where the x-dependence for the nth mode is given by

$$w_{p,n}(x) = A_{p,n} e^{-i\kappa_{2,n}x} + B_{p,n} e^{i\kappa_{2,n}x} + C_{p,n} e^{-\kappa_{1,n}x} + D_{p,n} e^{\kappa_{1,n}} \quad (4)$$

where $p = 1, 2, 3, 4$ and $\kappa_{1,n}, \kappa_{2,n} = (\kappa^2 \pm k_n^2)^{1/2}$ with the wavenumber κ and the eigenvalues $k_n = n\pi/L_y$ corresponding to the mode functions

$$\varphi_{p,n}(y) = \sin(k_n \cdot y) \qquad 0 \leq y \leq L_y \quad (5)$$

where the integers n are the mode numbers associated to $\varphi_{p,n}(y)$; they are called cross-mode numbers in opposition to the propagation mode numbers in the x-direction.

The time harmonic dependence $e^{i\omega t}$ is implicitly considered throughout this paper. To determine the coefficients $A_{p,n}$, $B_{p,n}$, $C_{p,n}$, $D_{p,n}$, boundary and continuity conditions at junctions $j = 1, 2, 3, 4$ and 5 are used: no displacement at each junction j, no bending moment at $j = 1$ and 5, angle and moment continuity at $j = 2$, 3 and 4. When all the coefficients of Eq. (2) are determined, the transfer matrix $[T_p]$, corresponding to plate p, is constructed from computation of angular displacements Θ_j and bending moments M_j calculated as derivatives of w_p. In addition, a perturbation vector (P_p) is defined to take into account any external excitation F, such as concentrated or distributed force, exerted on plate p.

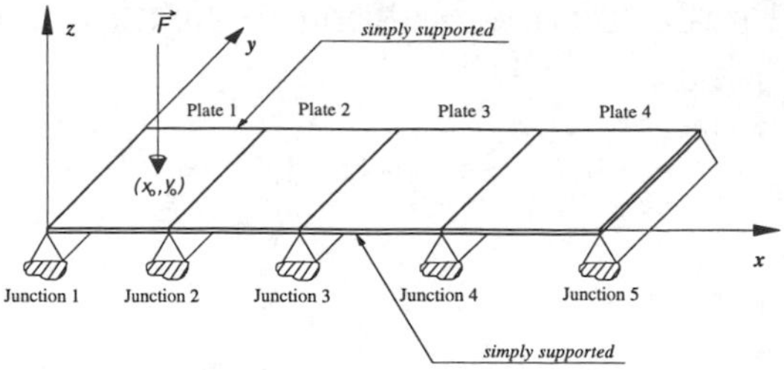

Figure 1. A coupled plate structure.

Finally, the equation relating angular displacements and exciting moments at the coupling junctions of plate p, is, in the most general case, written as

$$\begin{pmatrix} \Theta_{j+1} \\ M_{j+1} \end{pmatrix} = \begin{bmatrix} c_{11} & c_{12} \\ c_{21} & c_{22} \end{bmatrix} \cdot \begin{pmatrix} \Theta_j \\ M_j \end{pmatrix} + \begin{pmatrix} b_1 \\ b_2 \end{pmatrix} \cdot F = [T_p] \cdot \begin{pmatrix} \Theta_j \\ M_j \end{pmatrix} + (P_p) \cdot F \quad (6)$$

where $j = 1,...,4$.

If there is no external excitation on plate p, ie F equal to zero, Eq. (4) simplifies to

$$\begin{pmatrix} \Theta_{j+1} \\ M_{j+1} \end{pmatrix} = \begin{bmatrix} c_{11} & c_{12} \\ c_{21} & c_{22} \end{bmatrix} \cdot \begin{pmatrix} \Theta_j \\ M_j \end{pmatrix} = [T_p] \cdot \begin{pmatrix} \Theta_j \\ M_j \end{pmatrix} \quad (7)$$

Note that, for the structure considered in Fig. 1 , Eq. (7) applies to plates $p = 2,...,4$ and Eq. (6) to plate $p = 1$, respectively. The point force F exerted on plate 1 at (x_0, y_0) is assumed to be harmonic and expressed by

$$\vec{F} = F_0 \cdot e^{i\omega t} \cdot \delta(x - x_0) \cdot \delta(y - y_0) \cdot \vec{e}_z \quad (8)$$

driving plate 2 and subsequently plates 3 and 4, causing bending moments at the junctions.

The various procedures and calculations mentioned above are detailed in [4].

2.2 *Power Calculation by the Junction Method*

The plate displacement, velocity or acceleration in any point of the system is well defined for a given excitation. The space and time average of the plate velocity can consequently be calculated directly as a function of the angular displacement and bending moment at the plate junctions. Therefore,

$$\langle v_p^2 \rangle = \frac{1}{2L_x L_y} \int_0^{L_x} \int_0^{L_y} \mathrm{d}x\, \mathrm{d}y \left| v_p \right|^2 \quad (9)$$

with $v_p = i\,\omega\, W_p(x,y)$ (10)

where W_p is expressed by the series (3).
However, the average velocity is obtained more easily by considering the power balance of the system. For a stationary process, it yields

$$\Delta P_p = \omega\eta_p\, E_{\text{total}} \quad (11)$$

where ΔP is the power dissipated in plate p and E_{total} is the total energy which thus can be approximated for a 'sufficiently resonant' system as twice the kinetic energy E_k.

$$E_{\text{total}} = L_x L_y \mu \langle v_p^2 \rangle \quad (12)$$

where $\langle v_p^2 \rangle$ is the time and space average velocity of plate p and μ is the mass per unit area.
The power dissipated in the plates is obtained from a calculation of the power at the plate junctions where the needed parameters, such as angular displacements and bending moments, are known. Applying the boundary conditions and taking the temporal and plate-width average, the power at the plate junction is

$$P_j = \frac{\omega L_y}{4} \cdot \text{Real}\left\{ M_j \cdot (i\Theta_j)^* \right\} \quad (13)$$

where j = 1,...,5 and the superscript * indicates the complex conjugate.
The calculation of the power input into the system is determined by

$$P_{\text{input}} = \frac{\omega}{2} \cdot \text{Real}\left\{ F \cdot (iW_0)^* \right\} \quad (14)$$

where W_0 is the displacement at the excitation point (x_0, y_0).
The power dissipated, ΔP_p, in the pth plate is given by

$$\Delta P_p = P_j - P_{j+1} \quad (15)$$

where $j = p$ for p = 2, 3 and 4.

Note the particular case for plate 1 where the power dissipated is written

$$\Delta P_1 = P_{\text{input}} - P_2 \quad (16)$$

where P_{input} is the power fed into the system as given in Eq. (8).
The time and space-average of the velocity for plate p is finally given by

$$\langle v_p^2 \rangle = \frac{\Delta P_p}{\omega\eta_p \mu L_x L_y} \quad (17)$$

This is valid for any of the plates composing the system.
The junction method permits the rather cumbersome calculation of the integral (9) to be avoided. It is also well adapted for utilisation with the transfer matrix technique.

3. MEASUREMENTS

3.1 *Experiment objectives*

To validate the calculation procedures, measurements are performed on a model, sketched in Fig.3, consisting of a set of five thin rectangular steel plates coupled in the width by welding. In addition, the plate ensemble rests on a frame consisting of welded ribs.

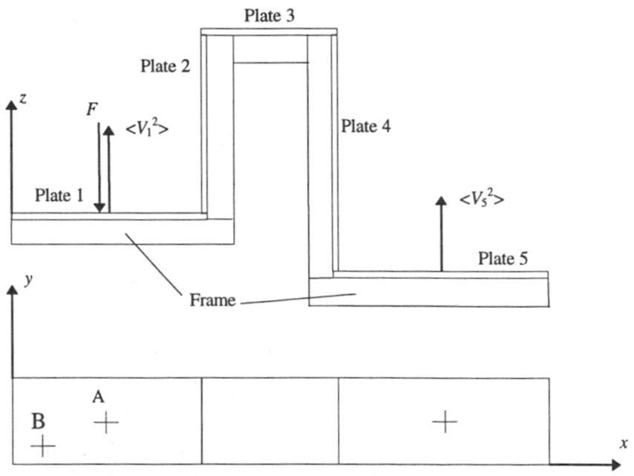

Figure 2. Sketch of the five plate model.

Table 1. Material and geometric properties of the plates.

Property	Value
Young's modulus	210 GN·m^{-2}
Poisson's ratio	0.3
Density	7800 kg·m^{-3}
Thickness	0.003 m
Width	0.30 m

Table 2. Plate length.

Plate	1	2	3	4	5
Length	0.70 m	1 m	0.80 m	1.20 m	0.90 m

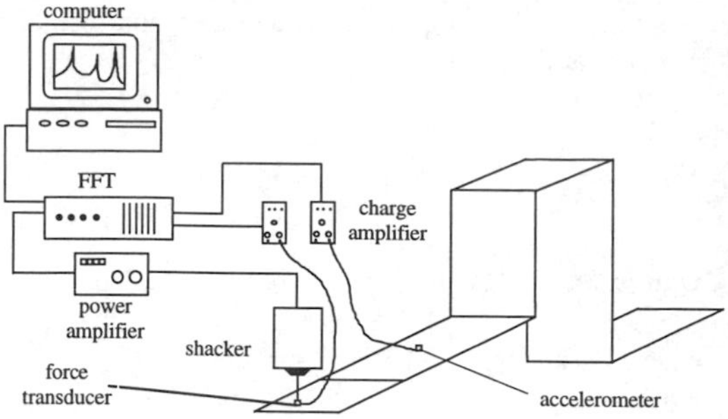

Figure 3. Sketch of the set-up used to measure plate mobilities on a plate model (frame omitted).

All the plates have the same widths and thicknesses but different lengths. The entire structure is made of the same type of common construction steel. The various material and geometric data of the structure are indicated in Tables 1 and 2.

The plate-plate junctions are made by line-welding whereas the plate-frame coupling is achieved by spot-welding. The weldings were made with special care to avoid any effect of plate buckling on the vibration path.

The structure under test is also used to investigate the general properties of a waveguide system. The influence of the point excitation location as well as the effects of added damping are observed under several measurements. Two different types of coupling for the plate elements are examined: line-welded and uncoupled junctions, ie plate elements connected only via the frame. The respective coupling behaviours toward vibration transmission are then compared. The plate velocity levels are obtained by measuring mobilities defined by the ratio between velocity and excitation.

3.2 *Description of the experiment*

The task of the set-up, sketched in Fig.2, is to measure mobilities in order to then deduce the plate velocity levels.

For measurement quality and practical reasons, the test structure is suspended from the frame area.

To excite the structure, an electrodynamic shaker (*B&K* 4809) is attached to plate 1. In both cases, the exciting signal is generated by a Tektronix 2630 signal analyser, ie Fast Fourier Transform (FFT) analyser. The excitation at the shaker attachment point is recorded by a force transducer (*B&K* 8200) screwed on plate 1 and whose output signal is amplified by a charge amplifier (*B&K* 2635) before analysis by the Tektronix 2630 analyser. To measure the response signals, a low weight accelerometer (*B&K* 4393) is fixed with bees wax to the surface of the test structure. To record the response signals, the accelerometer is not to be mounted too close to the boundaries, to avoid as much as possible uncontrolled nearfield influences. The response signal is also sent via a charge amplifier (*B&K* 2635) to the Tektronix 2630 analyser where this signal is processed and combined with the excitation signal to finally obtain the mobility function.

The excitation is situated on plate 1 and the measurements are made in the narrow band, with a step of 2.5 Hz, and in a frequency range up to 8500 Hz. In the low frequency range, ie up to 1000 Hz, a resolution of 1.25 Hz is used. Furthermore, to ensure a good accuracy, 40 averages are made for each measurement point. The number of points measured on each plate depends on the plate dimension and the distance from the exciting source.

3.3 *Utilisation of the measurement data*

As mentioned above, mobilities are measured and plate velocity levels are to be obtained. The conversion process is completed in three steps. Firstly, the plate mobilities are normalized either by a unit force or by a spectrum recorded at the excitation source. Secondly, the squared norms of the normalized mobilities are calculated, which yields the square velocities in several points on the plate under test. An arithmetic average is finally made to obtain the plate square velocity denoted $\langle v_p^2 \rangle$.

In addition, to interpret and compare the results in an easy way, the plate square velocities, functions of the frequency, can be plotted in third octave bands, where the velocity level in a frequency band is determined from the expression

$$L_{v_p} = 10\log\left\{\int_{f_1}^{f_2} \frac{\langle v_p^2 \rangle}{\langle v_{\text{ref}}^2 \rangle} \mathrm{d}f\right\} \tag{18}$$

where v_{ref} is the reference velocity and f_1 and f_2 are the limits of the frequency band. The integration is here replaced by a summation yielding

$$L_{v_p} = 10\log\sum_{f_1}^{f_2}\left\{\frac{\langle v_p^2 \rangle}{\langle v_{\text{ref}}^2 \rangle}\Delta f\right\} \tag{19}$$

where Δf is the frequency resolution.
The velocity level difference between two plates ΔL_{ij} is in the main utilised. This is defined as

$$\Delta L_{ij} = L_{v_i} - L_{v_j} \tag{20}$$

where L_v is the velocity level as given in Eq. (19) with indices i and j referring to the plate number.

4. RESULTS

4.1 *Comparison theory-experiment*

To make a comparison between theory and experiment, the structure, sketched in Fig. 2, is mathematically described by applying the transfer matrix method to a five plate system excited by a point force on plate 1. The calculation method permits the plate displacements, velocities and accelerations to be obtained at any plate location. In the experimental model, the excitation is produced by the shaker set in the centre of plate 1 and the resulting squared velocities are measured in the centres of plates 1 and 5, as indicated in Fig.2. Experimental and theoretical results are then plotted as a function of the frequency. The resolution used for the calculation is the same as the one obtained for the measurements. The material constants, such as Young's modulus and Poisson's ratio, and the structure's dimensions, employed in the computer simulations, are given in Tables 1 and 2. The plate loss factors are determined *in situ* through reverberation time measurements made on the third octave band. They are, thus, the sum of the internal and coupling losses. For the basic plate structure, the loss factor is 15×10^{-4} in the 125 Hz third octave band and decreases with increasing frequencies to 2×10^{-4} in the 10000 Hz third octave band.
In order to increase the damping within the structure, damping layers consisting of a visco-elastic self-adhesive material are clad to plates 2 and 5. In this case, measurements indicate that the added damping also affects the adjoining undamped plates. Therefore, on the average, the loss factors are found to be 0.01 for the plates with added damping and 0.003 for the other plates. These values are only indicative, more accurate results are used in the computer simulations.
The predicted squared velocities, displayed in Figs. 4-6, are calculated for the case in which the plate elements are assumed to be simply supported along the frames, as expressed in Eqs. (3) and (5). There also exists an approximate solution describing the clamped condition. This one was also used for making some prediction calculations, but a comparison with experimental results revealed a fairly large shift of the resonance peaks.
In Fig. 4, the squared velocity, as defined in Eq. (10), predicted and measured in the middle of plate 1 at the excitation point, is plotted in a frequency range up to 1000 Hz with a resolution of 1.25 Hz. In this frequency range, the calculation is made by using only the first cross-mode, ie the first mode in the y-direction.
In Figs. 5 and 6, the squared velocity, predicted and measured in the middle of plate 5, is plotted, first, in a frequency range up to 1000 Hz with a resolution of 1.25 Hz, then in a frequency range from 4500 Hz to 8500 Hz with a resolution of 2.5 Hz.
In the lower range, ie up to 1000 Hz, only the first cross-mode is needed. The use of higher cross-modes, in this rather low frequency range, does not result in any change of the squared velocity level. This remark is also valid for the results of plate 1, shown in Fig. 4. The curve, displayed in Fig. 5,

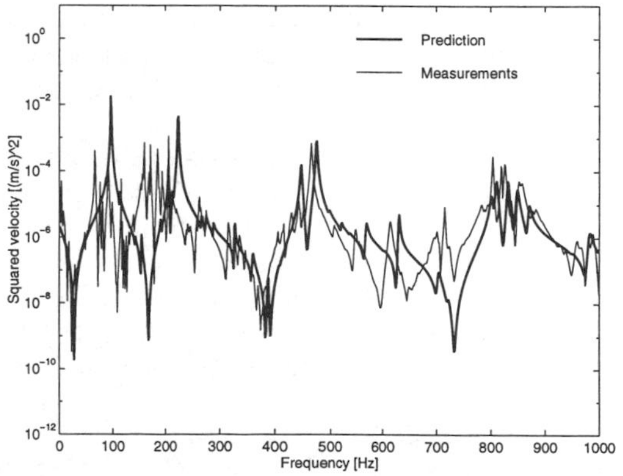

Figure 4. Velocity level predicted and measured in the middle of plate 1.

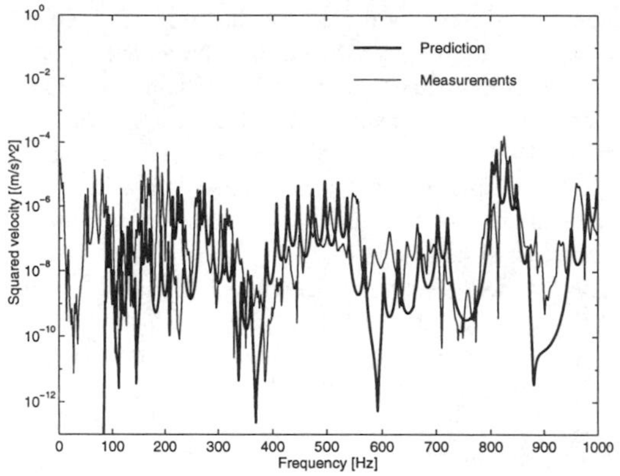

Figure 5. Velocity level predicted and measured in the middle of plate 5.

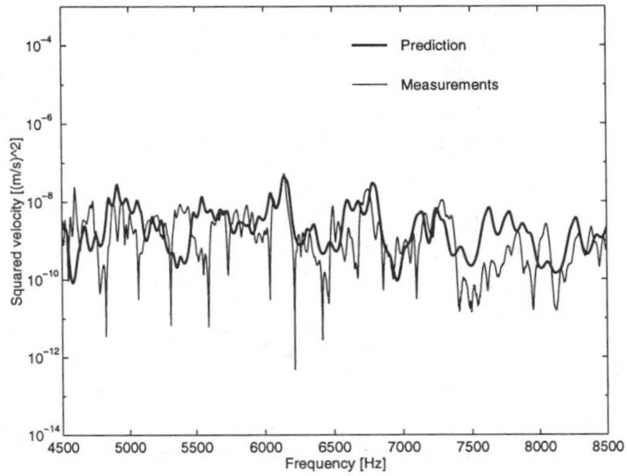

Figure 6. Velocity level predicted and measured in the middle of plate 5.

reveals the so-called cut-on frequency, at about 80 Hz for the theoretical model and around 50 Hz for the measurements. This frequency indicates the lower limit of the waveguide approach, below which the bending waves do not propagate.

In the upper range, the squared velocity is determined, from Eq (3), by using the three first cross-modes. The utilisation of solely the first cross-mode leads to underestimations of the squared velocity at certain frequencies. The corresponding calculations for higher cross-modes than the third one result in predicted squared velocities which are too large.

The plate velocity level, as defined in Eq. (17), can be recalculated according to Eq. (19) so that the velocity level difference between plates 1 and 5 is obtained in third octave bands, as displayed in Fig. 7.

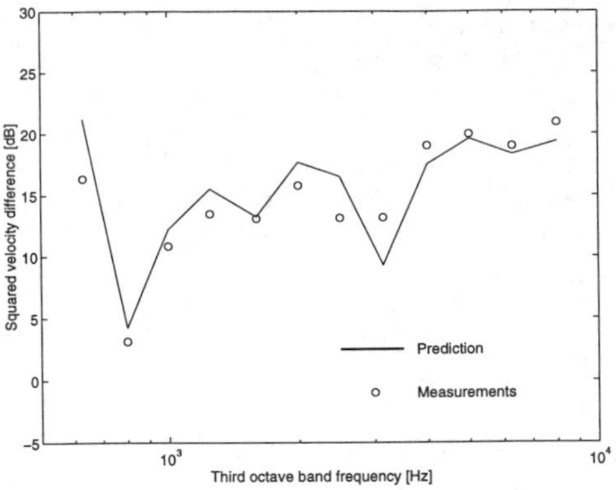

Figure 7. Velocity level difference between plates 1 and 5, ΔL_{15}.

The quantity ΔL_{15}, in the following figures, is equal to the velocity level between plates 1 and 5 as determined in Eq. (20). The velocity level difference, shown in Fig. 7, is calculated by taking into account only the first cross-mode in the whole frequency range. The corresponding calculations for higher modes yield velocity level differences which are too large compared to measurements. The discrepency is of the order of 5 dB in some regions.

4.2 *Influence of the force location*

The basic five plate structure, sketched in Fig. 2, is used here to determine the influence of the point force location on the vibration transmission. To achieve this, the excitation, produced by the shaker, is set in two different positions on plate 1, named A and B, as illustrated in Fig. 2. Point A is at the center with coordinates (x_o = 0.15 m, y_o = 0.35 m) and point B is located at the coordinates (x_1 = 0.12, y_1 = 0.05). The velocity level differences between plates 1 and 5 are measured for both excitation locations and compared in Fig. 8. The curves present level differences which can reach 5 dB, especially in low frequencies. In higher octave bands, the curves seem to come together, especially when admitting a potential error of 2dB. However, there is no obvious trend, in the studied frequency range, permitting one to conclude that a location is always more suitable than an other to obtain a decrease of the vibration transmission. This may be due to the plate-frame coupling of the model, which is not as stiff as in a common built-up plate structure.

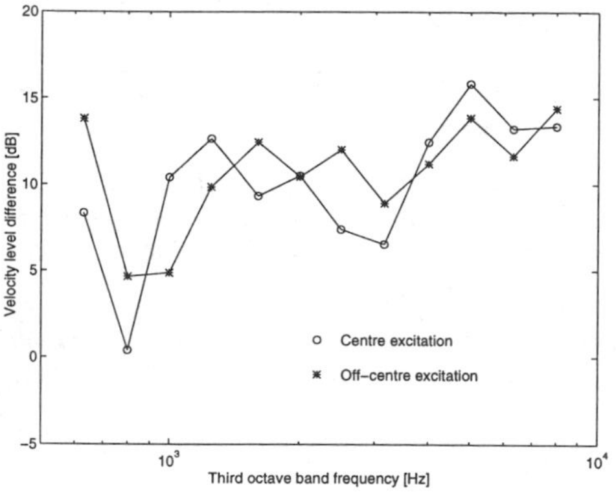

Figure 8. Velocity level difference between plates 1 and 5, ΔL_{15}, for two different locations of the point force.

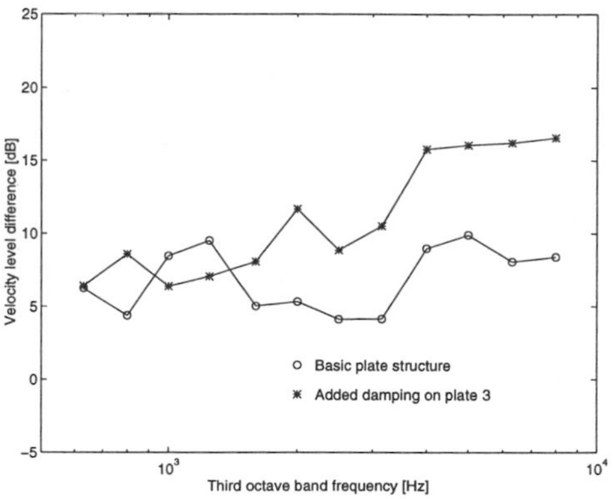

Figure 9. Velocity level difference between plates 1 and 3, ΔL_{13}, with and without added damping on plate 3.

4.3 *Influence of added damping*

To investigate the influence of damping layers on the model, the basic five plate structure, sketched in Fig. 2, is used to carry out measurements, first, without modifications, then, by adding damping layers on plate 3. In both cases, the velocity level differences between plates 1 and 3, then between plates 1 and 5, are measured and compared in Figs. 9 and 10, respectively. The damping layers used consisted of a visco-elastic self-adhesive material. The layers are clad to cover the whole surface of the plate. The weight per unit area of the layer is 1.6 $kg{\cdot}m^{-2}$.

The curves, in Fig. 9, display the velocity level difference between plates 1 and 3 before and after application of damping material. They reveal level differences from, at least, 5 dB up to 10 dB for frequencies over 1600 Hz. This indicates a major decrease of the vibration level of plate 3.

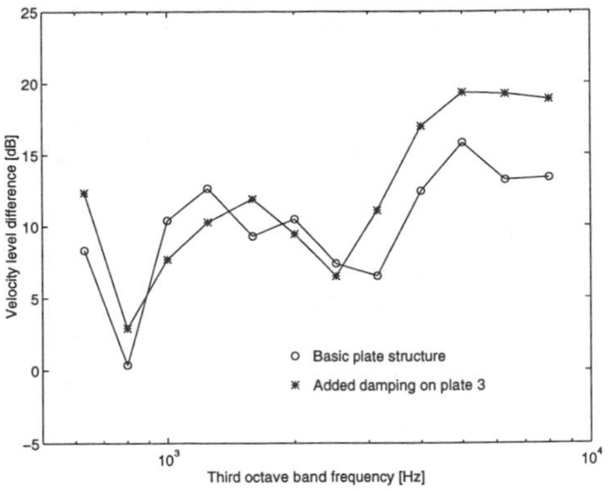

Figure 10. Velocity level difference between plates 1 and 5, ΔL_{15}, with and without added damping on plate 3.

Level differences, observed in Fig. 10, are, at most, 5 dB for frequencies over 3150 Hz. There is, thus, a decrease of the vibration level of plate 5, but much more limited in extent than that one observed for plate 3.

This shows that damping introduced locally may have only limited effect in reducing the vibration level of plates situated further away in a plate structure, even if major effects are locally observed. It could be argued that the undamped frame acts as a short cut for the vibration propagation, but it is shown, in section 4.4, that, compared to the plate elements, little vibrational power propagates throughout the frame.

4.4 *Changed coupling conditions*

The waveguide model is based on the concept that vibrational power propagates mainly as flexural waves in the plate elements. It is, therefore, expected that the vibrational power, flowing through the structure, may be reduced if the plate elements are uncoupled. Based on this assumption, two configurations of a system are examined.

In this section, a plate model, derived from the five plate structure, is used. It is the same model as before with an additional short thick plate welded to plate 1, which makes the original model a six plate structure. This new plate is referred to as plate 0; other denominations remain the same as defined in Fig. 2.

First, all the couplings between the plate elements are made by line welding. Then, junctions 2 to 5 are disconnected so that the plate elements are connected only via the frame. Junction 1 is kept unchanged, ie line-welded. The velocity level differences between

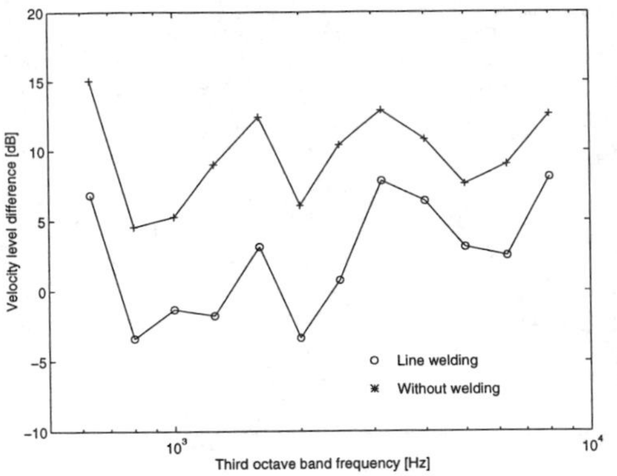

Figure 11. Velocity level difference between plates 1 and 5, ΔL_{15}, for line-welded and uncoupled plate elements.

plates 1 and 5 are measured in both cases and displayed in Fig. 11 The results show a level difference of 5 dB, at least, up to 10 dB. This tends to prove that vibrational power does propagate mainly in the plate elements and that the frame carries a minor part of this power. In addition, it indicates that the incorporation of dissipative elements in frames may not be useful in decreasing vibration transmission since the ribs composing the frame do not act as short-circuits for the vibration path.

5. CONCLUSION

The calculation procedure, developed according to a waveguide model for describing the vibrational behaviour of built-up plate structures, is easily adaptable to similar structures with various boundary conditions and excitation types. The transfer matrix technique permits point or averaged quantities, such as velocity and vibrational power, to be computed. Comparing with experimental data, the method provides satisfactory results. It is a computationally efficient tool for a systematic analysis of vibration propagation in built-up plate structures, enabling convenient parametric studies in acceptable computation time.

Measurements show that the force location may have some effects on the vibration propagation, but these are not systematically predictable. They also reveal that additional damping has to be applied where the vibration level is to be reduced.

6. REFERENCES

[1] A. C. Nilsson 1977. *Attenuation of structure-borne sound in superstructure on ships.* J. of Sound and Vibration 55(1), 71-91.

[2] B. M. Gibbs and Y. Shen 1987. *The predicted and measured bending vibration of an L-combination of rectangular thin plates.* J. of Sound and Vibration 112(3), 469-485

[3] J. L. Guyader, C. Boisson and C. Lesueur 1982. *Energy transmission in finite coupled plates, parts 1 and 2: theory and application to an L-shaped structure.* J. of Sound and Vibration 81(1), 81-105

[4] R. Haettel 1994. *Prediction of Vibration propagation in plate structures*, TRITA-FKT report 9427, MWL, KTH.

[5] A. C. Nilsson 1978. *Reduction of structure-borne sound in simple ship structures: results of model tests.* J. of Sound and Vibration 61(1), 45-60.

Modern Practice in Stress and Vibration Analysis, Gilchrist (ed.) © *1997 Balkema, Rotterdam, ISBN 90 5410 896 7*

Dynamic modeling of a turbo-prop aircraft using the U-vector Expansion Method

M.O.Gustavsson
Division of Structural Mechanics, Lund University, Sweden

ABSTRACT: Dynamic modeling of a furnished turbo-prop aircraft is performed by the use of a recently introduced technique called the U-vector Expansion Method (UEM). Transfer functions derived from test data are used as input for the modeling and the resulting model consists of a similar, but significantly larger, set of transfer functions to be used for design of noise reducing measures.

1. Introduction

Noise and vibration problems in turbo-prop aircraft are often characterized by the high contribution of tonal components generated by the rotating propeller blades. Typically the noise level in the passenger cabin of a turbo-prop aircraft, with no special noise reducing measures, is in the range 80-90 dB(A), totally dominated by the propeller Blade Passage Frequency (BPF) component. Reducing the tonal noise at the BPF normally results in a reduction of the total noise level by 5-10 dB(A), bringing the noise level down to a more acceptable level.

The noise control methods used for this type of low frequency noise are typically tuned vibration absorbers or/and active noise control. Both of these two methods require models with a large number of possible 'source positions' i.e the mounting points for the tuned vibration absorber units, or loudspeaker positions for the case of using active noise control. It is also desirable to be able to predict the response at a large number of positions, including the passenger ear positions. To get the dynamic properties of the fuselage structure-cabin cavity system for this number of driving points and response locations requires extensive testing, with a high cost associated.

The U-vector expansion method may be an attractive way to reduce the amount of testing to be performed, or to extend the test data base to get a model with more 'driving points' than actually used in the test.

2. The U-vector expansion method

The U-vector expansion method (UEM) [1,2,3] is based on measured transfer function data and is used to model dynamic systems. The method was developed as a result of poor performance of Experimental Modal Analysis (EMA) for heavily damped systems in frequency ranges with fairly high modal density, typical properties for vibro-acoustic problems in turbo-prop aircraft.

UEM is essentially based on the concept of singular value decomposition of a transfer function matrix, and may be seen as a natural step after the introduction of principal component analysis and the complex modal indicator function. Both principal component analysis and the complex modal indicator function are, as well as UEM, based on the concept of singular value decomposition of a matrix [4].

From a mathematical point of view UEM is nothing but a method to re-construct a symmetric, rank deficient matrix, from a set of vectors spanning the total vector space of the matrix. The rank deficiency makes it possible to use a sub-set of a matrix to find a set of space vectors spanning the space of the complete matrix. This is the central part of the UEM. The symmetry property of a matrix means that the same space vectors can be used to span both the row space of the matrix and the column space of the matrix. This is the second key characteristic on which the UEM is founded.

By using the concept of singular value decomposition a symmetric matrix, $\boldsymbol{A}$, may be decomposed in

$$\boldsymbol{A} = \boldsymbol{U}\boldsymbol{s}\boldsymbol{U}^{T} \qquad \textbf{[eq. 1]}$$

$\boldsymbol{U}$: is a matrix with orthogonal column vectors

$\boldsymbol{s}$: is a diagonal matrix

$[*]^{T}$: denotes the 'ordinary' transpose of a matrix (simply interchanging rows and columns).

The column vectors in U are called the U-vectors, and the diagonal elements in **s** are the singular values of A.

The number of non-zero singular values in **s** is equivalent to the rank of A and consequently if A is rank deficient only some of the U-vectors are required to generate the complete matrix A.

Consider if only some of the columns in the matrix A are available. This part of A is denoted $\bar{A}$. If the number of columns in $\bar{A}$ is equal to, or higher, than the rank of the complete matrix A it possible that the columns of $\bar{A}$ span the column space of the complete matrix A. The more columns in $\bar{A}$, compared to the rank of A, the more likely this situation becomes.

If the U-vectors of $\bar{A}$ span the same column space as the U-vectors of the complete matrix A, it is possible to re-construct the complete matrix A using the U-vectors of $\bar{A}$. A similar relation as [Eq. 1] can be formed, but since now we do not necessarily have the same U-vectors as would be found from the complete matrix A, a general matrix Σ, has to be used in place of the diagonal matrix **s**.

$$A = U(\bar{A})\Sigma U(\bar{A})^T \qquad \textbf{[eq. 2]}$$

The matrix Σ can be determined by using $\bar{A}$,

$$\bar{A} = U(\bar{A})\Sigma \bar{U}(\bar{A})^T \qquad \textbf{[eq. 3]}$$

with $\bar{U}(\bar{A})$ being a sub matrix of $U(\bar{A})$ with the entities to get the columns of $\bar{A}$. The matrix Σ is typically non-diagonal, but symmetric to ensure the symmetry of A.

Relating the mathematical description above to the case of modeling dynamic systems, the matrix A could be a transfer function matrix $H(f)$ at a particular frequency f. The entities in the transfer function matrix $H_{ij}(f)$ gives the response at location i for a unit input at j. By using complex notation both the amplitude and phase of the relation between input and response (output) is given by $H_{ij}(f)$.

The symmetry requirement for the complete matrix $H(f)$ is equivalent to assuming reciprocity for the dynamic system. This means that interchanging driving point and response point would give the same relation between excitation and response - i.e the transfer functions $H_{ij}(f)$ and $H_{ji}(f)$ are identical.

The assumption of a rank-deficient matrix is, for the case of transfer function matrices, related to the number of modes active at the frequency of interest. For the dynamic systems considered here the modal density is high and, consequently, several significant 'principal response components' (U-vectors) must be used to get accurate models. This implies that several driving points have to be used in the test.

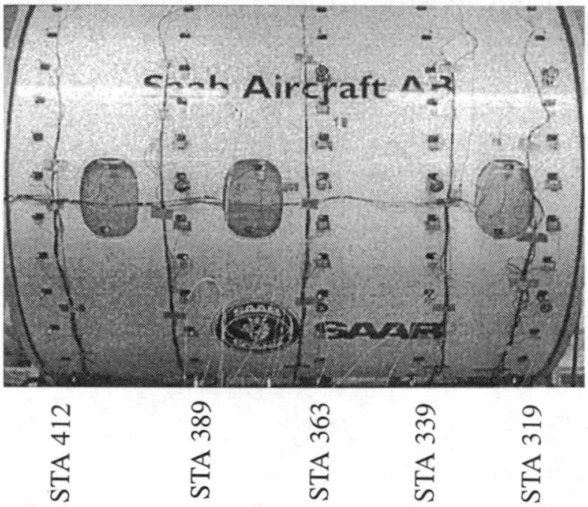

Figure 1: The Saab 340 Acoustic Test Section.

3. Test Description

The test case used to evaluate the performance of UEM is to model the dynamic behavior of a turbo-prop aircraft. Transfer function measurements on a 2.5 m long Acoustic Test Section of the Saab 340 are used as the input data for the UEM.

In the test 140 accelerometers and 105 microphones were used to register the acceleration response of the structure (Figure 1,2) and the acoustic response in the cabin cavity (Figure 3). The locations are identified by the cross-section name ('STA' in figure 1) and the cross-sectional position (figure 2,3).

What makes this test quite unique is the high number of excitation positions used: 108 structural excitation locations and 65 acoustic excitation positions.

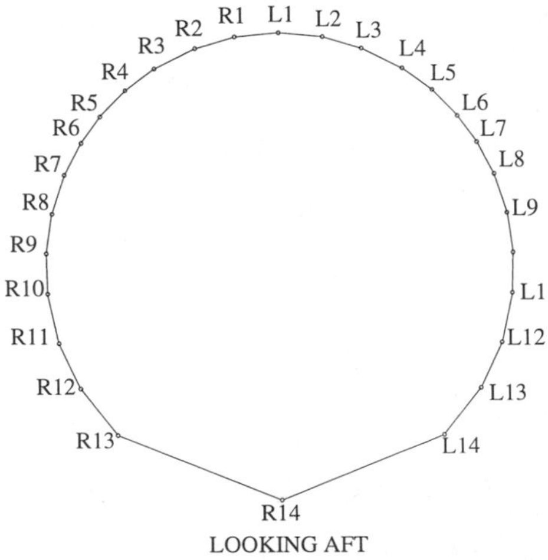

Figure 2: Positions used for vibration measurements and structural exciters.

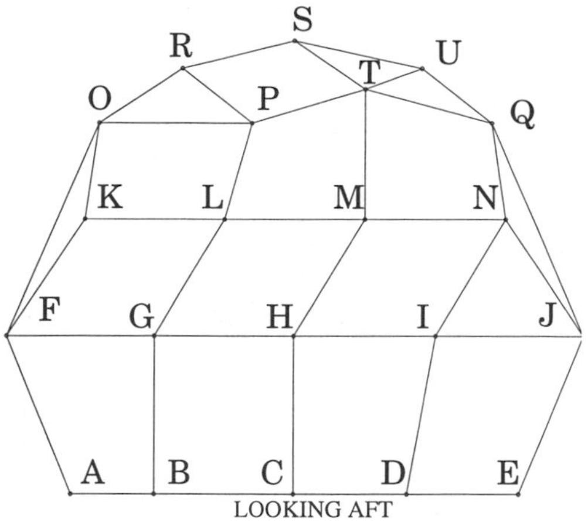

Figure 3: Positions used for acoustic pressure measurements and loudspeakers.

For the structural excitation electro-dynamic inertia shakers were used, and for the acoustic excitation small loudspeakers were used.

4. Transfer function data

In order to get reliable models with the U-Vector Expansion Method it is essential to remember the assumption of a:

- Rank deficient

and

- Symmetric

transfer function matrix.

Working with test data, and quantities derived from test data, one has to accept some deviations from the two assumptions above in the original data. It is, however essential to force the transfer function matrix to be symmetric before applying UEM.

Forcing symmetry is equivalent to eliminating factors that cause non-reciprocal data. In the current data set the response measurements were taken not exactly at the point of the excitation but at a distance of approximately 5cm from the excitation point for the structural locations and at about 10 cm from the 'source centre' for the loudspeaker excitation. This could not be avoided due to installation restrictions (i.e the size of the exciter/loudspeaker and the actuators). It is believed that this misalignment is more serious in some positions than others, in particular for the structure, explaining the difference in reciprocity characteristics.

Another reason for the in some cases poor reciprocity is the fact that the response level differs significantly for the two types of excitation (structural vs. acoustical). For the structural excitation the structural response was about 0.05 m/s^2 and the acoustic response was about 20 mPa. For the case of acoustic excitation the response levels were about 0.02 m/s^2 and 500 mPa respectively. With this quite large difference in response levels it is likely that non-linearity effects are influencing the results. In particular the attachment of the interior trim panel is a source of non-linear behavior.

Table 1: Reciprocity correlation

Driving point/ Response point	*sac*
Structure/Structure	0.87
Acoustic/Acoustic	0.90
Structure/Acoustic (and vice versa)	0.64

To estimate the validity of the reciprocity assumption a spatial correlation coefficient, ***sac***, for two vectors ***u*** and ***v*** is used.

$$sac = \frac{uv^H}{\|u\|\|v\|} \qquad \text{[eq. 4]}$$

Table 1 give the average correlation for structural/ structural acoustic/acoustic and structure/acoustic (or visa versa) degrees of freedom (dof's) at the frequency line of most interest (82 Hz).

The propeller Blade Passage Frequency of the Saab 340 is 82 Hz. That is the reason why the work presented in this paper is focused on 82 Hz.

In principle UEM can be applied to combined vibration/acoustic data, but with the quite low correlation for the structural/acoustic reciprocity found in the current data set, separate models will be derived for the structure and the acoustics respectively.

5. Test data Preparation

A first step, before the UEM is applied, is to 'prepare' the transfer function matrix to obey the symmetry requirement. Several methods may be used to force symmetry. Probably the simplest method is to use the average of $H_{ij}(f)$ and $H_{ji}(f)$. In matrix notation

$$H(f)_{sym} = (H(f) + H(f)^T)/2 \qquad \text{[eq. 5]}$$

However a method based on response rank reduction might be a better choice. To fulfill the assumption of a rank deficient and symmetric transfer function matrix we may use some of the U-vectors of $[\mathbf{H}(f)\, H^T(f)]$ and let

$$H(f)_{sym} = U\bar{s}V_{avg} \quad \textbf{[eq. 6]}$$

U: is the U-matrix of [H H^T]

$\bar{s}$: is a diagonal matrix with a selected set of the singular values of [H H^T]

$\bar{V}$: is the average of the part of V related to the H and H^T respectively

Leuridan et al. [5] describe a similar method to force reciprocity of transfer functions, although their method uses data for a frequency band rather than spatial data.

The rank deficiency requirement does not have to be handled separately, but will be a natural part of the UEM.

6. Transfer function matrix expansion

With as many as 108 structural and 65 acoustical driving points available in the test data set the performance of the UEM can be evaluated starting with different sub-sets of the complete test data. In this case the transfer functions derived with a model, based on typical test data, can be compared with the actual (measured) transfer functions.

Data for 30 frequency lines is available but, again, the major interest is to derive models valid at the fundamental Blade Passage Frequency (BPF) of the Saab 340, which is 82 Hz.

7. Results

Starting with a transfer function matrix containing all structural responses and all but one of the driving points (i.e using data for 107 of the 108 driving points) the 'missing' transfer function is derived by UEM. Comparing this result to the actual (measured) transfer function give an estimate of how well unmeasured transfer functions can be modeled with the existing data. Figure 4, 5 show two comparisons between measured and UEM-modeled transfer functions at 82 Hz using 25% of the U-vectors for the UEM.

Figure 6,7 show a 'snap-shot' of the measured and UEM derived structural response respectively.

The result in Figure 4-7 is for the case of using 26 of the 107 available U-vectors (25%) for the UEM, which is equivalent to assuming a transfer function matrix with rank 26. The average of the correlation coefficient, defined by [eq. 4], for the measured and modeled transfer function is 0.82.

The effect of using different portions of the U-vectors is given by table 2.

For this particular set of data it appears to be optimal to use about 25% of the 107 U-vectors.

For the acoustic data similar calculations were made. In this case there are data for 65 driving points

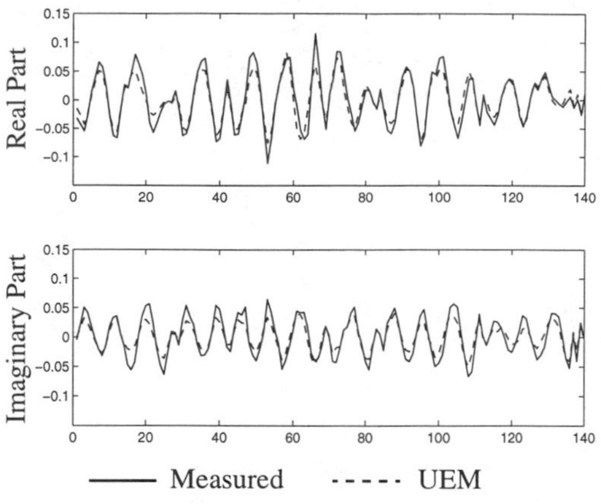

Figure 4: Structural transfer function response. Excitation at STA 363 pos. R10.

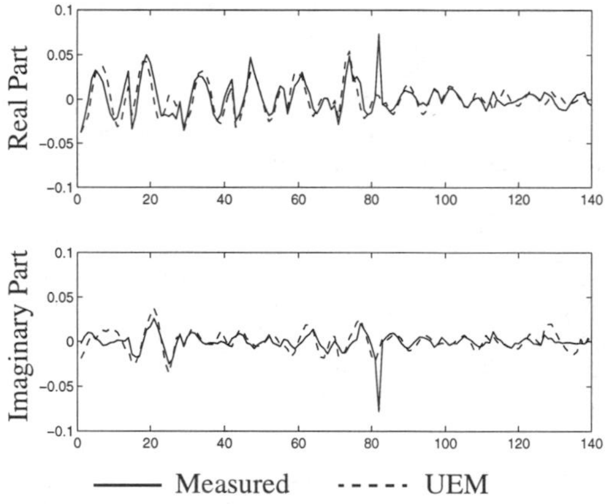

Figure 5: Structural transfer function response. Excitation at STA 339 pos. R12.

Table 2: Correlation for the structural model (107 of 108 driving points used).

U-vectors	10%	25%	50%	100%
Correlation	0.77	0.82	0.78	0.69

(loudspeaker locations), and using the same portion of the U-vectors (25% 50% and 100%) is equivalent to using 16, 32 and 64 U-vectors. The results in figure 8,9 are for the case of using 25% of the U-vectors.

From table 3 one gets the impression that the results for the acoustic data are not as good as for the structural data, but taking a closer look at Figure 8,9 this is

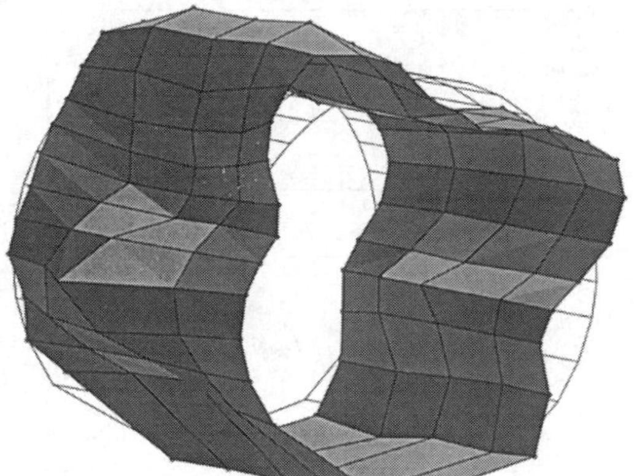

Figure 6: Measured transfer function response. Excitation at STA 363 pos. R10.

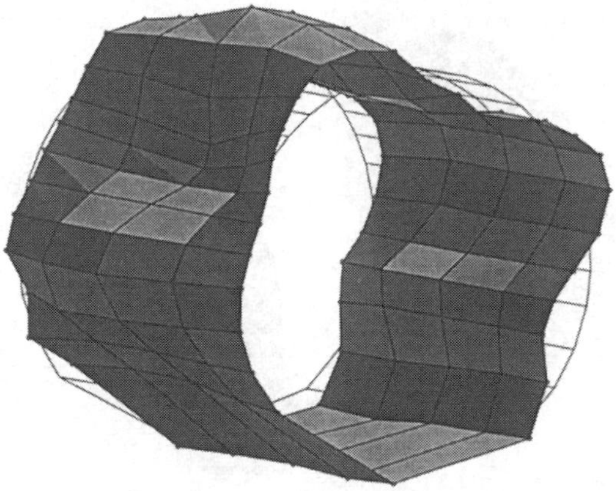

Figure 7: UEM transfer function response. Excitation at STA 363 pos. R10.

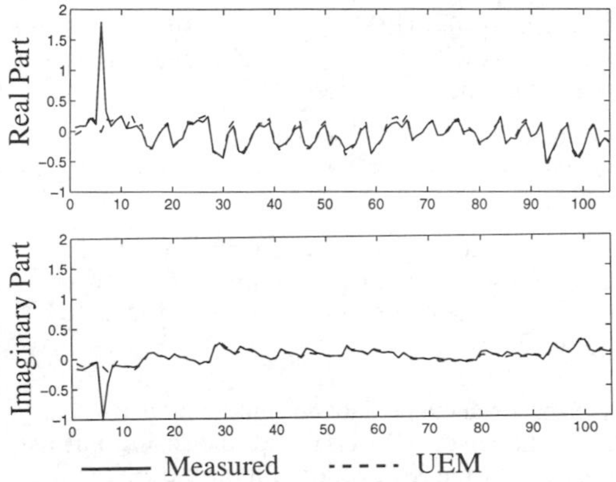

Figure 8: Acoustic transfer function response. Excitation at STA 319 pos. A.

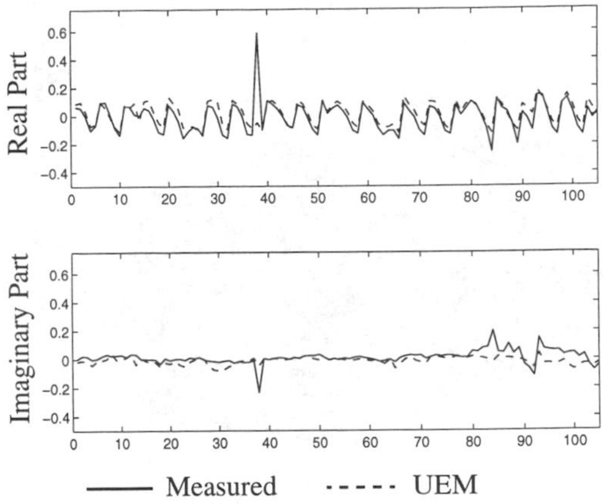

Figure 9: Acoustic transfer function response. Excitation at STA 389 pos. E.

Table 3: Correlation for the acoustic model (64 of 65 driving point used).

U-vectors	10%	25%	50%	100%
Correlation	0.57	0.58	0.58	0.30

Table 4: Correlation for the acoustic model excluding the driving point.

U-vectors	10%	25%	50%	100%
Correlation	0.80	0.79	0.84	0.42

found to be a consequence of the expected poor performance of the UEM for the driving point response. This is typical for expansion methods based on principal response vectors (UEM), or modal vectors (EMA), if the driving point response is high compared to the response at other points in the system. Excluding the driving point response, the result is completely different (Table 4).

In figure 10, 11 show the measured and modeled transfer function response respectively for acoustic excitation at STA 319 pos. A.

The correlation between the measured and modeled response is quite good, although the measured response close to the driving point differs.

The results above are for the case of taking a very large part of the data and using it to derive a model. For the structural excitation case 108 of the 140 columns of the transfer function matrix were available, and in the acoustic data 65 of the 105 columns were present. A more realistic situation would be to use data for significantly fewer driving points; however the number of driving points has to be large enough to allow for

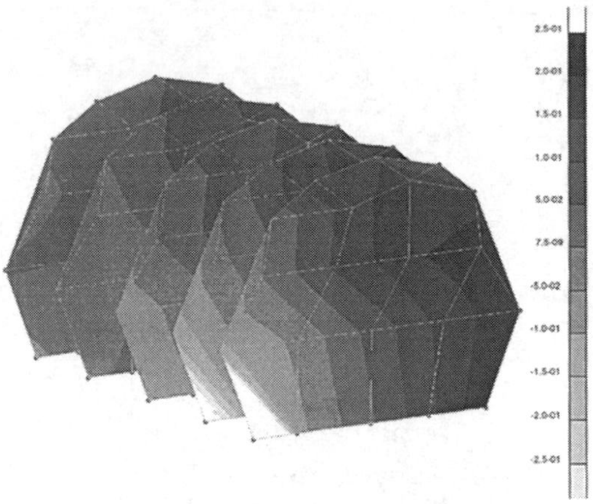

Figure 10: Measured transfer function response. Excitation at STA 319 pos. A.

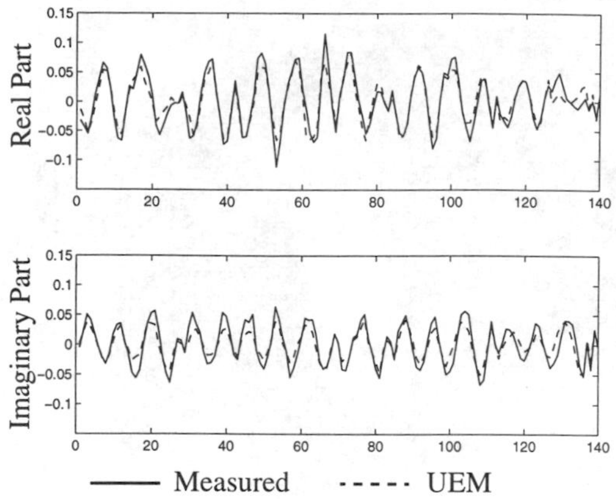

Figure 12: Structural transfer function response. Excitation at STA 363 pos. R10.

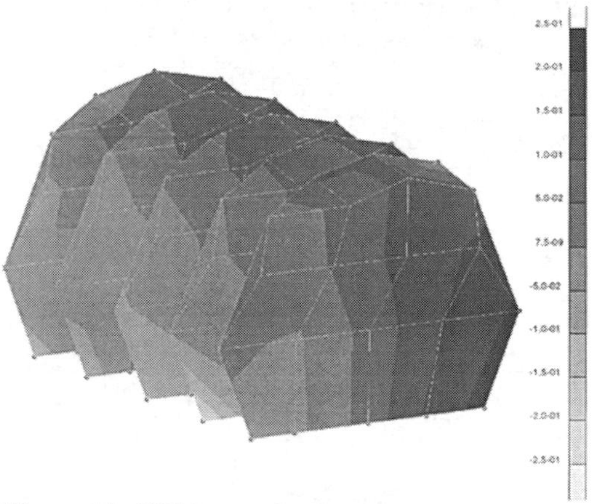

Figure 11: UEM transfer function response. Excitation at STA 319 pos. A.

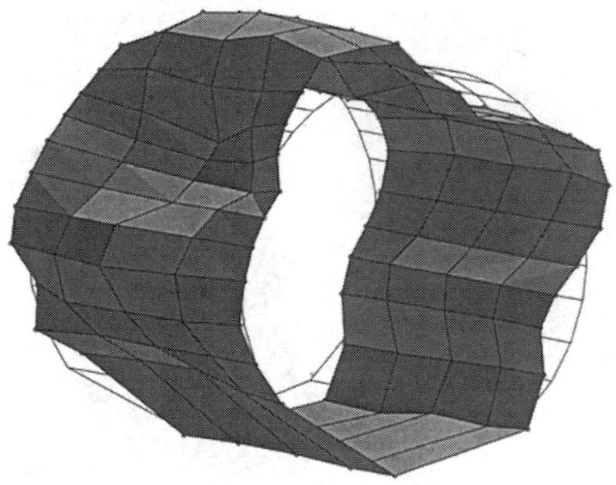

Figure 13: UEM transfer function response. Excitation at STA 363 pos. R10.

Table 5: Correlation of the structural model using 50% of the transfer functions.

U-vectors	10%	25%	50%	100%
Correlation	0.71	0.80	0.77	0.76

finding a set of U-vectors spanning the space of the complete transfer function matrix to be derived by UEM.

A 50% reduction of the number of driving points would be a significant reduction in test time, both for the data acquisition and for the installation time for structural exciters/loudspeakers. To simulate this situation only 50% of the transfer function data available was used for the UEM. Again we can compare the UEM results with the 'true' transfer functions as they were determined by the test.

For the structural data almost the same results as when using all but one transfer function were obtained (Table 5.).

Not only the average correlation but also the individual responses appear to by only slightly influenced by the quite substantial reduction in input for the UEM (figure 12,13).

Also for the acoustic data no significant performance degradation of the UEM is found when using only half the existing transfer functions for the modeling (Table 6, Figure 14,15).

Table 6: Correlation for the acoustic model excluding driving point response and using 50% of the transfer functions.

U-vectors	10%	25%	50%	100%
Correlation	0.74	0.80	0.79	0.61

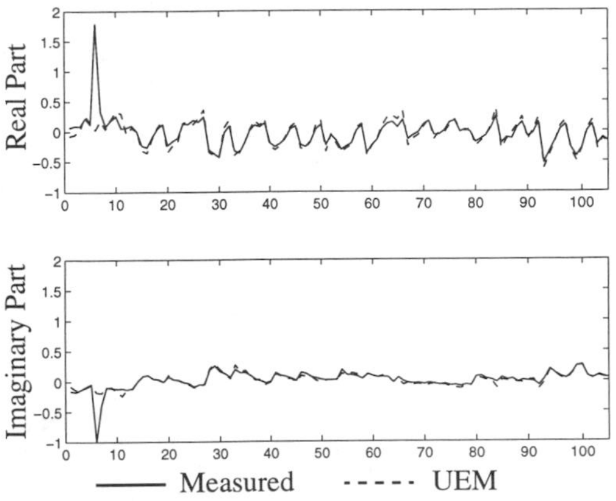

Figure 14: Acoustic transfer function response. Excitation at STA 319 pos. A.

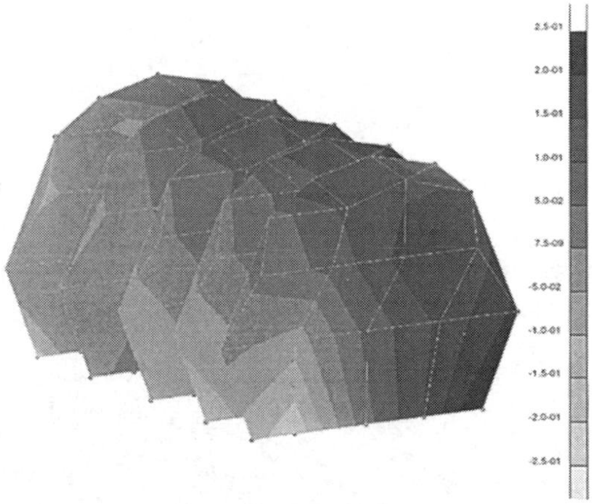

Figure 15: UEM transfer function response. Excitation at STA 319 pos. A.

The good result for the UEM using 50% of the transfer function data encourages one to try using fewer of the available transfer functions. The results will, however, now become very sensitive to which transfer functions (driving points) are used to model the response of a certain driving point.

8. Conclusions

The U-vector Expansion Method (UEM) has been used to model the structural vibrations and the acoustic response of a turbo-prop aircraft. The results show that quite accurate models of the structural vibration and the acoustic response can be achieved by UEM. Almost no degradation in model performance was found using 50% of the available test data compared to the case of using practically all the test data available for the modeling.

9. Acknowledgments

The work presented in this paper is financed by the National Swedish Aeronautics Research Program (NFFP). All test data used for the analysis were made available by Saab AB and are a result of the BRITE/EURAM project ASANCA II.
The financial support by NFFP, and the permission for using test data from the ASANCA II project are gratefully acknowledged.

10. REFERENCES

[1] Gustavsson M.
A method to generate dynamic models and its application to aircraft noise.
SVIB Conference Riksgränsen, Sweden 1991.

[2] Halvorsen W, Barney P, Brown D.
The U-vector expansion method for modelling structural/acoustic systems.
Journal of Sound and Vibration, July 1991.

[3] Otte D.
Development and evaluation of singular value analysis methodologies for studying multivariante noise and vibration problems
Doctoral dissertation,
Katholieke Universiteit Leuven, Belgium

[4] Golub G.H, Van Loan C. F.
Matrix Computations, Second Edition (1989).
The John Hopkins University Press.

[5] Leuridan J. et al.
Coupling of Structures Using Measured FRF's: Some Improved Techniques.
Proc. 13th Int. Seminar on Modal Analysis, Leuven Belgium.

Modern Practice in Stress and Vibration Analysis, Gilchrist (ed.) © *1997 Balkema, Rotterdam, ISBN 90 5410 896 7*

Diagnostic analysis of vibrations in a turbine generator support structure at Kilroot Power Station, Northern Ireland

P. Fanning
Department of Civil Engineering, University College Dublin, Ireland (Formerly: Structures & Computers Ltd)

P. Dowdie
NIGEN, Kilroot Power Ltd, UK

D. Sweeney
IDAC (Ireland) Ltd, Ireland (Formerly: Structures & Computers Ltd)

ABSTRACT: Kilroot Power Station is located on the north shore of Belfast Lough, east of Carrickfergus County Antrim. A dual fired station able to burn either coal or heavy fuel oil, it comprises two Turbine Generators each capable of producing 300 Megawatts (when firing oil). During normal operating conditions significant levels of vibration were observed to be induced in the Turbine Generator support structure. As part of a diagnostic analysis, levels of vibration on the turbine generator support structure were monitored over time and Structures and Computers Ltd, a firm of finite element consultants, were commissioned to undertake a series of analyses, with a view to assessing alternatives suitable for reducing or preventing these vibrations. These analyses and measures taken to reduce the levels of vibration are discussed in this paper.

1. INTRODUCTION

The turbine generators at Kilroot Power Station were manufactured, installed and commissioned by GEC Ltd (now GEC Alsthom) between 1976 and 1982. Steam at 163 bar and 540°C flows from the reheat Boiler to the Turbine which is mounted on a structural steelwork platform and supported on a number of steelwork columns. The rotors are coupled in a train to the Generator and rotate at 3000 revolutions per minute (50 Hz).

During normal operating conditions significant levels of vibration were observed to be induced in both the mounting platform and its supporting columns. Although not detrimental to the performance of the turbines these vibrations were perceptible to maintenance staff working on the platform and also caused partial cracking of the tiled floor at the platform level. A series of investigations were undertaken with a view to reducing or eliminating these vibrations.

2. THE SUPPORT STRUCTURE

All three sections of the turbine are mounted in series on a steel framed platform which is in turn supported on steel columns approximately 10.0m high. A sketch of the structural configuration is shown in Figure 1.

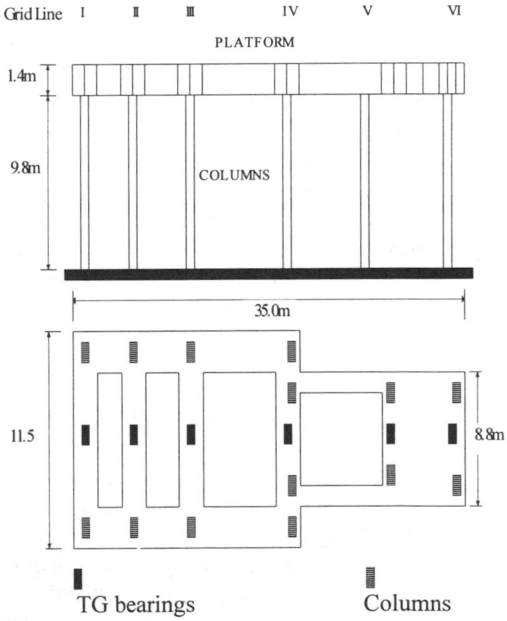

Figure 1 (Elevation & Plan View on Platform)

Both column and platform members are fabricated boxed sections. Columns are offset approximately 1.0m from the perimeter of the platform. There are two columns on grid lines I, II, III, V and VI and four along grid line IV. The turbine generator units are fixed at points to the platform as indicated.

The columns are typically 600mm x 600mm square sections and are split into approximately 600mm segments by internal stiffeners. Plate thicknesses between 6mm and 20mm are variously used in the fabrication of the platform and column members. Moment connections were designed for the column bases and the column/platform interfaces.

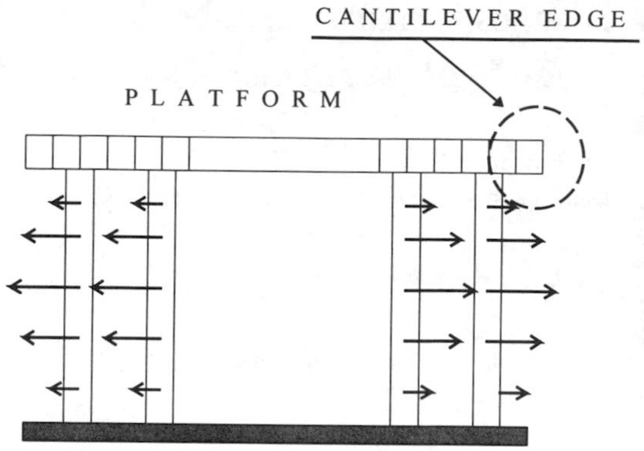

Figure 2: Vibration of Columns on Grid Line IV

3. DYNAMIC RESPONSE OF THE SUPPORT STRUCTURE DURING OPERATION

After commissioning of the power station and during normal operation significant levels of vibration were observed on the platform and along the supporting columns. Readings of displacement and associated phase angles at the turbine mounting locations indicated that these vibrations were not detrimental to the performance of the turbine. However the vibrations were such that maintenance staff were uncomfortable when working on the platform and more significantly cracking of the tiled working surface was initiated.

As part of an initial investigation into the extent of the levels of vibration accelerometers were mounted on the platform structure and also along the length of the columns.

The accelerometers mounted on the platform indicated that the maximum vibrations levels were present where the platform cantilevers beyond its supporting columns.

With respect to the columns the maximum levels of vibration were witnessed in the columns on Grid Line IV. These columns were observed to be excited in their first mode of vibration with each pair of outer columns being in phase but out of phase with the pair on the opposite side of the platform, refer to Figure 2.

Following these in-situ measurements additional mass, by means of sandbags, was added to the cantilever edges in an attempt to reduce the levels of vibration. Vibration levels on the platform were reduced and neoprene pads installed under the tiled floor surface prevented further cracking and deterioration. The additional mass however had little effect on the response of the supporting columns and

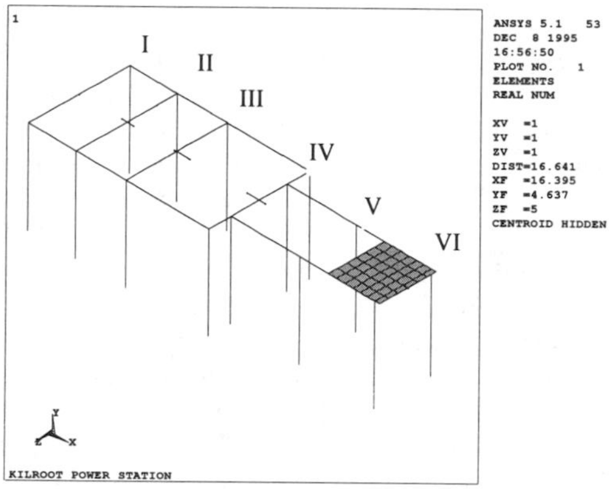

Figure 3 The Finite Element Model

Structures and Computers Ltd, a firm of finite element consultants, were commissioned to undertake a series of finite element analyses on the support structure aimed at reducing these vibrations.

4. FINITE ELEMENT ANALYSIS OF THE DYNAMIC RESPONSE OF THE SUPPORT STRUCTURE

4.1 *The Finite Element Model*

The ANSYS V5.1 finite element software (ANSYS, 1995) was used to construct a numerical model of the platform and its supporting columns. The finite element model is shown in Figure 3.

In constructing the finite element model it was assumed that the dynamic response of the structure was linear, that the turbines do not add significant stiffness to the complete assembly and that a value of 5% of critical damping would be used in examining the dynamic response. Furthermore the in-situ measurements had demonstrated that the flexural modes of the structural members were of primary importance indicating that a relatively simple three dimensional 'beam model' of the complete assembly might be appropriate.

BEAM4 elements, two-noded three-dimensional isoparametric elements, were specified for the platform plate girders and boxed column sections. Appropriate cross section data was calculated from section sizes specified on construction drawings available from NIGEN.

Between Grid Lines V and VI, Figure 1, the extent of the plated sections on the platform was such that SHELL99 elements, eight noded isoparametric shell elements, were preferred. Three layers were specified for these shell elements. The outer two layers modelled the top and bottom plates and the webs were catered for by a deep middle layer.

The accurate prediction of the dynamic response of a structure requires that both its stiffness and mass be appropriately specified (Hitchins,1992) and to this end the densities of the various elements were adjusted to cater for web stiffeners, flange stiffeners and any ancillary items attached to the platform and column elements in order that the complete mass of the structure be captured in this greatly simplified model.

The turbine was not modelled explicitly but was accounted for by a series of lumped mass elements, MASS21 elements, located at its bearing points. Where required short very stiff beam elements were used to offset these masses from the centrelines of the platform elements.

4.2 *Boundary Conditions & Loading*

The column footings were fully restrained against displacement and rotation.

The structure is subjected to gravitational loading, some live loading on the platform level and dynamic excitation at the turbine mounting points. The excitation at the mounting points is harmonic and occurs at a frequency of 50Hz, the running speed of the rotors. The displacement at the mounting points is routinely monitored by maintenance staff and amplitudes and phases of displacements at these locations were available from NIGEN for assignment in the finite element model.

4.3 *Numerical Analysis*

The ANSYS finite element software offers the analyst the option of using modal or direct integration methods (Craig, 1981 and Bathe, 1982) for determining the dynamic response of structures. Additionally harmonic analysis routines are available for harmonically varying loads such as encountered in this case. Harmonic response analysis routines are extremely efficient for harmonic loading but in ANSYS the facility to account for prestressing effects, for example the effect of loads in members due to gravitational loading, is not available. Modal techniques enable prestress effects to be accounted for but imposed displacements, to model the dynamic excitation at the turbine mounting points, are not possible. Direct integration techniques are the most flexible but also the most numerically intensive. Given the above a number of modal analyses were undertaken to assess the effect of prestressing effects on the natural frequencies and modes of vibration of the structure (and hence its dynamic response) prior to undertaking a series of harmonic response analysis to investigate mechanisms for reducing or eliminating current vibration levels.

In a first analysis gravitational loading and additional point loading at the column heads and uniformly distributed loading along the cross members, to account for any live loading, were applied in an initial prestressing linear static loadcase. A modal analysis of the structure in its stressed state was subsequently undertaken.

A second modal analysis was undertaken without the initial prestressing stage. Resonant frequencies for modes of vibration 1-4 and 43-50 for both analysis are tabulated in Table 1.

Table 1

Mode	**Frequency (Hz)**	
--	*Prestressed*	*Unstressed*
1	1.16	1.18
2	1.21	1.23
3	1.52	1.54
:	:	:
43	49.43	49.56
44	49.97	50.11
45	50.01	50.16
46	50.45	50.58
47	50.55	50.69
48	50.68	50.89
49	51.27	51.43
50	51.32	51.49

The effect of the prestress induced in the structure due to gravitational and live loading is clearly small, less than 2%. In view of such a negligible impact on the resonant frequencies and modes of vibration the decision to proceed with a series of harmonic response analysis instead of a more numerically intensive direct integration techniques is justified.

Modes of vibration for resonant frequencies 50.11Hz and 50.16Hz are plotted in Figures 4 and 5. These modes of vibration are system modes and typically involve vibration of single or several of the support columns. It is likely that the real modes of the structure will be variants of these calculated mode shapes due to inevitable biases in the as built structure. The fundamental behaviour of the structure exhibiting column vibrations at 50Hz however still holds true. These analyses demonstrated that the structure had numerous modes of vibration in and around 50.0Hz, the running speed of the generator, (roughly 20 modes centred on 50Hz were predicted in the range 50Hz±5%). It was expected that harmonic excitation at the turbine bearing points at 50 Hz would cause several, or a combination, of these modes to be excited.

In a third analysis the movement at the turbine mounting points were imposed on the model and a harmonic response analysis carried out for the frequency range 47Hz to 52Hz. Snapshots of the structure at different phase angles at 50Hz are plotted in Figures 6, 7,8 and 9. It is seen clearly that the columns of the support structure tend to vibrate at the running speed of the generator. Additionally the pairs of columns on line IV are observed to be out of phase with each other. This is consistent with on site measurements.

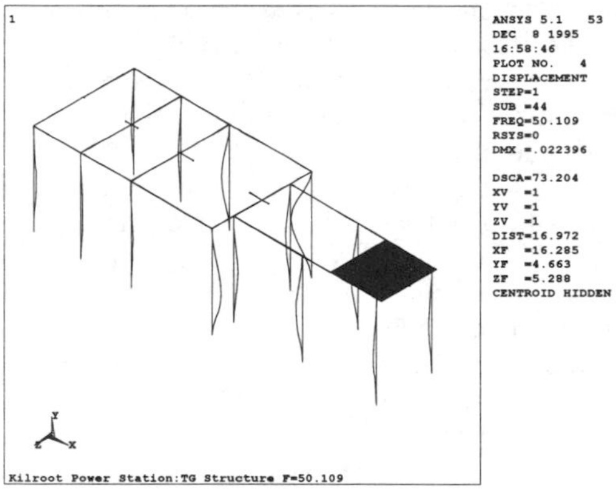

Figure 4 Mode 44, Frequency 50.11 Hz

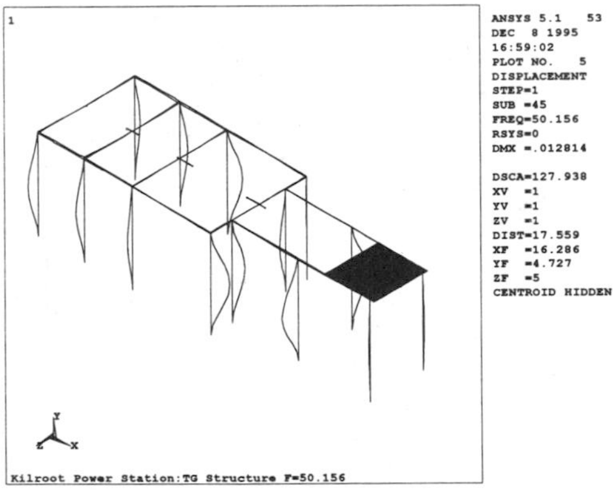

Figure 5 Mode 45, Frequency 50.16 Hz

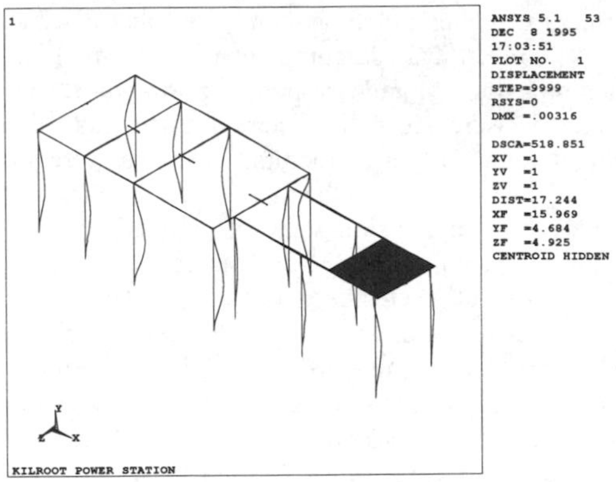

Figure 6 Snapshot 1 - Structural Response to Harmonic Excitation at Turbine Bearing Points

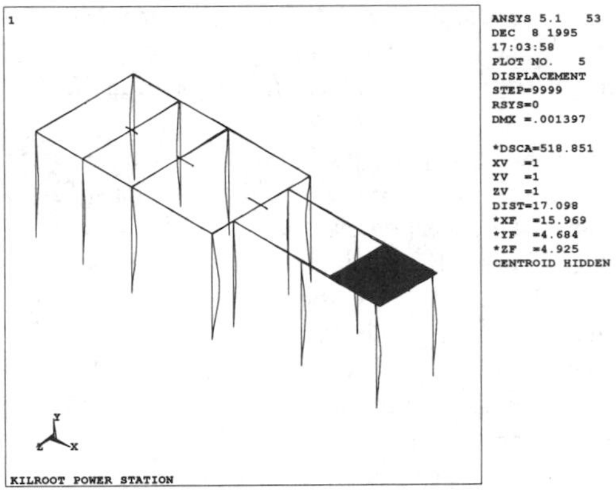

Figure 7 Snapshot 2 - Structural Response to Harmonic Excitation at Turbine Bearing Points

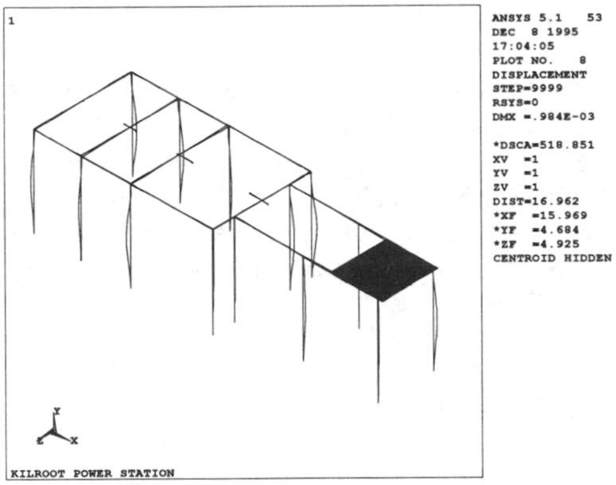

Figure 8 Snapshot 3 - Structural Response to Harmonic Excitation at Turbine Bearing Points

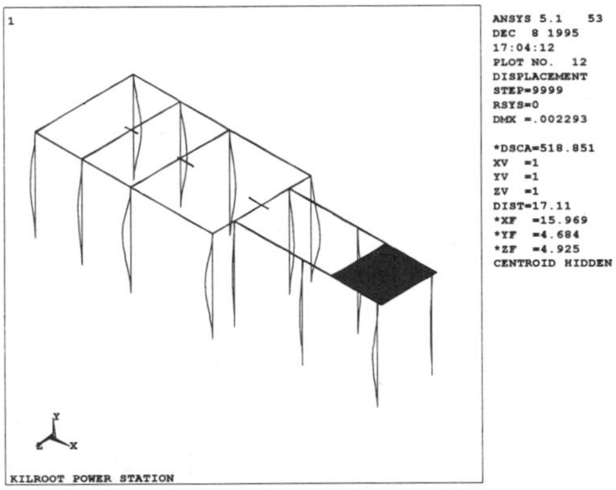

Figure 9 Snapshot 4 - Structural Response to Harmonic Excitation at Turbine Bearing Points

Column vibrations are excited because the running speed of the generator corresponds to resonant frequencies of the support structure characterised by vibration of one or more columns. The running speed of the generator could not be altered and hence various alternatives to move the critical frequencies of the structure away from the running speed of the generator were considered.

The extent to which the structure could be altered was limited by both time and logistical constraints. Bracing the column legs and decreasing the effective length of the columns by local stiffening at the column bases and column/platform interfaces were both demonstrated numerically to be successful solutions. Bracing the column legs was discounted due to congestion beneath the platform and between the column legs. The time frame available for physical implementation of a solution was limited to a two week annual maintenance when the turbine is shut down. Local stiffening of column ends could not be made in that time frame.

An attempt to change the dynamic response of the columns by the addition of mass to these members was identified as a viable and practical solution.

Three lumped masses were added to all column legs in the finite element model to model the effect of filling the central three segments of the boxed column members with ballast. Internal stiffeners in the column members facilitated this solution. The system modes characterised by vibration of the columns were now predicted around the 30Hz region and it was anticipated that the levels of column vibration would be significantly reduced.

This was confirmed by a further harmonic analysis on this altered structure. Levels of vibration were reduced by 80%.

In summary, the finite element model identified that the platform on its supporting columns had a significant number of modes of vibration occurring at and around the running speed of the generator. These modes, characterised by vibration of single or several columns, were excited causing noticeable vibration in the columns. Furthermore it was identified that these system modes could be shifted by the addition of extra mass at the centres of each of the columns and the levels of vibration significantly reduced. It was proposed that this solution should be implemented in practice.

5. PROVISION OF ADDITIONAL MASS TO PLATFORM SUPPORT COLUMNS

The columns are fabricated boxed sections with internal stiffeners at approximately 600mm centres. These internal stiffeners split the column cavities into nine segments which could conveniently be filled with a suitable material thereby adding additional mass to the columns without altering their stiffnesses. The amount of mass added and its distribution along the column length would determine the extent by which the natural frequencies of the column legs would be shifted.

Following due consideration and consultation it was decided that a trial mass would be added to the central segment of each column. This consisted of filling the middle segment of each column with ballast, in this case cast shot ballast treated with preservative oil. In-situ vibration measurements of the columns showed that this reduced the column vibrations on average by 57%. Encouraged by this success, and increasingly confident in the finite element results, further masses were added to the segments above and below the central segment on all columns, (as per the finite element model). The total mass added to the structure was approximately 20 tonnes. The result was to reduce on average the level of column vibrations by 74% of the original values.

This compared very favourably with the 80% reduction predicted by the finite element analyses. The recommended solution proved both simple to implement and effective.

6. SUMMARY

The manufacture, installation and commissioning of the turbine generators at Kilroot Power Station started in 1976. Numerical modelling techniques have advanced significantly in the past twenty years and areas of structural response that were once difficult, or even often not possible, to solve are now routinely and cost effectively analysed. The importance of understanding the dynamic response of a structure or system subjected to loading that may vary with time is highlighted by this case.

A finite element model of a turbine generator support structure subjected to harmonic excitation has been successfully constructed and used to examine the potential for reducing existing vibration levels. The numerical model was demonstrated to be representative of the in-situ dynamic response.

A relatively simple solution, which could be implemented in a short time scale, to reduce column vibrations, was identified and successfully implemented on site.

The ability of the numerical models to capture the on-site response gave maintenance engineers the confidence to proceed with the structural alterations and encouraged future development of the possibility of further reduction of vibrations on the platform level without compromising the integrity of the structure.

The time frames available for undertaking structural alteration are limited and infrequent. To date the effort has been concentrated on reducing the levels of vibration experienced by the columns. Vibration of the platform area is currently restrained by strategically located sand bags and floor mounting pads. In view of the success of column analysis further work is proposed to achieve a more aesthetic solution in the platform area.

7. REFERENCES

ANSYS 1995, ANSYS Revision 5.1 Theory Manuals, ANSYS Inc., Pittsburgh, USA.

Bathe, K J 1982. Finite Element Procedures in Engineering Analysis, Published: Prentice Hall

Craig, R 1981. Structural Dynamics: An Introduction to Computer Methods, Published: John Wiley & Sons.

Hitchins, D 1992. A Finite Element Dynamics Primer, Published: NAFEMS, Birniehill, East Kilbride, Glasgow G75 0QU.

Modern Practice in Stress and Vibration Analysis, Gilchrist (ed.) © *1997 Balkema, Rotterdam, ISBN 90 5410 896 7*

Impact modelling in lateral vibrations of cantilevered beams including friction dampers

D. Szwedowicz
Mechanical Department, CENIDET, National Centre for Research and Technological Development, Cuernavaca, Morelos, Mexico

ABSTRACT: In the paper, a numerical technique for steady-state vibrations of rotating turbine blades including lateral impact and dry friction effects is proposed. A linear model of elastic impact forces between the cover band and airfoil tenon is discussed in details. The presented model is verified experimentally and numerically. Good agreements between numerical and empirical results of impact force characteristics are obtained.

1 INTRODUCTION

Applying the finite element method, steady-state vibration responses of rotating turbine blades may be computed with a good reliability. Considering a possible spectrum of harmonic resonance conditions, a concept of circumferentially coupled blades through 360° of the turbine stage has more benefits in relation to uncoupled or packeted blades (Ewins and Imregun, 1984). Moreover, the blades shrouded at their airfoil tips enclose the flow. Thus, the shrouded turbine stage has higher flow efficiency in comparison with other types of the blade coupling.

In the turbine technique, cover bands can be mounted to the blade tip by riveting or screw attachments. Also, a shroud can be integrally machined at the blade tip. Due to corrosive processes or manufacture failures, a cover band or tenon head at the airfoil tip can be damaged. Hence, the blade may lost its designed connection in the disc assembly. Then, the blade can be excited by elastic impacts at shroud clearances (Swain et al., 1992, Yang and Griffin, 1995). A generated impulsive energy propagates along the blade reflecting at the blade attachment and airfoil tip. Often, friction dampers under platforms of rotating blades are mounted as an additional passive damper reducing vibrations of the blade (Griffin, 1980). Thus, the propagated impulsive wave can be dissipated by friction forces either at the friction damper or the blade attachment (Jones and Muszynska, 1978).

In the paper, a numerical technique for forced vibration analyses of cantilevered beams including lateral elastic impact and dry friction effects is proposed. At the free end of the cantilevered beam, an elastic restraint with the initial clearance is applied. The dry friction forces are considered in the discrete model. Hence, the analysed beam can be treated as a reference model of the turbine blade with a damaged connection at the shroud. For steady-state excitations, vibration responses of the beam are analysed numerically and computationally. Characteristics of the empirical and numerical impact forces are found with good agreements.

2 LINEAR VIBRATIONS OF DISC ASSEMBLIES

Applying finite elements, linear steady-state vibrations of the elastic structure are given as

$$[M]\{\ddot{q}\} + [D]\{\dot{q}\} + [K]\{q\} = \{F_o\}\exp(j\upsilon t),\quad j=\sqrt{-1} \tag{1}$$

where [M], [D], [K] refer to mass, damping and stiffness matrices, $\{\ddot{q}\}, \{\dot{q}\}, \{q\}$ relate to Cartesian acceleration, velocity and displacement vectors of the vibration response, υ and $\{F_o\}$ are an excitation frequency and amplitudes, respectively.

Due to the stress stiffness matrix $[K(x,\sigma)]$, the matrix [K] of the rotating disc assembly is a nonlinear term in Eq. (1) and it can be expressed by

$$[K] = [K(x)] + [K(x,\sigma)] + [K(x,\Omega^2)] + [K(x,\tau)] \tag{2}$$

where $[K(x)]$ denotes a linear, conventional stiffness matrix in terms of geometry x, the stress stiffness matrix $[K(x,\sigma)]$ depends on stresses σ caused by centrifugal loads, the linear stiffness matrix of centrifugal effects $[K(x,\Omega^2)]$ and the linear stiffness matrix $[K(x,\tau)]$ due to temperature τ. For a nominal rotation speed Ω of the turbine, the static equilibrium state of the disc assembly is obtained by a numerical iteration process. Finally, free or steady-state vibration analysis of the rotating disc assembly can be performed.

In practical analyses, Rayleigh's (proportional) damping model gives good agreements in comparison with experimental results. In the Rayleigh definition, the damping matrix [D] is a linear combination of the mass α_M[M] and stiffness α_K[K] matrices, where constants α_M and α_K are evaluated empirically.

Because of different manufacturing irregularities of the turbine casing, non-symmetric flow distributions in the turbine inlet along the circumferential direction are generated after the vane stage. By entering into and moving out of zones of different static flow pressures, rotating blades are stimulated to vibration. For the rotating blades, the non-uniform circumferential pressure distribution is a periodic excitation function with the period of T=1/Ω. Thus, for a chosen rotational excitation harmonic k, the linear steady-state vibration of the rotating disc assembly is given by

$$[M]\{\ddot{q}\} + [D]\{\dot{q}\} + [K]\{q\} = \{F_{o,k}\}\exp(\,j\,k\Omega t\,), \quad j=\sqrt{-1} \tag{3}$$

For getting a good relation between numerical and experimental results, fine finite element meshes of the structure have to be used. Therefore, the modal vibration analysis of the rotating disc assembly is an adequate and efficient solution technique and it can be performed using a conventional commercial finite element system. In the modal domain, linear equations of steady-state vibrations of the rotating disc assembly are expressed as

$$\ddot{\eta}_i + 2\,\xi_i\,\omega_i\,\dot{\eta}_i + \omega_i^2\,\eta_i = \frac{1}{m_i}\{\phi_i\}^T\{F_o\}\exp(\,j\,k\Omega t\,), \quad i = 1, 2, 3, \ldots \quad , k= 1, 2, \ldots, \tag{4}$$

where m_i, ξ_i are the modal mass and damping ratio, $\ddot{\eta}_i$, $\dot{\eta}_i$, η_i denote the modal acceleration, velocity and displacement response of mode i where its natural frequency ω_i and mode shape $\{\phi_i\}$ are obtained form

$$(\,[K] - \omega_i^2[M]\,)\{\,\phi_i\,\} = \{\,0\,\} \tag{5}$$

Using the known analytical formulas on a harmonic response of a damped system with single degree of freedom (Eq. (4)) and the superposition technique, the steady-state response of the disc assembly can be calculated in the Cartesian system.

3 NON-LINEAR VIBRATIONS OF DISC ASSEMBLIES

If the tenon head at the airfoil tip is damaged, the blade loses its designed connection to the shroud. Then, the damaged blade vibrates independently on oscillations of the disc assembly. Considering small clearances between the shroud and airfoil tenon, the elastic impacts act periodically on the blade during each single response cyclic 1/kΩ. Then, the vibration of the rotating blade can be written by

$$[M]\{\ddot{q}\} + [D]\{\dot{q}\} + [K]\{q\} + [R]\{q\} = \{F_o\}\exp(\,j\,k\Omega t\,) \tag{6}$$

where the stiffness matrix [R] represents stiffness properties of the restraints.

Considering much higher stiffness of the shroud along the circumferential direction in comparison with the bending stiffness of the airfoil tenon, the restraint stiffness matrix [R] can be simplified into a diagonal form containing equivalent stiffness' C of a local elasticity at the restraint. Finally, the stiffness term [R]{q} in Eq. (6) can be substituted by a vector of impact forces F_r acting periodically on the vibrating blade (Fig. 1). Then, the blade vibration motion in the modal domain is described by

$$\ddot{\eta}_i + 2\,\xi_i\,\omega_i\,\dot{\eta}_i + \omega_i^2\,\eta_i = \frac{\{\phi_i\}^T}{m_i}\{F_o\}\exp(\,j\,k\Omega t\,) + \frac{\phi_{r,i}}{m_i}F_r(t,q_r) \tag{7}$$

where

$$F_r = 0 \quad \Leftrightarrow \quad -\delta < q_r < \delta \tag{8}$$

and δ indicates a clearance between the airfoil tenon and shroud (Fig. 1).

Considering the friction damping either at the blade attachment (Jones and Muszynska, 1978) or the mounted friction damper (Griffin, 1980), the vibration energy of the blade can be dissipated by friction forces. In the discrete model, the friction phenomena is represented by the non-linear dry (Coulomb) friction forces (Ostachowicz and Szwedowicz 1986, Szwedowicz 1986) shown in Fig. 1. Finally, the modal dynamic equations of the vibrating blade including non-linear impact and dry friction forces are given by

$$\ddot{\eta}_i + 2\,\xi_i\,\omega_i\,\dot{\eta}_i + \omega_i^2\,\eta_i = \frac{\{\phi_i\}^T}{m_i}\{F_o\}\exp(\,jk\Omega t\,) + \frac{\phi_{r,i}}{m_i}F_r(t,q_r) + \frac{\phi_{s,i}}{m_i}\mu N\,\mathrm{sgn}(\,\phi_{s,i}\,\dot{q}_s\,) \tag{9}$$

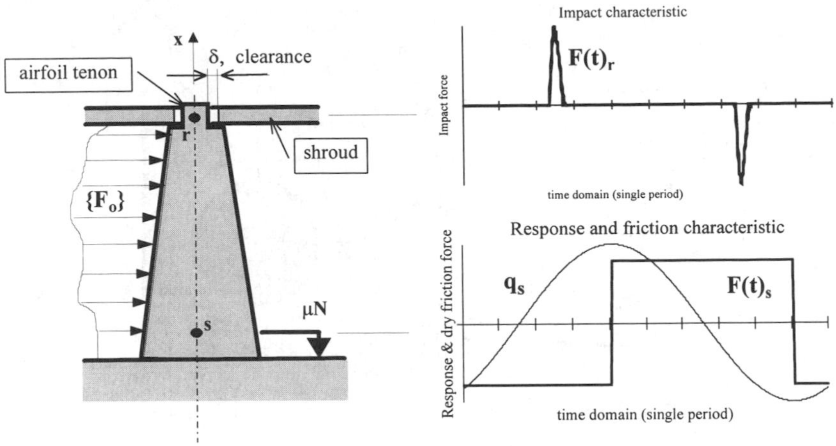

Fig. 1 Non-linear impact $F(t)_r$ and dry friction $F(t)_s$ forces acting on the vibrating blade during single response cyclic

where μ, N, $\dot{q}_s$ denote the friction coefficient, normal load and Cartesian tangential oscillation velocity at the node s on the contact plane, respectively. For the considered rotational speed Ω, the normal load N on the contact can be obtained numerically from the static finite element analysis.

In literature, a general impact force F_r is not given in the analytical form. The characteristic of the impact force has to be evaluated from experimental tests referring to the analysed structure.

4 DEFINITION OF IMPACT FORCES

For the translation nodal degrees of freedom, a non-linear Fricker's definition of the impact force can be applied as (Fricker, 1984)

$$F_r = C_r \sqrt[3/2]{q_r - \delta} \tag{10}$$

where q_r indicates the translation displacement, C_r and δ denote the unknown equivalent local stiffness and initial clearance between the shroud and the airfoil tenon, respectively. However, Moorthy et al. suggested a linear representation of the impact force (Moorthy et al., 1993) by

$$F_r = C_r \left(q_r - \delta\right) \tag{11}$$

that gives better agreements to Monn and Shaw experimental Poincare plot (Moon and Shaw, 1983) than Eq. (10). Considering the typical rectangular or circular tenon geometry, the equivalent normal stiffness C_r of the elastic contact can be determined from the Hertz's theory (Roark and Young, 1983). Applying the Hertz's formulas for a pressed cylinder (radius r) in a cylindrical socket (radius r + δ), the normal deformation q_r is given by

$$q_r = \sqrt[3]{\frac{\delta}{r\left(r+\delta\right)}\left(\frac{3F_r\left(1-\nu^2\right)}{2E}\right)^2} \tag{12}$$

where $E = E_{airfoil} = E_{shroud}$ indicates Young's modulus, $\nu = \nu_{airfoil} = \nu_{shroud}$ Poisson's ratio, F_r normal resultant force, r, δ the tenon radius and clearance, respectively (Fig. 2).

Using Eq. (12), a relation between the contact normal deformation q_r and force F_r is calculated in terms of the clearance δ and presented in Fig. 2. For small clearances up to 0.5 mm, the impact force F_r can be good approximated by a linear spring of the stiffness C_r (Fig. 2). For bigger clearances than 0.5 mm, the impact force F_r should be represented by a non-linear system of two springs (a soft spring between 0 and 20 N and a hard spring between 20 and 200 N, Fig. 2).

5 EXPERIMENTAL VERIFICATION OF THE DEFINITION OF IMPACT FORCES

Vibration tests were performed on a cantilevered beam of the total length of 700 mm and made of the steel 1018. Along the distance of 178 mm, one end of the beam is screwed between two plates (thickness of 20 mm) to the heavy steel frame (detail 1 in Fig. 3; where additional heavy plates mounted on the frame sides are not presented). Hence, the length of the vibrating cantilever beam equals 522 mm, where its thickness and width are equal to 5 and 50 mm, respectively (Fig. 1). The complete experimental set-up with a system of the data acquisition is presented in Fig. 3.

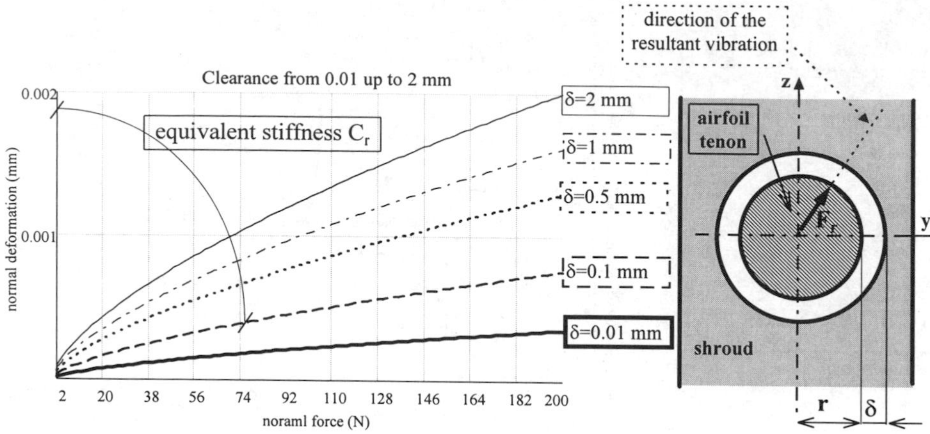

Fig. 2 Relationship between the normal contact deformation q_r and force F_r in terms of the clearance δ (Eq. 12) for the Young's modulus of 2.1×10^5 N/mm^2, Poisson's ratio of 0.3 and radius of 20 mm

A light electric motor of 0.35 kg with a rotating mass eccentric mounted at one end of the revolving shaft is used as the vibration exciter of the cantilevered beam. For getting better amplifications of the excitation, the vibration exciter is attached at the free end of the cantilevered beam. A rotational speed of the electric motor can be varied and measured by the electronic control system designed in CENIDET (National Centre for Research and Technological Development). The measured rotational speed relates directly to the excitation frequency Ω of the cantilevered beam.

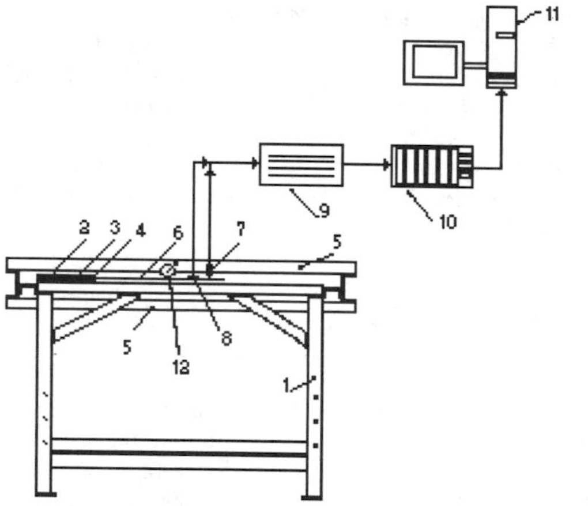

Fig. 3 Experimental set-up with indicated details: the carrying frame (1), assembly lower/upper plates (2/4), the tested beam (3, 6), the support (5) of the mounted stopper (7), strain-gauges for bending measurements (8), power amplifier (9), Hewlett Packard spectrum analyser type 3566A (10), Hewlett Packard PC486 computer (11) and vibration exciter (12)

If oscillations of the vibrating specimen are bigger then the clearance δ (Fig. 4), the impact occurs in the system. In the test set-up, characteristics of the impact force are measured by a load sensor designed in CENIDET and shown in Fig. 4. One side of the load sensor is mounted to the frame (detail 5 in Fig.3). At the other end of the sensor, the screw M4 is attached at which impacts of the cantilever beam occur. Axial deformations of the load sensor are measured by four strain gauges of Measurements Group type EA-06-062AQ-350Ω. The strain gauges are glued on the sensor sides and connected in Wheatstone bridge. The measured signal is transferred to Hewlett Packard spectrum analyser type 3566A (detail 10 in Fig. 3). After attaining a stationary state of the vibration, measured signals are stored on the personal computer (detail 11 in Fig. 3).

The cantilevered beam is stimulated nearly at the fundamental eigenfrequency (ω_1 = 12.95 Hz, ω_2 = 90.3 Hz) with the constant excitation frequency Ω of 12.9 Hz and excitation load amplitude F_O of 4 N. For the clearance δ of 2 mm, the experimental analyses were performed. For stationary states of the vibration, the measured impact forces and its duration are presented by a solid line in Fig. 4.

6 NUMERICAL SIMULATIONS OF THE MEASURED CANTILEVERED BEAM

The beam is made of steel 1018. The beam material is linear elastic with the tension (Young's) modulus of 2.1×10^5 MPa, the shear (Kirchhoff's) modulus of 8.1×10^4 MPa, Poisson's ratio of 0.3 and the density of 7860 kg/m^3. The beam length L, thickness H and width W are equal to 522, 5 and 50 mm, respectively. The ratio between the length (522 mm) and

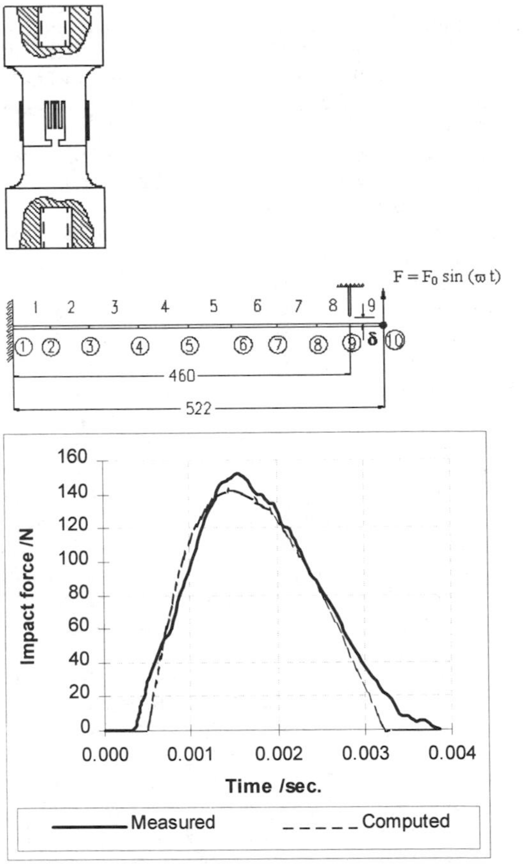

Fig. 4 The load sensor designed in CENIDET (National Centre for Research and Technological Development) for measurements of impact forces, the finite element model of the measured cantilevered beam, the experimental and numerical characteristic of the impact force

depth (5 mm) of the beam is bigger than 104. Thus, shear effects are negligible and the Bernoulli-Euler beam finite element formulation is used for generating a discrete model of the measured beam.

Because the cantilever beam is screwed to the experimental set-up (Fig. 3), the friction damping at the beam clamping occurs during the measurements. For bolted or riveted steel, Newmark and Hall (1982) suggested the damping ratio ξ between 0.03 and 0.05 that are used as equivalent damping ratios ξ (material and friction) in the numerical analyses.

For non-linear vibration analyses, the ADVV computer code is applied. The ADVV computer code was elaborated in CENIDET (Szwedowicz and Sotelo, 1995). The cantilevered beam is represented by nine beam finite elements (Fig. 4). The first two computed natural frequencies of the weak bending vibration are equals 12.95 and 90.3 Hz. The measurement of specimen eigenfrequencies was not made.

In the discrete model, a real contact geometry (elliptical area) between the impacting bodies is simplified to single contact point. Hence, the equivalent stiffness of the load sensor (Fig. 4), referring to the local stiffness of the restraint, is calculated from

$$C_r = \frac{E\,A}{B} \qquad (13)$$

and it is equal to 1410 N/mm, where B, A and E are the depth, minimal area of the cross section and Young's modulus of the load sensor, respectively.

Recommended for impact problems (Moorthy et al. (1993), Bathe and Wilson (1978)), the Newmark numerical integration with the constant acceleration scheme (Newmark, 1959) is used in the numerical analyses. Taking into account smallest damping ratio ξ of 0.03 and smallest size of the applied beam finite element (Szwedowicz, 1995), the stability condition is obtained for the time increment step of 0.5 ms (Szwedowicz and Sotelo, 1995). Because a real damping of the cantilevered beam was not measured, few numerical analyses are performed for different equivalent damping ratios. For the equivalent damping ratio ξ of 0.036, the computational characteristic of the impact force is shown by a dashed line in Fig. 4.

7 FINAL CONCLUSIONS

Using the linear representation of the impact force F_r (Eq. (11) and (13)), a good agreement between the numerical and empirical results are obtained (Fig. 4). The difference between the maximal measured (152 N) and computed (143 N) impact peak equals 6%. The variance between the measured (0.31 ms) and computed (0.28 ms) impact duration is equal to 11%. In general, a qualitative agreement (Fig. 4) is found with a good reliability. On the other hand, some improvements of the test rig should be done for getting a better quantitative relation between the numerical and experimental results (Szwedowicz and Sotelo, 1996).

Due to the high damping ratio ξ of 0.036 (material and friction damping), the participation of higher modes in the characteristic of impact forces (a not regular solid line in Fig. 4) cannot be obtained in the numerical simulation (a dashed line in Fig. 4). Thus, the computed impact force refers mainly to the response due to the fundamental frequency (a dashed line in Fig. 4) and higher modes of the beam do not participate in the built-up of the impact force (differences between the solid and dashed lines in Fig.4).

Using the Hertz contact theory, for clearances bigger than 0.5 mm, the non-linear representation of the impact force was suggested according to the results shown in Fig.2. In the presented experimental results (a solid line, Fig. 4), an increase of the impact force up to 20 N lasts very short. At the end on the impact process, a decreasing of the impact force from 20 N to zero takes more time in comparison with the start-up of the impact (Fig. 4). According to the numerical simulations, the linear representation of the impact force gives a good agreement in relation to the measurements. It seems that a soft stiffening effect in modelling of the impact force (forces between 0 and 40N in Fig.2) can be omitted. Hence, the linear model describing only a strong stiffening effect of the impact force (bigger than 20 N in Fig. 2) is a fundamental aspect of elastic impact problems. Thus, the proposed technique can applied for modelling of vibrations of turbine blades including elastic impact and dry friction effects.

Acknowledgements

This work is partially supported by CONACYT (the National Council for Science and Technology, Mexico) under contract No. 0831P and COSNET (Council of the National System for Technological Education, Mexico) under contract No. 206.95-P.

REFERENCES

Bathe, K. J. & E.L. Wilson 1978. *Numerical Methods in Finite Element Analysis*. New York: Prentice-Hall.

Ewins D.J. & M. Imregun 1984. Vibration Modes of Packeted Bladed Disks, *Journal of Vibration, Acoustic, Stress and Reliability in Design.* 106:175-180.

Fricker, A.J. 1984. The analysis of impacts in vibrating structures containing clearances between components, Proceedings of International Conference on Numerical Methods for Transient and Coupled Problems, Edited by Lewis W., R., Hinton E., Bettes P., Venice.

Griffin, J.H. 1980. Friction Damping of Resonant Stresses in Gas Turbine Engine Airfoils. *Transactions of the ASME, Journal of Eng. Power.* 102:329-333.

Jones, D.I.G. & A. Muszynska 1978. Vibrations of a Compressor Blade with Slip at the Root. *Shock Vibration Bull.* 48:53-61.

Moon, F.C. & S.W. Shaw 1983. Chaotic vibrations of a beam with non-linear boundary conditions. *International Journal of Non-linear Mechanic.* 18:465-477.

Moorthy, R.I.K., A. Kakodkar, H.R. Srirangarajan & S. Suryanarayan 1993. Finite element simulation of chaotic vibrations of a beam with non-linear boundary conditions. *Computers and Structures* 49(4):589-596.

Newmark, N.M. & W.J. Hall 1982. *Earthquake spectra and design*. Earthquake Engineering Research Institute, Berkeley, U.S.A.

Newmark, N.M. 1959. A method of computation for structural dynamics. *Journal of Engineering Mechanics Division, Proc.ASCE.* 85:67-94.

Ostachowicz, W. & D. Szwedowicz 1986. Vibrations of beams with elastic contact. *Computer and Structures.* 22(5):763-771.

Roark, J.R. & C.W. Young 1983. *Formulas for Stress and Strain.* ISBN 0-07-085983-3, Tokyo: McGraw-Hill International Book Company, Kosaido Priniting Co., Japan.

Swain, E., G.M. Chapman & M. Yang 1992. Vibration characteristics of integrally-shrouded blades. *Proceedings of the Institution of Mechanical Engineers, International Conference: Vibrations in Rotating Machinery*, Bath, Published for IMechE by Mechanical Engineering Publications Limited, Septmeber 1992: 73-82

Szwedowicz, D., 1986. *Application of Finite Element Method in Contact Problems of Mechanical Systems* (in Polish), Ph.D. Thesis, Technical University of Gdansk, Gdansk, Poland.

Szwedowicz, D. 1995. Droplet impact modelling in steam turbine blades. *Proceedings of the ASME International Joint Power Generation Conference*, Minneapolis, Minnesota, 8-12 October 1995: Volume 3, PWR-Vol.28, pp.331-340, U.S.A.

Szwedowicz, D. & C. Sotelo November 1995. Visualisation of the numerical model for impacts in mechanical vibrating structures (in Spanish). *Proceeding of the 1st National Congress of Mechanical Systems*, Puebla, Mexico, 15-17 November 1995:1-8

Szwedowicz, D. & C. Sotelo 1997. Updating of impact models for vibrating cantilever beams, The ASME Design Engineering Technical Conference, 16th Biennial Conference on Mechanical Vibration and Noise, Sacramento, California, 14-17 Septemebr 1997, U.S.A., (in issue)

Yang, M.-T., J,.H. Griffin, 1995. Exploring how shroud constraint can affect vibratory response in turbomachinery. *Transactions of the ASME, Journal of Engineering for Gas and Turbines and Power.* 117:198-206

8 Optimisation and design

Modern Practice in Stress and Vibration Analysis, Gilchrist (ed.) © *1997 Balkema, Rotterdam, ISBN 90 5410 896 7*

Approximate static and dynamic reanalysis techniques for structural optimization

P.B. Nair
Department of Mechanical Engineering, University of Southampton, UK

ABSTRACT

This paper presents an approach based on reduced basis approximation concepts for static and dynamic reanalysis of structural systems. In the presented approach, scaling parameters are introduced to increase the range of applicability of local approximation techniques based on Taylor or matrix perturbation series. The terms of the local approximation series are used as basis vectors for constructing an approximation of the perturbed response quantities. The undetermined scalar quantities are then estimated by solving the perturbed equilibrium equations in the reduced basis. This approach was earlier proposed in the context of statics by Kirsch (1991). This paper presents in brief the reanalysis procedure for statics and a new method based on a similar line of approach is proposed for approximate dynamic reanalysis. The method is applied to approximate dynamic reanalysis of a cantilevered beam structure. Preliminary results for this example problem indicate that high quality approximation of the natural frequencies and mode shapes can be obtained for moderate perturbations in the stiffness matrix elements of the order of ±40%.

1. INTRODUCTION

The development of approximate response models of structural systems has been a topic of extensive research over the last few decades. This has been mainly due to the ever increasing requirement of efficiently designing large scale structural systems using variable complexity analytical models. Applications also exist in the area of structural identification involving the reconciliation of finite element models with experimental data.

A detailed review of static reanalysis techniques can be found in Topping (1987). More recent reviews of the field have been presented by Barthelemy and Haftka (1991) and Grandhi (1993). It is important to mention here that many of the approximation concepts reported in the literature are valid only for small perturbations in the structural parameters. The literature review included in this paper is restricted to reanalysis techniques which are valid for moderate to large perturbations in the structural parameters.

Approximate reanalysis techniques can be broadly classified in to global and local methods. Global approximation methods either make use of polynomial regression (response surface analysis) or reduced basis approaches. The reduced basis approaches (see, for example, Kapania and Byun, 1993) are also commonly referred to as model reduction in the literature. These methods essentially seek to approximate the response throughout the design space of interest. In contrast, the local methods are based on Taylor or matrix perturbation series around a nominal design point. Hence the range of applicability is local in nature since the series may not converge for moderate to large perturbations in the structural parameters.

Kirsch (1991, 1995) proposed a method based on reduced basis approximation concepts to static reanalysis. This method can be thought of as an combined approximation technique since it attempts to give global characteristics to the conventional local approximation. It was shown that this approach can be used to compute high quality approximation of the static response quantities for very large perturbations in the structural parameters. More recently, Kirsch and Liu (1997) presented a formulation based on a similar line of approach to static reanalysis of structures undergoing topological modifications. The approach developed in this paper has been primarily motivated by the research of Kirsch.

Very few studies in the literature have approached the structural dynamic reanalysis problem for large perturbations in the structural parameters. High (1990) proposed an iterative modal method to compute the perturbations in the frequencies and mode shapes. This method was implemented in version 66 of MSC NASTRAN. Later studies by Eldred *et. al.* (1992) indicated that difficulties may arise in the convergence of High's method for moderate perturbations. They developed an improved scheme for normalization of the eigenvector perturbations in order to improve the convergence properties.

An interesting approach based on interpreting the eigen parameter perturbation equations as differential equations in terms of the perturbation parameters was proposed by Inamura (1988). It was shown via demonstration examples that this procedure could lead to improvements over the conventional local approximation. Pritchard and Adelman (1991) developed a similar procedure using the sensitivity equations of the eigenvalues and eigenvectors.

An iterative procedure using the nonlinear form of the eigenproblem perturbation equations was developed by Eldred et al. (1992). It was shown that this method converges to the exact solution for moderate to large perturbations in the stiffness matrix of the order of 150%. An exact method based on the block Lanczos algorithm was proposed by Carey *et. al.* (1994). Even though both these procedures can provide exact results, the computational effort involved is substantial compared to conventional techniques based on first order Taylor or matrix perturbation series approximation.

More recently, Balmes (1996) presented a novel approach in which the finite element model is represented as a parametric family of reduced order. The full order order finite element model is reduced using a transformation matrix composed of Ritz vectors evaluated at different points in the design space. Excellent results were obtained for approximate static and dynamic reanalysis of a cantilevered box beam structure.

The approach developed in this paper is in spirit similar to the approach of Balmes (1996). However the present method differs in the choice of basis vectors used to construct the transformation matrix. In the present approach, the basis vectors are chosen to be an implicit function of the parametric perturbations. In particular, the terms of the conventional Taylor or matrix perturbation series are chosen as basis vectors. Hence the present approach can also be viewed as an improved local approximation procedure. In contrast, Balmes's method uses a constant transformation matrix which is invariant with the parametric perturbations. Furthermore, the approach presented in this paper aims at building a reduced basis approximation of each eigenmode independently.

The long term objective of the present study is to develop techniques which can be used for obtaining high quality approximation of the dynamic response quantities for moderate to large perturbations in the structural parameters.

This paper is organized as follows. Section 2 briefly describes Kirsch's formulation for approximate static reanalysis. Based on Kirsch's method for statics, the extension of the theory to approximate dynamic reanalysis is proposed. Some comments on the computational aspects of the proposed procedure and analogies with existing approximation concepts are briefly discussed in section 3. Section 4 presents results for approximate dynamic reanalysis of a cantilevered beam. Approximations are sought for the natural frequencies and mode shapes for perturbations in the flexural rigidities. The effects of both global and local perturbations in the structural parameters on the proposed procedure is studied. Results indicate that a reliable approximation of the natural frequencies and mode shapes can be obtained for moderate perturbations in the stiffness matrix of the order of ±40%. Comparison studies with the conventional first order Taylor series approximation shows that the accuracy has been considerably improved with a relatively small computational effort. Section 5 summarizes the present work and future areas of investigation are outlined.

2. THEORETICAL DEVELOPMENT

The equations of motion of a multi-degree of freedom linear structural system can be written as

$$\mathbf{M}\ddot{\mathbf{x}} + \mathbf{C}\dot{\mathbf{x}} + \mathbf{K}\mathbf{x} = \mathbf{F} \tag{1}$$

where $\mathbf{M}$, $\mathbf{C}$ and $\mathbf{K} \in \Re^{n\times n}$ are the structural mass, damping and stiffness matrices respectively, $\mathbf{x}$ is the vector of displacements corresponding to the analytical degrees of freedom and $\mathbf{F}$ is the vector of external forces.

2.1 Approximate Static Reanalysis

Consider the case of static analysis, wherein the derivatives of the displacement vector $\mathbf{x}$ with respect to time are zero. For this case, if the stiffness matrix is perturbed by $\Delta\mathbf{K}$, the displacement at the perturbed design point can be expressed as a matrix series of the form

$$\mathbf{x} = (\mathbf{I} - \mathbf{B} + \mathbf{B}^2 - \ldots)\mathbf{x}^o \quad (2)$$

where $\mathbf{B} = \mathbf{K}^{o-1}\Delta\mathbf{K}$. $\mathbf{K}^o$ and $\mathbf{x}^o$ are the stiffness matrix and displacement vector at the nominal design point respectively. Note here that the above series will converge to the exact value of the perturbed displacement vector only when $(\mathbf{K}^{o-1}\Delta\mathbf{K})^k \to 0$ as $k \to \infty$. In general, for moderate to large perturbations in the stiffness matrix, the series will not converge.

Kirsch (1991) proposed the use of terms of the above series as high quality basis vectors for constructing an approximation to the perturbed displacement vector. It was shown via various demonstration examples that high quality approximation of the static response quantities can be obtained for very large perturbations in the stiffness matrix of the order of 900%. In a sequel paper, Kirsch (1995) showed that the use of Taylor series terms can lead to similar results. More implementation specific details including results for various benchmark structures can be found in the afore mentioned references.

In the next section, extension of Kirsch's method to approximate dynamic reanalysis is described. In particular, it is sought to develop a procedure for approximation of the eigenvalues and eigenvectors of perturbed eigensystems on a similar line of approach.

2.2 Approximate Dynamic Reanalysis

Typically, modal analysis is used for computing the dynamic response of large scale finite element models of structural systems. This procedure essentially involves computing the first few natural frequencies and the corresponding mode shapes and transforming the original equations of motion to the modal coordinates. This permits dynamic response analysis of the finite element model in a computationally efficient fashion. This paper is restricted to approximate reanalysis of the eigen parameters for perturbations in the stiffness and mass matrices. Once the eigen parameters have been approximated for the perturbed system, computation of the dynamic response for arbitrary time varying loading conditions is relatively straight forward.

The free vibration undamped natural frequencies and mode shapes of a structural system, whose equilibrium equations can be expressed in the form of equation (1), can be computed by solving the algebraic eigenvalue problem posed below.

$$\mathbf{K}\phi = \lambda\mathbf{M}\phi \quad (3)$$

where ϕ denotes the mode shape of the structural response and λ is the eigenvalue which is the square of the natural frequency.

The original/baseline system matrices and the response quantities are denoted by the superscript o. The set of structural parameters or design variables are denoted by the vector $\mathbf{X}$. Let $\Delta\mathbf{X}$ denote the perturbation in the structural parameters. Then the corresponding perturbation in the system matrices and the response quantities can be expressed as

$$\begin{aligned}
\lambda_i &= \lambda_i{}^o + \Delta\lambda_i \\
\phi_i &= \phi_i^o + \Delta\phi_i \\
\mathbf{K} &= \mathbf{K}^o + \Delta\mathbf{K} \\
\mathbf{M} &= \mathbf{M}^o + \Delta\mathbf{M}
\end{aligned}$$

The index i is used to denote the eigenmode number. Typically, the eigenvalue and eigenvector perturbations are calculated using first order sensitivity information, i.e., the perturbation in the eigenvalue and eigenvector for mode i can be expressed as

$$\Delta\lambda_i = \sum_{j=1}^{p} \frac{\partial\lambda_i}{\partial x_j}\Delta x_j \quad (4)$$

$$\Delta\phi_i = \sum_{j=1}^{p} \frac{\partial\phi_i}{\partial x_j}\Delta x_j \quad (5)$$

where $\frac{\partial\lambda_i}{\partial x_j}$ and $\frac{\partial\phi_i}{\partial x_j}$ are the sensitivities of the eigenvalues and eigenvectors with respect to the structural parameters denoted by $\mathbf{X} = \{x_1, x_2, \ldots, x_p\}$. For many cases of practical interest, the eigenvalue and eigenvector sensitivities may not be easy to evaluate. Hence it may be more convenient to compute the eigenvalue and eigenvector perturbations using a first order matrix perturbation series. Note here that the first order approximation of eigen parameters using matrix perturbation series are equivalent to the first order Taylor series terms. A first order approximation of the eigenvalues and eigenvectors using a matrix perturbation approach (Brandon, 1990) can be written as

$$\Delta\lambda_i = \frac{\phi_i^{oT}[\Delta\mathbf{K} - \lambda_i{}^o\Delta\mathbf{M}]\phi_i^o}{\phi_i^{oT}\mathbf{M}\phi_i^o} \quad (6)$$

$$\Delta\phi_i = \sum_{j=1,j\neq i}^{n} \alpha_{ij}\phi_j^o \qquad (7)$$

where

$$\alpha_{ij} = \frac{\phi_j^{oT}[\Delta\mathbf{K} - \lambda_i{}^o\Delta\mathbf{M}]\phi_i^o}{(\lambda_i{}^o - \lambda_j{}^o)\phi_i^{oT}\mathbf{M}\phi_i^o} \qquad (8)$$

Using these equations, a first order approximation of the eigenvalues and eigenvectors of the perturbed system can be calculated. Note here that in order to compute the perturbation in the eigenvector, a modal summation approach is used which requires all the eigenvectors of the original/baseline system to be evaluated. This can prove to be computationally prohibitive for large scale systems. In order to circumvent this requirement, a truncated set of baseline eigenvectors could be used. This could potentially lead to reduced accuracy for large perturbations in the structural parameters. The other alternative could be to make use of iterative schemes (see for example; Zhang and Zerva, 1997) or Nelson's method (Nelson, 1976) in order to compute the eigenvector perturbations in a computationally efficient fashion.

The approximation procedure developed in the present study is based on the proposition stated below.

Proposition : The eigenvector of the perturbed system can be approximated in the subspace spanned by $\phi_i{}^o$ and $\Delta\phi_i$. i.e., an approximation to the perturbed eigenvector can be written as

$$\hat{\phi}_i = \zeta_1\phi_i^o + \zeta_2\Delta\phi_i \qquad (9)$$

where ζ_1 and ζ_2 are the undetermined scalar quantities in the approximate representation of the perturbed eigenvector. The assumption implicit in the proposition is that even for moderate to large perturbations in the structural parameters, the first order approximation yields a $\Delta\phi_i$ vector which gives a reasonable indication of direction of change of the baseline eigenvector, although the magnitude of change may be erroneous. It can also be seen that for $\zeta_1 = \zeta_2 = 1$, the proposition reduces to the conventional first order approximation. Equation (9) can be expressed in matrix form as

$$\hat{\phi}_i = \mathbf{T}\mathrm{Z} \qquad (10)$$

where $\mathbf{T} = \{\phi_i^o, \Delta\phi_i\} \in \Re^{n\times 2}$ and $\mathrm{Z}^T = \{\zeta_1, \zeta_2\} \in \Re^{1\times 2}$

Substituting equation (10) in to equation (3) and premultiplying by $\mathbf{T}^T$, the resulting set of equations can be expressed as

$$\mathbf{K}_T\mathrm{Z} = \lambda\mathbf{M}_T\mathrm{Z} \qquad (11)$$

where $\mathbf{K}_T = \mathbf{T}^T\mathbf{K}\mathbf{T} \in \Re^{2\times 2}$ and $\mathbf{M}_T = \mathbf{T}^T\mathbf{M}\mathbf{T} \in \Re^{2\times 2}$ are the reduced stiffness and mass matrices. Hence using the present approach, the original $n \times n$ eigensystem is represented by a reduced 2×2 eigensystem for each eigenmode to be approximated.

A non-trivial solution to Z can be obtained only when λ is an eigenvalue of the matrix pair $(\mathbf{K}_T, \mathbf{M}_T)$. Hence an approximation to the eigenvalue of the perturbed system $(\hat{\lambda}_i)$ can be computed by solving for the roots of the quadratic given below.

$$a\hat{\lambda}_i^2 + b\hat{\lambda}_i + c = 0 \qquad (12)$$

where $a = m_{11}m_{22} - m_{12}^2$, $b = 2k_{12}m_{12} - k_{11}m_{22} - m_{11}k_{22}$ and $c = k_{11}k_{22} - k_{12}^2$, k_{ij} and m_{ij} are the elements of the reduced stiffness and mass matrices ($\mathbf{K}_T$ and $\mathbf{M}_T$) respectively. The elements of $\mathbf{K}_T$ and $\mathbf{M}_T$ are given below as

$$\begin{aligned} k_{11} &= \phi_i^{oT}\mathbf{K}\phi_i^o & m_{11} &= \phi_i^{oT}\mathbf{M}\phi_i^o \\ k_{12} &= \phi_i^{oT}\mathbf{K}\Delta\phi_i & m_{12} &= \phi_i^{oT}\mathbf{M}\Delta\phi_i \\ k_{22} &= \Delta\phi_i^T\mathbf{K}\Delta\phi_i & m_{22} &= \Delta\phi_i^T\mathbf{M}\Delta\phi_i \end{aligned}$$

Solution of the above quadratic give two values for the perturbed eigenvalue. Since the transformed matrices $\mathbf{K}_T$ and $\mathbf{M}_T$ are real and symmetric, the roots of equation (12) will be real. Now the question arises regarding which root to choose as the best approximation for the perturbed eigenvector. For the demonstration example considered, it was confirmed via numerical experiments that the root with the lowest magnitude gives the best approximation. The mathematical proof of this is involved and is beyond the scope of this paper. Once an approximation to the eigenvalue has been computed, the approximate eigenvector can be evaluated by calculating the values of c_1 and c_2 (i.e., the eigenvectors of the reduced eigensystem given by equation (11)) and using equation (9).

3. COMMENTS

It can be observed from the formulation that an approximation to the eigenvalues and eigenvectors of the perturbed system can be calculated by solving for the roots of an quadratic for each eigenmode of interest. The coefficients of the quadratic equation can be easily calculated after the first order approximation of the perturbed eigenvector is computed. Hence the proposed procedure involves only a few additional computations when compared to the conventional first order local approximation.

For many large scale structures, the response is dominated by the first few eigenmodes (typically 10-20 even for a structure with 10,000 degrees of freedom). Hence using the proposed procedure, the perturbations in the eigenvalues and eigenvectors can be approximated by solving a few number of quadratic equations which could lead to substantial savings in the computational time required for dynamic response synthesis.

In order to improve the accuracy of the present procedure, it may be desirable to make use of second order approximation terms (i.e, three basis vectors) at the cost of increased computations. It will be shown in the subsequent section that using the first order terms alone, a reliable approximation of the perturbed frequencies and mode shapes can be computed for moderate perturbations in the structural parameters.

Variants of the proposed approach, for example; using a common set of basis vectors for each eigenmode and solving a single reduced eigensystem also merits consideration. This could potentially lead to better accuracy in the approximation of eigen parameters for many modally complex structural systems. However, due to constraints on data presentation, results for these cases will not be presented in the present paper.

It is well known that the use of Rayleigh's quotient yields a better approximation to the lowest natural frequency as compared to the conventional first order approximation. The lowest eigenvalue is typically approximated using the equation

$$\lambda_{rqa} = \frac{\phi^{oT}\mathbf{K}\phi^{o}}{\phi^{oT}\mathbf{M}\phi^{o}} \tag{13}$$

The assumption made here is that the mode shapes are invariant to the parametric perturbations. Earlier studies (see for example; Canfield, 1990) have conclusively shown that the modal strain and kinetic energy (i.e., the numerator and denominator of Rayleigh's quotient) can be used as intervening variables to approximate the natural frequency with better accuracy as compared to the local first order approximation.

It can be checked that if $\Delta\phi_i$ is considered to be a very small quantity, the approximation for the first eigenmode using the present procedure tends to the Rayleigh quotient approximation.

4. DEMONSTRATION EXAMPLE, RESULTS AND DISCUSSION

The proposed procedure is applied to approximate structural dynamic reanalysis of a cantilevered beam structure. The baseline values of the structural parameters are taken as : flexural rigidity $EI = 1.286 \times 10^4 Nm^2$, mass per unit length $m = 2.73kg/m$ and length of the beam $L = 2.5m$. The beam is modeled using five finite elements with each node constrained to have only two degrees of freedom (translational and rotational). For this example, the structural parameters which are perturbed correspond to the flexural rigidities of the five elements, i.e, only perturbation in the stiffness matrix is considered.

Results are presented for two cases. In the first case, the approximate natural frequencies and mode shapes are evaluated for simultaneous (global) perturbations in the structural parameters. In the second case, the effect of local perturbations on the approximation procedure is studied.

In order to evaluate the accuracy of the approximation, two error indices are defined. The first error index is the Frobenius norm of the difference between the exact mode shape (obtained using exact reanalysis) and the approximate mode shape using the present approach. The mode shape error index for eigenmode i is defined below as

$$MSE_i = ||\hat{\phi}_i - (\phi_i)_{exact}||_f \tag{14}$$

where $||.||_f$ denotes the Frobenius norm of the vector $\{.\}$. The second index, henceforth referred to as FE_i is the percentage error in approximation of the natural frequency for eigenmode i.

Case 1 : For this case five different parameter sets were considered with perturbations in the flexural rigidities ranging from ±10% to ±50%. The percentage perturbations in the flexural rigidities for the five parameter sets are given in table 1.

Table 1: **Percentage Perturbation in Flexural Rigidities of the Elements for the Parameter Sets**

Parameter	Perturbation in Parameters(%)				
Set	1	2	3	4	5
PS1	+5	+8	-9	+10	-5
PS2	+15	-18	-19	+20	-12
PS3	+15	-25	+30	-27	+22
PS4	-35	-25	+25	-40	+37
PS5	+40	-50	+45	-30	+47

For each case, the flexural rigidities of the five elements were perturbed from the baseline values and the approximate frequencies and mode shapes were evaluated using the proposed procedure. The approximate results are compared with results obtained via exact eigensolution of the perturbed system. Table 2 summarizes the results for all the parameter sets. The results using the present approach is compared with those obtained a first order Taylor series approximation denoted by **TS1**. Results are presented only for approximation of the first three eigenmodes. The accuracy of the approximation is evaluated using the two error indices defined earlier.

It can be seen from the results that excellent improvements have been obtained over the conventional first order approximation for moderate perturbations in the structural parameters of the order of $\pm 30\%$. Deterioration in the approximation can be observed only when the perturbations approach the order of $\pm 50\%$. It can also be seen that for the range of perturbations considered in this demonstration example, the errors in approximation using first order Taylor series are substantial.

The mode shape error index defined in this study does not give a good indication of the accuracy of the approximation. In general, it was found that the mode shape can be approximated with better precision as compared to the natural frequency. Figure 1 compares the approximate and exact mode shape of the third eigenmode for parameter set PS5. Note here that the transverse displacement and rotation quantities alternate in the mode shape vector. It can be seen from the figure that using the present approach, the first order approximation of the perturbed eigenvector has been improved substantially.

Case 2 : The effect of perturbations in the flexural rigidity of element 3 on approximation of the natural frequencies and the mode shapes are studied. Perturbations in the flexural rigidity of this element is studied in the range of $10 - 100\%$.

Figure 2 depicts the variation of errors in approximation of the first three natural frequencies for increasing perturbation in the flexural rigidity of element 3. It can be seen from the figure that even for large local perturbation in the flexural rigidity of the order of 100%, the maximum error in approximation of the first three frequencies are of the order of 5%. It can also be noted that the maximum error occurs in approximation of the second eigenmode. Numerical experiments on cases involving perturbation of other local parameters

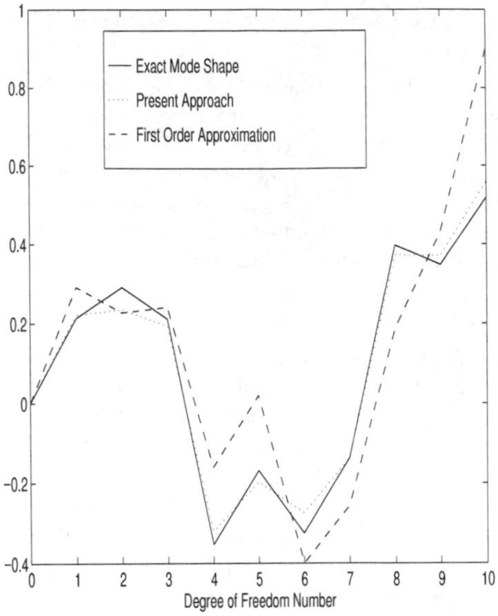

Figure 1: **Comparison Between the Approximate and Exact Mode Shape of the Third Eigenmode for Parameter Set PS5**

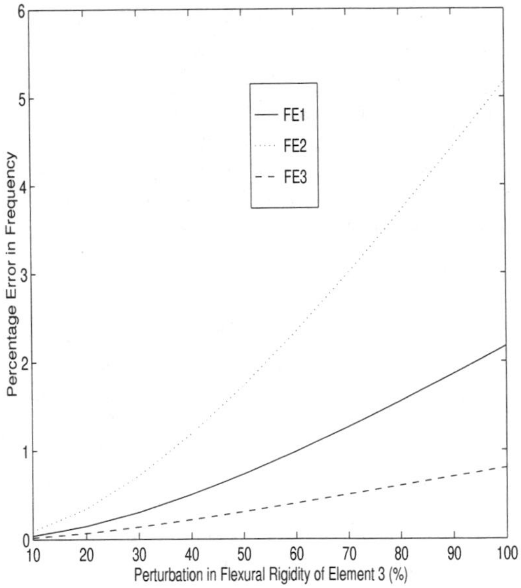

Figure 2: **Variation of Frequency Errors with Local Perturbation in the Structural Parameters**

indicated that the eigenmode which is difficult to approximate may change. It was found that when

Table 2: **Comparison of Results using the Present Approach with Taylor Series Approximation for the Five Parameter Sets**
*Negative Eigenvalue Approximation

Parameter Set	Method	FE1 (%)	FE2 (%)	FE3 (%)	MSE1	MSE2	MSE3
PS1	Present Approach	0.11	0.31	0.17	2.9×10^{-5}	4.9×10^{-4}	6.7×10^{-4}
	TS1	4.27	1.39	-3.16	1.2×10^{-4}	1.6×10^{-3}	8.8×10^{-3}
PS2	Present Approach	1.09	1.05	0.9	1.6×10^{-4}	1.9×10^{-3}	9.5×10^{-3}
	TS1	34.3	16.8	1.4	1.2×10^{-2}	6.1×10^{-3}	9.6×10^{-2}
PS3	Present Approach	2.01	2.86	2.03	2×10^{-4}	5.3×10^{-3}	5.7×10^{-3}
	TS1	28.5	10.86	19.8	1.2×10^{-2}	1.6×10^{-2}	0.12
PS4	Present Approach	1.85	4.74	3.51	4.7×10^{-4}	6.9×10^{-3}	9×10^{3}
	TS1	*	-3.92	3.7	5.2×10^{-3}	0.05	0.26
PS5	Present Approach	9.78	5.82	5.15	1×10^{-3}	8×10^{-3}	0.01
	TS1	69.8	27.97	33.23	0.09	0.05	0.3

the flexural rigidity of element 3 is perturbed by 150%, the percentage errors in approximation of the first three natural frequencies were 3.9, 9.2 and 1.3 respectively. Hence it can be concluded that for this particular example, high quality approximation of the first eigenmode can be obtained for large local perturbations in the structural parameters.

5. CONCLUDING REMARKS

An improved first order approximation procedure for reanalysis of eigenvalues and eigenvectors of modified structural systems was proposed. It has been shown via a simple demonstration example that the present approach can be used to estimate a reliable approximation of the natural frequencies and mode shapes for simultaneous perturbation in the structural parameters of the order of $\pm 40\%$. It was also demonstrated that the proposed procedure can be used for arriving at high quality approximation of the eigen parameters for large local perturbations in the structural parameters.

It is expected that the present formulation may find applications in the area of structural optimization and identification. It is important to note here that extension of the present approach to approximate eigensensitivity is relatively straight forward and could lead to computationally efficient procedures for structural design with dynamic response constraints.

In the form presented in this paper, the formulations lack mathematical rigor. It would be more useful if a formal theoretical background could be established which would enable one to study the numerical characteristics of the approximation procedure and possibly arrive at bounds on the estimates of the perturbed eigen parameters.

Further studies are also required to examine issues related to the effect of mode swapping on the proposed approximation procedure. It is expected that for modally complex structures, moderate perturbations in the structural parameters could potentially lead to switching of the eigenmodes. It may be required to make use of more basis vectors in order to capture the mode switch accurately.

ACKNOWLEDGMENTS

This work was partly carried out during the author's stay at the Indian Institute of Technology, Bombay as a postgraduate student funded by a Government of India Scholarship. The author would like to thank Prof. P. M. Mujumdar at IIT Bombay for valuable discussions and suggestions on this topic and for his encouragement. The author would also like to acknowledge the Faculty of Engineering and Applied Sciences at the University of Southampton for financial and computational support towards this work.

REFERENCES

Abu Kassim, A. M., and Topping, B. H. V., 1987, "Static Reanalysis of Structures : A Review," *Journal of Structural Engineering*, ASCE, Vol. 113, pp. 1029-1045.

Balmes, E., 1996, "Parametric Families of Reduced Finite Element Models. Theory and Applications," *Mechanical Systems and Signal Processing*, Vol. 10, pp. 381-394.

Barthelemy, J. -F. M., and Haftka, R. T., 1991, "Recent Advances in Approximation Concepts for Optimum Structural Design," *Proceedings of NATO/DFG ASI on Optimization of Large Structural Systems*, Berchtesgaden, Germany, pp. 235-256.

Brandon, J.A., 1990, "*Strategies for Structural Dynamic Modification*," Wiley, New York.

Canfield, R. A., 1990, "High Quality Approximation of Eigenvalues in Structural Optimization," *AIAA Journal*, Vol. 28, pp. 1116-1122.

Carey, C. M. M., and Golub, G. H., and Law, K. H., 1994, "A Lanczos-Based Method for Structural Dynamic Reanalysis Problems," *International Journal of Numerical Methods in Engineering*, Vol. 37, pp. 2857-2883.

Eldred, M. S., Lerner, P. B., and Anderson, W. J., 1992, "Higher Order Eigenpair Perturbations," *AIAA Journal*, Vol. 30, pp. 1870-1876.

Grandhi, R.V., 1993, "Structural Optimization with Frequency Constraints - A Review," *AIAA Journal*, Vo. 31, pp. 2296-2303.

High, G. D., 1990, "An Iterative Method for Eigenvector Derivatives," *Proceedings of 1990 MSC World Users Conference*, Paper 17, Los Angeles, CA.

Inamura, T., 1988, "Eigenvalue Reanalysis By Improved Perturbations," *International Journal of Numerical Methods in Engineering*, Vol. 26, pp. 167-181.

Kapania, R. K., and Byun, C., 1993, "Reduction Methods based on Eigenvectors and Ritz Vectors for Nonlinear Transient Analysis," *Computational Mechanics*, Vol.11, pp. 65-82.

Kirsch, U., 1991, "Reduced Basis Approximation of Structural Displacements for Optimal Design," *AIAA Journal*, Vol. 29, pp. 1751-1758.

Kirsch, U., 1995, "Improved Stiffness-Based First-Order Approximations for Structural Optimization," *AIAA Journal*, Vol. 29, pp. 143-150.

Kirsch, U., and Liu, S., 1997, "Structural Reanalysis for General Layout Modifications," *AIAA Journal*, Vol. 35, pp. 382-388.

Nelson, R. B., 1976, "Simplified Calculation of Eigenvector Derivatives," *AIAA Journal*, Vol. 14, pp. 1201-1205.

Pritchard, J. I., and Adelman, H. M., 1991, "Differential Equation Based Method for Accurate Modal Approximations," *AIAA Journal*, Vol. 29, pp. 484-486.

Zhang, O., and Zerva, A., 1997, "Accelerated Iterative Procedure for Calculating Eigenvector Derivatives," *AIAA Journal*, Vol. 35, pp. 340-348.

Modern Practice in Stress and Vibration Analysis, Gilchrist (ed.) © *1997 Balkema, Rotterdam, ISBN 90 5410 896 7*

Development of a finite element software design aid for automotive rear-view mirrors

B.B.McCarthy & P.E.McHugh
Department of Mechanical Engineering, University College, Galway, Ireland
P.M.Heslin
Donnelly Mirrors Ltd, Naas, Co. Kildare, Ireland
M.P.O'Grady
Donnelly Corporation, Holland, Mich., USA

ABSTRACT: All automotive mirrors undergo very strict vibration tests before being installed in an automobile. The aim of this project was to use the finite element method to predict the vibration performance of a mirror prior to the tool being manufactured, thus reducing or eliminating all expensive and time consuming modifications to the tool. The finite element predictions were compared with equivalent experimental results. Good correlation was achieved between the predicted and measured natural frequencies of the mirrors. An investigation of the damping properties of the materials was carried out in order to predict the magnitude of displacement of the mirror. It was concluded that the most accurate way to model damping was using the damping ratio. Damping ratios of certain models were calculated and used to perform harmonic analyses. In certain cases good correlation was achieved between the predicted and measured displacements.

1 INTRODUCTION

Prior to an automobile rear-view mirror being launched onto the market it has to undergo an extensive series of vibration tests. The aim of this project is to use the finite element method to predict the vibration performance of a mirror. If the simulation does not conform to the vibration specifications laid down by the customer then the design of the mirror can be modified prior to the tool being manufactured. This will lead to a reduction in expensive and time consuming testing and tool modifications. As a result of this it is hoped that the lead time from concept to production will be reduced.

Initially a mirror is designed using a 3D CAD package. The CAD data is then transferred, via IGES, to the ANSYS finite element package. A finite element model is then created. This model is meshed in the appropriate manner and a modal analysis is carried out. A modal analysis yields the natural frequencies of the structure. Following this a harmonic analysis is carried out yielding the amplitude of displacement over a range of frequencies.

Analyses have yielded an accurate prediction of the natural frequencies when they are compared with experimental results. However in order to carry out a harmonic analysis the damping behaviour of the model has to be known. Initially it was decided to investigate the damping properties of the materials from which the mirrors are manufactured. Next it was decided to look at the overall damping characteristics of the mirror.

2 MODELS

All components are modelled in a 3D CAD package. In order for a dynamic analysis to take place they have to be modelled within ANSYS. The models are exported from the CAD package to an IGES file and then imported into ANSYS. When the models are imported into ANSYS surfaces often become untrimmed. Due to this the models that are imported into ANSYS cannot be meshed directly. A number of Boolean operations have to be carried out in order for the model to be suitable for meshing. Once this is done the model can be meshed.

The following models were created and meshed

Bracket and 80g block
Bracket and 315g block
Bracket and toggle
Bracket, case and toggle
Bracket and complete mirror assembly

Figures 1 and 2 show the bracket and 80g block and the bracket and complete mirror assembly

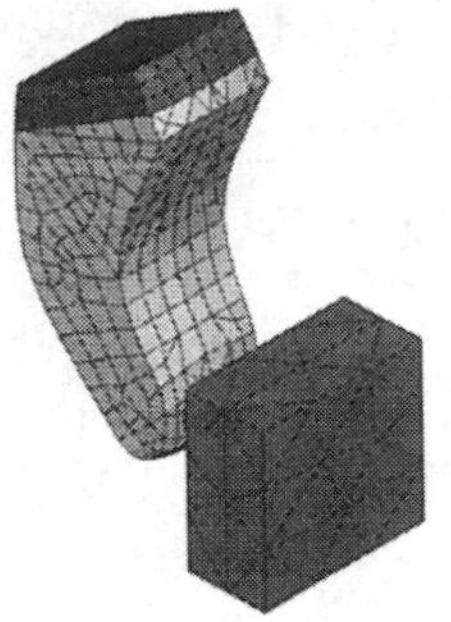

Figure 1. Bracket and 80g block

respectively. They were modelled using Solid and Shell elements.

The metal block was used to simulate the mass of the case, toggle and prism at the beginning of the project as it was simpler to model.

Generally in European automobiles when the complete mirror is set in position in an automobile a die cast metal button is attached to the windscreen with an adhesive. The bracket is then attached to the button with an interference fit. Instead of modelling this clip region, the geometry of which is quite complex, it was decided to experimentally determine the equivalent properties of the region. The clip area was modelled as a volume with these properties. When the modal analysis was performed displacement constraints were applied to all nodes of the clip region that are in contact with the windscreen.

3. ASSEMBLY EXPERIMENTS

In order to evaluate the success of the ANSYS predictions it was necessary to compare them to experimental results. A constant acceleration test was used to determine the natural frequencies. The test was run in a predetermined direction over a range of frequencies. The test yields the natural frequencies and the direction of displacement.

In order to measure the displacement it was decided to place an accelerometer at a certain point on the structure and place a master degree of freedom at the corresponding point on the model. In this way the model would correspond to the experimental set-up.

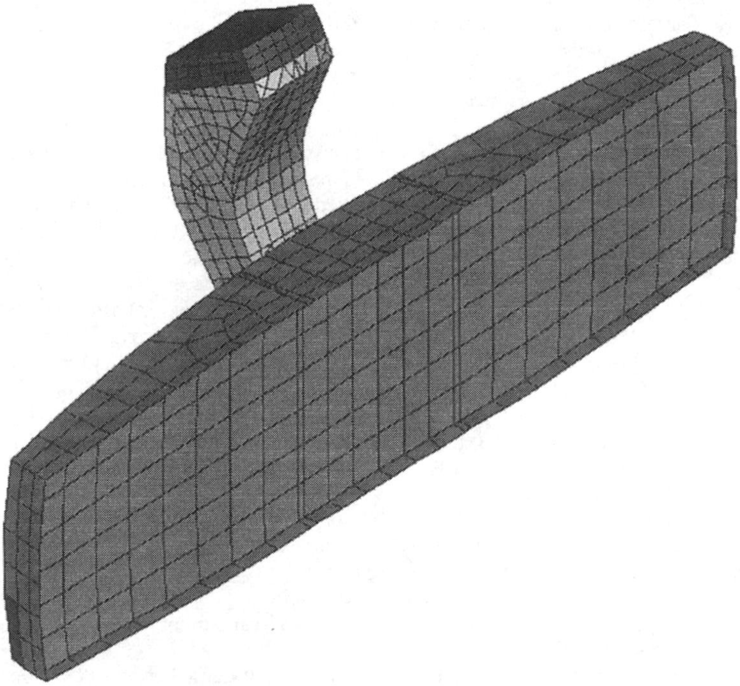

Figure 2. Bracket and complete mirror assembly

Table 1. Experimentally determined natural frequencies V natural frequencies determined by ANSYS for bracket and 80g block.

Experimental (Hz)	Direction	ANSYS (Hz)	Direction
123	Y	114.14	Y
180	Z	159.78	Z
		302.59	Torsional
345	Z	345.43	Z
		556.39	Torsional

Table 2. Experimentally determined natural frequencies V natural frequencies determined by ANSYS for bracket and 315g block.

Experimental (Hz)	Direction	ANSYS (Hz)	Direction
52.5	Y	52.9	Y
77.5	Z	76.48	Z
		90.66	Torsional
155	X	152.39	X
		187.98	Y

4. MODAL ANALYSES

Once the models were created modal analysis were carried out to determine the natural frequencies. These natural frequencies were then compared with the experimentally determined natural frequencies. Tables 1 and 2 show the experimental natural frequencies being compared with the natural frequencies determined by ANSYS. Excellent correlation has been achieved between the experimental and ANSYS determinations.

5. HARMONIC ANALYSIS

The next stage of the project was to predict the displacement that was experienced at the natural frequency. In order to do this the theory of damping was studied.

5.1 Theory of damping

For a single degree of freedom system the damping ratio is defined as the amount of damping that a system experiences divided by the critical damping of the system and is defined by equation 1

$$\zeta = c/c_{cr} \tag{1}$$

where ζ is the damping ratio, c is the damping coefficient and c_{cr} is the critical damping.

The logarithmic decrement describes the relationship that successive oscillation amplitudes have with one another in a viscously damped harmonic system, as shown in figure 3, and is defined in equation 2

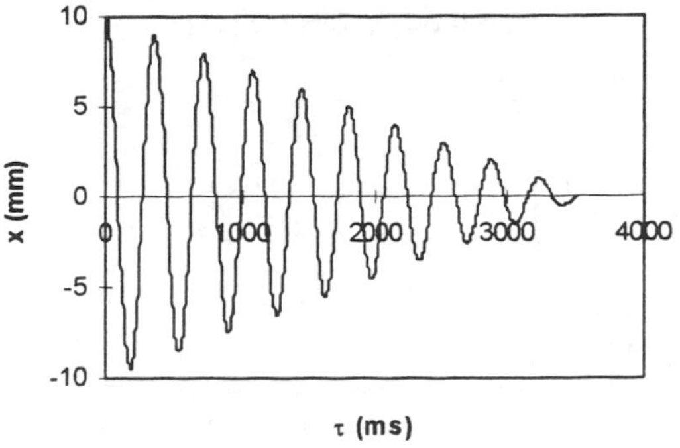

Figure 3. Response of viscously damped system

$$\xi = 1/n\,(\ln\,(X_0/X_n)) \tag{2}$$

where ξ is the logarithmic decrement, n is the number of complete cycles that the system has experienced, X_0 is the initial displacement of the system and X_n is the displacement after n cycles.

The logarithmic decrement and damping ratio can be related by equation 3

$$\xi = c/2m(2\pi/\omega_d) = 2\pi\zeta/(1-\zeta^2)^{1/2} \tag{3}$$

The damped natural frequency, ω_d, and the undamped natural frequency, ω_n, are related to one another through the dimensionless damping ratio, ζ, the viscous damped natural frequency always being less than the undamped. Equation 4 defines the relationship

$$\omega_d = \omega_n\,(1-\zeta^2)^{1/2} \tag{4}$$

5.2 Damping in ANSYS

Damping can be modelled in ANSYS in any of the following ways, the ANSYS commands are in brackets,

Rayleigh constants (ALPHAD, BETAD)
Material dependent damping (MP,DAMP)
Constant damping ratio, (DMPRAT)
Modal damping (MDAMP)

It may also be modelled using a combination of the above methods.

Rayleigh constants are used when a system is proportionately damped. This means that the damping of the system can be related to the mass and stiffness by the equation 5

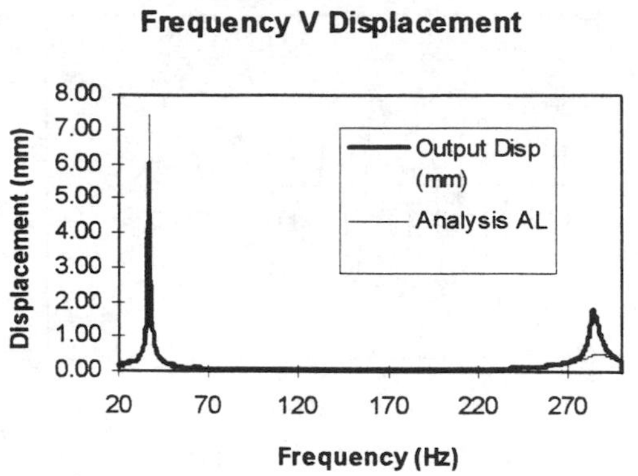

Figure 4. POM beam No.1, run No. 24, mass of accelerometer = 0.2g and analysis AL ($\alpha = 2.166427$ and $\beta = 3.12301*10^{-5}$)

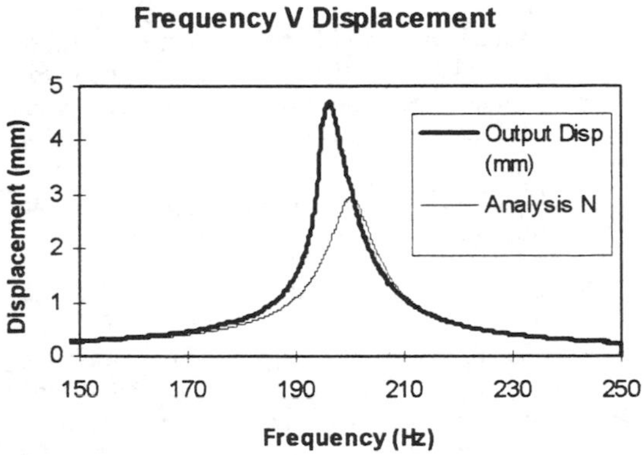

Figure 5. POM beam No.2, run No. 60, mass of accelerometer = 0.2g and analysis N ($\alpha = 2.166427$ and $\beta = 3.12301*10^{-5}$)

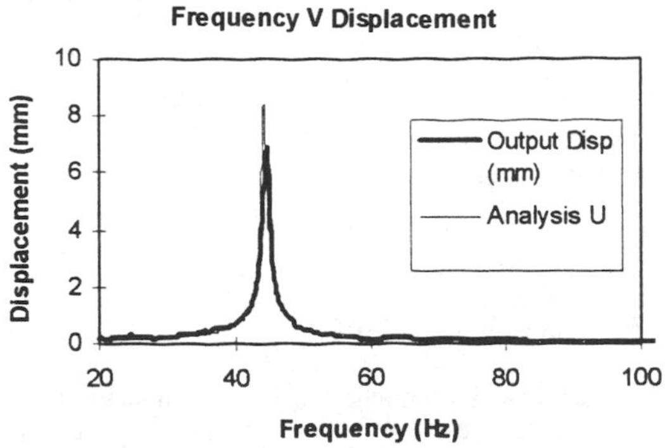

Figure 6. POM beam No.3, run No. 64, mass of accelerometer = 0.2g and analysis U ($\alpha = 2.166427$ and $\beta = 3.12301*10^{-5}$)

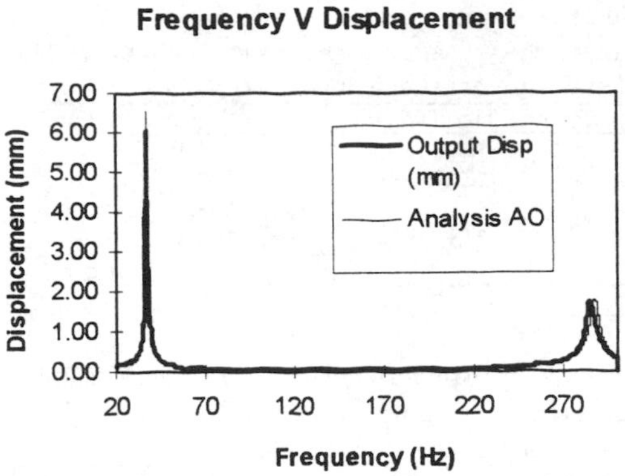

Figure 7. POM beam No.1, run No. 24, mass of accelerometer = 0.2g and analysis AO ($\alpha = 4$ and $\beta = 0.8*10^{-5}$)

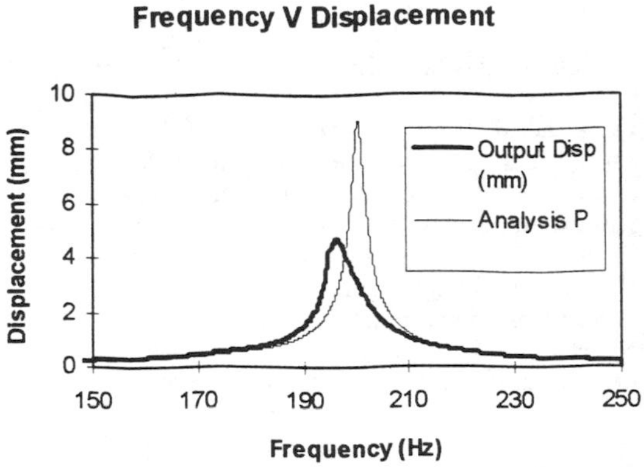

Figure 8. POM beam No.2, run No. 60, mass of accelerometer = 0.2g and analysis P ($\alpha = 4$ and $\beta = 0.8*10^{-5}$)

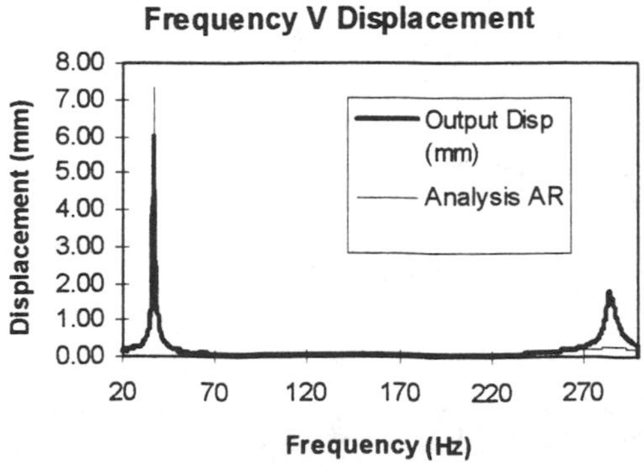

Figure 9. POM beam No.1, run No. 24, mass of accelerometer = 0.2g and analysis AR (MP,DAMP = $7.054619533*10^{-5}$)

$$[C] = \alpha [M] + \beta [K] \quad (5)$$

where [C] is the overall damping matrix, [M] is the mass, [K] is the stiffness, and α and β are dimensionless parameters known as Rayleigh constants (Swanson 1993). However this method of simulating damping is mainly used in transient analyses and values for α and β are calculated for a particular structure at a particular frequency.

The idea of this method is to give a prediction of displacement that is of the same order of magnitude of the actual displacement. This method of damping is used by ANSYS because the equations decouple and as a result of this they are simpler to solve. The main disadvantage with this method is that the system is supposed to be proportionately damped and in practice no real system is proportionately damped.

Material dependent damping allows one to specify damping as a material property. In the case of an automobile mirror this is quite tedious because a complete mirror maybe manufactured from several materials.

The damping ratio is the traditional method of quantifying damping. It represents the ratio of actual damping to critical damping and is inputted into ANSYS for the whole of the system and not different values for individual parts.

Modal damping allows you to specify different damping ratios for different modes of vibration. Modal damping should give the same results as the damping ratios as it is these damping ratios that are being inputted at various modes.

6 MATERIAL EXPERIMENTS

In order to assess the damping properties of the materials from which mirrors are manufactured a series of tests had to be performed. The following tests were carried out on simple beam structures. All tests were carried out in a controlled environment of 23°C and 50% relative humidity.

6.1 Free oscillation test

The free oscillation test yields the logarithmic decrement and from that it is possible to calculate the damping ratio. The main disadvantage of this test is that it only yields the first natural frequency of the beams.

6.2 Sweep sine test

The sweep sine test is used to assess the dynamic performance of a simple beam. A constant sine displacement is applied to a beam over a frequency range. These experiments are simulated in ANSYS. The results are compared in order to assess how well the finite element method has predicted the dynamic response of the beam. It is also possible to derive a damping ratio from the results of the sweep sine test. One of the advantages of the sweep sine test is that you are not limited to the first natural frequency.

7 MATERIAL DAMPING

Analyses were carried out on Polyoxymethylene (POM), commercially known as Acetal, in order to investigate its damping properties. Sweep sine tests were carried out on a series of simple beam specimens and these beam specimens were also modelled in ANSYS.

POM was investigated first because it is a very stable material, relative to Polyamide, and is not adversely affected by varying atmospheric conditions. The grade of POM that was investigated was an unfilled variety therefore it would not be affected by fibre orientation.

It should be noted that all analyses were assigned a letter in order to keep a record of them, i.e. Analysis AL, while the research work was carried out. In the course of this paper all the information for a particular analysis is given.

7.1 Rayleigh constants (α and β)

When Rayleigh constants were used to model the damping they did not yield an accurate prediction of the displacements that were experienced by the simple beams as can be seen from figures 4, 5, and 6.

Even when the Rayleigh constants were chosen, and not experimentally determined, they were only able to predict the magnitudes of displacement for one structure and not for different structures of the same materials as can be seen from figures 7 and 8.

As a result of this it was decided that Rayleigh constants were not a suitable method of modelling the damping characteristics of materials.

7.2 Material Damping (MP,DAMP)

When the MP,DAMP command was used to model

the material damping it was found that the damping was frequency dependent as can be seen from figure 9.

The difference between the results, for the first natural frequency, that were obtained experimentally and the results that were obtained by finite element analysis were not very significant as can be seen in figure 9. This was very encouraging and it was felt that it may be possible to obtain a MP,DAMP value over a range of frequencies.

Figure 10 shows the results of a sweep sine test similar to figure 9 the only difference being the mass of the accelerometer used.

From figures 9 and 10 it can be seen that the natural frequencies occur at approximately the same frequency and that the mass of the accelerometer has little effect on the natural frequencies. However the same value for MP,DAMP is inputted and as can be seen from figure 10 an accurate prediction of the displacement is not given. This leads us to conclude that the mass of the accelerometer has an effect on the damping of the system.

Looking at figure 11 an accurate prediction of the displacement is given however the natural frequency is different from the specimen whose results are in figure 9.

This would indicate that the results are certainly dependent on operating conditions, in this case the mass of the accelerometer, but may not be as dependent on frequency or structure as was initially thought.

7.3 Constant Damping Ratio (ζ)

The damping ratio is the traditional method of measuring damping. It represents the ratio of actual damping to critical damping and is inputted into ANSYS for the whole of the system and not different values for individual parts.

It is assumed that the damping ratio varies with frequency and temperature for a given structure. The damping ratio for a particular structure can be obtained, experimentally, using the sweep sine test and the free oscillation test. Once these values were obtained they were used in ANSYS. However the main drawback with this method is that the system has to be tested in order to determine the damping ratio. This is counter productive to us as the aim of our project is to predict the vibration performance of a mirror prior to it being manufactured. Figure 12

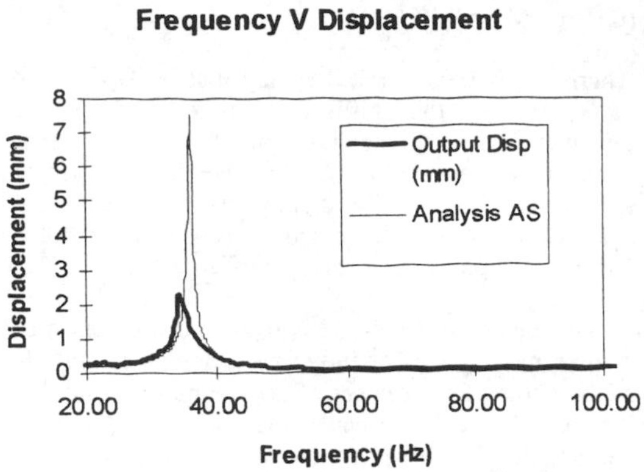

Figure 10. POM beam No.1, run No. 10, mass of accelerometer = 0.64g and analysis AS (MP,DAMP = $7.054619533*10^{-5}$)

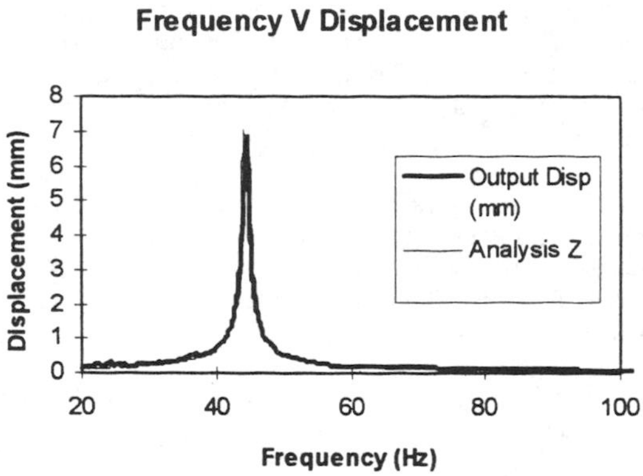

Figure 11. POM beam No.3, run No. 64, mass of accelerometer = 0.2g and analysis Z (MP,DAMP = $7.054619533*10^{-5}$)

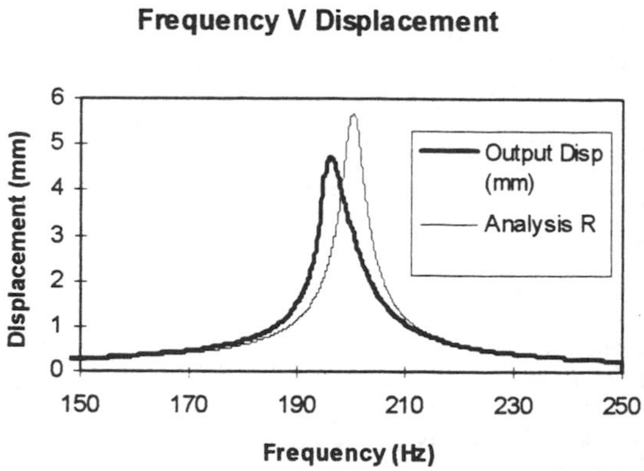

Figure 12. POM beam No.2, run No. 60, mass of accelerometer = 0.2g and analysis R (DMPRAT = 0.01061)

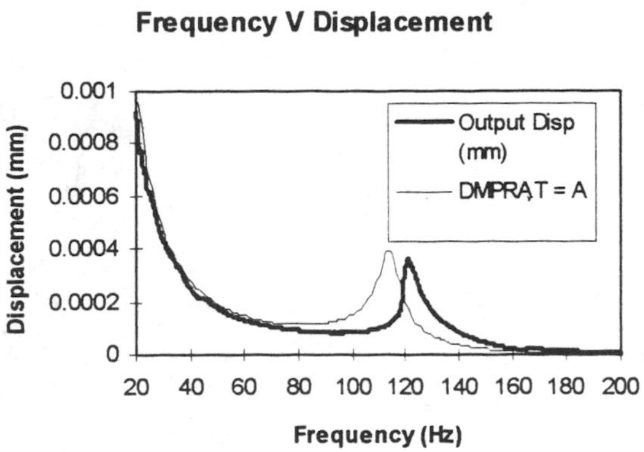

Figure 13. Bracket and 80g block in Y direction with constant input acceleration of 1.5*g* compared with finite element analysis equivalent with a damping ratio of A.

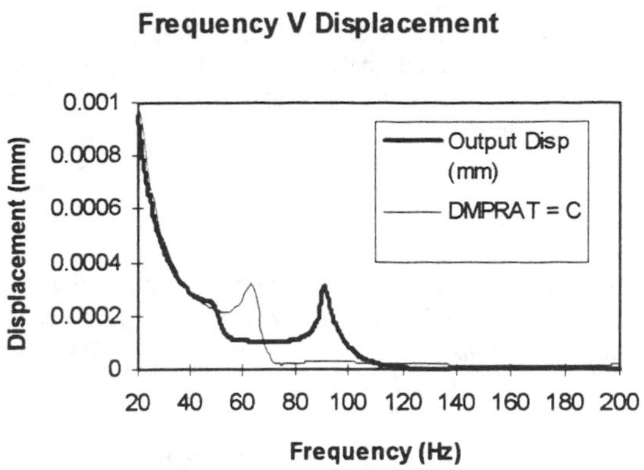

Figure 15. Complete mirror in Z direction with constant input acceleration of 1.5*g* compared with finite element analysis equivalent with a damping ratio of C.

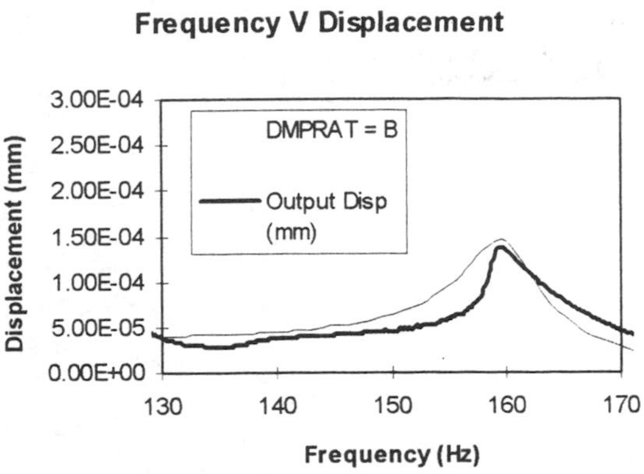

Figure 14. Bracket and 80g block in Z direction with constant input acceleration of 1.5*g* compared with finite element analysis equivalent with a damping ratio of B.

shows a sweep sine test being compared to a harmonic analysis. The prediction is quite accurate when compared with the equivalent experimental result and this similarity was seen throughout all analyses where the damping ratio was used to simulate damping.

8 COMPLETE ANALYSIS

Once the various methods of modelling damping in ANSYS had been investigated it was decided to perform some harmonic analysis on complete mirrors. The main aim of this section is to assess the overall damping properties of the complete mirror. The method of testing the complete mirrors is described in section 3.

The overall aim of this approach is to build up a data base of the damping ratios that are involved in the analyses of mirrors and establish a pattern that may be dependent on frequency or mode shape. A number of analysis using Rayleigh constants were carried out. However no pattern could be established from the Rayleigh constants and it was decided to look at the damping ratio as the method of modelling damping. Once all the materials that make up a mirror are investigated harmonic analyses using material damping to model damping will be carried out.

As can be seen from figure 13 the damping ratio once again gives an accurate prediction of the amount of displacement when it is compared with the experimental equivalent.

A series of analyses were carried out on the three structures and it was concluded that the damping ratios of the three structures range between two values. It was also possible to determine a narrower range for each of the three structures.

9 CONCLUSIONS

An accurate prediction of the natural frequencies is critical. The damping ratio is the most accurate and convenient way of modelling damping.

It is hoped that it will be possible to establish a much more detailed pattern of the damping ratios of complete mirrors in the future taking into account the frequency and mode shape. Also it is planned to carry out a detailed analysis of the material damping properties of all the materials that make up a complete mirror.

REFERENCES

Mc Carthy, B. 1995. *Experimental Investigation of the Button/Windscreen Support of an Automobile Internal Rearview Mirror*, Final Year Project, University College Galway. pp. II.

Swanson Analysis Systems Inc., *Dynamics, User Guide for Revision 5.0,* Updl DN-S211:50, 1993. pp. 1.12 - 1.26.

O'Grady, M. *Finite Element Software Design Aid for Automobile Rearview Mirrors*, Masters Project, University College Galway, 1995, pp. 33 - 37.

ACKNOWLEDGEMENTS

The authors would like to express their thanks to *Donnelly Mirrors Ltd.* for their support in the course of this project.

Modern Practice in Stress and Vibration Analysis, Gilchrist (ed.) © *1997 Balkema, Rotterdam, ISBN 90 5410 896 7*

Vibration of engine foundations

L. Feng, A. Nilsson & L. Kari
The Marcus Wallenberg Laboratory for Sound and Vibration Research, KTH, Stockholm, Sweden

ABSTRACT: Vibration transmission through an engine foundation is investigated. Two new designs, which are much lighter than the standard one, are suggested. Numerical estimation of transmission loss is made by using a simplified transmission line model. Acoustic properties of the three different types of foundations are thoroughly compared experimentally. Three parameters are used: vibration level at the excitation point; vibration level at the reference point; and averaged vibration level at the bottom plate. It is found that the new designs have much better acoustic properties than the standard one.

1 INTRODUCTION

Sandwich structures are widely used in marine constructions, with engine foundation as a very important application. The cross section of a typical sandwich engine foundation is of the shape like "╨". The thickness of the vertical panel is usually about a few times of that of the bottom panel. A marine engine is mounted on the top of the structure and the vibration is transmitted to the bottom of the structure, which is usually a part of the hull or is directly connected to the hull. An important task in designing an engine foundation is how to reduce the vibration level at the bottom panel, or ship hull, when it is excited by the engine.

2 DIFFERENT TYPES OF FOUNDATIONS

A typical sandwich engine foundation (hereafter it is referred as "standard" foundation) is shown in Fig. 1. It is mainly constructed of sandwich panels. A steel structure, which is usually thick and heavy, stands on the top of that. A marine engine is mounted on the steel panel.

In order to reduce the vibration at the bottom panel, two new types of engine foundations are designed. They are also shown in the same figure and are referred as "type 2" and "type 3". In both designs, a steel panel, which is smaller and thinner than that in the standard design, is embedded inside

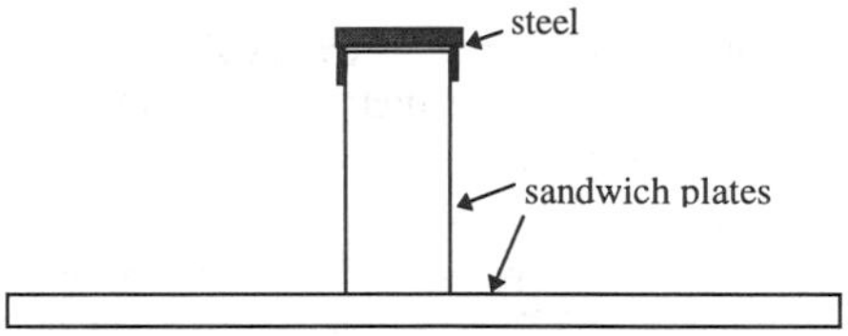

Type 1: a typical sandwich engine foundation (standard design)

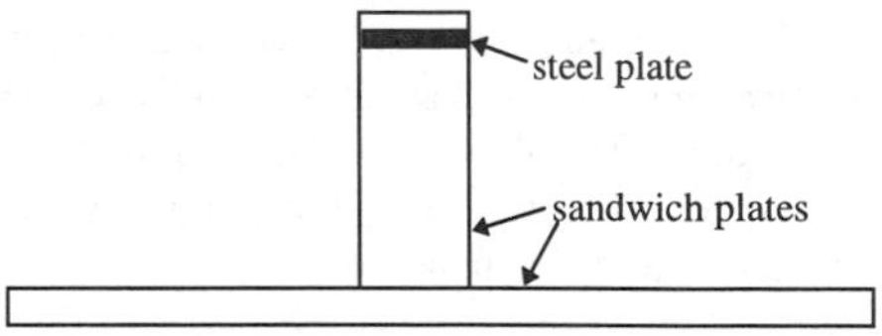

Type 2: with embedded steel plate

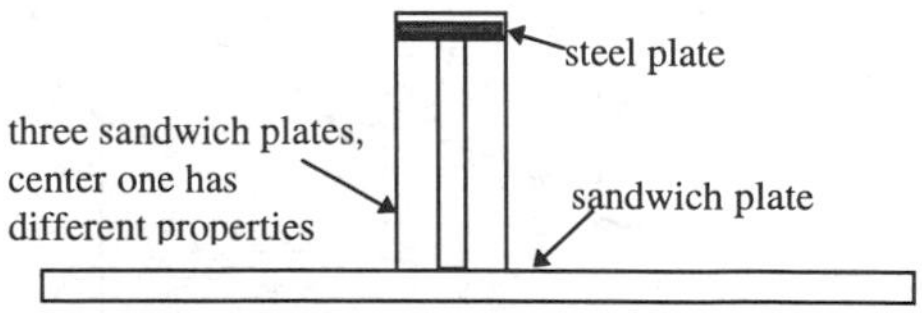

Type 3: with embedded steel plate and vertical sandwich plates

Fig. 1: Schematic diagram of foundations

the vertical sandwich panel. In design type 3, the vertical sandwich panel under the steel plate is replaced by three thinner sandwich panels. The middle one has different acoustic properties than the side panels, in order to smoothen the frequency characteristics of the structure.

3 ESTIMATION OF TRANSMISSION LOSS

Two factors may influence the vibration at the bottom panel when the top panel is excited: input mobility and transmission loss of the foundation. In this section we try to estimate and compare the vibration transmission loss of the three different foundations.

A two-dimensional problem, i.e., the foundation is infinite long, is considered. As a first order approximation, a transmission line model is used. The whole foundation is divided into several elements with different material parameters and cross sections. Low frequency approximation is used to convert all elements into equivalent uniform ones. Each element is represented by a transmission matrix and a reflection matrix, which include both bending wave and longitudinal wave and wave type interchanges. Those elements are joined together in a straight line or at a right angle. The transmission and reflection coefficients of each element are calculated after Cremer and Heckl (1988). Estimated results are shown in Fig. 2, where the transmission loss stands for the difference between the vibration levels at the input plate and at the bottom plate.

From the estimation one may expect that at most frequencies, both new designs should have higher vibration transmission loss than the traditional one. This is especial the case for design type 3, which shows the best acoustic behavior.

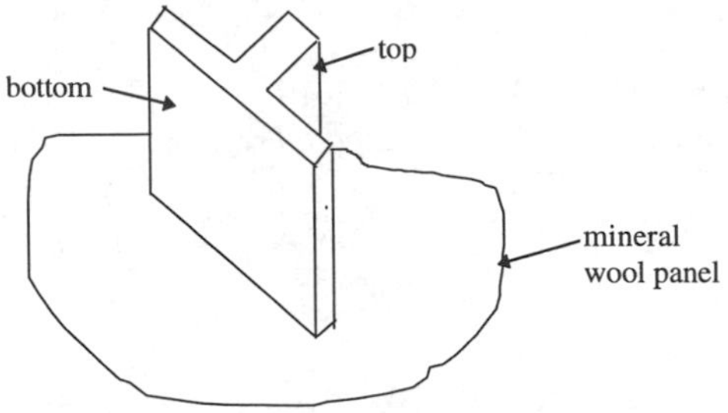

Fig. 3 Arrangement of the engine foundation

4 EXPERIMENTS

The numerical model shown above estimates only the transmission loss of a foundation. Another factor, input mobility, also influences the characteristics of a foundation to a great extent. Furthermore, the above model is an over simplified one and can only be used to estimate the acoustic properties qualitatively. In order to make comparisons quantitatively, one has to make a more delicate model or make experiments. The purpose of the experiment shown below is to evaluate the overall acoustical properties of the structures. That is, under a certain force excitation, the vibration level at the bottom panel of the different foundations.

The three types of foundations are tested in real scale. All foundations are of the same size and same material. The bottom panel of the foundation is of the size 2200 x 1500 mm and the top one is of the size 260 x 1500 mm. The height is about 572 mm (It varies a little because of different constructions).

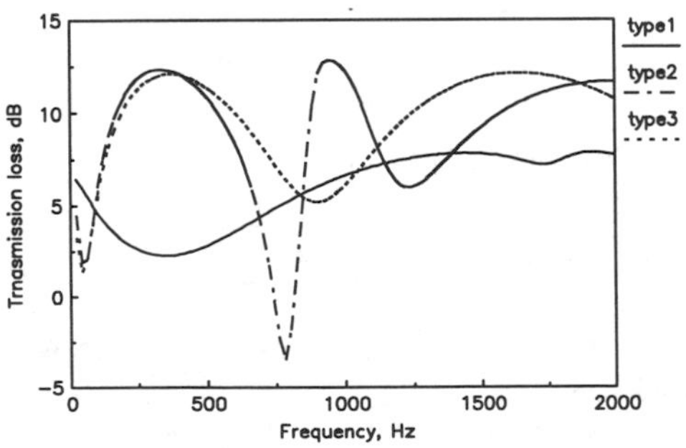

Fig. 2 Transmission loss of different foundations

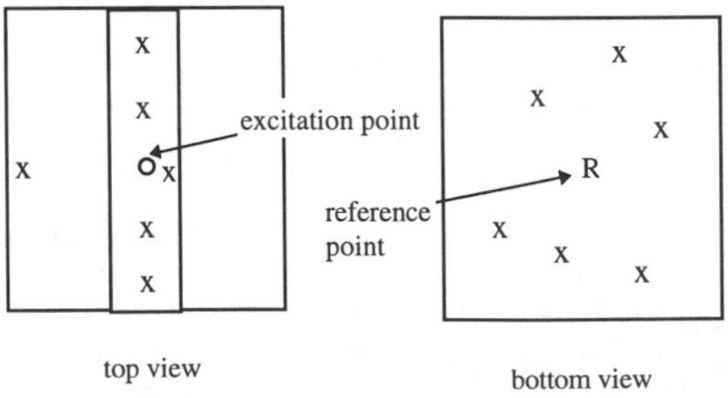

Fig. 4 Measurement and excitation points
O: excitation point; R, x: measurement point.

The foundations under test are standing on mineral wool panels in vertical direction, as shown in Fig. 3. The thickness of the mineral wool panel is 50 mm and the density is 135 kg/m³. A shaker (Ling Dynamic type 409) is connected to the foundation at the center of the top plate via a steel bar and a force transducer (B&K type 8200). Vibration signals are obtained via an accelerometer (B&K type 4367) and a charge amplifier (B&K type 2635). There are totally eight measurement points at the bottom plate, with seven points at the back side and one at the front side. Point R is at the center of the back side of the bottom plane and is used as the reference point. Those points are shown in Fig. 4. For the standard design, there are also five measurement points at the top plate to check the vibration level at the input plane.

The measurements are performed by using a computer-controlled four-channel Fourier Analyzer (Tektronix type 2630). White noise excitation and narrow band mobility measurements are applied, with frequency resolution 1.25 Hz and frequency range 0 - 2 kHz. With the information of mobility, we can calculate power spectrum of vibration velocity under any force excitation by using the formula

$$S_{XX}(\omega) = |H(\omega)|^2 \cdot S_{FF}(\omega) \qquad (1)$$

where $S_{FF}(\omega)$ is the power spectrum of input force and $H(\omega)$ is mobility. $S_{XX}(\omega)$ is power spectrum of velocity at the input point if $H(\omega)$ is input mobility and is power spectrum at the reference point if $H(\omega)$ is transfer mobility.

In order to make comparison easier, the power spectrum of the applied force is set to be unit ($1\ N^2 / Hz$) everywhere. Hence the velocity power spectrum under a unit force excitation is

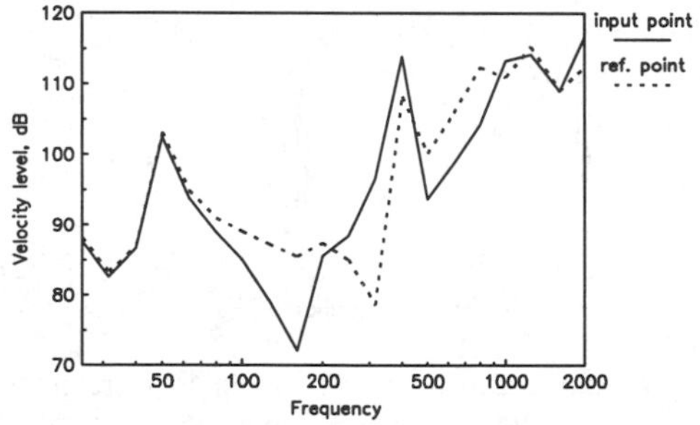

Fig. 5 Vibration level at input and reference points. standard foundation, unit force excitation

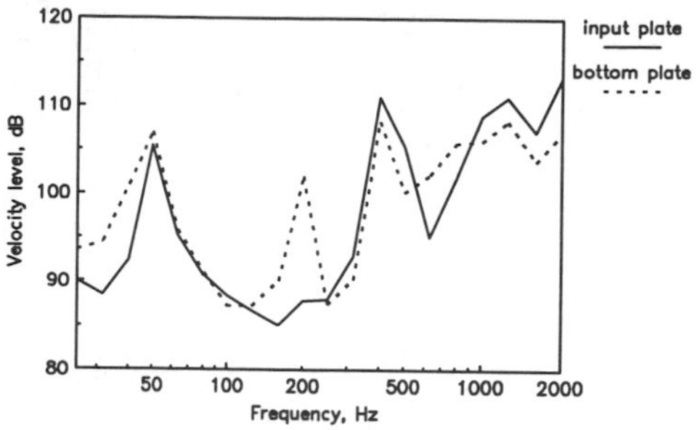

Fig. 6 Averaged vibration level. standard foundation, unit force excitation

$$S_{xx}(\omega) = |H(\omega)|^2 \qquad (2)$$

The third octave band velocity level is calculated from the power spectrum as

$$L_{vi} = 10 * \log(\sum_i |H(\omega_i)|^2) + 10 * \log(v_{ref}^2) \qquad (3)$$

where the summation is over the respective third octave band. v_{ref} is reference velocity (1e-9 m/s).

Loss factors are also calculated from the same transfer function measurements.

5 RESULTS ON STANDARD ONE

5.1 *Vibration level*

The first measurement is performed on the standard engine foundation. Fig. 5 shows the vibration level at the input point and at the reference point when there is a unit force excitation at the center of the top plate. The values are shown in 1/3 octave band. The low vibration level of the input velocity at 160 Hz band is due to the vibration pattern, since the measurement point is close to node of the vibration mode at that frequency. The same reason explains the low level at the reference point in 315 Hz band. Besides these two frequency bands, the vibration level at the input point and at the reference point is roughly the same. This is due to the fact that the impedance of the steel panel (input plate) is much higher than that of the sandwich panel (bottom plate). From the steel panel to the sandwich panel, the amplitude of vibration is actually *amplified*. One should not only look at the difference of vibration

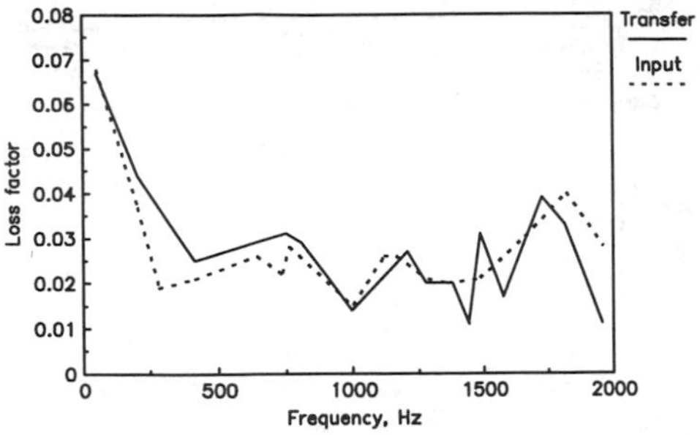

Fig. 7 Loss factor of standard foundation. Averaged value: 0.028

levels between the top plate and the bottom plate to see the characteristics of the foundation. Previous *in situ* measurements on ship engine foundations show the same vibration level as we measured on the test sample.

For the vibration measurement of a large structure at low frequencies, one point measurement is usually not enough, since the modal density is low and the vibration level changes much from point to point. The 163 Hz band at the input point and the 315 Hz band at the reference point are examples of this. Fig. 6 shows averaged vibration levels at the top and at the bottom plates of the engine foundation. The average of the input plate is taken over the five measurement points. For the bottom panel, the average is taken over all the eight points (seven at the back side and one at the front side). The low level frequency points mentioned before disappeared after the average. It is clearly seen from the figure that there is a resonance at the bottom panel at 200 Hz band. At this frequency band, vibration at the bottom panel is about 15 dB higher than that at the top panel. This resonance is due to the size of the bottom panel of the foundation. The same peak appears also for the newly designed foundations, since the bottom panels are of the same size and material.

Comparing Fig. 5 and Fig. 6 we see that there are two types of peaks for the vibration at the bottom panel. The first group consists of resonant peaks at 50 Hz and 400 Hz. For this group, vibrations at both input and bottom plates are at their resonances. The strong vibration at the bottom panel is caused by the strong vibration at the input panel, or high input mobility.

Another type of peak is the one occurs at around 200 Hz, where the vibration at the input plane is low but at the bottom plane is high. This is caused by the bottom panel resonance and is dependent on the size of the sample under test.

It should also be noticed that the vibrations at the both input and bottom panels are high when frequency is higher than about 400 Hz. Though there are certain fluctuations, the vibration velocity level keeps high.

5.2 *Loss factor*

Loss factors are calculated via Nyquist circle method. Both input mobility measurements and transfer mobility measurements (when the shaker is connected) are utilized for the calculation. They give roughly the same results. They are shown in Fig. 7. The frequency-averaged value of the loss factor is about 0.028.

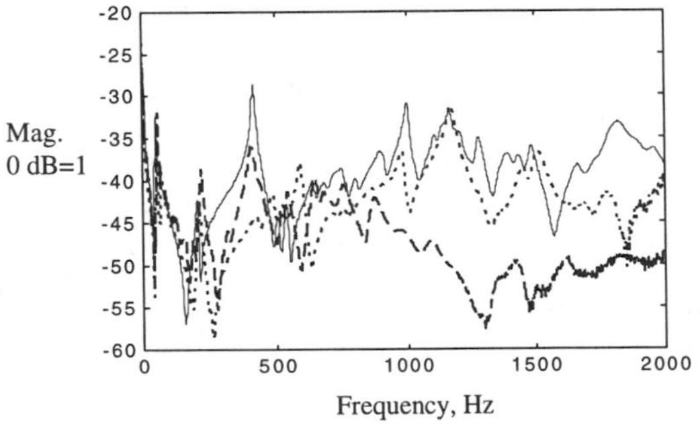

Fig. 8 Comparison of input mobility. Solid: standard foundation; dashed: foundation type 2; dotted: foundation type 3.

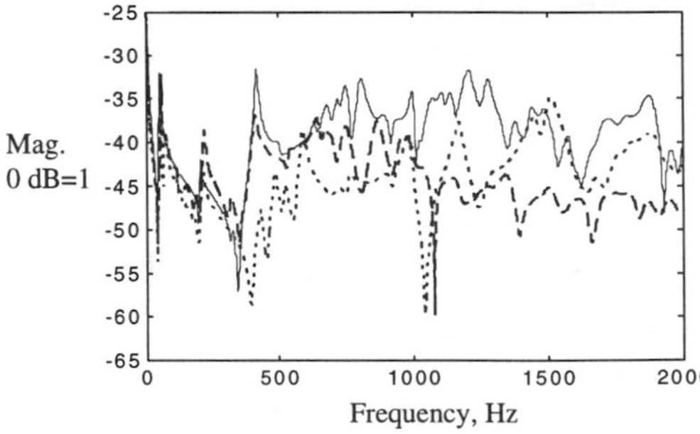

Fig. 9 Comparison of transfer mobility at reference point. Solid: standard foundation; dashed: foundation type 2; dotted: foundation type 3.

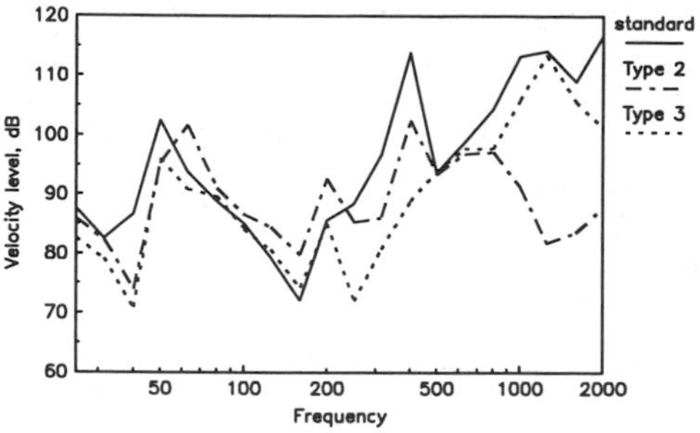

Fig. 10 Comparison of vibration level at input point

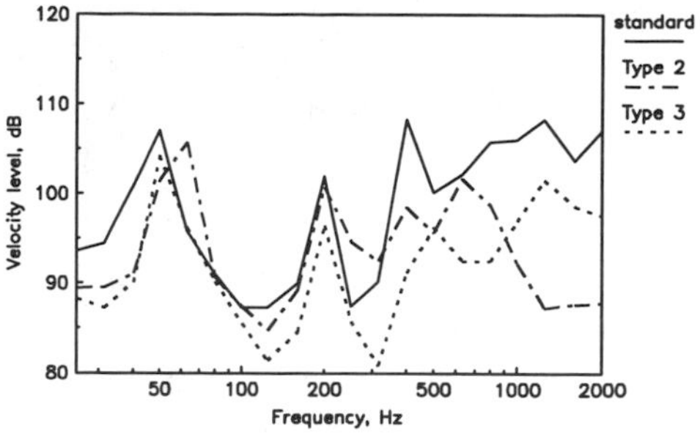

Fig. 12 Averaged vibration level at the bottom panel

6 COMPARISONS

The same measurements are performed for the other two foundations. Since the top steel plate of the new foundation is embedded into the sandwich panel, a special hole (20 x 40 mm) is made at the top of the foundation, in order to excite and measure at the steel plate. Except that, the excitation and measurement situations are exactly the same as those for the standard one.

Comparisons are made for three different quantities: input mobility; transfer mobility at the reference point; and averaged vibration level at the bottom panel. Fig. 7 shows the comparison of the magnitudes of the input mobilities. It is seen that when frequency is under about 200 Hz, the differences among the three designs are not very big. Above that frequency, the input mobilities of the two new designs are lower than that of the standard one. Foundation type 3 has the best performance at frequencies between about 200 Hz and 800 Hz. One interesting thing is that the highest resonance of input mobility, which occurs at about 400 Hz, disappeared in this type of foundation. When frequency is even higher, foundation type 2 shows much lower input mobilities than those of the others. This is because that for that foundation, the steel panel is only supported by the core material, which has low Young's modulus and hence the system has a low resonant frequency. As we know, the vibration of the plate is very difficult to be excited when frequency is high above the resonant frequency of the corresponding spring-mass system.

The two new designs show even better acoustic performance for the transfer mobilities at the reference points (Fig. 9), this is especially the case for the foundation type 3. This foundation keeps low vibration at a wider frequency range (200-1k Hz). Above 1 kHz, the difference between the two new designs is also smaller than that of the input mobility, since the transmission loss of type 3 is higher than that of type 2 at high frequencies, as we expected from the simple numerical model. Since the ultimate goal is to reduce the vibration at the bottom panel, the vibration level at the reference point is more important than the properties of the input mobility.

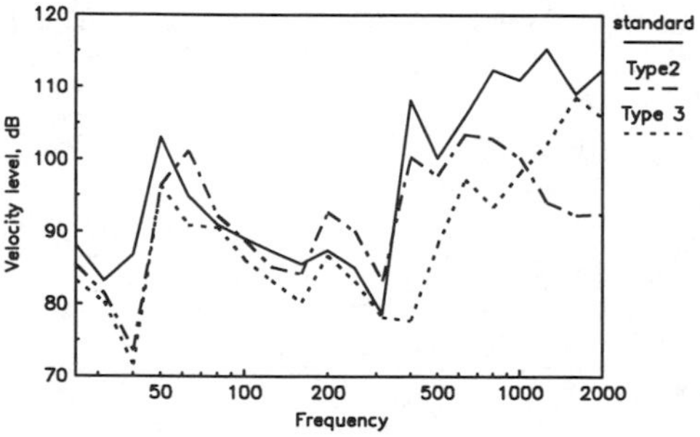

Fig. 11 Comparison of vibration level at the reference point

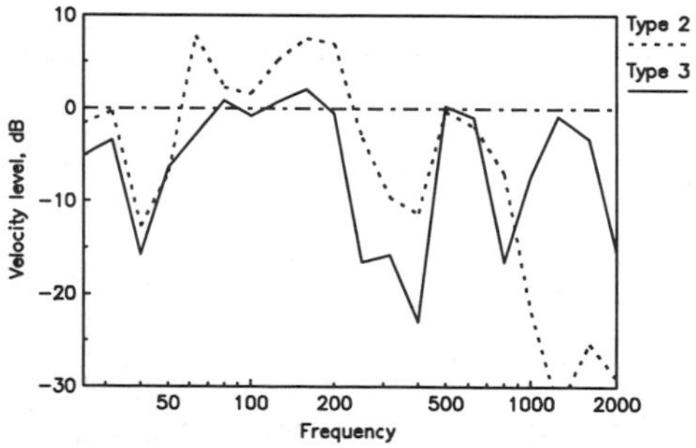

Fig. 13 Difference of vibration level at input point. The value for standard one is taken as zero.

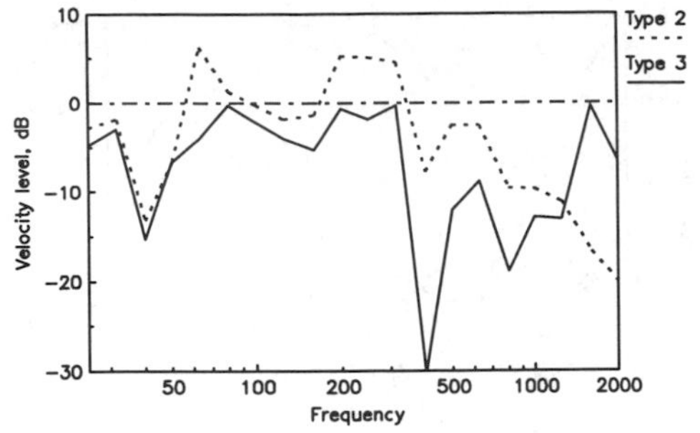

Fig. 14 Difference of vibration level at the reference point

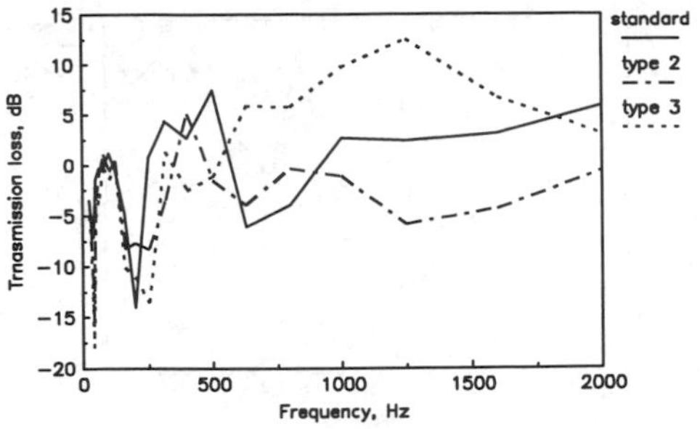

Fig. 16 Transmission loss at the reference point

Fig. 10 and Fig. 11 show vibration levels at these two points under a unit force excitation, which are actually the 1/3 octave band correspondents of Fig. 8 and 9. Fig. 12 is the comparison of the averaged vibration level of the bottom panel. The average is taken over the eight measurement points on the bottom panel of the each foundation. Similar as the vibration at the reference point, foundation type 3 shows the lowest vibration when frequency is below 1 kHz. After that, foundation type 2 has the lowest vibration. As mentioned before, the resonance at 200 Hz, which occurs for all foundations, may be due to the resonance of the bending vibration of the bottom panel.

Fig. 13 - Fig. 15 show the difference of vibration levels between new designs and the standard one. In those figures, the vibration level of the standard one, under the same excitation, is taken as zero. The values below or above the zero line show the advantages and disadvantages of the new designs. Although the input mobilities of foundation type 3 is a little higher than the standard one at a few frequency bands, the bottom panel vibration of this foundation is always lower than that of the standard one in the whole frequency range. Foundation type 2 has its best performance only when frequency is above 1 kHz. In the frequency range about 50 - 500 Hz, the vibration of foundation type 2 is somewhat comparable with the standard one.

It is also interesting to compare the measured transmission losses with the predicted one. Fig. 16 shows the measured transmission loss of the three types of foundations at the reference point. It seems the frequency scale in Fig. 3 should be adjusted. Otherwise the general tendency of Fig. 16 looks similar to Fig. 3, although the numerical model is for infinite bottom panel and the measurement is for a finite one.

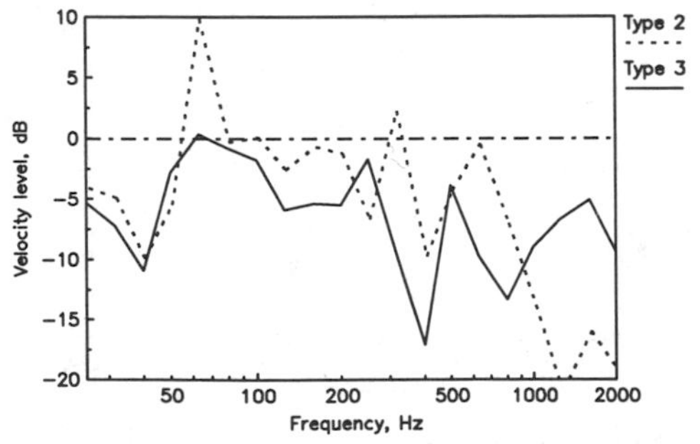

Fig. 15 Difference of averaged vibration level at the bottom panel

7 CONCLUSIONS

The acoustic properties of the new designs are superior to those of the standard one. They are also much lighter.

When frequency is below about 1 kHz, foundation type 3 has the best acoustical performance. When frequency is even higher, foundation type 2 is better.

8 REFERENCES

Cremer, L. and Heckl, M. 1988 *Structure-Borne Sound* (Springer-Verlag Berlin), chapter V.

Modern Practice in Stress and Vibration Analysis, Gilchrist (ed.) © 1997 Balkema, Rotterdam, ISBN 90 5410 896 7

Design for vibration characteristics

A.J. McMillan
Department of Engineering Science, University of Oxford, UK (Currently: Transmissions Department, Rolls Royce plc, Derby, UK)

A.J. Keane
Department of Mechanical Engineering, University of Southampton, UK

ABSTRACT: In recent years, with continued development in computer hardware and computational methods, engineering design has become increasingly driven by computational analysis. Designs are routinely optimised for cost and mass; in this work, optimisation for vibrational characteristics is investigated. The modes of vibration of a rectangular plate are manipulated by adding a number of small masses to the plate, with the objective of controlling vibration in a frequency band width corresponding to the 37th to 42nd modes. It is shown that the mechanism by which the vibration suppression is achieved is highly dependent on the optimisation strategy.

1 INTRODUCTION

In this work we assess the possibility of making significant modifications to the vibrational modes and frequencies of a structure in a pre-determined way by making small modifications to the structural design. Such a technique might be used to create a frequency band over which the vibrational response is minimal, to coincide with the operating frequency band of some connected machinery. Alternatively, one might wish to fine-tune some of the modes and frequency, for example, to improve the tuning and tone colour of musical instruments (Schoofs *et. al.*, 1987).

As this is an inverse problem, it is not possible to simply give a list of frequencies and deduce the shape of the structure directly. In many cases there will be no corresponding physical structure, and in the cases where a solution is found it may not be unique (Gordon *et. al.*, 1992). It is clear that the fundamental and first few frequencies are strongly determined by the size and shape of the structure - and this will often be a constraint. Conversely, the very high eigen-number frequencies will tend to be very regularly spaced, regardless of the shape of the structure (Kac, 1966). There will be further constraints which depend on the particular structure considered and the types of modification that may be considered.

Inverse problems involving a large number of degrees of freedom are rather difficult to solve. One approach is to use an optimisation technique. Although there are many such techniques they all involve the generation of a range of structural designs and giving each a rank depending on its closeness to the required solution. This ranking is determined by an objective function, however the choice of this is in itself a non-trivial matter and can have a profound effect on the nature of the optimised solution. For example, to create a frequency band over which the vibrational response is minimal, one might try to reassign the resonant frequencies which fall in that band to frequencies outside the band. In effect the objective function would be a count of the number of resonances falling within the band, and (possibly) a measure of the total differences in frequency between each reasonance and the edge of the band. Alternatively, one might take the view that it does not matter if a resonance does occur in this band so long as the modal displacement is low over a given region of the structure (it is hard to excite a mode close to one of its nodes). In this case the objective function would be the area under the frequency response curve over the frequency band in question.

In any practical problem, the computational time involved is a major issue, and this will strongly influence decisions concerning objective function and optimisation technique (McMillan and Keane, 1996, 1997).

2 THE CASE STUDY

We consider the free vibration of a rectangular plate of length a and breadth b, simply supported on all four sides, and carrying a single concentrated mass. This problem is relatively simple in that the eigenfunctions of the plate and mass system may be constructed (Amba-Rao, 1964) from the well known plate eigenfunctions (Bishop, 1960). In view of the power of numerical techniques, particularly finite element methods, to solve such problems, the reader may question the value of such an approach. However, while the finite element method gives robust solutions to particular engineering problems, it can not give more general insights. For example, in this case study we are not so much concerned with finding the eigenvalues and eigenfunctions for the plate and mass system, but with understanding how they are affected by the position and size of the mass.

The equation of motion is

$$D\nabla^4 w + \{\rho h + M\delta(x-u)\delta(y-v)\}\frac{\partial^2 w}{\partial t^2} = 0, \quad (1)$$

where $w = w(x,y,t)$ is the transverse deflection of the plate, $D = Eh^3/12(1-\nu^2)$ is its flexural rigidity, ρ is the density of the plate material, h is the plate thickness, and the concentrated mass has a mass of M and coordinates (u,v). The solution of (1) is given in the form of a summation over normal modes,

$$w(x,y,t) = \sum_{r=1}^{\infty} a_r \Psi_r(x,y) e^{i\omega_r t}, \quad (2)$$

where the eigenfunctions are

$$\Psi_r(x,y) = \frac{4M\omega_r^2}{Dab}\Psi_r(u,v)\times \sum_{m=1}^{\infty}\sum_{n=1}^{\infty}\frac{\sin(\frac{m\pi u}{a})\sin(\frac{n\pi v}{b})\sin(\frac{m\pi x}{a})\sin(\frac{n\pi y}{b})}{\{[(\frac{m\pi}{a})^2+(\frac{n\pi}{b})^2]^2-\omega_r^2\frac{\rho h}{D}\}} \quad (3)$$

and the eigenvalues are the roots of

$$f(\omega_r) = \frac{4M\omega_r^2}{Dab}\times \sum_{m=1}^{\infty}\sum_{n=1}^{\infty}\frac{\sin^2(\frac{m\pi u}{a})\sin^2(\frac{n\pi v}{b})}{\{[(\frac{m\pi}{a})^2+(\frac{n\pi}{b})^2]^2-\omega_r^2\frac{\rho h}{D}\}} - 1. \quad (4)$$

Although some of the eigenvalues are given as the roots of equation (4), some care has to be taken when the mass lies on a node line or when there are degenerate eigenvalues. Clearly, if the mass lies on a node line, then the plate acts as if the mass were not present. In this case the (un-normalised) eigenfunction is simply

$$\Psi_r(x,y) \equiv \Psi_{mn} = \sin\left(\frac{m\pi x}{a}\right)\sin\left(\frac{n\pi y}{b}\right) \quad (5)$$

and the corresponding eigenvalue is

$$\omega_r \equiv \omega_{mn} = \sqrt{\frac{D}{\rho h}}\left[\left(\frac{m\pi}{a}\right)^2+\left(\frac{n\pi}{b}\right)^2\right]. \quad (6)$$

The mass will lie on a node line of some eigenfunction if either of u/a or v/b are rational numbers. Furthermore, if the plate has degenerate eigenvalues then the corresponding eigenfunctions may be recombined to give node lines through any point on the plate. In this case, by adding a mass, the symmetry is broken, and for the plate and mass system, one eigenvalue is given by the root of expression (4) and its corresponding eigenfunction from equation (3). The other eigenvalue remains that of the plate without the mass and the eigenfunction is the linear combination of the two degenerate eigenfunctions which is orthogonal to the eigenfunction already determined.

It is simple to prove that the plate and mass system does not exhibit degeneracy, except where the plate without the mass is degenerate. Degenerate eigenvalues exist if and only if $f(\omega) = f'(\omega) = 0$. Differentiating (4) we obtain

$$f'(\omega) = \frac{8M\omega}{Dab}\times \sum_{m=1}^{\infty}\sum_{n=1}^{\infty}\frac{\sin^2(\frac{m\pi u}{a})\sin^2(\frac{n\pi v}{b})[(\frac{m\pi}{a})^2+(\frac{n\pi}{b})^2]^2}{\{[(\frac{m\pi}{a})^2+(\frac{n\pi}{b})^2]^2-\omega^2\frac{\rho h}{D}\}^2} \quad (7)$$

which is clearly greater than zero for all $\omega > 0$.

The demoninator of equation (4) tends to zero as the frequency approaches the eigenvalues of the plate without the mass. Expression (4) is discontinuous at these points, but from (7) it is monotonically increasing in between. Therefore $f(\omega) = 0$ for exactly one value of ω between each pair of consecutive eigenvalues for the plate without the mass, with ω_1 lying between zero and the first.

In conclusion, the effect of adding a mass is to reduce the eigenvalues, although the gap between each will remain roughly constant and is strictly bounded by the eigenvalues of the plate without the mass. If the eigenvalues of the plate are sorted into ascending order, then the nth eigenvalue of the plate and mass system lies between the $(n-1)^{th}$ and the nth eigenvalues of the plate alone.

There is also an effect on the eigenfunctions of the plate; the node lines are distorted towards the mass (an infinite mass would would be equivalent to clamping the plate at that point). Since this will distort the node line pattern over the whole plate, judicial positioning of the mass could cause a node line of a given mode shape to appear at a given point on the plate, ie at $x = x_o$, $y = y_o$. The effect of this would be to dramatically reduce the frequency response at this point for the frequencies around the resonant frequency for this mode shape.

3 ADDING SOME MORE MASSES

To obtain more control over the eigenvalues, we might simply try adding more masses. We could proceed inductively; constucting the set of eigenfunctions for a plate with n masses from the eigenfunctions of the plate with one fewer mass. Although this would be numerical folly, it is easy to see, analytically, that as each mass is added, the eigenvalues will be reduced, as was the case when the first mass was added. Thus we are now able to provide bounds for the eigenvalues of a plate carrying a number of masses;

$$\omega_r^{[0]} < \omega_{r+n}^{[n]} \le \omega_{r+n}^{[0]}, \tag{8}$$

where the subscript refers to the eigennumber, and the superscipt referes to the number of masses added. In other words, to shift n resonances from a given frequency band, at least n masses are necessary. Even if there were no practical limitation on the size of the masses and where they can be put, it is likely that more masses would be required.

Although we have demostrated that n prechosen resonances may be shifted downwards, we have not yet addressed the problem of the higher resonances also shifting down, into the frequency we wish to clear. It is obvious from the reasoning above that, if further masses are not positioned on nodelines, then the order of the eigenvalues will remain the same. If all the masses lie very close to the node lines of a given eigenfunction then the masses will have little effect on that eigenfunction or the corresponding eigenvalue. This eigenvalue then is reduced very gradually, and all the higher numbered eigenvalues must be greater, by equation (8).

As more or larger masses are added, the effect on that eigenfunction will become greater, and it may change form suddenly with a corresponding sharp drop in eigenvalue. It would seem that the best that can be obtained is that the masses are placed exactly on the node lines, although the theorem then does not apply and the eigenvalues may re-order.

The vibration of a plate carrying a number of point masses may be solved directly using a method analogous to that given by Amba-Rao (1964) for a single point mass. For a plate carrying n point masses, where the kth mass is given by M_k, and its location is given by (u_k, v_k), the partial differential equation is

$$D\nabla^4 w + \left\{\rho h + \sum_{k=1}^{n} M_k \delta(x-u_k)\delta(y-v_k)\right\}\frac{\partial^2 w}{\partial t^2} = 0. \tag{9}$$

Writing $w(x,y,t)$ as a sum over normal modes

$$w(x,y,t) = \sum_{r=1}^{\infty} A_r \Psi_r(x,y) e^{i\omega_r t}, \tag{10}$$

we obtain

$$D\nabla^4 \Psi_r - \omega_r^2 \left\{\rho h + \sum_{k=1}^{n} M_k \delta(x-u_k)\delta(y-v_k)\right\}\Psi_r = 0. \tag{11}$$

Taking the finite sine transform of equation (11), it may be shown that

$$\int_0^b \int_0^a \Psi_r(x,y)\sin\left(\frac{m\pi x}{a}\right)\sin\left(\frac{n\pi y}{b}\right) dx dy = \frac{\frac{\omega_r^2}{D}\sum_{k=1}^{n} M_k \Psi_r(u_k,v_k)\sin(\frac{m\pi u_k}{a})\sin(\frac{n\pi v_k}{b})}{\{[(\frac{m\pi}{a})^2 + (\frac{n\pi}{b})^2]^2 - \omega_r^2 \frac{\rho h}{D}\}} \tag{12}$$

Re-inverting gives

$$\Psi_r(x,y) = \frac{4\omega_r^2}{Dab}\sum_{k=1}^{n} M_k \Psi_r(u_k,v_k) \times \sum_{m=1}^{\infty}\sum_{n=1}^{\infty}\left\{\sin(\frac{m\pi x}{a})\sin(\frac{n\pi y}{b}) \times \frac{\sin(\frac{m\pi u_k}{a})\sin(\frac{n\pi v_k}{b})}{\{[(\frac{m\pi}{a})^2 + (\frac{n\pi}{b})^2]^2 - \omega_r^2\frac{\rho h}{D}\}}\right\}. \tag{13}$$

Substituting (u_l, v_l) for (x,y) gives n equations which can be written as a matrix eigenvalue equation of the form

$$(\mathbf{A} - \lambda \mathbf{I})\mathbf{x} = 0 \tag{14}$$

where $x_l = \Psi_r(u_l, v_l)$, $\lambda = \frac{Dab}{4\omega_r^2}$ and

$$A_{k,l} = M_k \sum_{m=1}^{\infty} \sum_{n=1}^{\infty} \left\{ \sin(\frac{m\pi u_l}{a}) \sin(\frac{n\pi v_l}{b}) \times \frac{\sin(\frac{m\pi u_k}{a}) \sin(\frac{n\pi v_k}{b})}{\{[(\frac{m\pi}{a})^2 + (\frac{n\pi}{b})^2]^2 - \omega_r^2 \frac{\rho h}{D}\}} \right\}.$$

Equation (14) is true if and only if $|\mathbf{A} - \lambda \mathbf{I}| = 0$.

To solve for ω_r evaluate the determinant for a range of values and use those either side of a sign change as a starting point for a bisection root search. N.B. the reader must not confuse the eigenvalues of the matrix equation, λ, with those of the plate and masses system, ω_r. In this context λ and the matrix $\mathbf{A}$ both contain the eigenvalue ω_r.

As a test case, a plate with the dimensions and material properties given in Table 1 was considered. Five 10 kg masses were added individually to the plate and at each stage the eigenvalues lying between 100 Hz and 110 Hz and the corresponding eigenfunctions were determined. The masses were constrained to lie on the node lines of the 42nd eigenfunction, which has an eigenvalue of about 110 Hz, so that this eigenfunction would be unaffected by the addition of the masses. For the purpose of choosing the $n+1^{th}$ mass position, the eigenfunctions for the plate carrying n masses were inspected by eye, and the mass placed at a point of large deflection coinciding with the node lines of the 42nd eigenfunction.

With the addition of five masses there is an encouraging degree of reduction in eigenvalues 37 to 41, although it is not really sufficient to be useful; it is clear that significantly more masses must be applied to the plate to have the desired effect.

The eigenfunctions for the un-loaded plate, which are illustrated in Figure 1, exhibit a regular grid structure. The effect on the eigenfunctions of the addition of further masses is then shown in Figures 2 to 6, where the crosses indicate the positions of the added masses. Although the eigenfunctions are distorted by the addition of masses, the nodelines are still similarly spaced. The spacing of the nodelines gives an indication of the modal amplitude across the plate. It is possible that addition of further masses may concentrate high amplitude deflection to localised regions of the plate.

Table 1. Plate Parameters.

Young's modulus (E)	206.8	G Pa
Poisson's ratio (ν)	0.29	
Density (ρ)	7820	Kg m^{-3}
Plate thickness (h)	0.01	m
Length (a)	7	m
Breadth (b)	2	m

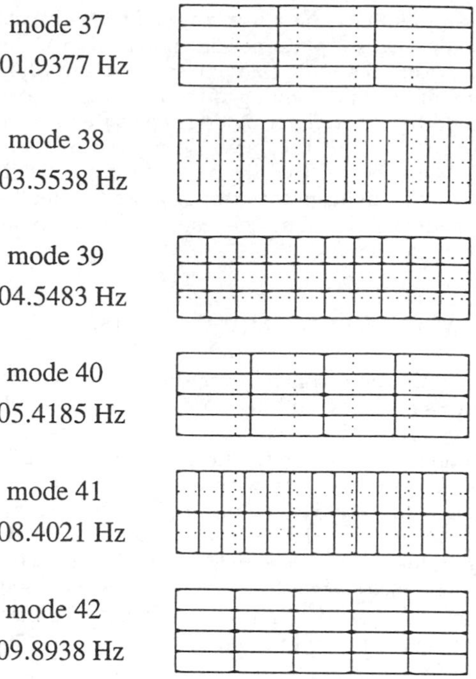

Figure 1. Eigenvalues and functions for the un-loaded plate.

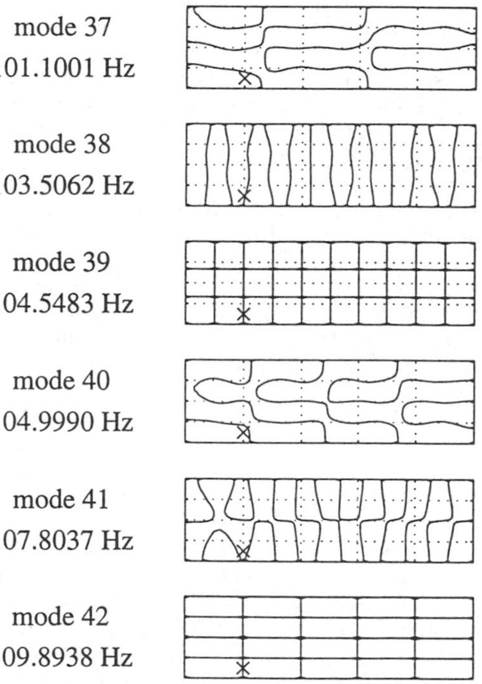

Figure 2. Eigenvalues and functions for the plate carrying one mass.

4 OPTIMISATION OF MASS POSITION

Suppose that a larger number of masses were positioned on the plate. To keep within the philosophy of making *small* modifications to the structure, it would be necessary to reduce the size of these masses. Then, the problem is to decide how to optimise the locations of the masses. Although more than one mass may be allowed at any one location, it is intuitive that the masses will tend to separate.

There are two main approaches to the problem; to optimise the location of each mass *sequentially*, or to optimise the locations of all the added masses *simultaneously*.

4.1 *Sequential optimisation*

The basis of the sequential optimisation method is to start with a given structure (the plate without any added masses) and improve it slightly (by adding just one mass). This new improved structure may be further improved (by the addition of another mass), *etc.*. Although this methodology can be followed until all available masses are placed, this solution may not be the absolute optimum.

To make the problem more tractable, the possible mass positions were discretised to lie at 0.05 m intervals along the node lines of the 42nd mode. The objective function was calculated for the plate carrying just one mass, for each allowable position. The optimum position of that mass was then taken to be the position which gave the lowest objective function. A second mass was added, and its optimum position found, and so on.

4.2 *Simultaneous optimisation*

A Genetic Algorithm optimisation of the mass positions has been performed utilising the same objective function and also restricting the masses to lie on the node lines of the 42nd mode. Unlike the method described above, the GA controls the full set of 50 mass positions simultaneously. Since there were 50 variables, it was necessary to use a large generation size (499). The number of generations was restricted to 20, so that the total computational expense of the two different approaches would be comparable. The control parameters used by the algorithm are those found to be optimal settings by Keane [4], and have been tested for many different optimisation problems.

4.3 *Objective function*

When applying an optimiser to the question of the optimum mass positions, some serious consideration must be given to the objective function, *i.e.* the function by which each set of mass positions is graded. In this case, we calculated the sum of frequency responses at frequencies 100.5, 101.5, 102.5,.... 109.5 over three points on the plate due to white noise forcing at three other points. These forcing and response points were chosen to be at the corners of equilateral triangles with the distance between them being approximately half the wavelength of a travelling wave of frequency between 100 and 110 Hz, Table 2. This would make it impossible for the optimiser to choose mass positions which led to all the forcing or response points occuring at nodes.

Table 2. Forcing and Response Points.

Forcing points (m)	0.859	1.141
	1.193	1.052
	0.948	0.807
Response points (m)	6.141	0.859
	5.807	0.948
	6.052	1.193

The response of the plate due to harmonic forcing can be found by treating the masses as equivalent frequency dependent point forces, so that the partial differential equation we now solve is

$$D\nabla^4 w + \rho h \frac{\partial^2 w}{\partial t^2} = F(x, y, t), \qquad (15)$$

where $F(x, y, t)$ is the sum of all forces applied to the plate. This includes both the real force applied at (x_i, y_i) and the forces associated with the point masses.

The deflection of the plate at (x_o, y_o) due to a single force of magnitude f_i applied at (x_i, y_i) is given by

$$\begin{aligned} w(x_o, y_o, \omega) &= f_i g(x_i, y_i, x_o, y_o; \omega) \\ &= f_i \sum_{m=1}^{\infty} \sum_{n=1}^{\infty} \frac{\Psi_{mn}(x_i, y_i)\Psi_{mn}(x_o, y_o)}{b_{mn}\{\omega_{mn}^2 - \omega^2\}} \end{aligned} \qquad (16)$$

The eigenfunctions $\Psi_{mn}(x, y)$ and the eigenvalues ω_{mn} are given in equations (5) and (6) respectively, and $b_{mn} = \rho hab/4$.

Now, if there are n masses with positions and values of (x_k, y_k) and M_k, respectively, then each mass provides a point force on the plate of

$$\begin{aligned} f_k(x, y) &= -M_k \frac{\partial^2 w}{\partial t^2}\bigg|_{(x=x_k, y=y_k)} \\ &= M_k \omega^2 w(x_k, y_k). \end{aligned} \qquad (17)$$

mode 37
99.6801 Hz

mode 38
102.6834 Hz

mode 39
104.5483 Hz

mode 40
104.6155 Hz

mode 41
107.4578 Hz

mode 42
109.8938 Hz

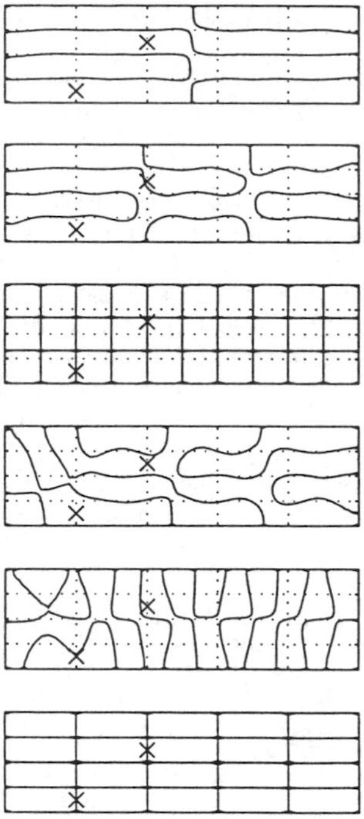

Figure 3. Eigenvalues and functions for the plate carrying two masses.

mode 37
99.6667 Hz

mode 38
102.3307 Hz

mode 39
103.1882 Hz

mode 40
104.6155 Hz

mode 41
107.4566 Hz

mode 42
109.8938 Hz

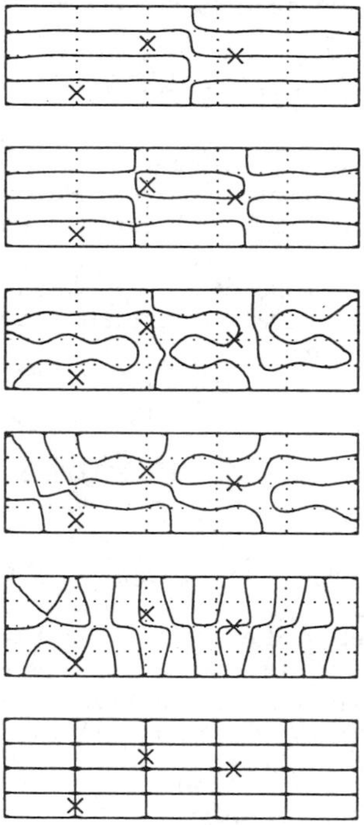

Figure 4. Eigenvalues and functions for the plate carrying three masses.

mode 37
99.0232 Hz

mode 38
102.2485 Hz

mode 39
103.1146 Hz

mode 40
103.6750 Hz

mode 41
107.0980 Hz

mode 42
109.8938 Hz

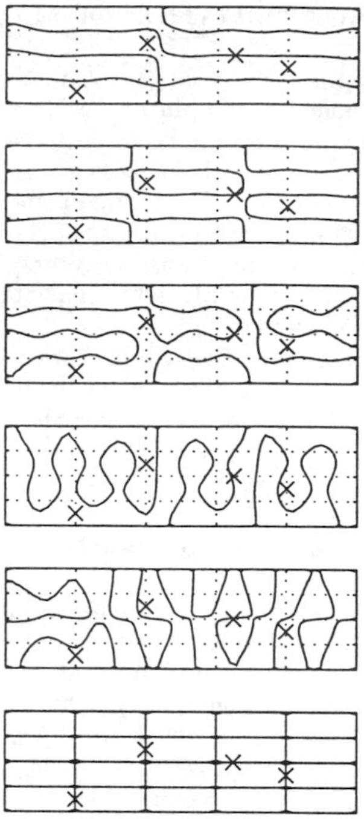

Figure 5. Eigenvalues and functions for the plate carrying four masses.

mode 37
96.7913 Hz

mode 38
102.0956 Hz

mode 39
103.0951 Hz

mode 40
103.6670 Hz

mode 41
106.1106 Hz

mode 42
109.8938 Hz

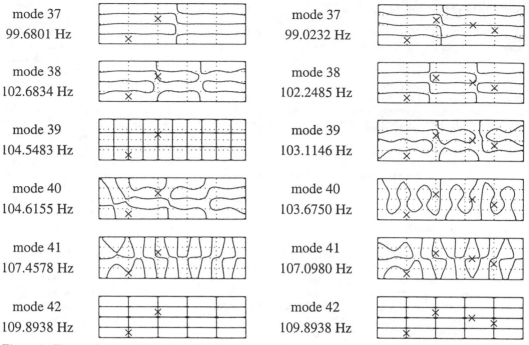

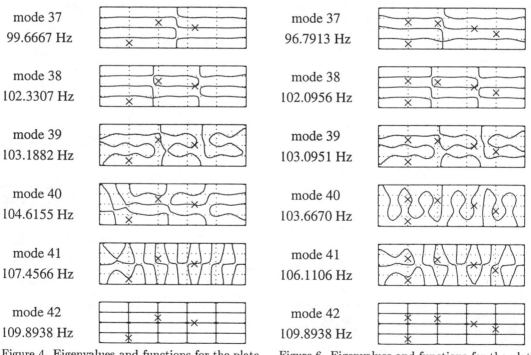

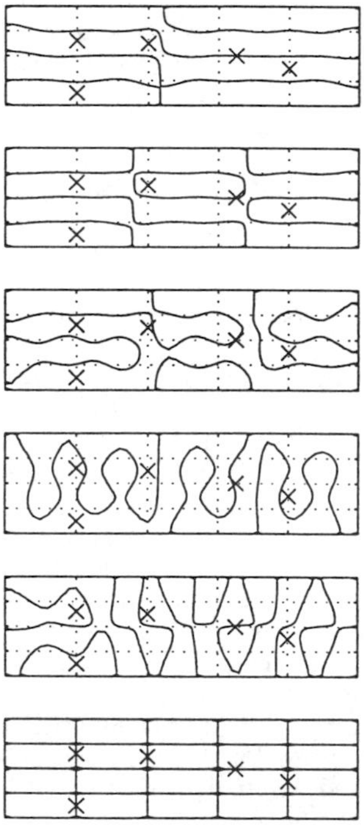

Figure 6. Eigenvalues and functions for the plate carrying five masses.

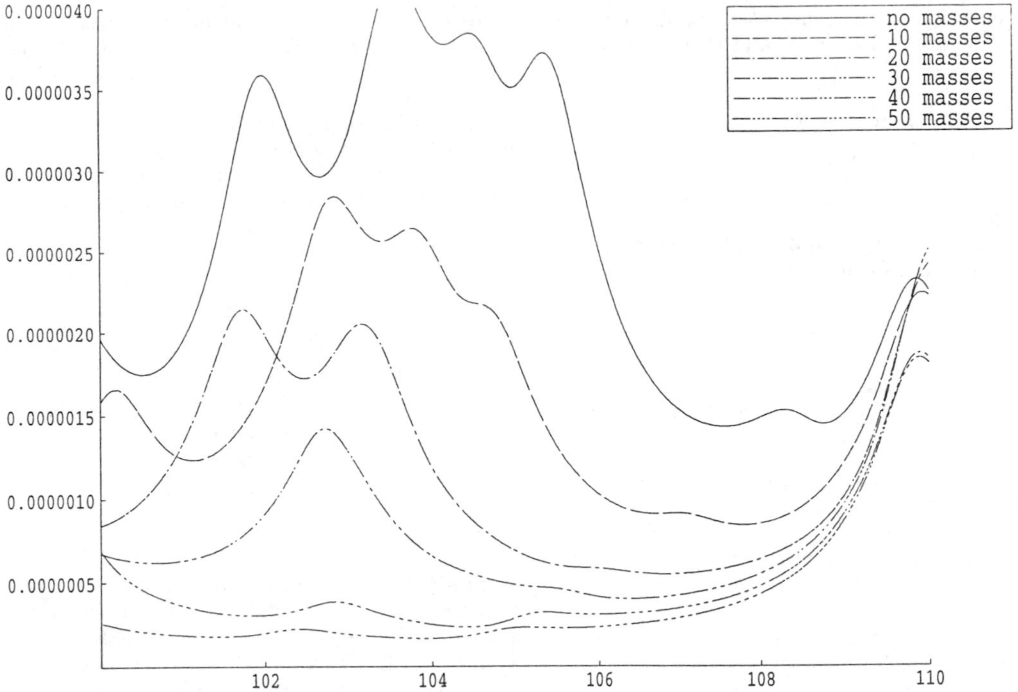

Figure 7. Response of plate with mass positions chosen by sequential positioning.

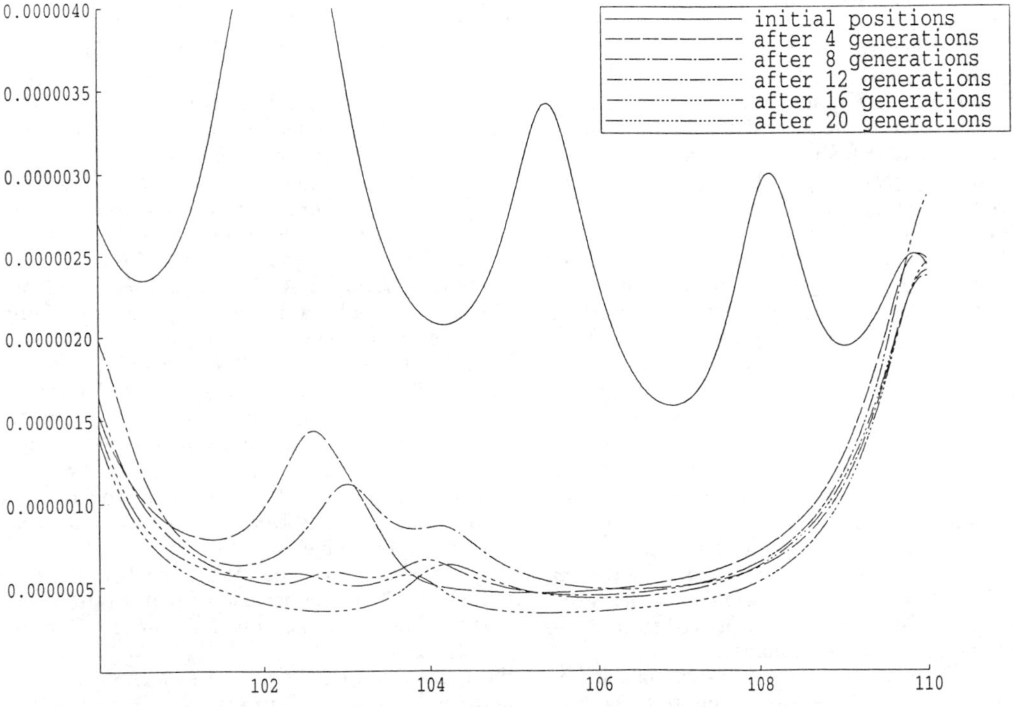

Figure 8. Response of plate with mass positions chosen by Genetic Algorithm.

The deflection of the plate at any point (x_o, x_o) due to the applied force f_i and those due to the n masses is then

$$w(x_o, y_o, \omega) = f_i g(x_i, y_i, x_o, y_o; \omega) + \sum_{k=1}^{n} f_k g(x_k, y_k, x_o, y_o; \omega). \quad (18)$$

In particular, using equation (17), we obtain n equations of the form

$$f_j = M_j \omega^2 \left[f_i g(x_i, y_i, x_j, y_j; \omega) + \sum_{k=1}^{n} f_k g(x_k, y_k, x_j, y_j; \omega) \right], \quad (19)$$

for n positions (x_j, y_j) on the plate corresponding to the n masses, which can be solved for the n unknown frequency dependent forces f_k. Then, by equation (18), it is possible to calculate the deflection of the plate at any given point (x_o, y_o).

5 DISCUSSION

The sequential mass positioning method achieved a slightly greater vibration suppression than the Genetic Algorithm, which achieved the bulk of its improvement within the first few generations. However, as the computational time for evaluating the objective function increases with the square of the number of masses in use, the degree of noise isolation achieved by either method per unit of computer effort would be rather similar. Figures 7 and 8 show the objective function (the frequency response of the plate, with multiple forcing and response points), at various stages of the optimisation process.

Perhaps the greatest difference between these methods is in the type of results that are obtained. The sequential positioning method achieved its objective by steadily pushing back the resonance near 108 Hz to around 105 Hz, and by making the response to the lower resonances low. However, a higher mode crossed over the 42nd, with a frequency of 109 Hz. Although this mode is not significantly excited at the response or forcing points chosen for the optimisation process, it would be significant if other locations on the plate were considered. The modeshapes of this plate system are characterised by localised areas of and low high deflection maxima, so it seems that noise isolation is achieved in part by the sculpting of the modeshapes.

The GA results are quite diffrerent in nature. After just 8 generations, there seem to be no resonances between about 104.5 and 110 Hz, and there seems to be no evidence of higher mode crossover. However, the frequency responses for all generations beyond the 8th seem to peak at around 104 Hz, and this might suggest that to clear a 10 Hz frequency band would require more masses. The modeshapes are generally much more uniform — exhibiting many peaks of a similar height — than those for the plate carrying the sequentially optimised masses.

ACKNOWLEDGEMENTS

The work reported here has been supported by the EPSRC under grant number GR/J06856. The determinant of a matrix was found using the LU decomposition routine given in "Numerical Recipes" (Press *et. al.*, 1986).

REFERENCES

Amba-Rao, C. L. 1964. On the vibration of a rectangular plate carrying a concentrated mass. *Journal of Applied Mechanics* 31:550-551.

Bishop R. E. D. & D. C. Johnson 1960. *The mechanics of vibration.* Cambridge University Press.

Goldberg D. E. 1989. *Genetic algorithms in search, optimization and machine learning.* Addison-Wesley.

Gordon, C., D. L. Webb, & S. Wolpert 1992. One cannot hear the shape of a drum. *Research Announcements, bulletin of the* American Mathematical Society 27:134-138.

Kac, M. 1966. Can one hear the shape of a drum? *Amer. Math. Monthly* 73:1-23.

Keane A. J. 1995. Genetic algorithm optimization of multi-peak problems — studies in convergence and robustness. *Artificial* Intelligence in Engineering 9(2):75-83.

McMillan, A. J. & A. J. Keane 1996. Shifting resonances from a frequency band by applying concentrated masses to a thin rectangular plate. *J. Sound Vib.* 192(2):549-562.

McMillan, A. J. & A. J. Keane 1997. Vibration isolation in a thin rectangular plate using a large number of optimally positioned point. masses. *J. Sound Vib.* to appear.

Press W. H., B. P. Flannery, S. A. Teukolsky & W. T. Vetterling 1986. *Numerical recipes.* Cambridge University Press.

Schoofs, A. G., F. van Asperen, P. Maas & A Lehr 1987. Computation of bell profile using structural optimisation. *Music Perception* 4:245-254.

Siddall J. N. 1982. *Optimal engineering design: priciples and applications.* New York: Marcel Dekker, Inc.

Modern Practice in Stress and Vibration Analysis, Gilchrist (ed.) © *1997 Balkema, Rotterdam, ISBN 90 5410 896 7*

Assessing the effects of viscoelastic inserts by experimental reanalysis

E.M.O.Lopes
Federal University of Santa Catarina, Florianópolis, Brazil & University of Wales Cardiff, UK

J.A.Brandon
University of Wales Cardiff, UK

J.J.de Espíndola
Federal University of Santa Catarina, Florianópolis, Brazil

ABSTRACT: In troubleshooting situations, it is by no means unusual to predict effects of design changes in mechanical systems via models constructed only from experimental data. For a large number of problems in which specific unacceptable response characteristics of a structure should be treated, response models can be a very convenient alternative. This is particularly true when an ordinary design strategy for structural response attenuation, the use of viscoelastic materials, is followed. The current paper is concerned with the extension of methods of experimental response reanalysis in order to assess the addition of polymeric inserts into structures. Numerical results with a simple mechanical system support the mathematical representations and provide data for preliminary conclusions.

1 INTRODUCTION

In troubleshooting situations, it is by no means unusual to predict effects of design changes in mechanical systems via models constructed only from experimental data. As a matter of fact, experimentally based approaches have played a fundamental role in guiding and assessing structural modification efforts for purposes of vibration control. This is particularly observed when, concomitantly, design data related to those mechanical systems are not available.

For a large number of problems in which specific unacceptable response characteristics of a structure should be treated, response models (as opposed to modal models) can be a very convenient alternative (Brandon 1990). Assisted by such models, it is possible to appraise an intended (proposed) structural modification for a certain system, given the existing (measured) receptance matrix and the acceptable (established) levels of response for the system thus modified. It is usually understood that those undesirable response characteristics would present themselves as conditions restricted either in frequency or location.

The use of viscoelastic inserts is an ordinary design strategy for structural response attenuation (Sun and Lu 1995). If its effects are to be realistically described via mathematical models, a detailed characterisation of the material properties of the inserts is a fundamental prerequisite. As is well-known, these properties are, to a lesser or greater extent, dependent on frequency, temperature, and loading (Nashif et al 1985).

Although there have been many research studies over the years in the estimation of material properties, standard techniques still depend on a formulation which does not provide appropriate boundary conditions. For this reason, the authors have concentrated on the development of a more representative formulation for the standard test configuration. In tackling this problem, numerical techniques for the inverse problem have been developed based on tracking the zeroes of the singular value decomposition rather than standard root-finding techniques (Espíndola et al 1995).

Much of the research in experimental structural dynamic modification has been done by adopting the modal approach rather than by employing direct methods for response analysis. As a particularly elegant example, it is mentioned a fairly new optimisation procedure presented by Bucher and Braun (Bucher and Braun 1994), which aims at minimising responses of structures using truncated modal data.

On the non-modal front, a substantial bulk of the original literature has already been surveyed by one of the authors (Brandon 1990), contextualizing contributions of his own along with fellow co-workers. Of the same period is a procedure for optimal structural modification of viscously-damped systems, developed by Sestieri and D'Ambrogio (Sestieri and D'Ambrogio 1989). Tahtali and Özgüven, following a pseudo-force approach considered previously (Özgüven 1987), have proposed a method specially advantageous for handling major structural modifications (Tahtali and Özgüven 1994). More recently, Heylen and Mas (Heylen and Mas 1996) have discussed the applicability of two optimisation methods directed at reducing frequency-limited undesired behaviour through optimal structural modifications at specific locations.

In a broader sense, related topics have been addressed by a number of researchers. Srivastava and Kundra (Srivastava and Kundra 1993), on the assumption that complete models are available, have proposed an algorithm for structural dynamic modification of structurally-damped systems with damped beam elements. From a primarily theoretical point of view, Level et al (Level et al 1994, 1996) have manifested particular interest in the computational efforts required by methods of calculating the modified response of systems under periodic excitation.

Problems involving viscoelastic materials have also been tackled. Ravi et al (Ravi et al 1995) have developed a perturbation method for response reanalysis of beams with unconstrained and constrained viscoelastic damping layers, via theoretical modal analysis. Likewise they have investigated clamped and simply supported plates, with free viscoelastic layers applied in various configurations (Ravi et al 1996). From a perspective of model correlation, Lin and Ling (Lin and Ling 1996) have devised a method for identifying damping properties of viscoelastically- damped structures, based solely on an incomplete set of frequency response functions.

This paper is concerned with the extension of methods of experimental response reanalysis in order to assess the addition of polymeric inserts into structures. Special emphasis is placed towards the suitability of response models in those cases. Numerical results with a simple mechanical system will support the mathematical representations and provide data for preliminary analyses.

2 THEORETICAL BACKGROUND

The matricial equation of motion of a linear multi-degree of freedom system of order n can be expressed in the frequency domain as

$$\left[-\omega^2[M]+\left[\overline{K}(\omega)\right]\right]\{X(\omega)\}=[S(\omega)]\{X(\omega)\}=\{F(\omega)\} \tag{1}$$

where $[M]$ = n x n square, symmetric, positive definite, constant mass matrix, $\left[\overline{K}(\omega)\right]$ = n x n square, symmetric, frequency-dependent, complex stiffness matrix, $\{X(\omega)\}$ = n x 1 generalised displacement vector, $[S(\omega)]$ = n x n square, symmetric, frequency-dependent dynamic stiffness matrix, $\{F(\omega)\}$ = n x 1 generalised force vector, and ω = angular frequency.

Usually the complex stiffness matrix is represented by

$$\left[\overline{K}(\omega)\right]=[K(\omega)]\left([I]+i[\eta(\omega)]\right) \tag{2}$$

where $[K(\omega)]$ = n x n square, symmetric, frequency-dependent stiffness matrix, $[I]$ = identity matrix of order n, and $[\eta(\omega)]$ = n x n square, frequency-dependent loss factor matrix. The loss factor matrix equals $[K(\omega)]^{-1}[H(\omega)]$, $[H(\omega)]$ being a n x n square, symmetric, frequency-dependent damping matrix.

From equation (1) it results that

$$\{X(\omega)\}=[S(\omega)]^{-1}\{F(\omega)\}=[R(\omega)]\{F(\omega)\} \tag{3}$$

The frequency-dependent matrix $[R(\omega)]$, which is the inverse of the dynamic stiffness matrix $[S(\omega)]$, is known as the receptance matrix. The receptance matrix contains all the input-output relationships for the whole system under investigation. It constitutes what is called a response model and may, within certain practical limits, be approximated experimentally.

When changes $\Delta[M]$ and/or $\Delta\left[\overline{K}(\omega)\right]$ are introduced into the mass and/or complex stiffness matrices of the system, respectively, a generic modifying matrix $\Delta[S(\omega)]$, representing these changes, can accordingly be written as

$$\Delta[S(\omega)]=\left[-\omega^2\Delta[M]+\Delta\left[\overline{K}(\omega)\right]\right] \tag{4}$$

yielding, hence, a modified receptance matrix $[R^*(\omega)]$ as follows

$$\left[R^*(\omega)\right]=\left[[S(\omega)]+\Delta[S(\omega)]\right]^{-1} \tag{5}$$

If the modifying matrix $\Delta[S(\omega)]$ may be factorised as

$$\Delta[S(\omega)] = [U(\omega)]_{nxr}[V(\omega)]_{rxn} \tag{6}$$

i.e., a matricial product of rank r (≤ n), it is then known that the receptance matrix $[R^*(\omega)]$ is alternatively given by (Householder 1953, Noble and Daniel 1988)

$$[R^*(\omega)] = [R(\omega)] - [R(\omega)][U(\omega)][W(\omega)]^{-1}[V(\omega)][R(\omega)] \tag{7}$$

with

$$[W(\omega)] = [I] + [V(\omega)][R(\omega)][U(\omega)] \tag{8}$$

where $[R^*(\omega)]$ = modified receptance matrix, $[R(\omega)]$ = original receptance matrix, $[U(\omega)]$ = n x r frequency-dependent matrix, $[V(\omega)]$ = r x n frequency-dependent matrix, $[W(\omega)]$ = r x r frequency-dependent matrix, and $[I]$ = identity matrix of order r.

It is immediately verified that equation (7) is tailor-made for experimental reanalysis of slightly modified systems (r much smaller than n) since it entails only information about the original receptance matrix and the modifying matrix. It may also be conveniently adapted in a number of ways, in accordance with its particular applications (Brandon 1990, Level et al 1996).

Apart from the condition stated by equation (6), it is important to note that equation (7) holds only if the matrix $[W(\omega)]$ is non-singular. A sufficient (but not necessary) condition is that any matrix norm of the product $[V(\omega)][R(\omega)][U(\omega)]$ be smaller than 1. This point has already been analysed before for similar circumstances (Brandon 1990).

By way of complementary information, it is noticed that there are other alternatives to equation (7) (Özgüven 1987, Brandon 1990, Heylen and Mas 1996). They, nevertheless, will not be pursued herein (nor will any comparisons), given their equivalence to equation (7).

Although the factorisation to be applied to the modifying matrix may seem a severe restriction (see equation(6)), that is not so. In fact, a great number of modifications likely to be made in practice can be represented in that form. Some simple examples, as mass changes and viscoelastic springs, will be discussed in the next section. More elaborate inserts, as damping layers and neutralisers, will be deferred to a subsequent paper (Lopes et al 1997).

3 NUMERICAL RESULTS

A three-degree-of-freedom system, as displayed in Figure 1, has been chosen for preliminary numerical investigations of the formulations presented in the preceding section. For such a simple system, it is feasible to introduce localised modifications and compare results over a significant frequency range. It is also expected that both changes in mass and stiffness can be assessed in a wholly controlled fashion.

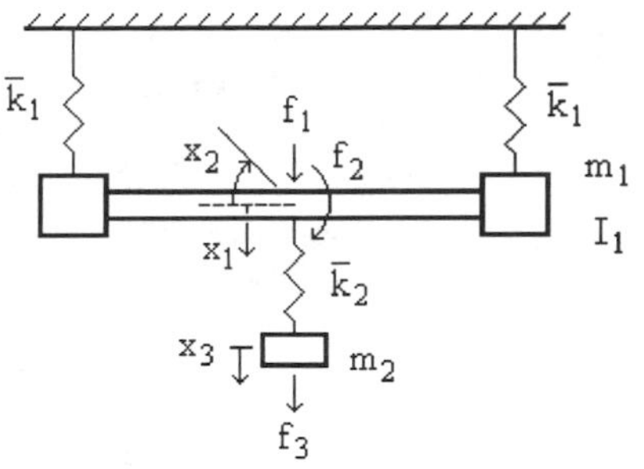

Figure 1. Three-degree-of-freedom system

By expressing its equations of motion in the frequency domain as

$$\left[-\omega^2\begin{bmatrix} m_1 & 0 & 0 \\ 0 & I_1 & 0 \\ 0 & 0 & m_2 \end{bmatrix} + \begin{bmatrix} 2\bar{k}_1(\omega)+\bar{k}_2(\omega) & 0 & -\bar{k}_2(\omega) \\ 0 & 2\bar{k}_1(\omega)L^2 & 0 \\ -\bar{k}_2(\omega) & 0 & \bar{k}_2(\omega) \end{bmatrix}\right]\begin{Bmatrix} X_1(\omega) \\ X_2(\omega) \\ X_3(\omega) \end{Bmatrix} = \begin{Bmatrix} F_1(\omega) \\ F_2(\omega) \\ F_3(\omega) \end{Bmatrix} \tag{9}$$

where $\bar{k}_j(\omega) = k_j(\omega)(1 + i\eta_j(\omega)), j = 1,2$, it follows that

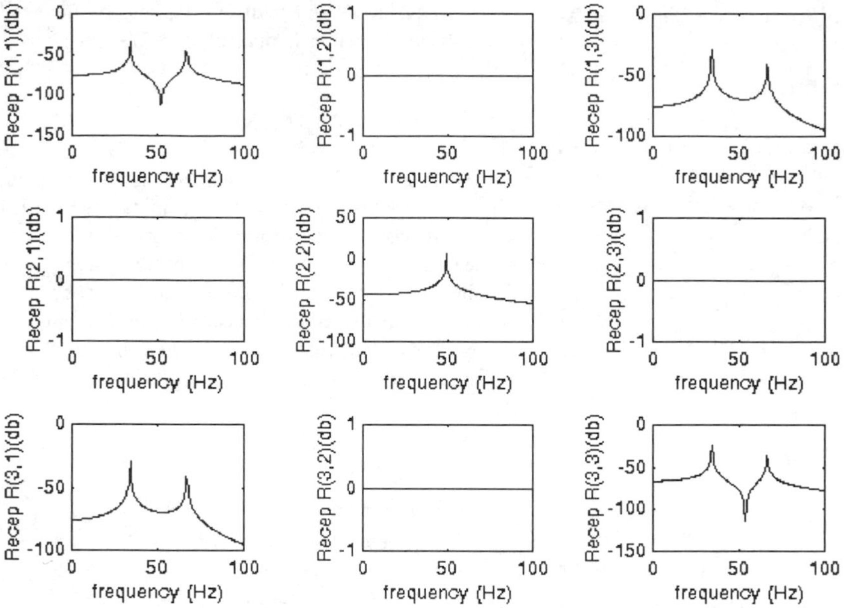

Figure 2. Pictorial representation of the original receptance matrix

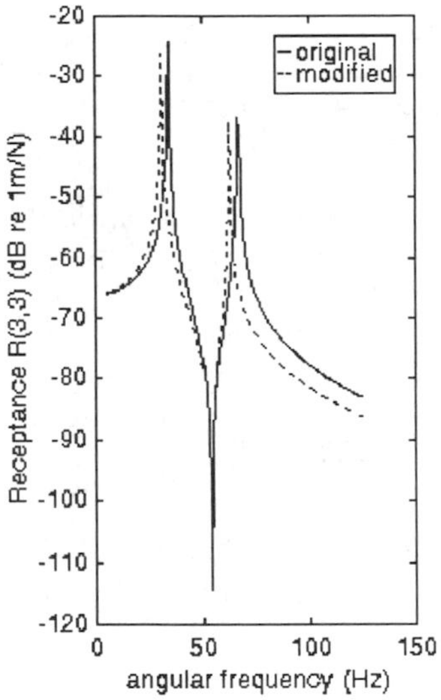

Figure 3 - Case I: original and modified $R_{33}(\omega)$

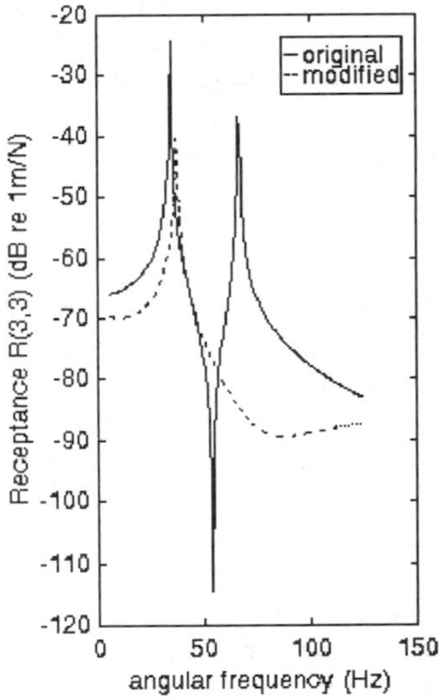

Figure 4 - Case II: original and modified $R_{33}(\omega)$

$$[S(\omega)] = \left[-\omega^2 \begin{bmatrix} m_1 & 0 & 0 \\ 0 & I_1 & 0 \\ 0 & 0 & m_2 \end{bmatrix} + \begin{bmatrix} 2\bar{k}_1(\omega) + \bar{k}_2(\omega) & 0 & -\bar{k}_2(\omega) \\ 0 & 2\bar{k}_1(\omega)L^2 & 0 \\ -\bar{k}_2(\omega) & 0 & \bar{k}_2(\omega) \end{bmatrix} \right] \tag{10}$$

and

$$[R(\omega)] = \left[-\omega^2 \begin{bmatrix} m_1 & 0 & 0 \\ 0 & I_1 & 0 \\ 0 & 0 & m_2 \end{bmatrix} + \begin{bmatrix} 2\bar{k}_1(\omega) + \bar{k}_2(\omega) & 0 & -\bar{k}_2(\omega) \\ 0 & 2\bar{k}_1(\omega)L^2 & 0 \\ -\bar{k}_2(\omega) & 0 & \bar{k}_2(\omega) \end{bmatrix} \right]^{-1} \tag{11}$$

Figure 2 shows a pictorial representation of equation (11), taking $m_1 = 3120$ kg, $I_1 = 0.059$ kg.m^2, $m_2 = 1120$ kg, $k_1 = 3015$ N.m^{-1}, $k_2 = 3054$ N.m^{-1}, $\eta_1 = \eta_2 = 0.015$ and $L = 0.153$ m. It has to be stressed here that the complex stiffnesses were only taken as constants for the sake of simplicity. As is well known, this assumption implies a system with structural damping and is not acceptable under generic excitations. In practice, however, the original receptance matrix (or parts of it) is supposed to be approximated experimentally.

From equation (11) it is seen that the displacement at mass m_2 due to a single excitation at the same mass is

$$X_3(\omega) = \{R_{31}(\omega) \quad R_{32}(\omega) \quad R_{33}(\omega)\} \begin{Bmatrix} 0 \\ 0 \\ F_3(\omega) \end{Bmatrix} = R_{33}(\omega)F_3(\omega) \tag{12}$$

If the spectrum of the excitation is particularly pernicious in the regions around the resonances, design changes may be required in order to avoid undesired responses.

For the system under investigation, a change in mass m_A at mass m_2 (hereafter called case I), is accordingly represented as

$$\Delta[S(\omega)] = -\omega^2 \begin{bmatrix} 0 & 0 & 0 \\ 0 & 0 & 0 \\ 0 & 0 & m_A \end{bmatrix} = \begin{Bmatrix} 0 \\ 0 \\ -\omega^2 m_A \end{Bmatrix} \{0 \quad 0 \quad 1\} = -\omega^2 m_A \begin{Bmatrix} 0 \\ 0 \\ 1 \end{Bmatrix} \{0 \quad 0 \quad 1\} \tag{13}$$

whereas the addition of a viscoelastic spring $\bar{k}_A(\omega) = k_A(\omega)(1 + i\eta_A(\omega))$ between masses m_1 and m_2 (case II) is given by

$$\Delta[S(\omega)] = \begin{bmatrix} \bar{k}_A(\omega) & 0 & -\bar{k}_A(\omega) \\ 0 & 0 & 0 \\ -\bar{k}_A(\omega) & 0 & \bar{k}_A(\omega) \end{bmatrix} = \bar{k}_A(\omega) \begin{Bmatrix} 1 \\ 0 \\ -1 \end{Bmatrix} \{1 \quad 0 \quad -1\} \tag{14}$$

The factorisations required in each case by equation (6) are illustrated by equations (13) and (14), respectively. Both modifications are of unit rank.

In case I, according to equations (7), (8), and (13), the new displacement will be

$$X_3^*(\omega) = R_{33}^*(\omega)F_3(\omega) \tag{15}$$

where

$$R_{33}^*(\omega) = \left(R_{33}(\omega) + (\omega^2 m_A / (1 - m_A R_{33}(\omega)))R_{33}^2(\omega)\right) \tag{16}$$

Equation (16) indicates that, apart from the proposed modification, just one element of the original receptance matrix is needed for computing the modified frequency response function of interest.

For case II, mutatis mutandis, it yields

$$R_{33}^*(\omega) = \left(R_{33}(\omega) - (\bar{k}_A(\omega) / (1 + \bar{k}_A(\omega)(R_{11}(\omega) - 2R_{13}(\omega) + R_{33}(\omega))))(R_{13}(\omega) - R_{33}(\omega))^2\right) \tag{17}$$

a formula which involves the complex stiffness of the additional viscoelastic spring and three elements of the original receptance matrix.

Figures 3 and 4 portrait both the original and the modified frequency response functions for cases I and II, as per equations (11), (16) and (17). In the simulations, the mass m_A equalled 0.408 kg, while the viscoelastic spring had, for the frequency range of concern (1Hz £ f (=ω/2p) £ 20 Hz, at 15 °C), the following characteristics:

$$k_A(\omega) \cong 3.05x10^3 f^{0.446} (N/m) \; ; \; \eta(\omega) \cong 0.60 f^{0.169}$$

Those characteristics correspond to springs made of ISODAMP C-1002, a viscoelastic material manufactured by EAR Specialty Composites.

4 DISCUSSION

Equation (7) is an exact expression. Therefore it is expected to yield the same results as the more time-consuming direct inversion indicated in equation (5). And that was exactly what happened in Figures 3 and 4, with the simulated frequency response functions shown in Figure 2 taking the part to be played by the measured ones.

The direct use of measured frequency response functions is arguable (Ewins 1984). In the absence of data smoothing procedures, inconsistent models can easily be generated. Moreover, for a number of reasons, the measured receptance matrix constitutes only an approximated model (often incomplete!) to the actual structure. It is however the authors' point of view that those are challenges worth facing.

The evaluation of new response properties at each frequency of interest is usually listed as a major weakness of response reanalysis. Nevertheless, when it comes to studying the effects of viscoelastic inserts in structures, it is an imperative. After all, viscoelastic materials are inherently frequency-dependent! Usual modal approaches do not account for such behaviour, which places directly-obtained response data in evidence (Lin and Ling, 1996).

The modifications presented in section 3 are very simple and just illustrate the general formulation. In order to consider more elaborate design changes, as the addition of damping layers or dynamic neutralisers, some refinements in modelling are required. As already mentioned, this point is going to be addressed in a near future.

Another very important feature, which was not addressed in this paper, is that of optimal selection of design changes. It is supposed that modifications will not be introduced in a trial and error basis but rather in systematic and realistic fashion. Attempts in this direction have already been reported (Sestieri and D'Ambrogio 1989, Heylen and Mass 1996).

5 CONCLUSIONS

A method of experimental response reanalysis has been extended in order to assess the addition of viscoelastic inserts into mechanical systems.

Numerical simulations for a simple mechanical system have illustrated the mathematical representations and provided results for preliminary discussions.

The evaluation of new response properties at each frequency of interest is an imperative when viscoelastic materials are introduced into a structure. It has therefore been argued that response models may play a major part in that area of experimental reanalysis.

REFERENCES

Brandon, J. A. 1990. *Strategies for structural dynamic modification*. Research Studies Press.

Bucher, I. & S. Braun 1994. Efficient optimization procedure for minimizing vibratory response via redesign or modification, part I: theory. *Journal of Sound and Vibration* 175(4):433-453.

Bucher, I. & S. Braun 1994. Efficient optimization procedure for minimizing vibratory response via redesign or modification, part II: examples. *Journal of Sound and Vibration* 175(4):455-473.

Ewins, D. J. 1984. *Modal Testing: Theory and Practice*. Research Studies Press.

Heylen, W. & P. Mas 1996. Optimisation of structural modifications based on frequency response functions. In P. Sas (ed.), *Proceedings of ISMA 21 Noise and Vibration Engineering, Leuven, 18-20 September 1996*, III:1843-1850.

Householder, A. S. 1953. *Principles of Numerical Analysis*. McGraw-Hill.

Level, P., D. Moraux, P. Drazetic & T. Tison 1996. On a direct inversion of the impedance matrix in response reanalysis. *Comm. in Num. Meth. in Engineering* 12(3):151-159.

Lin, R. M. & S-F Ling 1996. Identification of damping characteristics of viscoelastically damped structures using vibration test results. *Proc. Instn. Mech. Engrs., Part C: Journal of Mech. Engineering Science* 210:111-121.

Espíndola, J. J., E. M. O. Lopes & J. A. Brandon 1995. Numerical conditioning in the inverse problems of heterogeneous sub-structures. In L. Jezequel (ed.), *Proc. of the Intnl. Conference MV2 New Advances in Modal Synthesis of Large Structures: Non-linear, Damped and Non-deterministic Cases (pre-prints), Lyon, 5-6 October 1995*, 1:125-136.

Lopes, E. M. O., J. A. Brandon & J. J. de Espíndola 1997. Modelling of viscoelastic inserts for experimental reanalysis. *University of Wales Cardiff, Division of Mechanical Engineering and Energy Studies, Report No. 2268* (submitted for publication).

Moraux, D., P. Level & Y. Ravalard 1994. Comparison of forced response reanalysis techniques - their potentials and limitations. In P. Sas (ed), *Proceedings of ISMA 19 Tools for Noise and Vibration Analysis, Leuven, 12-14 September 1994*, II:887-897.

Nashif, A. D., D. I. G. Jones & J. P. Henderson 1985, *Vibration damping*, John Wiley & Sons.

Noble, B. & J. W. Daniel 1988. *Applied linear algebra*. Prentice-Hall.

Özgüven, H. N. 1987. A new method for harmonic response of non-proportionally damped structures using undamped modal data. *Journal of Sound and Vibration* 117(2):313-328.

Ravi, S. S. A., T. K. Kundra & B. C. Nakra 1995. A response re-analysis of damped beams using eigenparameter perturbation. *Journal of Sound and Vibration* 179(3):399-412.

Ravi, S. S. A., T. K. Kundra & B. C. Nakra 1996. Reanalysis of plates modified by free damping layer treatment. *Computers and Structures* 58(3):535-541.

Sestieri, A. & W. D'Ambrogio 1989, Why be modal: how to avoid the use of modes in the modification of vibrating systems. *The Intnl. Journal of Anal. and Experim. Modal Analysis* 4(1): 25-30.

Srivastava, R. K. & T. K. Kundra 1993. Structural dynamic modification with damped beam elements. *Computers and Structures* 48(5):943-950.

Sun, C. T. & Y. P. Lu 1995. *Vibration damping of structural elements*. Prentice Hall.

Tahtali, M. & H. N. Özgüven 1994. Vibration analysis of dynamic structures using a new structural modification method. In *Proc. of the 6th Intnl. Machine Design and Production Conference, Ankara, 21-23 September 1994*, 511-520.

E. M. O. Lopes is sponsored by CNPq, Brasília, Brazil.

Modern Practice in Stress and Vibration Analysis, Gilchrist (ed.) © *1997 Balkema, Rotterdam, ISBN 90 5410 896 7*

Development of a vibration control strategy for flexible robot arms

D.W. Lang & W.J.O'Connor
Mechanical Engineering Department, University College Dublin, Ireland

ABSTRACT: A novel control strategy for flexible robot arms is presented which combines vibration absorption and position control. It is based on mechanical wave propagation and absorption. The arm is modelled by a lumped-parameter mass-spring system with an actuator at one end and a load mass at the other. The actuator is required to position the remote load and, simultaneously, to provide active vibration damping. It does so by propagating mechanical waves through the system and absorbing reflected waves. Only the first two masses and springs in the system need to be characterised and observed to determine the required actuator movement. The control algorithm is robust and compares very favourably with the time-optimal performance of bang-bang control. It is also inherently adaptive. Finally the extension of the concept to the control of arms modelled as Euler-Bernoulli beams is investigated and some practical difficulties are identified and discussed.

1 INTRODUCTION

Flexible robot arms offer a number of significant advantages over more conventional stiff and heavy robot arms. They consume less energy, exhibit safer operation due to a lighter and 'softer' structure, and for a given input power can respond more rapidly. Certain applications in the aerospace and nuclear power industries require the use of light, long reach (and therefore very flexible), robot arms. Even in industries traditionally dominated by stiff and heavy robot arms the drive for increased throughput and reduced costs has brought about a renewed interest in flexible robot arms, with their inherent advantages.

Their main disadvantage, and the reason they are not widely used, is precisely their flexibility, which may give rise to serious oscillations of the arm after any movement, resulting in a loss of accuracy in positioning the arm tip. Furthermore, internal (material) damping is typically light and the addition of passive external dampers would impede the dynamic response of the arm, offsetting the advantages gained by the light structure. Thus any oscillations of the arm may take unacceptably long to decay naturally. Hence, some form of active vibration absorption, ideally using the existing arm actuator, is necessary to achieve acceptable performance.

There are two principal approaches to control of flexible robot arms. The first, 'open-loop', approach attempts to move the arm in such a way as to avoid exciting resonances of the structure. There is no 'feedback' of the state of the arm, i.e. no attempt is made to compare the state of the arm with the desired state and hence this approach relies on previous knowledge of the arm dynamics. In general, however, the dynamics of the arm are difficult to predict as it may be manipulating a load-mass of unknown magnitude and so this approach has limited practical use. The second, 'closed-loop', approach aims to move the arm as quickly as possible and subsequently remove residual vibration from the structure by using the arm actuator as a vibration absorber. This approach generally relies on feedback of some or all of the variables describing the state of the arm. A complication associated with such a control strategy for flexible robot arms is that the precise location of the arm tip is frequently difficult to ascertain, so information regarding the state of the arm may have to be based on the output of sensors located close to the actuator. (This difficulty is one further reason why traditionally robots have been stiff and heavy.)

Much research has been carried out in recent decades on the problem and many different strategies exist for control of flexible robot arms. Book (1993) reviews a number of approaches including feedback and feedforward control, passive damping enhancement, and structural design strategies. Meckl and Seering (1985) develop an open-loop forcing function designed to avoid exciting resonances of the system under control. This is found to compare favourably with the time-optimal performance of 'bang-bang' control. Abduljabar et al. (1992) develop and compare several control strategies for a single-link flexible robot arm modelled as an Euler-Bernoulli beam with an actuator located at the base. A controller design robust to variations in the natural frequency of the flexible beam under control is proposed by Korolov and Chen (1989), while Feliu et al. (1990) propose a simple and elegant adaptive control scheme for a single-link flexible robot arm which is modelled as having only one mode of vibration.

Despite the large amount of research that has gone into the problem in recent years, for most practical applications of flexible robot arms, the control technique continues to depend on moving the joints slowly and then waiting for residual oscillations to decay naturally. Thus, the challenge is to develop a practical control strategy for flexible robot arms which, with minimum knowledge of the dynamics of the arm, and with minimum necessity for sensory input, can move the arm in such a way so as to minimise the degree of vibration excited in the arm by the movement, and which can then absorb these (and any other) vibrations which are excited in as short a time as possible.

2 WAVE-ABSORPTION CONTROL

Consider the multiple degree-of-freedom (MDOF) lumped parameter system shown in Figure 1. Such a system serves as a good initial model of a flexible robot arm. Damping is assumed to be negligible as this presents the 'worst-case' scenario in the context of vibration control.

To understand how such a system may be controlled using wave absorption techniques, consider initially that the system is completely uniform, i.e. $k_1 = k_2 = \ldots = k_n = k$ and $m_1 = m_2 = \ldots = m_n = m$. Any movement of the actuator will set-up a transient wave that will initially propagate rightwards through the system. If the actuator undergoes a net displacement as a result of its movement, then the wave propagated by that movement will cause the same net displacement of any mass as the wave passes that mass. When the wave reaches the final mass, it will reflect and start to travel back towards the actuator. Somewhat surprisingly, as the returning wave passes any given mass, it will cause the mass to displace by the same amount *again* (in the same direction as the original displacement). Thus if the actuator propagates a wave with a net displacement of one unit then, after one reflection, the end-mass will have undergone a net displacement of two units. Now, if a zero reflection condition can be established at the actuator for the returning wave, then the actuator will effectively absorb the returning wave, and the entire system (including the actuator) will come to rest having undergone a net displacement of two units.

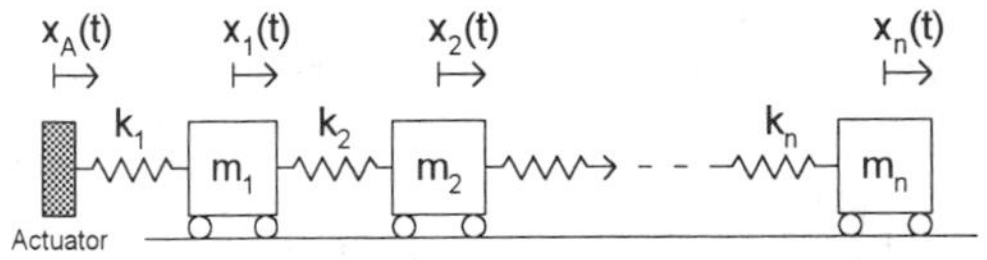

Figure 1. Multiple degree-of-freedom lumped parameter mass-spring system representing robot arm.

In other words, to move the end-mass a desired amount, the actuator must initially move half this amount, and then subsequently move the remaining distance in the course of absorbing the returning wave. This has the effect of efficiently absorbing all vibration energy out of the system while leaving it at rest at the desired new position. Thus the two requirements of position control and vibration absorption are combined, independently of the load mass. This concept of wave propagation by the actuator, reflection at the end-mass, and detection and absorption of reflected waves, forms the basis of the control strategy for multiple degree-of-freedom systems such as that shown in Figure 1.

The concept may be extended to systems that are not uniform. Any non-uniformity in the system (differing masses or stiffnesses along the arm) will cause an incident wave to be partially reflected and partially transmitted at each non-uniformity. Nevertheless, if all reflected waves returning to the actuator are absorbed, then it may be shown that the end-mass will ultimately come to rest at the correct position.

2.1 *Absorption of Reflected Waves*

For the actuator to present a zero reflection (or total absorption) condition to any returning wave two requirements must be met. Firstly, from the motion of the masses, *returning* transient waves need to be detected, and distinguished from outgoing waves. This is not a trivial problem and is considered below. Secondly, there should be no dynamic mismatch between the first mass-spring unit and the actuator. To achieve this the concept of dynamic impedance might be considered as a first approach. This, however, assumes continuous sinusoidal waves whereas in the present problem transient effects are dominant. However, if the actuator can be made to behave as if it is the first mass in an *infinitely long string* of uniform mass-spring units, it will effectively absorb both transient and steady-state waves, as such a system will absorb all waves without reflection.

In order to develop a control law for the actuator that will allow it behave as if it is the first mass in an infinitely long string of uniform mass-spring units, it is necessary to determine how such a system responds to a unit impulse input, as once the unit impulse response has been determined the response to an arbitrary input may be obtained by convolution. Consider the semi-infinite (i.e. extending to infinity in one direction only) uniform mass-spring system shown in Figure 2.

If $x_0(t)$, the motion of the actuator and $x_1(t)$, the motion of the first mass in the infinite mass-spring system are transformed to the Laplace domain to yield $X_0(s)$ and $X_1(s)$ respectively, then it is possible to define a transfer function in the Laplace domain, $G(s)$, which describes the response of the first mass in the system to any actuator movement by:

$$X_1(s) = G(s)X_0(s) \tag{1}$$

Essentially $G(s)$ describes how a wave propagates through the system. For any rightwards propagating wave:

$$X_{n+1}(s) = G(s)X_n(s) \tag{2}$$

where $X_n(s)$ is the Laplace transform of the motion of the nth mass in the system and similarly $X_{n+1}(s)$ is the Laplace transform of the motion of the $(n+1)^{th}$ mass in the system. Thus (for a rightwards propagating wave):

$$X_n(s) = G^n(s)X_0(s) \tag{3}$$

To derive an expression for $G(s)$, two approaches may be taken. The first approach results in a convenient closed-form expression for $G(s)$, which when converted to the time domain yields an infinite-series type expansion for $g(t)$, the time-domain impulse response. The second approach, which is more useful in practice, approximates the response of the infinite mass-spring system by that of a long, but finite, mass-spring system.

To determine the exact form of $G(s)$, consider the uniform infinite-infinite (i.e. extending to infinity in both directions) mass-spring system shown in Figure 3.

The equation of motion for mass (2) is:

$$m\ddot{x}_2 = k(x_1(t) - 2x_2(t) + x_3(t)) \tag{4}$$

Assuming all initial conditions are zero, and letting $\omega_n = \sqrt{k/m}$, this may be transformed to the Laplace domain to yield:

$$s^2 X_2(s) = \omega_n^2(X_1(s) - 2X_2(s) + X_3(s)) \tag{5}$$

Suppose there is a wave travelling from left to right. Then:

$$X_2(s) = G(s)X_1(s) \tag{6}$$

$$X_3(s) = G^2(s)X_1(s) \tag{7}$$

Substituting equations (6) and (7) into equation (5), dividing across by $X_1(s)$, and rearranging, yields:

$$\omega_n^2 G^2(s) - (s^2 + 2\omega_n^2)G(s) + \omega_n^2 = 0 \tag{8}$$

This is a quadratic equation in $G(s)$, and thus, with some simplification, the solution is given by:

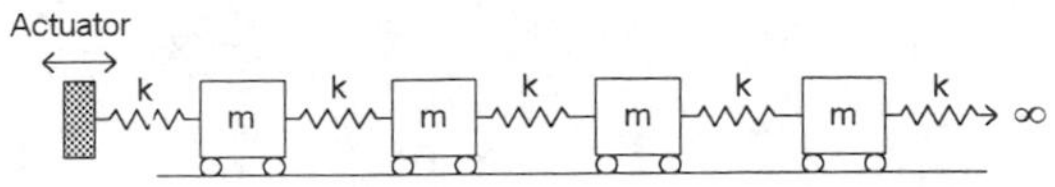

Figure 2. Semi-infinite uniform mass-spring system.

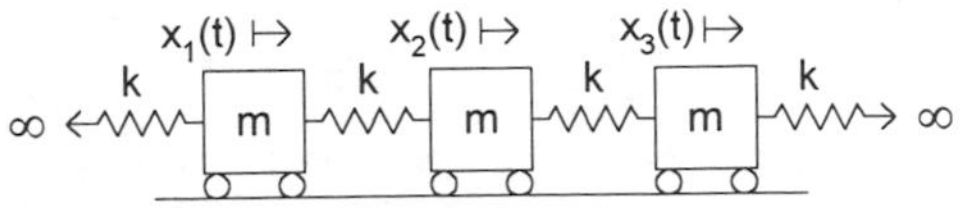

Figure 3. Infinite-infinite uniform mass-spring system.

$$G(s)=\frac{1}{2\omega_n^2}\left((s^2+2\omega_n^2)\pm s\sqrt{s^2+4\omega_n^2}\right) \qquad (9)$$

This gives two solutions for $G(s)$. Only one of these tends to zero as s tends to infinity (that with the negative sign before the radical), and thus is the only solution which may be transformed to the time-domain. The other solution corresponds to wave propagating from right to left, which contradicts the initial assumption. Hence, the closed form solution for $G(s)$ is given by:

$$G(s)=\frac{1}{2\omega_n^2}\left((s^2+2\omega_n^2)-s\sqrt{s^2+4\omega_n^2}\right) \qquad (10)$$

Expanding the term under the radical using the binomial theorem and simplification yields an infinite series in s, the mth term of which is given by:

$$T_m(s)=-\frac{1}{2}\binom{1/2}{m+1}\left(4^{m+1}\right)\left(\frac{\omega_n}{s}\right)^{2m} \qquad (11)$$

Transforming back to the time-domain yields the mth term in the infinite series expansion of $g(t)$:

$$T_m(t)=-\frac{1}{2(2m-1)!}\binom{1/2}{m+1}\left(4^{m+1}\right)\omega_n^{2m}t^{2m-1} \qquad (12)$$

The closed form solution for $G(s)$ given in equation (10) is useful for determining properties of the MDOF control algorithm (such as stability, settling position etc.), however for practical purposes a more useful approximation to $G(s)$ is obtained by truncating the response of a long, but finite, mass-spring system as shown in Figure 4.

An approximation to the infinite mass-spring system impulse response can be made by taking the first section of the waveform shown in Figure 4 (up to the cut-off time, T, which is given in terms of normalised time, $\omega_n t$), and thereafter setting the response identically equal to zero.

The response of the first mass in a system consisting of m mass-spring units to a unit impulse input at $t=0$, is of the form:

$$x_1(t)=\sum_{i=1}^{m}A_i\sin f_i\omega_n t \qquad (13)$$

where A_i are the amplitudes of the m frequencies present in the response and f_i are the ratios of these frequencies to ω_n. Hence, the approximated infinite

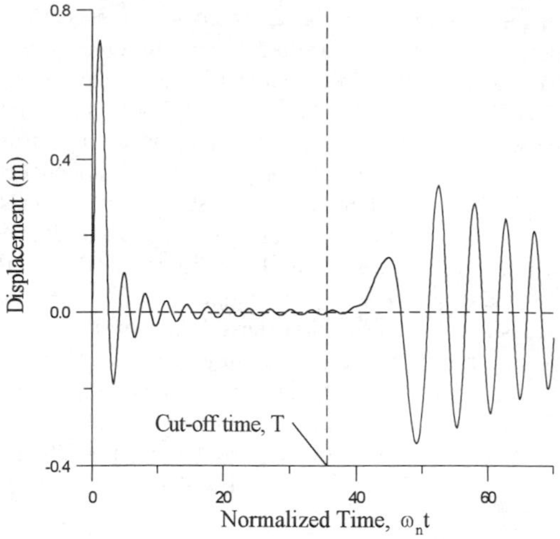

Figure 4. Unit impulse response of a finite length mass-spring system. If the response is set to zero after the cut-off time, T, then a good approximation to the impulse response of the infinite mass-spring system is obtained.

mass-spring system impulse response may be defined as:

$$g(t)=\begin{cases}\sum_{i=1}^{m}A_i\sin f_i\omega_n t & (0\le t\le \frac{T}{\omega_n})\\ 0 & (t>\frac{T}{\omega_n})\end{cases} \qquad (14)$$

To absorb reflected waves, it is necessary to be able to distinguish between rightwards and leftwards propagating waves in the mass-spring system under control. Consider again the infinite-infinite uniform mass-spring system shown in Figure 3, and suppose there are both leftwards and rightwards propagating waves in the system. As the system is linear, the motion of mass (1) may be regarded as consisting of two components, motion due to rightwards propagating waves, $R_1(s)$, and motion due to leftward propagating waves, $L_1(s)$. Hence:

$$X_1(s)=R_1(s)+L_1(s) \qquad (15)$$

Similarly,

$$X_2(s)=R_2(s)+L_2(s) \qquad (16)$$

Now,

$$R_2(s)=G(s)R_1(s) \qquad (17)$$

$$L_1(s) = G(s)L_2(s) \tag{18}$$

Substituting equations (17) & (18) into equations (15) & (16) and solving between the two resulting equations for $R_1(s)$ yields:

$$R_1(s) = \frac{X_1(s) - G(s)X_2(s)}{1 - G^2(s)} \tag{19}$$

Solving between this result and equations (15) and (18) gives:

$$L_2(s) = \frac{X_2(s) - G(s)X_1(s)}{1 - G^2(s)} \tag{20}$$

Now, the component of motion of the mass to the left of mass (1) which is due to a leftward propagating wave, will be given by:

$$L_0(s) = G^2(s)L_2(s) \tag{21}$$

Hence:

$$L_0(s) = \frac{G^2(s)\left(X_2(s) - G(s)X_1(s)\right)}{1 - G^2(s)} \tag{22}$$

If $X_1(s)$ and $X_2(s)$ are the positions of the first two masses in the MDOF mass-spring system to be controlled (i.e. the system shown in Figure 1), then $L_0(s)$ is the movement that the actuator must execute in order to completely absorb any incoming waves, and so may be used to form part of the control signal to the actuator. Equation (23) may be implemented in the time-domain using a type of positive feed-back loop to determine this control signal.

Thus, if it is possible to observe $X_1(s)$ and $X_2(s)$, it will be possible to differentiate between incoming and outgoing waves, and hence, to instruct the actuator to absorb any incoming (returning) waves. To do this, it is sufficient to have knowledge of the nature of the system close to the actuator (i.e. the nature of the first two spring-mass units which should be uniform), and to be able to observe the position of the first two masses. No knowledge of the nature of the system beyond the first two mass-spring units is required.

2.2 *Implementation of the Control Algorithm*

Figure 5 shows, in block diagram form, how the wave-absorption based MDOF control algorithm may be implemented in practice. The various $G(s)$

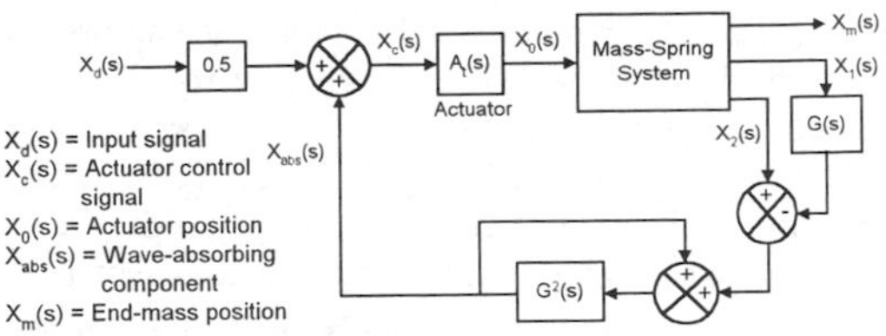

Figure 5. Block diagram representation of MDOF control system.

blocks may be implemented in the time domain by simply convolving the appropriate signal with the time-domain approximation to $G(s)$ given in equation (14).

3 PERFORMANCE EVALUATION

To evaluate the performance of the MDOF control algorithm, a computer-based simulation of the control system as shown in Figure 5 has been developed. Figure 6 shows a typical simulated response of the end-mass of a uniform three-mass system to a unit step positional input at t = 0.05s, which the actuator modelled as a dc servomotor.

It is found experimentally that the control system is capable of both controlling and effectively absorbing vibrations from any multiple degree-of-freedom mass-spring system provided that the nature of the first two units in the mass-spring system has been accurately characterised, and that all vibration can propagate back to the actuator. This is a significant point as the first two mass-spring units in the lumped-parameter system act as a low-pass filter and thus may not allow high frequency vibrations to propagate back to the actuator. In practice however, if the multiple degree-of-freedom system is being used to model a flexible robot arm, this last point is unlikely to present a problem as such an arm is likely to be of uniform stiffness or else will stiffen towards the actuator.

The performance of the control algorithm compares very favourably to the time-optimal performance of the open-loop 'bang-bang' control algorithm (for a full description of the 'bang-bang' control strategy see, for example, Meckl and Seering (1985)), and also compares favourably with other full state-variable feedback designs (despite the fact that the wave-absorption control algorithm *does not* require full state-variable feedback).

Other factors affecting the performance of the MDOF control algorithm include the presence of damping and the nature of the actuator being used to control the system. When considering the effects of damping, two situations must be considered: (a) material or internal damping, which may be modelled by viscous dampers between adjacent masses in the mass-spring system, and (b) environmental or external damping (i.e. viscous friction as a result of the arm being moved through air, for example), which may be modelled by viscous dampers between individual masses and 'ground'. Material damping causes a slower response with less residual vibration but does not however affect either the final end-mass position nor the ability of the control law to control the mass-spring system effectively.

Environmental damping also slows down the propagation of wave energy through the system, and has the additional effect of removing momentum from the propagating wave, resulting in less net displacement as the propagating wave passes a given mass. This will cause an error in the final settling position of the system. However, this problem may be overcome by monitoring the settling position of the actuator, and correcting for any discrepancy between it and the desired settling position. In a practical case, the length of time taken for the actuator to settle after propagating a wave could be determined by initially performing a test movement with the system.

The use of a non-ideal actuator does not impair the performance of the control system unduly provided that the bandwidth of the actuator is sufficiently high relative to the bandwidth of the system under control. In fact, as the section of the system nearest the actuator acts as a low-pass filter, the actuator bandwidth merely needs to be sufficiently high relative to that section of the system. This further reduces the need to characterise the system beyond the first two mass-spring units.

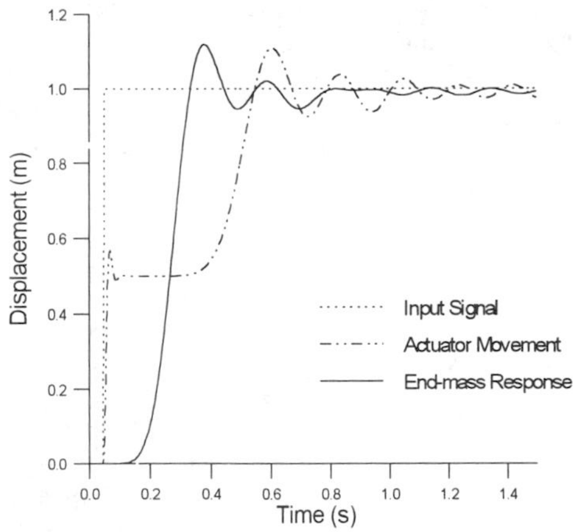

Figure 6. Simulated response of end-mass of uniform three-mass system to unit step input at t = 0.05s. The actuator is modelled as a dc servomotor of high bandwidth relative to that of the system under control.

For any control system, a primary concern is to determine what conditions, if any, give rise to unstable operation. By considering the frequency response of an open-loop equivalent system derived from the system shown in Figure 5 and applying the Nyquist stability criterion, it is possible to determine conditions for stable operation. It is found that the nature of the system under control (i.e. the presence/absence of damping and the nature and number of mass-spring units in the system under control) does not adversely affect the stability of the control system. However, it may be shown that the actuator bandwidth is a significant factor in ensuring the stability of the control system. This must be sufficiently large relative to the bandwidth of the system under control to ensure stability.

4 EXTENSION TO CONTINUOUS SYSTEMS

The wave-absorption control algorithm offers significant advantages over more conventional control algorithms for the control of multiple degree-of-freedom elastic-inertial systems. Principal amongst these advantages is the ability of the control algorithm to deal with any number of degrees of freedom, and thus any number of modes of vibration, i.e. there is theoretically no limit on the number of modes of vibration that may be controlled. Thus it should be possible to extend the concept of wave-absorption control to continuous systems in which the inertial and elastic properties are distributed in a continuous way throughout the system. Such systems have an infinite number of vibratory modes. This development would allow the control of flexible robot arms modelled as continuous systems, which is a significantly more realistic model than the discrete lumped-parameter model used so far.

The Euler-Bernoulli beam model provides a relatively realistic model of a single-link flexible robot arm. This model assumes that all deflection is due to bending and neglects the effects of shear deformation and rotary inertia which are only significant for the higher vibratory modes (which are unlikely to be encountered in practice). The standard form of the Euler-Bernoulli beam equation is given by:

$$\frac{\partial^2 y(x,t)}{\partial t^2} + \frac{EI}{m}\frac{\partial^4 y(x,t)}{\partial x^4} = 0 \qquad (23)$$

where EI is the flexural rigidity of the beam and m is the mass per unit length.

In order to apply the wave-absorption control technique (as developed for MDOF systems) to flexible robot arms modelled as Euler-Bernoulli beams, it is necessary to derive the transfer function that describes the propagation of transient flexural waves along semi-infinite length Euler-Bernoulli beams. (This transfer function is analogous to $G(s)$, which describes the propagation of transient waves along semi-infinite discrete mass-spring systems.) Equation (23) may be transformed to the Laplace domain to give:

$$\frac{d^4 Y(s,x)}{dx^4} - q^4 Y(s,x) = 0 \qquad (24)$$

$$\text{where } q^4 = -\frac{s^2 m}{EI} \qquad (25)$$

The general solution to equation (24) is of the form:

$$\begin{aligned} Y(s,x) &= C_1 \cosh qx + C_2 \sinh qx \\ &+ C_3 \cos qx + C_4 \sin qx \end{aligned} \qquad (26)$$

The value of the four arbitrary constants C_1 to C_4 must be determined from the boundary conditions. For the semi-infinite length beam the first two boundary conditions are given by the deflection and slope of the beam at $x = 0$ ($Y_0(s)$ and $Y_0'(s)$ respectively). The remaining 'boundary' conditions may be derived from the fact that effectively no other boundary exists, as the beam is of semi-infinite length. Substitution of the four constants thus obtained back into equation (26) yields:

$$\begin{aligned} Y(s,x) &= Y_0(s)\left[\frac{e^{-qx} + e^{-jqx} - je^{-qx} + je^{-jqx}}{2}\right] \\ &+ Y_0'(s)\left[\frac{-e^{-qx} + e^{-jqx} - je^{-qx} + je^{-jqx}}{2q}\right] \end{aligned} \qquad (27)$$

Note that the overall response $Y(s,x)$, is a linear combination of two individual transfer functions, one acting on a *deflection* input at $x = 0$, and the other acting on a *slope* input at $x = 0$. Note also that both deflection and slope are necessary to completely define the state of the beam at any given point.

From equation (25) it can be seen that there are four solutions for q, each with a corresponding solution for equation (27). However, two pairs of these solutions for q give the same solution for $Y(s,x)$, and thus there are only two distinct solutions. Furthermore one of these solutions corresponds to leftwards propagating waves, which contradicts the initial assumption. Hence the following equation is obtained which describes the propagation of transient flexural waves along semi-infinite length Euler-Bernoulli beams:

$$\begin{aligned} Y(s,x) &= Y_0(s)\left[e^{-ax\sqrt{s}} \cos ax\sqrt{s} + e^{-ax\sqrt{s}} \sin ax\sqrt{s}\right] \\ &+ Y_0'(s)\left[\frac{1}{a\sqrt{s}} e^{-ax\sqrt{s}} \sin ax\sqrt{s}\right] \end{aligned} \qquad (28)$$

where $a = \sqrt[4]{\frac{m}{4EI}}$ (a can be thought of as a parameter indicating the inertia to stiffness ratio for the beam). Equation (28) may be transformed back to the time-domain to give:

$$\begin{aligned} y(x,t) &= y_0(t) * \left[\frac{ax}{\sqrt{\pi t^3}} \sin\frac{(ax)^2}{2t}\right] \\ &+ y_0'(t) * \left[\frac{1}{a\sqrt{\pi t}} \sin\frac{(ax)^2}{2t}\right] \end{aligned} \qquad (29)$$

where ' * ' represents convolution. As the state of the beam at any point x is defined by both deflection and slope, it is convenient to represent this using vector notation, defined by:

$$\vec{Y}(s,x) = \begin{Bmatrix} Y(s,x) \\ Y'(s,x) \end{Bmatrix} \qquad (30)$$

The response of the beam to a deflection and/or slope input may then be written:

$$\vec{Y}(s,x) = [\theta]\vec{Y}_0(s) \qquad (31)$$

where $\vec{Y}_0(s)$ is the vector describing the state of the beam at $x = 0$ and $[\theta]$ is a 2×2 matrix of transfer functions. The first row elements of $[\theta]$ are the two transfer functions contained in equation (28), while the second row elements are the spatial derivatives of the corresponding first row elements.

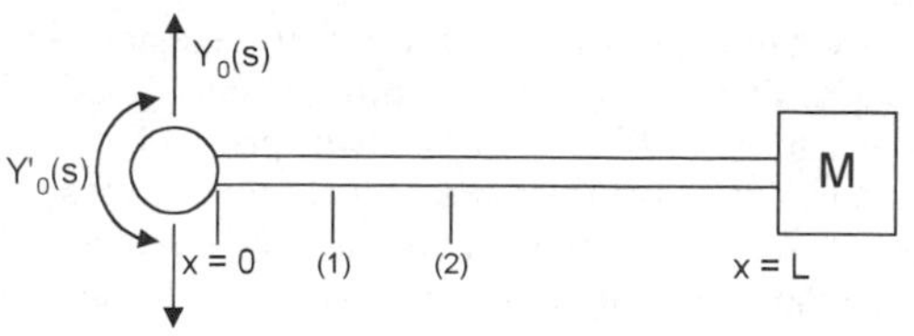

Figure 7. Model of a flexible robot arm based on Euler-Bernoulli beam with an actuator at the left-hand end capable of giving a slope and deflection input, and an inertia load at the right-hand end. Information on the state of the arm is fedback from points (1) and (2).

With the transfer functions describing transient wave propagation along Euler-Bernoulli beams established, it is possible to develop a control strategy based on the same principle of wave propagation, followed by reflection at the end-mass, then detection and absorption by the actuator, as has been proposed for MDOF systems. Such a control system would require feedback from two observation points close to the actuator as shown in Figure 7.

Practical implementation of this control algorithm requires the ability to convolve signals with the time-domain version of the transfer functions contained in $[\theta]$ (these time-domain functions are essentially responses to unit deflection and unit slope impulse inputs). A difficulty arises here as these functions contain infinitely high frequency components (due to the dispersive nature of the Euler-Bernoulli beam) and thus cannot be discretised without loss of information. Further work is required to allow implementation of the control algorithm on a digital controller.

5 CONCLUSIONS

Control of flexible robot arms aims to take advantage of their improved dynamic response and reduced energy consumption to increase throughput and drive down costs. A novel control technique has been developed which combines the dual requirements of position control and active vibration absorption using wave control concepts. The control strategy has been simulated for flexible robot arms modelled as discrete, lumped-parameter mass-spring systems. Its performance compares favourably with both the time-optimal performance of bang-bang control and with other existing control strategies. Furthermore, the control algorithm requires only a limited amount of sensory input and limited knowledge of the dynamics of the system under control. The concept has been extended to the control of flexible robot arms modelled as continuous systems, although additional work is required before practical implementation is possible.

REFERENCES

Abduljabbar, Z., ElMadany, M. M., Al-Dokhiel, H. D., "Controller Design of a One-link Flexible Robot Arm," *Computers and Structures,* Vol. 49, No. 1, 1993, pp. 117-126.

Book, W. J., "Controlled Motion in an Elastic World," ASME *Journal of Dynamic Systems, Measurement, and Control,* Vol. 115, June 1993, pp. 252-261.

Feliu, V., Rattan, S. R., and Brown, H. B., "Adaptive Control of a Single-Link Flexible Manipulator," IEEE *Control Systems Magazine*, Feb. 1990, pp. 29-33.

Jayasuriya, S., Choura, S., "On the Finite Settling Time and Residual Vibration Control of Flexible Structures," *Journal of Sound and Vibration,* Vol. 148, No. 1, 1991, pp. 117-136.

Korolov, V. V., Chen, Y. H., "Controller Design Robust to Frequency Variation in a One-link Flexible Robot Arm," ASME *Journal of Dynamic Systems, Measurement, and Control,* Vol. 111, March 1989, pp. 9-14.

Meckl, P., Seering, W., "Minimizing Residual Vibration for Point-to-Point Motion," ASME *Journal of Vibration, Acoustics, Stress, and Reliability in Design,* Vol. 107, Oct. 1985, pp. 378-382.

9 Modelling of damage and fracture

Modern Practice in Stress and Vibration Analysis, Gilchrist (ed.) © *1997 Balkema, Rotterdam, ISBN 90 5410 896 7*

Development and implementation of an application phase analysis feature to simulate rapid crack propagation in a gas pipeline

P.E.O'Donoghue
University College Dublin, Ireland

ABSTRACT: An analysis based methodology to predict crack propagation or crack arrest in gas pipelines is implemented here. This is achieved using a unique fluid/structure/fracture interaction program called PFRAC. This paper describes a modification to PFRAC such that the crack location as a function of time can be determined for a given fracture toughness of the pipe material (application phase analysis). Example problems are presented to illustrate these concepts.

1 INTRODUCTION

There have been several catastrophic instances where cracks have propagated axially in a rapid manner in gas pressurised pipelines. This has been particularly true for steel gas transmission pipelines but it has also been identified as a problem in the gas distribution industry. Clearly it is important that these rupture events must be prevented. This problem is recognised as one of the most challenging issues in dynamic fracture mechanics and it has attracted significant attention from the research community over the last forty years.

The development of an analysis model for dynamic crack propagation in a pipeline has been studied extensively over the years (see reference [O'Donoghue et al. 1997] for a more detailed list of relevant publications). The recent development of a coupled fluid/structure/fracture analysis procedure, PFRAC (Pipeline Fracture Analysis Code) at Southwest Research Institute [O'Donoghue et al. 1991] has overcome many of the difficulties associated with earlier models. The primary features of this package are outlined in Section 2.

Up to now, the PFRAC code has been used primarily to simulate fracture experiments where the crack speed (location) is measured as a function of time. In this case the input is the crack speed as a function of time and PFRAC then computes the crack driving force. In this paper, the methodology is extended to predict the crack speed for a given material fracture toughness. While this crack speed is not of great interest to the utility engineer, more importantly, the predictive feature can also be used to determine whether crack arrest will occur.

2 FEATURES OF PFRAC

Due to the complexities of the rapid crack propagation event, such as gas escaping from the opening breech, the calculation of the crack driving force is a non-trivial task. Fortunately, the PFRAC package allows this calculation. There are three basic segments to this code: a structural mechanics unit, a fluid mechanics unit and a fracture mechanics unit. These three modules are fully coupled together and are used to analyse crack propagation in a pipeline.

The primary requirements of the structural dynamics code are that it must model the large deformations of cylindrical structures. The program incorporates a Lagrangian finite element description with four node quadrilateral elements that allow for geometric nonlinearities. An explicit finite difference scheme is used to march forward in time. This code is ideally suited for shell like structures undergoing large deformations such as the flap opening experienced in ruptured pipes. A three dimensional finite difference scheme is used to model the complex highly transient flow that takes place when the pipeline is fractured and the gas is escaping. Finally, a node release algorithm is implemented to numerically simulate crack propagation in the finite element module.

The shell finite element module and the finite difference module are linked such that the gas pressures are used to calculate the forces on the opening pipe wall and this in turn defines the confinement for the gas. Validation of the fluid/structure interaction program has been accomplished through comparison with full scale data both for steel gas transmission pipelines and polyethylene gas distribution pipelines. The program has been successfully used to predict cases of crack propagation and arrest. In addition to the gas pipeline applications, PFRAC has also been used to make predictions for fracture in aircraft and to assess the integrity of space stations to debris impact.

3 FRACTURE MECHANICS CONCEPTS

Central to the successful implementation of a fracture analysis methodology is the requirement for a valid crack-tip characterising parameter. For the inelastic dynamic behaviour that is observed during fracture of line pipe steel, the crack tip opening angle (CTOA) is found to be the most appropriate for the present application [Venzi et al. 1980]. This is the driving force measure that is used here. Crack propagation then takes place when this driving force equals the critical CTOA for the material of interest. This governing equation for fracture can be expressed as

$$CTOA = (CTOA)_c \tag{1}$$

where $(CTOA)_c$ is the material toughness which can be a function of velocity. If the driving force is less than the toughness, then crack propagation will not take place.

The CTOA can be calculated very easily using PFRAC. This quantity is of course dependent on the particular pipeline operating conditions. Measurements of the $(CTOA)_c$ have been successfully made using a two specimen drop weight tear test (DWTT). In this technique, two specimens are used with different ligament lengths [Venzi et al. 1980].

4 APPLICATION PHASE METHODOLOGY

The approach that was adopted here is based on the procedure that was introduced in reference [Jung et al. 1981]. It assumes that the crack travels in a straight line in the longitudinal direction of the pipe and that this path is coincident with the boundaries of the elements in PFRAC. This procedure is outlined in the following steps.

(a) When the crack is at the beginning of a particular element, the edge of that element in the direction of crack advance is subdivided into a number of equal sub-intervals. During the crack propagation, the crack is permitted to advance in discrete increments, each of which is equal to the size of a sub-interval. The number of sub-intervals is based on the maximum crack velocity, V_{max}, that is allowed in the analysis. For a given value of V_{max}, the maximum distance, Δx, that a crack can propagate in an integration time step Δt is

$$\Delta x = V_{\max} \Delta t \tag{2}$$

For a given element boundary length, l, the number of sub-intervals, N, is taken as

$$N = \frac{\Delta x}{l} + 1 \tag{3}$$

where N is truncated to an integer. This ensures that the crack will not travel faster than one subdivision per time-step. It also implies that the maximum crack speed is slightly less than V_{max}. Note that in most problems of practical interest, it will generally take a crack many time steps to propagate one element length.

(b) When the crack is at the start of an element, it is also necessary to determine the value of the nodal force at the node that is being released to simulate crack growth in the finite element model. As four node elements are used in PFRAC, this is the node on the crack path at the beginning of the element. As the crack propagates through the element, this force is gradually released to simulate the creation of a stress free surface as the crack tip advances. The magnitude of this force is assumed to decrease linearly with the advance of the crack tip through the element. This is simply expressed as

$$F = F_0\left(1 - \frac{x}{l}\right) \tag{4}$$

where F is the current value of the nodal force, F_0 is the initial value of the nodal force to be released and x is the distance that the crack tip has advanced in the element. While other nodal force reduction procedures have been used, numerical studies have demonstrated that a linear decrease of the nodal force produces the optimal results.

(c) At each time step, the crack driving force CTOA is calculated. This is based on the average displacement normal to the direction of crack propagation at the two nodes immediately behind the crack tip.

(d) The crack speed, if the crack was to propagate at this time step, is now computed. This speed can be viewed as a pseudo-crack speed, V_p, as the crack will only propagate at this speed if the crack driving force equals the fracture toughness at that instant. The pseudo-speed is given by

$$V_p = \frac{\Delta x}{n\Delta t} \qquad (5)$$

where Δx is the length of the current sub-interval. The quantity n is the number of time steps it takes before the crack driving force equals the fracture toughness.

(e) Based on the current value of the pseudo-speed, the material fracture toughness, $(CTOA)_c$, is determined. This toughness may or may not be a function of velocity. If the driving force is less than the toughness, then there will be no crack advance at this time step. The time step is incremented by Δt, another integration is performed and a new crack driving force is computed as in step (c) and the process is continued.

(f) If the driving force does indeed equal the toughness, as given by Equation (1), then the crack tip is advanced by one sub-division at that time. Note that the computed driving force may be marginally greater than the toughness. The crack speed of this advance is given by Equation (5) and the nodal force is adjusted according to Equation (4).

(g) Finally, a check is made to determine if the crack has reached the end of the current element. If it has, then steps (a) - (g) are repeated for crack growth through the next element. If it has not, then steps (c) - (g) are repeated.

In this manner, the crack position-time history is determined. This should be viewed at an element level by considering the time that the crack tip reaches each nodal location. The crack speed can be determined from this. The crack speeds from Equation (4) are not a true reflection of the actual propagation process. Crack arrest can be considered to occur when the crack does not advance for an arbitrarily long time period.

5 RESULTS

A number of examples are presented to illustrate the crack propagation mechanisms. The first of these illustrates the convergence of the results by considering two different meshes to simulate propagation in a 760 mm diameter steel pipe. In Mesh A, the elements were 127 mm in the axial direction and 120 mm in the circumferential direction while the elements in Mesh B were 63 mm by 66 mm. Figure 1 illustrates the crack growth time history computed using each model. The results indicate that the velocity is essentially constant in each case (180 m/s for Mesh A and 174 m/s for Mesh B) and that the results are insensitive to mesh size.

A second set of computations was carried out for a 1.42 m diameter pipe with a 18.7 mm wall thickness and an initial line pressure of 124 bars. In these parametric computations, different values of $(CTOA)_c$ were used, ranging from 7° to 14°. However in each analysis, the $(CTOA)_c$ was assumed to be independent of velocity. These results are presented in Figure 2 where the crack length time histories are plotted. As expected, the cracks propagate more or less with a constant

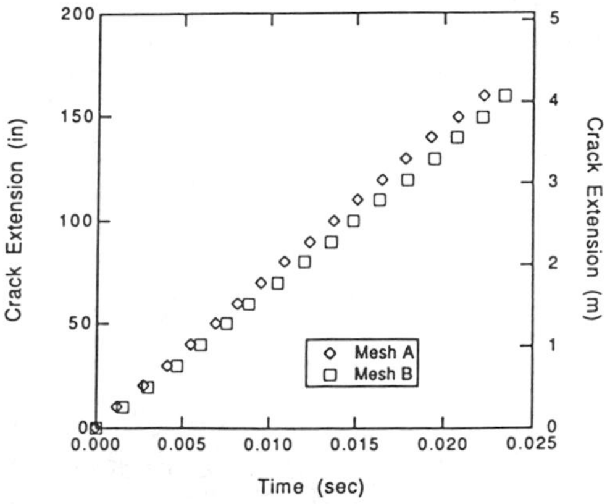

Figure 1. Comparison of Computed Crack Extension with Time for Different Mesh Sizes

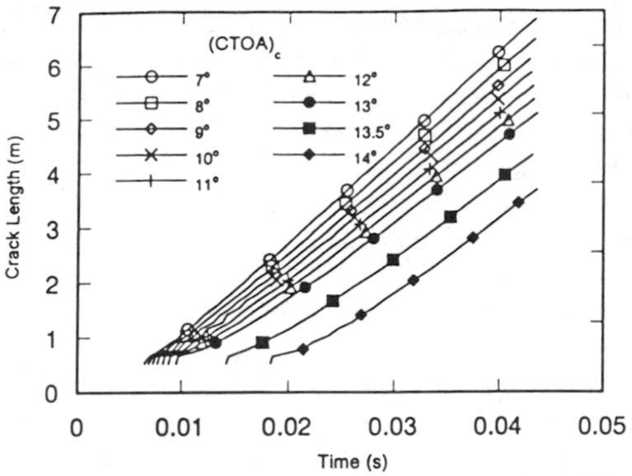

Figure 2. Crack Length Time Histories for Different Constant Values of $(CTOA)_c$

velocity, after some initial transients. The higher the toughness, the lower the crack speed. It can be seen that the cracks initiate at different times and this is again as a result of the different toughnesses.

It is frequently of interest to determine what happens to a crack as it propagates in a pipe that has a varying toughness. This is illustrated in Figure 3 where a number of composite pipe scenarios (sections with different toughnesses) are analysed. The first section is 2.5 m long and has a toughness of 8°. Foe the remainder of the pipe three different toughnesses are considered; 8°, 13° and 18°. Crack length versus time computations for these three scenarios are presented in Figure 3. When the toughness changes in the second section of pipe, a noticeable reduction in the velocity is also observed, with the lowest velocity for the toughest pipe section. This shows that the application phase analysis methodology can be used to pipe sections with different toughnesses.

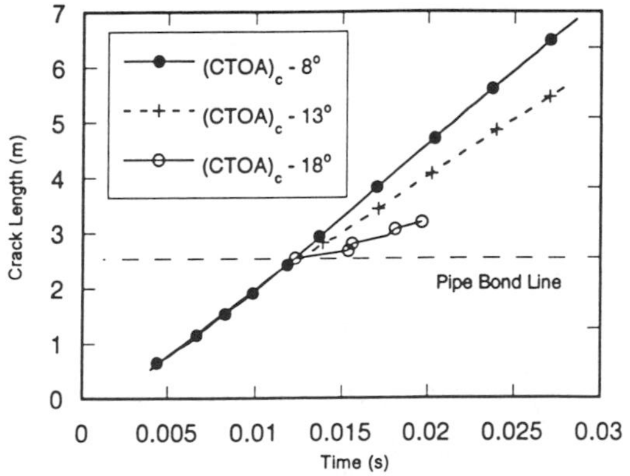

Figure 3. Computed Crack Length Time Histories for Pipe Sections with Different Toughnesses

6 DISCUSSION

In all the examples presented here, the $(CTOA)_c$ is assumed to be independent of crack velocity. This is probably not always true in practice but it is a reasonable approximation here in the absence of more detailed experimental data. However, the above application phase analysis approach can be used easily when the $(CTOA)_c$ is piece-wise linear function of the velocity.

The importance of the application phase analysis feature is that it can now be used to directly predict the likelihood of crack arrest such as when the crack propagates into a tougher section of pipe material. This will be of significant benefit to the gas industry.

ACKNOWLEDGEMENT

This research project was supported by a grant from Southwest Research Institute to University College Dublin. The encouragement and assistance provided by Professor C. H. Popelar and Mr. T. S. Grant are greatly appreciated.

REFERENCES

Jung, J., Ahmad, J., Kanninen, M. F. & Popelar, C. H. 1981. Finite Element Analysis of Dynamic Crack Propagation. *Proceedings of the ASME Failure Prevention and Reliability Conference, Hartford, CN*: 7-12.

O'Donoghue, P. E., Kanninen, M. F., Leung, C. P., Demofonti, G. & Venzi, S. 1997. The Development and Validation of a Dynamic Fracture Propagation Model for Gas Transmission Pipelines. *International Journal of Pressure Vessels and Piping*, 70: 11-25.

O'Donoghue, P. E., Green, S. T., Kanninen, M. F. & Bowles, P. K., 1991. The Development of a Fluid/Structure Interaction Model for Flawed Fluid Containment Boundaries with Applications to Gas Transmission and Distribution Piping. *Computers and Structures*, 38 (5/6): 501-514.

Venzi, S., Martinelli, A. & Re, G., 1980., Measurement of Fracture Initiation and Propagation Parameters from Fracture Kinetics. In G. C. Sih & M. Mirabile (eds.), *Conference on Analytical and Experimental Fracture Mechanics*: 737-756.

Modern Practice in Stress and Vibration Analysis, Gilchrist (ed.) © *1997 Balkema, Rotterdam, ISBN 90 5410 896 7*

Computational modelling of penetration of biomaterials by a sharp knife

J.Ankersen, A.Birkbeck & R.Thomson
Ballistics and Impact Group, University of Glasgow, Scotland

P.Vanezis
Department of Forensic Medicine and Science, University of Glasgow, Scotland

ABSTRACT: This study aims to quantify the force needed to penetrate human tissue in order to reduce the subjectivity of expert opinion in stabbing incidents. The results will also help clinicians to assess the severity of injury and help prevent the unnecessary deaths which can occur when this is not fully appreciated. Video footage of simulated knife attacks with a stationary victim and assailant showed blade velocities of the order of 3 m/s. Quasi-static penetration experiments were then performed in which knife-blades were pushed into a target of skin and flesh simulants. The blade profiles were chosen to show the effect of sharpness and to admit finite element simulation. The key computed quantity to be compared with experiments is the reaction force on the knife. Both experiments and analysis showed a drop in reaction force as skin perforation occurs but only by refining the mesh in the contact area can quantitative agreement be achieved.

1 INTRODUCTION

In the UK the ownership of firearms is tightly controlled and even today their criminal use is unusual. Stabbings therefore account for more than 50% of the recent homicides in the Strathclyde Region of Scotland and by 1993 the frequency of knife attacks was sufficient to merit a focused campaign by the police (Op BLADE) to reduce the incidence of knife-carrying in the streets. This met with some success but knife attacks remain a problem, with yet another police officer stabbed to death in July 1994. However most knife attacks are not directed against police officers but, and increasingly, against civilians; the murders of a medical practitioner in his surgery and a physiotherapist in her clinic highlight the vulnerability of particular groups of individuals.

The lack of quantitative knowledge of the force needed for a sharp instrument to penetrate human tissue (and clothing) makes it impossible to model stabbings mathematically with any degree of confidence. This presents a problem in forensic pathology, since it is impossible to infer, after the event, whether a specific wound was necessarily the result of a deliberate blow. The common defence "he fell upon the knife" then becomes difficult to challenge in cases where evidence suggests otherwise. The requirement is for a quantitative method of assessing the force used in any stabbing incident. This will reduce the margin of doubt, both in single stabbing "alleged accidental" cases and in multiple stabbings where the effort required and hence the speed of repetition of the blow, is often a critical factor.

A quantitative force assessment method should therefore lead to speedier resolution of cases and, by reducing to subjectivity of expert opinion, improve the soundness of the verdict. In addition, sounder knowledge of the forces involved in the production of stab wounds, in relation to location on the body and likely damage to underlying structures, would enable clinicians to make a more rapid and appropriate assessment of the severity of injury. This is important since unnecessary deaths still occur from time to time in cases where the stab wound track length and the severity of trauma to the underlying structures had not been fully appreciated.

2 PREVIOUS WORK

The engineering mechanics of the penetration of biomaterials such as human tissue has received little attention in the literature. This may reflect the biomedical background of most of the professionals who come into contact with the problem. It may alternatively reflect the difficulty of developing a quantitative dynamic model for the penetration of biomaterials by a sharp object. Knight (1975) recorded penetration forces by making incisions on cadavers using a simple, instrumented knife blade and showed the overriding factor to be the sharpness of the blade. He also noted that pretensioning of the skin reduced the level of force needed to cause penetration and so the intercostal spaces were penetrated more easily than the looser abdominal skin. A further significant result was that penetration was usually by the full length of the blade, the force of penetration reaching a maximum just before initial penetration of the skin. However his attempts to model human tissue using a polyethylene "epidermis" over a synthetic foam "dermis" met with little success. Indeed, the spring-loaded instrumentation used may well have influenced the results.

From experiments in which cadavers were laid on top of a force table, Green (1978) confirmed Knight's earlier findings regarding blade tip sharpness. Unlike Knight though, Green (1978) also tested clothed bodies and recorded a tenfold increase in the force of penetration. Both workers noted that lower forces were required at higher impact velocities.

3 CASE STUDIES

To establish the physical characteristics which should be included in laboratory simulations of knife attacks, pathologist's reports and press files relating to real stabbing incidents were examined. From the 32 cases studied in detail, it was concluded that:

1. knife attacks can be usefully classified as either slashings or stabbings
2. stabbings are more likely to prove fatal than slashings
3. in stabbings, penetration is usually by the full length of the blade
4. there is no "critical length" below which a blade is safe; even 3" blades have proved fatal
5. blades often break during an attack and the most dangerous are those with strength and stiffness in bending, such as spike bayonets and sharpened screwdrivers.

Observation 3 supports Knight's (1975) conclusion that a force sufficient to break the skin on a victim is usually more than enough to allow continued penetration into the flesh. This implies that it may not be necessary to mathematically model flesh in detail and that effort might be more productively focused on the modelling of skin, bone and, where appropriate, clothing. However, while bone has a significant effect in determining the wound track, it plays no role in the initiation of penetration unless it is near the surface and supports the skin. This is indeed the case in the chest region, a common site of injury and studies of bone are in progress. Here attention is restricted to the penetration of skin.

4 LABORATORY EXPERIMENTS

Measurements taken from video footage of simulated knife attacks showed the velocity of the blade at impact to be of the order of 3 m/s. This is low in comparison with the 8 m/s, occasionally 14 m/s, found by Parker (1995). However this difference may simply reflect particular styles of attack. Those recorded here were close-quarter events involving an essentially stationary victim and assailant. In such cases, the inertia of the participants and of the weapon, which must be included in simulations of running attacks, will be less significant and even quasi-static penetration experiments might be expected to yield useful information. Such experiments were conducted on a Lloyd's Instruments LR30K Universal Testing Machine. The knife-blade was attached to the crosshead and forced into a target of flesh simulant contained in a plastic tub, as shown in Figure 1.

A force transducer fitted into the load train allowed the force of both penetration and withdrawal to be measured. The crosshead was moved under displacement control at 16 mm/s and plots of force against displacement captured and displayed on an on-line PC. This loading rate was high enough to avoid the effects of relaxation in the viscoelastic skin but low enough to avoid, for the present, the controversial issue of whether energy, momentum or force is the controlling parameter at higher velocities (Hetherington (1995). Indeed, the authors believe that many knife incidents involve relatively low velocities with high follow-through forces, i.e. low energy but high momentum, and are not well simulated by the gas-gun apparatus specified for the UK Home Office tests on stab-resistant body armours (Parker (1993)). This conclusion has also apparently been reached by Calvano (1993).

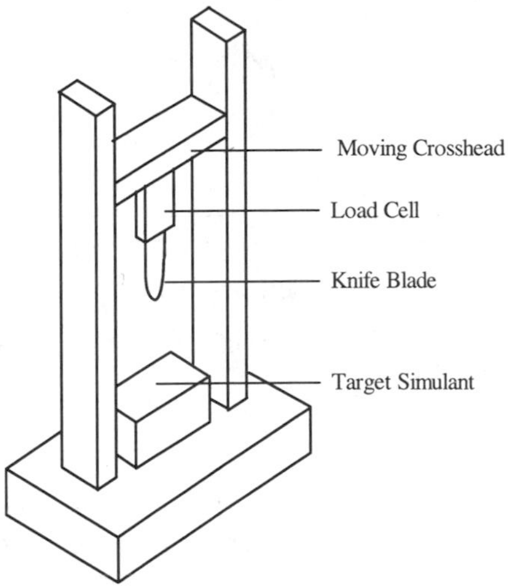

Figure 1. Experimental apparatus.

No	Description	Figure
1A	Square-ended blade with blunt edge	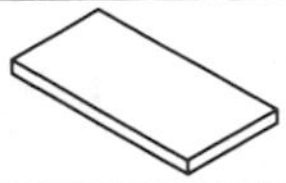
1B	Square-ended blade with sharp edge	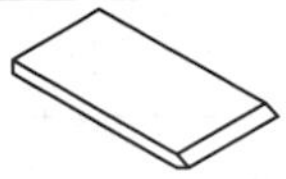
2A	Blunt-tipped blade with blunt edges	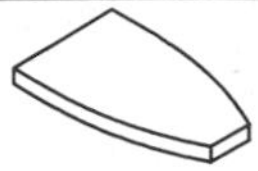
2B	Blunt-tipped blade with sharp edges	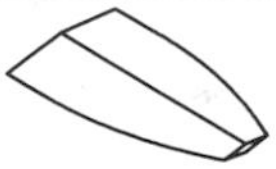
3A	Pointed blade with blunt edges	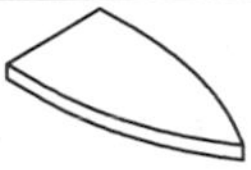
3B	Pointed blade with sharp edges	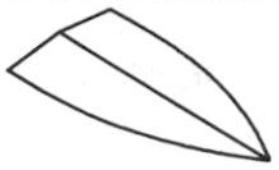

Figure 2. The model knife profiles

4.1 *Knife profiles*

The commercially manufactured knife blades described by Parker (1993) are rather too complicated to admit simple finite element modelling. A number of simplified model blades were therefore devised as in Figure 2. These were ground from high-quality steel, readily available as broken sections of power-hacksaw blades. Such blades typically break in service at lengths of about 20 cm with no damage to the microstructure of the remaining metal. The profiles were designed to highlight any effect of the sharpness of the blade tip and edge and to admit computational prediction of tip stresses.

4.2 *Tissue simulants*

While the eventual need is for a skin simulant and, to a lesser extent, a flesh simulant, commissioning tests on the apparatus were performed using a single component, Lewis "Newplast" as the target. This is a readily available modelling clay but it is not a good flesh simulant for a number of reasons. It is, for example, much denser than human flesh, although this is not a drawback in the quasi-static tests performed here. More seriously, it absorbs the work done during penetration primarily by friction between the material and the blade flanks. The force needed to continue penetration then increases strongly with depth. This is in stark contrast to the results of tests on raw meat and to the observations of Knight (1975). The response of modelling clay-based materials is also very sensitive to temperature and good temperature control is essential if consistency is to be achieved.

These criticisms are indeed applicable to all modelling-clays, including the Roma "Plastilina" recommended by Parker (1993), and their use has been criticised by Jason and Fackler (1990). These materials were included in the current study for completeness but a gelatine mix was also prepared to support the skin. For ease of handling, commercially prepared gelatine blocks were used, with 50 ml of hot water added to each 142 g gelatine block to give a firm consistency similar to flesh.

The ease with which a knife penetrates live flesh is also enhanced by the lubricating effects of body fluids, largely blood. Indeed this lubricant can make it difficult to withdraw the blade and combat knives are designed with a "fuller", a longitudinal groove in the blade which breaks the suction effect. To investigate any such effect, a soap-like lubricant was injected into the modelling clay for one series of experiments.

Finally, and most significantly in the current work, a knife must first pass through a surprisingly tough layer of skin and, perhaps, clothing before reaching the more easily damaged underlying flesh. Computational simulation of the energetics of

Table 1. The maximum force recorded during penetration

	Force on Blade [N]					
Target	1A	1B	2A	2B	3A	3B
Roma Plastilina	229	240	162	137	178	145
Roma Plastilina + Lubricant	173	202	126	123	135	143
Softened Roma Plastilina	183	202	102	115	127	148
Gelatine	58	50	25	23	24	19
Gelatine and Chamois	187	137	88	90	33	26

wounding by complete or partial penetration of the blade must admit the drop in force and the strain energy release consistent with penetration of a tough, elastic skin. Barbarel (1994) has recommended synthetic chamois as a consistent, readily available skin simulant. This was used in initial experiments but it suffers from two problems. It is weak in comparison with human skin and, more significantly, it is a non-woven fabric made from short fibres laid down in a 0/90° pattern. The material properties are then markedly different in different directions, in contrast to the almost in-plane isotropy of skin. Experiments were therefore also done using pigskin, which is also relatively easy to obtain and was found from uniaxial tensile tests to match such properties of human skin as could be found in the literature (Jansen and Rottier (1958), Tregear (1966)).

4.3 *Experimental results*

The force/displacement results for the curved blades are in agreement with intuition (Table 1). In such a small experimental sample there are necessarily statistical variations, e.g. due to machining differences in nominally identical blades, but the trend is that a sharper tip and, to a lesser degree, a sharp edge both reduce the force of penetration in the "better" flesh simulants. With the skinless samples, the force increases monotonically with depth of penetration up to the maximum value. The increase is small with the gelatine, in accord with the experiments on raw meat, but considerable with the Plastilina.

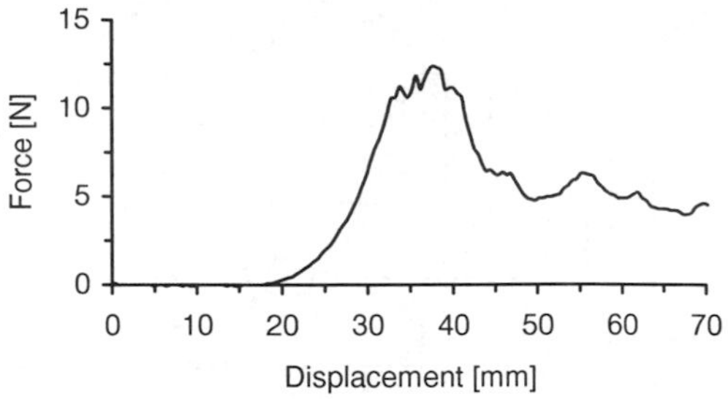

Figure 3. Force-displacement behaviour of pigskin in stab-penetration.

An unusual feature of the Plastilina behaviour is that the blunt squared blade (1A) has a lower force of penetration than does the chisel-edge blade (1B), i.e. the sharp front is ineffective as a weapon feature. This result is counterintuitive and the effect is not shown with the gelatine targets. It may well result when a "bow wave" of plastic clay, which fails to re-bond to the blade, is pushed aside during penetration by the blunt tip.

When skin is present, the force peaks just before penetration and drops off rapidly after penetration, as in Figure 3. This peak is more marked with blunt blades. These stretch the elastic skin more before penetration is achieved and so, potentially, release more elastic energy on penetration. A similar effect is observed in traditional engineering materials containing an advancing crack, which suggests that perforation of the skin might be better modelled by fracture mechanics than by solid mechanics. However there is little in the literature on tearing of flexible materials by compressive forces on the crack flanks.

5 FINITE ELEMENT ANALYSES

The authors regard mathematical modelling, whether using the well-developed methods of solid mechanic, fracture mechanics, or the less-well developed continuum damage mechanics, as an essential component of future forensic pathology. However a large number of factors influence the force needed to cause a stab wound and the significance of each factor varies in each incident. A single model which included all of them would be too complicated to be of practical use - if it could be built at all - but simplified models with restricted applicability seem feasible. Even such restricted models are likely to be computer based and finite element analysis (FEA) offers some promise. FEA has been applied to many different types of problem in areas as diverse as engineering, geology, mathematics, biology and medical research. Fundamentally, it is a technique for solving partial differential equations whose domains are geometrically too complicated to admit analytic solutions. The domain is divided into subdomains, into a "mesh" of "finite elements" connected at "nodes". Each element is small enough for the field variable, commonly displacement or temperature, to be approximated by a simple function of position. In principle then, FEA can be used to solve almost any problem which can be modelled by differential equations and it has indeed been described as the most powerful analysis technique ever made available to engineers. In practice, the FEA of penetration is however numerically intensive and is still at the threshold of possibility. Nevertheless its obvious potential in the design of weapons and armour has been noted (Birkbeck and Thomson (1995)) and it has potentially far-reaching implications for forensic pathology.

5.1 *Finite element mesh*

The finite element analyses were carried out using Abaqus/Explicit, a commercially available package specifically tailored to solve non-linear contact and penetration problems and with particular features of use when modelling damage. A quarter model with symmetry on two planes, as in Figure 4, was used to model the knife blade and the target flesh.

Linearly interpolated, isoparametric elements were used since the package does not support quadratic elements for penetration analyses. Initially, the target comprised a fairly coarse mesh of solid elements but numerical problems associated with excessive distortion of the elements were encountered and the skin layer was later modelled with membrane elements, which have no bending stiffness. Given the experimental observation that the peak force developed in the blade is not strongly influenced by the gelatine flesh-simulant, the authors felt justified in modelling the skin only. Membranes were used in single layer 60-element meshes while solid elements were used in 60-element meshes but with five layers in the thickness direction.

To reduce the processing time in analyses where attention is to be focused on the response of the human tissue, a type-3B knife blade, which is pointed with sharp edges, was modelled with rigid elements.

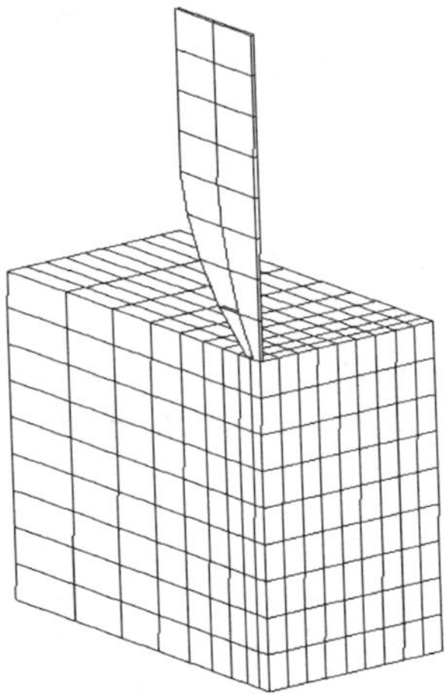

Figure 4. Finite element mesh.

However these rigid elements cannot return blade-tip stresses and to allow comparisons between the different blade geometries, several analyses were performed with the knife itself represented by solid elements.

As a general rule, the finite element model of a contact problem should use a more refined mesh for the more flexible material. Better results are also obtained when the elements actually in contact are of broadly similar sizes. A reliable model of the contact between a rigid, sharp point and a softer target material then requires a very fine mesh in the contact zone. Furthermore, large angles between element faces in the two contact surfaces are known to cause problems.

5.2 *Material modelling*

In this study the computational difficulties are compounded by the paucity of data on the mechanical properties of both human tissues and pigskin, both of which are non-linear elastic anisotropic materials. A number of constitutive models have been proposed but failure criteria have not been addressed to any extent. Since the Abaqus material library does not contain a suitable generic material type, a Fortran90 subroutine incorporating material damage was written to describe the pigskin and linked to the main Abaqus program using the VUMAT feature. This used constants derived from uniaxial tensile tests on specimens 60 mm long by 13 mm wide and 2 mm thick. A typical uniaxial load-deflection curve is shown in Figure 5.

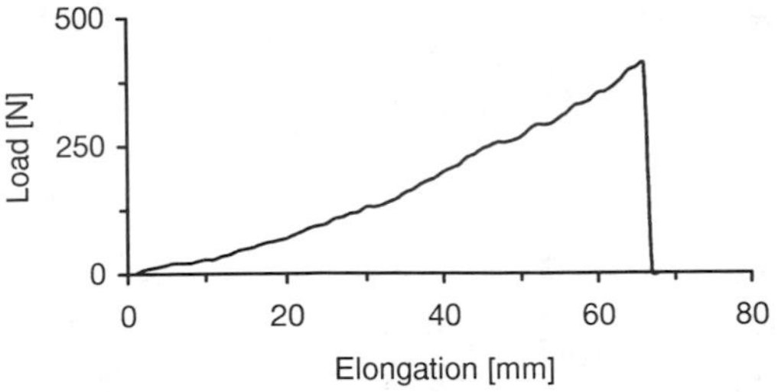

Figure 5. Stress-strain behaviour of pigskin.

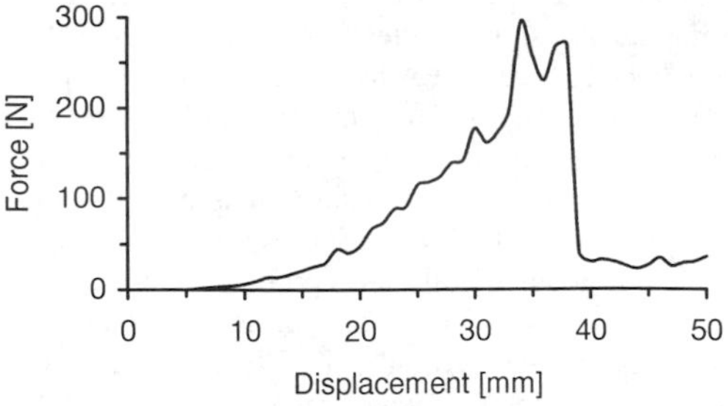

Figure 6. Computed reaction force at the knife.

Schneider (1982) has noted that a ten-fold increase in strain rate can produce a doubling of the force needed to penetrate a viscoelastic material but assuming perfect elasticity and that strain-rate and hysteresis effects are negligible over the range of velocities considered here, the material may be described by a bilinear curve up to failure. The skin is almost isotropic in the plane of the membrane but the through-thickness modulus is quite different. However values for this were not available at the time and so the stresses were calculated from the strains on the assumption of 3-D isotropy. Failure of the material was simulated computationally by deleting the element from the analysis once the Mises stress had reached the experimentally determined uniaxial failure stress of 25 MPa. This element erosion technique was originally designed to avoid the numerical difficulties associated with heavily distorted elements but it is commonly used as a means of admitting penetration of a target. Extensive tests were performed on the material model to ensure that its performance matched the experimental observations.

5.3 *Boundary conditions*

Since the material model contains no strain rate dependence, the velocity of the blade when forced into the target is of little consequence, provided it is small enough for inertial forces to be insignificant. This velocity-independence of the force was confirmed experimentally for speeds from 16 mm/s to 500 mm/s. The velocity of the knife does however have an influence on the computational time, which decreases as the knife speed increases. A velocity of 500 mm/s was therefore chosen for the computational simulations. The results were computed on a Sun workstation and displayed using the Abaqus/Post graphical postprocessor.

5.4 *Results*

The key quantity to be extracted from the analysis is the reaction force developed in the blade as penetration occurs. This is essentially the force required to effect the stabbing action. Figure 6 shows a typical reaction force on the knife blade as it penetrates the target which in this case is formed by solid elements. These plots have a similar form to the experimental data but the computed values are somewhat higher.

The penetration force was found to be dependent on mesh refinement and was calculated for a range of meshes with different bias in the element sizes. A bias value of one corresponds to a uniform mesh while a lower value indicates a high mesh concentration around the blade tip i.e. smaller elements in the area of the penetration. The resulting penetration force as a function of bias is shown in Figure 7.

Both membrane elements and solid elements show a decrease in penetration force as the bias decreases (i.e. at higher mesh density). This tendency only applies in the bias range 0.65 to 1.00. Below this range the in-plane measures of the elements are probably too small compared to the thickness. Interlocking of the membrane elements due to bending are thought to cause the inconsistency in penetration force below 0.6 bias. Penetration forces are found much higher for membrane elements than for solid elements but the models are not directly comparable as the membrane model has a single element layer and the solid model has five elements through the thickness.

6 DISCUSSION

The computated peak force just before penetration is strongly dependent on size of the finite elements.

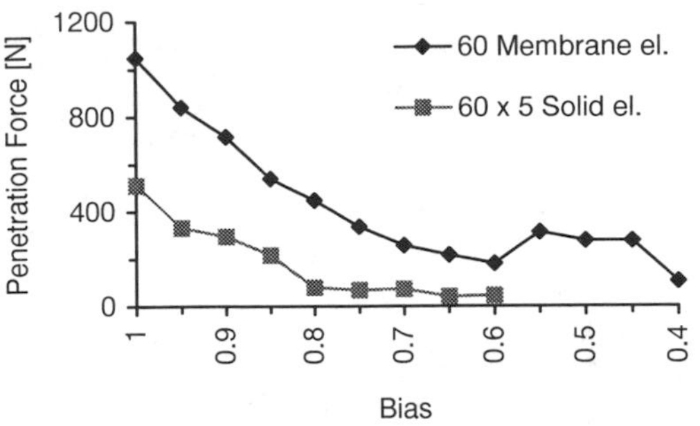

Figure 7. Reaction force with bias.

With the current blade mesh the tip is effectively a point load until penetration of the target initiates, after which time area contact between the blade and the target is established. Reducing the element size in the target will then always result in a lower penetration force and the computed peak force will always be an upper bound on the the experimental value, especially for the sharper blades. It may then be necessary to model the detailed geometry of the blade tip, perhaps using a boundary formulation, to achieve good agreement. Such work is in progress.

Since there is a limit to the scope for mesh refinement using membrane elements of a given thickness, solid elements are the preferred option when high mesh density is required. The use of multiple elements in the target thickness direction presents no difficulty in principle but will require material properties in the thickness direction to be known. The through-thickness mechanical properties of skin have not however been extensively discussed in a literature which generally adopts membrane theories, where such properties are of no importance. This aspect is also being addressed.

A fracture mechanics model, incorporating a stress intensity factor to describe the near-elliptical "crack" which develops in the skin and the flesh after initial penetration might be more appropriate for the later stages of the simulations. However, fracture mechanics can only be employed once a crack has been established and cannot model the approach to penetration or allow assessment of the peak force.

7 CONCLUSIONS

In contrast to ballistic experiments, simulations of stabbing incidents must include a skin simulant and, potentially, clothing. Pigskin supported over a layer of gelatine is a good combination.

Finite element analysis is a useful tool for the analysis of stabbing but the computational and materials engineering problems are daunting. The potential benefits to forensic pathology make this a worthwhile aspiration.

A finite element mesh which is fine enough to model the finite geometry of the blade tip may be needed to return accurate peak forces. Solid elements remain the preferred solution.

8 REFERENCES

Barbarel (1994), University of Strathclyde, Private communication.

Birkbeck A.E. and Thomson R.D. (1995), *Ballistic Impact and Penetration Resistance of Engineering Materials, Computational Plasticity - Fundementals and Applications*, Proceedings of the 4th International Conference.

Calvano N.J. (1993), *A Study to Determine the Most Important Parameters for Evaluating the Resistance of Soft Body Armor to Penetration By Edged Weapons*, U.S. Department of Justice.

Green M.A. 1978, *Stab Wound Dynamics - A Recording Technique for Use in Medico-legal Investigations*. J. Forens. Sci. Soc., 18, 161-163

Hetherington J.G. (1995), *Energy and Momentum Changes During Ballistic Perforation*, Int. J. Impact Engng 18, 319-337.

Jansen L.H. and Rottier P.B. (1958), *Some Mechanical Properties of Human Abdominal Skin Measured on Excised Strips*, Dermatologica 117, 65-83.

Jason A. and Fackler M.L. (1990), *Body Armor Standards - A Review and Analysis*, Center for Ballistic Analysis, Final report - second edition.

Knight B. (1975), *The dynamics of stab wounds*, Forensic Science 6, 249-255.

Parker G. (1993), *Stab Resistant Body Armour Test Procedure*, UK Home Office Police Scientific Development Branch publication 10/93.

Schneider D.C. (1982), *Viscoelasticity and Tearing Strength of the Human Skin*, PhD Thesis, University of California.

Tregear (1966), *Physical Functions of Skin*, Academic Press, London.

Modern Practice in Stress and Vibration Analysis, Gilchrist (ed.) © 1997 Balkema, Rotterdam, ISBN 90 5410 896 7

The analysis of stress concentrations in components: A modified fracture-mechanics approach

D.Taylor & G.Wang
Mechanical and Manufacturing Engineering Department, Trinity College, Dublin, Ireland

J.Devlukia, A.Ciepalowicz & W.Zhou
Rover Group, Gaydon Test Centre, Warwick, UK

ABSTRACT: This paper describes a new technique known as "crack modelling" which is intended for the analysis of fatigue failure in engineering components. Early work has been done on the application of this method to components of complex geometry, which were subjected to bending loads. The present paper presents new data on the effect of loading mode and material type. It is shown that the method works equally well when applied to situations in which the dominant loading is torsion. In these cases the approach for finding the equivalent K value is to examine stresses on the plane normal to the maximum principal stress at the stress-concentration point. New data are presented from an analysis of components made from a cast aluminium/silicon alloy, to show that the method works equally well in this material. Finally, consideration is given to the initiation and growth of fatigue cracks from stress concentrations. An understanding of the mechanism of crack growth in the early stages of fatigue is essential to the development of the crack-modelling approach. It is shown that, in some materials, it is necessary to modify the calculation of K to take account of this developing crack.

1 INTRODUCTION

Fatigue usually initiates at stress concentrations: geometric features in components such as holes, grooves and corners. Traditional methods of fatigue analysis, some of which are now incorporated into finite element (FE) code, show poor accuracy when applied to high stress concentrations (e.g. sharp notches) or to materials of low notch-sensitivity (e.g. cast irons, mild steels). The crack modelling approach deals with this problem by modelling the notch as a crack. This allows the calculation of an equivalent stress-intensity factor (K), enabling standard fracture mechanics methodology to be used. Fatigue is predicted to occur if the cyclic value of this equivalent stress intensity exceeds the crack propagation threshold.

Previous papers (Taylor etc. 1996,1997) have reported the application of this method to components of complex geometry, which were subjected to bending loads. In this paper the method is extended to the different load and different materials.

2 MATERIALS AND MICROSTRUCTURE

The components chosen for study were a C-shaped bracket and a camshaft which are in commercial use in Rover vehicles. C-shape brackets, which are made from the cast aluminium alloy LM25TF, were tested under tension/compression fatigue loading. Camshaft components which are made from grey iron were tested experimentally under bending and torsion fatigue loads at R=-1. The mechanical properties of the materials are given in Table 1.

Table 1. Mechanical properties

Material	Grey iron grade 17	LM25TF
Young `s Modulus	170 GPa	71 GPa
Poisson's ratio	0.29	0.34
Yield stress	202 MPa ($\sigma_{0.2}$)	230 MPa
Ultimate strength	249 MPa	275 MPa

The microstructure of LM25TF contains a high percentage of pores which reach 7.17% in the worst condition. On machined specimens large shrinkage pores were found to have very irregular three-dimensional shapes and sizes varying from 86 to 373 μm. These pores resulted from insufficient metal flow into the space between connected dendrites during solidification. Under fatigue tests failure was found from a crack initiated at this kind of pore. Unlike machined specimens, as cast specimens were found to fail mainly from a micro-crack initiating from

Table 2. Chemistry of LM25TF (% Weight)

Cu	Fe	Mn	Si	Mg	Zn	Ti
0.01	0.01	0.01	7.25	0.36	0.01	0.02

Table 3. Chemistry of grey iron grade 17 (% Weight)

C	S	Mn	Si	Ni	Cu	P	Cr
3.3	0.09	1.5	1.8	0.07	0.2	0.03	0.05

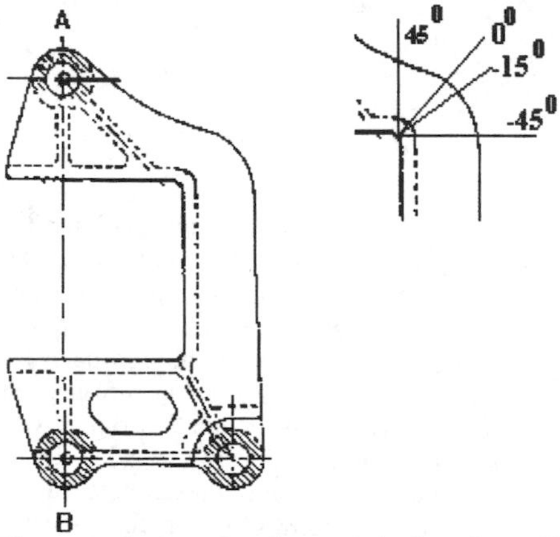

Figure 1. C-shape bracket and the location of *S-D* curves

hollows which were 100-300 μm in length, much larger than the other defects (10-50 μm), e.g. inclusions, micro-pores.

The microstructure of grey cast iron, grade 17, contains carbon in the form of graphite flakes which is embedded in a matrix of either ferrite or pearlite, or a ferrite-pearlite mixture. From the as-cast surface to the centre, the amount of pearlite decreases and the amount of graphite increases. At the surface the graphite is spherical in shape, becoming flake-like with longer flakes inside. The grain size also increases with distance.

Tables 2 and 3 give the composition of LM25TF and grey iron.

3 COMPONENT EXPERIMENTS AND FINITE ELEMENT ANALYSIS

The experimental results for the C-shape component (as-cast) under tension/compression load along the *AB* line (see Figure 1) and camshaft components, under bending and torsion load, are shown in Table 4.

The IDEAS software was employed for FE analysis. For the C-shape bracket shown in Figure 1, the geometry is approximately symmetrical in the plane of the diagram. The hot spot stress concentration is at the top right-hand corner. Therefore, a fine mesh was made for the volume around the root and a coarse mesh elsewhere. Three elements were placed along the curve of the root whose radius is 5 mm.

Because the hot spot is just around the corner where the other two principal stress components can be predicted, the difference between maximum principal stress and the Von Mises stress is very small, about 3%. Therefore, for the fatigue prediction discussed below both of them can be used without affecting the result.

The distribution of the maximum principal stress (S11) near the corner is shown in Figure 2. The stress-distance (*S-D*) curves are measured at different angles from the hot spot (see Figure 2). The effect of the angle is very small.

Most parts of the camshaft are axisymmetric except the cam lobes. The influence of the cam lobes on the stress outcome at the notch from using axisymmetric analysis is negligible. Therefore an axisymmetric model was used for the camshaft stress analysis. Away from the notch where detailed stress

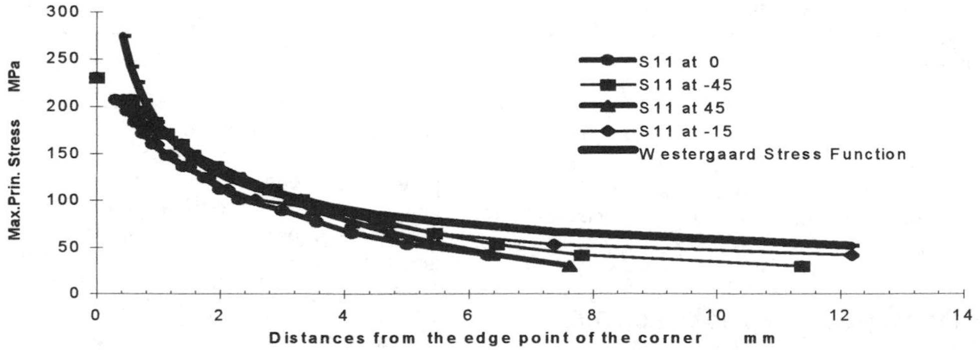

Figure 2. *S-D* curves on the corner of C-shape bracket

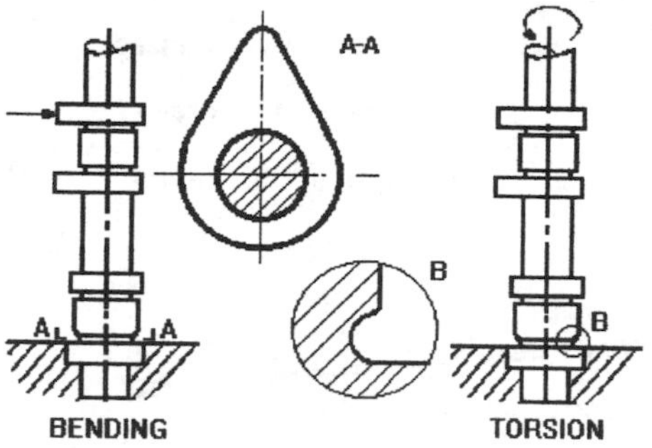

Figure 3. Camshaft loading condition

was expected, the influence of geometrical change, such as another notch, cam lobes or a diameter change is very small and can be also negligible.
Two kinds of load conditions were taken for FE analysis, bending and torsion (see Figure 3). For torsion a specific plane was examined to which the local maximum principal stress direction is perpendicular, about 45° to the bearing axis. Figure 4 shows the distribution of maximum principal stress and Von Mises stress (SEQU) near the hot spot on the plane. *S-D* curves were measured at different angles in the same way as in the C-shape bracket. The effect of the angle is still very small.
For bending, the plane containing the hot spot and bearing axis was examined. Figure 5 shows the stress distribution near the hot spot at different angles.

4 PREDICTION OF THE FATIGUE FAILURE

The crack modelling method (Taylor 1996) is based on the Westergaard function which gives σ as a function of distance r for a crack of length a_w in tension at a stress σ_w:

$$\sigma = \sigma_w \Big/ [1-(a_w/(a_w+r))^2]^{1/2} \quad (1)$$

The values of σ_w and a_w in equation (1) can be varied to obtain a best fit with the *S-D* curve (see Figure 1, Westergaard stress function). Then the appropriate K value is given by

$$K = \sigma_w \sqrt{\pi a_w} \quad (2)$$

This method assumes that if the equivalent K, found by applying equation (2), is less than the threshold value of the material, fatigue failure will not occur.

The stress intensity threshold of materials at a certain R ratio is needed for the method described above. For the cast aluminium alloy LM25TF, the value is 10 MPa √m at R=-1. At the same R ratio, the threshold value was obtained from the literature (Taylor, Hughes and Allen 1996)for the grey iron grade 17, which is 15.94 MPa √m.

The results of the prediction using crack modelling are given in Table 4 which shows predicted and experimental values of the fatigue limit load range. All experimental data are amplitude values at the ratio, R = -1.

Table 4. Comparison of experiments and predictions

Component name	Experimental Fatigue Limit	Predicted Fatigue Limit	Error %
C-bracket in tension	0.76 kN	0.643 kN	15.39
Camshaft in bending	0.75 kN	0.68 kN	9
Camshaft in torsion	255 Nm	263 Nm	-3

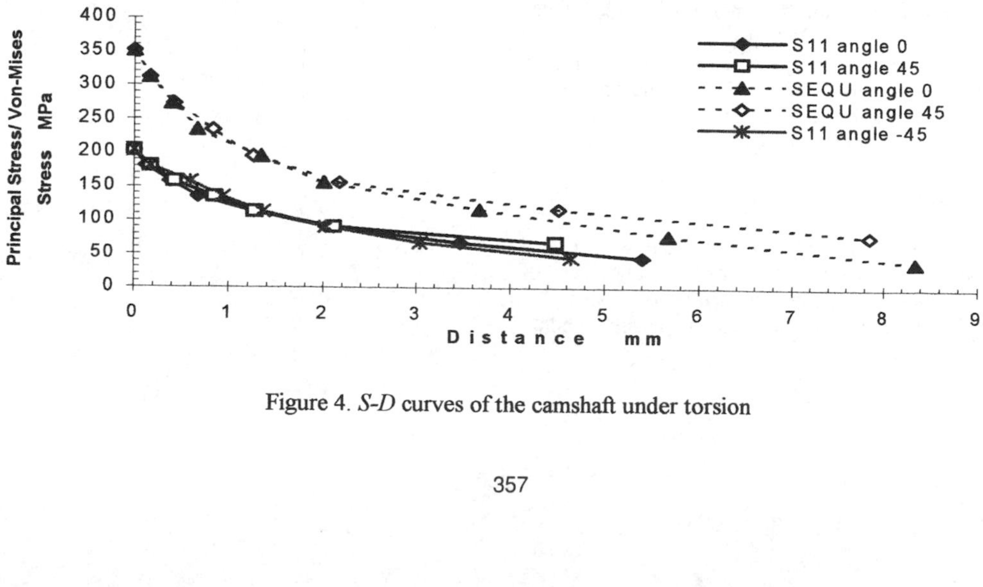

Figure 4. *S-D* curves of the camshaft under torsion

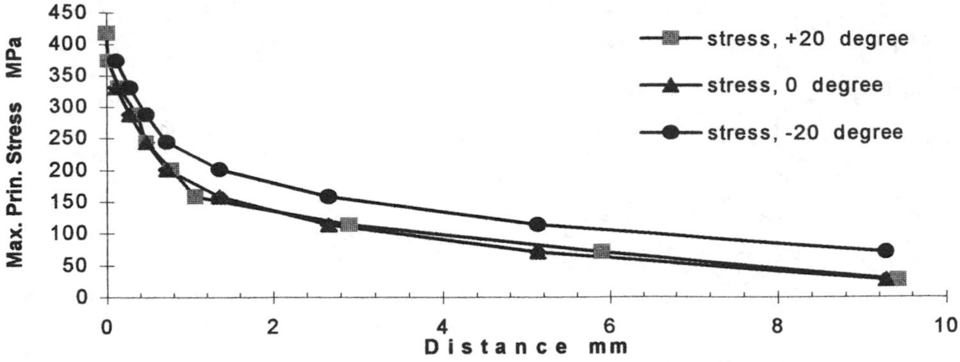

Figure 5. *S-D* curves of the camshaft under bending

5 DISCUSSION

Good accuracy was achieved in all three predictions, showing that the method is capable of estimating the fatigue limits of components containing stress concentration. Two of the predictions were conservative by 9% and 15%, which is useful in design calculations.

The method adapted for analysing tensile loading refers to stresses on the principal tensile plane. This implies that cracks grow perpendicularly to the maximum principal stress. Since the method works for tensile and shear loading, it will also work for any combination of tension and shear.

This approach was successful for both an aluminium alloy and a cast iron. Previous work (Taylor and Lawless 1996) has shown that other materials such as mild steel can also be analysed. The method works best for low-strength materials which normally face low notch sensitivity.

Only an elastic Finite Element is required, and the results are not strongly sensitive to mesh density. As Figure 2-4 show, the result varies only slightly with angle: normally the line would be drawn normal to the local contours but the choice of angle is not critical.

Further work is being carried out to investigate the effect of surface condition and of the size of the stress concentration and to study the growth of fatigue cracks from the hot-spot.

6 CONCLUSIONS

1. The technique of crack modelling is useful method of estimating the fatigue behaviour of components in which stress concentrations exist.

2. The technique can be used for tensile, bending and shear loading.

3. For some materials, the surface-layer effect should be considered when using the technique in order to predict accurately.

REFERENCES

Taylor, D. 1996. Crack modelling: A technique for the fatigue design of components. *J. Engineering Failure Analysis,* Vol. 3, No. 2, pp. 129-136

Taylor, D., Hughes, M. and Allen, D. 1996. Notch fatigue behaviour in cast irons explained using a fracture mechanics approach, *Int. J. Fatigue* Vol. 18, No. 7, pp. 439-445

Taylor, D. and Lawless, S. 1996. Prediction of fatigue behaviour in stress-concentrators of arbitrary geometry, *Eng. Fracture Mechanics,* Vol. 53, No. 6, pp. 929-939

Taylor, D., Ciepalowicz, A. J., Rogers, P. and Devlukia, J. 1997. Prediction of fatigue failure in a crankshaft using the technique of crack modelling, *Fatigue Fract. Engng Mater. Struct.* Vol. 20, No. 1, pp. 13-21

10 Experimental stress analysis techniques

Modern Practice in Stress and Vibration Analysis, Gilchrist (ed.) © 1997 Balkema, Rotterdam, ISBN 90 5410 896 7

New optical interferometric technique for stress analysis

S. Yoshida, Muchiar, I. Muhamad, R. Widiastuti, B. Siahaan, M. Pardede & A. Kusnowo
Research and Development Center for Applied Physics, Indonesian Institute of Sciences, Tangerang, Indonesia

ABSTRACT: Using an optical interferometric technique we have visualized a band structure in aluminum-alloy specimens under a tensile load. By monitoring the temporal and spatial behavior of the visualized band structure it is possible to judge if the specimen is close to a fracture and where the fracture will occur. Our analysis based on a recently developed general theory of plastic deformation indicates that the band structure is caused by stress concentration. We have applied this technique to welded specimens and visualized the stress concentration caused by heat.

1 INTRODUCTION

We investigate an optical, nondestructive test method for solid-state materials under external loads. Our particular interest is to develop a technique capable of predicting the location and timing of a fracture at an early stage of deformation. Recently we have found that when deformation develops to be a certain stage a shear-band-like band structure can be visualized on an interferometric fringe-pattern, and that by monitoring the temporal and spatial behavior of this optical band structure it is possible to diagnose if the material is close to a fracture and where the fracture will take place (Yoshida et al. 1996a). Since this optical band structure appears to be conspicuously whiter than the bright peak of the surrounding interferometric fringes we call it the white band (WB). Our subsequent investigation has revealed that the WB represents the band structure caused by stress concentration, that strain is localized at the location of the WB, and that when such strain localization develops it is responsible for a fracture (Yoshida et al. 1997a).

Theoretically, the above-mentioned band structure can be well explained by a recently developed general theory of plastic deformation called mesomechanics (Panin 1995). Based on a solid physical basis, and without relying on empirical formula, mesomechanics can describe all the stages of deformation including the fracture in the same theoretical system. This enables mesomechanics to define criteria of the fracture in a unique way, which has been verified experimentally (Yoshida et al. 1994). We have found that several features of the WB that appear toward the end of deformation show good accordance with these mesomechanical fracture-criteria (Yoshida et al. 1996a). This justifies to use those features of the WB for diagnosing a fracture. It is the purpose of this paper to explain the WB in connection with stress concentration and to report on our recent experiment in which we investigated specimens containing various levels of stress concentration caused by weld. It will be shown that the observation of the WB is a convenient way to visualize the stress concentration responsible for a fracture.

2 THEORETICAL BASIS

In mesomechanics (Panin 1995), the plastic deformation is defined as the loss of shear stability. It is an energy dissipation process in which stress concentration relaxes. The plastic deformation develops through hierarchy of scale levels, called micro-, meso- and macro-level. In this process, the translational and rotational modes of deformation operate synergetically; spatial inhomogeneity in translational-mode deformation causes rotational-mode deformation in a higher level, which in turn induces secondary translational-mode deformation. Because of this interaction, the plastic deformation possesses wave characteristics and the mechanical field of the deforming material becomes vortical. The

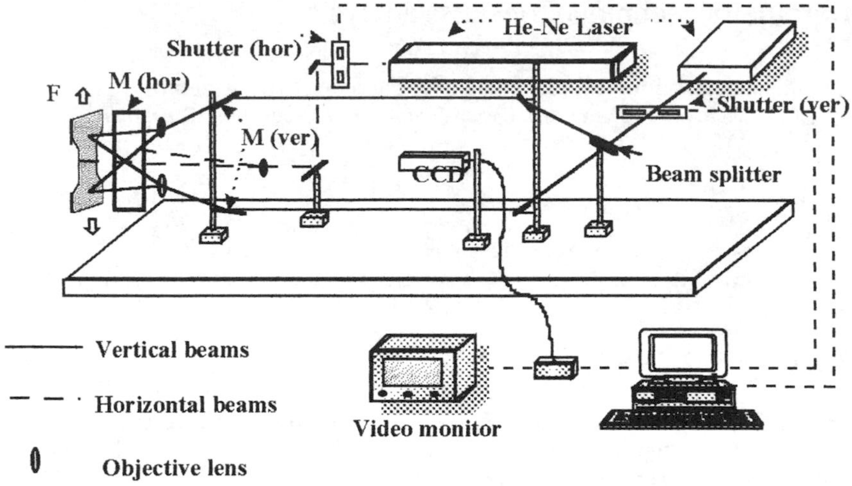

Fig. 1 Experimental arrangement

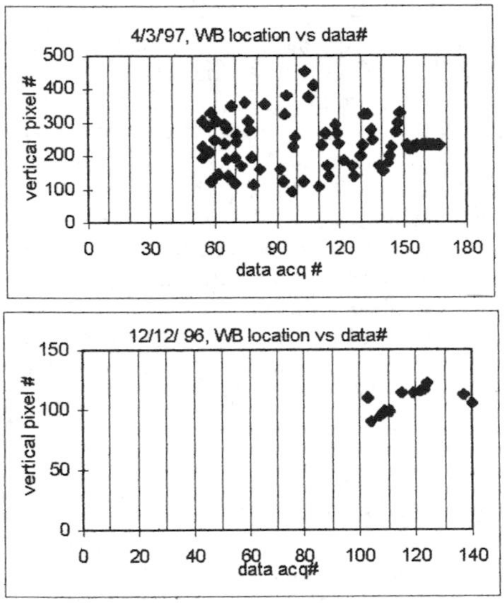

Figure 2 Dynamic characteristics of MS type (upper) and LS type (lower) WBs. The fracture occurs at the final location of the WB.

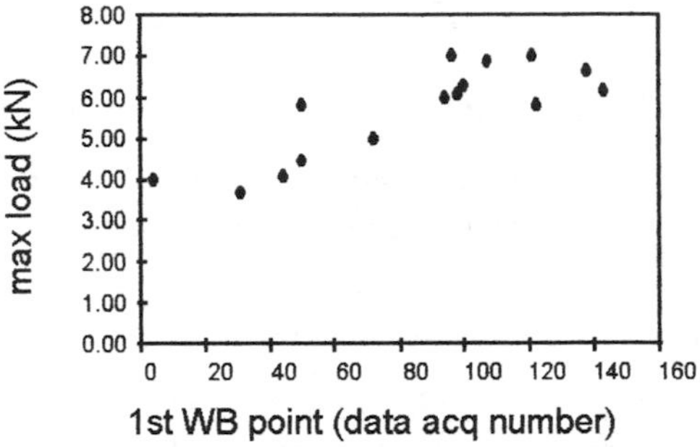

Figure 3 Maximum load vs appearance of the first WB. The material is AA6063. The maximum load is defined as the highest load measured in the stress-strain curve.

former is referred to as the plastic deformation wave and the latter is referred to as the translatonal-rotational vortex.

As the deformation develops, the plastic deformation wave and the translational-rotational vortex also develop in their scale levels. When the deformation grows to be the final stage, the plastic deformation wave tends to have a wavelength comparable to the specimen size and a peak of it becomes stationary. At the same time, the translational-rotational vortexes are assembled to be two vortexes mutually rotating in the opposite directions. Mesomechanics defines these features manifested in the plastic deformation wave and translational-rotational vortex as the fracture-criteria (Yoshida et al. 1994). In this situation, intensive strain localization takes place at the boundary of the developed vortexes, causing a bend moment over the specimen. The material tends to be discontinuous at the boundary of the vortexes. If the stress somehow relaxes without further intensifying this localized strain, the material at this location can recover from the discontinuous situation. If the strain localization is further intensified, the discontinuity leads to a macro-scale crack and, eventually, a fracture.

The band structure is formed when defects propagate as a consequence of stress relaxation in any scale level. In the micro-level, the dynamic range of defects is within the deformation structural element and so is the dimension of the band structure. As the scale level grows, a large band structure is formed. In the macro-level a band running across the specimen transversely is formed and at that location a large scale strain localization occurs. We have experimentally confirmed that such a developed band structure appears at the boundary of the above-mentioned two developed vortexes (Yoshida et al. 1997a). It is thus reasonable to relate the characteristics of the WB and the mesomechanical explanation of the evolution of deformation (Yoshida et al. 1996b).

3. EXPERIMENTAL

3.1 *Experimental arrangement and procedure*

Figure 1 illustrates the experimental arrangement. We placed the specimen on a standard tensile machine and gave a load at the head speed of 0.4 mm/min. The specimen was illuminated by an interferometer consisting of a 30 mW He/Ne laser as the light source and an inplane sensitive interferometric setup (Løkberg 1993). We took pairs of speckle images of the specimen by a CCD (Charge Coupled Device) camera at an interval of fifteen seconds (called data acquisition interval) and stored the images in a computer memory by a frame grabber. Each pair of images were taken with an interval of five seconds (called the deformation interval). This was repeated until the specimen fractured. Along with the images, we recorded the stress-strain characteristics of the specimen.

To form interferometric fringe-patterns, we made subtraction between the two images taken in every data acquisition interval. This subtraction was made by a second frame-grabber/computer set that was connected to the master computer by a net work. In this fashion we could monitor the WB by a TV camera on a real time basis.

3.2 *Specimens*

We used two types of aluminum alloys. The former was a low grade material whose chemical composition was conformable to the standard material AA6063. The post casting process of this material was unknown, but judging from the versatility of the observed tensile characteristics we speculated that it was not standardized. The other type of aluminum alloy was conformable to the standard material A5052H32 both in the chemical composition and the post casting process. We call the former the AA6063 material and the latter the A5052 material in this paper. For both materials, the effective specimen size was 20 mm wide, 100 mm long and 2 mm thick. To increase the brightness of the image, all the specimens were painted white.

In order to introduce different levels of stress concentration, we welded specimens of the A5052 material by a cw (continuous wave) carbon dioxide laser under various welding conditions shown in Table 1. In this table, the entry "graphite coated" means that the material was coated by graphite before it was welded. It is well known that if A5052 is graphite-coated the void density greatly increases when welded.

Table 1. Welding condition

Type of welding	Welding method	Welding speed	Laser power
Shallow welding	Bead-on-plate	6.0 m/min	3.75 (kW)
Deep welding	Bead-on-plate	2.0 m/min	3.75 (kW)
Butt-welding	Butt-welding	2.0 m/min	3.5 (kW)
Graphite coated	Bead-on-plate	2.5 m/min	2.5 (kW)

4. RESULTS AND DISCUSSIONS

In this section we will analyze the results of the tensile experiment by examining the dynamic characteristics of the WB and the stress-strain curve. Here the dynamic characteristics of the WB is defined as the time-historical trace of the location of the WB on the specimens.

4.1 *AA6063 material*

The special feature of this material is that the stress condition varies among species because of the unstandardized post-casting process. Consequently all the specimens showed different types of the WB that can be classified as the move-and-stay (MS) type WB, stay-and-stay (SS) type WB, or Late-start-stationary (LS) type WB (Yoshida et al. 1996b). Figure 2 shows the dynamic characteristics of typical MS and LS type WBs observed in AA6063 specimens. In both types, the fracture occurs at the location where the WB finally becomes stationary. This is how we can diagnose the location of a fracture (Yoshida et al. 1996b).

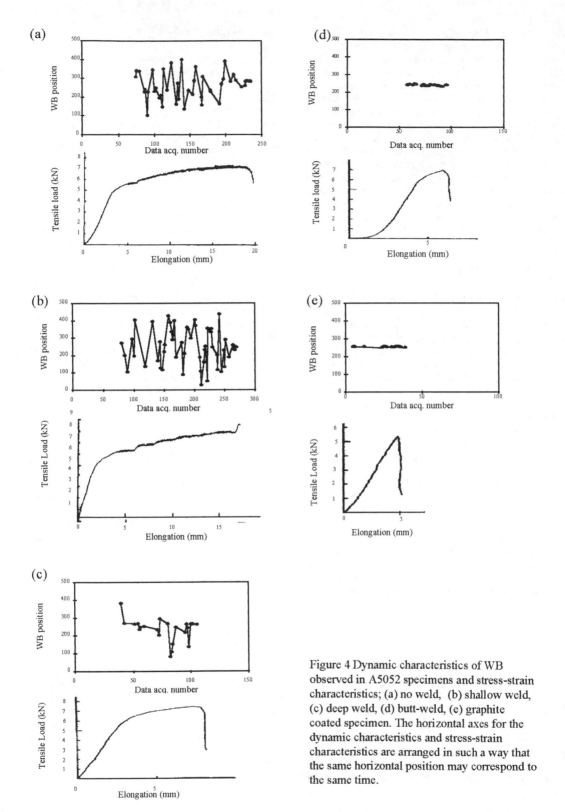

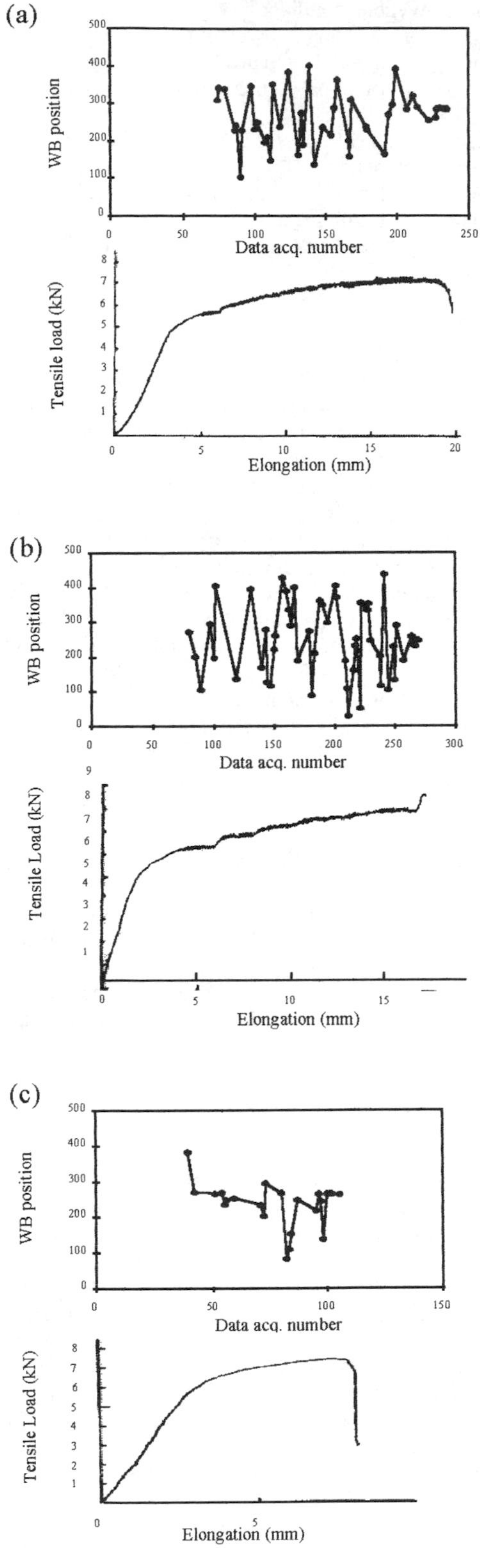

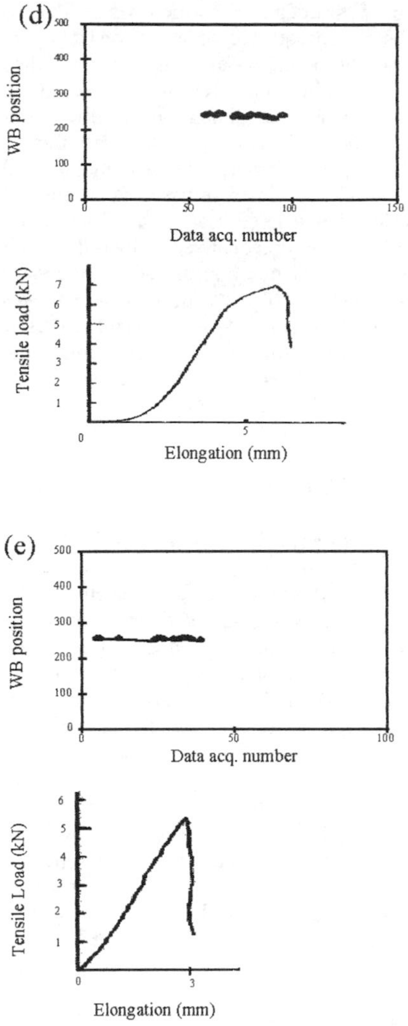

Figure 4 Dynamic characteristics of WB observed in A5052 specimens and stress-strain characteristics; (a) no weld, (b) shallow weld, (c) deep weld, (d) butt-weld, (e) graphite coated specimen. The horizontal axes for the dynamic characteristics and stress-strain characteristics are arranged in such a way that the same horizontal position may correspond to the same time.

Figure 3 shows the relationship between the maximum load and the data acquisition number when the first WB appears. A positive slope is seen indicating that the earlier the first WB appears the smaller the work hardening is. With the help of mesomechanics, this can be explained as follows. According to mesomechanics, the work hardening is a self action of the gauge field that takes place in the process of stress relaxation (Panin 1995). We have found that the stress-strain curve of a specimen of the MS type WB shows a zigzag character and that the beginning of this zigzag character coincides with the appearance of the first WB (Yoshida et al. 1997a). This indicates the following mechanism of stress relaxation in connection with the formation of the WB. During plastic deformation, the stress does not relax apparently on the stress-strain curve until it reaches a critical value. On reaching this critical value the stress relaxes intensively, and this is accompanied by the formation of a WB. In other words, the formation of the WB is a way of intensive stress relaxation. As soon as this stress relaxation is completed the work hardening resumes. When the stress rises back to the critical value, the intensive stress relaxation takes place again being accompanied by the formation of a WB at a different location. Thus the appearance of a WB basically suppresses work hardening, and this is why the work hardening is more intense in the cases where the WB appears later.

It should be noted that in a different experiment we have observed that every time a WB appears at a different location the plastic deformation wave becomes temporally discontinuous (Yoshida et al. 1997b). This implies that the propagation of the plastic deformation wave is a way of stress relaxation that occurs while the stress is below the critical value. As soon as the stress reaches the critical value, and hence a WB is formed, the current plastic deformation wave decays and a new plastic deformation wave is generated and propagated as work hardening resumes.

4.2 *A5052 material*

Figure 4 (a) through (e) shows the dynamic characteristics and stress-strain curve of the A5052 specimens with no weld, shallow weld, deep weld butt-weld and graphite coating, respectively. We examined the cross-section of the fused zone for each welding condition and observed that the void density increased in going from (b) to (e). Thus it can be said that the prestrain level increases in this order.

When not welded, the A5052 material always shows the MS type WB. (Note that this is a standard material). Figure 4 indicates that as the prestrain level increases in going from (a) to (e), the dynamic characteristics of the WB show two tendencies: (i) the first WB appears prior to the yield point, and (ii) the WB tends to be stationary from the beginning. Of these two tendencies, (i) is observed at a lower degree of the prestrain than (ii). Thus, it can be said that the first sign of prestrain is that the WB begins to appear in the elastic (linear) region of the stress-strain curve. If the degree of the prestrain is higher, the WB tends to be stationary from the beginning. The former indicates that if a prestrain exists strain localization can occur locally before the bulk material yields. The latter indicates that if the prestrain level is higher the strain localization tends to be intensive from the beginning. This is in good accordance with the mesomechanical explanation concerning the evolution of deformation to a fracture, i.e., as the deformation develops, first strain starts to be localized, and if the deformation develops further a large strain localization occurs intensively at a certain point of the specimen, causing discontinuity there leading to a fracture (Panin 1995).

5. CONCLUSIONS

The results of this investigation confirms our previous conclusions that the WB represents stress relaxation, and that the stationary WB indicates intensive strain localization leading to a fracture. When the material is prestrained, the WB can appear prior to the yield point. If the prestrain is intensive, the WB is stationary from the beginning. These features can be used to evaluate the stress condition of a weld.

ACKNOWLEDGMENT

The laser welded specimens were prepared in collaboration with the Institute of Research and Innovations. We are grateful to K. Takahashi and S. Sato for this collaboration. This investigation was supported by Science and Technology Corporation of Japan.

REFERENCES

Løkberg, O. J. 1993. Recent development in video speckle interferometry. In R. S. Sirohi (ed), Speckle Metrology, Vol. 38 of Optical Engineering: 163-166. New York: Marcel Dekker.

Panin, V. E., 1995. Physical basis of mesomechanics of plastic deformation and fracture of solid-state materials. In V. E. Panin (ed), Physical mesomechanics and computer-aided design of materials, Vol.1: 7-49. Novosibirsk: Nauka.

Yoshida,S., Y. Hida, V. E. Panin and L. B. Zuev 1994. Application of the Wave Theory of Plastic Deformation to a Novel Scheme of Non-Destructive Analysis of Mechanical Behavior of Solid-State Object. Proc. 10th Int. Conf. on the Strength of Materials: 295. Sendai: Japan Inst. of Metals.
Yoshida, S., Suprapedi, R. Widiastuti, M. Pardede, S. Hutagalong, J. Marmaung, A. Faizal and A. Kusnowo 1996a. Direct observation of Developed Plastic Deformation and Its application to Nondestructive Testing. Jpn., J. Appl. Phys. Vol. 35, Pt. 2, No. 7A: L854-L857.
Yoshida, S., Suprapedi, R. Widiastuti, Marincan, Septriyanti, Julinda, Faizal and A. Kusnowo 1996b. Study of plastic deformation and fracture based on a new optical-interferometric technique. To be published in the SEM VIII International Congress Post Conference Proceedings.
Yoshida, S., I. Muhamad, M. Pardede, R. Widiastuti, Muchiar, B. Siahaan and A. Kusnowo 1997a. Optical Interferometry Applied to Analyze Deformatio and Fracture of Aluminum Alloys. To be published from TAFM Journal.
Yoshida, S., M. H. Pardede, B. Siahaan, N. Sijabat, I. Muhamad, H. Simangunsong, T. Simbolon, Adlin and A. Kusnowo 1997b. New optical interferometric technique for deformation analysis. To be presented at International Conference on Advanced Technology in Experimental Mechanics, July 25-26, Wakayama, Japan.

Modern Practice in Stress and Vibration Analysis, Gilchrist (ed.) © 1997 Balkema, Rotterdam, ISBN 90 5410 896 7

A comparison of analytical, finite element and photoelastic results for a cylinder in contact with an elastic half-space

R.L. Burguete & E.A. Patterson
Department of Mechanical Engineering, University of Sheffield, UK

ABSTRACT: The methods of finite element (F.E.) analysis and stress frozen photoelasticity were used to investigate the stresses resulting from the contact of an infinite cylinder with an elastic half-space for comparison with the well known theoretical solution. Two load cases were studied, that of a load normal to the plane and another where a normal and tangential load are applied to the cylinder. The coefficient of friction was controlled in the photoelastic experiment and the same value used in the F.E. analysis. Plots of the Cartesian stresses in the plane of symmetry, the Cartesian shear stress along a line below the surface and the full field of the Cartesian shear stress, were obtained and compared. The correlation between the experimental, numerical and theoretical results is very close and suggests that combining the experimental and numerical techniques provides a reliable method of stress determination in complex components and structures.

NOMENCLATURE

a	- semi-contact width
$u_{y,x}$	- normal and tangential displacements
p_0	- maximum value of the normal traction distribution
E	- Young's modulus
$F_{y,x}$	- normal and tangential nodal forces
$K_{N,T}$	- normal and tangential stiffness
P,Q	- total normal and tangential forces ($\Sigma F_{y,x}$)
μ	- coefficient of friction
ν	- Poisson's ratio
$\sigma_{x,y}$	- Cartesian stress components
τ_{xy}	- Cartesian shear stress

1 INTRODUCTION

A comparative study of a classical contact geometry was performed using the finite element method and photoelasticity. The two situations considered were firstly a cylinder on its side in contact with a half-plane subject to normal loading and the same with normal and tangential loading. In both cases the coefficient of friction was controlled in the photoelastic experiment and duplicated in the F.E. analysis. The resulting stress distributions along a line in the half-space and for the full field were compared to a well known analytical solution.

One of the aims of this study was to validate the F.E. code for use in simple contacting bodies so that it could be used for more complex geometries. The F.E. code for a non-linear contact element had been validated previously (Vaillancourt et al. 1996) but this was for complete contact cases where the contact area remained constant as the applied load was varied. A different class of contact is considered here, which is incomplete or advancing contact, where the contact area increases with the applied load. Although some F.E. studies of these types of contact have been considered before (Shih et al. 1992), none have considered the effect of friction.

The classical cases studied in the past have also been limited by the fact that analytical formulae do not exist for the case of a cylinder loaded with normal and tangential forces, where $Q \neq \mu P$ (P and Q are the normal and tangential loads respectively). In this study in addition to the cases where $Q = \mu P$ the stress field for loading cases where $Q \neq \mu P$ were obtained numerically and compared with experimental and numerical data.

The comparison of experimental, numerical and analytical analyses provides a source of confidence in the different methodologies. This is desirable when validating F.E. anlysis against photoelastic tests and theoretical analysis so as to investigate the stresses in engineering components and structures.

2 THEORY

The surface tractions and sub-surface stresses for a cylinder on an elastic half-space are described by a well defined theory. The first investigation of this problem was by Hertz in 1882 (Johnson 1994) who amongst other things obtained the contact pressure distribution for two parallel cylinders in contact. This analysis has been greatly developed and for the comparisons between experimental results, numerical data and theoretical models the work of Hills et al. 1993 has been used. This gives simple, closed form solutions for all the components of stress which result from tangential and normal loads. In the case of normal and tangential loading the subsurface stress field is given for a static or sliding line contact by Hills et al. 1993 and will not be repeated here.

When the tractive force transmitted through the interface is less than the limiting friction force, a condition known as incipient sliding occurs. This creates two separate regions, one of stick and the other of slip. Assuming that the normal traction distribution is Hertzian and that the Coulomb laws of friction apply, the distribution of the tangential surface tractions is given by (Johnson 1994):

$$q(x) = q'(x) + q''(x) \tag{1}$$

$$q'(x) = \mu p_0 \left(1 - x^2 / a^2\right)^{\frac{1}{2}} \tag{2}$$

$$q''(x) = -\frac{c}{a}\mu p_0 \left[1 - (x+d)^2 / c^2\right]^{\frac{1}{2}} \tag{3}$$

where μ is the coefficient of static friction, p_0 is the maximum value of the normal traction distribution, c is half the length of the stuck region and d is the distance from the centre of contact to the centre of the stuck region. When the total normal and tractive forces, P and Q, are known then the dimensions of the stick and slip regions can be found:

$$\frac{d}{a} = 1 - \frac{c}{a} = 1 - \left(1 - {Q}/{\mu \mathrm{P}}\right)^{\frac{1}{2}} \tag{4}$$

since the area under the curve defined by *q(x)* is equal to the total tractive force. The subsurface stress distributions produced by the normal and tangential loads can be obtained by integrating the distributions of normal and tangential tractions. These integrals are given below:

$$\sigma_x = -\frac{2z}{\pi}\int_{-a}^{a}\frac{p(s)(x-s)^2 ds}{\{(x-s)^2 + y^2\}^2} - \frac{2}{\pi}\int_{-a}^{a}\frac{q(s)(x-s)^3 ds}{\{(x-s)^2 + y^2\}^2} \tag{5}$$

$$\sigma_y = -\frac{2z^3}{\pi}\int_{-a}^{a}\frac{p(s) ds}{\{(x-s)^2 + y^2\}^2} - \frac{2z^2}{\pi}\int_{-a}^{a}\frac{q(s)(x-s) ds}{\{(x-s)^2 + y^2\}^2} \tag{6}$$

$$\tau_{xy} = -\frac{2z^2}{\pi}\int_{-a}^{a}\frac{p(s)(x-s) ds}{\{(x-s)^2 + y^2\}^2} - \frac{2}{\pi}\int_{-a}^{a}\frac{q(s)(x-s)^2 ds}{\{(x-s)^2 + y^2\}^2} \tag{7}$$

For the case of incipient sliding these integrals do not have a closed form solution. Hence to obtain the stress distributions these functions must be integrated numerically.

3 METHOD

The photoelastic modelling procedure is described in detail elsewhere (Burguete and Patterson 1997). A brief outline is given below due to considerations of the available space. The photoelastic models were made of epoxy resin MY750 and prepared so that control of the coefficient of friction could be achieved. This was done by altering the surface roughness and using a powder lubricant. These models were loaded and stress frozen and thin slices were cut from the middle plane for photoelastic analysis using an automated system based on the method of phase stepping (Carazo et al. 1994). The dimensions and loading of the models were similar to those used in the F.E. analysis.

The model used for the F.E. analysis is shown diagramatically in Figure 1. The cases modelled using the F.E. method were subjected to the same conditions as the photoelastic models, that is the same material constants were used and the same friction coefficients applied. In the case where there is only normal loading, the semicircular part of the model was deflected in the negative y-direction only by 0.3 (for the finite element analyses all the values used are in consistent units, i.e. if one measurement is in metres, all others must also be, this includes compound units such as N/m^2). The rectangular block which models the infinite half-space was given

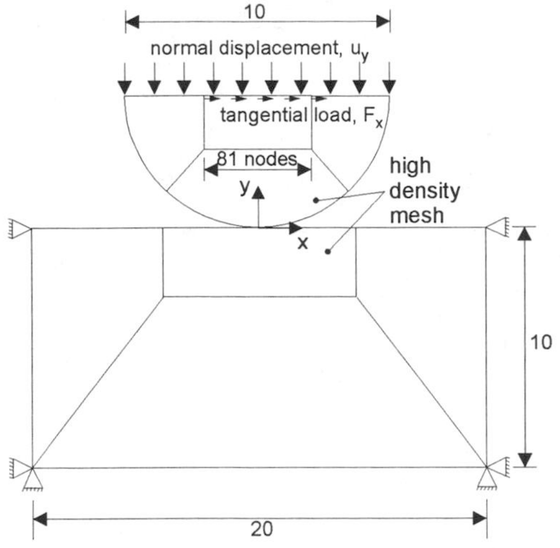

Figure 1: Diagram of the F.E. model of a cylinder on its side on an infinite half-plane showing dimensions, constraints and loading conditions.

Table 1. Material and model properties for the finite element models

Material/Model Property	Normal loading	Normal and tangential loading
E	35	35
μ	0.2	0.9
ν	0.3	0.3
K_N	350	350
K_T	-	35

Table 2. Loading Parameters for the finite element models

Loading Parameter	Normal loading	Normal and tangential loading
ΣF_x	-	0.66
ΣF_y	3.38 (Resultant)	3.34 (Resultant)
u_x	-	-
u_y	0.3	0.3

constraints in the x- and y-direction along the bottom edge and in the x-direction along the two sides.

For normal and tangential loading the constraints were the same apart from a tangential force applied to the 81 nodes in the central section of the semicircular part of the model. The sum of all these nodal forces is equivalent to the total tangential force. The normal load was applied as a displacement because the upper half of the model underwent rigid body motion and the solution did not converge if a force was used. To obtain the correct value of deflection for a desired normal load, an iterative procedure was performed where the reaction forces for a small deflection was found first. The sum of all the vertical reaction forces on the horizontal edge of the semicircular part of the model were then taken to be the applied normal load. This procedure was repeated until the desired load was achieved.

The finite element analysis software ANSYS 5.0 was used for all the work described. The elements used to model this geometry were 4-node planar structural solid elements, PLANE42 (ANSYS User's Manual 1992). Symmetry conditions exist out of the plane so as to simulate the plane strain conditions of an infinitely long half cylinder and half-space. The other material properties and dimensions not detailed above are given in Table 1 and Figure 1. The applied loads and constraints are given in Table 2.

The interface was modelled using CONTAC48 two-dimensional point-to-surface contact elements. These elements allow sliding and sticking so that a contact where incipient sliding occurs can be modelled. For these elements to provide good results various controlling parameters have to be carefully selected. The two critical parameters are K_N and K_T, which are the normal and tangential contact stiffness respectively. Dealing with the normal contact stiffness first, this value is related to the penalty function which determines the restoring force if a node penetrates a surface. If a large value of K_N is selected then the compatibility between the surfaces is well maintained but problems can arise with the convergence of the solution. It is therefore desirable to select the largest value of K_N which allows convergence.

The selection of K_T is only necessary when sliding is going to take place and differential tangential forces act across the interface. The physical significance of this value is shown in Figure 2. The value of K_T needs to be large enough to allow shear forces to be transmitted with fidelity but not so high that no incipient sliding can occur and difficulties with convergence are experienced. If this value is of the same order of magnitude as the Young's modulus, then a reliable solution will be achieved. In all the cases tested the static coefficient of friction was used because no rigid body motion or sliding takes place. The values used are given in Table 1.

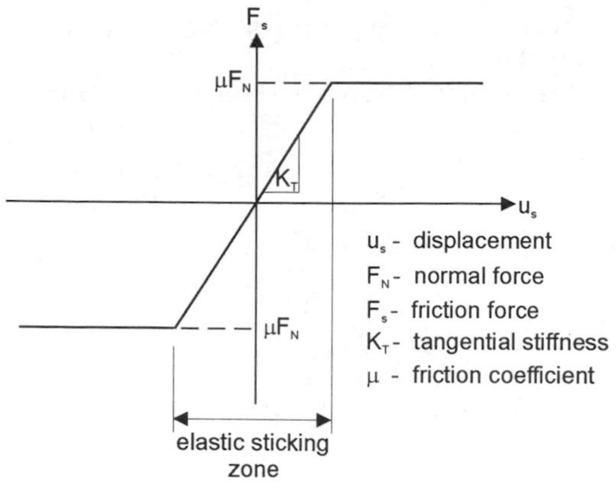

Figure 2: Graph showing the significance of the tangential contact stiffness parameter, K_T.

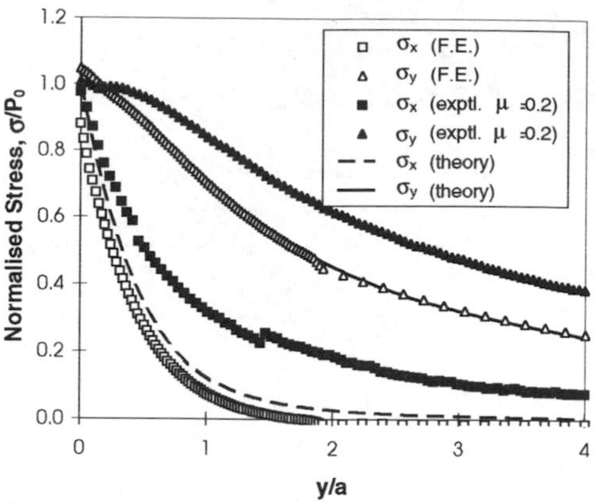

Figure 3: Plots of the cartesian components of stress, σ_x and σ_y along the axes of symmetry for the half-plane with the cylinder on its side subject to normal load only. Finite element, photoelastic and theoretical results are shown.

The mesh density used was arrived at by performing a series of convergence tests. Convergence of the model was determined by comparing one of the stress components in a selection of arbitrarily chosen points in the area of interest. If the stress for two different levels of mesh refinement differed by less than 1% the coarser mesh was used for the analysis.

As a result of the analyses performed above, a methodology for using F.E. methods when analysing contact problems was produced. This meant that further studies of more complex contacts could be carried out with a high level of confidence in the results.

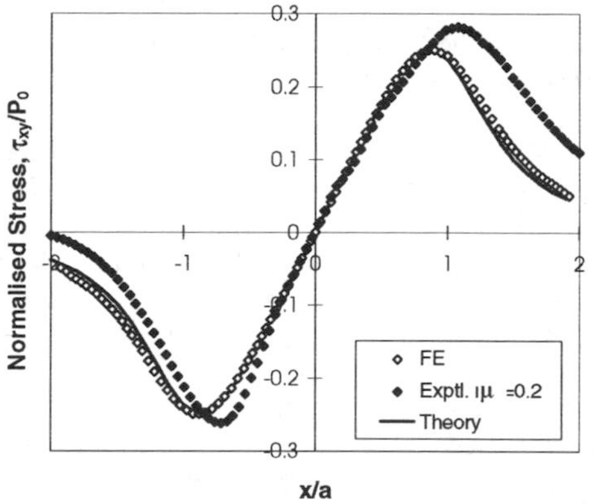

Figure 4: Cartesian shear stress, τ_{xy} along a line y = 0.5a below the contact interface, where a is the semi-contact width. The data shown is for the half plane with the cylinder on its side subject to normal load.

4 RESULTS AND DISCUSSION

The results from classical contact tests where the cylinder has been normally loaded are shown in Figures 3 and 4. Figure 3 shows the cartesian stress components σ_x and σ_y along the line of symmetry. The photoelastic results and the theoretical values are from Burguete and Patterson 1997. Close correlation was found between theory and the finite element results. The lesser quality of fit between the experimental and theoretical results arises from a number of possible causes which are detailed in Burguete and Patterson 1997. The most significant cause is due to errors in the stress separation routines. These errors arise from incorrect seeding of the stress separation routines as a result of marks or scratches on the specimen.

Figure 4 shows the cartesian shear stress τ_{xy} along a line y=0.5a below the contact interface. Here again the correlation between the finite element results and theory are excellent. Full field plots of the theoretical and finite element analysis results are given in Figure 5 which shows the cartesian shear stress, τ_{xy}.

Figures 6 and 7 show the results from the cylinder with normal and tangential applied loads. As for the previous results the correlation between the

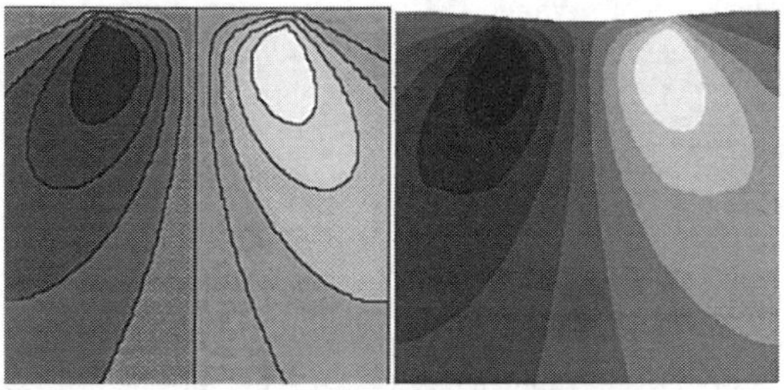

Figure 5: Maps of the cartesian shear stress, τ_{xy}, in the half-plane with the cylinder on its side subject to normal load. The results from theory (left) and finite element analyses (right) are shown.

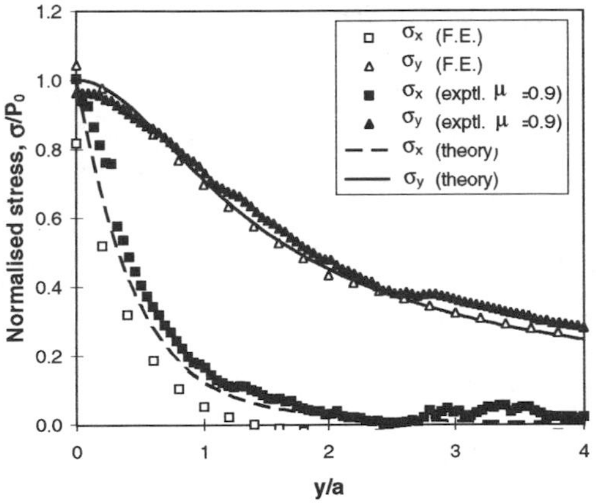

Figure 6: Plots of the cartesian components of stress, σ_x and σ_y along the line of symmetry for the half plane with the cylinder on its side subject to normal and tangential loads.

experimental, numerical and analytical results is excellent. The full field data is given in Figure 8.

The accuracy and reliability of the automated photoelastic system have been considered by Haake et al 1993 and by Carazo et al 1994. They found that the readings of isochromatic fringe order and isoclinic angle were repeatable to ± 0.007 fringes and ± 0.75° respectively. This is comparable to other automatic systems and manual analysis. The accuracy of the stress separation procedures were investigated by Haake et al 1995 and they concluded that σ_x and σ_y would be obtained with errors of no more than 5%.

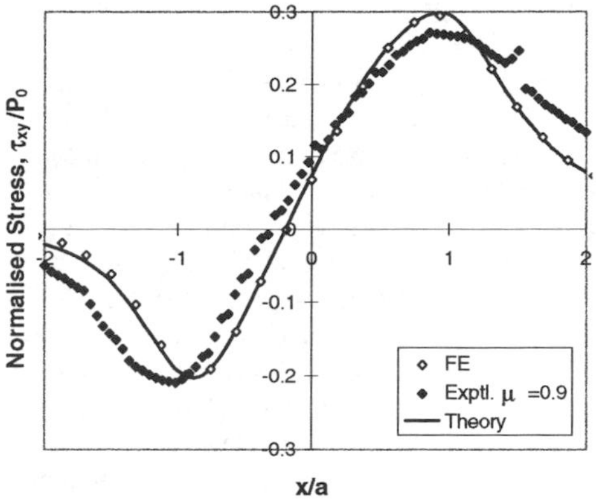

Figure 7: Cartesian shear stress, τ_{xy}, along a line y = 0.5a below the contact interface, where a is the semi-contact width. The data shown is for the half plane with the cylinder on its side and subject to normal and tangential loads.

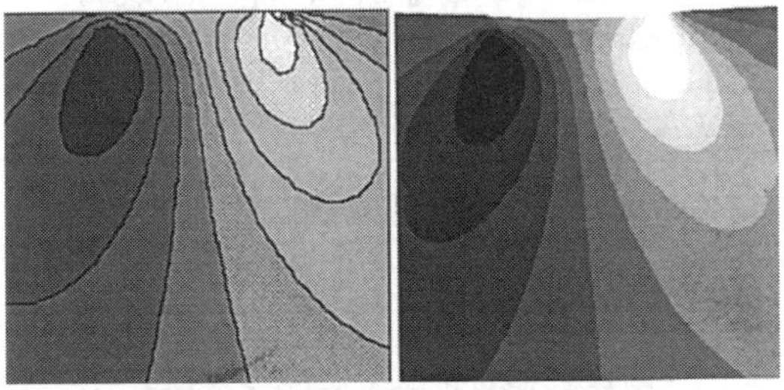

Figure 8: Maps of the cartesian shear stress, τ_{xy}, in the half plane with the cylinder on its side subject to normal and tangential loads. The results from theory (left) and finite element analyses (right) are shown.

Sources of error in the F.E. analysis arise from the selection of the values of the contact stiffness as these determine the deformations of the model particularly at the interface. Other errors may have resulted from the fact that the F.E. mesh is an approximation to the infinite half-space.

The close correlation between the results from the finite element method, photoelasticity and theoretical analysis in these studies, gives a high degree of confidence in both the F.E. and photoelastic modelling indicating that the methods used are reliable. Ideally the two approaches should be used in tandem for maximum effectiveness and efficiency when analysing stress in engineering components.

5 CONCLUSIONS

Two loading cases for a cylinder in contact with an elastic half-space have been analysed using the finite element method and photoelasticity. These compared very favourably with well known analytical solutions. The photoelastic results were obtained using the automated polariscope and included the distribution of cartesian shear stresses for large areas around the contact interface. The shear difference method was employed to obtain the cartesian components of stress in the plane of symmetry of the half-space. Significant 'tuning' of the finite element models was necessary, but would not need to be repeated for similar geometry and conditions. Good correlation was obtained between the results from experiment, numerical analyses and analytical theory.

The methods of automated stress-frozen photoelasticity and finite element analysis have been shown to be reliable for the analysis of a simple engineering component. The combined use of these methods would lead to a high level of confidence in the results of the analysis of more complex engineering structures.

ACKNOWLEDGEMENTS

The authors would like to thank Mr R Kay and Mr J Driver for their careful manufacture and preparation of photoelastic specimens, and Mr I D Bruce for his assistance in running the F.E. models. Financial support under EPSRC grant number GR\J 09413 is gratefully acknowledged.

REFERENCES

ANSYS User's Manual for Revision 5.0, Volume III—Elements. Swanson Analysis Systems, Inc. 1992.

Burguete, R.L. and Patterson, E.A. 1997. A photoelastic study of contact between a cylinder and a half-space. *Experimental Mechanics* in press.

Carazo-Alvarez, J., Haake, S. J. and Patterson, E. A. 1994. Completely automated photoelastic fringe analysis. *Optics and Lasers in Engineering* 21:133-149.

Haake, S. J., Wang, Z. F. and Patterson, E. A. 1993. Evaluation of full field automated photoelastic analysis based on phase stepping. *Experimental Techniques* 17(6):19-25.

Haake, S. J., Wang, Z. F. and Patterson, E. A. 1995. 3D separation of stresses - 40 years on. Proc. of SEM Spring Conference, June 12-14, Grand Rapids, Michigan, USA, pp.183-190.

Hills, D. A., Nowell, D. and Sackfield, A. 1993. *Mechanics of Elastic Contacts.* Oxford: Butterworth Heinemann.

Johnson, K.L. 1994. *Contact Mechanics.* Cambridge: Cambridge University Press.

Shih, C.W., Schlein, W.D. and Li, J.C.M. 1992. Photoelastic and finite element analysis of different size spheres in contact. J. Mater. Res. 7(4):1011-1017.

Vaillancourt, H., McCammond, D. and Pilliar, R.H. 1996. Validation of a non-linear two-dimensional interface element for finite-element analysis. *Experimental Mechanics* 36(1):49-54.

Modern Practice in Stress and Vibration Analysis, Gilchrist (ed.) © 1997 Balkema, Rotterdam, ISBN 90 5410 896 7

Photoelastic analysis of stress intensity factors in notches

S.J. Haake & J.R. Yates
Department of Mechanical Engineering, University of Sheffield, UK

ABSTRACT: Simulated cracks from three notch geometries were analysed using two dimensional photoelasticity. The specimens were manufactured from epoxy-resin sheet (CT1200) and the notches machined using a diamond impregnated saw. The notches were semi-circular, U-shaped and V-shaped and the crack length was varied up to 0.6 of the specimen width. The specimens were subjected to pure Mode I loading and the fringe patterns used to calculate the Mode I stress intensity factor. A previously suggested notch correction term gives the ratio of the stress intensity factor of a notched to an unnotched specimen. The correction term was determined for all three notch geometries and all crack lengths and the experimental data compared to a previously suggested empirical equation and Finite Element (FE) results. The three sets of results compared well although the FE results were closer to the empirical equation than the photoelastic results. The notch correction factor provides a simple but effective way of calculating stress intensity factors within notches.

NOMENCLATURE

a crack length in un-notched specimen
D notch depth
E Young's modulus
F notch correction term
J far-field J-integral
K_I mode I stress intensity factor
K_T stress concentration factor, based on gross section stress
l length of crack from notch root
w width of gross section of specimen
Y_c non-dimensional stress intensity factor of a crack in an un-notched specimen
Y_n non-dimensional stress intensity factor of a crack in a notched specimen
ν Poisson's ratio
ρ radius of root of notch
σ nominal gross section stress

1 INTRODUCTION

Fabricated structures are susceptible to fatigue crack growth arising from the interaction of non-steady loads and the local geometry. Common sites for the start of structural failure by fatigue are the localised stress or strain concentrations arising from changes in shape. The historical approach to understanding the behaviour of notched components has been to estimate the elevation of the elastic stress at the notch root. However, many industries are now following the lead set by aerospace and nuclear power in using linear elastic fracture mechanics (LEFM) methods in assessing the integrity of flawed structures.

A fracture mechanics assessment of a cracked structural member requires knowledge of the stress intensity factor for the particular crack and geometrical configuration. Although handbooks (Murakami 1987) of stress intensity factors provide solutions for a wide range of cracked geometries, the number of examples for cracks at notches is more limited.

The application of linear elastic fracture mechanics to cracks at notches has been progressing for more than twenty years (Fuhring 1973, Miller 1973, Smith and Miller 1973, 1977, 1978, Murakami and Okazaki 1976, Karlsson and Backlund 1978). A common approach has been to formulate stress intensity factors from the remote applied stress, the crack length and a non-dimensional correction term based on the shape of the notch (Schijve 1982, Lukas 1987, Yates 1991, Yates *et al.* 1993, Yates and Cardew 1996).

In general terms, stress intensity factors for cracks at notches may be described by

$$K_I = Y_n \sigma \sqrt{\pi a} \quad (1)$$

where $a = D + l$ and there are several ways of estimating Y_n. Schijve (1982) proposed that

$$Y_n = C K_T \sqrt{\frac{l}{a}} \quad (2)$$

where

$$C = 1.1215 - 3\left(\frac{l}{\rho}\right) + 4\left(\frac{l}{\rho}\right)^{1.5} - 1.7\left(\frac{l}{\rho}\right)^{2} \quad (3)$$

was found by fitting to numerical data on elliptical holes in infinite sheets (Newman 1971, Nisitani 1978). Lukas (1987) proposed a method similar to that of Schijve, differing only in the expression for C

$$C = \frac{1.12}{\sqrt{1 + 4.5\left(\frac{l}{\rho}\right)}} \quad (4)$$

A simple alternative would be to take

$$C = Y_c \quad (5)$$

and re-arrange equations 1 and 2 to give

$$K_I = K_T Y_c \sigma \sqrt{\pi l} \quad (6)$$

Yates (1991) suggested that a correction factor based on the un-notched geometry and a general notch correction term may be suitable:

$$Y_n = F Y_c \quad (7)$$

$$K_I = F Y_c \sigma \sqrt{\pi a} \quad (8)$$

where $0 \le F \le 1$ as the crack length increases. Studies of various notch geometries and loading configurations (Yates 1991, Yates *et al.* 1993, Cardew and Yates 1996) suggest that

$$F = 4.0\left[\frac{l}{\sqrt{D\rho}}\right]^{0.5} - 5.25\left[\frac{l}{\sqrt{D\rho}}\right] + 2.25\left[\frac{l}{\sqrt{D\rho}}\right]^{1.5} \quad (9)$$

for $l/\sqrt{(D\rho)} \le 0.4$ and $F = 1$ for $l/\sqrt{(D\rho)} > 0.4$ is a reasonable estimate of the notch correction term for many situations.

This paper presents the results of photoelastic analyses of cracks in notched bodies subjected to tensile loading. The results are presented in terms of stress intensity factors and notch correction terms and are compared with finite element solutions for similar geometries.

2 METHOD

2.1 *Experimental method*

Three 6 mm thick two-dimensional epoxy resin (Araldite CT1200) models were, manufactured with a 45° V-shaped notch, a U-shaped notch and a semi-circular shaped notch machined in one side of the specimen to the dimensions shown in Figure 1. A simulated crack of length *a* was cut into the specimen along the horizontal axis of each notch as shown using a diamond-impregnated circular saw. The simulated cracks produced had a nominal width of 0.4 mm, and the lengths of which were measured using a microscopic vernier gauge. A tensile load was applied to the specimens using two clamps, to ensure a uniform distribution of the load across the sheet, which were pin-jointed to the testing machine.

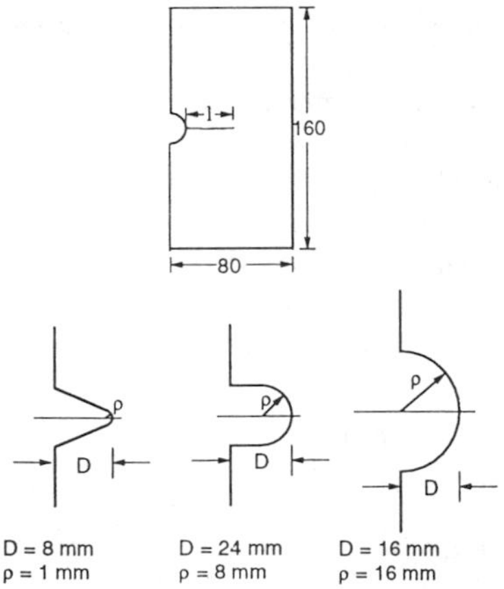

Figure 1. Dimensions of epoxy resin models and notch geometries.

The specimen was viewed under load in a circular polariscope and using a sodium light source. The fringes patterns around the crack tip were recorded using a CCD camera and a video printer and the images analysed using a digitising tablet (Nurse and Patterson 1993). Data from the pattern were input to an algorithm for calculating stress intensity factors (Nurse and Patterson 1990). Data were collected outside a zone of 10 times the crack tip radius and within a zone 0.4 times the crack length in order to reduce the effects of plasticity at the tip and to ensure that data were from the singularity dominated region.

A method of producing a natural crack, suggested by Post (1954) and Wells and Post (1958) in a photoelastic specimen was evaluated, the apparatus for which is shown in Figure 2. A steel ball is dropped down a tube onto an anvil attached to a razor blade. The impact causes a crack to develop in the specimen. It was found that the crack lengths produced were a linear function of drop height and cracks of lengths 0.4 to 0.7 mm could be made. It was estimated that the cracks had a width of approximately 10^{-2} mm.

Cracks were propagated using this technique by creating the crack, analysing the fringe pattern, using the circular saw to machine out the crack and then extending the crack using the drop weight razor apparatus.

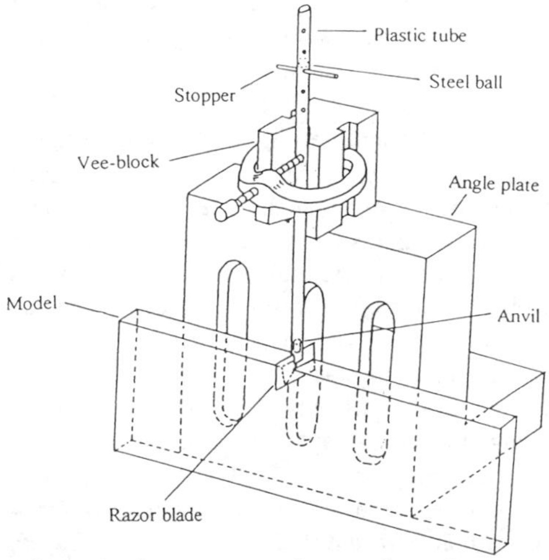

Figure 2. Apparatus used to produce sharp cracks in epoxy resin models. A steel ball drops down the tube from a set height on to an anvil attached to a razor blade, causing a crack to propagate in the specimen.

2.2 *Numerical method*

The TOMECH finite element program (Goldthorpe 1986) was used for the analysis of notch crack stress intensity factors. Two-dimensional plane strain finite element meshes of single edge cracked and edge notched plates were set up using 8-noded isoparametric elements with quarter point singular elements in the vicinity of the crack tip. The material properties selected were a non-dimensional modulus, E/σ_y, of 325, unit yield stress and Poisson's ratio of 0.3.

Stress intensity factors were obtained through using results for the J-integral (Rice 1968) determined on contours remote from, but surrounding, the crack tip. Under conditions of plane strain the stress intensity factor is related to J as follows,

$$K_I = \left[\frac{JE}{1-\upsilon^2}\right]^{0.5} \tag{10}$$

Three profiles of edge notch were used in the solutions: a semicircular notch $D/\rho=1$, $D/w=0.2$; a U-shaped notch $D/\rho=2$, $D/w=0.3$ and a V-shaped notch $D/\rho=5$, $D/w=0.1$ with an included angle of 45°. Loading was achieved by prescribed axial displacement of the end of the mesh remote from the notch. The notch profiles were chosen to represent typical engineering shapes over a range of notch acuity.

3 RESULTS AND DISCUSSION

3.1 *Effect of crack width*

Figure 3 shows $Y_c = K_I / K_o$ against overall crack length for the 45° V-shaped notch, the U-shaped notch and the semi circular notch with results for the crack produced with the circular saw and the crack from the drop weight razor apparatus. Cracks were not produced with the razor apparatus for the semi-circular notch. The solid line shows results for Y_c from Murakami (1987). The 45° V-shaped notch had a depth of 8 mm, i.e. 0.1 $(D+\ell)/w$ and it can be seen that K_I / K_o reduces with crack length in line consistent with published values (Murakami 1987).

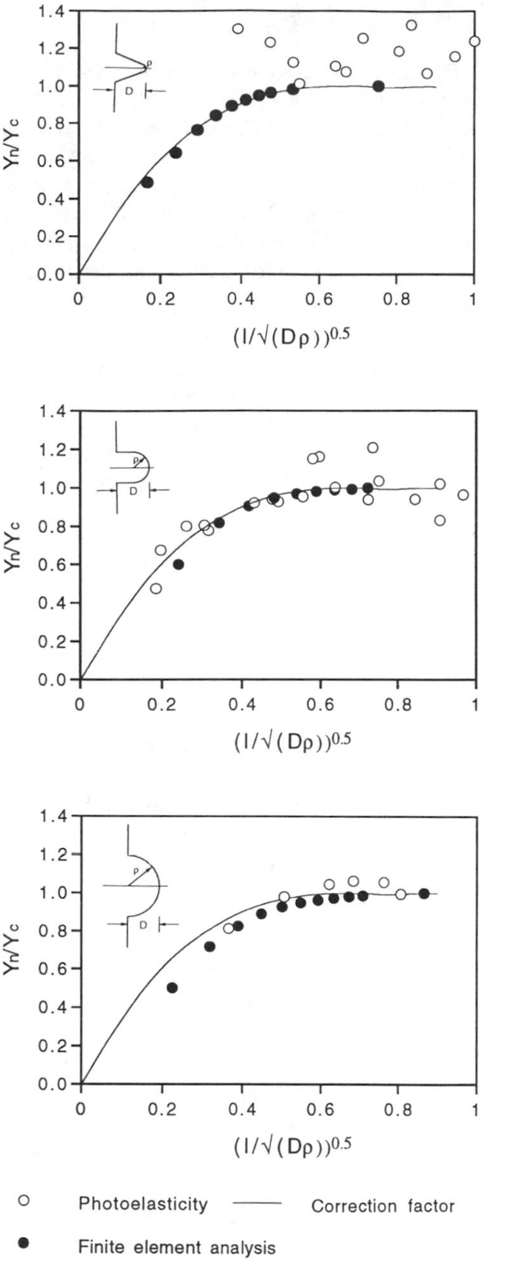

Figure 3. Normalised stress intensity factors $Y_c = K_I / K_o$ versus overall crack length for edge cracks emanating from a 45° V-shaped notch, a U-shaped notch and a semi-circular notch for cracks produced using a circular saw and a razor.

The depth of the U-shaped notch was 24 mm giving a minimum overall crack length $(D+\ell)/w$ of 0.3. It can be seen that at 0.3 $(D+\ell)/w$ the normalised stress intensity factor drops off sharply, tending towards zero. A similar effect is seen for the semi-circular shaped notch which had a minimum of 0.2 $(D+\ell)/w$.

There appears to be little difference between the results for the crack produced with a saw and those for the crack produced with a razor blade. The results for the V-shaped notch show considerable scatter at the short crack lengths close to the notch. Results for the U-shaped notch show scatter for longer cracks. Inspection of the cracks made by the razor show that the cracks are not always perpendicular either along the crack or through the thickness. This twisting of the crack plane in three dimensions makes determination of the crack tip difficult and hence increases scatter. There appears to be no advantage of the razor method over using a conventional saw in Mode I loading. It may be that the 'razor' cracks produce greater realism in mixed mode loading due to crack closure effects.

3.2 *Stress intensity factors close to the notch*

Figure 3 shows clearly for the case of the U-shaped notch and the semi-circular notch that the normalised stress intensity factor decreases at short crack lengths close to the notch simply due to the closeness of the boundary of the specimen. At longer crack lengths the normalised stress intensity factor equals Y_c (Murakami, 1987).

Figure 4 shows all the photoelastic results compared to Y_c and results from finite element analysis. At long crack lengths the finite element results compare well with the Murakami solution. At short crack lengths the FE results, show the same decrease in normalised stress-intensity factor as shown by the photoelastic results. It can be seen that the normalised stress intensity factor decreases rapidly at very short crack lengths for the V-shaped notch over a range of only 0.02 $(D+\ell)/w$ while the change for the U-shaped and semi-circular notch occurs over ranges 0.08 and 0.15 $(D+\ell)/w$ respectively. These ranges correspond to crack lengths of 0.2, 0.27 and 0.75 mm for the V-shaped, U-shaped and semi-circular notches respectively. It appears, therefore, that, for short sharp notches, such as the V-shaped notch, the stress field around the notch has very little effect on the crack tip unless the crack is very short. The corollary is that the wider,

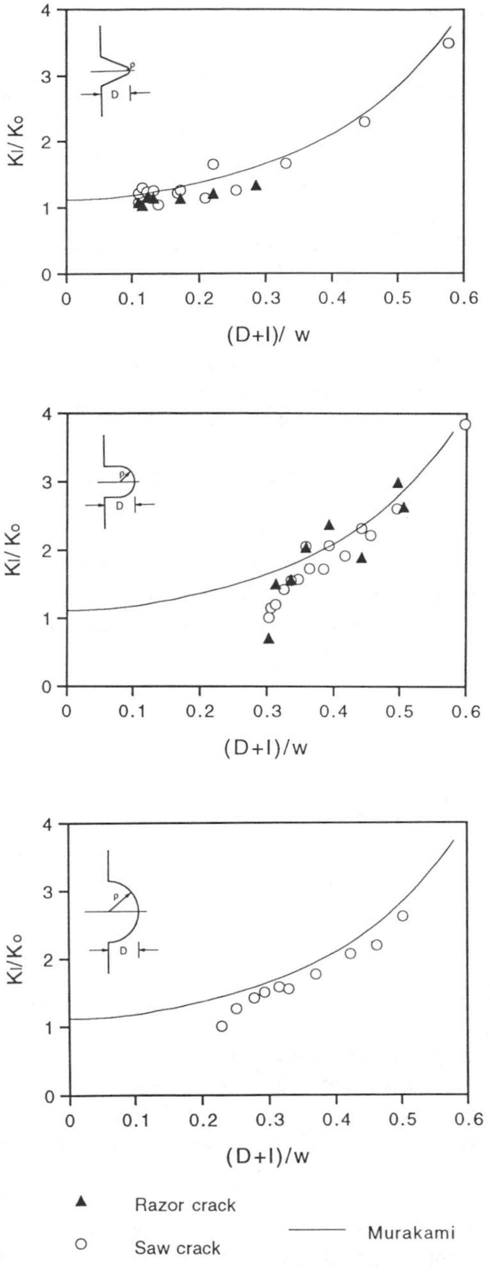

Figure 4. Normalised stress intensity factors $Y_c = K_I / K_o$ versus overall crack length for edge cracks emanating from a 45° V-shaped notch, a U-shaped notch and a semi-circular notch using photoelasticity and finite element analysis.

deeper semi-circular notch affects cracks three times as long.

Clearly, the geometry of the notch has an effect on the reduction in stress intensity factor for a particular crack length. A correction factor suggested by Yates (1991) has been used to take account of the influence of D, ρ and the crack length ℓ within the region of the notch.

3.3 *A notch correction factor*

Figure 5 shows the stress intensity factors Y_n from Figures 3 and 4 normalised with respect to Y_c versus $\left(\ell/\sqrt{D\rho}\right)^{0.5}$. The solid line shows the correction factor given by equation 9. The correction factor is the same for each notch, however the x-axis limit of $\left(\ell/\sqrt{D\rho}\right)^{0.5} = 1$ for each notch represents crack lengths of approximately 3, 14 and 16 mm for the V-shaped, U-shaped and semi-circular notches respectively.

The finite element results compare well with the correction factor over the full range of ℓ, although the correction factor slightly overestimates the value of Y_n / Y_c for the larger notches. The finite element results indicate that the normalised stress intensity factor with a notch Y_n equals Y_c the stress intensity factor with an un-notched specimen at long crack lengths.

The experimental results from the photoelastic experiment show variability. The values of Y_n / Y_c for the V-shaped notch are consistently above the values of finite element analysis and the correction term. They also show a large amount of scatter. This is due to the very short crack lengths involved (less than 3 mm) and the practical difficulties associated with measuring fringe patterns around cracks of this size.

Results for the U-shaped and semi-circular notch show less scatter; the latter does not contain data from the razor cracks which have been shown to contain more variability.

The photoelastic results certainly show the same trend as the finite element results suggesting that the correction factor F is a good empirical solution for the calculation of stress intensity factors for cracks emanating from notches.

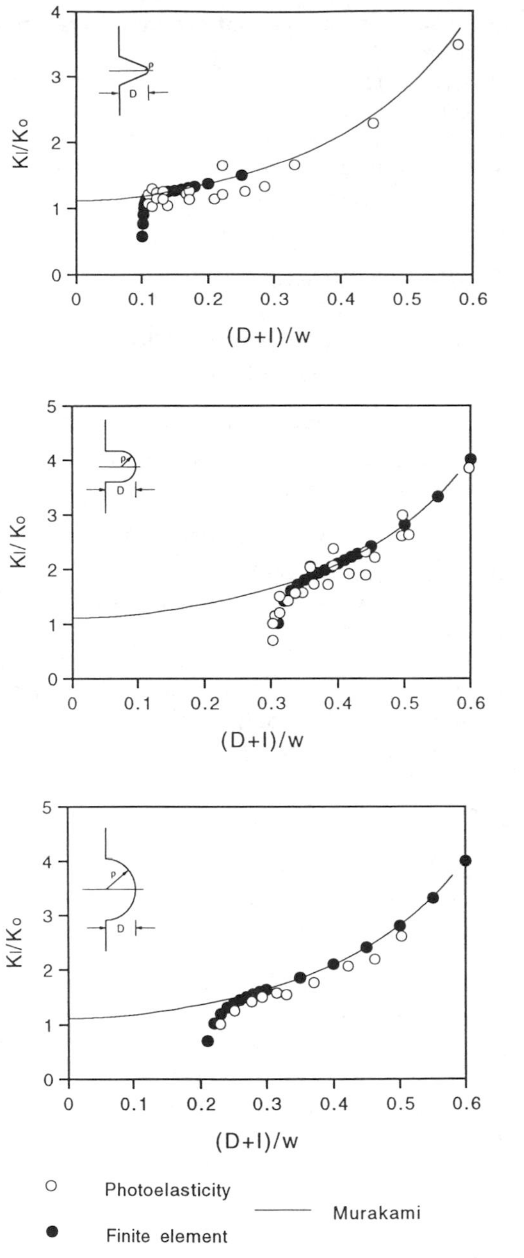

Figure 5. Y_n / Y_c versus $(\ell / \sqrt{D\rho})^{0.5}$ for edge cracks emanating from a 45° V-shaped notch, a U-shaped notch and a semi-circular notch. The solid line shows the correction factor given by equation 9.

4 CONCLUSIONS

It was found that for short crack lengths in the region of a notch, the normalised stress intensity factor was considerably reduced in comparison to that given by the published solution for a total crack length $(D+\ell)/w$. The effect was more noticeable for wide, deep notches.

Normalised stress intensity factors for cracks in epoxy-resin models calculated using photoelasticity compared well with results from finite element analysis.

The technique of using a razor blade to produce realistic cracks appeared to show more scatter than results from cracks produced using a diamond impregnated circular saw.

The results from both the photoelastic analysis and the finite element analysis showed good comparison with the suggested correction factor. The photoelastic results for the 45° V-shaped notch showed considerable scatter due to the difficulties associated with measuring fringe patterns at the small scale required.

ACKNOWLEDGEMENTS

The authors acknowledge access to TOMECH, the research finite element code of SIRIUS, and support in its use by the TOMECH group. Acknowledgement also goes to the technicians of the Experimental Stress Analysis Laboratory.

REFERENCES

Cardew G.E. and Yates J.R., 1996. A local grid refinement method for determining stress intensity factors for cracks at notches. Fatigue and Fracture of Engineering Materials and Structures, **19**, 523-528.

Fuhring H., 1973, Approximation functions for K-factors of cracks in notches, *Int. J. Fract.*, Vol. 9, pp. R328-R329.

Goldthorpe M.R., 1986, *An elastic-plastic finite element program with applications to cracked bodies,* Ph.D. thesis, Department of Mechanical Engineering, University of Sheffield.

Karlsson A. and Backlund J., 1978, Summary of SIF graphs for cracks emanating from circular holes, *Int. J. Fract.,* Vol. 14, pp. 585-596.

Lukas P., 1987, Stress intensity factor for small notch-emanated cracks, *Engng Fract. Mech.,* Vol. 26, pp. 471-473.

Miller K.J., 1973, Application of fracture mechanics to fatigue at notches, *Int. J. Fract.,* Vol. 9, pp. R326-R328.

Murakami Y., 1987, *Stress Intensity Factors Handbook,* Pergamon Press, Oxford.

Murakami Y. and Okazaki Y., 1976, A simple procedure for the accurate determination of stress intensity factors for a round bar with a circumferential crack at notch root, *Trans. Japan. Soc. Mech. Engrs,* Vol. 42, No. 364, pp. 3679-3687.

Newman J.C., Jr., 1971, An improved method of collocation for the stress analysis of cracked plates with various shaped boundaries, Technical Report NASA TN D-6376.

Nisitani H., 1978, Solutions of notch problems by body force method, *Mechanics of Fracture*, Vol. 5, ed. G.C.Sih, Noordhoff 1978, pp. 1-68.

Nurse, A.D and Patterson, E.A. 1990. Photoelastic determination of stress intensity factors for edge cracks under mixed mode loading. *Proc. 9th Int. Conf. Expt. Mechs, Copenhagen*, **2**, 948-957.

Nurse, A.D and Patterson, E.A. 1993. Determination of predominantly mode II stress intensity factors from isochromatic data. *Fatigue and Fract. in Engng. Maters. and Structures,* **16** (12), 1339-1354.

Rice J.R., 1968, A path independent integral and the approximate analysis of strain concentration by notches and cracks, *J. Appl. Mech.*, Vol. 35, pp. 379-386.

Schijve J., 1982, The stress intensity factor of small cracks at notches, *Fatigue Engng Mater. Struct.*, Vol. 5, pp. 77-90.

Smith R.A. and Miller K.J., 1973, The growth of fatigue cracks from circular notches, *Int. J. Fract.,* Vol. 9, pp. R101-R104.

Smith R.A. and Miller K.J., 1977, Fatigue cracks at notches, *Int. J. mech. Sci.*, Vol. 19, pp. 11-22.

Smith R.A. and Miller K.J., 1978, Prediction of fatigue regimes in notched components, *Int. J. mech. Sci.*, Vol. 20, pp. 201-206.

Yates J.R., 1991, A simple approximation for the stress intensity factor of a crack at a notch, *J. Strain Anal.,* Vol. 26, pp. 9-13.

Yates J.R., Kiew H.T. and Goldthorpe M.R. (1993) Stress intensity factors for cracks at the root of blunt notches. Proc. 12th Int. Conf. on Offshore Mechanics and Arctic Engineering, Glasgow, June 1993, ASME, Vol III, Part B, pp. 855-861.

Modern Practice in Stress and Vibration Analysis, Gilchrist (ed.) © *1997 Balkema, Rotterdam, ISBN 90 5410 896 7*

Some design considerations for a thermoelastic strain gauge

J.M. Dulieu-Smith, S.Quinn & P.Davies
University of Liverpool, Department of Mechanical Engineering, UK

P.Stanley
University of Manchester, School of Engineering, UK

ABSTRACT: Design studies are described for a thermoelastic strain gauge consisting of a thin strip of material bonded to a specimen at its ends, with the central portion unattached. Separate values of the direct stresses in the specimen parallel and normal to the length of the strip are obtained from the thermoelastic signals from the strip and the adjacent specimen surface. Factors considered in this work include the gauge material, the gauge-to-specimen adhesive and the use of the gauges on standard test specimens.

1. INTRODUCTION

An adiabatic stress change in an elastic body causes a small change in the temperature of the body (Kelvin 1853). This phenomenon - the thermoelastic effect - is the basis of the thermoelastic stress analysis technique (Harwood and Cummings 1991) in which the stress changes in a cyclically loaded elastic body are derived from measurements of the temperature changes induced by the cyclic loading. Since these temperature changes are of the order 0.01° C to 0.1° C, the technique requires a very sensitive measuring system. Two such systems are available. The SPATE system (Harwood and Cummings 1991) has been in use since the early 80s. It incorporates a highly sensitive radiometric infra-red detector (CdHgTe type) and operates in a scanning mode. In practice a selected area (or "frame") of the cyclically loaded specimen is scanned and the received signal is processed and displayed on a video monitor as a contour plot of the stress data over the scanned area. The time required for a scan is typically of the order of 1 hour. Hard copy contour plots or line plots or point values can be readily obtained from the stored data. The DELTATHERM system (Boyce and Lesniak 1994) is a recent development in which a 128 x 128 InSb detector array replaces the single detector of the SPATE system. As a result, scanning is no longer necessary and considerable reductions in the data acquisition time are attained.

The essential "working equation" relating the received signal (S) to the stress changes in an isotropic solid is

$$\sigma_x + \sigma_y = AS \qquad (1)$$

in which

$$A = \frac{DGR}{TeK2048} \qquad (2)$$

and

$$K = \alpha/\rho C \qquad (3)$$

In these equations:

σ_x and σ_y are orthogonal direct stresses at a point on the specimen surface,
A is a calibration factor,
T is the absolute temperature,
e is the surface emissivity,
D and G are system parameters (Dulieu-Smith 1995),
R is a correction factor (approximately equal to unity) (SPATE Manual 1989),
K is the "thermoelastic constant" of the specimen material (Harwood and Cummings 1991),
α, ρ and C are the coefficient of thermal expansion, density and specific heat, respectively, of the specimen material.

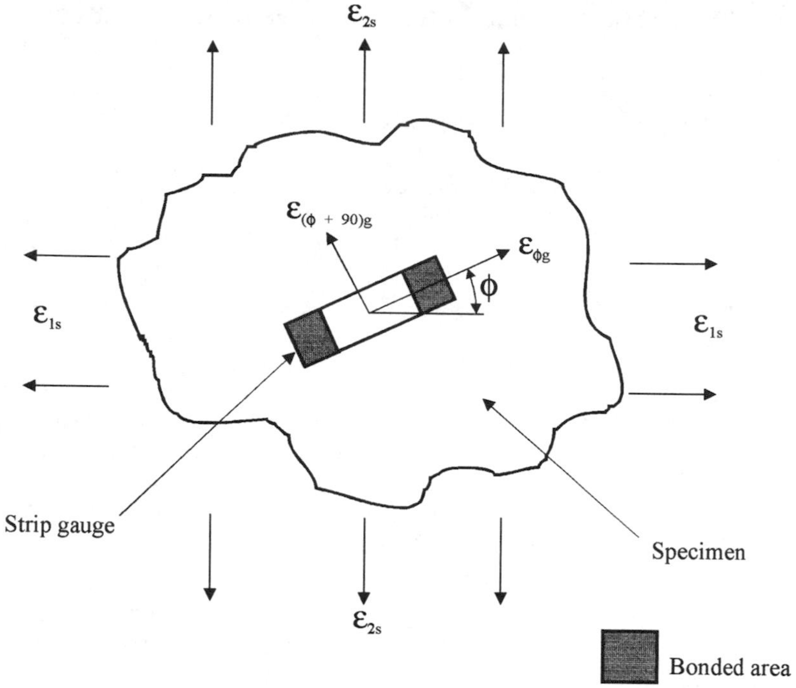

Fig.1 Strip gauge

Calibration techniques for the determination of the factor A have been described by Dulieu-Smith (1995).

It is a feature of the technique that the measured quantity is proportional to the sum of the orthogonal direct stresses on the specimen surface (i.e. the first stress invariant). The process of determining individual stresses from the stress-sum data is referred to as "stress separation". Stanley and Dulieu-Smith (1996) have reviewed numerical stress separation methods and have described an experimental approach in which one of three different devices, bonded to the specimen surface, can be used to provide further thermoelastic data which allows separation of the specimen stresses. These devices are referred to as thermoelastic strain gauges.

One of the devices (the "strip gauge") takes the form of a thin strip of material bonded to the specimen at its ends but unbonded over the central portion. The specimen strain in the direction of the length of the strip is transferred into the strip and by combining the thermoelastic signals from the strip and the adjacent specimen surface, values of the individual direct surface stresses in the specimen can be obtained. This paper describes design work with this form of gauge which includes the choice of gauge material and bonding adhesive, and some evaluation work on standard specimens. The specimen material was mild steel.

2. THEORY OF STRIP GAUGE

The ends of the strip gauge shown in Figure 1 are bonded to the specimen in which the principal strains are ε_{1s} and ε_{2s}. The length of the gauge is inclined at an angle ϕ to the greater principal strain ε_{1s} as shown. The axial strain developed in the gauge ($\varepsilon_{\phi g}$) is equal to the strain in the specimen in the ϕ direction, i.e.

$$\varepsilon_{\phi g} = \varepsilon_{1s} \cos^2\phi + \varepsilon_{2s} \sin^2\phi \qquad (4)$$

The gauge is in a state of uniaxial stress and hence the thermoelastic signal from the gauge S_g (see equation (1)) is given by

$$S_g = E_g \varepsilon_{\phi g} / A_g \qquad (5)$$

where E_g is the Young's modulus of the gauge material and the subscript g signifies the gauge.

It follows therefore that the gauge signal is directly proportional to the gauge strain, and that for any pair of gauges (i and j)

$$\frac{S_{gi}}{S_{gj}} = \frac{\varepsilon_{gi}}{\varepsilon_{gj}} \qquad (6)$$

Equation (6) is used in the appraisal of the test data presented later in the paper. Further details of the interpretation of the gauge signal for the purpose of stress separation are given by Stanley and Dulieu-Smith (1996).

3. CHOICE OF GAUGE MATERIAL

For a given specimen material it is evident from equation (5) that the gauge signal is directly proportional to the quantity E_g/A_g and therefore to the product E_gK_g (see equation (2)). This latter quantity has been evaluated for a large number of materials in a recent review (Dulieu-Smith and Stanley (1996)) and, amongst metals readily available in strip form, aluminium and copper emerge as attractive possibilities for the gauge.

It is also preferable that the elastic strain range of the gauge material should be large and that the possible local reinforcing effect of the strip gauge on the specimen should be minimal. The elastic strain range is given by the quantity σ_{yg}/E_g (where σ_{yg} is the yield stress or limiting elastic stress of the gauge material); the reinforcing effect of the gauge is proportional to E_g and a relatively low modulus value is therefore favoured. On both of these counts aluminium and copper are attractive, and these two materials were chosen as alternatives for the strip gauge.

4. PRELIMINARY TESTS

Following an extensive survey of commercially available products two adhesives were selected for evaluation: Permabond E04 (a two part epoxy adhesive) and Permabond Flexon (a toughened acrylic). The test specimen for this evaluation consisted of a mild steel bar (40 mm wide x 6 mm thick x 176 mm long), with two copper and two aluminium strip gauges fully bonded to the surface along the central line and aligned with the length of the bar. The strips were 6 mm wide × 0.5 mm thick × 25 mm long. One strip of each material was bonded with E04, the other with Flexon. Strain gauges were bonded to each strip and strain readings were taken with the bar loaded in axial tension. An axially aligned control gauge was also bonded directly to the specimen. The test procedure consisted of taking strain readings at static load intervals of 1 kN as the load was increased from zero up to 20 kN (83 MPa) and then decreased back to zero. This was followed by cyclic load testing in which strain ranges were obtained for load ranges of ± 1 kN to ± 9 kN in increments of 1 kN about a mean value of 10 kN. The load cycle frequency was 20 Hz throughout.

The results for the static and dynamic load testing are shown in Figures 2 and 3 respectively. It can be seen from Figure 2 that the strip strains followed the control strains tolerably well for both adhesives, but whilst the strains in the E04 bonded strips "bracketed" the control values, those from the Flexon bonded strips tended to fall below the control values. (The cause of the slight non-linear trend in the Figure 2 data was not further pursued.) Again, in general, the dynamic strain values (Figure 3) compared well with the control readings; the best match was from an E04 bonded strip. Overall, these comparisons resulted in a marginal preference for the E04 adhesive and this was used in all subsequent work.

This work on fully bonded strips was followed by a similar test series with strips bonded only at the ends. The test specimen is shown in Figure 4. Two gauge orientations were studied i.e. parallel to the bar centre line and inclined at 30° to it ($\phi = 0$, $\phi = 30°$, see Figure 1). The static load results showed some features possibly associated with the bond shear stress enhancement occurring in the partially bonded gauges. These results required further study, the outcome of which will be presented elsewhere. Figure 5 shows the dynamic test results which are directly relevant to thermoelastic stress analysis. It can be seen that whilst the $\phi = 0$ strip strain values are somewhat less than the control strain for both materials, in general the correspondence is acceptable. The ratios of the $\phi = 0$ and $\phi = 30^{o}$ strip strains were compared with the theoretical strain ratio obtained using equation (4) assuming a Poisson's ratio of 0.3 for the steel specimen. The experimental values of the ratio (obtained for an applied load range of 16 kN) were 0.632 for the aluminium and 0.685 for the copper, the theoretical value was 0.675. The theoretical strain ratio changed by

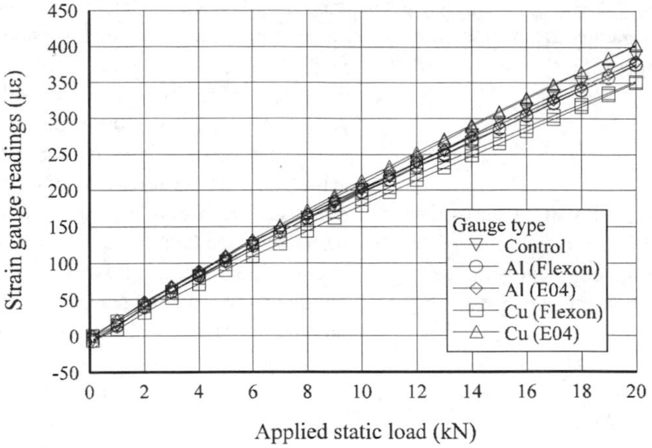

Fig. 2 Static load testing of the fully bonded gauges

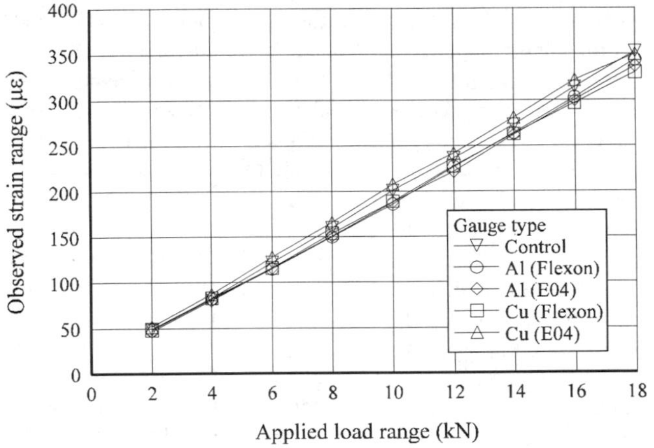

Fig. 3 Dynamic load testing of the fully bonded gauges

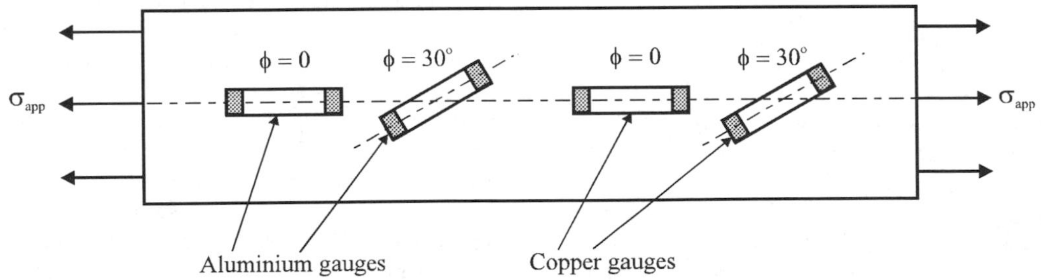

Fig. 4 Tensile specimen with aluminium and copper strip gauges

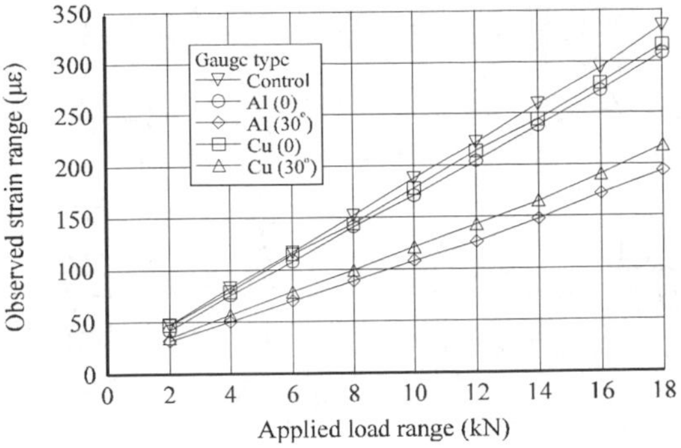

Fig. 5 Dynamic load testing of the partially bonded gauges

0.003 for a change of 0.01 in Poisson's ratio, and by 0.020 for a 1° change in the angle ϕ. These results confirmed that the E04 adhesive was satisfactory for use with the partially bonded strip gauge under dynamic loading.

5. THERMOELASTIC TESTS

Thermoelastic tests were carried out using the standard SPATE 9000 equipment, with a cyclic load frequency of 20 Hz. Both tensile specimens and Brazilian disc specimens (Dulieu-Smith 1995) were used, and partially bonded strip-gauges in both aluminium and copper were studied. The test procedure was standard in all respects. The specimens were sprayed with RS matt-black aerosol paint to enhance and standardise the thermal emissivity (see the term e in equation (2)). Signal scans were taken over a square area around each strip gauge.

5.1 *Tensile tests*

The tensile specimen was that used in the preliminary test work and is shown in Figure 4. The applied load cycle was ± 8 kN about a mean value of 10 kN giving a cyclic stress range of 67 MPa. The "working distance" (i.e. the distance from the SPATE detector to the specimen) was 390 mm and the "spot size" on the specimen from which the thermal emissions were received was 0.64 mm in diameter. The sampling time (or "dwell time") for each spot was 2 sec. Total scan times were 40 min and 90 min for the 0° and 30° gauges respectively. The received signals were smoothed using a median filter and average values, with standard deviations, were obtained from line scans parallel and transverse to the length of each gauge. The test was repeated (a) with the cyclic load range reduced by 50% and (b) at a cyclic load frequency of 10 Hz. The results are summarised in Table 1. The standard deviations of the several reading sets ranged from 3 to 10, giving a typical coefficient of variation of the order of 2%. The difference between the signals from the two line scans for each gauge never exceeded the combined standard deviations and was usually much smaller.

The Table 1 data confirms that the signal is linearly related to the cyclic load range and is not greatly dependent on frequency over the 10-20 Hz range. (It is suspected that the 10% change for the Cu (30°) gauge is anomalous.) The ratios of the signals from corresponding $\phi = 0$ and $\phi = 30^o$ gauges were also appraised; values are given in Table 1. The theoretical value of this ratio for a Poisson's ratio for 0.30 is readily obtained from equations (4) and (6) as 0.675 (see also Section 4 above). The experimental values are consistently greater than this figure; detailed consideration had produced no explanation for this.

Table 1.
Average signal values from strip gauges - tensile tests.

Gauge	±8kN, 20 Hz Signal (ratio)	±4kN, 20 Hz Signal (ratio)	±8kN, 10 Hz Signal (ratio)
Al(0)*	397	201	-
Al(30°)	316 (0.796)	156 (0.776)	317
Cu(0)	304	154	312
Cu(30°)	227 (0.747)	108 (0.701)	248 (0.798)

* The angle ϕ (see Figure 1) is given in brackets

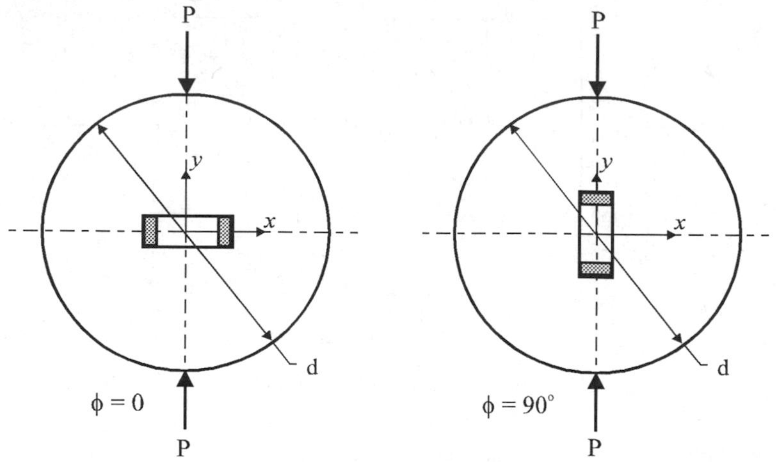

Fig. 6 Strip gauge orientation on the Brazilian disc

Table 2.
Average signal values from strip gauges - Brazilian disc tests

Gauge	Signal	Standard dev	Ratio
Al (0)*	132	4	
Al (90°)	-310	5	- 2.35
Cu (0)	67	5	
Cu (90°)	-172	5	- 2.55

* (0) denotes a horizontal gauge, (90°) denotes a vertical gauge.

5.2 *Brazilian disc tests*

The Brazilian disc specimen consists of a circular disc subjected to two diametrically-opposed in-plane radial forces at the periphery (see Figure 6). The specimen used was 79.2 mm in diameter and 5.0 mm thick; the material was mild steel. A partially bonded strip gauge, with an unbonded length of *ca* 20 mm, was applied symmetrically about the disc centre; an aluminium strip gauge was attached to one face of the disc, a copper gauge to the other, the dimensions of which were the same as before. The disc was subjected to a cyclic load of 7 ± 5 kN in the test machine first with the strip gauges horizontal ($\phi = 0$), then with the strip gauges vertical ($\phi = 90^\circ$). A SPATE scan was taken around each gauge (working distance 430 mm, sampling time 2 sec, scan time 80 min). Again, average signals were obtained from line scans along and transverse to the strip length. The results are summarised in Table 2. The standard deviations were of the same order as those obtained in the tensile tests.

A theoretical treatment of the Brazilian disc (Den Hartog 1952) provides the following expressions for the coordinate stresses along the horizontal and vertical diameters respectively:-

$$\sigma_x = \frac{2P}{\pi dt}\left[1 - \frac{16x^2d^2}{(d^2+4x^2)^2}\right]; \quad \sigma_y = \frac{2P}{\pi dt}\left[1 - \frac{4d^4}{(d^2+4x^2)^2}\right] \tag{7}$$

$$\sigma_x = \frac{2P}{\pi dt}; \quad \sigma_y = \frac{2P}{\pi dt}\left[1 - \frac{4d^2}{\left(d^2-4y^2\right)}\right] \tag{8}$$

where t is the disc thickness and the remaining symbols are defined in Figure 6. Substituting from equations (7) and (8) into the generalised Hooke's Law relationship, the following expressions are obtained for the strain in the x direction (ε_x) along the horizontal diameter

$$\varepsilon_x = \frac{2P}{\pi dtE}\left[(1-\nu) - \frac{4d^2(4x^2 - \nu d^2)}{(d^2+4x^2)^2}\right] \tag{9}$$

and for the strain in the y direction (ε_y) along the vertical diameter

$$\varepsilon_y = \frac{2P}{\pi dtE}\left[(1-\nu) - \frac{4d^2}{\left(d^2 - 4y^2\right)}\right] \qquad (10)$$

where E and ν are the Young's modulus and Poisson's ratio of the specimen material.

Three forms of strain ratio were obtained from equations (9) and (10) for comparison with the signal ratios obtained from corresponding horizontal and vertical strip gauges (see Table 2; see also equation (6)). First the strain ratio ($\varepsilon_y/\varepsilon_x$) was obtained by putting x = y = 0. The resulting expression is

$$\varepsilon_y/\varepsilon_x = -(3+\nu)/(1+3\nu) \qquad (11)$$

For a Poisson's ratio of 0.30 this becomes -1.74, which is considerably smaller than the values given in Table 2. This value would be appropriate for the "zero length" strip gauge but it is clear that account must be taken of the finite length of the gauge and the fact that the points of attachment were not at the disc centre.

Secondly, the strains in the specimen at the ends of the unbonded portion of the strip gauges were obtained by substituting x = y = 10 mm into equations (9) and (10). The derived strain ratio ($\varepsilon_y/\varepsilon_x$) was -2.33 which compared very well with the experimental signal ratios (see Table 2). However, it was recognised that this ratio value was as "unreal" as that derived from the centre strain values.

Finally, therefore, the strain variations in the specimen beneath the unbonded part of the strip gauge were considered in detail. Strains and thence displacement increments were obtained from equations (9) and (10) at 1 mm intervals from the disc centre to the x = ±10 mm and y = ±10 mm positions. The displacement increments were then summed to give the nett relative displacement of the bonded ends of the strip gauge in the two orientations; these values were divided by the unbonded gauge lengths to give the strip gauge strains. The resulting strain ratio was -1.91; this differed significantly from the experimentally derived signal ratios.

The disparity in these ratios has not been satisfactorily explained. Any tendency of the gauge to reinforce the specimen would affect the two gauge strains (ε_x and ε_y) proportionately and would not therefore affect the signal ratio. The possibility of buckling in the unbonded part of the gauge under a compressive strain (e.g. ε_y) has been mentioned previously (Stanley and Dulieu-Smith 1996). If the unbonded portion is treated as a column with built-in ends, it is readily shown the critical compressive strain (ε_c) is given by

$$\varepsilon_c = \pi^2 t^2 / 3\ell^2 \qquad (12)$$

where t and ℓ are the thickness and length of the unbonded part of the strip gauge. The derived critical strain is 20 x 10^{-4} $\mu\varepsilon$. The maximum compressive strain in the strip gauges in the y orientation was about 18% of this value. It has to be mentioned also that simple buckling would result in a lessening of the compressive surface strain at the centre of the gauge. Clearly the buckling possibility is not realistic and does not offer an explanation of the differences between the experimental signal ratios and the theoretical strain ratio.

6. CLOSING REMARKS

The authors' intention in presenting this "warts and all" account of the current state of development of the thermoelastic strip gauge is three-fold. First, it is considered important to demonstrate that work in this area is being actively pursued. The development of numerical stress separation techniques continues and these will doubtless have a prominent role in the future; nevertheless, it would be a matter of regret if the parallel development of experimental alternatives were neglected. Secondly, a wider awareness of the work presented may encourage other workers to discuss the ideas and problems involved, and possibly to take up related studies themselves. New insights, new solutions, and new proposals which may be forthcoming could prove invaluable in advancing the technology. Finally the work illustrates once again how the successful implementation of basically simple ideas often demands a degree of persistence and rigour greatly in excess of what may have been assumed at the beginning.

Development work on the partially bonded thermoelastic strip gauge (and on the other proposed versions) will continue. A particularly important aspect requiring further study is the nature and distribution of the stresses in the adhesive bond resulting from the cyclic load, and the response of the bond material to these stresses. (Clearly, compared with a fully bonded gauge (e.g. the conventional strain gauge) there is a considerable increase in the bond stresses associated with a particular specimen strain.) Moreover, it is essential for the validity of equation (1) that the thermal response of the gauge/bond specimen combination is adiabatic. Future work will also include further consideration of the thermal behaviour of the system subjected to cyclic temperature changes.

REFERENCES

Boyce, B.R. and Lesniak, J.R. 1994. A high-speed differential thermography camera. *Proc. SEM Spring Conf. On Exp. Mech.* Baltimore: 491-497.

Den Hartog, J.P. 1952 *Advanced Strength of Materials*. New York: McGraw-Hill.

Dulieu-Smith, J.M. 1995. Alternative calibration techniques for quantitative thermoelastic stress analysis. *Strain.* 31: 9-16.

Dulieu-Smith, J.M. and Stanley, P. 1996. On the interpretation and significance of the Grüneisen parameter in thermoelastic stress analysis. *Proc. 5th International Scientific Conference on Achievements in Mechanical and Materials Engineering (AMME '96).* Gliwice-Wisla, Poland: 121-124.

Harwood, N. and Cummings, W.M. (Eds).1991 *Thermoelastic Stress Analysis.* Bristol: IOP Publishing Limited (Adam Hilger).

Stanley, P. And Dulieu-Smith, J.M. 1996. Devices for the experimental determination of individual stresses from thermoelastic data. *Journal of Strain Analysis.* 31: 53-63.

SPATE 9000 Manual. 1989. London: Ometron Ltd.

Thomson, W. (Lord Kelvin) 1853. On the dynamical theory of heat, *Trans. Roy. Soc. Edinburgh.* 20: 261-283.

Modern Practice in Stress and Vibration Analysis, Gilchrist (ed.) © 1997 Balkema, Rotterdam, ISBN 90 5410 896 7

Thermoelastic stress analysis of oblique holes in flat plates

J.M. Dulieu-Smith & S. Quinn
University of Liverpool, Department of Mechanical Engineering, UK

ABSTRACT: The stresses in the region of an oblique hole in a flat plate are analysed using the thermoelastic stress analysis technique. Particular attention is paid to the obliquity of the hole and to the line of action of the applied load in relation to the hole axes. The effects of varying plate thickness are also studied. The data from around the rim and along the internal axis of the hole are presented as a series of stress factors. The results are compared to those from previous studies and effects such as variations in Poisson's ratio and loading frequency are discussed.

1. INTRODUCTION

In the past attention has focused on the problems of a central circular hole in a flat plate (e.g. Green 1948). In an infinitely large plate the stresses at the hole are dependent only on the plate thickness to hole diameter (h/d) ratio. It is well known that the plane stress solution (i.e. at the central plane of the plate) results in a stress concentration factor of 3.

The stresses that are developed in the region of an oblique hole in the centre of an infinitely large flat plate are more complex and are dependent on the following parameters:-

(i) the hole obliquity, α, i.e. the angle between the hole axis and a plane normal to the plane of the plate (see Figure 1),
(ii) the line of action of the applied load, ϕ, i.e. the angle between the major axis of the hole and the direction of the load (see Figure 1),
(iii) the h/d ratio.

The solution of the stresses at an oblique hole in a flat plate can be seen as the first step in understanding the stresses at oblique intersections in cylindrical bodies. The practical application being a solution for the stresses at a nozzle/vessel interface in a pressurised system. In view of the importance of the problem very little work has been carried out on oblique holes in flat plates. Experimental work has centred on the use of the photoelastic "frozen stress" technique (e.g. Stanley and Day 1990) to provide full-field stress information from both the rim and the interior of the hole. Rau (1971) used reflection photoelasticity and Moiré interferometry to obtain the stresses at the rim of the hole. Ellyin et al (1966) and Ellyin and Sherbourne (1968) used a plane elasticity approach in which the solution was based on separating the plate into thin membranes parallel to the plane of the plate. Abdul-Mihsein et al (1979) and Talfreshi and Thorpe (1995) have used boundary integral equation and finite element approaches. In general the results from these findings are (i) as α increases the maximum stress increases and (ii) as ϕ increases the maximum stress increases. However much of the data on the magnitude and position of the maximum stress is conflictory. Many researchers conclude that the maximum stresses are positioned at the surface of the plate (Abdul-Mihsein et al 1979, Tafreshi and Thorpe 1995, Leven 1970 and Daniel 1970). Ellyin and Sherbourne's (1968) results show that the maximum stress is at the surface of the plate until h/d exceeds 0.6. Stanley and Day (1990) state that the maximum stresses occur in the hole interior; the actual position being dependent on α, ϕ or h/d. McKenzie and White (1968) note that the maximum stress moves from the mid-plane to the surface with increasing α. Perhaps the most interesting comparison is between Stanley and Day's (1990) work and that of Tafreshi and Thorpe (1995). In these two cases identical α, ϕ and h/d ratios are used. A close inspection of the results shows that there are differences in the maximum stress values of up to 23%. This clearly warrants further investigation. In comparing frozen-stress

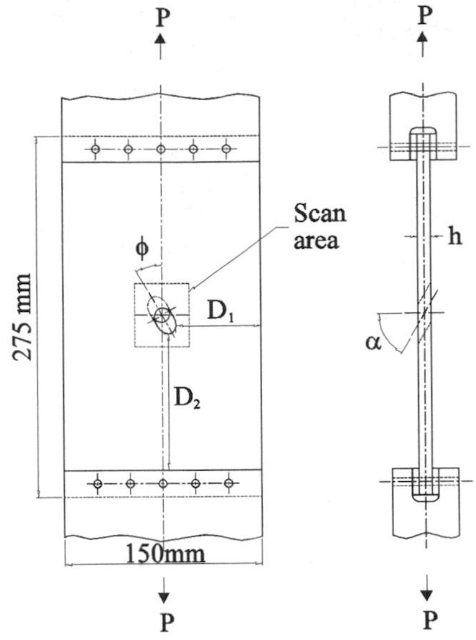

Fig. 1 Typical test specimen and notation

photoelastic work with numerical work there is a danger of neglecting the possible effects of differing Poisson's ratio (i.e. around 0.3 for steel or aluminium alloy and 0.5 for epoxy resin at the stress freezing temperature). Ellyin and Sherbourne (1968) consider this and show that the maximum stress is 6.83% less when Poisson's ratio is 0.5.

The work described in this paper uses the well established experimental technique of thermoelastic stress analysis (TSA) (Harwood and Cummings 1991) to determine the stresses around an oblique hole in a flat plate loaded in uniaxial tension. The thermoelastic technique is based on the measurement of the small temperature change that occurs in a material when it is subjected to elastic cyclic stresses. In a linear elastic, isotropic, homogeneous material the temperature change is directly related to the sum of the principal stresses (Stanley and Chan 1985). Currently the standard equipment for TSA is the SPATE (Stress Pattern Analysis by Thermal Emissions) system. The SPATE system incorporates a highly sensitive infra-red detector that permits the measurement of the small temperature change.

It is readily shown that the output from the detector, i.e. the thermoelastic signal, S, is related to the principal stress changes on the surface of the material, $(\sigma_1 + \sigma_2)$, by the following

$$(\sigma_1 + \sigma_2) = AS \quad (1)$$

Table 1. Summary of Plate Parameters

Plate Number	h (mm)	h/d	α (degrees)	ϕ (degrees)
1	10	0.67	0	90
2	10	0.67	30	90
3	10	0.67	60	90
4	10	0.67	45	0
5	10	0.67	45	30
6	10	0.67	45	60
7	10	0.67	45	90
8	15	1.00	45	90
9	20	1.33	45	90

where A is a calibration constant.

The SPATE detector operates in a scanning mode so that a full-field representation of the stresses over a pre-defined area are developed on the SPATE computer monitor over a period of 1 to 2 hours. The SPATE technique has the advantage that it is non-contact, the actual component material is used and only minimum surface preparation is required.

In developing equation (1) two assumptions are made. Firstly that the temperature change occurs adiabatically; this is achieved by cycling the specimens at such rates so as no heat conduction takes place. The second is that the elastic constants are independent of temperature; in most materials at room temperature, this accounts for differences of the order of only 1 to 2% over the entire elastic range.

The purpose of the present study is to apply TSA to the problem of oblique holes. The SPATE data obtained from the rim of the hole directly provides stress concentration factor data. The remaining data is proportional to the sum of the principal stresses (see equation 1) and can only therefore be seen as stress factors. A detailed investigation of a variety of hole configurations is carried out thus providing further insight into this difficult stress analysis problem.

2. SPECIMEN DESIGN AND TEST PROGRAMME

A diagram of a typical test specimen and the loading jig is shown in Figure 1. In all nine plates were manufactured so that a parametric study of α, ϕ and h/d could be achieved. A summary of the plate parameters is given in Table 1; in all cases the hole diameter, d, was 15 mm. Plates 1 to 7 are 10 mm thick giving an h/d ratio of 0.67. Plates 8 and 9 are 15 and 20 mm thick respectively (i.e. h/d = 1 and h/d = 1.33). Four values of α are studied (0, 30°, 45° and 60°) and four values of ϕ are studied (0, 30°, 60° and 90°). When varying α, ϕ is always 90° and h/d = 0.67. When varying ϕ, α is always 45° and h/d = 0.67. When h/d is varied $\alpha = 45°$ and $\phi = 90°$.

The plates were manufactured from 5083 annealed condition aluminium alloy. The choice of dimensions was such that the edges of the plate had no influence on the stress at the hole. Stanley and Day (1990) showed that if the distance between the edge of the hole and the edge of the plate, D_1 (see Figure 1), is greater than 3.3d and the distance between the edge of the hole and the end of the plate, D_2 (see Figure 1), is greater than 6.7d then infinite plate conditions can be assumed. For convenience the plates are all 275 mm long by 150 mm wide. This results in a minimum D_1 value of 3.42d and a minimum D_2 of 6.66d.

The plates are loaded *via* five 6 mm silver steel pins at each end of the plate (see Figure 1); this ensures an even load distribution across the plate. The plates are mounted between two mild steel spacing plates. This assembly is then mounted in special end-grips that minimise any bending of the plates.

3. EXPERIMENTAL WORK

The mechanical properties of the aluminium used to manufacture the test specimens were established from a series of tensile tests. (The tensile test specimens were produced from the same billet of material as that used to manufacture plates 1 to 7 (see Table 1).) The specimen material had a yield strength of 114.0 MPa, a Young's modulus of 73.8 GPa and a Poisson's ratio of 0.33. It is essential in TSA that the yield strength of the material is not exceeded. In Stanley and Day's (1990) study the maximum stress concentration factor was 5.64. Therefore each plate was loaded so that a nominal stress of one sixth of the yield strength was developed in the plates; a small factor of safety was also introduced by taking the yield strength to be 100 MPa. The applied load levels were 12 ± 11.5 kN, 18 ± 17 kN and 23 ± 22.5 kN for the 10, 15 and 20 mm thick plates respectively; in each case the loading frequency was 10 Hz. Prior to testing each of the specimens was coated with two passes of RS matt black paint.

The first section of the SPATE work concentrated on the surface of the specimen close to the hole. SPATE scans were taken from a rectangular area with boundaries approximately 10 mm away from the edge of the holes; the scan area is indicated in Figure 1. The interior of the hole was blanked off by filling the hole with plasticine so that the rim of the hole could be easily identified in the SPATE data. The second part of the study investigated the stresses developed on the interior surface of the hole. The limiting factor is possible signal attenuation due to the obliquity of the hole interior so only obliquities of 30° or greater were studied. In these tests the flat surface was covered with plasticine. In both sets of tests the SPATE detector was positioned at a working distance of 450 mm from the surface of the specimen, so that the scanning spots were around 0.7 mm in diameter. Each scan took approximately 2 hours to complete. To obtain detailed data the resolution of the scanning spots was about half the spot diameter.

The SPATE data was calibrated using a calibration factor, A, (see equation 1) of 0.050 MPa U^{-1} (U = uncalibrated signal). This was obtained by taking SPATE readings from two tensile test specimens manufactured from the same grade of aluminium as the plates. The effects of the grain of the aluminium were investigated as one of the tensile specimens was cut along the grain and the other transverse to the grain. The values were determined using a technique described in Dulieu-Smith (1995) and were found to be within 1% of each other. A further factor was any possible attenuation of signal due to the oblique angle of the hole interiors. Calibration studies on the tensile specimens oriented at angles corresponding to the hole obliquities showed that this had no discernible effect on the SPATE output.

The calibrated SPATE data was normalised by dividing it by the nominal applied stress so that the SPATE output was expressed in terms of stress factors. Figure 2 shows a SPATE contour plot from the surface of plate 3 (i.e. $\alpha = 60°$, $\phi = 90°$). (The oblique side of the hole is to the right of the figure, the acute is to the left.) There are two clear positive stress factor maxima either side of the hole, the largest is at the acute side. The negative maxima are skewed towards the acute side of the hole and are not on the centre line of the hole as would be expected in the case of a central hole. Figure 3 shows a contour plot from the interior of the hole in plate 3. There is a maximum stress factor located on the hole generator through the major axis of the surface ellipse close to the acute intersection with the surface.

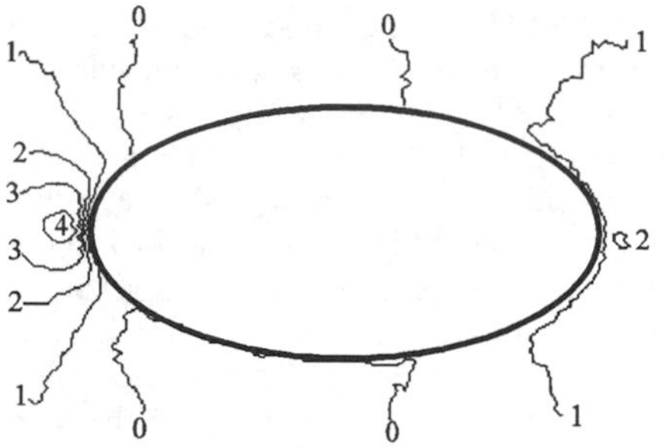

Fig. 2 SPATE contour plot from the surface of plate 3

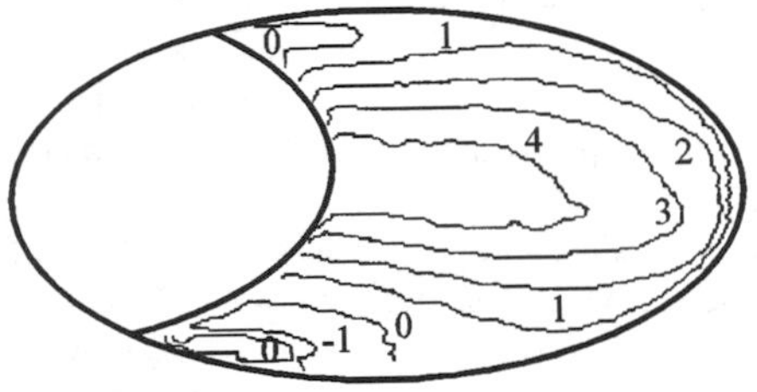

Fig. 3 SPATE contour plot from the interior of the hole in plate 3

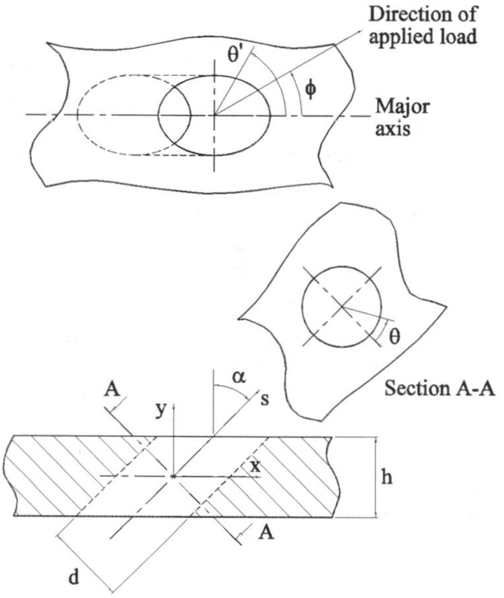

Fig. 4 Oblique hole geometry and notation

4. ANALYSIS OF THE SPATE DATA AND RESULTS

To compare the results from each of the different hole geometries it is important to relate the data to the circular hole generator that produces the elliptical intersection with the surface. Figure 4 shows a complete representation of all of the geometrical factors involved. (This is consistent with Stanley and Day's (1990) notation so that a straightforward comparison can be made between photoelastic and thermoelastic results.) The linear coordinate parallel to the hole axis, measured from the mid-plane of the plate is s, while y is the linear coordinate normal to be mid-plane of the plate. The angular coordinate in the plane normal to the hole axis is θ while the angular coordinate parallel to the mid-plane of the plate is designated θ'. (Note that $\theta' = \theta$ only at the major and minor axis of the surface ellipse.) The relationship between θ' and θ is

$$\tan \theta' = \tan \theta \cos \alpha \tag{2}$$

The SPATE data shown in Figure 2 is related to θ' and not θ. Therefore for any direct comparisons of the position of the maximum stress with data from other plates it is necessary to transpose the data to the θ-plane. Clearly this manipulation will have no effect on plate 1 (the circular hole) and will be most pronounced when varying ϕ. Readings were taken from the SPATE area scans in 10° intervals around the rim of each hole. From these polar plots were constructed so that the stress concentration factor at the rim, I_R, was plotted against θ'. A second polar plot was then produced of I_R against θ. The two polar plots for plate 3 are shown in Figure 5; the datum for the plots is the hole and negative values are shown with a broken line. There are two positive maxima in the plot the largest being at the acute edge of the hole, I_{Rmax}, the secondary maxima being at the oblique edge of the hole, I_{Rs}. It is a straightforward matter to locate the position of the maxima in the plots, i.e. θ_{max} and θ_{maxs}.

Figure 6 shows plots of I_{Rmax}, I_{Rs}, θ_{max} and θ_{maxs} for varying α, ϕ and h/d. Figure 6a shows I_{Rmax} and I_{RS} plotted against α for a constant ϕ of 90° and h/d = 0.67. When $\alpha = 0$ I_{Rmax} and I_{Rs} have virtually equal values giving an average of 2.5. As α increases I_{Rmax} increases to a maximum of 4.3. I_{Rs} decreases slightly to a minimum of 2.0. At $\alpha = 30°$ the maxima differ by only a few percent from the $\alpha = 0$ case; there is a sharp change in I_R values between $\alpha = 30°$ and $\alpha = 45°$. Figure 6b shows the corresponding position of θ_{max} for the I_R values given in Figure 6a. From this it is clear that changes in hole obliquity have practically no effect on the position of the maximum stresses at the rim of the

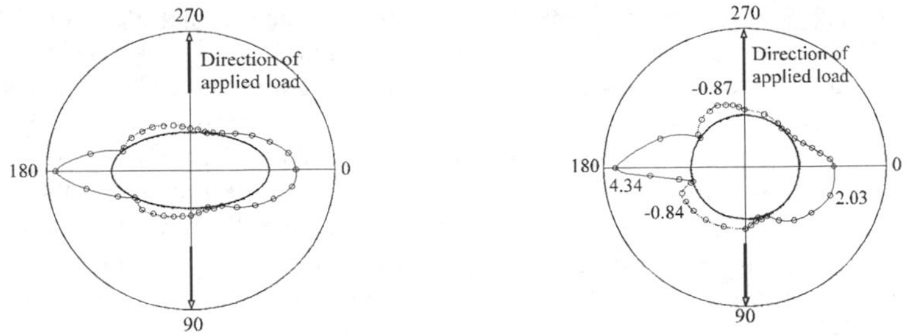

(a) I_R against θ' (b) I_R against θ

Fig. 5   Polar plots of I_R for plate 3

(a) I_R against α (b) θ against α

(c) I_R against ϕ (d) θ against ϕ

(e) I_R against h/d (f) θ against h/d

Legend for (a), (c) and (e)
- ● I_{RMAX} (TSA)
- ■ I_{RS} (TSA)
- ○ I_{RMAX} (PE)
- □ I_{RS} (PE)

Legend for (b), (d) and (f)
- ● θ_{MAX} (TSA)
- ■ θ_{MAXS} (TSA)
- ○ θ_{MAX} (PE)
- □ θ_{MAXS} (PE)

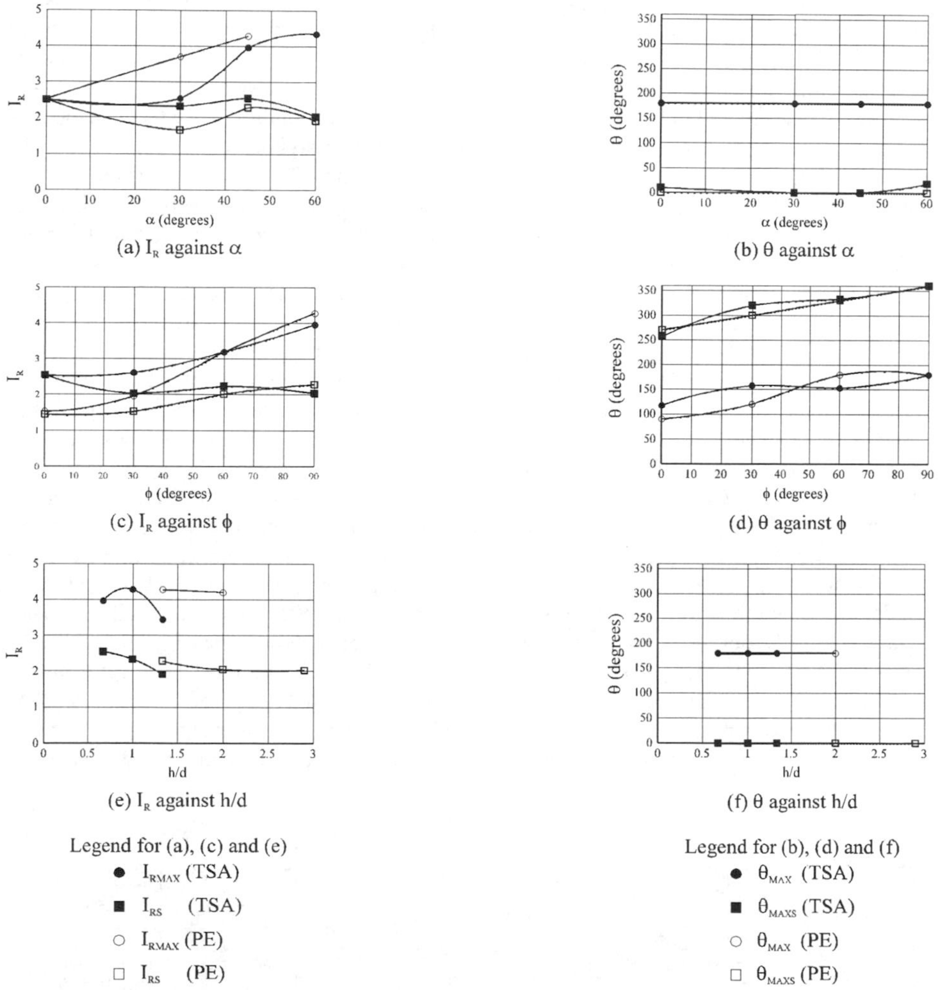

Fig. 6 I_{RMAX}, I_{RS}, θ_{MAX}, and θ_{MAXS} for varying α, ϕ, and h/d

hole. Figure 6c shows I_{Rmax} and I_{Rs} plotted against ϕ for a constant α of 45° and h/d = 0.67. When $\phi = 0$ I_{Rmax} and I_{Rs} are equal at 2.5. As ϕ increases I_{Rmax} increases monotonically to a maximum of 4.0 at ϕ = 90°. I_{Rs} decreases sharply between $\phi = 0$ and ϕ = 30° and then gradually increases back to the $\phi = 0$ value of 2.5. Figure 6d shows the corresponding position of the maximum values shown in Figure 6c. Changing the direction of the applied load has a significant effect on the position of the maximum stresses. When $\phi = 0$ the maximum stress is at θ = 117° and when ϕ = 90° the position of the maximum stress is at θ = 180°. Likewise the secondary maxima rotates from 256° when $\phi = 0$ to 360° when ϕ = 90°. Figure 6e shows the effect of varying h/d on I_{Rmax} and I_{RS} for a constant α of 45° and a constant ϕ of 90°. There is an overall decreasing trend in I_R as h/d increases, (although there is a slight increase in I_{Rmax} for h/d = 1) thus indicating a departure from the plane stress condition for the thicker plates. For completeness Figure 6f shows that the position of the maximum stress is unaffected by the changes in h/d.

To analyse the date from the interior of the hole SPATE line plots were taken along the hole generator aligned with the major axis of the surface ellipse. The data from the line plots provides information about the magnitude of the principal stress sum and the position of the maximum principal stress sum in relation to the x-axis (see Figure 4). To compare the data from each hole stress factors, I, were plotted against s (see Figure 4) so that the variation in the stresses along the true length of the hole could be examined. Figure 7 shows plots of I against s for varying α and h/d. From Figure 7a the maximum stress factor occurs in the plate with the largest obliquity and decreases from around 4.5 for α = 60° to 3.0 for α = 30°. When α = 60° the maximum stress factor is at the acute edge of the plate (i.e. s = 0), there is a plateau between 3.0 and 11.0 mm were the stress factor is constant and then a sharp decrease. When α = 45° the maximum I is at s = 1.5 mm; this remains constant for 5 mm and then reduces sharply. When α = 30° the maximum I is 4 mm into the plate, there is a small plateau region where I is at a maximum and then a sharp decline in I to the oblique edge of the hole. Figure 7b shows that as h/d increases the maximum I value is unaffected. However the position of the maximum I moves from s = 1.5 mm to s = 7 mm.

5. DISCUSSION

Table 2 gives I_{Rmax}, I_{Rs}, θ_{max} and θ_{maxs} for the TSA work and compares this with the results of the previous photoelastic frozen stress study carried out by Stanley and Day (1990). (Stanley and Day's results are also indicated in Figure 6.) A good indication of the validity of the comparison between the TSA and the photoelastic work is provided by the circular hole (i.e. α = 0) where the data is practically identical. The agreement between the rest of the hole configurations is only tolerable and warrants further discussion. When α is varied and ϕ remains constant (see Figure 6a) the photoelastic results are greater than the thermoelastic results for I_{Rmax} and less for I_{Rs}. With the exception of α = 30° this is only by a few percent. The position of the maximum I_R values from both techniques shows a good correspondence. Figure 6c shows that for the two higher values of ϕ the agreement is good. However when $\phi = 0$ and 30° the agreement is poor. It is interesting to note that when $\phi = 0$ Stanley and Day's (1990) maximum value of I was observed at the surface of the plate as 2.52. However, their rim data of 1.52 is 40% less. Tafreshi and Thorpe's (1995) data also showed that I_{max} also occurs at the surface and was 2.59 almost identical to that given in the present study. When ϕ = 30° Stanley and Day's (1990) maximum value was 2.5 mm away from the surface, Tafreshi and Thorpe's (1995)

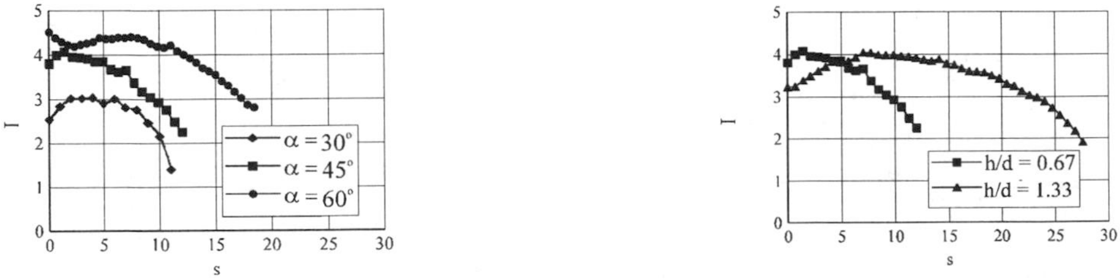

(a) α (b) h/d

Fig. 7 I against s for varying α and h/d

value is 3.20 and at the surface. As ϕ increases Stanley and Day (1990) show that the maximum stress moves further into the hole. Once again the θ_{max} positions show a good correspondence.

The differences between the thermoelastic and photoelastic data could be attributed to the differences in the plate thickness i.e. 10 mm in the thermoelastic work and 25.4 mm in the photoelastic work. This would result in the TSA always producing higher values of I_R; Figure 6e shows that as the plate thickness increases I_R reduces. A more likely cause is the difference in the Poisson's ratio value between the two materials used in each study. The higher Poisson's ratio value in the Araldite specimens used in the photoelastic work would have no effect on the circular hole. As the hole obliquity increases the effects would be more pronounced. Another consideration would be the larger displacement experienced by the Araldite specimens. A further factor is the assumption of adiabatic conditions in the thermoelastic theory. Non-adiabatic behaviour is more likely to occur when α is large because of the very thin sections at the hole edge.

Figure 7a shows that as the hole obliquity increases the maximum I is located closer to the surface agreeing with McKenzie and White (1968). However Stanley and Day (1990) indicate that as the hole obliquity increases the position of the maxima is further into the hole. An important point to note is that the thermoelastic data from the hole interior is proportional to the sum of the principal stresses. Figure 7b shows that as the plate thickness increases the maximum I moves away from the surface. A further interesting feature of the work is the plateau that is evident in the plots, this was also noted by Adbul-Mihsein et al (1979).

6. CONCLUSIONS

The principal conclusions arising from this investigation of the stresses at an oblique hole in an infinite flat plate are the following:

1. For a given plate thickness, hole diameter and load direction the maximum principal stress at the rim of the hole increases with hole obliquity.
2. For a given hole obliquity the maximum principal stress increases as the load direction rotates from 0 to 90°.
3. For plate thickness (h) to hole diameter (d) ratios of $0.67 \leq h/d \leq 1.33$ the maximum stress at the rim of the hole can reduce by up to 15%.

Table 2. Comparison of TSA Results with Photoelastic (PE) Work (Stanley and Day (1990))

α (°)	ϕ (°)	I_{Rmax}		θ_{max} (°)		I_{Rs}		θ_{maxs} (°)	
		TSA	PE	TSA	PE	TSA	PE	TSA	PE
0	90	2.50	2.51	180	180	2.49	2.49	10	0
30	90	2.53	3.70	180	180	2.31	1.65	0	0
60	90	4.34	-	180	-	2.03	1.91	19	0
45	0	2.54	1.52	117	90	2.54	1.44	256	270
45	30	2.61	1.95	158	120	2.02	1.52	320	300
45	60	3.19	3.19	153	180	2.23	2.01	333	330
45	90	3.96	4.28	180	180	2.54	2.28	360	360

4. The position of the maximum stresses at the hole rim is independent of hole obliquity and plate thickness, and only dependent on the line of action of the load.
5. For a given load direction and h/d ratio the maximum principal stress sum position moves from the hole interior to the surface as the hole obliquity increases.
6. For a given hole obliquity and load direction when h/d varies from 0.67 to 1.33 the maximum principal stress sum moves from a position close to the edge of the plate to a position approaching the mid-plane of the plate.

REFERENCES

Abdul-Mihsein, M J, Fenner, R T and Tan, C L (1979). Boundary integral equation analysis of elastic stresses around an oblique hole in a flat plate. *J Strain Analysis* 14: 179-185.

Daniel, I M (1970). Photoelastic analysis of stresses around oblique holes. *Exp Mech* 10: 467-473.

Dulieu-Smith, J M (1995). Alternative calibration techniques for quantitative thermoelastic stress analysis. *Strain* 31: 9-16.

Ellyin, F, Lind, N C and Sherbourne, A N (1966). Elastic stress field in a plate with a skew hole. *J Eng Mech*, 92: 1-10.

Ellyin, F, and Sherbourne, A N (1968). Effect of skew penetration on stress concentration, *J Eng Mech* 94: 1317-1336.

Harwood, N and Cummings, W M (1991). *Thermoelastic stress analysis*. Bristol: IOP Publishing Ltd.

Leven, M M (1970). Photoelastic determination of the stresses at oblique openings in plates and shells. *WRC Bulletin* 153: 52-80.

McKenzie, H W and White, D J (1968). Stress concentration caused by an oblique hole in a flat plate under uniaxial tension. *J Strain Analysis* 3: 98-102.

Rau, C A (1971). Elastic-plastic strain concentrations produced by various skew holes in a flat plate under uniaxial tension. *Exp Mech* 11: 133-141.

Stanley, P and Chan, W K (1985). Quantitative stress analysis by means of the thermoelastic effect. *J Strain Analysis* 20: 129-137.

Stanley, P and Day, B V (1990). Photoelastic investigation of stresses at an oblique hole in a thick flat plate under uniaxial tension. *J Strain Analysis* 25: 157-175.

Tafreshi, A and Thorpe, T E (1995). Numerical analysis of stresses at oblique holes in plates subjected to tension and bending. *J Strain Analysis* 30: 317-323.

11 Composites and anisotropic materials

Modern Practice in Stress and Vibration Analysis, Gilchrist (ed.) © *1997 Balkema, Rotterdam, ISBN 90 5410 896 7*

Sound transmission through honeycomb panels

E. Nilsson & A. Nilsson
The Marcus Wallenberg Laboratory for Sound and Vibration Research, KTH, Stockholm, Sweden

ABSTRACT: Sound transmission through light weight orthotropic honeycomb panels is discussed. For the prediction of the sound transmission through an orthotropic plate it is shown that the plate can be substituted by an isotropic plate with frequency dependent bending stiffness, loss factor and dimensions. The bending stiffness along the two main axes of the plate can be determined by means of simple measurements. Models for the prediction of the sound reduction index of finite isotropic plates can be modified for calculating the sound reduction index of orthotropic plates. Predicted and measured sound reduction indices for orthotropic honeycomb panels are compared. The prediction model can be used to determine the effects of changing material parameters, plate dimensions and boundary conditions.

1 INTRODUCTION

During the last decade, various types of light-weight structures have been introduced in the vehicle industry. This trend is dictated by demands for higher load capacity for civil and military aircrafts, reduced fuel consumption for passenger cars, increased speed for passenger and navy vessels of catamaran type, and increased acceleration and retardation for trains to increase average velocity.

Due to reduced fuel consumption and increased load capacity, the environmental impact of light-weight vehicles could be considerable. However, there are also certain constrains like passenger comfort, safety and costs for new types of vehicles. Passenger comfort requires low noise and vibration levels in any type of vehicle.

Light weight honeycomb structures can be used as floorpanels in aircrafts. The sound transmission from turbo prop engines through fuselage and floor panels and into the cabin can be considerable for small aircrafts. It is therefore essential to increase the sound transmission loss of floor panels while maintaining or reducing the weight.

The sound transmission loss of a composite panel is often estimated by means of numerical models. This makes it sometimes difficult to make simple parameter studies. An alternative approach is to model a composite plate as an equivalent single leaf panel. Traditional methods are thereafter used to determine the sound transmission through the structure.

2 LATERAL DEFLECTION OF HONEYCOMB PANELS.

A honeycomb panel consists of a light weight honeycomb core with thin laminates bonded to the core as shown in Figure 1. The core is primarily acting like a spacer between the laminates to give the required bending stiffness of the panel. The cells in the core and the fibres in the laminates are usually oriented in such a way that the plate is orthotropic.

The lateral deflection or apparent bending stiffness of a honeycomb panel is caused by bending as well as shear. In the low frequency region pure bending dominates. Shear effects in the core are of importance for higher frequencies. For typical aircraft panels shear tends to dominate in the frequency range from 100 to 1000Hz. For higher frequencies the apparent bending stiffness is almost completely determined by the bending stiffness of the laminates.

For a symmetric honeycomb beam - identical laminates - the apparent bending stiffness D_x of the beam along its main axis can according to Kurtze and Watters (1969) and E.Nilsson (1996) be written as:

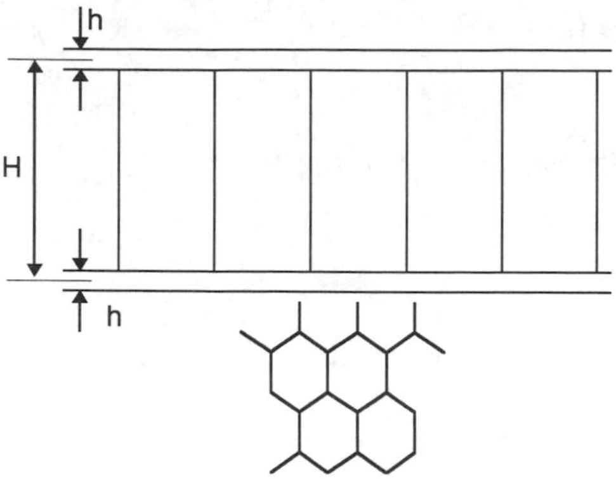

Figure 1. Honeycomb panel and core.

$$\left(\frac{GH}{\omega\mu^{1/2}}\right)\left[\frac{D_x^{3/2}}{D_1}-D_x^{1/2}\right]+D_x-2D_1=0 \qquad (1)$$

The bending stiffness corresponding to pure bending of the entire structure is D_1. The bending stiffness of just one of the two identical laminates is D_2. The shear stiffness of the core is G and the total thickness H. The mass per unit area of the structure is μ. The angular frequency is ω. If all the material parameters are known to apparent bending stiffness Dx can be calculated from equation (1). This is generally not possible. However if the apparent bending stiffness can be measured at a minimum of three different frequencies the bending stiffness at any other frequency can be determined. Equation (1) can bewritten in a more general way as:

$$A_x D_x^{3/2}/f - B_x D_x^{1/2}/f - D_x + C_x = 0 \qquad (2)$$

where A_x, B_x and C_x are constants and f is the frequency. The constants or rather bending stiffness D_x can be determined through measurements of the first few resonance frequencies for the beam. The bending stiffness D_x at the resonance frequency $f_{//}$ is for a freely suspended beam, length L and mass per unit area μ, given by

$$D_x = \mu 4\pi^2 f_n^2 (L_x/\alpha_n)^4 \qquad (3)$$

For a freely suspended beam the parameter α_n is 4.73, 7.85, 11.00 etc. for n=1,2,3... For a beam with

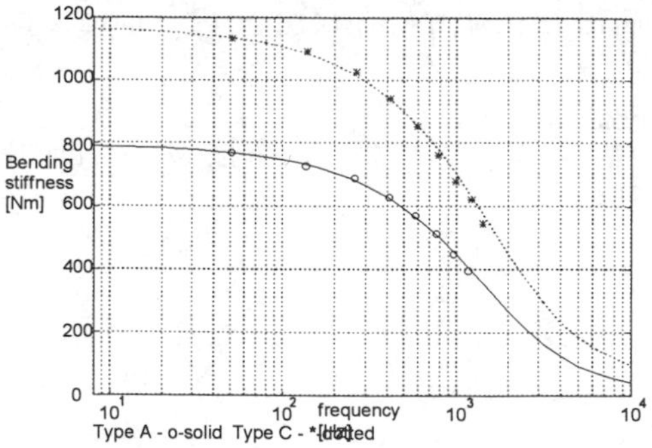

Figure 2. Bending stiffness for two beams.

a length of 1.2m, six to ten resonances can be determined. The unknown parameters A_x, B_x and C_x in equation (2) can be determined from measured data using the least square method. The result of this type of regression analysis is shown in Figure 2 for two beams. The elements are taken from the same rectangular plate. The beams represent the two main directions, parallel and perpendicular to the length of the plate. For typical honeycomb panels used by the aircraft industry the bending stiffness can be strongly dependent on frequency. The bending stiffness along the two main axes of the plate can be very different.

3 RESPONSE OF PLATE

According to Szilard (1974) the lateral displacement w of an orthotropic plate oriented in the x-y- plane is governed by the equation

$$D_x\frac{\partial^4 w}{\partial x^4}+2B\frac{\partial^4 w}{\partial x^2\partial y^2}+D_y\frac{\partial^4 w}{\partial y^4}+\mu\frac{\partial^2 w}{\partial t^2}= \\ = p_1-p_2=p \qquad (4)$$

The bending stiffness in the x- and y-directions are denoted D_x and Dy respectively. B is the effective torsional rigidity and can be approximated by $B=\sqrt{D_x D_y}$. The plate is exposed to an acoustical pressure p_1 on one side and p_2 on the other side. The pressure difference is p. For a simply supported rectangular plate with one corner at origo and with the sides a and b along the two coordinate axes the plate displacement can be expanded along the eigenvectors

$$f_{mn}=\sin\left(\frac{m\pi x}{a}\right)\cdot\sin\left(\frac{n\pi y}{b}\right) \qquad (5)$$

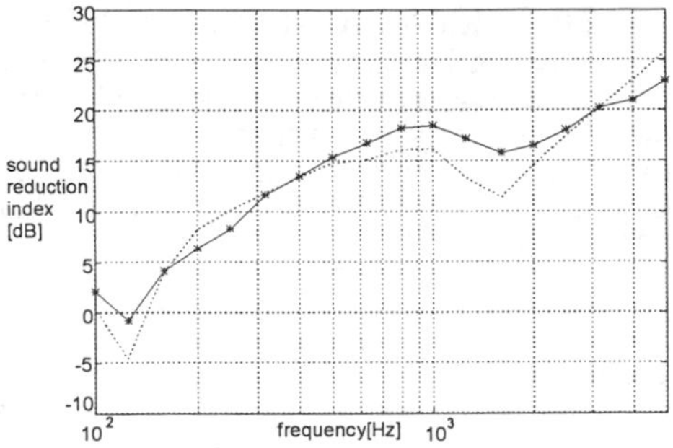

Figure 3 Measured and predicted sound reduction index. Solid - measured, * - measurement points, dotted - predicted.

For a time dependence exp(iωt) the displacement w of the plate is obtained as

$$w = \sum_{m,n} \frac{4}{ab} \cdot \frac{\langle p | g_{mn} \rangle}{Q_{mn}} \cdot g_{mn} \cdot \exp(i\omega t) \qquad (6)$$

$$Q_{mn} = \left(\frac{m\pi}{a}\right)^4 \cdot D_x + 2\sqrt{D_x D_y}\left(\frac{m\pi}{a}\right)^2\left(\frac{n\pi}{b}\right)^2 + \left(\frac{n\pi}{b}\right)^4 D_y - \mu\omega^2 \qquad (7)$$

This expression is modified by making the standard substitution

$$\alpha^4 = a^4\sqrt{D_y / D_x} \quad ; \beta^4 = b^4\sqrt{D_x / D_y} \qquad (8)$$

The parameters D_0, κ and κ_{mn} are defined as

$$D_0 = \sqrt{D_x D_y}$$

$$\kappa^4 = \mu\omega^2 / D_0$$

$$\kappa_{mn}^2 = \left(\frac{m\pi}{\alpha}\right)^2 + \left(\frac{n\pi}{\beta}\right)^2 \qquad (9)$$

The expression Q_{mn} defined in equation (7) is, based on the equations (8) and (9), reduced to

$$Q_{mn} = D_0\left[\kappa_{mn}^4 - \kappa^4\right] \qquad (10)$$

The product ab in equation (6) is unchanged since ab=αβ.

The result represented by the equations (10) and (6) would also be obtained for a simply supported and homogeneous plate with the bending stiffness D_0, mass per unit area μ and dimensions α and β.

The result implies that a model for predicting the sound reduction index for a homogeneous, single leaf panel could be used to estimate the sound transmission through an orthotropic panel. This could be done by using suitable transforms for stiffness and geometry. In addition the bending stiffness and loss factor should be allowed to depend on frequency.

4 SOUND REDUCTION INDEX

One model describing the sound transmission through a single leaf and homogeneous panel is presented by A.Nilsson (1974). The sound reduction index for the panel is found to be a function of mass per unit area, bending stiffness and loss factor of the plate as well as the frequency. In addition, the boundary conditions and the geometry of the plate and the dimensions of the baffle the plate is mounted in are parameters which influence the sound reduction index.

The results presented by A.Nilsson (1974) are valid for typical building elements with certain weights, dimensions etc. A good acoustical coupling between a vibrating panel and the driving and induced sound fields requires that the wave number k in air should satisfy $k \geq \kappa_{mn}$ where κ_{mn} is given in equation (9). This implies that the quantity Q_{mn}, defined in equation (19), can be approximated by $Q_{mn} \approx -D_0\kappa^4$ for $\kappa >> k$ or in the frequency range well below coincidence. For light-weight, stiff and small panels, the first plate resonance frequency f_{11} can be fairly high. κ_{11} can therefore not be neglected as compared to κ in equation (10). At the resonance frequency f_{11} the sound transmission through the panel has a maximum.

The sound reduction index of a panel can according to A.Nilsson (1974) and considering the arguments above, be written as

$$R = 20\log\mu + 10\log\left\{\left[(f_{11}/f)^2 - 1\right]^2 + \eta^2\right\} + 20\log f - 10\log[\Gamma\Lambda + G] - 48 \text{ [dB]} \qquad (11)$$

for $f < f_c$

$$R = 20\log\mu - 10\log f_x + 30\log f + 10\log\eta + 5\log(1 - f_x / f) - 47 \text{ [dB]} \quad \text{for } f > f_c \qquad (12)$$

$$f_x = \frac{c^2}{2\pi}\sqrt{\frac{\mu}{D_0(f)}} \tag{13}$$

where $D_0(f)$ is the frequency dependent generalised bending stiffness defined in equation (9). At coincidence $f_x=f=f_c$. The function Γ in equation (11) depends on frequency and the dimensions of panel and baffle. The function G accounts for the resonant transmission through the panel. Λ is a function depending on the boundary conditions of the plate. For a simply supported plate Λ is

$$\Lambda = 1 + \frac{3\cdot 10^4}{4\eta f^{1/2} f_x^{3/2}}\left(\frac{1}{\alpha^2}+\frac{1}{\beta^2}\right) \tag{14}$$

The parameters α and β are defined in equation (8). If the boundary conditions are changed from simply supported to clamped, the function Λ is increased. Consequently the sound reduction index is reduced below coincidence. Since the plate is very light the fluid loading on the plate contributes to the apparent mass of the panel. This will influence the first plate resonance frequency f_{11} of the plate. At the first resonance the apparent mass μ_0 of the panel is approximately equal to

$$\mu_0 = \mu + \frac{2\rho}{\kappa_{11}} \tag{15}$$

where ρ is the air density. The first resonance frequency can be calculated by means of the Rayleigh-Ritz method.

The total loss factor of the panel not only depends on material parameters but also on radiation losses and can be written as

$$\eta_{tot} = \eta_{panel} + \frac{2(\rho c)\cdot\sigma}{\mu\omega} \tag{16}$$

where η_{panel} is due to internal damping and transmission losses at the edges of the panel. The sound radiation ratio is defined as σ. The wave impedance in air is equal to (ρc).

5 COMPARISON BETWEEN PREDICTIONS AND MEASUREMENTS

A honeycomb panel with the dimensions 1.2 x 0.42m^2 and area weight 2.5 kg/m^2 was mounted between a reverberant and an anechoic room. The panel was excited by loudspeakers in the reverberant room. The acoustical power radiated into the anechoic room was measured by means of the intensity technique. The loss factor for the mounted plate was determined. The bending stiffness of two beams representing the two main axes of the plate were measured. The predicted and measured sound reduction indices are shown in Figure 3. The first resonance frequency of the plate is within the 125Hz 1/3 octave band. The coincidence frequency is within the 1600Hz octave band.

6 DISCUSSION

Models developed for the prediction of the sound reduction index for homogeneous single leaf panels can be used for estimating the sound transmission through honeycomb panels. The model used has to be improved in the frequency range just below coincidence and at the first plate resonance frequency.

Below coincidence the sound reduction index of a panel can be improved by adding mass. Due to weight problems this is not an acceptable solution. The addition of damping material only slightly improves the sound reduction index at the first plate resonance frequency and just below and above coincidence. A relaxation of the boundary conditions improves the sound reduction index below coincidence and decreases the first resonance frequency of the plate. This is often advantageous.

Damping can be increased without unduly increasing the weight by mounting a frictional layer between the edge of the plate and its supporting structure. The sound reduction index depends on the plate dimensions as well as the orientation of the stiffness axes of the plate. The coincidence frequency can be shifted by changing the dynamical properties of the core.

REFERENCES

Kurtze G. & Watters B.G. 1969. New wall design for high transmission loss or high damping. *JASA 31(6)*; 739-748.

Nilsson E. 1996. Acoustic properties of honeycomb panels, *Report 96/36*, MWL, KTH, Stockholm.

Nilsson A. 1974. Sound transmission through single panels, *Report 74-01*, Applied Acoustics, Chalmers, Gothenburg.

Szilard R. 1974. Theory and analysis of plates, *Prentice Hall Inc.*

Modern Practice in Stress and Vibration Analysis, Gilchrist (ed.) © 1997 Balkema, Rotterdam, ISBN 90 5410 896 7

Rapid detection of defects in thin plates using ultrasonic Rayleigh-Lamb waves

M. D.Gilchrist
Mechanical Engineering Department, University College Dublin, Ireland

ABSTRACT: It is shown in this paper how the presence in thin plates of horizontally symmetric crack-like defects can be detected rapidly using longitudinal ultrasonic waves. Using longitudinal waves has the potential to be significantly faster for non-destructively detecting defects than conventional ultrasonic techniques, which rely on transverse waves propagating through the thickness of a plate. Computational and analytical methods are used to predict reflection coefficients due to the attenuation of a longitudinal ultrasonic wave by physically small defects.

1 INTRODUCTION

A wide range of physical defects can exist in metallic and composite material systems, and many engineering fractures tend to initiate at the site of naturally-occurring material imperfections such as voids, inclusions and crack-like defects. Within metals such defects could be porosities, impurities, imperfections, slag, etc., while delaminations, matrix cracks, fibre fracture and splitting can often be the source of failure of composite systems. Defects such as delaminations along ply interfaces in composites, impurities along the rolling direction of hot- or cold-rolled metals, or surface machining scratches are orientated in particular directions. Other defects can be unevenly aligned and randomly distributed throughout a material. Many such defects are often microscopically small and will propagate due to applied forces, for example, in-flight loads on a rotorcraft, or internal pressure in a filament-wound composite pressure vessel. It is only when a defects has grown to a certain size that it will be detected by means of some NDE technique. At that stage, maintenance scheduling is used to monitor the development and growth rate of defects before they reach the critical size at which catastrophic failure of the structure will occur.

Conventional non-destructive evaluation (NDE) by ultrasonic C-scanning, for example, uses high frequency waves (of the order of MHz) which propagate transversely through the thickness of a plate. A single transducer can be used both to emit incident waves into a plate and to receive reflected waves whilst a double transducer system will have a receiving transducer located on the opposite side of the plate from the emitting transducer. Such systems can characterise the position and dimensions of a defect with good accuracy. However, to assess an area of plate requires that the transducer(s) scans the complete surface; this is particularly time consuming for large plates. The method proposed in this paper is for longitudinal waves, rather than transverse waves, to be propagated along the length of a plate and to monitor the attenuation of such a wave. Because guided waves (i.e., Rayleigh-Lamb waves) travel large distances without attenuation (in a non-lossy material) they are ideally suited for NDE of large structures and components that are difficult to inspect. It is not expected that this procedure will identify a defect as accurately as conventional transverse waves, although it should provide some estimate of the position of a defect in a plate. The primary advantage of this procedure is that the need to scan the surface of a plate is avoided and consequently, the time required to assess the integrity of a plate is reduced by at least an order of magnitude. Should this method identify the presence of a defect above some critical size, conventional transverse ultrasonic waves could subsequently be used to characterise the defect more accurately.

The scattering of Rayleigh-Lamb waves by an anisotropic weldment in a metal plate has been analysed by Datta and co-workers (Al-Nasser et al., 1991) using a hybrid finite element method. These results showed that maximum sensitivity during NDE (reflection coefficients of up to 70%) could be achieved by operating in certain frequency ranges ($3 < \Omega < 8$) and by probing with certain modes (symmetric S_0 mode). The same method has also been used to predict reflection coefficients associated with S_0 (symmetric) and A_0 (antisymmetric) waves scattered by transverse cracks in laminated composite plates (Datta et al., 1988, 1990; Bratton et al., 1991; Karunasena et al., 1991a, b; Ju & Datta, 1992). However, the main disadvantages of this hybrid method is that it is limited to sub megahertz frequencies and that it is not immediately compatible with commercially available finite element software.

Guo & Cawley (1993) showed that the amplitude of the reflection of the S_0 Rayleigh-Lamb wave from a delamination in a composite plate is strongly dependent upon the position of the delamination through the thickness of the laminate. The delamination locations corresponding to the maximum and minimum reflectivity correspond to the locations of maximum and minimum shear stress across the interface in the S_0 mode. Delaminations of 10- to 20-mm diameter (= O(wavelength) in size) were detected at a range of up to 500mm in 8-ply laminates (Guo & Cawley, 1993). However, delaminations could not be detected at interfaces through the thickness of a composite laminate that were free of shear stress. Rokhlin (1979) has similarly examined delaminations in rolled sheets where the length of the delaminations was greater than the wavelength of the Rayleigh-Lamb waves. Negligible reflection coefficients were predicted using an analytical method. The most promising method for long-range inspection (Guo & Cawley, 1994) was a pulse-echo configuration in which the S_0 mode is generated by a transmitting transducer and reflections of the same mode are monitored returning to a receiver placed close to the transmitter.

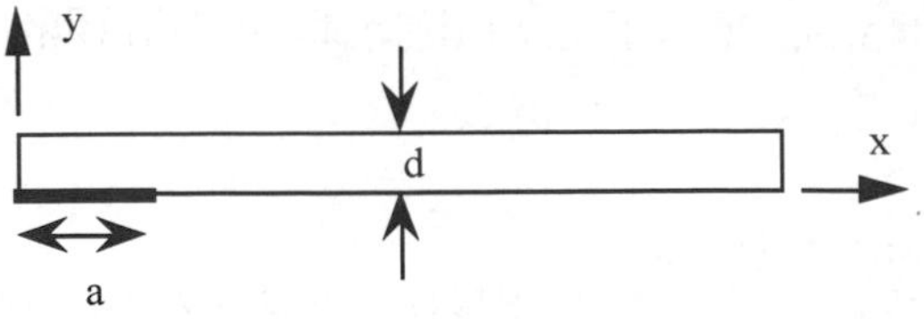

Figure 1. Symmetric crack of length *2a* embedded in a thin plate of thickness *2d*. One quarter of this plate is modelled due to the presence of two planes of symmetry ($x = 0$and $y = 0$ planes).

A simple example of an aligned defect, which is amenable to mathematical analysis, is that of a two-dimensional horizontal crack, symmetrically embedded in a large metal sheet (infinite thin plate). This paper will consider the scattering and attenuation of longitudinal waves by such a crack, oriented parallel to the plane of a sheet, as identified in Figure 1. It is physically small defect sizes that are considered in order that the initial stages of crack propagation may be detected. A combined analytical and computational method is used to predict the reflection coefficients of Rayleigh-Lamb waves by such defects.

2 THEORETICAL ANALYSIS

For an S wave, the general formulae for the complex displacements in the *x, y* directions, namely, *u, v*, are given (Viktorov, 1967) by

$$u = +ik\left[\frac{\cosh qy}{\sinh qd} - \frac{2qs}{k^2+s^2}\cdot\frac{\cosh sy}{\sinh sd}\right]e^{i(kx-\omega t)}$$

$$v = +q\left[\frac{\sinh qy}{\sinh qd} - \frac{2k^2}{k^2+s^2}\cdot\frac{\sinh sy}{\sinh sd}\right]e^{i(kx-\omega t)}$$

where

$$q = \sqrt{k^2 - k_l^2} \quad \text{and} \quad s = \sqrt{k^2 - k_t^2}$$

and

$$k = 2\pi/\lambda \qquad k_l/k = c/c_l \qquad k_t/k = c/c_t$$

The displacement of the surface $y=0$ is given by

$$u(y=0) = ik\left[\frac{1}{s_1} - \frac{2qs}{k^2+s^2}\cdot\frac{1}{s_2}\right]e^{i(kx-\omega t)}$$

$$s_1 = \sinh qd, \qquad s_2 = \sinh sd$$

The formula for the normal stress, τ_{yy}, is given by

$$\tau_{yy} = \lambda_1\left[\frac{\partial u}{\partial x} + \frac{\partial v}{\partial y}\right] + 2\mu_1\frac{\partial v}{\partial y}$$

Restricting this analysis to the case of Poisson ratio $\nu=1/3$, $\lambda_1 \approx 2\mu_1$. The above formula for the normal stress then reduces to

$$\tau_{yy} = 2\mu_1\left[\frac{\partial u}{\partial x} + 2\frac{\partial v}{\partial y}\right]$$

Calculating the out of plane stress on the plane $y=0$ gives that

$$\left(\frac{\partial u}{\partial x}\right)_{y=0} = -k^2\left[\frac{1}{s_1} - \frac{2qs}{k^2+s^2}\cdot\frac{1}{s_2}\right]e^{i(kx-\omega t)}$$

$$2\left(\frac{\partial v}{\partial y}\right)_{y=0} = 2q\left[\frac{q}{s_1} - \frac{2k^2 s}{k^2+s^2}\cdot\frac{1}{s_2}\right]e^{i(kx-\omega t)}$$

$$\left|\frac{\tau_{yy}}{2\mu_1}\right| = \frac{1}{s_1}\left[-k^2+2q^2\right] + \frac{1}{s_2}\left[\frac{2qsk^2}{k^2+s^2} - \frac{4qsk^2}{k^2+s^2}\right]$$

$$= \frac{1}{s_1}\left[-k^2+2q^2\right] + \frac{1}{s_2}\left[-\frac{2qsk^2}{k^2+s^2}\right]$$

$$= \frac{1}{s_1}\left[-k^2+2k^2-2k_l^2\right] + \frac{1}{s_2}\left[-\frac{2qsk^2}{k^2+s^2}\right]$$

$$= \frac{1}{s_1}\left[k^2-2k_l^2\right] - \frac{1}{s_2}\left[\frac{2qsk^2}{k^2+s^2}\right]$$

where the amplitude of an S_0 wave, which produces a pressure of 1Pa on the crack face, is given by $|u_{(y=0)}|/|\tau_{yy}|$ (Crane et al., 1997). Three different frequencies have been considered and the corresponding amplitudes of incident S_0 waves have been calculated and are given in Table 1.

Table 1. Amplitudes of in-plane displacements of incident S_0 waves.

Ω	incident wave amplitude [m]
0.534	49.44E-14
1.000	12.54E-14
1.600	2.308E-14

The reflection coefficients associated with the defects in the various cracked plates are calculated as the ratio of the amplitude of the reflected wave to that of the incident wave. Specifically, the amplitudes of the reflected waves are calculated computationally by modelling the propagation of a sinusoidal wave, of pressure 1Pa, as described below in Section 3.

3 COMPUTATIONAL ANALYSIS

A series of two-dimensional finite element models of cracked plates were created using commercial software (ABAQUS, 1996). Because the cracks are all symmetrically located within the plates there are two planes of symmetry and it has only been necessary to model one quarter of each plate, as shown in Figure 1. The same plate dimensions have been assumed in all the numerical analyses: a half-thickness of 0.5mm and a half-length of 130.0mm. Such a large ratio of plate length to thickness is necessary to ensure that the wavelength of the high frequency waves is a small fraction of the plate length.

All finite element models were meshed with regular plane strain isoparametric quadrilateral elements with only vertex nodes and 3x3 integration points. Eight elements were used to model the plate half-thickness which provided 24 Gaussian integration points to represent the model variables through the 0.5mm plate half-thickness. Ten elements were used to model the half-length of the crack and 520 for the remaining uncracked half-length of the plate. Consequently, each model consisted of 4240 elements. The boundary conditions appropriate to the various planes of symmetry and the cracked region of the plates were enforced and material properties were specified for aluminium (Young's modulus = 70.7GPa, Poisson ratio = 0.34, density = 2700.0kg/m^3). Model symmetry within the various cracked plates was defined by restraining the vertical (y-axis) displacements of nodes on the y=0 plane and by restraining the in-plane (x-axis) displacements of uncracked nodes on the x=0 plane. Merely by changing the material properties of elements within the model to those of steel, or indeed another metallic material, would have provided a finite element model of a crack in a different plate.

However, since the initial application for this work is primarily aimed at light-weight high performance engineering structures, it is only aluminium that has been modelled. Subsequent research will perform similar investigations for fibre reinforced polymer matrix composites.

A dynamic sinusoidal load, of amplitude $|P_n|$ and angular frequency w, identical to that of the incident wave, was applied normally outwards from the plane of the crack to simulate the propagation of the reflected wave. The sinusoidal reflected wave for the case Ω=1 was defined by

$$P_n = \cos(\omega t)$$

Reflection coefficients were calculated by normalising in-plane displacements (*x*-axis components) throughout the cracked plates due to the reflected wave against the amplitude of the incident S_0 wave (defined by Table 1) at identical positions in the cracked plates.

4 RESULTS

The reflection coefficients are calculated by normalising the amplitudes of the in-plane displacements against the amplitudes of the incident wave, which are given in Table 1. The normalised amplitudes of the in-plane displacements have been calculated at different distances away from the crack at both the midplane and plate surface positions; the variation with time of those along the midplane some 5mm away from the crack are given in Figure 2.

The normalised amplitude is seen to vary in an essentially sinusoidal manner (circular frequency, ω, = 10.0E6rads/sec). An initial transient phase is observed from 0 ≤ time ≤ 16μsec, and is caused by the initial stages of the reflected wave propagation. This is followed by a relatively long steady-state propagation phase. It is the maximum amplitude of the steady-state wave propagation that indicates the magnitude of the reflection coefficient, R. In the particular case of Figure 2 the reflection coefficient is some 14.2%. Such a value for a reflection coefficient is well above the signal noise levels that would be used with conventional digital signal processing techniques and it is therefore anticipated that this particular defect should be detectable in practice.

A further twenty analyses have been performed in similar fashion. Crack lengths ranging from 0.4 ≤ a/d ≤ 2.0 and three different frequencies

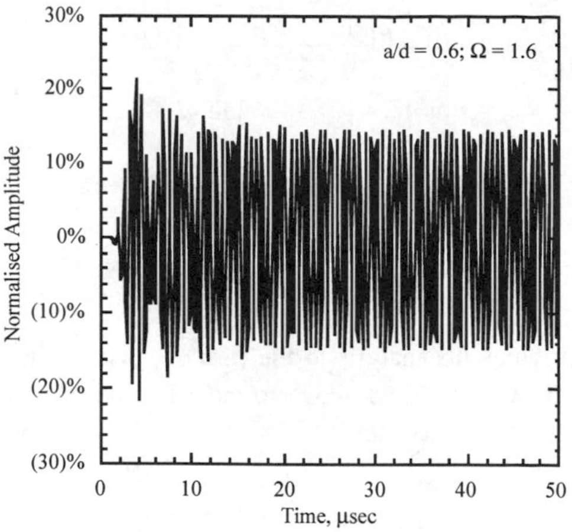

Figure 2. Normalised amplitude of the reflected wave calculated 5mm away from the end of the centrally embedded horizontal defect (a/d=0.6; Ω=1.6) at the midplane of the aluminium plate. The reflection coefficient is the magnitude of the normalised amplitude of the steady-state stage of the wave propagation (= 14.2% here).

Table 2: Reflection coefficients (units of %) predicted via finite element analyses for different horizontal defect lengths (2a) centrally embedded within aluminium plate (thickness 2d) given as a function of dimensionless frequency, Ω.

a/d	Ω = 0.534	Ω = 1.000	Ω = 1.600
0.4	0.0095	0.190	21.0
0.5	0.0165	0.373	25.5
0.6	0.028	0.740	14.2
0.8	0.06	5.25	10.5
1.0	0.2	4.75	10.0
1.5	4.5	2.5	11.0
2.0	1.0	2.2	6.5

were analysed. The reflection coefficients of these analysis are given typically in graphical form in Figures 2 and 3 and summarised in Table 2 and in Figure 4. The greatest reflection coefficients are generally associated with the highest dimensionless frequency, Ω = 1.6.

There is a transient stage associated with each analysis, the duration of which varies with defect length and, to a lesser degree, frequency. The transient stage of the analysis of Figure 2 (i.e., a/d=0.6 and Ω=1.6) is only some 16μsec whereas

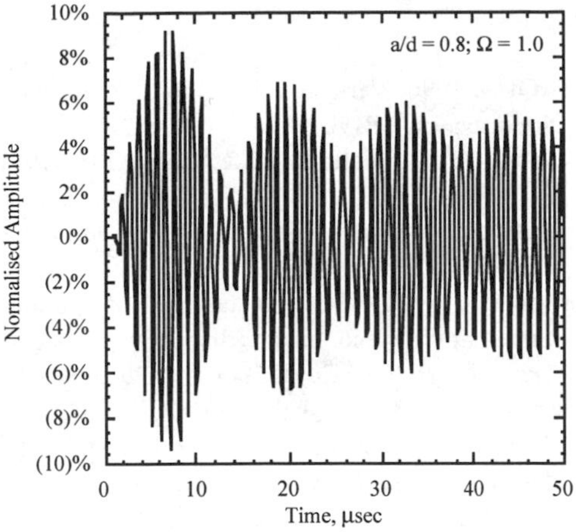

Figure 3. Normalised amplitude of the reflected wave calculated 5mm away from the end of the centrally embedded horizontal defect (a/d=0.8; Ω=1.0) at the midplane of the aluminium plate. The reflection coefficient is the magnitude of the normalised amplitude of the steady-state stage of the wave propagation (= 5.25% here).

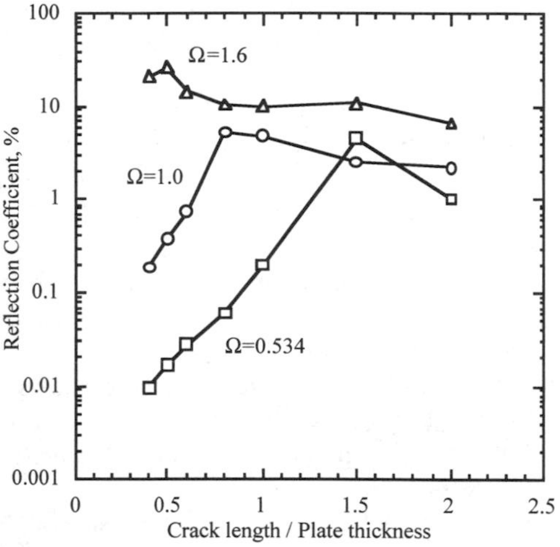

Figure 4. Predicted reflection coefficients for centrally embedded horizontal defects at the midplane of 1mm thick aluminium plate. cf Table 2.

that of the analysis of Figure 3 (i.e., a/d=0.8 and Ω=1.0) is greater than the 50μsec of the results that have been shown. While duration of the transient stage is a function of both the dimensionless frequency and the defect size, the nature of the transience does not appear to be systematically consistent. In other words, the degree of modulation that occurs in the transience of the analysis of Figure 3 does not occur in every analysis. It is possible that this is due to some underlying physical phenomenon (cf Section 6) rather than some numerical combination of defect size and wave frequency.

Figure 4 presents the reflection coefficients for all twenty-one analyses as a function of defect size and Rayleigh-Lamb wave dimensionless frequency. It is clear that measurable reflection coefficients (>10%) can realistically only be obtained by using high frequency waves (Ω=1.6) since the magnitude of the reflection coefficients that would otherwise be measured (O(1%)) would be difficult to distinguish from the noise levels during digital signal processing.

5 VIBRATION ANALYSIS

The vibration and natural frequencies of undamaged composite plates and plates containing midplane delaminations were predicted recently by Tenek et al., (1993) using finite element methods. Negligible change in the first few natural frequencies (up to the 8th natural frequency) of plates containing even large delaminations was predicted. Higher order natural frequencies, however, were decreased and resonating delaminations dissipated mechanical energy into heat which was experimentally detected using vibrothermography and thermoelastic emission.

Similar results were observed in the cracked plates of the present investigation. Natural frequencies were calculated for all the defects ($0.4 \leq a/d \leq 2.0$) via an eigenvalue analysis using two-dimensional finite element methods. The reduction of the first ten natural frequencies was not more than 0.2% of that of the corresponding natural frequencies of the uncracked plate. Natural frequencies up to 10^7 rads/sec were also obtained and the corresponding reduction was not more than 1%.

6 DISCUSSION

It is noteworthy to observe that the maximum amplitude of the reflected wave is not constant during the steady-state phase of the wave propagation, as shown typically in Figure 2. Some oscillation or degree of modulation of the wave is apparent. It is unclear at the present time whether this is due to numerical instability associated with the finite element analyses of the Ω=1.6 cases or to modulation of the reflected wave. However, the reflection coefficients have been taken as the average value of the maximum amplitude during the steady-state phase. A similar modulational instability was detected on finite gravity waves in deep water of arbitrary depth, h, (periodic wavetrains in nonlinear dispersive systems), which become unstable if the fundamental wavenumber k satisfies $kh > 1.363$. The greater the value of kh the more unstable the waves will be, in the sense that the modulational instability develops faster (Benjamin, 1967; Benjamin & Feir, 1967). Future work will attempt to establish whether the instabilities in this system are due to physical phenomena or to numerical time stepping increments.

7 CONCLUSIONS

The propagation of Rayleigh-Lamb waves within thin aluminium plates has been examined and the reflection of such waves by symmetrically embedded cracks has been modelled in order to predict reflection coefficients. It has been predicted that measurable values of reflection coefficient (>10%) can be obtained by using high frequency waves (Ω=1.6). Future work being done by the author is aimed at determining the reflection coefficients of delaminations within composite plates; this will provide useful information on the lower bound limits to the use of Rayleigh-Lamb waves for NDE of composite structures.

ACKNOWLEDGEMENTS

The author is grateful to Professor L. J. Crane (TCD and INCA) and to Dr G. Thomas (UCC) for useful discussions. The author would like to acknowledge the support provided by University College Dublin (President's Research Award), Forbairt (Basic Research Grant) and the European Commission (Contract no. BRE2-CT94-0990).

REFERENCES

ABAQUS, 1996, Version 5.5. Hibbitt, Karlsson & Sorensen, Inc., RI, USA.

Al-Nassar, Y. N., Datta, S. K. & Shah, A. H., 1991, Scattering of Lamb waves by a normal rectangular strip weldment. *Ultrasonics*, 29: 125-31.

Benjamin, T. B., 1967, Instability of periodic wavetrains in nonlinear dispersive systems. *Proceedings of the Royal Society*, A299: 59-75.

Benjamin, T. B. & Feir, J. E., 1967, The disintegration of wave trains on deep water. *Journal of Fluid Mechanics*, 27-3: 417-30.

Bratton, R., Datta, S. K. & Shah, A. H., 1991, Scattering of Lamb waves in a composite plate. *Review of Progress in Quantitative NDE*, 10B: 1507-14. D. O. Thompson & D. E. Chimenti, Eds., Plenum Press, New York.

Crane, L. J., Gilchrist, M. D. & Miller, J. J. H., 1997, Analysis of Rayleigh-Lamb wave scattering by a crack in an elastic plate. To appear in *Computational Mechanics.*

Datta, S. K., Shah, A. H. & Karunasena, W. M., 1990, Edge and layering effects in a multilayered composite plate. Computers & Structures, 37-2: 151-62.

Datta, S. K., Shah, A. H., Bratton, R. L. & Chakraborty, T., 1988, Wave propagation in laminated composite plates. *Journal of the Acoustical Society of America*, 83: 2020-26.

Guo, N. & Cawley, P., 1993, The interaction of Lamb waves with delaminations in composite laminates. *Journal of the Acoustical Society of America*, 94-4: 2240-6

Guo, N. & Cawley, P., 1994, Lamb wave reflection for the quick nondestructive evaluation of large composite laminates. *Materials Evaluation*, March, 404-11.

Ju, T. H. & Datta, S. K., 1992, Pulse propagation in a laminated composite plate and nondestructive evaluation. *Comp. Eng.*, 2: 55-66.

Karunasena, W. M., Shah, A. H. & Datta, S. K., 1991a, Plane-strain-wave scattering by cracks in laminated composite plates. *ASCE Journal of Engineering Mechanics*, 117: 1738-54.

Karunasena, W. M., Shah, A. H. & Datta, S. K., 1991b, Reflection of plane strain waves at the free edge of a laminated composite plate. *Int. J. Solids & Structures*, 27-8: 949-64.

Koshiba, M., Karakida, S. & Suzuki, M., 1984, Finite-element analysis of Lamb wave scattering in an elastic plate waveguide. *IEEE Transactions on Sonics & Ultrasonics*, SU-31-1: 18-24.

Rokhlin, S. I., 1979, Interaction of Lamb waves with elongated delaminations in thin sheets. *Int. Adv. in Nondestructive Testing*, 6: 263-85.

Tenek, L. H., Henneke, E. G., II & Gunzburger, M. D., 1993, Vibration of delaminated composite plates and some applications to non-destructive testing. *Composite Structures*, 23: 253-62.

Viktorov, I. A., 1967, *Rayleigh and Lamb Waves*. Plenum Press, New York.

Modern Practice in Stress and Vibration Analysis, Gilchrist (ed.) © *1997 Balkema, Rotterdam, ISBN 90 5410 896 7*

Interlaminar stress analysis of composite laminates with an open hole

F.Z. Hu & C. Soutis
Department of Aeronautics, Imperial College of Science, Technology & Medicine, London, UK

ABSTRACT: In this paper, a three-dimensional finite element analysis is performed to investigate the interlaminar stresses that develop in the ply interfaces, near a circular hole, in carbon fibre-epoxy composite laminates. The stress distributions are presented in three ways: through the thickness, along radial lines away from the hole and around the hole. Because an interlaminar stress singularity is expected between the plies at the hole, the computed stresses are presented near but not at the hole edge. The interlaminar normal and shear stresses are then averaged over a characteristic length from the hole boundary and employed in a quadratic stress criterion to predict delamination initiation load and delamination location.

1 INTRODUCTION

High-strength/low-density polymeric composite materials are being used extensively in aerospace structures. A large amount of work has been carried out to predict the strength of laminates containing holes (cut-outs) using in-plane stresses and stress based failure criteria, which has been reviewed by Soutis (1996) and Tan (1994). However, many of these laminates develop high through-thickness stresses near the traction free hole-edge regions due to stiffness discontinuity between plies although they are subjected to in-plane loading (Raju& Crews 1982, Herakovich 1989). Such stresses can be quite large and influence the laminate failure. A large interlaminar shear stress at the interface may produce matrix cracks at the free edge. These cracks then propagate into the laminate and initiate rupture, leading to a premature failure of the composite. The initial damage at the free edge is quite important for fatigue loading in which the ultimate failure may initiate at the edges.

As carbon fibre composites are considered for heavily loaded primary aircraft structures, increased attention is being devoted to the understanding and calculation of these through-thickness stresses. Over the past decade, there have been many experimental and analytical studies for the calculation of interlaminar stresses and the prediction of delamination initiation at the free edge (Pipes & Pagano 1970, Ko & Lin 1992). However, most of these studies have been performed for the straight free-edge and fewer studies have been devoted to a curved free edge (hole) due to its more complicated geometry and stress state. The analysis of a straight free edge may be assumed as a problem with two-dimensional stress and strain variations, while the stress state near a curved free-edge is a three-dimensional stress problem.

In this paper, a three-dimensional (3-D) finite element (FE) analysis is performed to investigate the interlaminar normal and shear stress distributions near a circular hole in carbon fibre-epoxy composite plates. A series of laminates, [90/0]s, [0/90]s, [±30]s, [±45]s and [±60]s, are examined in the current study. The FE77 finite element package, developed at Imperial College (Hitchings 1996), is used and the analysis is based on displacement formulation employing curved isoparametric 20-node elements; a very high mesh refinement near the intersection of the interface between plies and the hole boundary is required in order to capture the edge effect. The stress distributions are presented in three ways: through the thickness, along radial lines away from the hole, and around the hole. Because an interlaminar stress singularity is expected between the 0° and 90° plies at the hole, the computed stresses are presented near

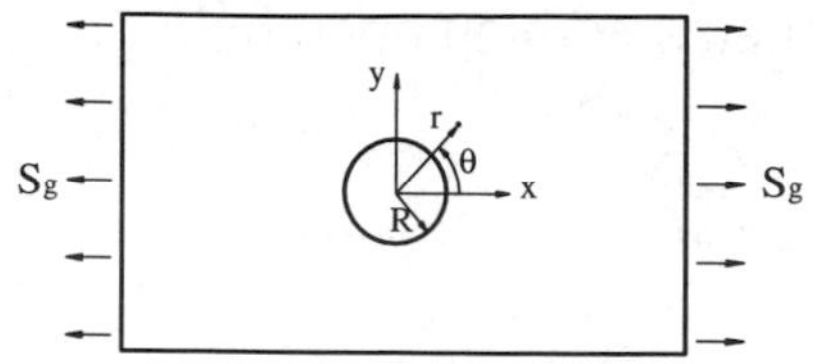

Figure 1. Laminate geometry and loading (the length of the laminate: 60 mm; the width: 30 mm; the thickness: 0.5 mm)

but not at the hole edge (r/R=1.0001). Finally, the interlaminar normal stress σ_z and the shear stress $\tau_{z\theta}$ are averaged over a characteristic length from the hole boundary and used with a quadratic stress criterion to determine the delamination location and delamination initiation load.

2 THREE-DIMENSIONAL STRESS ANALYSIS

Consider a symmetric laminate with a circular hole subjected to a uniaxial tensile load, *Sg*, as shown in Figure 1. The x-y plane of the Cartesian co-ordinate system lies in the mid-plane of the laminate and the origin is at the centre of the hole; a cylindrical co-ordinate system (r, θ, z) is also used. The total length of the laminate is 60 mm, the width 30 mm, the hole radius R is 2.5 mm and the ply thickness h is equal to 0.125 mm. Several four-ply laminates, [90/0]s, [0/90]s, [±30]s, [±45]s and [±60]s, are examined in the current study.

2.1 *Finite element model*

The FE77 finite element package, developed at Imperial College, is used and the analysis is based on displacement formulation employing curved isoparametric 20-node elements. Due to the symmetry of loading, hole location and lay-up, only one-eighth of the laminate is modelled by 4000 curved isoparametric 20-node elements, as shown in Figure 2. Because high interlaminar stress concentration is expected near the hole edge and between the plies, a high mesh refinement is required in this area. The smallest element size in the radial and thickness directions is 0.25 μm. Each ply is treated as a homogeneous, elastic and anisotropic material with the same elastic properties as those in the literature of Raju & Crews (1982):

$E_{11} = 138$ GPa
$E_{22} = E_{33} = 14.5$ GPa
$G_{12} = G_{23} = G_{31} = 5.86$ GPa
$\upsilon_{12} = \upsilon_{23} = \upsilon_{31} = 0.21$

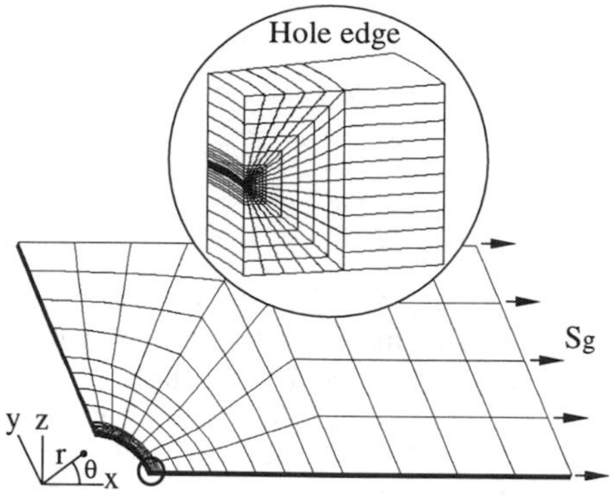

Figure 2. 3–D FE mesh for one–eighth plate

2.2 *In-plane stresses*

To evaluate the reliability of the current 3-D finite element analysis model, the in-plane stress distributions are compared with the exact solution obtained by Lekhnitskii's theory (1963). Figure 3 shows the laminate stress distribution σ_x normalised by the remote load S_g along the y-axis (θ =90°). The overall stress concentration factor for the $[90/0]_s$ laminate is 5.05 while the exact value is 5.02. The agreement between FE and theory is excellent. In Figure 3, the stress distributions developed in the 0° and 90° plies are also presented. The normalised stress σ_x/S_g in the 0°-ply is approximately 10 times of that developed in the 90°-ply due to the large difference between the longitudinal and transverse moduli.

2.3 *Interlaminar stresses*

This section presents the distributions of through-thickness stresses: σ_z , $\tau_{z\theta}$, τ_{zr} in the cylindrical co-ordinate system. Near the free hole-edge region, the shear stress component τ_{zr} is very small compared to $\tau_{z\theta}$ and can be neglected. The distributions of interlaminar stresses, σ_z and $\tau_{z\theta}$, are presented in three ways: through the thickness, along radial lines away from the hole and around the hole. As an

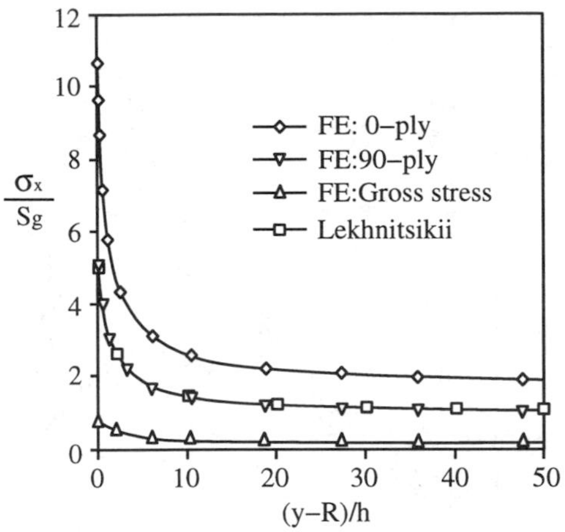

Figure 3. Normalised in–plane stress, σ_x/S_g, distributions along the y–axis of a [90/0]s laminate (Ply thickness: h=0.125 mm).

example, only the interlaminar stresses in the hole edge region of a [90/0]s laminate are presented in this section. However, the current FE model can be applied to any multidirectional laminate with any lay-up configuration and any number of plies under any loading condition.

Figure 4 shows the normalised interlaminar normal stress σ_z/S_g distributions through the thickness at four angular locations (θ = 0°, 30°, 60° and 90°) on the hole boundary in the [90/0]s laminate. Since an interlaminar stress singularity exists at the free edge between the 90° and 0° plies, the computed stresses are presented near but not at the hole boundary (at r-R = 0.002h = 0.00025 mm). At all angular locations, the curves have steep gradients at $z = h$ suggesting the existence of interlaminar stress singularity at the hole edge. Furthermore at θ = 0° and 90°, the σ_z stresses are discontinuous across the interface $z= h$. At the top face ($z=2h$), σ_z equals zero (free surface). Near the laminate mid-plane ($z=0$) the σ_z distributions show smooth gradients and there is no evidence of singularity. The interlaminar shear stress $\tau_{z\theta}$ also shows singularity at $z=h$ interface and equals zero at the top face ($z=2h$) and mid-plane ($z=0$). Therefore, it is found that the mid-plane, which was extensively studied by Rybicki & Schmueser (1978), Kim & Soni (1984) *et al*, is a less critical delamination location and the interlaminar stresses at $z=h$ interface should be examined.

Figure 5 shows the radial distributions of the normalised normal stress σ_z/S_g and shear stress $\tau_{z\theta}/S_g$ at the 90°/0° ply interface at θ = 45°. As the distance from the edge (r-R) increases, the interlaminar stresses are rapidly decreased. When (r-R)= 0.25 mm, *i.e.* two ply-thickness away from the hole boundary, σ_z and $\tau_{z\theta}$ become almost zero. The shear stress component τ_{zr} is very small in the hole edge region compared to $\tau_{z\theta}$ and can be neglected.

Figure 6 shows the circumferential distributions of the interlaminar normal stress σ_z at the 90°/0° ply interface around the hole ($z = h$ = 0.125 mm). Since an interlaminar stress singularity exists at the free edge between the 90° and 0° plies, the computed stresses are presented near but not at the hole boundary. The stresses closest to the hole are at r/R= 1.0001; that is, at a distance of 0.25 μm from the hole. As the distance from the edge (r-R) increases, the interlaminar stress σ_z is rapidly decreased. When (r-R)=0.1R, *i.e.* 0.25 mm or two ply-thickness (2h) away from the hole boundary, σ_z becomes almost zero. σ_z is compressive for most of the region around the hole with a small tensile region near θ = 90° when subjected to tensile load S_g. The largest compressive σ_z occurs at about 60° from the loading axis. For the [0/90]s lay-up, σ_z is again compressive for most of the region along the hole

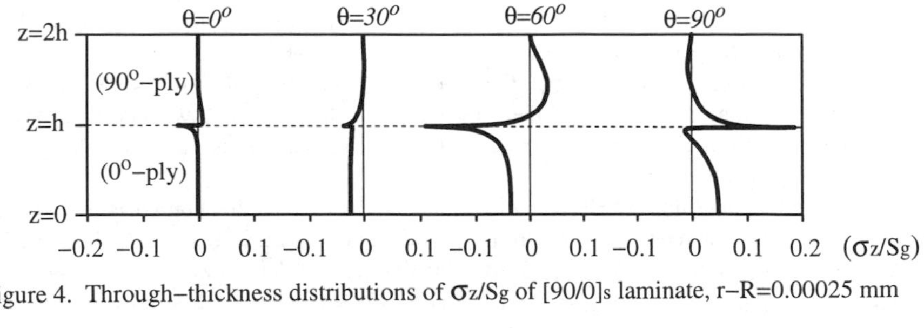

Figure 4. Through–thickness distributions of σ_z/S_g of [90/0]s laminate, r–R=0.00025 mm

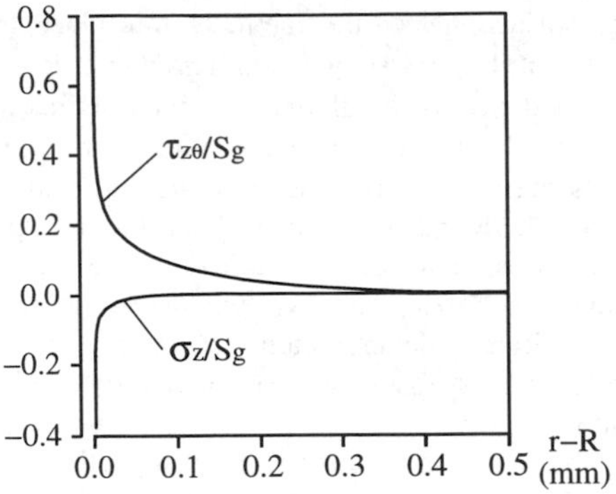

Figure 5. Radial distributions of σz/Sg and τzθ/Sg at the 90/0 interface of [90/0]s, θ=45°.

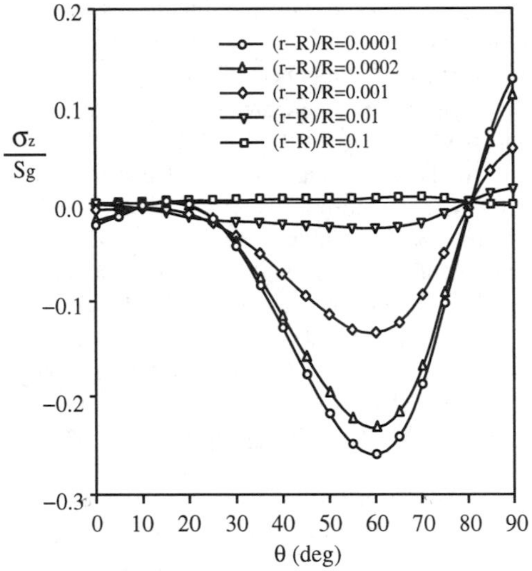

Figure 6. Normalised interlaminar normal stress distributions, σz/Sg, around hole at the 90/0 interface of [90/0]s laminate.

boundary but is tensile in the regions $10° \leq \theta \leq 28°$ and $80° \leq \theta \leq 90°$. Its largest compressive value occurs around θ =60°. The stress distributions in Figure 6 are having similar shapes at different (*r-R*) distances from the hole edge, indicating that the location in hoop direction is an important parameter for a curved free edge. This is the main difference from the straight free edge problem.

The circumferential interlaminar shear stress distributions $\tau_{z\theta}$ at various distances, *(r-R)/R*, from the hole boundary are shown in Figure 7. Similar to the normal stress σ_z , the interlaminar shear stress $\tau_{z\theta}$

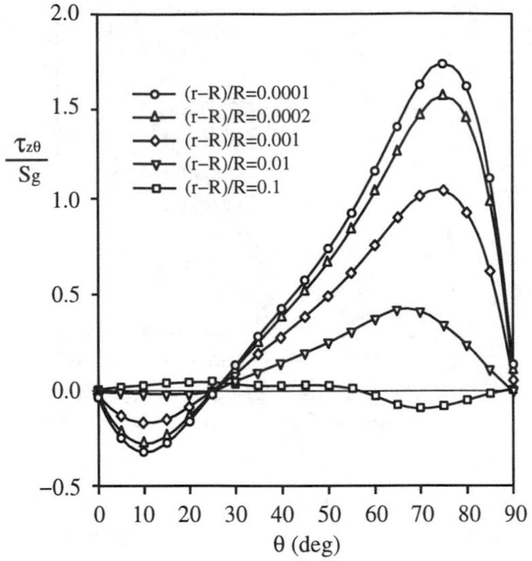

Figure 7. Normalised interlaminar shear stress distributions, τzθ/Sg, around the hole at the 90/0 interface of [90/0]s laminate.

decreases as the distance (*r-R*) from the hole boundary increases and becomes vanishingly small within two-ply thickness (0.25 mm) from the hole. Except for different signs, the $\tau_{z\theta}$ distributions are identical for the $[90/0]_s$ and $[0/90]_s$ laminates. Both of these distributions have their maximum values at approximately 75° from the loading axis. The largest value in Figure 7 is about $1.73S_g$, which is about seven times as large as the largest σ_z value (Figure 6) computed for the same distance *(r-R)/R* = 1.0001 from the hole. This comparison indicates that the $\tau_{z\theta}$ stress singularity is stronger than that for the σ_z interlaminar stress. It also suggests that the interlaminar shear stresses are mainly responsible for the delamination initiation in these cross-ply laminates. The shear stress component τ_{zr} is very small in the hole-edge region compared to $\tau_{z\theta}$ and can be neglected.

Comparison are also made between the interlaminar stress distributions of the current FE analysis and the FE results obtained by Raju & Crews (1982) at *(r-R)/R*=0.0001. The difference between the two finite element solutions is less than 2%.

However, the comparison between the FE results and the theoretical predictions by Ko & Lin's analytical model (1992) is in poor agreement (Hu&Soutis 1996). This is probably due to the zeroth-order approximation that Ko&Lin used when they solved the equilibrium equations in the boundary-layer region. Although their solution satisfied the equilibrium conditions at the boundary, the circumferential stress and the laminate in-plane shear stress gradients in the radial direction have been ignored. For the case of a hole, there are two stress gradients to be considered: the gradient in the interlaminar stresses near the hole, similar to the straight edge case, and also the gradient in the in-plane stresses due to the presence of the hole, which gives another contribution to the gradient in the through-thickness stresses. This suggests that their assumed boundary layer stresses need to be amended or employ higher-order approximation in their stress solution, which does not look easy judging by eqn.(5) of their paper. This is another indication of how much more complex the hole problem is, compared with straight free edges.

3 PREDICTION OF DELAMINATION INITIATION

After one has obtained the interlaminar stress distributions, the next step is to use these data to compute the delamination initiation load. Here, the delamination initiation is mainly attributed to interlaminar stress effects, so only the interlaminar shear stress $\tau_{z\theta}$ and the normal stress σ_z are required. Therefore, the Tsai-Wu criterion (Tsai&Hahn 1980) in polar co-ordinates can be simplified as,

$$\left(\frac{\sigma_z}{Z}\right)^2 + \left(\frac{\tau_{z\theta}}{S}\right)^2 = e^2 \quad \begin{cases} e < 1 & \text{no failure} \\ e \geq 1 & \text{failure} \end{cases} \tag{1}$$

where Z is the interlaminar normal strength and S is the interlaminar shear strength (ILSS); the term of τ_{zr} has been neglected since it is very small

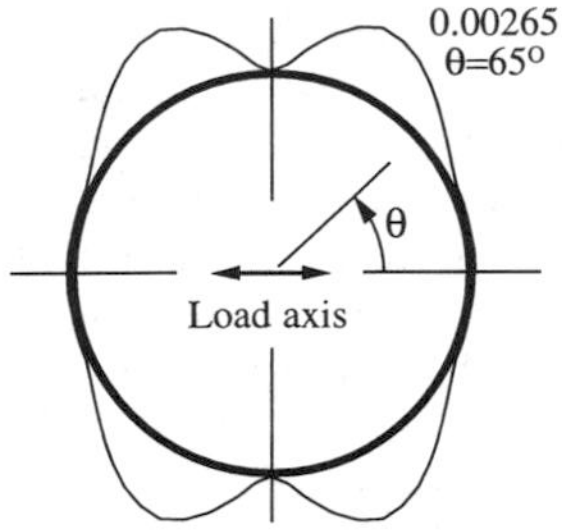

(a) [90/0]s: e_{max}=0.00265, $S_{g,f}$=377 MPa.

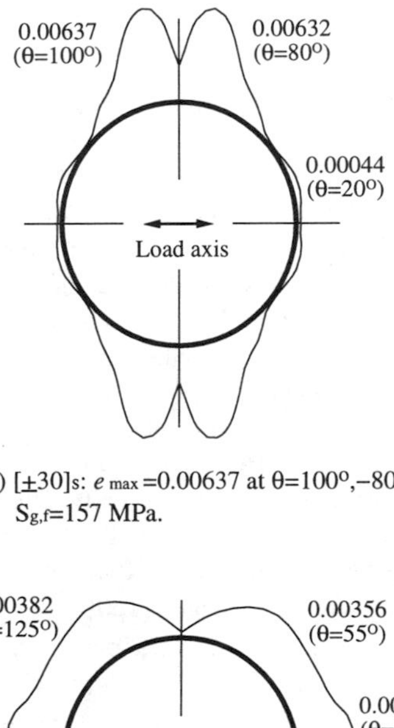

(b) [±30]s: e_{max}=0.00637 at θ=100°,–80°, $S_{g,f}$=157 MPa.

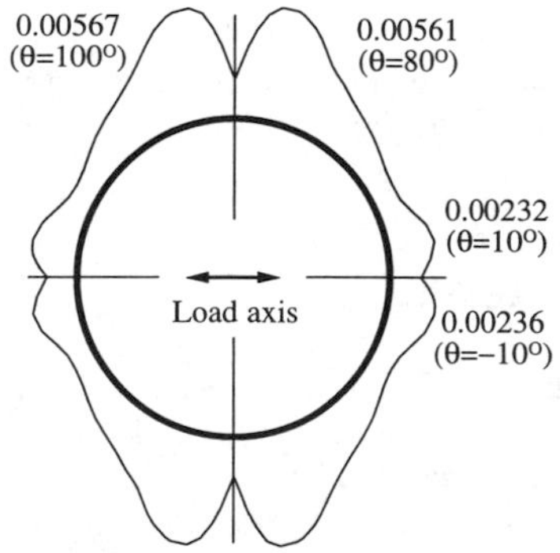

(c) [±45]s: e_{max}=0.00567 at θ=100°,–80° $S_{g,f}$=176 MPa.

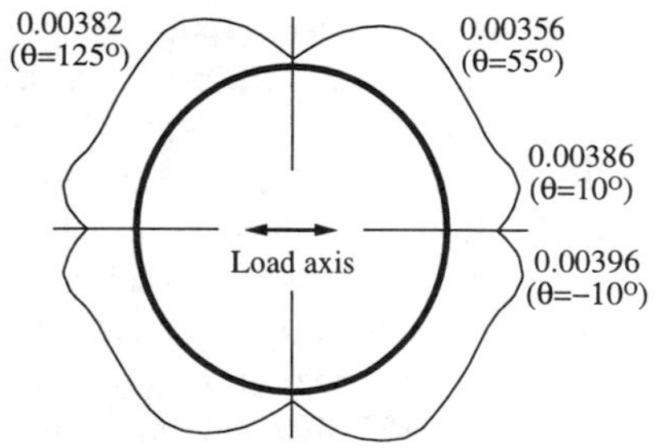

(d) [±60]s: e_{max}=0.00396 at θ=–10°, 170°, $S_{g,f}$=252 MPa.

Figure 8. ***e***–index distributions at hole edge interface determined by the average stress criterion (when S_g=1 MPa).

compared to $\tau_{z\theta}$ in the region near the hole boundary.

From the literature of Kim&Soni (1984), it can be seen that the approach for predicting failure in such laminates has been that of averaging the interlaminar stresses over a distance in from the hole edge. The average stress failure criterion (ASFC) assumes that delamination initiates when the stresses at a characteristic distance d_o from the discontinuity meet the failure criterion expressed by eqn.(1). The average of a stress component is defined as:

$$\overline{\sigma_{ij}} = \frac{1}{d_o}\int_R^{R+d_o} \sigma_{ij}\, dr \qquad (2)$$

The following strength properties:

$Z_t = 50.6$ MPa
$Z_c = 200$ MPa
$S = 103$ MPa

are used for the carbon fibre-epoxy laminates in eqn.(1) to determine the applied far field stress $S_{g,f}$ required to initiate delamination in the interface. Figure 8 shows the distributions of the *e*-index of eqn.(1) around the hole edge of [90/0]s, [±30]s, [±45]s and [±60]s laminates when the remote load S_g equals unity; d_o is assumed to be equal to 0.125 mm (one ply thickness). The maximum value of *e* occurs at $\theta = 65^0$ to the loading direction for the [90/0]s laminate, indicating that the delamination initiates from the hole edge at $\theta = 65^0$ angular location. The delamination load $S_{g,f}$ is predicted to be 377 MPa by assuming *e* equals unity. The *e*-index distributions are almost identical for the cross-ply laminates [0/90]s and [90/0]s. For the angle-ply laminates, [±30]s, [±45]s and [±60]s, the *e*-index distributions are not symmetric about the x- and y-axes, Figure 8(b), (c) and (d). The 1st ($\theta = 0$ to 90^0) and 3rd ($\theta = 180$ to 270^0) phases are the results of the one-eighth FE model (Figure 2) of a [+α/-α]s laminate while the 2nd and 4th phases are the results of [-α/+α]s. For the four lay-ups examined, the [±30]s laminate shows the lowest delamination initiation load, $S_{g,f} = 157$ MPa, indicating that large interlaminar stresses develop near the hole edge. However, the adjustable parameter d_o has not yet been related to micromechanical features and does not result from a mechanics analysis. Experimental evidence is required in order to verify these results.

4 DISCUSSION AND CONCLUDING REMARKS

The interlaminar stresses σ_z and $\tau_{z\theta}$ between the plies of different fibre orientations show high values near the hole edge due to the stiffness discontinuity. The σ_z has a relatively lower value in the mid-plane than that in the interface, indicating that the interface between plies of different orientations is more critical than the mid-plane for hole edge delamination.

The delamination is predicted by a quadratic stress failure criterion; the interlaminar stresses near the hole edge are averaged over a characteristic length, d_o, suggesting that the peak values of the stresses at the hole edge are not too important. In real laminates stress singularities at the hole edge may not occur owing to the local stress redistribution and material non-linearities.

In summary, the aim of the present work is to calculate the interlaminar stresses near the hole edge in composite laminates and to predict the delamination initiation. The FE method could be applied to any lay-up configuration (thin or thick) under any loading condition, tension or compression. The work does not consider matrix cracking, which may influence the delamination onset load and does not attempt to relate the adjustable characteristic length used in the quadratic failure criterion to any micromechanical feature. Further analytical and experimental work is required to resolve these issues.

ACKNOWLEDGEMENTS

The authors are grateful for funding from the Engineering and Physical Sciences Research Council (EPSRC, GR/K54892).

REFERENCES

Herakovich, C.T. 1989. *Handbook of Composites*. **2**. Structure and Design: 187-230. ed. C.T. Herakovich *et al.* Elsevier Applied Science.

Hitchings, D. 1996. *Finite Element Package FE77: User's Manual*. Imperial college.

Hu, F.Z. & C. Soutis 1996. Evaluation of the Ko-Lin model for interlaminar stresses in composite laminates with an open hole. *Advanced Composites Letters* **5**(5): 143-147.

Kim, R.Y. & S.R. Soni 1984. Experimental and analytical studies on the onset of delamination in laminated composites. *J. Comp. Mat.* **18**: 70-80.

Ko,C. C. & C.C. Lin 1992. Method for calculating the interlaminar stresses in symmetric laminates containing a circular hole. *AIAA Journal* **30**(1): 197-204.

Lekhnitskii, S.G. 1963. Theory of elasticity of an anisotropic elastic body. Holden-Dey Inc. San Francisco.

Pipes, R.B. & N.J. Pagano 1970, Interlaminar stresses in composite laminates under uniform axial extension. *J. Composite Materials* **4**: 538-548.

Raju, I.S. & J.H. Crews, Jr 1982. Three-dimensional analysis of [0/90]s and [90/0]s laminates with a central circular hole. *Comp. Tech. Rev.* **4**(4): 116-124.

Rybicki, E.F. & D.W. Schmueser 1978. Effect of stacking sequence and lay-up angle on free edge stresses around a hole in a laminated plate under tension. *J. Comp. Mat.* **12**: 300-313.

Soutis, C. 1996. Failure of notched CFRP laminates due to fibre microbuckling: a topical review. *J. Mech. Behaviour of Materials* **6**(4):309-330.

Tan, S.C. 1994. *Stress Concentrations in Laminated Composites*. Technomic Pub.

Tsai, S.W. & H.T. Hahn 1980. *Introduction to composite materials*. Technomic Pub.

Modern Practice in Stress and Vibration Analysis, Gilchrist (ed.) © 1997 Balkema, Rotterdam, ISBN 90 5410 896 7

On Saint-Venant's principle for composite materials under plane deformations

M. Kashtalian
Timoshenko Institute of Mechanics, National Academy of Sciences of Ukraine, Kiev, Ukraine

W.J. Stronge
Department of Engineering, University of Cambridge, UK

ABSTRACT: A self-equilibrated system of forces acting in a small region of an elastic solid results in a strain energy density that decreases rapidly with the distance. The strain energy density decreases more rapidly if the forces are in astatic equilibrium than if they are merely self-equilibrated. For an anisotropic solid, there is a small rate-of-decrease of strain energy density and a much more exaggerated effect of anisotropy on the distribution of radial displacements in the direction where Young's modulus is largest. These results are pertinent to applications of Saint-Venant's principle in the case of anisotropic solids, e.g. structures made from fibrous composite materials.

1 INTRODUCTION

The principle bearing his name was introduced by Saint-Venant (1855) who developed solutions for extension, torsion and flexure of prismatic and cylindrical bodies and found that "... the means of application and distribution of the forces toward the extremities of the prisms is immaterial to the perceptible effects produces on the rest of the length, so that one can always, in a sufficiently similar manner, replace the forces applied with equivalent static forces or with those having the same total moments and the same resultant forces...".

Boussinesq (1885) who obtained a solution for an unbalanced normal force acting at the boundary of elastic half-space, proposed the first universal statement of the principle as follows: "An equilibrated system of external forces applied to an elastic body, all the points of application lying within a given sphere, produces deformations of negligible magnitude at distances from the sphere which are sufficiently large compared to its radius". Another version of the principle was proposed by Love (1944): "The strains that are produced in a body by the application, to a small part of its surface, of a system of forces statically equivalent to zero force and zero couple (self-equilibrating), are of negligible magnitude at distances which are large compared with the linear dimensions of the part".

Examining tangential forces applied to the boundary of a half-space, von Mises (1945) noted that the stress did not decay faster when system of forces was simply self-equilibrated, and that the rate of decay depends also on whether the self-equilibrated system of forces is also in astatic equilibrium (the concept introduced by von Mises), i.e. it remains in equilibrium even when all the forces are turned through the same arbitrary angle. Von Mises provided an explicit form for rate of decay, a formal proof of which was later given by Sternberg (1954).

Later it was established also that end effects in composites which are anisotropic and inhomogeneous materials, decay more slowly than in isotropic and homogeneous ones. Investigations of Saint-Venant's principle for composite materials in 2-D anisotropic elasticity have focused mostly on the exponential decay of stresses resulting from self-equilibrated sets of tractions acting on the end of a semi-infinite strip (Everstein & Pipkin 1971, Horgan 1972, Choi & Horgan 1977, Arimitsu et al. 1995) and boundary of an infinite strip (Matemilola et al. 1995), see also reviews (Horgan & Knowls 1983, Horgan 1989, Horgan 1996).

In the present paper we investigate stress and displacement fields resulting from self-equilibrated sets of tractions acting on the boundary of an anisotropic elastic half-plane. Stress fields generated by different self-equilibrated systems of forces, each acting in a small region, can be categorised as follows,

(i) self-equilibrated forces:

$$\sigma_{ij} = \sigma_{ij}(\epsilon/r^2, \theta)$$

(ii) forces in astatic equilibrium:

$$\sigma_{ij} = \sigma_{ij}(\epsilon^2/r^3, \theta)$$

where r, θ are polar coordinates, and ϵ is the radius of the loaded region.

Fig.1 illustrates two sets of self-equilibrated tractions acting in a small region at the boundary of an elastic half-plane. System (i) is simply self-equilibrated while system (ii) is also in astatic equilibrium.

In this paper the distribution of strain energy as a function of θ around systems of self-equilibrated forces acting on the surface of an elastic orthotropic half-plane is examined and a system that is in astatic equilibrium is compared with one that is not. For both systems of self-equilibrated forces, there is additional complexity in the distribution of strain energy density if the material is highly anisotropic, e.g. a fiber-reinforced composite. The present paper demonstrates that although a self-equilibrated system of forces on the surface of an orthotropic half-plane always results in a displacement field that varies radially as a function of ϵ/r, nevertheless the displacements (and stresses) in some directions are much larger than they are in others.

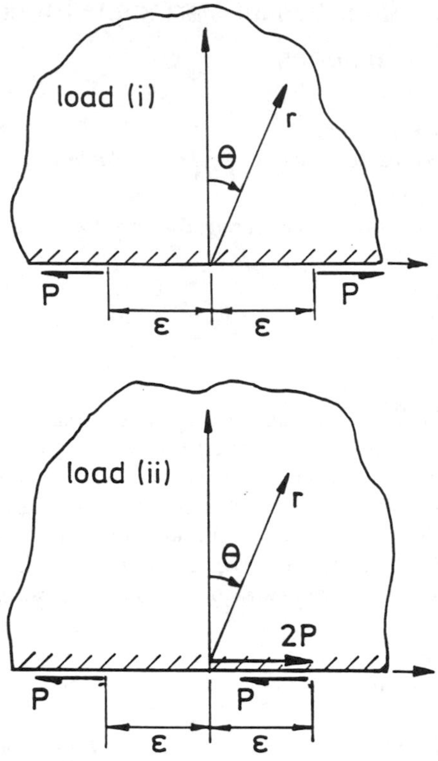

Figure 1: Self-equilibrated systems of forces (i) not in astatic equilibrium, (ii) in astatic equilibrium.

2 STRESS FIELDS

In the limit as $\epsilon/r \ll 1$, the self-equilibrated set of forces for problem (i) (Fig. 1) results in a stress field that can be expressed (Lekhnitskii 1963) as

$$\sigma_r = \frac{2P}{\pi}(u_1 + u_2)\beta_{11}\frac{\epsilon}{r^2}\phi_r(\theta)$$

$$\sigma_\theta = 0 \qquad (1)$$

$$\sigma_{r\theta} = \frac{2P}{\pi}(u_1 + u_2)\beta_{11}\frac{\epsilon}{r^2}\phi_{r\theta}(\theta)$$

where u_1 and u_2 are roots of characteristic equation

$$\beta_{11}u^4 - (2\beta_{12} + \beta_{66})u^2 + \beta_{22}u = 0 \qquad (2)$$

The parameters β_{11}, β_{12}, β_{22}, β_{66}, are reduced elastic constants that depend on elastic moduli of the material $c_{11}, c_{12}, c_{22}, c_{66}$. For plane stress state in particular they are defined as

$$\beta_{11} = \frac{c_{22}}{\Delta} \qquad \beta_{12} = -\frac{c_{12}}{\Delta} \qquad (3)$$

$$\beta_{22} = \frac{c_{11}}{\Delta} \qquad \beta_{66} = \frac{1}{c_{66}} \qquad (\Delta = c_{11}c_{22} - c_{12}^2)$$

Most composite materials have moduli giving roots u_1 and u_2 that are real and distinct but the roots can be also complex or, for isotropic materials, real and repeated.

The angular variations of components of stresses can be expressed as

$$\phi_r(\theta) = \frac{1}{L^2}\{L(2\sin^2\theta + 1) - \sin\theta\frac{\partial L}{\partial \sin\theta}\}$$

$$\phi_{r\theta}(\theta) = \frac{1}{L}\cos\theta\sin\theta \tag{4}$$

where

$$L = \beta_{11}\sin^4\theta + \beta_{22}\cos^4\theta + $$
$$+ (2\beta_{12} + \beta_{66})\sin^2\theta\cos^2\theta \tag{5}$$

For the problem (ii) (Fig. 1), i.e. a system of self-equilibrated forces in astatic equilibrium, the stress field can be expressed as

$$\sigma_r = \frac{2P}{\pi}(u_1 + u_2)\beta_{11}\frac{\epsilon^2}{r^3}\psi_r(\theta)$$

$$\sigma_\theta = \frac{2P}{\pi}(u_1 + u_2)\beta_{11}\frac{\epsilon^2}{r^3}\psi_\theta(\theta) \tag{6}$$

$$\sigma_{r\theta} = \frac{2P}{\pi}(u_1 + u_2)\beta_{11}\frac{\epsilon^2}{r^3}\psi_{r\theta}(\theta)$$

where

$$\psi_r(\theta) = \frac{1}{L^3}\{L^2(2\sin\theta + \sin^3\theta) -$$

$$-L(1 + 2\sin^2\theta)\frac{\partial L}{\partial \sin\theta} + \frac{L}{2}\sin\theta\frac{\partial L^2}{\partial \sin\theta^2} -$$

$$-\sin\theta(L\frac{\partial L^2}{\partial \sin\theta^2} - (\frac{\partial L}{\partial \sin\theta})^2\}$$

$$\psi_\theta(\theta) = \frac{1}{L}\cos^2\theta\sin\theta \tag{7}$$

$$\psi_{r\theta}(\theta) = \frac{1}{L^2}\cos\theta(L(1 + \sin^2\theta) - \sin\theta\frac{\partial L}{\partial \sin\theta})$$

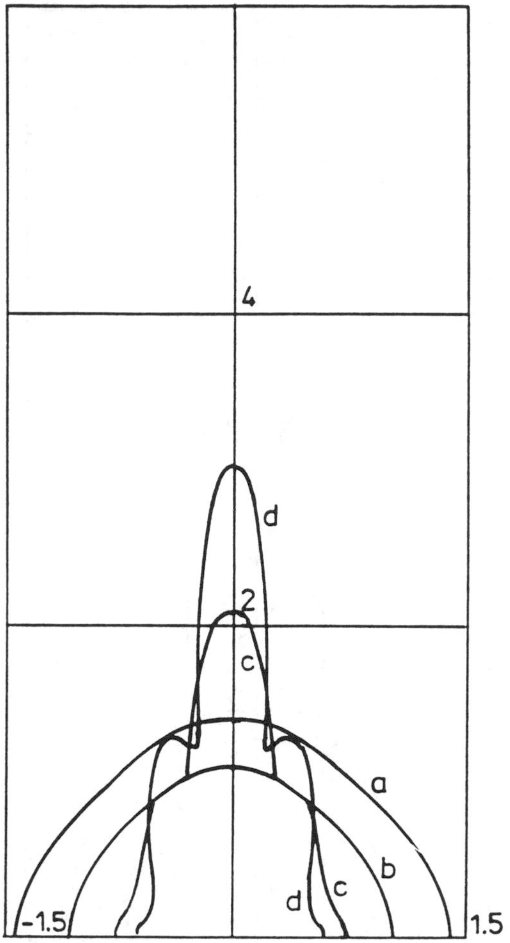

Figure 2: Contours of the constant strain energy for system (i)

Fig. 2 and 3 show polar plots of contours of constant strain energy density as a function of angle θ for (i) self-equilibrated and (ii) astatic self-equilibrated sets of surface tractions respectively. Contour *(a)* represents isotropic material with Poisson's ratio $\nu = 0.3$, contours *(b), (c), (d)* refer to the composite materials, the properties of which are given in Table 1 (Pira & Hasan 1996). The Steel-Aluminium composite has rather weak anisotropy, while Glass-Epoxy and Graphite-Epoxy composites are strongly orthotropic materials. The curves show the effect of anisotropy on the distribution of the strain energy density. For highly anisotropic composites there is pronounced ellipticity (or elongation) of the strain energy contours in the direction of larger stiffness. This is consis-

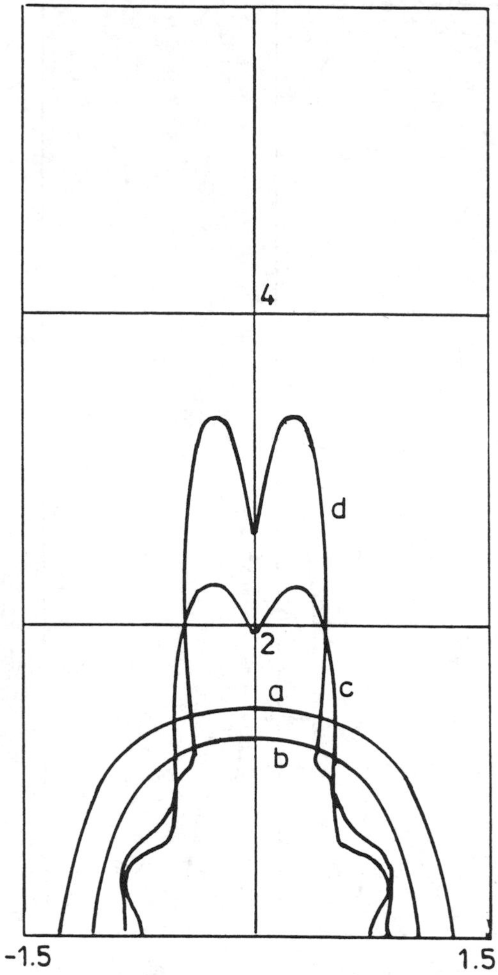

Figure 3: Contours of the constant strain energy for system (ii)

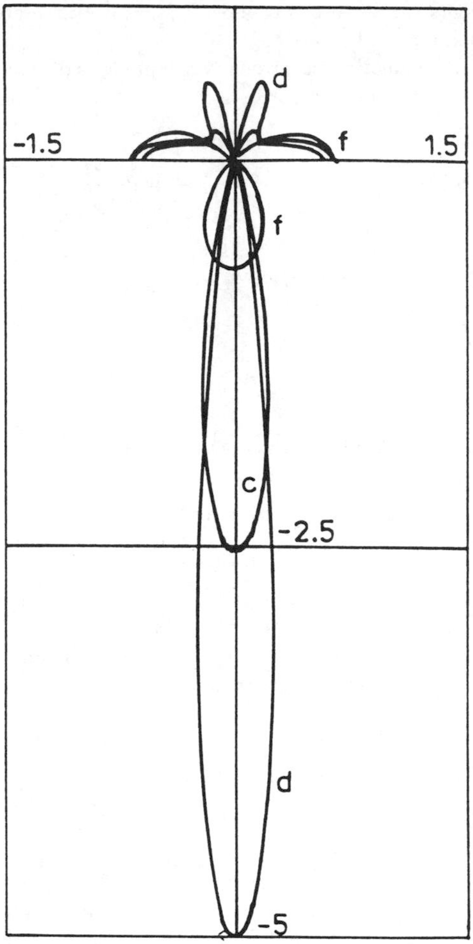

Figure 4: Contours of the constant radial displacement for system (i)

tent with ideas of stress channelling (Everstein & Pipkin 1971, Arimitsu et al. 1995). The ellipticity of contours is slightly more sensitive to anisotropy in case (i) than in case (ii).

High anisotropy brings also, increasing complexity of the shape of the strain energy contour as evidenced by the number of extrema on a contour curve.

3 DISPLACEMENTS

The displacement caused by self-equilibrated systems of forces can be obtained by integration of the strains. For each set of forces radial displacement $u_r(\theta)$ is

$$(i) \qquad u_r = -\frac{2F}{\pi E_x}\left(\frac{\epsilon}{r}\right)(u_1 + u_2)\phi_r$$

$$(8)$$

$$(ii) \qquad u_r = -\frac{2F}{\pi E_x}\left(\frac{\epsilon^2}{2r^2}\right)(u_1 + u_2)(\psi_r - \nu_{xy}\psi_\theta)$$

where functions ϕ_r and ψ_r, ψ_θ are given by formulas (4) and (7), respectively.

The effect of anisotropy on the distribution of radial displacement u_r is evident from Fig. 4 and

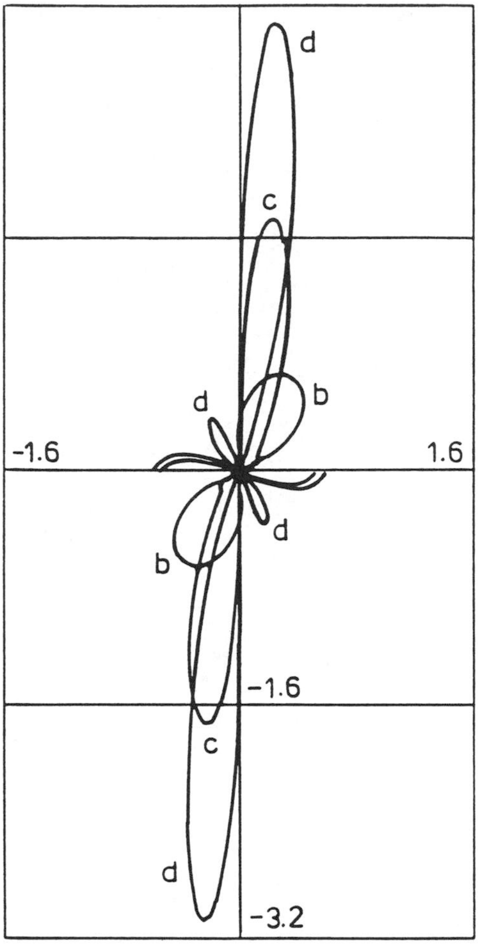

Figure 5: Contours of the constant radial displacement for system (ii)

5. They show polar plots of a constant value of radial displacement for different composite materials (Table 1). To make plots more readable, positive displacements are shown in the upper coordinate half-plane and negative ones in the lower half-plane. For astatic self-equilibrated forces (ii) some part of the displacement contour have very large curvature near the origin; consequently there are angular sectors where, at a given radius r, the radial displacement is negligibly small in comparison with the displacement at other angles θ. As against to isotropic materials, the curves of constant displacement for anisotropic materials are highly elongated in the direction of largest stiffness.

Table 1. Material parameters

Contour	Type of composite	c_{11}/c_{66}	c_{22}/c_{66}	c_{12}/c_{66}
b	Steel—Aluminium	3.952	4.155	1.959
c	Glass—Epoxy	3.139	12.190	1.155
d	Graphite—Epoxy	3.504	29.822	1.723

4 CONCLUSION

Near a set of self-equilibrated forces the stresses and strain energy have a rate-of-decrease that varies as a function of both radius r and angle θ. For an anisotropic material the rate-of-decrease also depends on the ratio of material properties.
Curves for equal values of rate-of-decrease of strain energy have the same variation with θ as the contours of equal strain energy (Fig. 2 and 3). The variation of strain energy with angle θ is especially strong in the case of an anisotropic material – there the slowest rate-of-decrease for strain energy occurs in the direction with the largest Young's modulus. The effect of anisotropy on the radial rate-of-decrease of strain energy is smaller if the set of forces is not only self-equilibrating but also in astatic equilibrium.

5 REFERENCES

Arimitsu, Y.,K. Nishioka & T. Senda 1995. A study of Saint-Venant's principle for composite materials by means of internal stress fields. *ASME J. Appl. Mech.* 62: 53-58.

Boussinesq, M.J. 1885. *Application des potentiels.* Paris: Gauthier-Villars.

Choi, I. & C.O. Horgan 1977. Saint-Venant's principle and end effects in anisotropic elasticity. *J. Appl. Mech.* 44: 424-430.

Eversteine, G.C. & A.C. Pipkin 1971. Stress channeling in transversely isotropic composites *ZAMP* 22: 825-834.

Horgan, C.O. 1972. Some remarks on Saint-Venant's principle for transversely isotropic composites *J. of Elasticity* 2: 335-339.

Horgan, C.O. 1989. Recent developments concerning Saint-Venant's principle: An update *Appl. Mech. Rev.* 42: 295-303.

Horgan, C.O. 1996. Recent developments concerning Saint-Venant's principle: A second update. *Appl. Mech. Rev.* 49: S101-S111.

Horgan, C.O. & J.K. Knowles 1983. Recent developments concerning Saint-Venant's principle. *Advances in Appl. Mech.* 23, 179-267.

Lekhnitskii, S.G. 1963 *Theory of Elasticity of an Anisotropic Body.* San Francisco: Holden-Day Inc.

Love, A.E.H. 1944. *A treatise on the mathematical theory of elasticity (4th ed.).* New York: Dover Publications.

Matemilola, S.A.,W.J. Stronge & D. Durban 1995. Diffusion rate for stress in orthotropic materials. *ASME J. Appl. Mech.* 62: 654-661.

Mises, R. von. 1945. On Saint-Venant's principle *Bull. Amer. Math. Soc.* 51: 555-562.

Pira, A. & W. Hasan 1996. Effect of orthotropy on the intersonic crack propagation. *ASME J. Appl. Mech.* 63: 933-938.

Saint-Venant, B. de. 1855. Mémoire sur la torsion des prismes. *Mém. Savants étrangers.* Paris.

Sternberg, E. 1954. On Saint-Venant's principle *Quart. Appl. Math.* 11: 393-402.

Modern Practice in Stress and Vibration Analysis, Gilchrist (ed.) © *1997 Balkema, Rotterdam, ISBN 90 5410 896 7*

Gyrotropic properties of rocks due to a dissymmetry of microstructure

T.I.Chichinina & I.R.Obolentseva
Institute of Geophysics of Russian Academy of Sciences (Siberian Branch), Novosibirsk, Russia

ABSTRACT: A new concept 'gyrotropy of rocks', introduced previously in the context of a phenomenological theory, is substantiated by means of constructing a model on a microlevel. A model is presented of a dissymmetric grainy medium possessing, on a macrolevel, gyrotropic properties because of the displacement of spherical grains in each column of cubic packing in the same azimuthal direction: clockwise or counter-clockwise. A mathematical modelling of stressed state has been performed to determine gyration constants of the model versus the parameters of dissymmetry of its microstructure. The results occurred to be in a good agreement with the seismic experimental data and in the theoretical limits for optically gyrotropic crystals. Thus, gyrotropy as well as anisotropy provide a principle possibility to recognize a microstructure of rocks through large scale observations of seismic waves.

1 INTRODUCTION

Anisotropy of rocks is a well known phenomenon. Recently, it was found (Obolentseva 1992, 1996) that rocks can be not only anisotropic but gyrotropic as well. Anisotropy and gyrotropy are similar in the sense that they arise when the microobjects (with linear dimensions much less than the wave length) such as grains, thin layers, microcracks etc. are situated in a particular way in space. An anisotropy appears if the microobjects are parallel to certain planes or lines. For a gyrotropy to appear, the condition is rather simple: microobjects must be arranged so that the rock has no centre of symmetry. In this case the polarization of shear waves will be elliptical. If, moreover, the microobjects form a structure with prevailing right or left orientations, the effect of rotation of polarization plane will be observed.

By now the phenomenological theory of elastic gyrotropy is developed (Obolentseva 1992, 1993, 1996; Chichinin 1993). The problem is to present a microscopic theory or at least to construct specific micromodels of real rocks. One of such models is put forward — the model of a dissymmetric grainy medium which is gyrotropic, namely, enantiomorphent (i.e. 'left' or 'right') as a result of the dissymmetry of its microstructure.

2 ON GYROTROPY

Gyrotropy is an exhibition of the spatial dispersion (of the first order). In a gyrotropic medium, Hooke's law,

$$\sigma_{ij} = c_{ijkl}\varepsilon_{kl} + b_{ijklm}\partial\varepsilon_{kl}/\partial x_m, \quad (1)$$

contains new terms describing the contribution of neighbouring points to the stressed state at the given point, i.e. the stresses are not local. The stress field in the vicinity of a given point must exist as a result of microheterogeneity of the medium.

The gyration tensor $(b_{ijklm}) = \boldsymbol{b}$ is a tensor of the fifth (odd) rank, and therefore it is not equal to zero only in the media without a centre of symmetry. The tensor $\boldsymbol{b}$ is invariant relatively a group of rotations. The inner symmetry of tensor $\boldsymbol{b}$ is

$$b_{ijklm} = b_{jiklm} = b_{ijlkm} = b_{jilkm}, \quad b_{ijklm} = -b_{klijm}.$$

Because of these symmetry properties, the number of independent components of the tensor $\boldsymbol{b}$ in the general case is equal to 45.

For seismic applications, two symmetry groups of the tensor $\boldsymbol{b}$ are the most important: $\infty\infty$ (gyrotropic properties are the same in all directions) and ∞ (they are the same in all planes passing through the symmetry axis of order ∞).

Non-zero components of the tensor $\boldsymbol{b}$ of the group symmetry $\infty\infty$ are

$$\begin{aligned} b_{12131} &= b_{23212} = b_{31323} = \eta, \\ b_{11123} &= b_{22231} = b_{33312} = 2\eta, \\ b_{11132} &= b_{22213} = b_{33321} = -2\eta, \end{aligned} \quad (2)$$

and their isomers.

In a medium of the group symmetry ∞ with a symmetry axis oriented along the axis X_3, non-zero components of the tensor $\boldsymbol{b}$ are as follows:

$$\begin{aligned} &b_{11333} = b_{22333} = \alpha, \\ &b_{13331} = b_{23332} = \beta, \\ &b_{13332} = -b_{23331} = \gamma, \\ &b_{13233} = \delta, \end{aligned} \qquad (3)$$

and their isomers.

The main features of elastic-wave propagation in gyrotropic media are an elliptical polarization of waves — in absence of attenuation — and rotation of displacement vectors (for attenuative gyrotropic media rotation of ellipses).

Seismic gyrotropy is introduced (Obolentseva 1992, 1993, 1996) by analogy with the well known optical gyrotropy (Landau & Lifshits 1992, Kisel & Burkov 1980) and acoustical gyrotropy (Sirotin & Shascolskaya 1979).

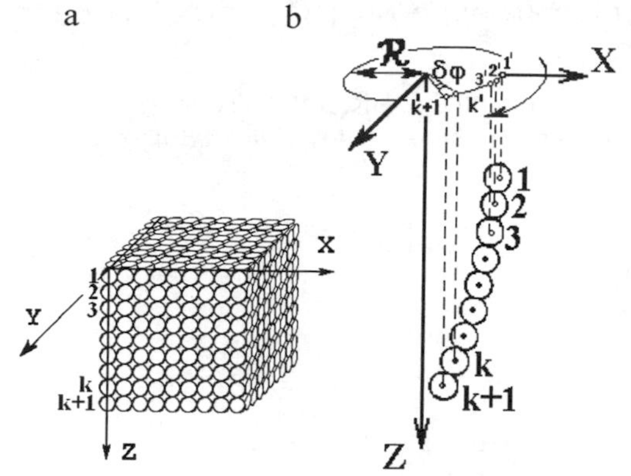

Figure 1. A regular cubic packing of spheres (a); a column of spheres from a disturbed cubic packing (b).

3 A GYROTROPIC MICROMODEL OF A DISSYMMETRIC GRAINY MEDIUM

Gyrotropy first has been revealed at borehole studying of the uppermost of the ground consisting of alluvial deposits, i.e. sandy – clayey rocks. Therefore, the first gyrotropic model, which we are going to represent, is a model of terrigeneous rocks consisting mainly of isometric particles not less than 10^{-3} cm, i.e. sand grains.

To construct the model, the Curie principle is used. According to this principle, when any phenomena exhibit a certain dissymmetry, the same dissymmetry necessarily presents in the causes that generated these phenomena. In the case considered, a phenomenon showing a dissymmetry is a rotation of a polarization plane of shear elastic waves, or a turn of a displacement vector; hence, the cause of this phenomenon — the dissymmetric microstructure of the medium — also must be somehow connected with turns. Therefore, it may be concluded that an essence of constructing of our model will be in an azimuthal rotation combined with translation. It will be demonstrated that such a medium rotates a displacement vector.

Let a medium be a cubic packing of spheres (see Figure 1, a). In further considerations it is called 'a regular model'. To construct a packing possessing gyrotropic properties, a spiral will be imitated for each column of the cubic packing drawn at Figure 1, a. Figure 1, b shows how it is made: centres of spheres (of radius R_0) are so displaced that their projections on a horizontal plane XY lie on an arc of the circle of radius $\mathcal{R}$, all being displaced in one direction: clockwise or counter-clockwise.

The central angles $\delta\varphi$ ought to be very small: $0.01 - 0.000001°$. This condition is needed to provide a similarity of model and real turn angles of shear-wave displacement vectors.

Dissymmetry parameters of the model are $\mathcal{R}$ and $\delta\varphi$ (at given R_0).

For a grain, the dissymmetry of the model looks as it is shown at Figure 2. Consider three sequential grains 1, 2, 3 from a column of the disturbed cubic packing shown at Figure 1, b. For the sphere 2, the top contact point (T) is situated in the middle of the interval $[O_1, O_2]$, and the bottom contact point (B) — in the middle of the interval $[O_2, O_3]$. Respectively, in XOY-plane the point T' is the middle point of $O'_1O'_2$ and the point B' is the middle point of $O'_2O'_3$. Both points, T' and B', can be considered to lie on a circumference of radius $\rho = \mathcal{R}\sin(\delta\varphi/2)$. The angle distance between the points T' and B' is $\pi - \delta\varphi$, i.e. $\angle T'O'_2B' = \pi - \delta\varphi$.

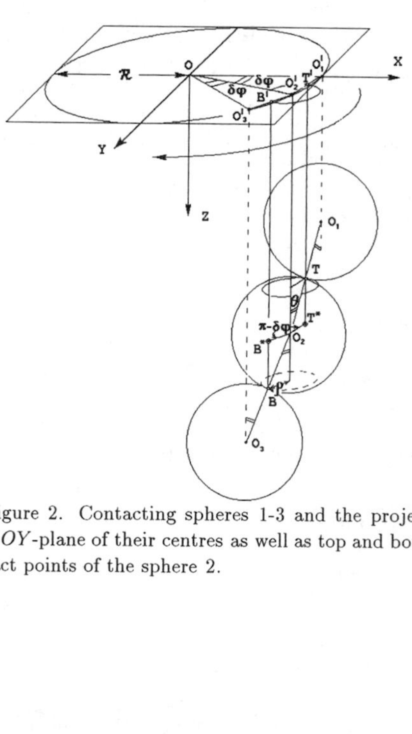

Figure 2. Contacting spheres 1-3 and the projections on XOY-plane of their centres as well as top and bottom contact points of the sphere 2.

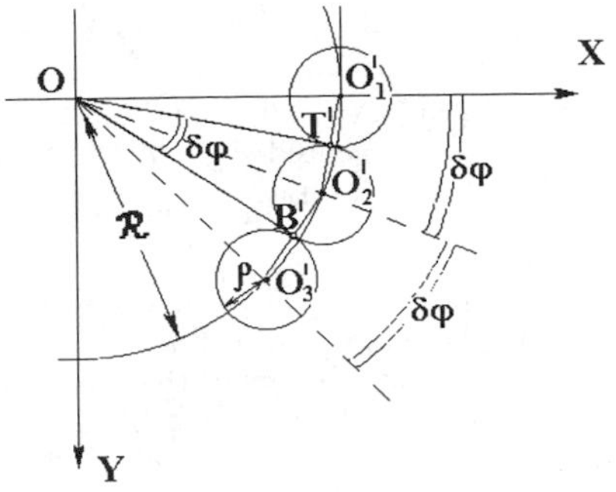

Figure 3. Projections of three grains on XOY-plane.

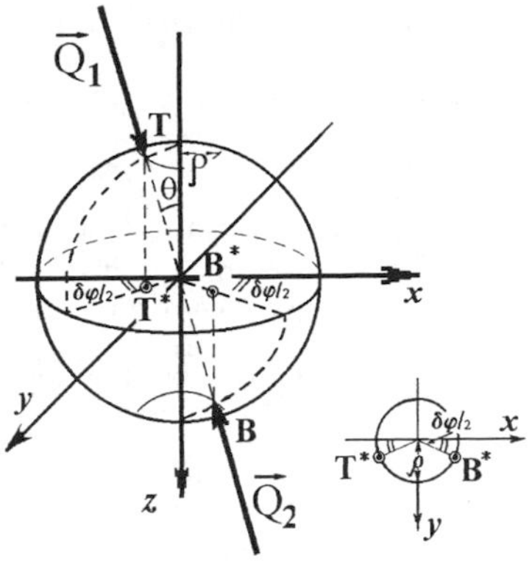

Figure 4. The forces $\vec{Q}_1, \vec{Q}_2$ applied to a grain's surface in the dissymmetric model.

(For a grain this angle is $\angle T^*O_2B^*$.) One can see it looking at the quadrangle $OT'O_2'B'$ drawn at Figure 3.

Projections of the circumference of radius ρ on the surface of the sphere 2 are drawn at Figure 2. One can see that $\rho = R_0 \sin\theta$, where $\theta = \angle TO_2O_2'$.

So the relative positions of contact points on the grain surface can be characterized by the azimuthal angle $\delta\varphi$ and the polar angle θ, $\sin\theta = \rho/R_0$.

The dissymmetry parameters for a grain of radius R_0 are related with the dissymmetry parameters for a model as a whole by formula

$$\frac{\rho}{\mathcal{R}} = \sin(\delta\varphi/2), \quad \text{or} \quad \mathcal{R} = \frac{R_0 \sin\theta}{\sin(\delta\varphi/2)}.$$

Thus, in building the dissymmetric model, a movement along a spiral is imitated (in reality, along a rather limited part of a spiral, and more exactly — along a part of a half-spire). In other words, the dissymmetric model is built in accordance with the principle 'an azimuthal turn plus translation'.

4 MODELLING OF ROTATION OF POLARIZATION PLANE

To simulate wave propagation in the direction of Z-axis, let the surface forces $\vec{Q}_1, \vec{Q}_2$, $|\vec{Q}_1|=|\vec{Q}_2|=Q$, be applied at the top (T) and the bottom (B) points of a sphere (Figure 4).

Since we want to model the phenomenon of rotation of shear-wave polarization plane, the forces must have components in a horizontal XY-plane, i.e. to be tangent to a grain's surface. However, for simplifying a solution of the problem, there is a good reason to consider the forces acting radially for their moments would be equal to zero. In this case, we ought to suppose that the forces $\vec{Q}_1, \vec{Q}_2$ are directed at some angles to Z-axis for they have horizontal components. For definitness, these forces are considered to be compressive.

A radius of a sphere is much less than a wavelength ($R_0 \ll \lambda$), therefore the problem of wave propagation can be reduced to the problem of statical equilibrium of a sphere, an element of the model.

In a local coordinate system xyz connected with a grain (see Figure 4), the forces $\vec{Q}_1, \vec{Q}_2$ have the components:

$$\vec{Q}_1 = \left\{ Q\sin\theta\cos\frac{\delta\varphi}{2},\ -Q\sin\theta\sin\frac{\delta\varphi}{2},\ Q\cos\theta \right\};$$

$$\vec{Q}_2 = \left\{ -Q\sin\theta\cos\frac{\delta\varphi}{2},\ -Q\sin\theta\sin\frac{\delta\varphi}{2},\ -Q\cos\theta \right\}.$$

Because of the dissymmetry of the model, equlibrium conditions are not satisfied:

$$\vec{Q}_1 + \vec{Q}_2 = -2Q\sin\theta\sin\frac{\delta\varphi}{2}\,\vec{e}_2 = -\vec{Q}_y \neq 0,$$

i.e. the model is not balanced one.

The compensative force $\vec{Q}_y$ can exist somewhere, since all grains are bounded. However, to have this force in the limits of one grain, it is necessary to change the model. The model can be transfered into balanced one (in the limits of one grain) in a number of ways. One of them is in placing a little particle (or particles) near the grain surface. (It is quite reasonable because the considered rock, sand, by definition, consists more than 50% of sand, but the rest are particles of less sizes, up to the smallest: 10^{-7}cm.) When the particles, for example clay plates, are situated near the surface of a given grain and, hence, far from the neighbouring grains, the elastic and Van der Waals forces applied to the grain can occur to be great enough to produce the force $\vec{Q}_y$.

The equilibrium equation for a sphere is

$$\mathrm{Div}\,\sigma_{ij} = (\lambda + \mu)\,\mathrm{grad\,div}\,\vec{u} + \mu\Delta\vec{u} = 0, \qquad (4)$$

where λ, μ are Lame constants of grain material.

A solution of equation (4) inside the sphere is searched for in Papkovich form (Lurye 1955)

$$\vec{u}(\vec{r}) = 4(m-1)/m \cdot \vec{B} - \mathrm{grad}\,(\vec{r}\,\vec{B} + B^0),$$

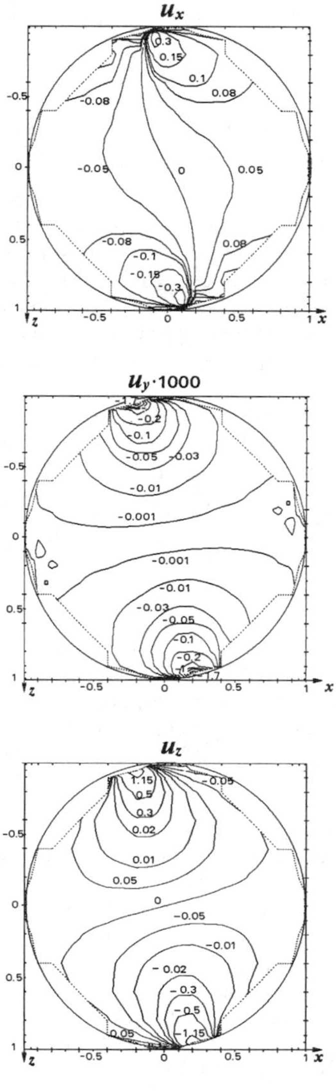

Figure 5. Isolines of the displacements u_x, u_y, u_z in xOz-section. (All values are to be multiplied by a factor $|Q|/(8\pi\mu R_0) \approx 4 \cdot 10^{-6}$ cm.)

where $m=2(\lambda+\mu)/\mu$ — Poisson number, $\vec{B}$ is a harmonic vector, and B^0 is a harmonic scalar ($\Delta B_x=0$, $\Delta B_y=0$, $\Delta B_z=0$, $\Delta B^0=0$). $\vec{B}$ and B^0 are presented in the form of series $\vec{B}=\sum_{n=0}^{\infty}\vec{B}_n$, $B^0=\sum_{n=0}^{\infty}B^0_{n-1}$, containing harmonic polynomials.

The computed displacements $\vec{u}(\vec{r})$ in xOz section of a sphere are shown in a form of isolines at Figure 5. The parameters of the model are as follows: $R_0=10^{-4}$ m, Lame constants $\mu=10^8$ N m^{-2}, $\lambda=2\mu$, the applied force $|Q|=10^{-2}$ N; dissymmetry parameters are $\theta=10°$, $\delta\varphi=0.06°$.

In a medium with gyrotropic properties, a rotation of S-wave-polarization plane is characterized by the turn angle Φ of a displacement vector $\vec{U} = U_X\vec{e}_1 + U_Y\vec{e}_2$: $\tan\Phi = U_Y/U_X$. (X, Y are global coordinates, see Figures 1, b and 6.)

The angle Φ accrues monotonously in the wave propagation from grain to grain, and the ultimate turn angle Φ is equal to the sum of turn angles of all spheres in an interval from $Z=0$ to $Z=2R_0 \cdot n$:

$$\Phi = \sum_{k=1}^{n}\phi_k = \sum_{k=1}^{n}\tan^{-1}(u_y/u_x)^{(k)}_m = n\tan^{-1}(u_y/u_x)_m,$$

where $(u_y/u_x)_m$ is a mean value of ratios u_y/u_x inside the individual grains.

It should be noted that

$$u_y/u_x = u_Y/u_X,$$

since the transition from the local coordinates to the global ones (Figure 6) is an orthogonal transformation and hence does not change a relation between x- and y-components of a vector.

Our calculations show that $\phi_k=0.025°$ (if the dissymmetry parameters $\theta=10°, \delta\varphi=0.06°$), and a polarization vector turns to an angle 1° at the depth $Z=0.8$ cm. Thus, to provide a specific rotation of shear-wave polarization plane equal 1°

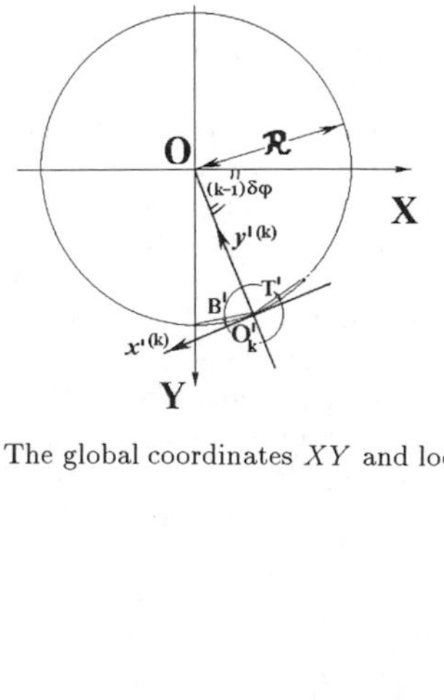

Figure 6. The global coordinates XY and local ones $xy^{(k)}$.

per metre, as it is observed in experiments (Obolentseva 1996), it is sufficiently to have a thin gyrotropic layer, in our case δh=0.8 cm (or fourty rows of grains 0.02 cm in diameter), in a layer 1 m thick. This means that, in the average, nearly every hundredth grain in the column ought to be displaced in the manner mentioned.

5 DETERMINATION OF GYROTROPY CONSTANTS

The main problem is to find the gyration constants of the model — components of the gyration tensor $\boldsymbol{b}$ in (1). This was done in the following way.

Two models: *symmetric* and *dissymmetric* ones — were considered. The dissymmetric model (Figure 4) has been characterized in details in the section 3. As for a symmetric model, it is necessary to choose it with a centre of symmetry to exclude a gyrotropy. Besides, to facilitate a comparison of symmetric and dissymmetric models and provide a possibility of correct determination of gyrotropy constants, it is appropriate to choose a symmetric model similar (as far as possible) to the dissymmetric one. Such a model is depicted at Figure 7.

In the *symmetric* model the forces $\vec{Q}_1$, $\vec{Q}_2$ have the following components:

$$\begin{aligned}\vec{Q}_1 &= \{Q\sin\theta,\ 0\ ,\ Q\cos\theta\},\\ \vec{Q}_2 &= \{-Q\sin\theta,\ 0\ ,\ -Q\cos\theta\}.\end{aligned} \tag{5}$$

To determine the gyration stresses (see (1)),

$$\Delta\sigma_{ij} = b_{ijklm}\partial\varepsilon_{kl}/\partial x_m,$$

the stresses in dissymmetric and symmetric models and then their differences have been computed $\delta\sigma_{ij} = \sigma_{ij}^{dissym} - \sigma_{ij}^{sym}$, thereafter an equation for determining b_{ijklm} was

$$b_{ijklm} = \delta\sigma_{ij}\,(\partial\varepsilon_{kl}/\partial x_m)^{-1}, \tag{6}$$

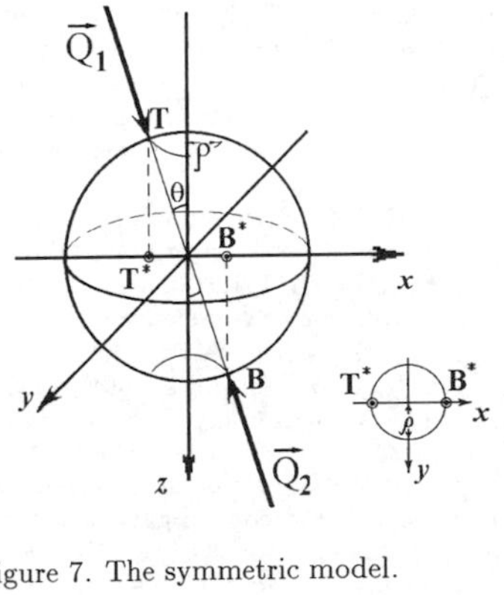

Figure 7. The symmetric model.

where $\varepsilon_{kl} = (1/2)\,(\partial u_k/\partial x_l + \partial u_l/\partial x_k)$.

Inside the grains, the values of $\partial\varepsilon_{kl}/\partial x_m$, i.e. of the second derivatives of displacements $\vec{u}(\vec{r})$, were computed numerically by means of bicubic spline interpolation. However, spline interpolation was failing near the surface of a sphere (at $r/R_0 > 0.95$) because of the quick growth of displacements $u(\vec{r})$. Therefore, the other way of approximation was found which occured to be rather effective.

Analysing the computed field of displacements, one can see (Figure 5) that the main contribution to the stress state inside the sphere is made by the regions in the vicinities of the applied forces. This fact leads to an idea to replace the problem considered by the Bousinesque problem for a point source at a half-space surface.

Bousinesque problem has a simple analytical solution (Rekach 1977):

$$u_x = \frac{Q}{4\pi}\left[\frac{zx}{\mu R^3} - \frac{x}{(\lambda+\mu)R(z+R)}\right],$$

$$u_y = \frac{Q}{4\pi}\left[\frac{zy}{\mu R^3} - \frac{y}{(\lambda+\mu)R(z+R)}\right],$$

$$u_z = \frac{Q}{4\pi}\left[\frac{z^2}{\mu R^3} - \frac{\lambda+2\mu}{\mu(\lambda+\mu)R}\right]; \tag{7}$$

$$\sigma_{11} = -\frac{Q}{2\pi}\left\{\left[\frac{3x^2z}{R^5} - \frac{\mu z}{(\lambda+\mu)R^3}\right] - \frac{\mu}{\lambda+\mu}\left[\frac{y^2+z^2}{R^3(R+z)} - \frac{x^2}{R^2(R+z)^2}\right]\right\},$$

$$\sigma_{22} = -\frac{Q}{2\pi}\left\{\left[\frac{3y^2z}{R^5} - \frac{\mu z}{(\lambda+\mu)R^3}\right] - \frac{\mu}{\lambda+\mu}\left[\frac{x^2+z^2}{R^3(R+z)} - \frac{y^2}{R^2(R+z)^2}\right]\right\},$$

$$\sigma_{33} = -\frac{Q}{2\pi}\frac{3z^3}{R^5},$$

$$\sigma_{12} = -\frac{Q}{2\pi}\left\{\frac{3xyz}{R^5} - \frac{\mu xy(z+2R)}{(\lambda+\mu)R^3(R+z)^2}\right\},$$

$$\sigma_{23} = -\frac{Q}{2\pi}\frac{3yz^2}{R^5}, \qquad \sigma_{13} = -\frac{Q}{2\pi}\frac{3xz^2}{R^5}, \tag{8}$$

where $R = (x^2+y^2+z^2)^{1/2}$, $\vec{r} = (x,y,z)$ is a point in a Cartesian coordinate system $x'y'z'$. This system has an origin $O'(x_0, y_0, z_0)$ at the point of applying the force $\vec{Q}_1$, and its z'-axis is normal to the sphere surface at this point.

Matrix of transition from the old local coordinate system xyz, which is connected with the regular model drawn at Figure 1,a to the new one

$x'y'z'$, connected with the direction of the applied force $\vec{Q}_1$, is as follows:

$$A = \begin{pmatrix} \cos\theta\cos\varphi & -\sin\varphi & -\sin\theta\cos\varphi \\ \cos\theta\sin\varphi & \cos\varphi & -\sin\theta\sin\varphi \\ \sin\theta & 0 & \cos\theta \end{pmatrix},$$

where θ, φ are polar and azimuthal angles of the applied force. The new origin $O'(x_0, y_0, z_0)$ has the coordinates of the point of applying the force $\vec{Q}_1$:

$$x_0 = R_0 \sin\theta\cos\varphi,$$

$$y_0 = R_0 \sin\theta\sin\varphi,$$

$$z_0 = R_0 - R_0\cos\theta.$$

For the symmetric model $\varphi = 0$, and for the dissymmetric one $\varphi = \delta\varphi/2$.

The second derivatives of displacements $\vec{u}$ involved into left-hand side of equation (6) were computed by double differentiation of expressions (7) using the program of analitical transformations Derive. Then these derivatives were transformed into the original coordinate system xyz:

$$\partial\varepsilon_{kl}/\partial x_m = A^T_{kk'} A^T_{ll'} A^T_{mm'} \partial\varepsilon_{k'l'}/\partial x_{m'}.$$

The stresses σ_{ij} (8) also were transformed into the coordinate system xyz:

$$\sigma_{ij} = A^T_{ii'} A^T_{jj'} \sigma_{i'j'}.$$

The tensor $\boldsymbol{b}$ has been determined for the considered model twice.

In the first case, we supposed that the model is of group symmetry ∞. For such a medium, one has four equations (Obolentseva 1993, 1996)

$$\Delta\sigma_{13} = \Delta\sigma_{31} = \beta\,\partial\varepsilon_{33}/\partial x_1 + \gamma\,\partial\varepsilon_{33}/\partial x_2 + 2\delta\,\partial\varepsilon_{23}/\partial x_3,$$

$$\Delta\sigma_{23} = \Delta\sigma_{32} = \beta\,\partial\varepsilon_{33}/\partial x_2 - \gamma\,\partial\varepsilon_{33}/\partial x_1 - 2\delta\,\partial\varepsilon_{13}/\partial x_3,$$

$$\Delta\sigma_{11} = \Delta\sigma_{22} = \alpha\,\partial\varepsilon_{33}/\partial x_3,$$

$$\Delta\sigma_{33} = -\alpha\,(\partial\varepsilon_{11}/\partial x_3 + \partial\varepsilon_{22}/\partial x_3) - 2\beta\,(\partial\varepsilon_{13}/\partial x_1 + \partial\varepsilon_{23}/\partial x_2) + 2\gamma\,(\partial\varepsilon_{23}/\partial x_1 - \partial\varepsilon_{13}/\partial x_2)$$

to determine four gyration constants: $\alpha, \beta, \gamma, \delta$, see (3). The constant $\delta = b_{31323}$ is responsible for rotation of polarization plane. The constants $\alpha, \beta, \gamma, \delta$ have been determined using averaging of the stresses $\Delta\sigma_{ij}$ and the second derivatives of displacements $\vec{u}$ inside the grain. The following values have been found (in $\mathrm{N\,m^{-1}}$):

$$\alpha = 60.3, \quad \beta = 71.6, \quad \gamma = 26.5, \quad \delta = 8.57. \tag{9}$$

Then we supposed that the model belongs to the group symmetry $\infty\infty$, i.e., its gyrotropic properties are the same in any direction. This suggestion is the less probable one, however, it is worthy of consideration. In this case, according to (Obolentseva 1993, 1996),

$$\Delta\sigma_{ii} = b_{iiklm}\partial\varepsilon_{kl}/\partial x_m = 4\eta\ \partial\varphi_i/\partial x_i,$$

$$\Delta\sigma_{ij} = b_{ijklm}\partial\varepsilon_{kl}/\partial x_m = 2\eta\ (\partial\varphi_i/\partial x_j + \partial\varphi_j/\partial x_i),$$

where

$$\varphi_i = -1/2\ e_{ijk}\omega_{jk},$$

$$\omega_{jk} = 1/2\ (\partial u_j/\partial x_k - \partial u_k/\partial x_j),$$

$\boldsymbol{\varphi} = (\varphi_1, \varphi_2, \varphi_3)$ is an axial vector, a vector of small rotations, (e_{ijk}) is a unit antisymmetric tensor, and $\eta = b_{31323}$, see (2). The constant η was determined from the following formula for $\Delta\sigma_{33}$:

$$\Delta\sigma_{33} = 2\eta\ (\partial/\partial x_3)(\partial u_1/\partial x_2 - \partial u_2/\partial x_1).$$

The calculated value is

$$\eta = 2.05\ \mathrm{N\,m^{-1}}. \tag{10}$$

Comparison of the value η (10) with the values $\alpha, \beta, \gamma, \delta$ (9) shows that the value (10) is by order less than the values (9), and $\eta/\delta \approx 1/4$.

Note that the gyration constants have the dimension of force moment per a unit area, or of force per a unit length.

The work was supported by Russian Fund of Fundamental Investigations (RFFI) grant 94-05-16738.

REFERENCES

Chichinin, I. S. 1993. Wave equation for shear-wave gyrotropic propagation: physical standpoint. In I. R. Obolentseva (ed.), *Elastic waves in gyrotropic and anisotropic media*: 23-34. Novosibirsk: Nauka.

Kisel, V. A. & V. I. Burkov 1980. *Gyrotropy of crystals.* Moscow: Nauka.

Landau, L. D. & E. M. Lifshits 1992. *Electrodynamics of solids.* Moscow: Nauka.

Lurye, A. I. 1955. *Three-dimensional problems of elasticity theory.* Moscow: Gostekhizdat.

Obolentseva, I. R. 1996. On seismic gyrotropy. *Geophys. J. Int.* 124: 415-426.

Obolentseva, I. R. 1993. On symmetry properties of the gyration tensor, characterizing spatial dispersion of elastic properties. In I. R. Obolentseva (ed.), *Elastic waves in gyrotropic and anisotropic media*: 5-23. Novosibirsk: Nauka.

Obolentseva, I. R. 1992. Seismic gyrotropy. In I. S. Chichinin (ed.), *Investigations of seismic-waves propagation in anisotropic media*: 6-45. Novosibirsk: Nauka.

Rekach, V. G. 1977. *Handbook for solution of elasticity theory problems.* Moscow: Higher School.

Sirotin, Yu. I. & M. P. Shascolskaya 1979. *Fundamentals of crystallophysics.* Moscow: Nauka.

Modern Practice in Stress and Vibration Analysis, Gilchrist (ed.) © *1997 Balkema, Rotterdam, ISBN 90 5410 896 7*

Computerized algebra in the analysis of composite plates and shells

I.A.Jones
Department of Mechanical Engineering, University of Nottingham, UK

ABSTRACT: A significant obstacle to the extension of classical isotropic plate and shell solutions and finite element formulations to cope with generally orthotropic, laminated and shear deformable structures lies in the greatly increased mathematical complexity of the derivations involved. This is due to the proliferation of independent material constants and the appearance of additional terms which would otherwise cancel or simplify. Four case studies examine the application and difficulties of using computerized symbolic algebra to manage this complexity: (i) the derivation of a Roark-like catalogue of homogeneous orthotropic circular plate solutions; (ii) the laminated orthotropic extension of an existing isotropic axisymmetric shell finite element; (iii) the further extension of this element to consider shear-deformable shells via the approach of Soldatos, requiring manipulation of shape functions of unspecified form; and (iv) the derivation of an eighth-order partial differential equation governing a laminated monoclinic thin cylindrical shell, by treating the differential operators as multiplicative constants. The considerable effort involved in putting the results into publishable form is noted.

1 INTRODUCTION

A considerable amount of work has been carried out worldwide in recent years on the analysis of anisotropic and laminated plates and shells for stiffness, stress, natural frequencies and mode shapes. This has included both the extension of classical shell theory to cope with anisotropy and laminated construction, and the relaxation of the classical (Kirchhoff-Love) assumption of non-deformable normals following the approaches of Soldatos and Timarci (1993) and of Reddy and Lui (1985). A significant obstacle to the practical application of laminated shell theories is the complexity of the algebra involved, for example in the application of practical boundary conditions or the derivation of finite element formulations based upon these theories. A practical and effective solution to this problem lies in computerized symbolic algebra, yet only recently have publications appeared regularly on this application of the technique and these have often concentrated upon the results rather than the details of the processes used to obtain them.

The derivation of almost any plate or shell solution involves a governing differential equation derived from a set of strain-displacement (compatibility) assumptions and a set of equilibrium equations. These are linked via a plane-stress version of Hooke's law (also incorporating transverse shear stiffnesses for thick plate/shell models). For isotropic materials, the Young's modulus and Poisson's ratio are the only independent material constants, and for homogenous shells various terms cancel when through-thickness integrations are performed between the limits of $\pm t/2$. For composite shells lacking isotropy the equilibrium and compatibility equations are unchanged but there are now at least four and (for off-axis or monoclinic layers in a thick shell) up to nine independent material constants. This clearly means that terms will not simplify to the same extent; furthermore, the lack of symmetry of a general laminate means that many additional terms persist in the final solution rather than cancelling.

This increase in complexity often means that an error-free solution is almost impossible to achieve manually and in a reasonable timescale. Computerized symbolic algebra provides the opportunity to make these derivations feasible. While the technique eliminates the possibility of direct human algebraic errors, the task of accurately programming the derivation (via a command batch file) is often non-trivial; it requires not only a very clear understanding of the derivation procedure but

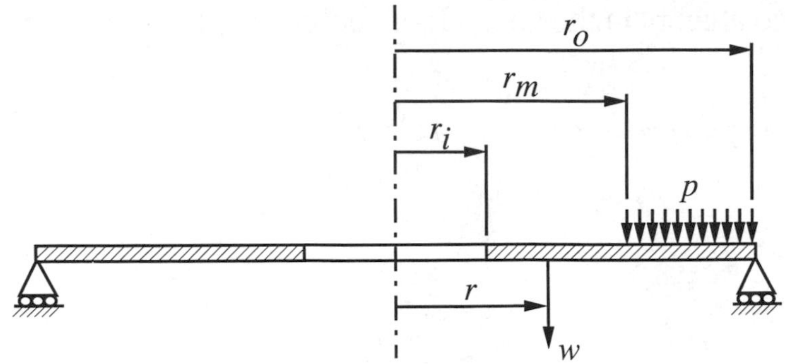

Figure 1. Typical loadcase for flat circular polar orthotropic plate.

often considerable ingenuity to enforce rules and identities not inherent in the computerized system.

Computerized symbolic algebra has been available for many years, during which time papers have been published rather infrequently on its application to composite plates and shells. For example, Wilkins (1973) described the static analysis of cylindrical shells using this technique, including a stage-by-stage commentary on the FORMAC coding of the solution method. Noor and Andersen (1979) surveyed the limited volume of literature on the technique's application to structural mechanics, including the vibration analysis of laminated composite elliptical plates. In addition to the advantages which included its reliability in evaluating integrals and derivatives, these authors identified several problems including the large size of intermediate expressions.

More recently, a greater number of relevant publications have appeared. Li *et al* (1995) describe work on cylindrical shells which utilised computerized symbolic algebra, and Han and Petyt (1996) mention its application to the vibration analysis of laminated rectangular plates, but neither paper gives many details of the technique's implementation. However, Argyris and Tenek (1996) present the formulation of a finite element for the study of temperature fields in laminated shells, and include in their publication the MACSYMA input file used in its derivation. Of particular relevance is a paper by Webber and Stewart (1991) who describe their use of MACSYMA to derive and solve the differential equations governing the buckling of sandwich panels, and also present the MACSYMA input file. Their paper shares with the fourth example in the present work the need to solve three simultaneous differential equations, although Argyris and Tenek used MACSYMA's own differential operators in contrast to the present author's use of multiplicative constants to represent them.

2 CASE STUDIES

This paper considers four plate/shell problems, most of which would have been difficult or impossible to solve without computational tools. The emphasis of the following descriptions is placed not on the algebra software itself nor on the results obtained (which will be published elsewhere) but on the derivation routes and the problems overcome in encoding the solution methods.

In all cases, the REDUCE system was used, together with the TAYLOR package for power series expansion and the RLFI postprocessor for generating output in LaTeX typesetting language (both supplied with REDUCE). REDUCE was initially chosen as the only system readily available to the author. It proved to be reasonably well suited to this kind of problem and has therefore been retained for the subsequent work. In common with other similar packages, it may be used interactively or using batch files of commands which for practical purposes constitute programs written in a Pascal-like high-level structured language.

2.1 *Polar Orthotropic Circular Plates*

This example relates not to laminated shells but to a range of homogeneous orthotropic circular plates, and was originally intended to form part of a Roark-like catalogue of solutions to assist in the design of composite components. It is presented as an example of a problem which would be messy and

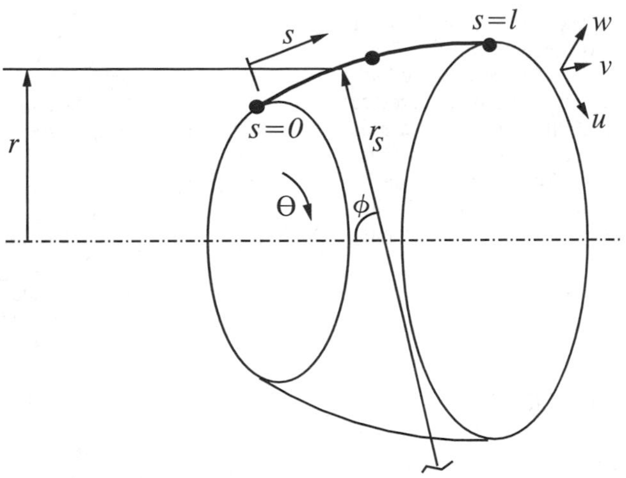

Figure 2. Axisymmetric shell element

tedious to solve manually but which involves few pitfalls in its solution using computerized symbolic algebra. The plate is assumed to be uniform and thin (i.e. shear-rigid) and to have radial and circumferential principal material directions. A variety of load cases were considered; a typical one is shown in Figure 1. The solution method (Threlfall, 1994) is broadly similar to that presented in standard texts e.g. that of Rees (1990) for isotropic problems, and leads to expressions such as that for the deflection w at radius r for the loaded region of the plate in Figure 1:

$$w = Fr^4 - Gr^2 + \frac{A_o r^{1+\beta}}{1+\beta} + \frac{B_o r^{1-\beta}}{1-\beta} + C_o \tag{1}$$

where F and G are constants dependent upon the pressure load and the plate's radial flexural rigidity, A_o, B_o and C_o are constants of integration which depend upon boundary conditions, and $\beta = E_\theta / E_r$ where E_θ and E_r are respectively the circumferential and radial moduli. The corresponding equation for the unloaded region omits the first two terms and contains three different constants A_i, B_i and C_i. In principle it is straightforward to find the constants from the boundary conditions; in practice this means enforcing continuity of deflection, slope and moment at radius r_m and imposing three further edge boundary conditions. This involves the solution of six linear simultaneous equations and is a trivial task for the SOLVE facility within REDUCE. The numerator and denominator of each of the six constants were typeset directly into LaTeX using the RLFI postprocessor, although much manual formatting was then required and some simplifications were possible. A total of 24 variations on loadcase and boundary conditions were examined. The resulting list of constants (notably the set of denominators) was rationalised to remove duplicate constants. The LaTeX code was then edited back to produce FORTRAN and REDUCE files, the latter being used for checking that the original differential equations and boundary conditions were satisfied. The results were presented in catalogue form (Jones and Threlfall, to be submitted).

2.2 *Laminated Orthotropic Axisymmetric Fourier Shell Element*

This example was actually the author's original application of computerized symbolic algebra to laminated shells, and involved extending an existing commercial implementation of a thin shell finite element based upon Flügge shell theory (Flügge, 1962) to cover laminated orthotropic shells.

The element derivation is based upon the following integration of strain energy over the element volume of an N-layered element:

$$U = \int_0^{2\pi}\int_0^{l}\sum_{k=1}^{N}\int_{z_{k-1}}^{z_k}\left[\frac{1}{2}\{\varepsilon\}^{\mathrm{T}}[\bar{Q}]_k\{\varepsilon\}\right] \times \frac{\sin\phi}{\bar{r}_s}(r_s+z)(r_\theta+z)\,dz\,ds\,d\theta \tag{2}$$

where $[\bar{Q}]_k$ is the orthotropic material stiffness

matrix of the kth layer, and where z is the through-thickness position, s and θ are the meridional and circumferential positions, r_s and r_θ are respectively the meridional and circumferential radii of curvature, and ϕ is the angle of the shell normal to the axis. The set of strains $\{\varepsilon\}$ for each harmonic order m are expressed in terms of the local circumferential, meridional and normal displacements and their derivatives using Flügge's strain-displacement relations:

$$\varepsilon_s = \frac{\partial v}{\partial s} + \frac{w}{r_s+z} - \frac{z}{r_s+z}\left(r_s \frac{\partial^2 w}{\partial s^2} + \frac{v}{r_s}\frac{\mathrm{d}r_s}{\mathrm{d}s} - \frac{\mathrm{d}r_s}{\mathrm{d}s}\frac{\partial w}{\partial s} \right)$$

$$\varepsilon_\theta = \frac{1}{r}\frac{\partial u}{\partial \theta} + \frac{v}{r_s}\cot\phi\left(\frac{r_s+z}{r_\theta+z}\right) + \frac{w}{r_\theta+z} - \frac{z}{r_\theta+z}\left(\frac{1}{r\sin\phi}\frac{\partial^2 w}{\partial \theta^2} + \cot\phi\frac{\partial w}{\partial s}\right)$$

$$\gamma_{s\theta} = \left(\frac{r_\theta+z}{r_s+z}\right)\left(\frac{r_s}{r_\theta}\frac{\partial u}{\partial s} - \frac{r_s}{r_\theta^2}\cot\phi\, u\right) + \left(\frac{r_s+z}{r_\theta+z}\right)\frac{1}{r_s\sin\phi}\frac{\partial v}{\partial \theta} - z\left(\frac{1}{r_\theta+z} + \frac{r_s}{r_\theta(r_s+z)}\right)\left(\frac{1}{\sin\phi}\frac{\partial^2 w}{\partial s\partial\theta} - \frac{\cot\phi}{r_\theta\sin\phi}\frac{\partial w}{\partial\theta}\right) \quad (3)(a\text{-}c)$$

where r is the radial position and u, v and w are the circumferential, meridional and normal displacements which vary sinusoidally with harmonic order m.

In practice, only the integrals through the shell thickness and around the circumference are evaluated explicitly, since the integral with respect to the meridional position s is evaluated numerically at run-time. The shell section modulus matrix for an N-layered laminate in this application

$$[D] = \sum_{k=1}^{N}\Big[[K]_k\Big]_{z_{k-1}}^{z_k} \quad (4)$$

is obtained by equating the strain energy calculated in equation 2 to the following:

$$U = \int_0^l \frac{1}{2}\{\varepsilon^0\}^{\mathrm{T}}\sum_{k=1}^{N}\Big[[K]_k\Big]_{z_{k-1}}^{z_k}\{\varepsilon^0\}\,\mathrm{d}s \quad (5)$$

where $\{\varepsilon^0\}$ is the vector of seven so-called "pseudostrains", which are the amplitudes of the sinusoidally-varying displacements u, v and w and their derivatives with respect to s.

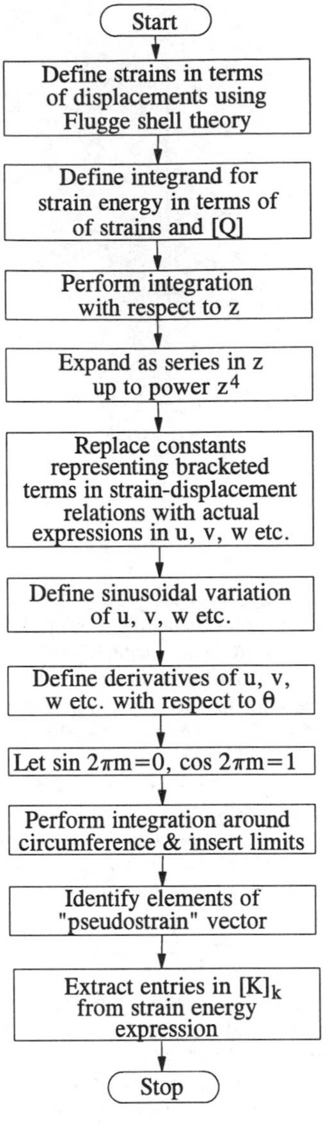

Figure 3. Derivation procedure for thin axisymmetric shell element

The derivation is given in detail in (Jones, 1996), closely follows that described in the literature (notably Henshell *et al*, 1972) for the existing isotropic element, and is illustrated by the flowchart in Figure 3. Flügge's strain-displacement expressions (equation 3) are entered into equation 2, the through-thickness integration is performed and the TAYLOR package is used to expand the result as a power series in the through-thickness co-ordinate

z. In order to optimise the derivation process, some bracketed terms in equation 3 were defined as constants which were not expanded until the series expansion had been performed. The circumferential integration was straightforward, but insertions of the integration limits involved the explicit definition of values of cos $2\pi m$ and sin $2\pi m$ ($m \in \mathbb{I}$) since these are not inherent in REDUCE. The result was an expression for strain energy per unit of meridional length s.

The individual elements of $[K]_k$ were extracted from the strain energy expression using a three-stage approach. Firstly, each of the leading diagonal terms K_{ii} was extracted directly as the coefficient of the second power of the ith pseudostrain. A subexpression was then found as the coefficient of the first power of each pseudostrain; the terms K_{ij} were then found by halving the coefficient of the jth pseudostrain in the ith subexpression.

While it was straightforward to output the elements of $[K]_k$ in the form of accurate and useful FORTRAN code, the grouping of the terms was somewhat verbose and unsuitable for publication in conventional mathematical notation. A further REDUCE program was therefore written to rearrange the results and the corresponding LaTeX code produced via the RLFI postprocessor was further simplified manually. This was then edited back to FORTRAN and REDUCE format for verification against the original solution.

Although the derivation process for this application is straightforward and requires few tricks to encode using REDUCE, earlier manual attempts at the derivation achieved little progress because of the very large amount of manipulation involved. Indeed, the impracticality of performing the task manually seems a plausible explanation of why this element was not previously extended to cover orthotropic laminates.

2.3 *Laminated thick axisymmetric shell element.*

Out of the four case studies, the greatest challenges in the application of computerized symbolic algebra to shells arose in the extension of the element described in Section 2.2 to take account of transverse shear deformations. The approach taken was that of Soldatos and Timarci (1993), whereby an additional displacement field representing shear deformations is superimposed onto the displacement field from a thin shell theory such as that of Flügge (1962). The displacements due to transverse shear (representing distortion of the shell normal) are extrapolated from the transverse shear strains u_1 and v_1 at the shell reference surface using functions $\psi_1(z)$ and $\psi_2(z)$. These are left undefined except that their first derivatives are unity at z=0 and (if the condition of zero free-surface shear stresses is enforced) zero at $z=\pm t/2$. In order to satisfy continuity of displacements and interlaminar shear stresses, functions such as piecewise cubics or piecewise hyperbolic functions are chosen. The resulting in-surface strain field was derived manually from the displacement field:

$$\begin{aligned}
\varepsilon_s &= \varepsilon_s^c + \frac{r_s}{r_s + z}\psi_1(z)\frac{\partial v_1}{\partial s} \\
\varepsilon_\theta &= \varepsilon_\theta^c + \frac{1}{(r_\theta + z)\sin\phi}\psi_2(z)\frac{\partial u_1}{\partial \theta} \\
&+ \frac{1}{r_\theta + z}\psi_1(z)\, v_1 \cot\phi \\
\gamma_{s\theta} &= \gamma_{s\theta}^c + \frac{r_s}{r_s + z}\psi_2(z)\frac{\partial u_1}{\partial s} \\
&- \frac{1}{r_\theta + z}\psi_2(z)\, u_1 \cot\phi \\
&+ \frac{1}{(r_\theta + z)\sin\phi}\psi_1(z)\frac{\partial v_1}{\partial \theta}
\end{aligned} \qquad \begin{matrix}(6)\\ \text{(a-c)}\end{matrix}$$

where ε_s^c etc. are given by Flügge's classical strain-displacement relations (equation 3). Furthermore, by differentiating the displacement field with respect to z, the following expressions for transverse shear strain were obtained:

$$\gamma_{sz} = v_1\frac{\partial \psi_1}{\partial z} \quad (a) \qquad \gamma_{\theta z} = u_1\frac{\partial \psi_2}{\partial z} \quad (b) \qquad (7)$$

The $[K]_k$ matrix used in evaluating the element's shell section modulus matrix was derived using an extension of the approach used in section 2.2, although some changes were made to the derivation sequence in order to minimise the processing time and memory required by REDUCE to perform the derivation. The derivation procedure is illustrated in Figure 3.

The particular challenge of the work lay in the need to manipulate $\psi_1(z)$ and $\psi_2(z)$, together with their squares and products, without ever specifying the actual form of these functions. Moreover, explicit through-thickness integration was retained in order to allow evaluation of the various options for $\psi_1(z)$ and $\psi_2(z)$ without any unwanted influence of the choice of numerical integration method. The following "tricks" were therefore employed in order to perform the required derivation.

1. In order to prevent the expansion of $\psi_1(z)$ and $\psi_2(z)$ as terms in the power series, REDUCE was not informed of the dependency of these functions upon z until the series expansion of their multiplicands had taken place. In practice, the definition of the

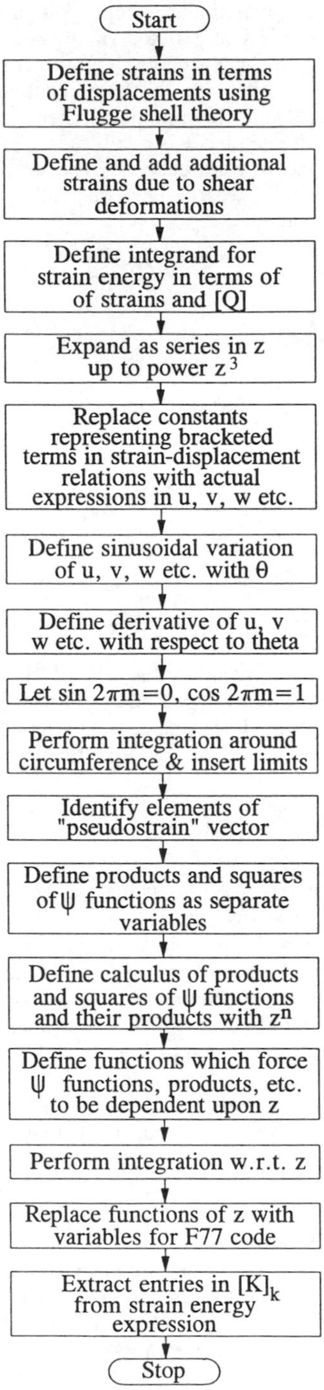

Figure 4. Derivation procedure for thick axisymmetric shell element

functions representing $\psi_1(z)$ and $\psi_2(z)$ (along with $\psi_1(z)\psi_2(z)$, $\psi_1'(z)\psi_2'(z)$ etc.) did not take place until much later, following the circumferential integration and the identification of the squares and products described below.

2. Squares and products such as $[\psi_1(z)]^2$, $\psi_1(z)\psi_2(z)$, $[\psi_1'(z)]^2$ etc. were separated out, noting that at this stage each of $\psi_1(z)$, $\psi_1'(z)$ etc. was represented as a variable not a function. Each of these products was later redefined as a function of z in its own right.

3. Multiple integrals (ignoring constants of integration) and the first derivatives of $\psi_1(z)$, $\psi_2(z)$ and the products identified in stage 2 were each defined as functions of z, and the integral and differential relationships between these were explicitly defined. Furthermore, integrals of quantities such as $z^n\psi_1(z)$, $z^n\psi_1(z)\psi_2(z)$ etc. (n=1,2...4) were each defined in terms of the multiple integrals defined above. The definitions were obtained manually by recursive use of integration by parts, leading to identities such as:

$$\int_a^b z^2[\psi_1(z)]^2 \mathrm{d}z = \left[z^2\int[\psi_1(z)]^2 dz\right]_a^b - \left[2z\int\left(\int[\psi_1(z)]^2 dz\right)dz\right]_a^b + 2\int_a^b\left(\int\left\{\int[\psi_1(z)]^2 dz\right\}dz\right)dz \tag{8}$$

It was verified that the constants of integration vanish either from all the expressions such as equation 8 or from the overall solution for the shell section modulus matrix [D] (equation 4) and hence their omission from the integrals is valid in this application.

4. Having established the rules governing the shape functions and their squares and products with each other and with z, it is straightforward to perform the through-thickness integration (i.e. that with respect to z). The earlier use of the series expansion is of influence here: by converting all multiplicands of $\psi_1(z)$, $\psi_2(z)$ etc. into power series in z, it becomes possible to carry out manipulations such as integration of the strain energy expression, having specified only the rules described in stage 3 and without prescribing the actual form of $\psi_1(z)$ and $\psi_2(z)$.

5. In practice, the result of the integration involves functions of z which include $\psi_1(z)$, $\psi_2(z)$, their products and various multiple integrals and first derivatives. In order to output these in machine-readable form (FORTRAN code), it was necessary to assign each of these to a simple variable. Each of

these variables is evaluated numerically within the finite element implementation at the limits (z_{k-1} and z_k) of the definite integral corresponding to the top and bottom of the kth layer, prior to the variable's use within the machine-generated FORTRAN code.

In order to allow implementation of the thick shell element, various modifications to the original element were carried out by Pafec Limited to include the shear deformations as additional degrees of freedom and hence allow incorporation of the new, larger $[K]_k$ matrix. It is also worthy of mention that a shell inertia matrix $[M]$ was derived using a very similar approach to that used for $[D]$, enabling the element to be used for natural frequency calculations. Excellent agreement with alternative FE and experimental results was obtained (Jones *et al*, 1996).

2.4 *Differential Equation for Laminated Monoclinic Cylinder*

A different set of challenges lay in the derivation of the eighth-order governing differential equation for a laminated monoclinic cylindrical shell, enabling a pinched-cylinder benchmark solution for orthotropic shells to be obtained. This derivation was based upon a method presented by Schwaighofer and Microys (1979) and took as its starting point Flügge's strain-displacement relations for a cylindrical shell (Flügge, 1962). These expressions involved partial derivatives with respect to meridional and circumferential positions. Flügge's equilibrium equations included more differentiations to obtain three simultaneous differential equations containing partial derivatives of axial, circumferential and normal displacements u, v and w and the distributed loads p_x, p_θ and p_r in the axial, circumferential and normal directions.

One problem encountered related to the through-thickness integrals for force and moment resultants, for example for the axial force per unit of shell width N_x for a shell of radius a:

$$N_x = \int_{-t/2}^{t/2} \sigma_x \left(1 + \frac{z}{a}\right) dz \tag{9}$$

Performing such an integration literally would have resulted in the neglection of shell coupling terms such as that for B_{11}:

$$B_{11} = \int_{-t/2}^{t/2} \bar{Q}_{11} z \, dz \tag{10}$$

since $\bar{Q}_{11}$ etc. vary with position through the laminate thickness. The approach actually taken was to omit from the derivation the explicit through-thickness integration of stresses and stress couples, and instead to pick out of the integrands for N_x etc. the integrands for shell stiffness terms such as B_{11}.

The facilities within REDUCE for dealing with partial derivatives were judged not to be appropriate for this problem, so in contrast to the approach of Webber and Stewart (1991), the solution of the three simultaneous differential equations was achieved by treating the differential operators $a\partial/\partial x$ and $\partial/\partial\theta$ throughout the whole problem as multiplicative constants (termed `DASH` and `DOT`, following the shorthand notation ()′ and ()˙ used by Schwaighofer and Microys for these differential operators). It was then noted that the three differential equations can be represented in the form:

$$\begin{aligned} \Delta_{11}u + \Delta_{12}v + \Delta_{13}w + p_x a^2 &= 0 \\ \Delta_{12}u + \Delta_{22}v + \Delta_{23}w + p_\theta a^2 &= 0 \\ \Delta_{13}u + \Delta_{23}v + \Delta_{33}w - p_r a^2 &= 0 \end{aligned} \tag{11}$$

where the differential operators Δ_{ij} are represented as rather complicated expressions containing various powers of `DASH` and `DOT`. It is then a relatively straightforward matter of linear algebra to obtain an eighth-order partial differential equation from which u and v had been eliminated, containing only w and the load terms and their derivatives. The coefficients of the individual differential operators on w were found by picking out the coefficients of the various powers of `DASH` and `DOT`. The retention of tension-shear coupling terms (i.e. the additional terms in off-axis or monoclinic behaviour) resulted in the appearance of odd derivatives in the eighth-order equation as well as the even ones in the much simpler equation for homogeneous orthotropic cylinders presented by Schwaighofer and Microys (1979). For the special case of a laminated cylinder with specially orthotropic properties (i.e. a cross-ply cylinder) the differential equation was solved manually using an approach similar to that of Yuan and Ting (1957) to give a variety of benchmark solutions (Jones, to be submitted). Virtually identical results for test problems were obtained as from the element described in Example 2.

3. DISCUSSION AND CONCLUSIONS

The problems described have covered the range from the very straightforward (if mathematically messy) to the rather complex. While no originality is claimed for the application of computerized symbolic algebra to this kind of problem, it has been demonstrated that a good deal of ingenuity is sometimes necessary to achieve the desired results. In addition to the actual results obtained from these analysis (which are presented in other publications, and are generally too voluminous to summarise here), the following conclusions can be drawn.

1. The use of computerized symbolic algebra can make feasible the algebraic derivation of plate and shell solutions which would be extremely tedious or quite impractical to achieve manually. Out of the four examples described, the author believes that only the first could reasonably be attempted by a patient and methodical analyst, the second and fourth examples would at best be exceedingly difficult and tedious to undertake manually and the third would probably be quite impractical to achieve without computer assistance. Thus the method makes it possible to obtain results such as finite element formulations and analytical benchmark solutions which would not otherwise be available.

2. Notwithstanding the usefulness of computer algebra in this application, the difficulty of presenting the output is worthy of mention. The machine-readable output (e.g. FORTRAN) can generally be used directly, and the facilities exist (via the RLFI postprocessor) to generate typesetting output (LaTeX code) directly. However, the resulting output often require considerable manipulation (either manually or by judicious use of commands within REDUCE prior to post-processing) and reformatting to make them acceptable for publication using reasonably concise mathematical notation. The author's experience is that this stage is generally more time-consuming than the actual generation of the results in machine-readable form.

ACKNOWLEDGEMENTS

The author wishes to thank Dr R.D. Henshell and Dr J. Platt (Pafec Limited) for their advice and practical assistance in implementing the orthotropic shell elements.

REFERENCES

Argyris, J. & L. Tenek 1996. Steady-state nonlinear heat transfer in stiffened composite plates and shells by exactly integrated facet triangular element. *Comput. Meths. App. Mech. Engng.* 129: 53-79.

Henshell, R.D., T. J. Bond & J.O. Makoju 1972. Ring finite elements for axisymmetric and non axisymmetric thin shell analysis. *Proc. Conf. Appl. Theor. Struct. Dyn., Southampton, 1972*: 4/44 - 4/58.

Elishakoff, I. & J. Tang 1988. Buckling of polar orthotropic circular plates on elastic foundation by computerized symbolic algebra. *Comput. Meths. App. Mech. Engng.* 68: 229-247.

Flügge, W. 1962. *Stresses in Shells* (2nd ed.). Berlin: Springer.

Han, W. & M. Petyt 1996. Linear vibration analysis of laminated rectangular plates using the hierarchical method – I. Free vibration analysis. *Comput. Struct.* 61(4): 705-712.

Jones, I.A. 1996. A curved laminated orthotropic axisymmetric element based upon Flügge thin shell theory. *Comput. Struct.* 60(3): 487-503.

Jones, I.A. (to be submitted). Flügge shell equations and solution for laminated composite cylindrical shells under pinching loads.

Jones, I.A., E.J. Williams & A. Messina 1996. Theoretical, experimental and finite element modelling of laminated composite shells. *Proc. DTA/NAFEMS/SECED Conf. Structural Dynamics Modelling, Cumbria, UK, 3-5 July 1996*: 267-279.

Jones, I.A. & A.L. Threlfall (to be submitted). Review and catalogue of solutions for flat circular polar orthotropic plates under transverse loads.

Li, Y.W., I. Elishakoff & J.H. Starnes Jr. 1995. Axial buckling of composite cylindrical shells with periodic thickness variations. *Comput. Struct.* 56(1): 65-74.

Noor, A.K & C.M Andersen 1979. Computerized symbolic manipulation in structural mechanics – progress and potential. *Comput. Struct.* 10: 95-118.

Reddy, J.N. & C.F. Lui 1985. A higher-order shear deformable theory of laminated elastic shells. *Int. J. Engng. Sci.* 23(3): 319-330.

Rees, D.W.A 1990. *Mechanics of Solids and Structures*. London: McGraw-Hill, 670-680.

Schwaighofer, J. & H.F. Microys 1979. Orthotropic cylindrical shells under line load. *J. App. Mech.* 46: 356-362.

Soldatos, K.P. & T. Timarci 1993. A unified formulation of laminated composite, shear deformable, five-degrees-of-freedom cylindrical shell theories. *Compos. Struct.* 25: 165-171.

Threlfall, A.L. 1994. *Database of Solutions to Orthotropic Problems.* University of Nottingham unpublished report.

Webber, J.P.H. & I.B. Stewart 1991. A theoretical solution for the buckling of sandwich panels with laminated face plate using a computer algebra system. *Comput. Meths. App. Mech. Engng.* 92: 325-341.

Wilkins, D.J. Jr. 1973. Applications of a symbolic algebra manipulation language for composite structures analysis. *Comput. Struct.* 3: 801-807.

Yuan, S.W. & L. Ting 1957. On radial deflections of a cylinder subjected to equal and opposite concentrated radial loads – infinitely long cylinder and finite-length cylinder with simply supported ends. *J. App. Mech.* 24: 278-282.

Modern Practice in Stress and Vibration Analysis, Gilchrist (ed.) © 1997 Balkema, Rotterdam, ISBN 90 5410 896 7

An inverse problem for identifying mechanical characteristics of composite materials

V. P. Matveyenko & N. A. Jurlova
Institute of Continuous Media Mechanics, Department of Russian Academy of Science, Perm, Russia

ABSTRACT: A numerical - experimental method is proposed to identify the effective mechanical constants of composite materials in shell structures. As experimental data, the results obtained by analysing the strain state of a material structure under different static loading conditions or the spectrum of natural frequencies and vibration modes are used. Numerical part of the method is based on the solution to the inverse elasticity problem involving the estimation of the model parameters to describe the strain state recorded experimentally. The mechanical behaviour of a material is considered in the framework of the models of elastic anisotropic body or with complex dynamic moduli in the case of vibrations. In this treatment, concurrent with the solution to the inverse problem, the sensitivity analysis techniques are applied to choose, for example, the required experiments, and the wanted data from a particular experiment. The potential of the method proposed is illustrated numerically.

1 INTRODUCTION

A distinguishing feature of fiber-reinforced composites is simultaneous production of the material and structure by continuous filament framework winding. As a result, the material structure is able to vary from point to point which implies that the properties of the structure material depend on the manufacturing technology, the size of an article, the fiber disorientation, the framework bending and tension, and the initial micro-and macroscopic stresses. Therefore, the standard test data on the specimens with a fixed structure or cut from an article workpiece may provide an inadequate assessment of the material behaviour within a structure.

One way to define the effective mechanical characteristics without recourse to a reference specimen is to use an experimental evidence for shells under load. This approach is particularly attractive for unique shell structures under non-destructive mechanical tests.

In this work, we develop further the above approach by constructing the algorithm for shells of complex geometry under different boundary conditions. The information required for identification of the mechanical characteristics can be gained from the experimental data on different static loading tests and on resonance regimes (the spectrum of natural frequencies and vibration modes). The algorithm should be also capable to estimate an informative value of the experimental evidence obtained.

2 PROBLEM FORMULATION

We use here the concept of the inverse problem suggesting that the equation coefficients are determined by the vector

$$a_p = \left(a_{p1},\ a_{p2},\ \ldots, a_{pq}\right)^{\mathrm{T}}, \tag{1}$$

where q is the number of the parameters to be found, in particular, the mechanical characteristics which provide a mathematical model for the structure strain state best fitted the experimental observations.

Assume that, for the structure under kth load, we have the data on the displacement $u_i^{эk}$, strains $\varepsilon_{ij}^{эk}$, and resonance frequencies $\omega_n^{э}$ ($n=1$, 2, ...). Here, the superscripts э and κ designate the experimental results at kth loading.

For numerical implementation, a parametrized formulation of the inverse problem is suggested, where the vector of coefficients of the state equations $\overline{a}_0$, providing within the assumed norm a minimal distance between the calculated $u_i^{rk}, \varepsilon_{ij}^{rk}, \omega_n^r$ and experimental $u_i^{эk}, \varepsilon_{ij}^{эk}, \omega_n^э$ results, must be found. As a norm, we take the following functional:

$$F(a_p) = \sum_{k=1}^{K} \int_V \left[\sum_{i=1}^{3} \alpha_i^k \left(u_i^{эk} - u_i^{rk}\right)^2 + \right. \qquad (2)$$

$$\left. + \sum_{i=1}^{3} \sum_{j=1}^{3} \beta_{ij}^k \left(\varepsilon_{ij}^{эk} - \varepsilon_{ij}^{rk}\right)^2 \right] dV + \sum_{n=1}^{N} \gamma_n \left(\omega_n^э - \omega_n^r\right)^2,$$

where N is the number of natural frequencies known from experiments, and $\alpha_i^k, \beta_{ij}^k, \gamma_n$ are the weight coefficients.

So, we arrive at the functional minimum problem (2)

$$F(a_0) = \min F(a_p), \qquad (3)$$

provided that the vectors a_p are in the range of admissible values.

As a rule, the experimental data on the displacement and deformation fields may be obtained at separate points of the region V occupied by the body considered. In this case, functional (2) becomes the function of several variables :

$$F(a_p) = \sum_{k=1}^{K} \left\{ \sum_{m=1}^{M} \sum_{i=1}^{3} \alpha_i^k \left[u_i^{эk}(x_m) - u_i^{rk}(x_m)\right]^2 + \right.$$

$$\left. + \sum_{l=1}^{L} \sum_{i=1}^{3} \sum_{j=1}^{3} \beta_{ij}^k \left[\varepsilon_{ij}^{эk}(x_l) - \varepsilon_{ij}^{rk}(x_l)\right]^2 \right\} + \qquad (4)$$

$$+ \sum_{n=1}^{N} \gamma_n \left(\omega_n^э - \omega_n^r\right)^2,$$

and the functional minimum problem reduces to searching for the minimum of the function of several variables. Here, (x_m) and (x_l) are the points for which the experimental data on displacements and deformations are known.

In the general case, oriented composite materials are anisotropic non-linear viscoelastic materials which, are, however, mostly exploited at temperature below the glass transition temperature. Therefore, in the first approximation these materials can be considered elastic solids, particularly, at small strains, and their mechanical behaviour can be described by the model of anisotropic elastic body with elastic constants

$$\sigma_{ij} = C_{ijkl} \varepsilon_{kl}. \qquad (5)$$

Dissipative properties of the material undergoing dynamic deformation are taken into account in the context of the model of visco-elastic body which for isotropic case can be written as

$$\sigma_{ij} - \sigma\delta_{ij} =$$

$$= 2G_0 \left\{ \varepsilon_{ij} - \frac{1}{3}\vartheta\delta_{ij} - \int_0^t R(t-\tau)\left[\varepsilon_{ij}(\tau) - \frac{1}{3}\vartheta(\tau)\delta_{ij}\right] d\tau \right\}, \qquad (6)$$

where

$$\sigma = K_0 \left[\vartheta - \int_0^t T(t-\tau)\vartheta(\tau) d\tau \right].$$

When the body is subjected to dynamic loading, the use can be made of the material model, where the dissipative properties are described by the complex dynamic moduli.

For isotropic body, the physical equations have the form

$$\sigma_{ij} = K\varepsilon_{ij}\delta_{ij} + 2\overline{G}\left(\varepsilon_{ij} - \frac{1}{3}\vartheta\delta_{ij}\right), \qquad (7)$$

where $\overline{G}$, $\overline{K}$ are the complex moduli:

$$\overline{G} = G_0[1 - \Gamma_c(\omega) - i\Gamma_s(\omega)],$$

$$\overline{K} = K_0[1 - K_c(\omega) - iK_s(\omega)],$$

$$\Gamma_c(\omega) = \int_0^\infty R(\tau)\cos\omega\tau\, d\tau,$$

$$\Gamma_s(\omega) = \int_0^\infty R(\tau)\sin\omega\tau\, d\tau,$$

$$K_c(\omega) = \int_0^\infty T(\tau)\cos\omega\tau\, d\tau,$$

$$K_s(\omega) = \int_0^\infty T(\tau)\sin\omega\tau\, d\tau,$$

where R, T are the relaxation kernels.

For anisotropic viscoelastic body, the physical relations (V.P. Matveyenko and E.P. Kligman , 1997) are

$$\sigma_{ij} = \overline{C}_{ijkl}\varepsilon_{kl}(t), \qquad (8)$$

where $\overline{C}_{ijkl}$ is the complex tensor of mechanical constants.

The formulated inverse problem is numerically implemented based on the results from the solution to direct problems on static or dynamic deformation. The last, in turn, is related to consideration of the following vibration problems:

1. *natural vibrations of elastic bodies*

Under uniform boundary conditions, we find the solution of the form

$$u_i(x,t) = \xi_i(x) \cos \omega t, \qquad (9)$$

where ω is the natural frequency, and $\xi_i(x)$ is the vibration eigenform .

2. *natural damping vibrations of viscoelastic bodies*

Damping vibrations of a viscoelastic body can be represented as

$$u_i(x,t) = \\ = e^{-\omega_I t}\left[u_i^c(x) \cos \omega_R t + i u_i^s(x) \sin \omega_R t\right] \qquad (10)$$

or in a complex form

$$u_i(x,t) = \bar{u}_i(x) e^{-i\omega t}, \qquad (11)$$

where $\bar{u}_i(x)$ are the complex components for the displacement vector amplitudes, and $\omega = \omega_R + i\omega_I$ is the complex frequency with the actual part being the frequency and with the imagine part being the damping ratio of natural damping vibrations.

In this case, under uniform boundary conditions, we find the solution of the form

$$u_i(x,t) = \bar{\xi}_i(x) e^{-i\omega t}, \qquad (12)$$

where $\bar{\xi}_i(x)$ is the complex eigenform for shell vibrations.

3. *steady-state forced vibrations*

We find here a periodic in time motion with the period equal the period of external action

$$u_i(x,t) = \bar{u}_i(x) e^{-ipt} \qquad (13)$$

When analysing shell structures, we use the relationships from the momentum theory of shells based on the Kirchgoff-Lyava hypotheses (V.L. Biderman , 1977).

Our study is concentrated on the shells of revolution. A numerical analysis represents a FEM semi-analytical variant (O. Zienkevich, 1975), according to which the components of displacement and load vectors, and the stress and strain tensors are given in Fourier series along the circumferential co-ordinate. As finite elements, the element suggested by Zienkevich has been used being a truncated cone with linear (axial and circumferential) and cubic approximations of normal components of the displacement vector for each harmonic of Fourier series.

The formulation proposed reduces to a classical problem of non-linear mathematical programming suggesting the minimisation of function (4) subject to constrained equalities and inequalities. The last define the range of admissible values of the vector a_p and result from the known constraints imposed on the anisotropic material parameters (S.S. Abramchuk and V.P. Buldakov, 1979.).

In deciding on a particular method to solve the problem on non-linear mathematical programming (the optimisation method), we take into account, among other factors, the error sensitivity of the optimisation methods (e.g. measurement errors). The methods (of simplex search type or barycentric co-ordinate method type), in which a nonlocal approximation of the objective function is constructed by its values at several points, proved to be least sensitive to measurement errors.

For numerical realisation of the optimisation problem with constraints, we use the simplex method of a sliding tolerance based on the Nelder-Mid method (D. Himmelblau, 1975).

The sensitivity analysis techniques (E.G. Houg et al, 1988) is applied to predict an informative value of the experimental data to identify the effective material constants and to estimate the possibility of defining any material constants from the same experiment by changing the form of the objective function (for example, on inserting the weight ratios),

When structures undergo static deformation, the finite element method reduces the problem considered to the following algebraic analogue:

$$[K]\{u\} = \{F\}, \qquad (14)$$

where $[K]$ is the stiffness matrix, $\{F\}$ is the vector of the external loads, and $\{u\}$ is the displacement vector.

If the vector $a_p = \left(a_{p1},\ a_{p2},\ \ldots, a_{pq}\right)^{\mathrm{T}}$ is taken as the design variables, the global stiffness matrix and the external load vector are then the functions of these design variables

$$[K]=[K(a_p)],\ \{F\}=\{F(a_p)\}. \qquad (15)$$

It is evident that the solution also depends on these variables

$$\{u\}=\{u(a_p)\}. \qquad (16)$$

In the problems on structure design, some objective function is minimised or maximised by satisfying the constraints imposed on the design variables.

Let us consider a generalised function

$$\Psi=\Psi(a_p,\ u(a_p)). \qquad (17)$$

Here, the objective function is determined by relations (4), the constraints are constants, and no consideration is given to them in the sensitivity analysis.

The sensitivity analysis is aimed at derivation of the full dependence of this function upon the design variables, that is at calculation of $d\Psi/da_p$.

Suppose that all values involved in $[K(a_p)]$ and $\{F(a_p)\}$ are s - times differentiable with respect to the design variables. Thus, according to the implicit function theorem, the solution to equation (14) are also s-times continuously differentiated.

The full derivative of the Ψ function with respect to a_p has been calculated by two methods (E. G. Houg et al, 1988). From the complex function differentiation rule, we obtain

$$\frac{d\Psi}{da_p}=\frac{\partial\Psi}{\partial a_p}+ \qquad (18)$$
$$+\frac{\partial\Psi}{\partial u}K^{-1}(a_p)\left[\frac{\partial}{\partial a_p}\{F(a_p)-K(a_p)\tilde{u}\}\right].$$

The second method of calculating the derivatives is based on the determination of a conjugate variable λ

$$\lambda\equiv\left[\frac{\partial\Psi}{\partial u}K^{-1}(a_p)\right]^T=K^{-1}(a_p)\frac{\partial\Psi^T}{\partial u}, \qquad (19)$$

in which case the sensitivity derivatives are found from

$$\frac{d\Psi}{da_p}=\frac{\partial\Psi}{\partial a_p}+$$
$$+\frac{\partial}{\partial a_p}\left[\tilde{\lambda}^T F(a_p)-\tilde{\lambda}^T K(a_p)\tilde{u}\right], \qquad (20)$$

where the symbol ~ designates that this parameter will be constant during differentiation.

Some computational problems on the sensitivity derivatives have been considered earlier (E. G. Houg et al, 1988 and G. S. Aurora et al, 1979).

The finite-element procedure, applied to calculate the natural vibrations, yields the following algebraic eigenvalue problem :

$$([K(a_p)]-\omega[M(a_p)])\{\xi\}=0, \qquad (21)$$

where the eigenvector $\{\xi\}$ is normalised as

$$\{\xi\}^T[M(a_p)]\{\xi\}=1. \qquad (22)$$

E. G. Houg and colleagues (1988) have proved the theorem suggesting that if the symmetric positively defined matrices $[K(a_p)]$ and $[M(a_p)]$ are continuously differentiable with respect to the design variables, and the eigenvalue ω is not multiple, then the eigenvalue and the eigenvector are also continuously differentiable with respect to the design variables. Reduce equation (21) to

$$\tilde{\xi}^T K(b)\tilde{\xi}=\omega\,\tilde{\xi}^T M(b)\tilde{\xi}. \qquad (23)$$

Hence, we can find the required expression for sensitivity derivatives

$$\frac{d\omega}{da_p}= \qquad (24)$$
$$=\frac{\partial}{\partial a_p}\left[\tilde{\xi}^T K(a_p)\tilde{\xi}\right]-\omega\frac{\partial}{\partial a_p}\left[\tilde{\xi}^T M(a_p)\tilde{\xi}\right].$$

When the variables characterising the eigenforms are taken as the state parameters, the dimensionality of the system identification problem increases significantly even for a small number of eigenforms (D.K. Kemmer et al, 1990).

3 RESULTS

3.1 *Determination of elastic constants using static deformation data for shells*

Numerical calculations have been done for cylindrical, cone and semispherical orthotropic shells under different boundary conditions on the shell edges. Static load was produced by twisting or tensile forces, and by the internal pressure.

In the numerical analysis made, instead of the full-scale experiments we have used the numerical and analytical results obtained at the prescribed values of mechanical characteristics. The calculated displacements and strains simulated the measurement data from the appropriate experiment, whereas the values of mechanical characteristics served as a reliability criterion for the inverse problem.

The calculation procedure involves the preliminary analysis of sensitivity coefficients. As would be expected, when the considered shells are loaded by twisting forces, only the sensitivity vector component $\partial\Psi/\partial G_{12}$ differs from zero. Thus, in solving the inverse problem such data are expected to provide the way of finding only the modulus G_{12}, as was calculated.

For the analysis, it is more convenient to use a normalised sensitivity vector

$$l_i^n = l_i \Big/ \sqrt{\sum_{i=1}^{4} l_i^2}\,, \tag{25}$$

where

$$l_1 = \frac{\partial\Psi}{\partial E_1},\quad l_2 = \frac{\partial\Psi}{\partial E_2},\quad l_3 = \frac{\partial\Psi}{\partial \nu_2},\quad l_4 = \frac{\partial\Psi}{\partial G_{12}}.$$

In this treatment, we determine the sensitivity to the four elastic constants for the case when the shell is made of orthotropic material. At each shell point all three principal directions of the material elasticity coincide with the directions of the corresponding co-ordinate lines. That is, we consider here the case of properly orthotropic shells (S.A. Ambartsumjan, 1974). In orthotropic shells, at each layer point one of the elastic symmetry planes is parallel to the midsurface of a shell, and the rest two are normal to the co-ordinate lines coinciding with the main curvature lines for the shell midsurface. For the anisotropy pattern considered, the relation $E_1\nu_2 = E_2\nu_1$ (S.G. Lehnitskii, 1977) is fulfilled.

Table 1 provides the sensitivity coefficients for a cylindrical shell of dimensions $L/R = 4$, $h/R = 0.02$ with the following values of elastic constants :

$$E_1/G_{12} = 5.0,\quad E_2/G_{12} = 2.3,\quad \nu_2 = 0.31.$$

Here, L, R, h are the length, radius, and the shell thickness, respectively. Calculations have been made for different values of the initial approximations in the function minimum problem (4), namely

1. - $E_1/G_{12} = 10.000\quad E_2/G_{12} = 3.053,\quad \nu_2 = 0.28$
2. - $E_1/G_{12} = 7.0\quad E_2/G_{12} = 2.35,\quad \nu_2 = 0.251$
3. - $E_1/G_{12} = 1.875\quad E_2/G_{12} = 1.25,\quad \nu_2 = 0.42$

The following patterns of shell loading were considered:

a) one shell end is fixed and the other is acted upon by the tensile force;

b) the shell is subjected to internal pressure, and no stresses are observed on its edges;

c) the shell is subjected to internal pressure, and its edges are fixed.

Table 1

Loading pattern	Initial condit.	l_1	l_2	l_3
	1	-0.764	0.104	-0.637
a	2	0.728	0.098	-0.678
	3	-0.993	-0.06	0.101
	1	0.000	0.231	-0.973
b	2	0.000	0.131	-0.991
	3	0.000	-0.512	0.859
	1	0.014	0.089	0.996
c	2	0.021	-0.437	-0.899
	3	-0.109	-0.293	-0.950

By comparing the sensitivity coefficients, we can estimate the possibility of seeking the corresponding elastic constants. For example, the sensitivity coefficient l_2, obtained when the shell is loaded according to the pattern a, is small as compared with the rest coefficients. From appropriate calculations followed that when the elastic constants were found based on the shell tensile test data, the values E_1 and ν_2 were obtained to be sufficiently accurate, and the value E_2 could be found, when a proper initial approximation was chosen.

In this problem, the sensitivity analysis techniques can be applied to find the most informative experiments to identify elastic constants. Consider, as an example, the shell with the stress free ends subjected to internal pressure. When this shell is reinforced at its midpart by the external rigid ring, the sensitivity coefficients (Table 2) prove to be more commensurable. The initial approximations are assumed the same as in the previous case.

Table 2

Initial conditions	l_1	l_2	l_3
Internal pressure			
1	0.000	0.231	-0.973
2	0.000	0.131	-0.991
3	0.000	-0.512	0.859
Internal pressure and the rigid ring frame at the shell midpart			
1	0.020	0.363	0.931
2	0.003	0.220	-0.976
3	-0.091	-0.978	-0.190

From the information about the natural frequency spectrum, we find those values of frequency which provide the determination of the appropriate elastic constants by analysing the sensitivity coefficients.

Numerical calculations lead to the following conclusions.

The experimental data on twisting, extension and internal pressure loading allows us, in the context of the approach proposed, to define the elastic characteristics of orthotropic shells of complex configuration. The results on shell extension can be used not only in solving the inverse problem but in checking calculations.

To identify the mechanical characteristics which vary linearly along the length of the shell, we can use a set of the experimental data on twisting and internal pressure loading of the shell reinforced with a rigid ring frame.

To find the elastic mechanical characteristics in terms of the natural frequency spectrums, we require the information about one of torsional vibration modes and one or two frequencies of untorsional vibration modes. The last were taken by estimating the sensitivity coefficients.

By modelling the experimental error, we come to the conclusion that the data on displacements provide the elastic moduli with an error commensurable with the experimental error, and the error for Poisson's ratios is, on the average, twice as large. When the strain values are used as the experimental data, the error for elastic constants is less than the measurement error.

To illustrate the above, we consider the problem of defining the effective elastic constants for a multilayer cylindrical shell obtained by continuous winding. In the work by V. D. Protasov and A. A. Fillipenko (1984), the experimental data are given for a shell subjected to internal pressure with one end fixed, and the other under the tensile force. Two variants of tensile forces have been implemented experimentally:

$$T = pR/2 \quad \text{and} \quad T = p(R - R_0)/2. \qquad (26)$$

The second corresponds to the case when the nonfixed end cap has the hole of radius R_0.

The shell geometrical parameters are: h_i =0.0023 m, R= 0.1 m, L= 0.35 m, R_0 = 0.025 m, where h_i is the thickness of a single layer. The shell is composed of 28 layers, of which circular 16 and spiral 6. The winding angle is 29^0. The shell is subjected to internal pressure, p = 0.85 MPa.

The displacements u_r and u_φ were measured on four cross-sections. For this, 6 gauges were mounted on each of these cross-sections. The inverse problem has been solved based on the data from four points along the structure length taken as the arithmetic mean of the gauge records for the section considered. The following values of elastic constants were found: E_1 = 2.012 MPa, E_2 = 4.826 MPa, ν_1 = 0.13, ν_2 = 0.3122.

The elastic characteristics given in this work for the unidirectional layer (E_1 = 7.10 MPa, E_2 = 0.245 MPa, ν_2 = 0.23 , G_{12} = 0.196 MPa) allow the calculation of the effective elastic characteristics for the multilayer shell material by the formulae commonly applied to obtain the mean (effective) elastic characteristics of an arbitrarily reinforced composite, see for example the work by I. F. Obraztsov and colleagues (1977).

These formulae give: E_1 = 1.932 MPa, E_2 = 4.216 MPa, ν_1 = 0.127, ν_2 = 0.277, which differ from the values found by solving the inverse problem by no more than 12 %.

3.2 *Determination of complex dynamic moduli in terms of shell vibration analysis data*

We define here the complex dynamic modulus components in the framework of the approach proposed. The body model with the complex dynamic modulus allows taking into account the material damping properties which can manifest themselves in different ways depending on the motion regime. Under forced vibrations, the dissipative properties show themselves in the values of finite amplitudes (displacements, deformations, and stresses) at the resonance frequencies. The damping properties can be also observed when the free vibrations attenuate in a finite time owing to damping.

In this work, to simulate the damping vibration rate, we suggest, instead of classical problem with initial conditions on free vibrations, a new mechanical spectral problem on natural vibrations in viscoelastic bodies This statement seems to be more convenient for numerical implementation of the algorithms developed to solve the inverse problems.

The first variant to define the dynamic moduli relates to the use of the experimental data on the amplitude-frequency characteristics of displacements or deformations at some shell points. These data could be obtained from tests on vibration-testing machines.

As in static problems, the above method is verified based on the numerical results (taken as experimental data) from the solution to the direct problem which is, in this particular case, the problem on steady-state forced vibrations in visco-elastic bodies.

Consider a shell with the geometrical dimensions $L / R = 2$, $h / R = 0.05$. Assume that the mechanical behaviour of the material is described by the relationships from a linear hereditary theory, and no rheological properties show themselves under volume deformation.

Thus, at steady-state forced vibrations the material behaviour is determined by the complex dynamic shear modulus

$$\overline{G} = G_R + iG_M \quad (27)$$

and by the volume modulus K_0.

The objective function can be written as

$$F = \sum_{m=1}^{M}\sum_{j=1}^{N}\sum_{i=1}^{3}\left[u_{ij}^{p}(z_m) - u_{ij}^{э}(z_m)\right]^2 . \quad (28)$$

Here, $u_{ij}^{p}(z_m)$, $u_{ij}^{э}(z_m)$ are the calculated and experimental values of displacements at z_m points of the shell at the first N resonance frequencies.

Three variants of forced steady-state vibrations of the shell are considered. Figure 1 schematically presents the way of applying the boundary disturbing displacements, changing by the harmonic law with the amplitude A_0.

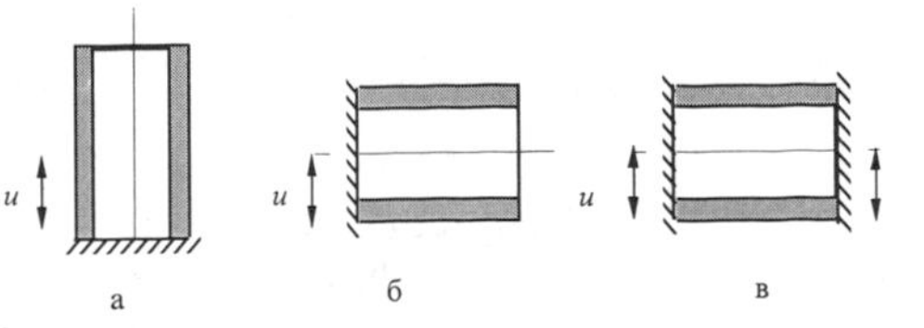

Figure 1. The schemes of applying the disturbing displacements.

The amplitude-frequency curves (AFC) for shell displacements and deformations result from the solution to the direct problems, simulating the test on a vibration-testing machine.

For calculation purposes, the following model material parameters were used:
$G_R / G_M = 0.2$, $K / G_R = 24.7$, $\rho = 1.0$.

Two types of polyurethane were considered:
$G_R = 22.802$ MPa, $G_M = 3.848$ MPa,
$K = 220.42$ MPa, $\rho = 0.1223$ kg / m^3
and
$G_R = 72.241$ MPa, $G_M = 18.413$ MPa,
$K = 698.33$ MPa, $\rho = 0.1295$ kg / m^3.

Calculations showed that, to determine the volume modulus and the components of the dynamic shear modulus with the accuracy of 1%, it is sufficient to have the amplitude values for one or two frequencies, which not unnecessarily be resonance.

Suppose that we have the test results on time variation of deformation and displacement produced by the initial impulse at different structure points. We further assume that these data have been treated to distinguish from the general picture the harmonic components which, by virtue of damping, are of attenuating character.

When the damping is described by the complex dynamic modulus, every appropriate eigenform can be written as

$$u(x,t) = \overline{u}(x)e^{-i\omega t} = \overline{u}(x)e^{-i\omega_R t}e^{\omega_I t}. \quad (29)$$

The problem on damping natural vibrations in the context of the considered approach - an inverse problem for identifying the material mechanical characteristics - can be used when the objective function has the form

$$F = \sum_{i=1}^{N} \left[\left(\omega_{Ii}^{p} - \omega_{Ii}^{э} \right)^{2} + \left(\omega_{Ri}^{p} - \omega_{Ri}^{э} \right)^{2} \right]. \qquad (30)$$

Numerical investigations have been made based on the results from the solution of the direct problem which we have used as experimental data. As a direct problem, we consider here the problem on damping natural vibrations of a cylindrical viscoelastic shell with one fixed and the other free ends. Dimensions and properties of the shell are :
$L = 0.1106$ m, $R = 0.00165$ m, $h = 0.002$ m, $G_R = 22.802$ MPa, $G_M = 3.848$ MPa, $K = 220.42$ MPa, $\rho = 0.1223$ kg / m^3.

Complex natural frequencies of the shell vibrations are as follows (sec^{-1}):

Table 3

harmonic number	frequency number	1	2	3	4
0	ω_I	17.059	28.048	51.283	80.526
	ω_R	203.61	347.11	612.07	974.05
1	ω_I	5.986	23.793	49.379	68.764
	ω_R	73.814	290.22	602.58	835.72

The obtained complex vibration frequencies have simulated the experimental values $\omega_R^{э}$ and $\omega_I^{э}$.

Numerical results show that the volume modulus and the complex dynamic modulus components can be found with an accuracy of 0.01% and, for this, we need only have knowledge of two natural frequencies at a single harmonic.

4 CONCLUSIONS

The approach proposed here opens up new possibilities to identify the effective mechanical constants for shell structures.

REFERENCES

Matveyenko V.P., Kligman E.P. , 1997. Numerical Vibration Problem of Viscoelastic Solids as Applied to Optimisation of Dissipative Properties of Constructions. *Int.J. of Vibr. and Control.* : 2, 87-102.

Biderman V.L. , 1977. *Mechanics of thinshelled structures.* M., Mashinostroyenie, 488.

Zienkevich O., 1975. *FEM in engineering.* M., Mir., 542.

Abramchuk S.S., Buldakov V.P., 1979. Admissible Values of Poisson's ratio for anisotropic materials. *Mech. Comp. Mater.,* 2: 235-239.

Himmelblau D., 1975. *Applied nonlinear programming.* M., Mir., 534.

Houg E.G., Choi K., Komkov V. , 1988, *Sensitivity analysis techniques in structure design,* M., Mir, 428.

Aurora G.S, Houg E.G., 1979. Sensitivity structure calculations in terms of design variables. *Rock. Eng. $ Cosm.,* V.17, 9: 52-58.

Kemmer D.K., Janson B.M., Meiso D.R. , 1990. Correlation of calculated and experimental data for a "Space Shuttel" KLAMI solid-propellant rocket engine central compartment. *Aerospace Eng.* 8: 138-149.

Ambartsumjan S.A. , 1974. *General theory of anisotropic shells,.* M., Nauka, 448.

Lehnitskii S.G. , 1977. *Elasticity theory of anisotropic body.* M., Nauka, 416.

Protasov V.D., Fillipenko A.A., 1984. Momentless laminate cylindric shells with variable elastic parameters obtained by continuous winding. *Mech. Comp. Mater.,* 3: 493-502.

Obraztsov IF, Vasileyv V.V., Bunakov V.A., 1977. *Optimal reinforcement of composite revolution shells.* M. Mashinostroeynie, 144.

12 Posters

Modern Practice in Stress and Vibration Analysis, Gilchrist (ed.) © 1997 Balkema, Rotterdam, ISBN 90 5410 896 7

Diagnosis of lubrication failure in gear systems by vibration analysis

D.P.S.Chauhan & K.N.Gupta
Department of Mechanical Engineering, Indian Institute of Technology, Bombay, India

ABSTRACT: The appropriateness of lubrication is critical in a gear system as regard to proper functioning and life of gears. In the present work it is shown that the vibration signal analysis can be effectively used for diagnosing the lubrication condition of gears. The spectral and statistical analysis of the gear vibration acceleration are investigated. Spectral analysis indicates that the deterioration in lubrication condition causes for overall increase in the amplitude of acceleration, but the relative increase in the amplitudes of higher harmonics of tooth meshing frequencies indicate the condition of lubrication. Difficulties in spectral analysis are also mentioned. In statistical analysis the probability density functions of the recorded time histories, along with their normal approximations are used. A *goodness of fit* parameter has been proposed for the normal approximation of the probability density function, which has been found to be a good indicator of the lubrication condition.

1 INTRODUCTION

In a gear system the transmission error is the major source of vibrational excitation (Kubo et al. 1986). This excitation is caused by the dynamic transmission error (Ozguven & Houser 1988, Ozguven 1991) and depends on gear rotational speeds (Umezawa et al. 1984, 1986). The damping offered by the lubricant film, present between the contacting tooth profiles controls the mesh excitation. Therefore, any change from the ideal lubrication, affects the excitation function and reflects in terms of gear system's vibratory response. Two approaches based on spectrum analysis and statistical analysis of the response vibration of gear system are investigated for diagnosing its lubrication condition.

1.1 *Spectrum analysis*

The important elements of the gear system vibrations are the tooth meshing frequency and its harmonics (Randall 1982). It is expected that the instantaneous effect of lubrication failure on the gear response vibration to be similar to, the effect of increased uniform wear of the tooth profiles. The wear of tooth profiles distorts the tooth meshing frequency of the response vibration signal, causing more pronounced effects on its higher harmonics.

1.2 *Statistical analysis*

The changes in the statistical parameters of gear vibration signature is another way of diagnosing the system behaviour. The normalized amplitude probability density function ($p(\bar{a}(j))$) is used for statistical analysis, which is computed from the histogram of acceleration time history:

$$p_\sigma(\bar{a}(j)) = \sigma \frac{100}{\delta(\bar{a})} q(\bar{a}(j)) \quad ...\%/\sigma \qquad (1)$$

where, $q(\bar{a}(j)$ is the histogram (Equation 2) for the signal whose amplitudes ($a(j)$) are normalized by its standard deviation (σ).

$$q(\bar{a}(j)) = \frac{1}{N}\sum_{n=0}^{N-1} k(\bar{a}(n)), \textit{ where:} \qquad (2)$$

$$k(\bar{a}(n)) = 1, \textit{ for } [\bar{a}(j) - \frac{\delta\bar{a}}{2} \le \bar{a}(n) \le \bar{a}(j) + \frac{\delta\bar{a}}{2}];$$
$$= 0, \quad \textit{otherwise}.$$
$$\textit{for } \ 0 \le n \le N-1, \textit{ and } \ 0 \le j \le J-1.$$

N is the number of available samples in time history, j is the amplitude group index and J is the total number of intervals into which the normalized amplitude range of the signal is divided. $\bar{a}(n)$ is the normalized amplitude of a time sample and $\bar{a}(j)$ is the fixed normalized amplitude value for each j. $\delta\bar{a}$ is the normalized amplitude interval and equals the normalized range of signal divided by J. The normal approximation to the time history is computed from Equation 3.

$$p_{nor}(\bar{a}(j)) = \frac{1}{\sqrt{2\pi}} \exp(-(\bar{a}(j) - \mu/\sigma)^2/2) \quad (3)$$

The value of mean (μ) and σ are same as those calculated from the time history. To quantify the goodness of normal approximation a *goodness of fit* parameter (λ) is defined:

$$\lambda = \sum_{i=1}^{J} \frac{(O_i - E_i)^2}{E_i} \quad (4)$$

where, O_i = observed frequency from observed histogram, E_i = expected frequency from normal approximation and J = number of intervals in histogram. $\delta\bar{a}$ (Equation 2) is kept as $\sigma/10$.

Use of the Beta distribution and its parameters for monitoring damage in gears has been earlier investigated. It is reported (Martin 1992) that the reciprocal of 4*th* Beta moment is a good indicator of tooth damage. Therefore, for comparison of insufficient lubrication cases, the reciprocal of 4*th* Beta moment is also computed and discussed in addition to the conventional statistical parameters.

2 EXPERIMENTAL INVESTIGATION

The experiments are conducted on a two stage compound spur-gear system driven by a variable speed *dc* motor. The details of gear system are shown in Table 1. The tooth meshing frequencies for the first stage (*tmf*-1) and second stage (*tmf*-2) respectively are 20.0 and 11.43 times the speed of the input shaft, in rotations per second (*rps*).

Table 1. Details of gear system.

Reduction stage	Number of teeth on	
	Pinion	Gear
First	20	35
Second	20	35

The vibration signals are picked-up by an accelerometer (B & K 4366) mounted below the bearing blocks. Charge output of the accelerometer is fed to a charge amplifier (B & K 2635) and finally stored in a digital signal analyser on a floppy diskette. These digital records are processed later in a personal computer.

The oil is drained out from the sump and the gear system is run for 30 minutes, to simulate the condition of failing lubrication. During this run one acceleration time history is recorded after every five minutes.

3 DISCUSSION

The spectrum analysis presented here is based on monitoring the gear vibration response spectrum, therefore a reference spectrum observed at ideal conditions is essential. The statistical analysis proposed is a diagnostic method, which does not require any reference or ideal signature of the gear system.

3.1 *Spectrum analysis*

The response amplitude of the gear system acceleration depends on the rotational speed of gears and applied load on the gear system. Therefore, to observe the effect of lubrication condition on the acceleration spectra two time histories are recorded at very close speeds (8.75 *rps*) and loads (4.5 *N-m*). The first time history belongs to properly lubricated gear system. The spectrum for properly lubricated case is shown in the Figure 1. The second time history recorded 15 minutes after draining out the lubricant from the sump, belongs to poor lubrication case. The spectrum for poor lubrication case is shown in Figure 2.

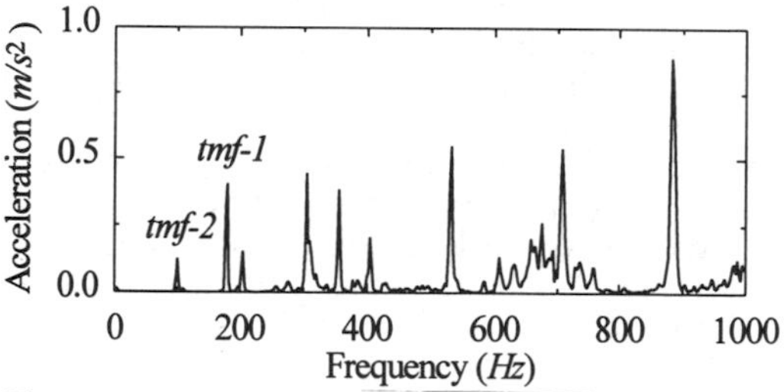

Figure 1. Spectrum for proper lubrication case.

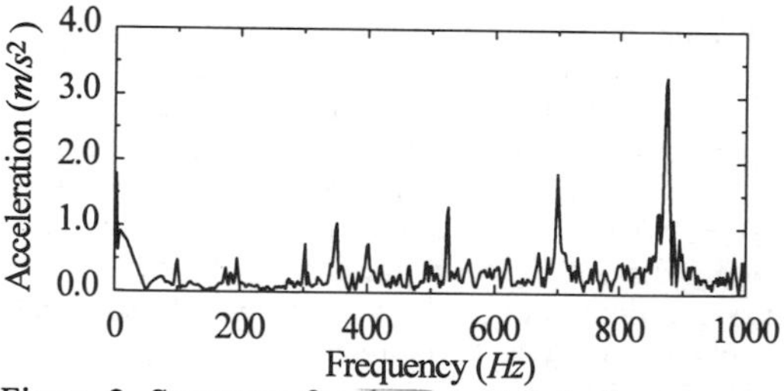

Figure 2. Spectrum for the poor lubrication case.

Comparing the spectra of proper and poor lubrication cases, following points are observed:
- In the poor lubrication case, overall acceleration is higher than the proper lubrication case.
- The carpet value of acceleration in the spectra is also higher in the poor lubrication case.
- In Figure 2, the higher frequencies in terms of harmonics of tooth meshing frequencies appear with increasing amplitudes.
- In Figure 1, the frequencies below *tmf*-1 (≈175 Hz) are found with very small amplitude. The same range of frequencies in Figure 2 has higher amplitudes and appear in form of the lobes.

Ratio of peak amplitudes at various tooth meshing frequencies, between those of Figure 2 and Figure 1 are shown in Table 2. Refer Table 2 for *tmf*-1 and its higher harmonics, belonging to the first stage gear pair operating at higher speed than the second stage, the ratio of peak amplitudes increases for higher harmonics. Therefore any spontaneous increase in the higher harmonics of a tooth meshing frequency can be attributed to failure of lubrication. In the second stage the ratio of peak amplitudes at various harmonics of *tmf*-2 do not fallow similar trend. This is due to the fact that in the poor lubrication case, an additive pulse caused by tooth hitting because of torsional mode of gear system's vibration is present in the time history. This pulse however very small, is reflecting at low frequencies in the form of a main lobe from zero *Hz*, alongwith a few visible side lobes. Hence, the presence of such pulses also affect the peak amplitudes of tooth meshing frequencies and pose difficulties in the spectrum comparison method. Another limitation of such a method is that, all the signatures to be compared should belong to the similar load and speed conditions for the following two reasons :
- The load and speed of the gear system affects the signal amplitude.
- Due to variation in speed the peaks at the tooth meshing harmonics may get affected at varying degrees by the transfer function of the signal path.

3.2 *Statistical analysis*

Five different cases of gear response acceleration signatures, recorded under different load, speed and condition of lubrication are analysed. Details of these acceleration signatures are shown in Table 3.

Table 2. Ratio of peak amplitudes at various *tmfs*.

Frequency	Ratio	Frequency	Ratio
tmf-1	0.91	*tmf*-2	3.91
2 x *tmf*-1	2.76	2 x *tmf*-2	3.37
3 x *tmf*-1	2.41	3 x *tmf*-2	1.66
4 x *tmf*-1	3.39	4 x *tmf*-2	3.61
5 x *tmf*-1	3.73		

Table 3. Various cases of gear response signature.

Case no	Input *rps*	load *N-m*	Lubrication condition
1	9.60	4.5	proper
2	8.75	4.5	poor*
3	14.07	5.5	very poor**
4	2.99	5.5	dry run

* 15 minutes after draining out the lubricant.
** 30 minutes after draining out the lubricant.

The acceleration signature of case-4 is recorded at relatively low *rps* to avoid any profile damage due to dry run. The acceleration time history of case-4 is shown in the Figure 3. In this time history a few pulses are seen, which are caused by separation and hitting of contacting tooth pair due to the torsional mode of gear system's vibration. Such pulses cause shift of the samples occurring within them.

The normalized probability density function (*NPDF*) and its normal approximation computed at 256 intervals for the time history of case-4 is shown in Figure 4. The effect of the impact pulses is seen

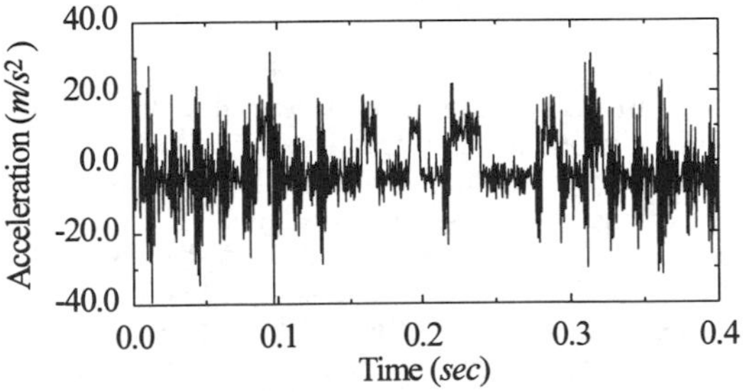

Figure 3. Acceleration time history of case-4.

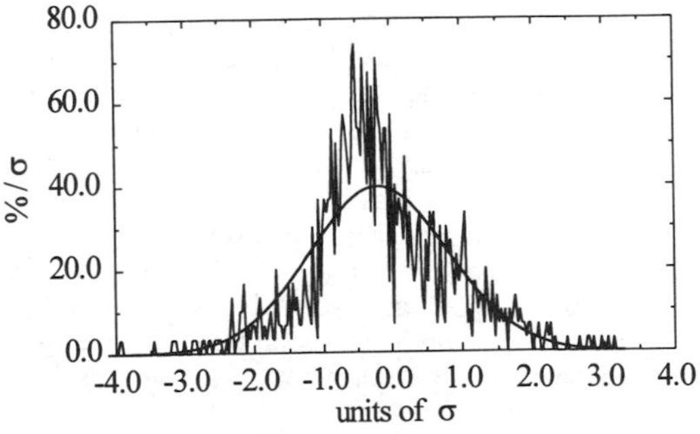

Figure 4. *NPDF* and normal distribution for case-4.

on the positive amplitudes in the range of 0.5σ to 1.5σ. Referring to the time history of Figure 3, the samples within the pulses are also found to shift on the positive amplitudes in the same range (σ =10.0 for case-4). Hence these pulses give rise to a mixed density function comprising of the events of samples within and outside them. Therefore a new record (case-5) is formed by eliminating the samples occurring within the pulses shown in Figure 3. This record is prepared to study the effect of impacting pulses on the *goodness of fit* parameter.

The statistical parameters of all the cases are shown in the Table 4.

Table 4. Statistical parameters for various cases.

	Cases of time histories				
Parameters	1	2	3	4	5
rms value	0.9	5.1	16.5	10.2	8.7
σ	0.9	5.1	16.5	10.0	8.7
Crest factor	3.0	3.6	3.7	3.9	4.7
*Kurtosis****	-.4	0.2	0.6	0.8	2.0
1/(4*th* Beta-moment)	32	54	61	57	113

*** *Kurtosis* is equal to the fourth normalized moment of signal minus three.

Compared with proper lubrication case, the increased root mean square (*rms*) value and σ for the case-2 is due to the dispersion of the signal at poor lubrication condition. In case-3 the *rms* value and σ are highest, because the corresponding signature is recorded at highest *rps* of the input shaft with very poor lubrication condition, compared to all other cases. The signature of case-4 is recorded at lowest *rps*, but due to dry run conditions its *rms* value is higher than the case-1 and 2. The increments in the *crest factor* indicate the relative increase in peak values as the lubrication condition deteriorates. The negative value of *kurtosis* for case-1, indicates presence of harmonic components in the time history. In other cases the *kurtosis* value steadily increases as the lubrication condition becomes more and more poor. It is interesting that the reciprocal of 4*th* Beta moment has upward trend for case-2 and case-3, but for the case-4 it falls. This reduction in the reciprocal of 4*th* Beta moment is due to the shift of samples by the impacting pulses, as its value further goes up for the case-5. Though the *kurtosis* seems to be independent of load and *rps*, it is rather difficult to remark on the lubrication condition of the gear system based on the statistical parameters listed in Table 4.

During the experimentation it has been observed on the averaged probability density functions (*PDFs*) of 100 acceleration records, computed for 256 intervals under identical range, that the peak of averaged *PDF* becomes more flat and stands apart from rest of the curve for the poor lubrication condition. The range of *PDF* or the signal amplitude also increases for poor lubrication condition. These observations lead to the need of quantifying the changes in *PDF* caused by insufficient lubrication. Therefore, the proposed method uses the normal approximation of *NPDF* to compute the *goodness of fit* parameter. The *goodness of fit* parameter is a measure of the resemblance of the actual *PDF* of the time history with a normal distribution. The *NPDFs* along with their normal approximations are shown in the Figure 5 for the cases-1,2 and 3. Similarly the cases-4 and 5 are shown in Figure 6.

In the Figure 5 and Figure 6, the actual *NPDF* of the signals are shown in bar diagram and their normal approximation computed on the same intervals are shown by continuous lines for clarity of comparison. Refer Figure 5, as the lubrication

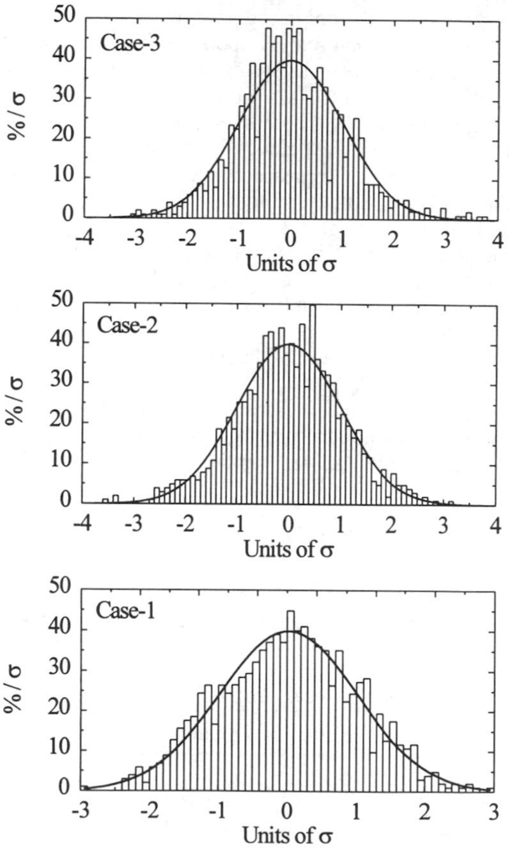

Figure 5. *NPDF* & Normal distribution(Case-1 to 3).

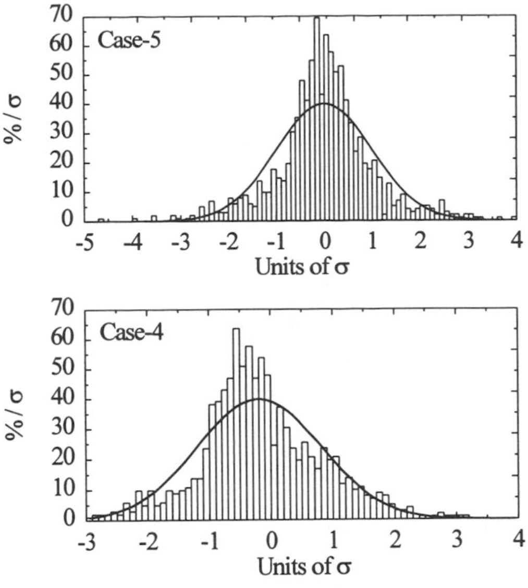

Figure 6. *NPDF* & Normal distribution(Case-4 & 5).

condition becomes poorer (from case-1 to case-3), difference between the actual *NPDF* and its normal approximation in the range of $\mp\sigma$ increases. This range of $\mp\sigma$ spreads towards the tails of the *NPDF*, when the lubricant is washed away, for the case-4 and case-5 (Figure 6).

The *goodness of fit* parameter computed for the entire signal range and partial signal ranges of $\mp\sigma$, $\mp 2\sigma$, and $\mp 3\sigma$ are shown in Table 5, for all the cases of acceleration time histories.

Table 5. *Goodness of fit* parameter.

Case no	*Goodness of fit* parameter			
	Range of signal			Full range
	$\mp\sigma$	$\mp 2\sigma$	$\mp 3\sigma$	
1	6.4	31.0	44.3	46.3
2	13.4	25.7	42.4	79.8
3	36.3	53.9	80.1	165.2
4	68.1	101.8	119.0	249.0
5	89.7	142.5	180.7	1564.9

Refer Table 5, the *goodness of fit* parameter for the range of $\mp\sigma$ increases almost 14 times from proper lubrication case to dry run of the gear system. In the range of $\mp 2\sigma$ and $\mp 3\sigma$ the similar increase is approximately 4 times. This parameter also shows an upward trend for the full range of signal also. But, the highest value of 1564.9 is contributed mainly by a few intervals located near the tails of the *NPDF* (Figure 6), for which the expected frequencies (Equation 4) are very low. From the repeated experimental observations it is confirmed that the *goodness of fit* parameter displays steady increments for the range of $\mp\sigma$ with deterioration in lubrication condition of the gear system.

4 CONCLUSIONS

1. The response acceleration of the gear system also depends on its lubrication condition. Under insufficient lubrication, the higher harmonics of the tooth meshing frequencies increase more.
2. In the spectrum comparison method the operating conditions such as speed and load must remain constant.
3. *Kurtosis* and reciprocal of the 4*th* Beta moment are found sensitive to the lubrication condition of the gears.
4. The proposed *goodness of fit* parameter is found to offer a reliable and steady index for the spur gear system, however more research needs to be carried out for other types of gear system also.

REFERENCES

Kubo, A., S. Kiyono & M. Fujino 1986. On analysis and prediction of machine vibration caused by gear meshing. *Bulletin of JSME*. 29(258): 4424-4429.

Martin, H. R. 1992. Detection of gear damage by statistical vibration analysis. *Proc. I Mech E, Int. Conf. on Vibrations in Rotating Machinery, 7-10 Sept., Univ of Bath*. c-432/006:395-401.

Ozguven, H. N., & D. R. Houser 1988. Dynamic analysis of high speed gears by using loaded static transmission error. *Journal of Sound and Vibration*. 125(1): 71-83.

Ozguven, H. N. 1991. A nonlinear mathematical model for dynamic analysis of spur gears including shaft and bearing dynamics. *Journal of Sound and Vibration*. 145(2): 239-260.

Randall, R. B. 1982. A new method of modelling gear faults. *Transactions of the ASME: Journal of Mechanical Design*. 104:259-267.

Umezawa, K., T. Sato & K. Kohino 1984. Influence of gear errors on rotational vibration of power transmission spur gears. *Bulletin of JSME*. 27(225): 569-575.

Umezawa, K., T. Ajima & H. Houjoh 1986. Vibration of three axis gear system. *Bulletin of JSME*. 29(249): 950-957.

Modern Practice in Stress and Vibration Analysis, Gilchrist (ed.) © 1997 Balkema, Rotterdam, ISBN 90 5410 896 7

Wavelet analysis applied to bearing vibration detection

H.R. Martin & S. Ziaei
Department of Mechanical Engineering, University of Waterloo, Ont., Canada

ABSTRACT: Machinery condition monitoring methods contribute significantly to an industries' competitiveness under the current economic conditions. Reducing the chances of a major equipment failure not only reduces repair or replacement costs, but also may prevent environmental damage. The effectiveness of such a monitoring system depends on the signal processing approach used. Experience shows that, as yet, there is no single approach that is satisfactory to cover the detection of all damage events.

The usual approach of using the Fourier transform to produce a frequency spectrum has the disadvantage that a whole range of features including those of short time duration, which disappear on averaging. A spectral line therefore represents the average of all the activity over the recorded time block, in that particular frequency window. Hence many features indicating early signs of failure could be lost.

An alternative approach to such non-stationary vibrational data is the application of the Wavelet Transform.

1. BACKGROUND

The most likely area of impending failure in machinery is where surfaces are moving in contact, hopefully in the presence of a lubricating film. Both bearings and gears fall into this category. Surface failures can result from excessive loading, the presence of abrasive particles, mishandling during assemble, to name a few sources of potential trouble. All these conditions produce vibrational signals that can be detected on the external body of the component, using a suitable sensor. The challenge is to develop signal processing methods that are sensitive enough to extract and categorise useful patterns that relate to the damage mechanisms in progress, while rejecting unwanted information.

A common approach to machine condition monitoring is to measure the vibration level using an accelerometer or velocity transducer. However this energy has passed through the structure from the source of interest, before reaching the sensor. As a result the data is more than likely to be contaminated with structural resonances and noise from other machinery sources in the vicinity. Some of the well established time domain methods currently used are, amplitude level monitoring, enveloping and tracking statistical parameters. The use of transforms such as the Fast Fourier and Hartley transforms can then be applied to filtered raw time domain data to obtain spectral distributions of various types. All these methods have been successful to one degree or another, but most signals emitted from machinery are a complex combination of sinusoidal and random energy.

2 INTRODUCTION TO THE WAVELET APPROACH

The Fourier method breaks a complex signal into a series of components parts, such that each component is a copy of a fundamental sinusoidal term scaled in amplitude and frequency. When these are all summed together, the original waveform can be reproduced. For example Figure 1 shows a base sinusoid together with copies of this, that is harmonics, which are halved in amplitude and doubled in frequency. The summation of all these components would then form a complex sinusoid. If $a = 1/k$ is called the 'scale factor', then the smaller this value is the more compressed is the original sine wave. The amplitude is reduced and the frequency increased by specific

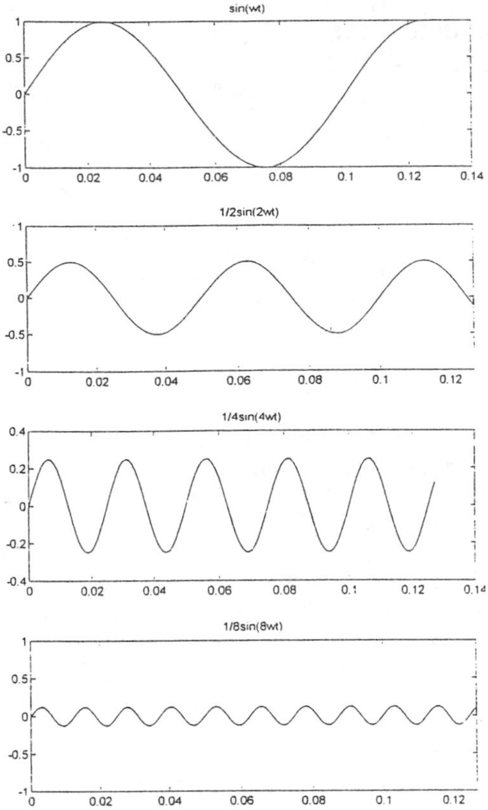

Fig 1. Sinusoidal Harmonics

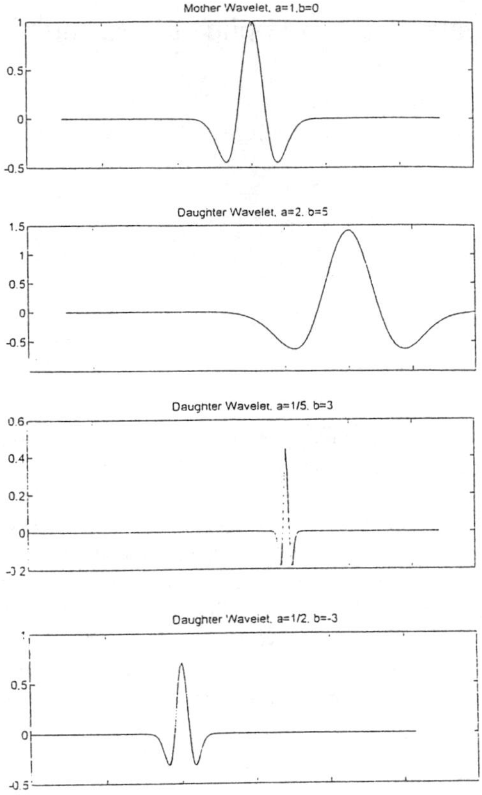

Fig 2. Wavelet Components

amounts. Hence in this particular case, we only need the base waveform and a series of values for 'k', to completely define the original waveform. If now the collected signal contains discontinuities such as that caused by a fault in a bearing outer ring surface, rather like an automobile going over a bumpy road, this transient event is difficult to detect. It is also impossible to tell when the event took place.

Wavelet analysis, on the other hand, is well suited to dealing with this type of situation and therefore has great potential for machine condition monitoring. This approach requires the selection of a wavelet prototype, called the mother wavelet. This is analogous to the fundamental sinusoid used in Fourier analysis, except that the fundamental and harmonics always start at $t = 0$ and carry on to $t = \infty$, wavelets start at different times (shifted or translated) and are waveforms of effectively limited duration that have an average value of zero. Also where sinusoid are smooth and predictable, the most useful wavelets for vibration analysis tend to be irregular and asymmetric.

To summarize, the smaller the scale 'a', the more compressed is the wavelet so the finer the resolution of detail. Hence small scales are related to the high frequency content of the data and vice versa.

The analysis of a signal f(t) can be carried out by constructing a base function, shifting it by a specific amount and changing its scale. The summation of copies of this base function are used in the analysis process. Repeat this process by taking the base function shifting it and scaling it repeatedly, then applying it to the same signal to get a new and hopefully better approximation. It turns out that this approach of scale analysis is less sensitive to noise than the Fourier approach, because it measures the average fluctuation of the signal at different scales.

The first and simplest wavelet is that conceived by Alfred Haar in 1909. Since that time numerous other types have been developed, for example Biorthogonal, Coiflets, Daubechies, Mexican Hat, Meyer, Morlet and Symlets Although there is a wide choice of wavelet functions available, the best for a particular application is an active research area The Daubachies wavelet family (Daubechies 1988) has been found to be useful in vibration analysis and has been actively pursued by Newland (1994).

The Continuous Fourier Transform is given by;

$$f(t) = \frac{1}{2\pi}\int_{-\infty}^{\infty} F(\omega)\, e^{j\omega t}\, d\omega \qquad (1)$$

where

$$F(\omega) = \int_{-\infty}^{\infty} f(t)\, e^{-j\omega t}\, dt \qquad (2)$$

= The sum over all time of the original signal weighted by a complex exponential

Each value of F(ω) represents the amplitude of each harmonic in f(t). The wavelet is defined in a similar way, except that the exponential term is replaced by,

$$\psi_{a,b} = \frac{1}{\sqrt{a}}\,\psi\!\left(\frac{t-b}{a}\right) \qquad (3)$$

where 'a' is the scale and 'b' is the position. Hence the corresponding Continuous Wavelet Transform (CWT) will be,

$$W_{a,b} = \frac{1}{\sqrt{|a|}}\int_{-\infty}^{\infty} f(t)\,\psi\!\left(\frac{t-b}{a}\right) dt \qquad (4)$$

= Sum over all time of the original signal weighted by scaled and shifted wavelet functions.

where ψ(t), for a = b = 0, is the basic or mother wavelet. This is illustrated in Figure 2. The term $1/\sqrt{a}$ is an energy normalizer which keeps the energy of the scaled wavelet equal to the original mother wavelet. Each value of $W_{a,b}$ represents the amplitude or wavelet coefficient for scale 'a' and position 'b'. The CWT can operate at every scale form the original signal up to the maximum capability of the computer used, since higher scales values generate a lot of calculations and storage requirements.

The Daubechies family of wavelets is derived from the dilation equation, meaning the spreading or expanding of a function. The basic dilation equation is a two-scale difference equation where;

$$\begin{aligned}\phi(t) &= c_o\phi(2t)+c_1\phi(2t-1)\\ &\quad + c_2\phi(2t-3)+c_3\phi(2t-3)\ldots\\ &= \sum c_k\phi(2t-k)\end{aligned} \qquad (5)$$

Beginning with this scaling function Daubechies reversed the order of c_k and applied alternate signs, giving for example;

$$\begin{aligned}\psi(t) &= -c_3\phi(2t)+c_2\phi(2t-1)\\ &\quad -c_1\phi(2t-2)+c_o\phi(2t-3)\end{aligned} \qquad (6)$$

which is now used in the wavelet function.

The particular shape of the wavelet is defined by the 'order' of the family. Hence for order 1, only c_o and c_1 are used while for order two all four coefficients are needed. In general, a wavelet of order p is produced by using r=2p coefficients. These coefficients are then evaluated using the following conditions;

For orthogonality criteria,

$$\sum c_b c_{b+2m} = 0 \quad m = 1,2,3,\ldots r/2-1$$
$$\sum c_b^2 = 2 \qquad (7)$$

For correct normalization,

$$\sum c_b = 2 \qquad (8)$$

To satisfy accuracy requirements,

$$\sum (-1)^b k^m c_b = 0 \qquad (9)$$

all over the range b=0 to (r-1).

The Daubechies D4 wavelet, used in this paper, has four coefficients, resulting from these condition equations, giving,

$$\begin{aligned}c_0 &= 0.683012\\ c_1 &= 1.183013\\ c_2 &= 0.316987\\ c_3 &= -0.183013\end{aligned} \qquad (10)$$

In general, the more coefficients used to develop

the wavelet, the smoother it is and this gives a better approximation to higher order polynomials. On the other hand, the computation time increases.

The Discrete Wavelet Transform (DWT) reduces this by defining scales and positions to powers of 2. Its purpose is to decompose any time domain signal f(t) over the range 0 to T, into a series of orthogonal wavelets of different scales and positions according to the expansion (Newland 1994),

$$f(t) = W_o + \sum_{a=0}^{\infty} \sum_{b=0}^{2^a-1} W_{2^a+b}\, \psi(2^a t - b) \qquad \textbf{(11)}$$

where

$$W_0 = \int_0^T f(t)\,dt = \text{mean amplitude value}$$

$$W_{2^a+b} = 2^a \int_0^T f(t)\, \psi(2^a t - b)\, dt \qquad \textbf{(12)}$$

In this form of the equation $\psi(2^a t - b)$ is one of the wavelets, all of them being developed from the mother wavelet by scaling and shifting. If n is the number of samples of f(t) taken, then the upper limit of 'a' is 'N', such that $2^N = n$. There will be (N+1) scale levels ranging from level -1 to level (N-1), against 'n' entries along the time axis.

Newland (1994) shows that a contour plot of the amplitude squared of each wavelet level plotted against scale and half the number of sample points gives a useful way of presenting the data collected from the sources of vibration. Half the number of points are used because of the convention used in drawing the map, so that the full time interval is covered with less points. The contour plot produced maps the distribution of the mean square of the function f(t) between wavelets of different level and different positions. Where there are significant contributions to a signals energy at particular frequencies and times, the mean square surface will show localised peaks. In other words the map show the distribution of energy against frequency as would the Fourier spectrum but also in relation to the time of occurrence.

3 SIMULATION RESULTS

To obtain a feel for the relationship between conventional and the wavelet approach to signal analysis consider Figure 3 which shows two

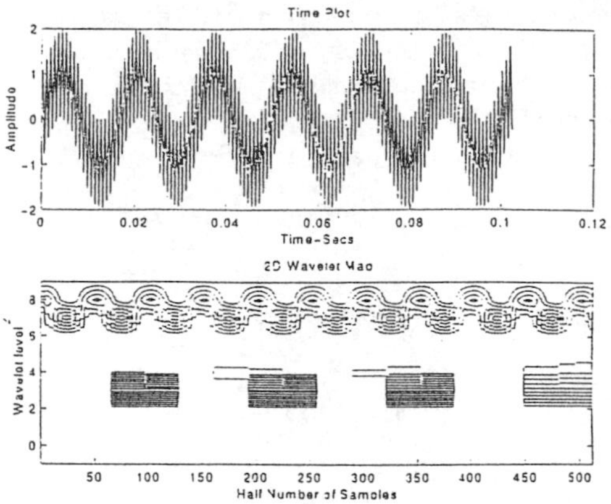

Fig 3. Comparing Time Domain with Wavelet Map

superimposed sine waves. The lower frequency is 60 Hz and the higher is at 1200 Hz. A sample of 1024 points was taken, in 0.1024 seconds. This time domain data is then analysed using the Daubechies D4 wavelet.

Referring to the second plot the wavelet level relates to the centre frequency of the octave band, a feature of the discrete wavelet transform given by $f = 2^j/T$, where j is the integer describing the wavelet level and ranges from -1 to (N-1). N is the power 2 is raised to for the number of points taken, and T is the sample block time, in this case 0.1024 seconds. The first row of activity shown on the 2D wavelet map is centred around level 3,

$$f_3 = \frac{2^3}{0.1024} = 78.125 \text{ Hz} \qquad \textbf{(13)}$$

while the second row of activity is centred around,

$$f_9 = \frac{2^7}{0.1024} = 1250 \text{ Hz} \qquad \textbf{(14)}$$

Figure 4 represents the simulation of data from a bearing with a mark on one of the balls. It is assumed that the bearing has 8 ball bearings and the shaft speed is 3000 rpm. This produces a ball passing frequency of 400 Hz. Superimposed on this are spikes, to represent the marked ball passing over the bearing surface and in addition there is a level of gaussian random noise. The latter represents the normal vibrations of two surfaces moving relative to each

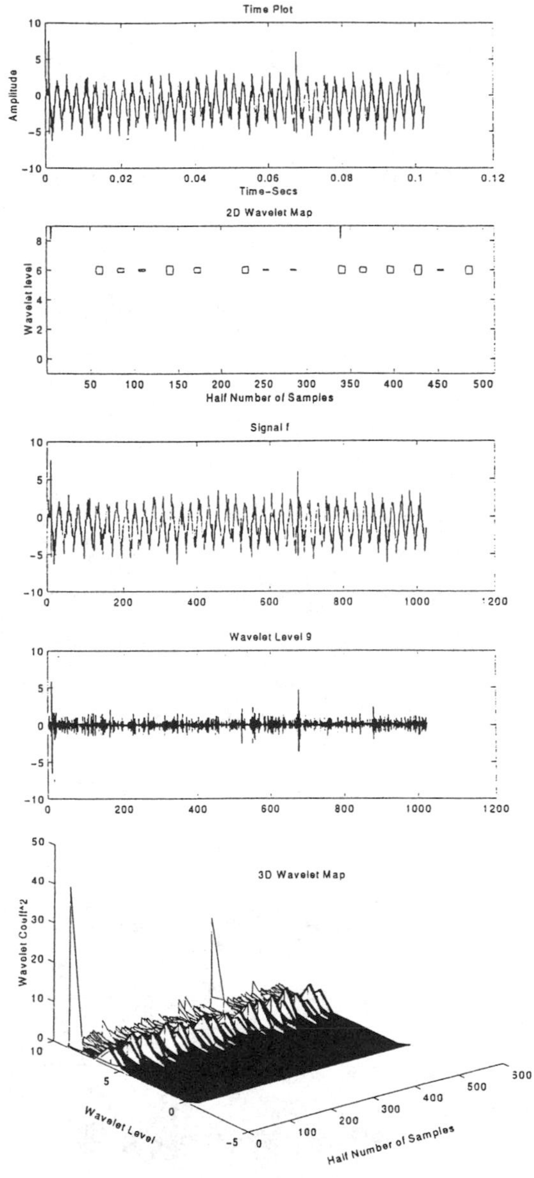

Fig 4. Simulation of a Bearing Fault

other. In the 2D wavelet map a spike shows up as a short line at sample number 340. This corresponds to 340x2x0.0001 = 0.068 s on the time domain plot. The row of activity occurring at wavelet level 6 for 1024 points corresponds to a centre band frequency of 625 Hz. Level 5 would correspond to 312.5 Hz, so the row of activity does fall in the region of the ball passing frequency. To improve this result more points would have to be taken in the sample.

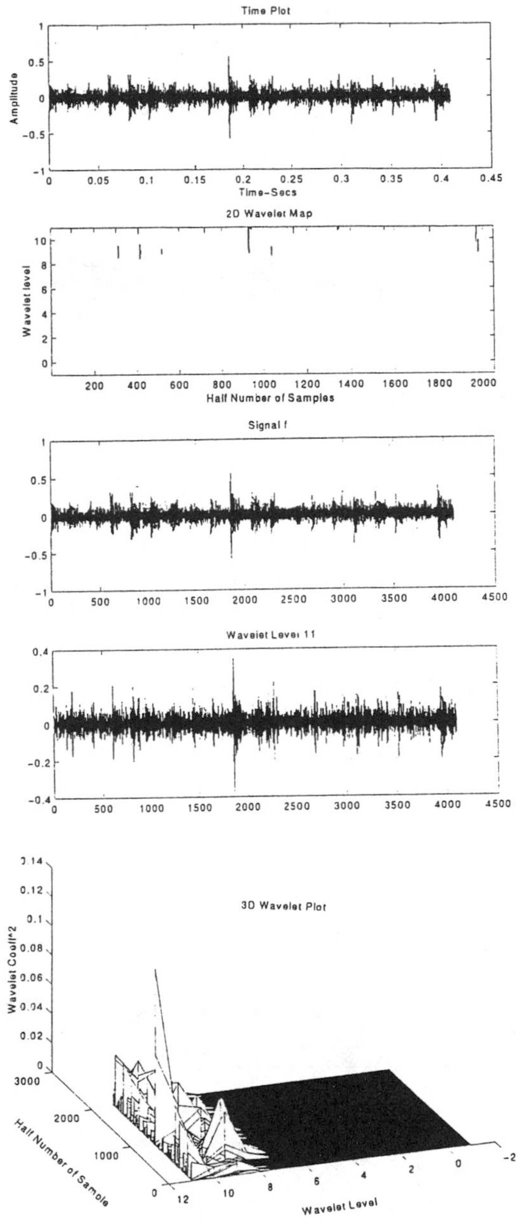

Fig 5. Results from an actual Faulty Bearing

Examining the separate wavelet levels, there will be N+1 = 11 of these, since $2^N = 2^{10} = 1024$, ranging from -1 to 9. In this case the 9th level shows the spike quite clearly at sample 680. The ball passing frequency has been and only the higher frequencies of the random component is present. This illustrates the filtering

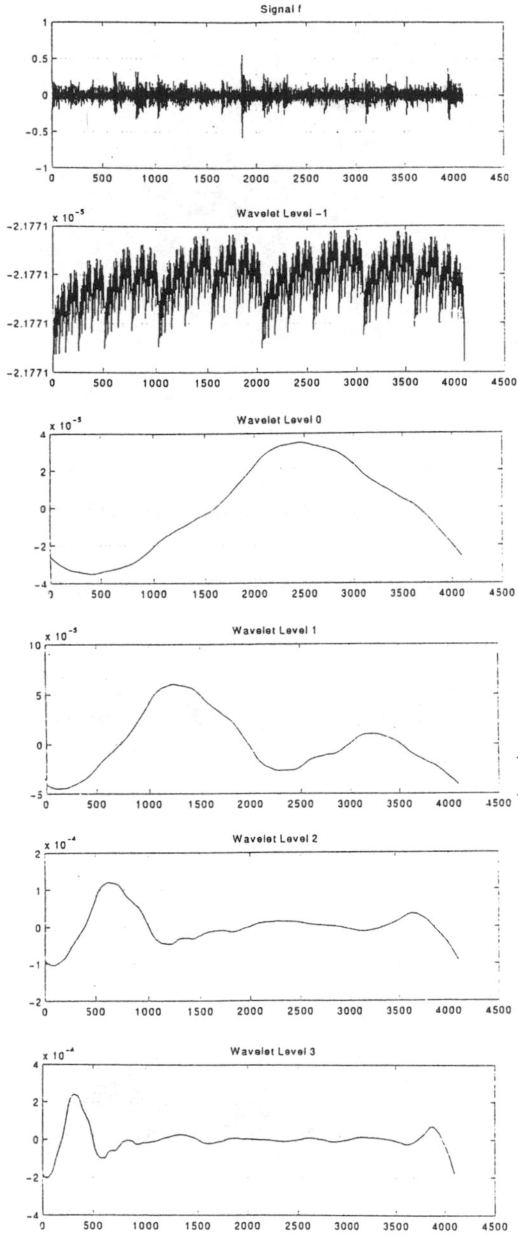

Fig 6. Wavelet Levels for Fig. 5

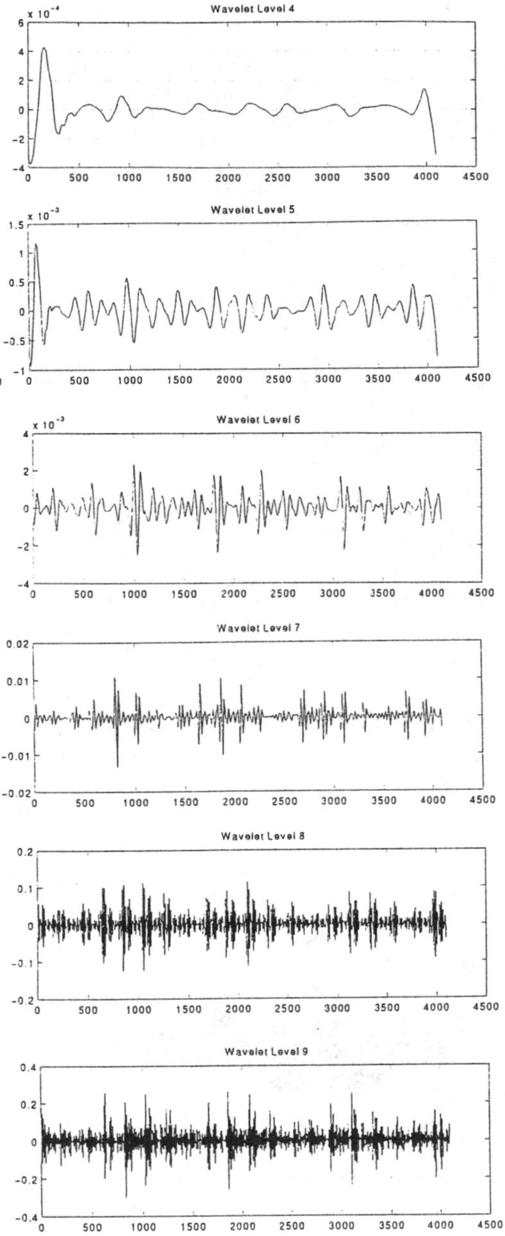

Fig 6. Continued

potential of the technique. The 3D wavelet map is produces from the 2D map for clarity.

This simulated data can then be compared with actual data taken from a test rig. With a shaft speed of 3000 rpm, one of the eight ball bearings of a RHP 6203 TBH EP7 bearing had a small defect made on the surface corresponding to about 8% of its diameter. The data was collected from the bearing housing using a PCB307A accelerometer, having a resonant frequency above 40 kHz. The shaft through the bearing was driven by a 1 HP permanent magnet dc motor with speed control to an accuracy of 2%. A Metrobyte Dash 16F A/D convertor in a PC microcomputer was used to collect 4096 points of

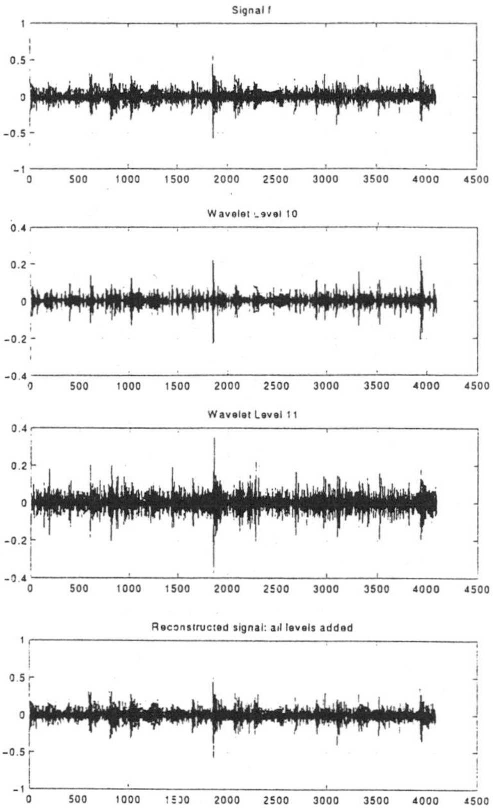

Fig 6. Continued

data. Analysis of the data was carried out using MATLAB and software based on reference (Newland 1993). At 3000rpm, 4096 points corresponded to 1.64 revolutions of the shaft.

Figure 5 shows the results of this test, giving very similar results to those from the simulation. Figure 6 shows the details of each wavelet level. Since there were $2^{12} = 4096$, data used, the number of wavelet level are now 13. In this analysis the Daubechies D8 wavelet was used.

4 CONCLUSIONS

The application of wavelets to the analysis of mechanical machine vibration is a relatively new area of investigation. These experiments show that the method has potential for detecting failure in machine moving parts. The next step is to determine if this approach offers additional advantages over more established methods.

The wavelet approach look at local activities in a signal, that is those events that cause transients. This type of phenomenon is often missed by Fourier analysis, due to its averaging nature. Many application of the technique have yet to be tried as well as the possibility of more effective orthogonal wavelets being found.

5 ACKNOWLEDGEMENTS

The authors wish to acknowledge the financial support provided by the Natural Sciences and Engineering Research Council of Canada, through Grant Number OGP0006468.

REFERENCES

Daubechies, I 1988. Orthonormal Bases of Compactly Supported Wavelets. *Communications on Pure and Applied Mathematics*. XLI:909-996.

Newland, D.E. 1994., Wavelet Analysis of Vibration Part 1: Theory. *Vibration and Acoustics J.* ASME. 116:409-416.

Newland, D.E. 1994. Wavelet Analysis of Vibration Part 2: Wavelet Maps. *Vibration and Acoustics J.* ASME, 116:417-425.

Newland, D.E. 1993. *An Introduction to Random Vibrations, Spectral & Wavelet Analysis*, (3rd ed), Longmans.

Modern Practice in Stress and Vibration Analysis, Gilchrist (ed.) © *1997 Balkema, Rotterdam, ISBN 90 5410 896 7*

A generalised transfer matrix method for free vibration of axisymmetric shells

X. Shang & J.J. Grannell
Department of Applied Mechanics, University College, Cork, Ireland

ABSTRACT: A generalised transfer matrix method is presented for analysing free vibration of axisymmetric shells. The general mathematical formulation is reduced to an eigenvalue problem for a system of first-order ordinary differential equations (ODE) based on the Reissner-Naghdi shell theory. The coefficient matrix occurring in the ODE system is derived for an arbitrary axisymmetric shell geometry and an unified form of frequency equation is given for general boundary conditions including arbitrary elastic restrictions. The transfer matrix method is developed by dividing the shell into segments and using a continuous piecewise matrix linear polynomial approximation for the coefficient matrix on each segment. The corresponding analytical solution for each segment transfer matrix is constructed by means of a truncated power series. The numerical examples confirm that the new method is simple and has excellent convergence, accuracy and numerical stability.

1 INTRODUCTION

Numerical methods for determining natural frequencies of free vibration of axisymmetric shells have been frequently discussed in the shell literature. Some methods yield an algebraic or generalised eigenvalue problem with large order matrices, e.g. finite difference method (Forsberg 1964), integral equation technique (Srinivasan 1989), boundary element method (Wang & Schweizerhof 1996), and various finite element methods (Luah & Fan 1989; Ozakca & Hinton 1994). An alternative approach, typified by the *transfer matrix method* involves reduction of the problem to calculation of the zeros of a determinant of fixed low order (depending only on the shell theory).

During the past three decades, transfer matrix methods have been studied and developed in various directions for the analysis of structures (Kalnins 1964a, 1964b, Uhrig 1966, Tottenham & Shimizu 1972, Horner & Pilkey 1978, Tavakoli & Singh 1989, Kayran et al. 1994). The transfer matrix corresponding to a segment of an axisymmetric shell relates the values of the kinematic and complementary force variables at one end of the segment to those at the other. Transfer matrix methods usually involve the division of the shell into segments - the global transfer matrix can be expressed as a product of element transfer matrices. Single segment transfer matrix methods are rarely used due to an instability which occurs if the shell is long and thin (Kalnins 1964a, 1964b). The basic idea is that the frequency equation may be expressed as the determinant of a low order matrix whose elements are related to the global transfer matrix, which is determined by solving, for each segment, a set of initial value problems for a system of first-order ordinary differential equations (ODE).

Runge-Kutta integration has been applied to calculate global transfer matrices for truncated conical shells (Irie et al. 1984) and for shallow spherical shells (Kobayashi & Yamada 1991) for single segment methods. Rung-Kutta (Sankar 1977) and Adams predictor-corrector methods (Kalnins 1964a, 1964b, Kraus 1967) have been used to calculate element transfer matrices for multisegment methods.

An alternative multisegment approach involves the approximation of the coefficient matrix occurring in the ODE system by a constant matrix on each segment. This enables the application of the matrix exponential to compute the element transfer matrix. For the case of a thin cylindrical shell, the matrix exponential in a segment has been evaluated by truncating the power series expansion (Tottenham & Shimizu 1972). For various thin shells, the Pade approximation of the matrix exponential has been employed together with the approximation of the coefficient matrix by its' integrated average in each segment (Tavakoli & Singh 1989).

The purpose of this paper is to generalise the transfer matrix method for free vibration of shells of revolution. The mathematical model of the problem is based on the Reissner-Naghdi shell theory which incorporates the effects of transverse shear deformation and rotary inertia (Kraus 1967) and is reduced to an eigenvalue problem for a system of first-order ODE. The coefficient matrix for the ODE system is derived for an axisymmetric shell of general shape including meridian curves with tangent jumps. Especially, owing to the introduction of two matrices describing general boundary conditions, it becomes possible to obtain a simple and unified form of frequency equation which is valid for various boundary conditions included arbitrary elastic restrictions. Moreover, matrix exponential methods are extended and developed in a new direction by using the approximation of continuous piecewise linear matrix polynomial for the ODE system coefficient matrix. The element transfer matrices are represented as truncated power series, the coefficients of which may be determined recursively. This makes it quite simple to develop a computer programme to implement the method. To illustrate the correctness and efficiency of the new method, numerical examples of implementation are given. The fundamental frequency for a thin spherical shell is computed and compared with the exact value. As an example of combination shell structures, natural frequencies of a thin cylindrical shell with two thin hemispherical endcaps are computed - the results agree well with those available in the literature. The numerical results show that the presented method has excellent convergence, accuracy and numerical stability.

2 THEORY

2.1 *Shell equations and boundary conditions*

Consider a general shell of revolution, the middle surface of which is generated by rotation of a curve (the meridian) about the symmetry axis. The arc-length along the meridian from one end is denoted by s. The shape of the meridian is specified by two coordinates; the distance, $r(s)$, from the axis of symmetry and the angle, $\phi(s)$, between the exterior normal to the meridian and the axis of symmetry. The following geometrical quantities

$$\kappa_s = \frac{d\phi}{ds}, \quad \kappa_\theta = \frac{\sin\phi}{r}, \quad \kappa = \frac{\cos\phi}{r}, \tag{1}$$

where κ_s and κ_θ are the two principal curvatures of the middle surface, appear in the shell equations below.

The shell model considered is the Reissner-Naghdi improved shell theory, in which the effects of transverse shear deformation and rotary inertia are included (Kraus, 1967). For axisymmetric and torsionless vibration, the kinematic variables are the middle-surface meridian displacement $u(s,t)$, transverse displacement $w(s,t)$ and the rotation of normal $\beta(s,t)$, where t is the time. The membrane stress resultants are N_s and N_θ, the moment resultants are M_s and M_θ, and the transverse shear stress resultant is Q_s. The constitutive relations are as follows:

$$\begin{aligned}
&N_s = K[(\frac{\partial}{\partial s} + \nu\kappa)u + (\kappa_s + \nu\kappa_\theta)w] \\
&-D(\kappa_s - \kappa_\theta)[\frac{\partial\beta}{\partial s} - \kappa_s\frac{\partial u}{\partial s} - \kappa_s^2 w)] \\
&N_\theta = K[(\nu\frac{\partial}{\partial s} + \kappa)u + (\nu\kappa_s + \kappa_\theta)w] \\
&+D(\kappa_s - \kappa_\theta)[\kappa(\beta - \kappa_\theta u) - \kappa_\theta^2 w)] \\
&M_s = D[(\frac{\partial}{\partial s} + \nu\kappa)\beta \\
&-(\kappa_s - \kappa_\theta)(\frac{\partial u}{\partial s} + \kappa_s w)] \\
&M_\theta = D[(\nu\frac{\partial}{\partial s} + \kappa)\beta \\
&+(\kappa_s - \kappa_\theta)(\kappa u + \kappa_\theta w)] \\
&Q_s = D_s[1 - \frac{h^2}{28}\kappa_s(\kappa_s - \kappa_\theta)] \\
&\times(\frac{\partial w}{\partial s} - \kappa_s u + \beta)
\end{aligned} \tag{2}$$

where $K = Eh/(1-\upsilon^2)$ is the membrane stiffness(E, ν and h are Young's modulus, Poisson's ratio and thickness of the shell, respectively), $D = h^2K/12$ is the bending stiffness and $D_s = \alpha K$ is the shear stiffness (the shear parameter, α , is given by $\alpha = \frac{1}{2}(1-\nu)k_s$ where $k_s = \frac{5}{6}$ is the shear coefficient).

The equations of motion for the free vibration are

$$\begin{aligned}
&\frac{\partial N_s}{\partial s} + \kappa(N_s - N_\theta) + \kappa_s Q_s \\
&= \rho h(c_1\frac{\partial^2 u}{\partial t^2} + \frac{h^2}{12}c_2\frac{\partial^2\beta}{\partial t^2}) \\
&\frac{\partial M_s}{\partial s} + \kappa(M_s - M_\theta) - Q_s \\
&= \frac{\rho h^3}{12}(c_2\frac{\partial^2 u}{\partial t^2} + c_3\frac{\partial^2\beta}{\partial t^2}) \\
&\frac{\partial Q_s}{\partial s} + \kappa Q_s - (\kappa_s N_s + \kappa_\theta N_\theta) \\
&= \rho h c_1\frac{\partial^2 w}{\partial t^2}
\end{aligned} \tag{3}$$

where ρ is the mass density of the shell and the coefficients c_1 , c_2 and c_3 are given by

$$
\begin{aligned}
c_1 &= 1+\frac{h^2}{12}\kappa_s\kappa_\theta \\
c_2 &= \kappa_s+\kappa_\theta \\
c_3 &= 1+\frac{3h^2}{20}\kappa_s\kappa_\theta .
\end{aligned}
$$

At each end of the shell, the boundary conditions considered have the form of *general elastic restrictions*

$$
\begin{aligned}
N_s &= -k_{i1}u, \\
M_s &= -k_{i2}\beta, \\
Q_s &= -k_{i3}w
\end{aligned}
\tag{4}
$$

where $0 \le k_{i1} \le +\infty$, $0 \le k_{i2} \le +\infty$ and $0 \le k_{i3} \le +\infty$ are the tension, bending and shear spring coefficients at the ends, respectively. The values *0* and *1* of the subscript *i* correspond to the ends $s=0$ and $s=L$, respectively (L is the total length of the shell meridian).

2.2 *Mathematical formulation of the eigenvalue problem*

The problem, which is now assumed to be time-harmonic, is formulated in terms of the following dimensionless variables and parameters:

$$
x=\frac{s}{L},\ \varepsilon=\frac{1}{12}(\frac{h}{L})^2,\ \lambda=\rho\omega^2\frac{L^2}{K},
$$

$$
[\kappa_1,\kappa_2,\kappa_3]=L[\frac{d\phi}{ds},\frac{\sin\phi}{r},\frac{\cos\phi}{r}],
$$

$$
\gamma_{ij}=\frac{1}{1+1/k_{ij}} \quad (i=0,1;j=1,2,3),
\tag{5}
$$

$$
\begin{aligned}
\mathbf{y}(x) &\stackrel{def}{=} [y_1,y_2,y_3,y_4,y_5,y_6]^{\top} \\
&= e^{i\omega t}[\frac{u}{L},\beta,\frac{w}{L},\frac{N_s}{K},\frac{M_sL}{D},\frac{Q_s}{\alpha K}]^{\top}
\end{aligned}
$$

where ω is the circular frequency and $0 \le \gamma_{ij} \le 1$.

It is straightforward to express (2)-(5) in the form of the following eigenvalue problem for a first order system of ordinary differential equations:

$$
\frac{d\mathbf{y}}{dx}=\mathbf{A}(x;\lambda)\mathbf{y}
\tag{6}
$$

$$
\mathbf{B}_0\mathbf{y}(0)+\mathbf{B}_1\mathbf{y}(1)=0
\tag{7}
$$

The elements of the coefficient matrix $\mathbf{A}$ are given by:

$$
\begin{aligned}
a_{11} &= -\nu\kappa_3/\Delta_2, a_{12}=-\varepsilon a_{21}, \\
a_{13} &= -(\kappa_1+\nu\kappa_2)/\Delta_2, \\
a_{14} &= 1/\Delta_2, a_{15}=\varepsilon(\kappa_1-\kappa_2)/\Delta_2, \\
a_{16} &= 0, a_{21}=\nu(\kappa_1-\kappa_2)\kappa_3/\Delta_2, \\
a_{22} &= -\nu\kappa_3\Delta_1/\Delta_2, \\
a_{23} &= -\nu(\kappa_1-\kappa_2)\kappa_2/\Delta_2, \\
a_{24} &= (\kappa_1-\kappa_2)/\Delta_2, \\
a_{25} &= \Delta_1/\Delta_2, a_{26}=0, a_{31}=\kappa_1, \\
a_{32} &= -1, a_{33}=a_{34}=a_{35}=0, \\
a_{36} &= 7/(10-3\Delta_1), \\
a_{41} &= \kappa_3^2\Delta_3/\Delta_2-\lambda(1+\varepsilon\kappa_1\kappa_2), \\
a_{42} &= \varepsilon a_{51}, a_{43}=\kappa_2\kappa_3\Delta_3/\Delta_2, \\
a_{44} &= \kappa_3(\nu/\Delta_2-1), \\
a_{45} &= \varepsilon\nu\kappa_3(\kappa_1-\kappa_2)/\Delta_2, \\
a_{46} &= -\alpha\kappa_1/\Delta_2, \\
a_{51} &= \kappa_3^2(\kappa_1-\kappa_2)(1-\nu^2/\Delta_2) \\
&\quad -\lambda(\kappa_1+\kappa_2), \\
a_{52} &= \kappa_3^2(1-\nu^2\Delta_1/\Delta_2) \\
&\quad -\lambda(1+\tfrac{9}{5}\varepsilon\kappa_1\kappa_2), \\
a_{53} &= \kappa_2\kappa_3(\kappa_1-\kappa_2)(1-\nu^2/\Delta_2), \\
a_{54} &= a_{21}, a_{55}=\kappa_3(\nu\Delta_1/\Delta_2-1), \\
a_{56} &= \alpha/(\Delta_2\varepsilon), \\
a_{61} &= a_{43}/\alpha, a_{62}=\varepsilon a_{53}/\alpha, \\
a_{63} &= [\kappa_2^2\Delta_3/\Delta_2 \\
&\quad -\lambda(1+\varepsilon\kappa_1\kappa_2)]/\alpha, \\
a_{64} &= (\kappa_1+\kappa_2\nu/\Delta_2)/\alpha, \\
a_{65} &= -\varepsilon a_{23}/\alpha, a_{66}=-\kappa_3
\end{aligned}
$$

where

$$
\begin{aligned}
\Delta_1 &= 1+\varepsilon\kappa_1(\kappa_1-\kappa_2), \\
\Delta_2 &= 1+\varepsilon\kappa_2(\kappa_1-\kappa_2), \\
\Delta_3 &= (1-\nu^2)-(\Delta_2-1)^2
\end{aligned}
$$

The boundary condition matrices appearing in equation (7) are defined as follows:

$$
\begin{aligned}
\mathbf{B}_0 &= \begin{bmatrix}\mathbf{\Gamma}_0 & \mathbf{I}-\mathbf{\Gamma}_0 \\ \mathbf{0} & \mathbf{0}\end{bmatrix}_{6\times6} \\
\mathbf{B}_1 &= \begin{bmatrix}\mathbf{0} & \mathbf{0} \\ \mathbf{\Gamma}_1 & \mathbf{I}-\mathbf{\Gamma}_1\end{bmatrix}_{6\times6}
\end{aligned}
$$

where $\mathbf{I}=diag(1,1,1)$ and $\mathbf{\Gamma}_i=diag(\gamma_{i1},\gamma_{i2},\gamma_{i3})$ ($i=0,1$)

Examples of boundary values are the following: (i) clamped end: $\mathbf{\Gamma}_i=diag(1,1,1)$, (ii) simple support: $\mathbf{\Gamma}_i=diag(1,0,1)$, (iii) free end: $\mathbf{\Gamma}_i=diag(0,0,0)$, (v) pole of a closed shell: $\mathbf{\Gamma}_i=diag(1,1,0)$. In the case of a toroidal shell, the ends $x=0$ and $x=1$ are coincident and $\mathbf{B}_0=-\mathbf{B}_1=\mathbf{I}$.

3. METHOD

Since the coefficient matrix $\mathbf{A}(x;\lambda)$ is dependent on x, in general no universally suitable approach is available to produce the exact solution of the eigenvalue problem (6)-(7). Here, the transfer matrix method is developed to seek an approximation to the solution.

The domain $[0,1]$ is partitioned into N segments, defining the mesh

$$\Delta : 0 = x_0 < x_1 < \cdots < x_N = 1$$

The ith segment is defined by $\Omega_i \stackrel{def}{=} \{x | \ x_{i-1} < x < x_i\}$, and the restriction of $\mathbf{y}$ to Ω_i is denoted by $\mathbf{y}_i$. Equation (6) may be rewritten, on Ω_i, as

$$\frac{d}{dx}\mathbf{y}_i = \mathbf{A}(x;\lambda)\mathbf{y}_i \tag{8}$$

The coefficient matrix $\mathbf{A}(x;\lambda)$ is replaced by the linear approximation

$$\mathbf{A}(x;\lambda) = \mathbf{A}_i^{(0)} + \mathbf{A}_i^{(1)}(x - x_{i-1}) \tag{9}$$

where

$$\begin{aligned} \mathbf{A}_i^{(0)} &= \mathbf{A}(x_i, \lambda), \\ \mathbf{A}_i^{(1)} &= \frac{1}{h_i}[\mathbf{A}(x_i, \lambda) - \mathbf{A}(x_{i-1}, \lambda)] \end{aligned}$$

with the step $h_i = x_i - x_{i-1}$. Next, the solution of equation (8) is approximated on Ω_i as a truncated power series:

$$\mathbf{y}_i(x;\lambda) = [\sum_{k=0}^{M} \mathbf{C}_i^{(k)}(x - x_{i-1})^k]\mathbf{z}_{i-1} \tag{10}$$

where

$$\mathbf{z}_{i-1} = \mathbf{y}_i(x_{i-1};\lambda) \tag{11}$$

Substitution of the approximations (9) and (10) into (8), yields the recursive relations which determine the coefficient matrix $\mathbf{C}_i^{(k)}$

$$\begin{aligned} &\mathbf{C}_i^{(0)} = \mathbf{I}, \mathbf{C}_i^{(1)} = \mathbf{A}_i^{(0)}\mathbf{C}_i^{(0)}, \\ &\mathbf{C}_i^{(k)} = \frac{1}{k}[\mathbf{A}_i^{(0)}\mathbf{C}_i^{(k-1)} + \mathbf{A}_i^{(1)}\mathbf{C}_i^{(k-2)}] \\ &(k = 2, \cdots, M) \end{aligned} \tag{12}$$

Continuity of the solution at each node, $x = x_i$, requires

$$\mathbf{y}_i(x_i;\lambda) = \mathbf{y}_{i+1}(x_i;\lambda) = \mathbf{z}_i \tag{13}$$

Letting $x = x_i$ in (10) and by using (11), it follows that the relationship between $\mathbf{z}_i$ and $\mathbf{z}_{i-1}$ is given by

$$\mathbf{z}_i = \mathbf{T}_i(\lambda)\mathbf{z}_{i-1} \quad (i = 1, 2, ..., N) \tag{14}$$

$\mathbf{T}_i(\lambda)$ is (an approximation to) the *segment transfer matrix* and is given by

$$\mathbf{T}_i(\lambda) = \sum_{k=0}^{M} \mathbf{C}_i^{(k)} h_i^k \tag{15}$$

Thus, from (13) and (14), the relationship between the boundary-values $\mathbf{y}(1)$ and $\mathbf{y}(0)$ is given by

$$\mathbf{y}(1) = \mathbf{T}(\lambda)\mathbf{y}(0) \tag{16}$$

where $\mathbf{T}(\lambda)$ is (an approximation to) the *global transfer matrix* and it may be expressed as a product of segment transfer matrices as follows:

$$\mathbf{T}(\lambda) = \mathbf{T}_N(\lambda)\ \mathbf{T}_{N-1}(\lambda)\ \cdots\ \mathbf{T}_1(\lambda) \tag{17}$$

Substituting (16) into (7) a system of linear algebraic equation given by

$$[\mathbf{B}_0 + \mathbf{B}_1\mathbf{T}(\lambda)]\mathbf{y}(0) = \mathbf{0} \tag{18}$$

is obtained for the initial value $\mathbf{y}(0)$. Since $\mathbf{y}(0) \neq \mathbf{0}$, the the *frequency equation* is

$$f(\lambda) \stackrel{def}{=} \det[\mathbf{B}_0 + \mathbf{B}_1\mathbf{T}(\lambda)] = 0 \tag{19}$$

The above method is an effective generalisation of methods based on the matrix exponential (Tottenham & Shimizu 1972; Tavakoli & Singh 1989) wherein it is assumed that the matrix $\mathbf{A}(x;\lambda)$ is constant on each segment. The new method can clearly be generalised to even higher order. In addition, by means of the general formulation of boundary conditions, the frequency equation is provided in the simple standardised form (19). It should be emphasised that it is not necessary to reorder the elements of $\mathbf{y}$ and to find a transformation of variables as is common in the literature (Kalnins 1964; Tavakoli & Singh 1989; Kayran et al. 1994).

4. EXAMPLES

4.1 *Comparison with an exact solution*

To illustrate the convergence and accuracy of the

method, an example of a closed spherical shell is considered. In this case, $\kappa_1 = \kappa_2 = \pi$, $\kappa_3 = \pi \cot x$. Comparison of the numerical results with the exact value of the lowest dimensionless frequency, $\Omega = \omega R \sqrt{\rho / E}$ ($h/R = 0.01, \nu = 0.3$) (Long 1969) is shown in Table 1. For fixed degree, M, of the truncated power series, excellent convergence is observed when the number of elements, N, is increased.

Table 1. Convergence of the lowest dismensionless frequency Ω

N	M			
	6	8	10	12
60	0.737	0.737	0.737	0.736
80	0.736	0.737	0.735	0.735
100	0.736	0.735	0.735	0.735
exact		0.735		

Many numerical methods typically encounter some inherent difficulties for thin shells. The difficulties originate in the stiffness of system (6) because the geometric parameter ε is small. To examine the numerical stability of the present method, the example of closed spherical shell ($\nu = 0.3$) is studied. The exact value (Long 1969) of the fundamental frequency is compared with the approximate value computed using the present method with $N = 300$ and $M = 10$. The results are recorded in Table 2. It can be seen that with decrease of the geometric parameter ε, the present method is still of the satisfactory accuracy for very small ε .

Table 2. Test of numerical stability $\Omega = \omega R \sqrt{\rho / E}$ ($\nu = 0.3$)

ε	present	exact
10^{-4}	0.73703	0.73695
10^{-5}	0.73509	0.73500
10^{-6}	0.73489	0.73481
10^{-7}	0.73486	0.73479
0	–	0.73478*

* The result is obtained in terms of membrane theory (Long 1969)

4.2 *Combined shell structures*

Shell structures consisting of a number of elementary segments (e.g. conical, spherical, cylindrical) joined together are widely used in engineering applications. The numerical method is applied to compute the eigenfrequencies of vibration of a thin hemi-spherically end-capped circular cylinder. The radii of the endcaps and cylinder all have the value $R = 114.3$ mm. All three segments have the same thickness $h = 2.03$ mm (so, $h / R \simeq 0.02$). The half length of the cylinder is $l_2 = 171.5$ mm ($\pi R/(2l_2) \simeq 1$). The material parameter values are as follows: $E = 207$ GPa, $\rho = 7800$ kg/ m^3 and $\nu = 0.3$. For the spherical shell segments $\kappa_1 = \kappa_2 = \pi/2$, $\kappa_3 = \pi \cot x/2$ and for the cylindrical segment $\kappa_1 = 0, \kappa_2 \simeq \pi/2, \kappa_3 = 0$. The computed values of the lowest three symmeric mode eigenfrequencies using $M = 8$ and $N = 300$, together with comparisons from the literature, are shown in the Table 3.

Table 3. Natural frequency, $f = \frac{1}{2\pi R}\sqrt{\frac{E}{\rho(1-\nu^2)}}\lambda$, for circular cylindrical shell with hemispherical endcaps.

	FEM*1	SMM*2	FEM*3	present
f_1	4005	4011	4004	4003
f_2	6297	6292	6292	6296
f_3	6808	6812	6804	6806

*1 Finite element method based on thin theory (Tavakoli & Singh 1989)
*2 State space method based on thin shell theory (Tavakoli & Singh 1989)
*3 C(0) Mindlin-Reissner finite element method (Ozakca & Hinton 1994)

5 CONCLUSION

A new and generalised transfer matrix method for computation of frequencies of free vibration of axisymmetric shells has been presented. A simple and unified form of frequency equation is given by introducing two matrices describing general boundary conditions including arbitrary elastic restrictions. The transfer matrix method has been developed by introducing a higher order and smoother approximation of the system coefficient matrix and by constructing an approximate analytical expression for the element transfer matrix in the form of a truncated power series. The numerical results have demonstrated that the present method is a simple and efficient computational technique with satisfactory numerical behaviour.

In addition, it should be pointed out that although the method has been derived for axisymmetric free vibration, it is straightforward to extend it to nonaxisymmtric vibration by means of Fourier expansion in the circumferential coordinate θ . The order of the coefficient matrix, $\mathbf{A}$, will become 10×10 in this case. Moreover, since the method of solution in this paper is discussed for a general eigenvalue problem for a system of first-order ordinary differential equations (6)-(7), it can be used for other similar mathematical problems with different applied backgrounds.

Acknowledgements - The first author is grateful for the opportunity of postdoctoral research afforded by the Department of Applied Mathematics, University College Cork. The authors would like to thank Dr. Michael J. A. O'Callaghan for his encouragement and support of this study.

REFERENCES

Forsberg, K. 1964. Influence of boundary conditions on the modal characteristics of thin cylindrical shells. *AIAA J.* 2: 2150-2157.

Horner, G. C. & W. D. Pilkey 1978. The Riccati transfer metrix method. *J. Mech. Design* 1 : 297-302.

Irie,T., G. Yamada & Y. Kaneko 1984. Natural frequencies of truncated conical shell. *J. Sound Vib.*92: 447-455.

Kalnins, A. 1964a, Free vibration of rotationally symmetric shells, *J. Acoust. Soc. Amer.* 36 (7): 1355-1365.

Kalnins, A. 1964b, Analysis of shells of revolution subjected to symmetrical and nonsymmetrical loads, *J. Appl. Mech. ASME* 31: 467-476.

Kayran, A., J. R. Vinson & E. S. Ardic 1994. A method for the calculation of natural frequencies of orthotropic axisymmetrically loaded shells of revolution. *J. Appl. Mech. ASME* 116: 16-25.

Kobayashi, Y. & G. Yamada 1991. Free vibration of a spinning polar orthotropic shallow spherical shell. *JSME Int. J.* 34 : 233-238.

Kraus, H. 1967. *Thin Elastic Shells*. New York: John Wiley.

Long, C. F. 1969. Frequency analysis of complete spherical shells. *J. Engng. Mech. Div. ASCE* 3: 505-517.

Luah, M. H. & S. C. Fan 1989. General free vibration analysis of shell of revolution using the spline finite element method. *Comput. Struct.* 33: 1153-1162.

Ozakca, M. & E. Hinton 1994. Free vibration analysis and optimisation of axisymmetric plates and shells--I. finite element formulation. *Comput. Struct.* 52 : 1181-1197.

Sankar, S. 1977. Extended transfermetrix method for free vibration of shell of revolution. *Shock and Vibr. Bulletin* 47: 121-133.

Srinivasan, R. S. 1989. Axisymmertric vibration of thick conical shells. *J. Sound Vibr.*135: 171-176.

Tavakoli, M. S. & R. Singh 1989. Eigensolution of joined / hermertic shell strucures using the state space method. *J. Sound Vib.* 130: 97-123.

Tottenham, H. & K. Shimizu 1972. Analysis of the free vibration of cantilever cylindrical thin elastic shells by the matrix progression method, *Int. J. Mech. Sci.* 14: 293-310.

Uhrig, R. 1966. The transfer metrix seen as one method of structural analysis among others. *J. Sound Vib.* 4: 136-148.

Wang, J. & K. Schweizerhof 1996. Boundary-Domain element method for vibration of moderately thick laminated orthotropic shallow shells. Int. J. Solids Structs. 33: 11-18.

Modern Practice in Stress and Vibration Analysis, Gilchrist (ed.) © *1997 Balkema, Rotterdam, ISBN 90 5410 896 7*

On 3-D stress analysis of laminated plates with perturbed interfaces

M. Kashtalian
Timoshenko Institute of Mechanics of National Academy of Sciences of Ukraine, Kiev, Ukraine

ABSTRACT: Three-dimensional stress-strain analysis of laminated plates with interfaces, the shape of which is slightly perturbed and may be described by sufficiently smooth function of two variables and small dimensionless parameter, has been fulfilled. Using the method of perturbation of boundary shape based on representing of stresses and displacements as series in respect to the small parameter which describes the shape of interfaces, the initial three-dimensional problem of theory of elasticity has been reduced to the sequence of problems for the plate with perfectly flat interfaces. The influence of the interface shape (amplitude of waving and value of the wave-forming parameter) on stresses and deflection of isotropic two-layered plates under bending with periodically waved in one direction interfaces has been examined. To obtain the numerical results with sufficient accuracy first three approximations for stresses and displacements have been proved to be enough.

1 INTRODUCTION

Mechanical failure of composites often results from stress concentration at the interfaces due to the interface defects, imperfections and inhomogeneities inherited from the manufacturing process (Piatti 1978). Imperfections of many varieties such as layers of uneven thickness, wavy layers or undulated interfaces in which nominally flat surfaces of the layers are developing during manufacturing, are present in laminated composite structures. The adhesion between the layers at the interfaces is not usually broken by the above-mentioned imperfections, and therefore, they may be considered as geometrical anomalies but not serious technological defects.

The nature of these imperfections reflects the nature of surface evolution dictated by the competition between surface and elastic energies. The nominally flat surface of an elastically stressed body is unstable with respect to the formation of surface undulations of wavelengths greater than some critical value (Srolovitz 1989). The amplitude of these perturbations is sufficiently small in comparison with structure length scales of the composite material structure. However, they could be so important in determining the response of composite materials that steps must be undertaken to understand their role.

For analytical description of deformation of macroscopically inhomogeneous solids like composite materials various kinds of continual theories have been developed and used in mechanics of composites, in which inhomogeneous material is modelled by the homogeneous anisotropic medium. Another approach to investigation of mechanical behaviour of composites is based on the structural model, i.e. the model of piecewise-homogeneous medium.

The first attempts to examine the role which play geometrical imperfections in the failure of composite material have been already undertaken within the both approaches (Guz & Akbarov 1995). They were concentrated on determination of self-equilibrated stresses in bulk composite material acting at the parts of the interfaces the size of which is comparable with or less than length scale of material structure. At this point the second approach is more advantageous, since structural model allows to determine the above stresses more precisely. However, no investigations of composite structures like plates or shells with geometrical interfacial imperfections has not yet been made.

The present paper is focused on investigation of stress distribution in laminated plates with perturbed interfaces and on examination of stress redistribution effects due to this kind of geometrical imperfections.

2 PROBLEM STATEMENT

The problem statement is as follows. Laminated plate of finite dimensions $0 \leq x \leq a$, $0 \leq y \leq b$, consisting of N layers with external surfaces S_0, S_N and interfaces S_i, $(i = 1, \cdots, N-1)$ has been considered (Fig. 1).
External surfaces and interfaces of the plate are being treated as slightly perturbed from the reference state in which they would be perfectly flat. Their geometry is assumed to allow description by shape function $f(x,y)$ - sufficiently smooth function of two variables which characterises the shape of perturbation, and a small dimensionless parameter ϵ which characterise the magnitude of deviation of the perturbed surface or interface $S_i (i = 0, \cdots, N)$ from its reference plate $z = h_i$. Shape equations for perturbed surfaces and interfaces are then

$$z = h_i + \epsilon f(x,y) \tag{1}$$

where $h_0 = 0$,$h_N = h$.

Suppose the bonding between the layers is perfect, and continuity conditions for stresses and displacements are fulfilled at the interfaces $S_k, (k = 1, \cdots, N-1)$

$$(\sigma_{xt,k+1} - \sigma_{xt,k}) n_{x,k} + (\sigma_{yt,k+1} - \sigma_{yt,k}) n_{y,k} +$$

$$+(\sigma_{zt,k+1} - \sigma_{zt,k}) n_{z,k} = 0 \tag{2}$$

$$u_{t,k} - u_{t,k+1} = 0 \tag{3}$$

where $\sigma_{jt,k}, u_{t,k}$ $(t = x, y, z)$ are stresses and displacements in the k-th layer $(k = 1, \cdots, N)$; $n_{t,i}$ $(t = x, y, z)$ are direction cosines for the surface or interface S_i $(i = 0, \cdots, N)$.

Given the shape equations (1), direction cosines $n_{t,i}$ may be expressed as follows

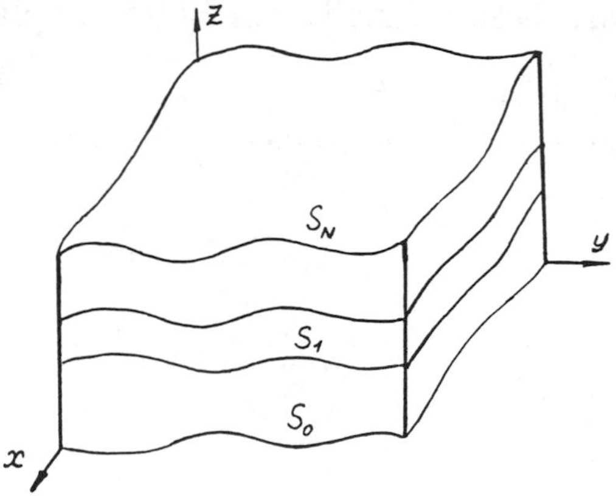

Figure 1: Laminated plate with perturbed external surfaces and interfaces

$$n_{x,i} = -\frac{\epsilon}{\Delta}\frac{\partial f}{\partial x} \quad n_{y,i} = -\frac{\epsilon}{\Delta}\frac{\partial f}{\partial y} \quad n_{z,i} = \frac{1}{\Delta} \tag{4}$$

where

$$\Delta = \pm(1 + \epsilon^2[(\frac{\partial f}{\partial x})^2 + (\frac{\partial f}{\partial y})^2])^{1/2} \tag{5}$$

Suppose the plate is loaded by transversal load $Q(x,y)$ at its top surface with bottom surface to be free. Boundary conditions at S_N, S_0 will be then

$$\sigma_{xt,N} n_{x,N} + \sigma_{yt,N} n_{y,N} + \sigma_{zt,N} n_{z,N} = Q n_{t,N} \tag{6}$$

$$\sigma_{xt,1} n_{x,0} + \sigma_{yt,1} n_{y,0} + +\sigma_{zt,1} n_{z,0} = 0 \tag{7}$$

On the edges of the plate $x = 0, a$ and $y = 0, b$ boundary conditions of Navier type are assumed to be fulfilled for each layer.

3 BOUNDARY SHAPE PERTURBATION METHOD

Due to the complexity of geometrical shape of interfaces the stated three-dimensional boundary-value problem (1)–(7) cannot be solved directly, and the method of perturbation of boundary shape (Nemish 1989) has been applied. According to it, stresses and displacements as well as components of load and direct cosines in boundary conditions are to be presented as series in respect to the small parameter ϵ , i.e.

$$[\sigma_{jt,k}, u_{t,k}, Q_t, n_{t,i}] = \sum_{p=0}^{\infty} \epsilon^p [\sigma_{jt,k}^{(p)}, u_{t,k}^{(p)}, Q_t^{(p)}, n_{t,i}^{(p)}] \quad (8)$$

and $\sigma_{jt,k}^{(p)}$ and $u_{t,k}^{(p)}$ are supposed to be expanded in Taylor series in vicinity of $z = h_i$, so that on S_i

$$[\sigma_{jt,k}, u_{t,k}]|_{S_i} =$$

$$= \sum_{p=0}^{\infty} \epsilon^p \sum_{q=0}^{p} \frac{f^q}{q!} \frac{\partial^q}{\partial z^q} [\sigma_{jt,k}^{(p-q)}, u_{t,k}^{(p-q)}]|_{z=h_i} \quad (9)$$

By means of this approach the initial boundary-value problem for the plate with perturbed surfaces and interfaces may be reduced to the sequence of recurrent boundary-value problems for plate with flat reference planes $z = h_i$.

In the zeroth-order approximation, i.e. for $p = 0$, the obtained boundary-value problem will coincide with one for a laminated plate with perfectly flat surfaces and interfaces $z = h_i$

$$\sigma_{zz,N}^{(0)}|_{z=h} = Q \qquad \sigma_{xz,N}^{(0)}|_{z=h} = \sigma_{yz,N}^{(0)}|_{z=h} = 0 \quad (10)$$

$$\sigma_{zz,1}^{(0)}|_{z=0} = \sigma_{xz,1}^{(0)}|_{z=0} = \sigma_{yz,0}^{(0)}|_{z=0} = 0 \quad (11)$$

$$\sigma_{zt,k+1}^{(0)} - \sigma_{zt,k}^{(0)}|_{z=h_k} = 0 \quad (12)$$

$$u_{t,k+1}^{(0)} - u_{t,k}^{(0)}|_{z=h_k} = 0 \quad (13)$$

In the higher-order ($p = 1, 2, 3, ...$) approximations boundary conditions at the reference planes $z = h_i$ which correspond to the boundary conditions at the perturbed surfaces and interfaces S_i will be as follows

$$\sigma_{zt,N}^{(p)}|_{z=h} = -Q_k^{(p)} + \quad (14)$$

$$+ \sum_{q=1}^{p} [N_{x,0}^{(q)} \sigma_{xt,N}^{(p-q)} + N_{y,0}^{(q)} \sigma_{yt,N}^{(p-q)} + N_{z,0}^{(q)} \sigma_{zt,N}^{(p-q)}]_{z=h}$$

$$\sigma_{zt,1}^{(p)}|_{z=0} = \quad (15)$$

$$= - \sum_{q=1}^{p} [N_{x,0}^{(q)} \sigma_{xt,1}^{(p-q)} + N_{y,0}^{(q)} \sigma_{yt,1}^{(p-q)} + N_{z,0}^{(q)} \sigma_{zt,1}^{(p-q)}]_{z=0}$$

$$\sigma_{zt,k+1}^{(p)} - \sigma_{zt,k}^{(p)}|_{z=h_k} = \quad (16)$$

$$= \sum_{q=1}^{p} [N_{x,k}^{(q)} (\sigma_{xt,k+1}^{(p-q)} - \sigma_{xt,k}^{(p-q)}) + N_{y,k}^{(q)} (\sigma_{yt,k+1}^{(p-q)} -$$

$$-\sigma_{yt,k}^{(p-q)}) + N_{z,k}^{(q)} (\sigma_{zt,k+1}^{(p-q)} - \sigma_{zt,k}^{(p-q)})]|_{z=h}$$

$$u_{t,k+1}^{(p)} - u_{t,k}^{(p)}|_{z=h_k} = \quad (17)$$

$$= \sum_{q=1}^{p} L_{t,k}^{(q)} (u_{t,k+1}^{(p-q)} - u_{t,k}^{(p-q)})|_{z=h_k}$$

Differential operators $N_{t,i}^{(q)}$ and $L_{t,i}^{(q)}$ in eqn (14)–(17) depend on the shape function f and its derivates. In the most general form they can be expressed as follows

$$N_{t,i}^{(q)} = - \sum_{s=1}^{q} \frac{f^s}{s!} n_{t,i}^{(q-s)} \frac{\partial^s}{\partial z^s} \quad (18)$$

$$L_{t,i}^{(q)} = f^q \frac{\partial^q}{\partial z^q} \qquad q = 1, 2, 3, ... \quad (19)$$

In the first and the second approximations, in particular, they are

$$N_{x,i}^{(1)} = \frac{\partial f}{\partial x} \qquad N_{y,i}^{(1)} = \frac{\partial f}{\partial y} \qquad N_{z,i}^{(1)} = -f \frac{\partial}{\partial z} \quad (20)$$

$$N_{x,i}^{(2)} = f \frac{\partial f}{\partial x} \frac{\partial}{\partial z} \qquad N_{y,i}^{(2)} = f \frac{\partial f}{\partial y} \frac{\partial}{\partial z} \quad (21)$$

$$N_{z,i}^{(2)} = -\frac{1}{2} [f^2 \frac{\partial^2}{\partial z^2} - (\frac{\partial f}{\partial x})^2 - (\frac{\partial f}{\partial y})^2]$$

For further solution of the obtained sequence of boundary-value problems (14)–(17) it is necessary to specify the shape function as well as the material of the plate.

Since application of the boundary shape perturbation involves only boundary conditions at the perturbed surfaces and interfaces, equilibrium equations and boundary conditions at the edges of the plate retain their initial form in all approximations. This peculiarity opens the possibility to use available general solutions of equilibrium equations for solving of recurrent boundary-value problems. In particular, for isotropic material well-known solution (Youngdahl 1969) has been successfully applied. For transversally isotropic and orthotropic materials the following representations (Green & Zerna 1954, Pagano 1972), for example, may be used.

4 NUMERICAL RESULTS

Numerical results for a two-layered free-supported plate under bending by sinusoidal load will be considered next to illustrate the influence of periodically waved in one direction interfaces on redistribution of stresses in the plate. The load and shape functions were taken respectively

$$Q(x,y) = -q\sin\frac{\pi x}{a}\sin\frac{\pi y}{b} \tag{22}$$

$$f(x,y) = f(y) = \pm h\cos\frac{\pi\omega y}{b} \tag{23}$$

where q and ω are load intensity and wave foundation parameter respectively.

The layers were considered to be isotropic with shear moduli G_1 and G_2 and Poisson's ratios ν_1 and ν_2. For solving of the obtained recurrent sequence of boundary-value problems (10)–(13), (14)–(17) Youngdahl's (1969) general solution of equilibrium equations

$$u_{x,k}^{(p)} = \frac{\partial\psi_{1,k}^{(p)}}{\partial x} - \frac{z}{4(1-\nu_k)}\frac{\partial^2\psi_{2,k}^{(p)}}{\partial x\partial z} + \frac{\partial\psi_{3,k}^{(p)}}{\partial y}$$

$$u_{y,k}^{(p)} = \frac{\partial\psi_{1,k}^{(p)}}{\partial y} - \frac{z}{4(1-\nu_k)}\frac{\partial^2\psi_{2,k}^{(p)}}{\partial y\partial z} - \frac{\partial\psi_{3,k}^{(p)}}{\partial x} \tag{24}$$

$$u_{z,k}^{(p)} = \frac{\partial\psi_{1,k}^{(p)}}{\partial z} - \frac{z}{4(1-\nu_k)}\frac{\partial^2\psi_{2,k}^{(p)}}{\partial z^2} + \frac{3-4\nu_k}{4(1-\nu_k)}\frac{\partial\psi_{2,k}^{(p)}}{\partial z}$$

has been used.

For the applied load (22) and the shape function (23) harmonic functions $\psi_{j,k}^{(p)} (j = 1,2,3)$ were obtained in the following closed form

$$\psi_{j,k}^{(p)}|_{j=1,2} = \sum_{s=1}^{p+1}(A_{(2j-1),k}^{(p,s)}\cosh\xi^{(p,s)} +$$

$$+A_{(2j),k}^{(p,s)}\sinh\xi^{(p,s)})\sin\frac{\pi x}{a}\sin\frac{\lambda^{(p,s)}\pi y}{b}$$

$$\psi_{3,k}^{(p)} = \sum_{s=1}^{p+1}(A_{5,k}^{(p,s)}\cosh\xi^{(p,s)} + \tag{25}$$

$$+A_{6,k}^{(p,s)}\sinh\xi^{(p,s)})\cos\frac{\pi x}{a}\cos\frac{\lambda^{(p,s)}\pi y}{b}$$

where

$$\lambda^{(0,1)} = 1$$

$$\lambda^{(1,1)} = \omega + 1 \qquad \lambda^{(1,2)} = \omega - 1 \tag{26}$$

$$\lambda^{(2,1)} = 2\omega + 1 \quad \lambda^{(2,2)} = 2\omega - 1 \quad \lambda^{(2,3)} = 1$$

$$\xi^{(p,s)} = ((\frac{\pi}{a})^2 + (\frac{\lambda^{(p,s)}\pi}{b})^2)^{1/2}\frac{z}{h} \tag{27}$$

Arbitrary constants $A_{j,k}^{(p,s)}$ $(j = 1,\cdots,6,\ s = 1,\cdots,(p+1),\ k = 1,\cdots,N)$ in the p-th order approximation $(p = 0,1,2,\cdots)$ are to be determined from the boundary conditions (10)–(13), (14)–(17) which build $p+1$ systems of linear algebraic equation of the $6N$-th order with respect to $A_{j,k}^{(p,s)}$.

Stress and displacements values were obtained using the first four approximations and are therefore of accuracy $O(\epsilon^3)$

$$\{\sigma_{jt,k}, u_{t,k}\} = \sum_{p=0}^{3}\epsilon^p[\sigma_{jt,k}^{(p)}, u_{t,k}^{(p)}\} + O(\epsilon^3) \tag{28}$$

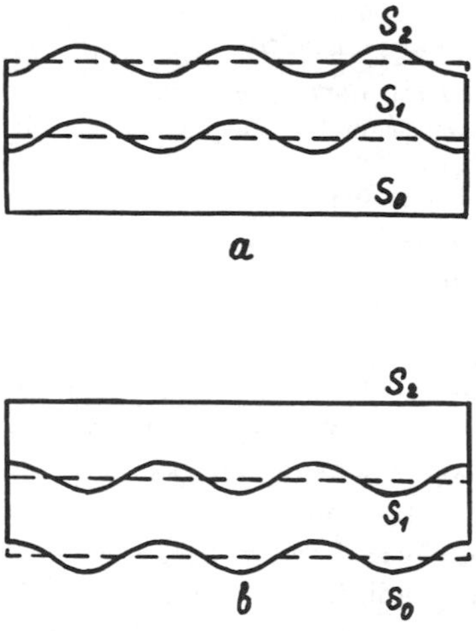

Figure 2: Types of plate geometry (profiles in $y0z$ plane).

Table 1.

ω	$\frac{\sigma_{xx,1}^{(0)}}{q}\|_{z=h_1}$	$\frac{\sigma_{xx,1}}{q}\|_{S_1}$	$\delta, \%$	$\frac{G_1}{G_2}$
		type a		
2	-8.085	-7.099	-12.2	
6	-8.085	-8.716	7.8	10
10	-8.085	-9.641	19.3	
2	0.820	0.660	-19.5	
6	0.820	0.630	-23.2	0.1
10	0.820	0.540	-34.1	
		type b		
2	-8.085	-6.620	-18.1	
6	-8.085	-6.204	-22.2	10
10	-8.085	-5.340	-33.8	
2	0.820	0.737	-10.2	
6	0.820	0.900	9.7	0.1
10	0.820	0.993	21.1	

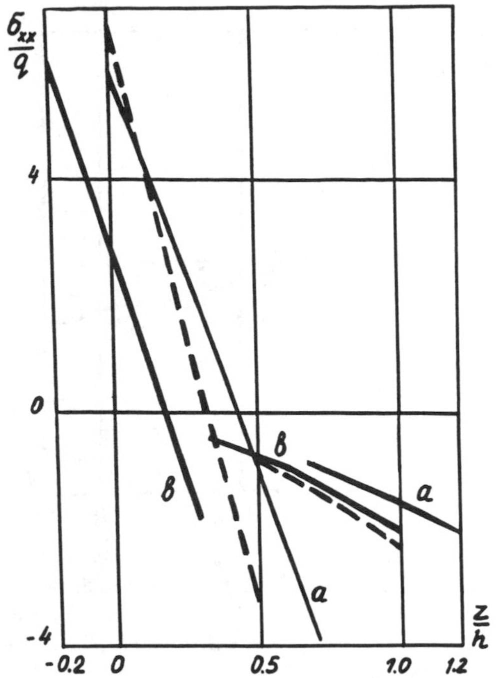

Figure 3: Distribution of σ_{xx}/q in plates with perturbed (solid line) and flat (hatched line) surfaces and interfaces: $h/a = 3$, $h/b = 6$, $h_1/h = 0.5$, $\nu_1 = \nu_2 = 0.3$, $G_1/G_2 = 5$; for perturbed interfaces: $\omega = 2$, $\epsilon = 0.2$.

$$\{\sigma_{jt,k}, u_{t,k}\}|_{S_i} = \tag{29}$$

$$= \sum_{p=0}^{3} \epsilon^p \sum_{q=0}^{p} \frac{f^q}{q!} \frac{\partial^q}{\partial z^q} \{\sigma_{jt,k}^{(p-q)}, u_{t,k}^{(p-q)}\}|_{z=h_i} + O(\epsilon^3)$$

Numerical results for the values of $\sigma_{xx,1}/q$ in the middle of the plate ($x = a/2, y = b/2$) at the perturbed interface S_1 with amplitude of perturbation $\epsilon = 0.1$ are given in Table 1. Two types of plates, the profiles of which in the co-ordinate plane $y0z$ are shown on Figure 2 (for $\omega = 6$), are considered Plate dimensions were taken $h/a = 3$, $h/b = 18$, and the layers had the same thickness, so that $h_1/h = 0.5$. Stress values at the perturbed interface have been are compared with the $\sigma_{xx,1}^{(0)}/q|_{z=h_1}$, i.e. a zeroth-order approximation, and the relative difference is expresses in percents. The results are presented for several values of wave foundation parameter ω and layers shear moduli ratio G_1/G_2

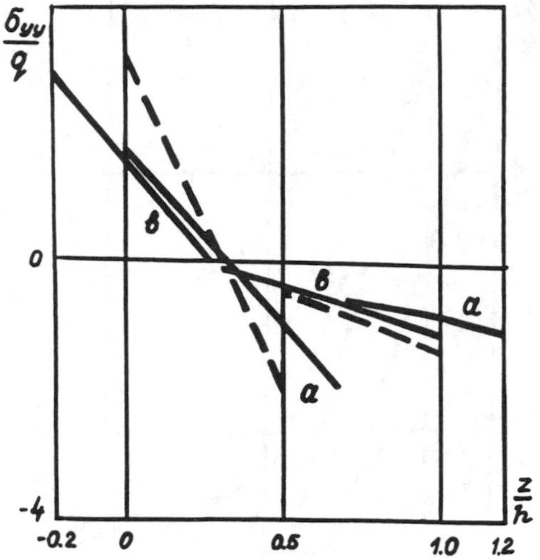

Figure 4: Distribution of σ_{yy}/q in plates with perturbed (solid line) and flat (hatched line) surfaces and interfaces: $h/a = 3$, $h/b = 6$, $h_1/h = 0.5$, $\nu_1 = \nu_2 = 0.3$, $G_1/G_2 = 5$; for perturbed interfaces: $\omega = 2$, $\epsilon = 0.2$.

and illustrate the influence of these parameters on stress value at the perturbed interface. Depending on the type of the plate and the shear moduli ration, stress value may increase as well as decrease in comparison with increasing of wave foundation parameter value.

Through-the-thickness distributions of σ_{xx}/q and σ_{yy}/q in the middle of the both types of plates are shown in the Figure 3 and 4 (solid lines). Plate dimensions were $h/a = 3$, $h/b = 6$, $h_1/h = 0.5$, geometrical characteristics of the perturbed surfaces and interfaces were $\omega = 2$, $\epsilon = 0.2$, and material properties of the layers were $G_1/G_2 = 5$,$\nu_1 = \nu_2 = 0.3$. As a reference, results of the zeroth-order approximation are given, which correspond to the plate with the same dimensions and all surfaces and interfaces being flat (hatched lines).

5 REFERENCES

Green, A.E. & W. Zerna 1954. *Theory of elasticity.* Oxford: Clarendon Press.

Guz, A.N. & S.D. Akbarov(eds.) 1995. *Mechanics of materials with curved structures.* In A.N. Guz (ed), *Mechanics of composites.* Kiev: Naukova Dumka.

Nemish, Yu.N. 1989. *Elements of the mechanics of piecewise-homogeneous bodies with non-canonical boundary surfaces.* Kiev: Naukova Dumka.

Pagano, N.J. 1970. Exact solution for rectangular bidirectional composites and sandwich plates. *J. Compos. Mater.* 4: 20-34.

Piatti, G.(ed.) 1978. *Advances in Composite Materials.* London: Applied Science Publishers.

Srolovitz, D.J. 1989. On the stability of surfaces of stressed solids.*Acta Metall.* 37: 621-625.

Youngdahl, C.K. 1969. On the completeness of a set of stress functions appropriate to the solution of elasticity problems in general cylindrical coordinates. *Int. J. Eng. Sci.* 7: 61-69.

Modern Practice in Stress and Vibration Analysis, Gilchrist (ed.) © 1997 Balkema, Rotterdam, ISBN 90 5410 896 7

Boundary element analysis of crack closure around inclusions embedded in dissimilar matrices

Y.H. Park & A.A. Becker
Department of Mechanical Engineering, University of Nottingham, UK
R.T. Fenner
Department of Mechanical Engineering, Imperial College of Science, Technology and Medicine, London, UK

ABSTRACT : A quadratic formulation of the Boundary Element (BE) method is used to analyze the fracture mechanics of spherical and circular cracks formed at the interfaces of second phase materials embedded in dissimilar matrices. Two types of geometries are considered; spherical inclusions and cylindrical fibres embedded in homogeneous dissimilar infinite matrices and subjected to tensile, compressive and biaxial remote loads as well as a thermal gradient. The fracture analysis covers the mixed-mode stress intensity factors and investigates the effect of the compressive crack-face contact stresses which cause crack closure. A range of different inclusion/matrix elastic material properties are considered in order to investigate the effect of increasing the rigidity of the inclusions. The effects of the coefficient of friction on the contact stresses and stress intensity factors are also investigated. For verification purposes, some of the BE solutions are compared to the corresponding FE solutions obtained using the ABAQUS FE code.

1. INTRODUCTION

Multi-phase materials such as ceramics and fibre-reinforced composites contain inclusions or cylindrical fibres embedded in dissimilar matrices. The study of the contact stresses and the fracture behaviour of these inclusions is important in assessing the structural integrity of such materials. Micro-cracks often nucleate in and around second phase materials due to the mismatch in elastic and thermal properties (see, for example, Hasselman and Fulrath, 1967, Evans, 1974 and Green, 1983).

Previous work on inclusions involving analytical approaches (see, for example, Wilson and Goree, 1967, and Kant and Bogy, 1980) have been limited to simple geometries and have ignored the effects of limited frictional slip in the interface between the dissimilar materials. Numerical techniques such as the Boundary Element (BE) and the Finite Element (FE) methods can be used to analyse the stresses around the inclusions, and are versatile enough to cover a range of geometries, loads and combinations of material properties. A limited amount of numerical modelling has been reported using the Finite Difference method (Ito et al, 1981) and the FE method (Ito and Nelson, 1983). More recently, the BE method has been used to analyze cracks in materials containing inclusions (see, for example, Tan and Gao, 1990 and Park et al, 1992, 1995). Previous work on this type of inclusion/matrix has included contact mechanics analysis and fracture but has not covered the effect of the crack faces pressing against each other thereby causing a "negative" stress intensity factor.

The BE method is particularly well suited for the analysis of dissimilar material contact problems and interface cracks due to the high resolution of the computed stresses and the lesser degree of mesh refinement needed to obtain comparable accuracy with the FE method. More significantly, the contacting surfaces can be modelled without special contact elements because the BE equations are coupled together to satisfy the compatibility and equilibrium conditions directly, whereas in most FE approaches, special contact elements are inserted between the contacting surfaces, and a very high degree of mesh refinement is needed around the contact/crack region.

A further advantage of the BE approach is realised when re-meshing is required around the edge of the contact area and the crack tips to obtain a better solution accuracy. A robust and reliable numerical algorithm must be devised to handle the sharp stress gradients around the crack tip as well as the possibility of slipping/sticking in the presence of friction. Previous applications of the BE method in contact problems have demonstrated its accuracy and versatility (see, for example, Andersson and Allan-Persson, 1983, Olukoko et al 1993, 1994).

In this paper, the BE method is used to analyze the crack closure occurring around second phase materials in the shape of spherical inclusions and cylindrical fibres embedded in dissimilar matrices. A range of different inclusion/matrix elastic material properties are considered in order to investigate the effect of increasing the rigidity of the inclusions. The analysis covers mixed-mode stress intensity factors and investigates the effect of the compressive crack-face contact stresses which cause crack closure. For verification purposes, some of the BE solutions are compared to the corresponding FE solutions obtained using the ABAQUS FE code (HKS, 1995).

2. BRIEF REVIEW OF THE BE FORMULATION

In the BE formulation used in this work, the partial differential equations of elasticity are transformed into integral identities relating the displacement at an interior point p to the displacements and tractions at a boundary point Q. Details of the formulation and theoretical background of the BE method can be found in several textbooks, e.g. Brebbia et al. (1983), Becker (1992) and Banerjee (1994).

To form a set of linear simultaneous algebraic equations, each point on the boundary is taken in turn as the load point p and the integrations performed numerically, using Gaussian quadrature formulae, over each element on the boundary, resulting in:

$$[A]\ [u] = [B]\ [t] \qquad \textbf{(1)}$$

where the matrices [A] and [B] are fully-populated matrices, and [u] and [t] are the displacement and traction vectors, respectively. To obtain a unique solution, the boundary conditions of prescribed displacements or prescribed tractions are then implemented, and the equations rearranged such that all the unknown variables, [x], are on the left hand side as follows:

$$[C]\ [x] = [d] \qquad \textbf{(2)}$$

where [C] is the solution matrix, and [d] contains all known quantities. The equations are then solved by Gaussian elimination since [C] is fully populated.

In contact problems, the elements on the contact interface need special treatment since neither the displacement nor traction values are given. At these points, the equilibrium conditions and compatibility relationships must be satisfied. Each domain is treated separately to form equation (1), and the resulting matrices [A] and [B] are directly coupled according to the relevant contact conditions, with the number of unknowns remaining equal to the number of equations. Therefore, the equilibrium and compatibility equations are directly incorporated, rather than approximately satisfied (as in many FE formulations).

In fracture problems, the stress singularity at the crack tip is represented by the established quarter-point node-shifted elements (see, for example, Barsoum, 1976). The stress intensity factors can be calculated by several methods, such as displacement and stress extrapolation or the J-contour integral (see, for example, Parker, 1981)

The BE solutions presented in this paper have been obtained using the BEACON software (Becker, 1989). The code uses quarter-point singularity elements to model the singularity at the crack tip, and directly implements the equilibrium and compatibility equations at the contacting surfaces without using special contact or interface elements. The FE results presented for verification purposes have been obtained by using the ABAQUS FE code (HKS, 1995), with special crack tip collapsed elements used to represent the singularity at the crack tip.

3. CONTACT AND FRACTURE ANALYSIS

Contact problems may be classified according to extent of the initial contact conditions and the progressive change of contact status. The geometries of the inclusions and the loading situations may be classified as non-Hertzian, conformal, receding contact problems, since the initial undeformed bodies fit exactly or closely together, and the contact region reduces as the load is applied. Friction slip is modelled by allowing the contacting nodes to slip if the tangential traction exceeds the value of the compressive normal traction multiplied by the coefficient of friction, μ.

Crack closure around inclusions can occur because of the mismatch of mechanical and/or thermal material properties as well as the loading conditions. Previous studies of similar geometries (see, for example, Park et al, 1994) have shown that crack closure occurs at spherical and circular cracks with angles over 70°. In most engineering design analysis, the compressive forces on closed crack faces, e.g. the compressive part of a fatigue load cycle, are often ignored and the mode-I stress intensity factor K_I is assumed to be zero.

A hypothetical "negative" stress intensity factor has been defined by some researchers, e.g. Cotterell and Rice (1980) and Heitzer and Matheck (1990), by reversing the compressive forces on the crack faces, i.e. allowing the crack faces to overlap. Such assumptions, however, do not take into account the compressive forces on the crack faces.

Examining the existing literature on crack closure, e.g. Paris and Tada (1975), Aksogan (1975), Gustafson (1976), Karami and Fenner (1986) and Comninou (1990), four categories of crack opening/closure configurations can be identified, as shown in Figure 1. Configuration (a) is the classical definition of K_I which can be calculated by using stress or displacement extrapolation techniques. In configuration (b), the crack faces are compressed against each other, but the region around the crack tip remains open, i.e. $K_I>0$. Configurations (c) and (d) represent partial and full closure of the crack faces, respectively, in which $K_I=0$.

In this paper, the contact stresses on the crack faces are calculated to establish the strength of the closure of the crack. Furthermore, the analysis also covers the shear mode of fracture (mode-II), which allows for a "negative" K_{II} since the crack faces open by a shearing action (see, for example, Comninou and Schmuser, 1979 and Gautesen and Dunders, 1988).

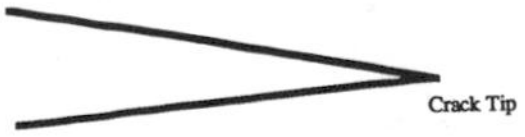

(a) Fully open crack face, $K_I>0$

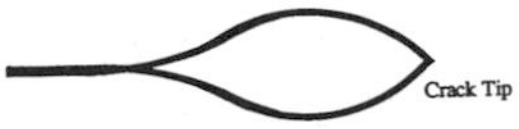

(b) Partially closed crack face, $K_I>0$

(c) Partially closed crack face, $K_I=0$

(d) Fully closed crack face, $K_I=0$

Figure 1 : Crack opening/closing configurations

4. CRACK CLOSURE AROUND SPHERICAL INCLUSIONS

This analysis concerns a spherical inclusion embedded in an infinite dissimilar matrix in which a spherical crack forms as a result of the separation of the inclusion from the matrix. The spherical inclusion is assumed to have material properties of Young's modulus E_i, Poisson's ratio ν_i and coefficient of thermal expansion α_i, whereas the corresponding material properties for the matrix are E_m, ν_m and α_m. In order to investigate the effect of the relative stiffness of the inclusion, a range of E_i/E_m values is used, together with a range of values of the coefficient of friction μ. For modelling purposes, the inclusion/matrix diameter ratio is taken as 0.2, which is considered adequate in representing an infinite matrix.

Three types of loading on the matrix are considered, (i) a tensile axial load, (ii) a compressive axial load, and (iii) a biaxial load consisting of a

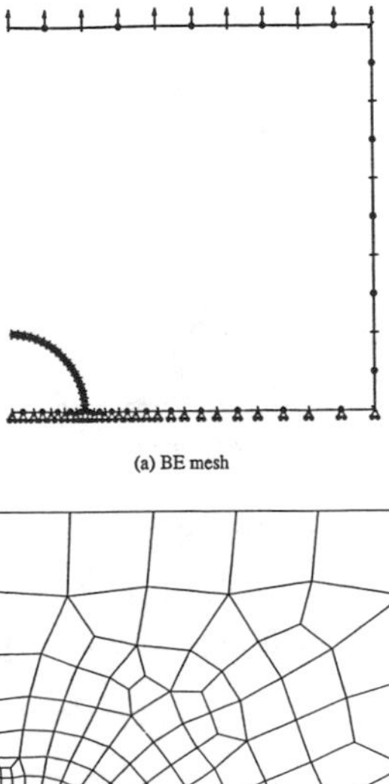

(a) BE mesh

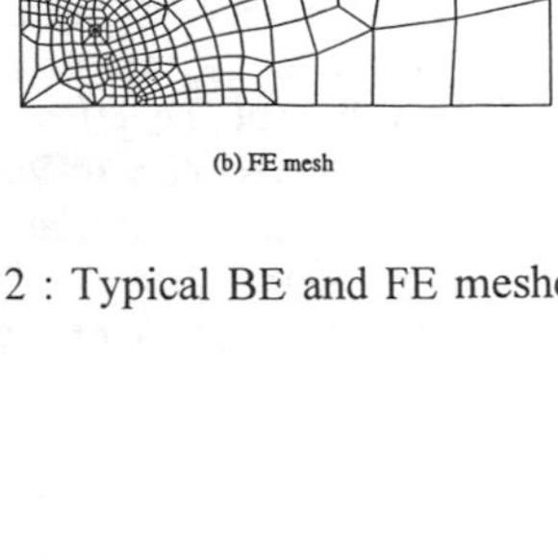

(b) FE mesh

Figure 2 : Typical BE and FE meshes used

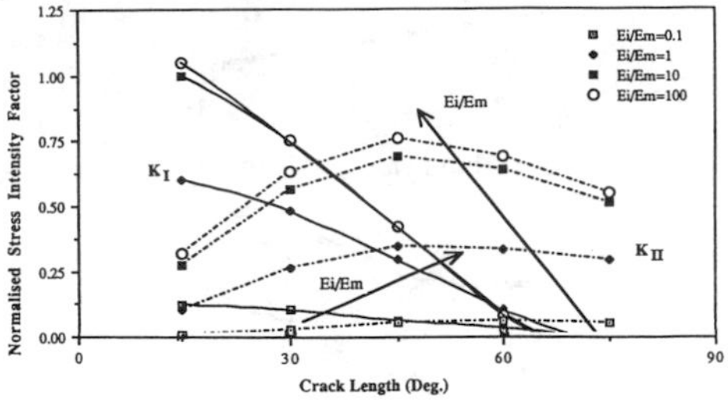

(a) Axial tensile load

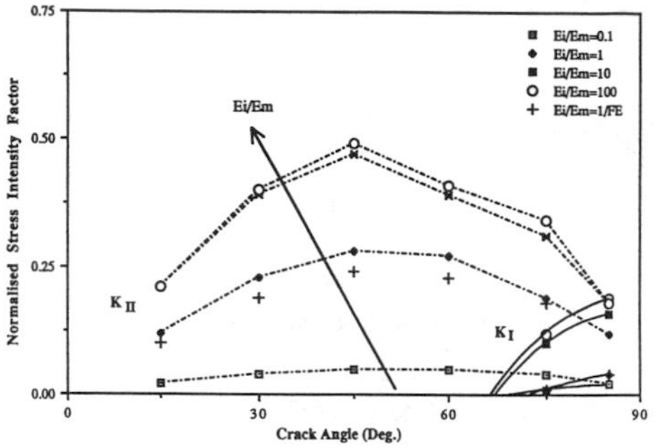

(b) Axial compressive load

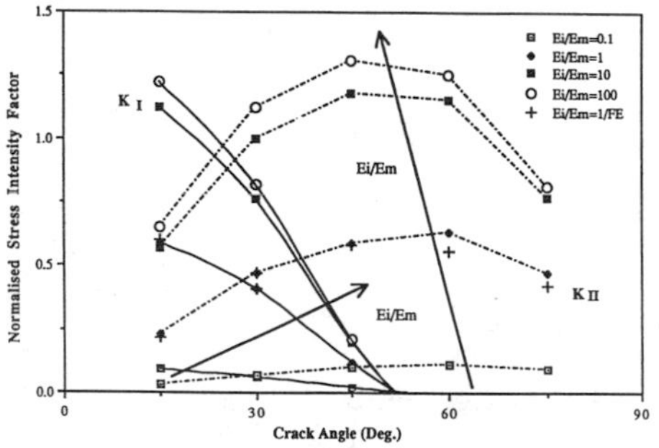

(c) Biaxial load

Figure 3 : Stress intensity factors around the spherical inclusion

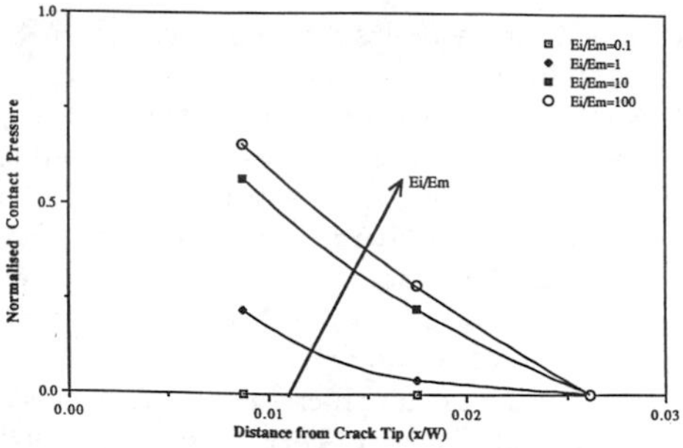

(a) Axial tensile load (crack tip at 75°)

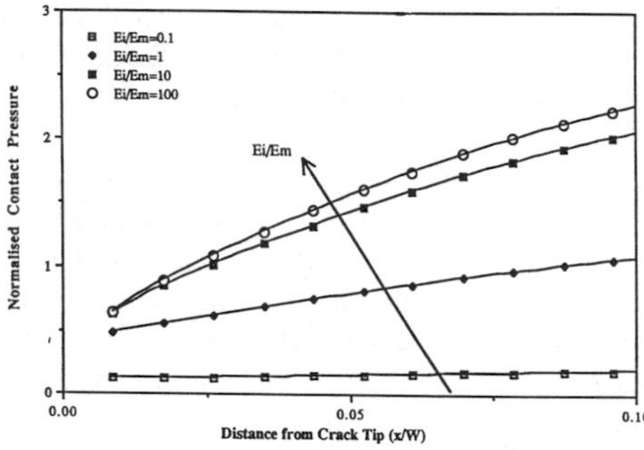

(b) Axial compressive load (crack tip at 30°)

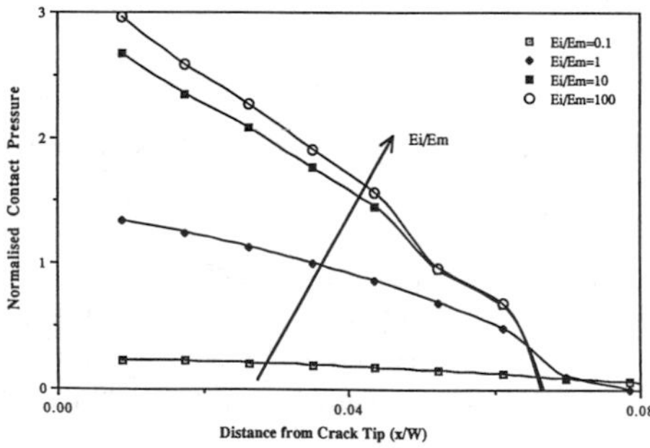

(c) Biaxial load (crack tip at 75°)

Figure 4 : Compressive stress on the closed crack surface around the spherical inclusion

tensile load in the axial direction and a compressive pressure on the outer surface of the matrix. The contact angle θ is measured clockwise from the vertical z-axis. To represent the effect of thermal properties mismatch, a temperature difference ΔT is also specified is some cases. The relative material properties used in this problem are as follows: E_i/E_m = 0.1, 1, 10 and 100, $\nu_i = \nu_m = 0.3$ $\alpha_i/\alpha_m = 2$. The applied load take the values +1 or -1, and ΔT = 0, -20 or -50 °C.

Figure 2 shows a typical BE mesh of 88 quadratic elements used for this axisymmetric problem, as well as the corresponding FE mesh of 358 quadratic elements, which have been carefully designed with smaller elements placed in the expected edge of the

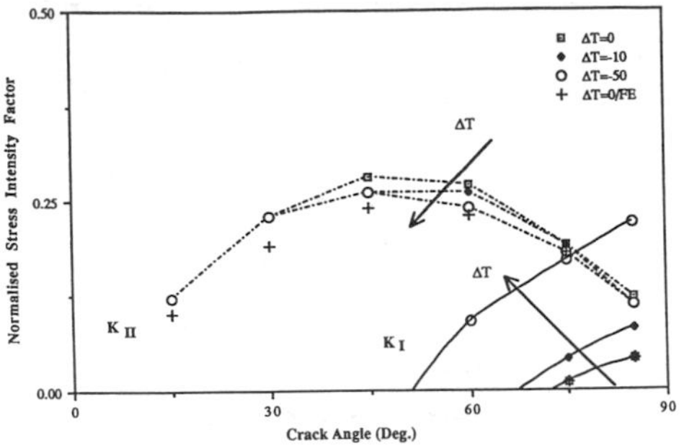

(a) Stress intensity factor

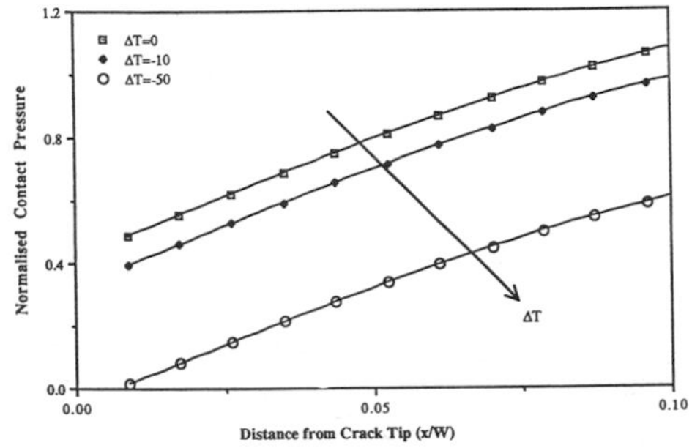

(b) Normal contact pressure (for crack tip at 30°)

Figure 5 : Effect of thermal cooling of the spherical inclusion under compressive axial load

contact/separation zone. It should be noted that the z-axis is not modelled in the BE program and the boundary curve of the matrix becomes an open one. This is due to the assumption of axisymmetry in which the z-axis is part of the interior (see, for example, Becker, 1992).

Figure 3 shows the stress intensity factors, K_I and K_{II}, normalised with respect to the stress intensity factor for a crack in an infinite homogeneous matrix, i.e. $\sigma_0(\pi a)^{1/2}$, for various crack angles and three loading configurations. The results show that the values of K_I and K_{II} increase as the E_i/E_m ratio increases, i.e. as the inclusion becomes more rigid. A good agreement is obtained with the corresponding FE solution for $E_i/E_m=1$. It is interesting to note that the K_{II} curves show a peak at crack angles around 45° for all three load configurations.

The degree of compressibility of the crack faces is demonstrated by the plots of the contact stresses shown in Figure 4 for closed cracks of angles 75° (tensile and biaxial loads) and 30° (compressive load). The contact stresses show a rapid increase near the crack tip for the tensile and biaxial loads,

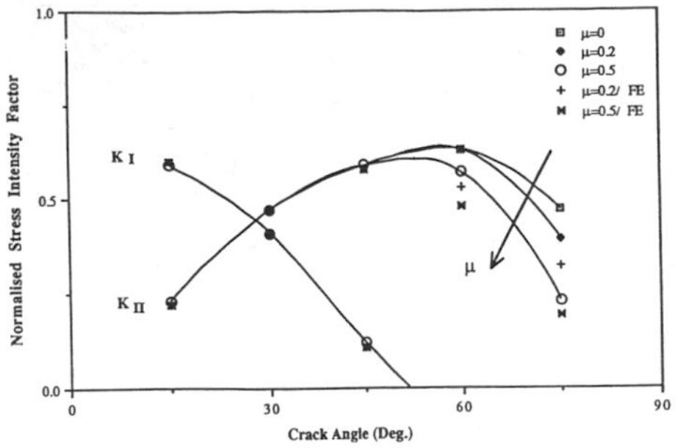

Figure 6 : Effect of surface friction on the spherical inclusion under biaxial load

whereas the compressive axial load causes a reduction in the contact stress near the crack tip. For all three loading cases, the compressive stresses increase as the inclusion becomes more rigid.

To demonstrate the effect of thermal cooling, Figure 5 shows the effect of changing ΔT on the stress intensity factors and the contact stresses for the axial compressive load case. Increasing ΔT causes a large increase in K_I but little effect on K_{II}, whereas the compressive contact stresses on the crack faces are reduced as a temperature difference is introduced.

To examine the effect of sticking/sliding of the crack faces, Figure 6 shows the stress intensity factors for different coefficients of friction for the biaxial load. As expected, the results show that K_I is almost unaffected by increasing μ since it has little effect on the normal contact stresses. However, increasing friction causes a decrease in K_{II}, particularly for large crack angles. Hills and Nowell (1989) presented solutions for closed crack faces in a cracked column under shear and direct loading, indicating the effect of Coulomb friction. Although their results were based on a different geometry, the trend in the change of K_{II} with the coefficient of friction is similar to that exhibited in Figure 6.

5. CRACK CLOSURE AROUND CYLINDRICAL FIBRES

This analysis concerns a long cylindrical fibre embedded in an infinite dissimilar matrix in which curved axial cracks are formed due to the separation of the fibres from the matrix. The fibres are considered long enough for plane strain conditions to be applicable, thus allowing two-dimensional modelling. For modelling purposes, the fibre/matrix diameter ratio is taken as 0.2, which is considered adequate in representing an infinite matrix. The

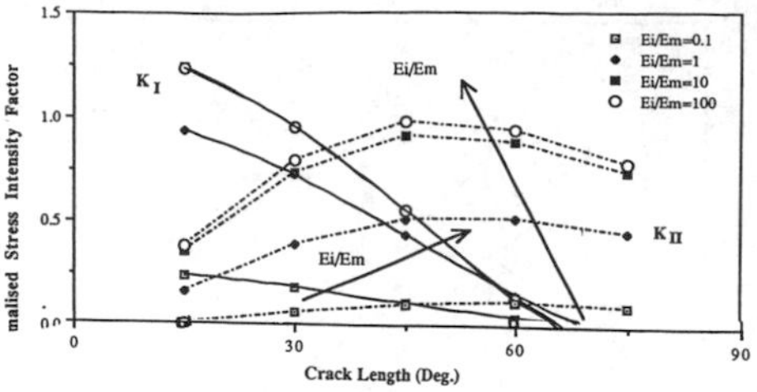

(a) Axial tensile load

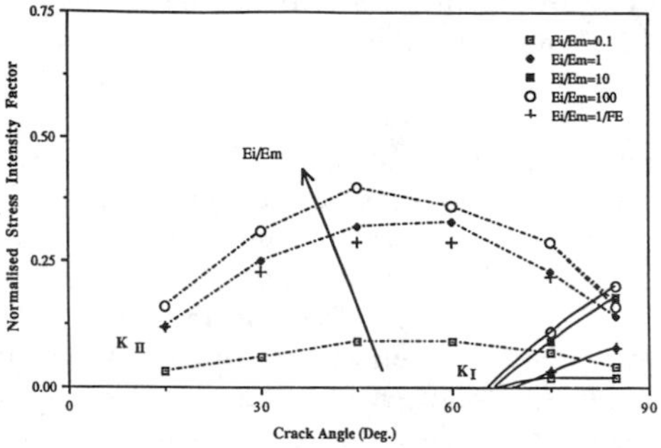

(b) Axial compressive load

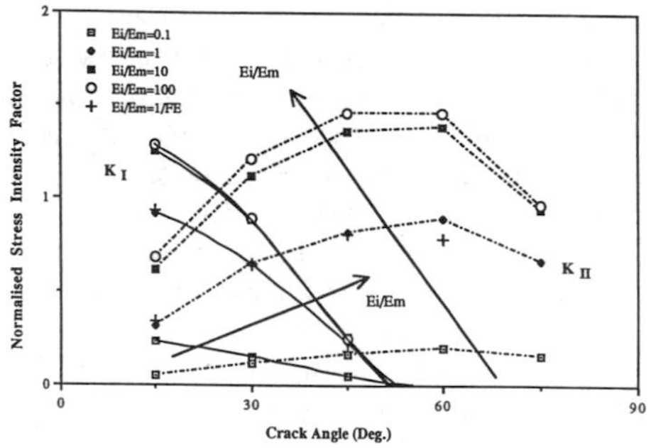

(c) Biaxial load

Figure 7 : Stress intensity factors around the cylindrical fibre

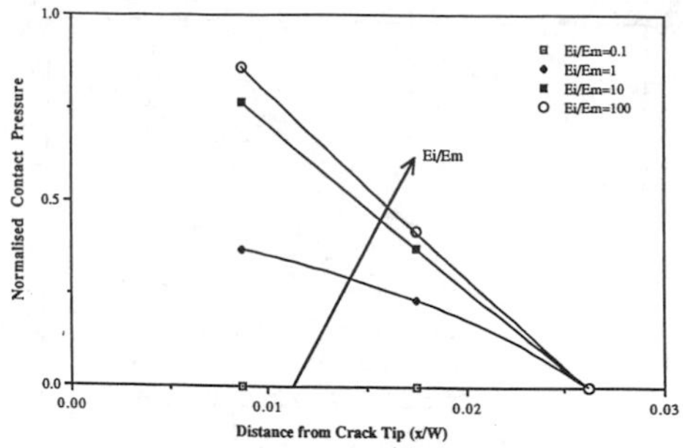

(a) Axial tensile load (crack tip at 75°)

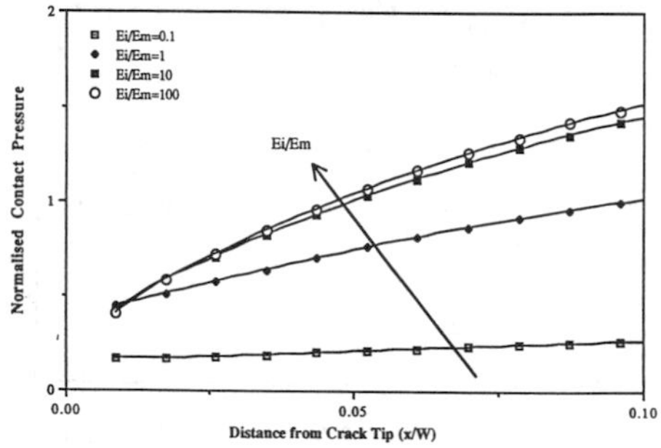

(b) Axial compressive load (crack tip at 30°)

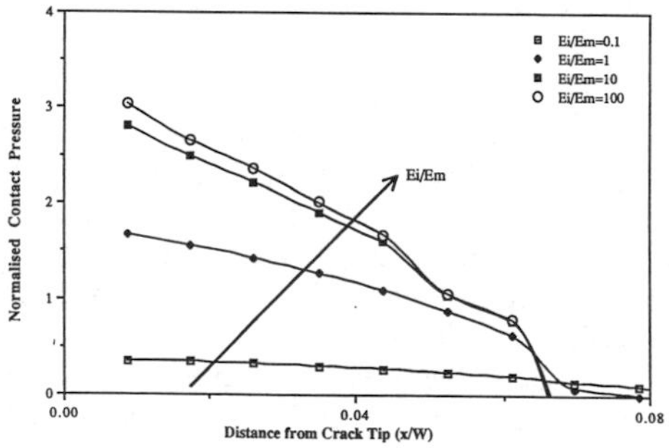

(c) Biaxial load (crack tip at 75°)

Figure 8 : Compressive contact stress on the closed crack surface around the cylindrical fibre

cylindrical fibre is assumed to have Young's modulus E_i, Poisson's ratio ν_i and coefficient of thermal expansion α_i, whereas the corresponding material properties for the matrix are E_m, ν_m and α_m. As in the spherical inclusion investigation, a range of material properties is considered to examine the effect of the relative stiffness of the fibre, as well as a range of coefficients of friction. Three loading configurations are applied to the matrix; the same as those applied to the spherical inclusion, i.e. tensile, compressive and biaxial loads. The contact angle θ is measured clockwise from the vertical axis. To represent the effect of thermal mismatch, a temperature difference of ΔT is also specified. The numerical values used in this problem are the same as those used in the spherical inclusion case.

The BE and FE meshes for this problem are very

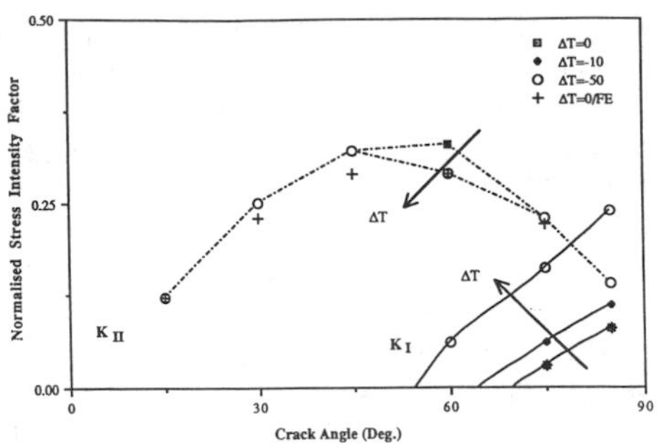

(a) Stress intensity factor

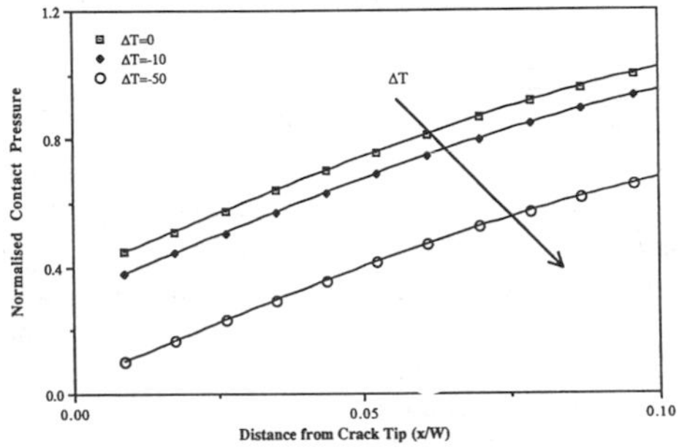

(b) Normal contact pressure (for crack tip at 30°)

Figure 9 : Effect of thermal cooling of the cylindrical fibre under compressive axial load

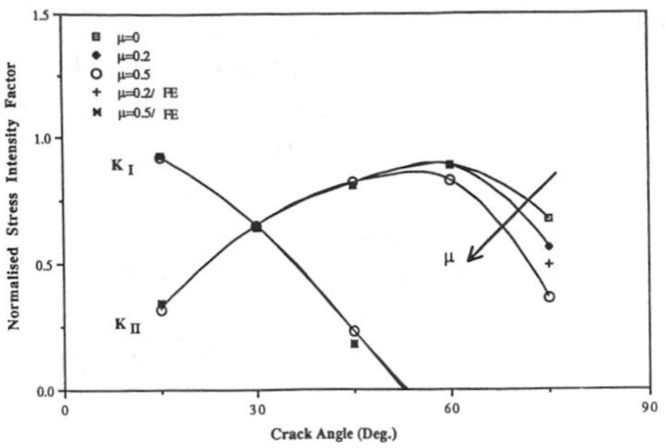

Figure 10 : Effect of surface friction on the cylindrical fibre under biaxial load

similar to those used in the axisymmetric spherical inclusion problem, as shown in Figure 2, except that in the BE mesh elements are placed on the vertical axis making the boundary curve a closed one.

Figure 7 shows the computed values K_I and K_{II}, normalised with respect to the stress intensity factor for a crack in an infinite homogeneous matrix, i.e. $\sigma_0(\pi a)^{1/2}$, for various crack angles and three loading configurations. The results show a similar trend to the spherical inclusions, in that the values of K_I and K_{II} increase as the inclusion becomes more rigid. A good agreement is obtained with the corresponding FE solution for E_i/E_m=1. A peak is exhibited in the K_{II} curves at crack angles of around 45° for the three load configurations.

Figure 8 shows the distribution of the contact stresses for closed cracks of angles 75° (tensile and biaxial loads) and 30° (compressive load) in order to assess the degree of compressibility of the crack faces. Similar to the spherical inclusion, the contact stresses show a rapid increase near the crack tip for the tensile and biaxial loads, whereas the opposite occurs for the compressive load. The contact stresses increase as the inclusion becomes more rigid, for all three load configurations.

The effect of thermal cooling is shown in Figure 9, where it can be seen that increasing ΔT causes a large increase in K_I but little effect on K_{II}, whereas the compressive contact stresses on the crack faces are reduced as a temperature difference is introduced.

Figure 10 shows the effect of changing the coefficient of friction on the stress intensity factors and the contact stresses for the biaxial load. As expected, and similar to the trend in the spherical inclusions, K_I is almost unaffected by increasing μ whereas increasing friction causes a decrease in K_{II}, particularly for large crack angles.

6. CONCLUSIONS

In this study, the crack closure effect of interface cracks formed by the separation of embedded inclusions or fibres from the dissimilar matrices is investigated using the BE method. A quantitative study of the stress intensity factors and the compressive contact stresses in closed crack faces is presented for spherical inclusions and long cylindrical fibres. The study includes the effect of the mismatch in the elastic and thermal material properties, as well as the effect of the interface surface roughness (Coulomb friction). In both types of geometries studied, under tensile, compressive and biaxial loads, a similar trend in obtained for the changes of the stress intensity factors with increasing crack angles and the compressibility of the crack faces in closed cracks.

In both geometries, the values of K_I and K_{II} increase as the inclusion becomes more rigid, with a peak exhibited in the K_{II} curves at crack angles of 45° for the tensile, compressive and biaxial loads. For closed cracks, the contact stresses show a rapid increase near the crack tip for the tensile and biaxial loads, whereas the opposite occurs for the compressive load. The normal contact stresses increase as the inclusion becomes more rigid, for all three load configurations. Introducing a temperature difference causes a large increase in K_I but little effect on K_{II}, whereas the compressive contact stresses on the crack faces are reduced. Increasing the surface roughness of the interface causes a decrease in K_{II}, particularly for large crack angles, but has negligible effect on K_I.

In this study, the BE method is shown to be an accurate and versatile technique in modelling the fracture and contact friction slip behaviour associated with inclusion problems. BE solutions, comparable to the corresponding FE solutions, have been obtained with a lesser degree of mesh refinement and much reduced data preparation time.

REFERENCES

Aksogan, O. 1975, Partial closure of a Griffith crack under a general loading, Int. J. Fracture, 11, 659-670.

Andersson, T. and Allan-Persson, B.G. 1983, The boundary element method applied to two-dimensional contact problems, *Progress in Boundary Element Methods- Vol 2*, edited by C.A. Brebbia, , pp. 136-157, Pentech Press, London.

Banerjee, P.K. 1994 *The Boundary Element Methods in Engineering*, McGraw-Hill, London.

Becker, A.A., 1989, A boundary element computer program for practical contact problems, *Modern Practice in Stress and Vibration Analysis*, edited by J.E. Mottershead, pp. 313-321, Pergamon Press, Oxford.

Becker, A.A. 1992, *The Boundary Element Method in Engineering*, McGraw-Hill, London.

Brebbia, C.A., Telles, J.C.F. and Wrobel, L.C. 1983, *Boundary Element Techniques - Theory and Applications in Engineering*, Springer-Verlag, Berlin.

Barsoum, R.S. 1976, On the use of isoparametric finite elements in linear fracture mechanics, *Int. J. Numerical Methods in Eng*, 10, 25-37.

Comninou, M. 1990, An overview of interface cracks, *Eng. Fracture Mechanics*, 37, 197-208.

Comninou, M. and Schmueser, D. 1979, The interface crack in a combined tension-compression and shear field, *J. Applied Mechanics*, 46, 345-348.

Cotterell, B. and Rice, J.R. 1980, Slightly curved or kinked cracks, *Int. J. Fracture*, 19, 155-169.

Evans, A.G. 1974, The role of inclusions in the fracture of ceramic materials, *J. Materials Science*, 9, 1145-1152.

Gautesen, A.K. and Dunders, J. 1988, The interface crack under combined loading, *J. Applied Mechanics*, 55, 580-586.

Green, D.J. 1983, Microcracking mechanisms in ceramics, *Fracture Mechanics of Ceramics - Volume 5*, edited by R.C. Bradt, et al., pp. 457-478, Plenum Press, New York.

Gustafson, C.G. 1976, Discussion : The stress intensity factors for cyclic reversed bending of a single edge cracked strip including crack surface interference by P.C. Paris and H. Tada, *Int. J. Fracture*, 12, 460-462.

Hasselman, D.P.H. and Fulrath, R.M. 1967, Micromechanical stress concentrations in two-phase brittle-matrix ceramic composites, *J. American Ceramic Society*, 50, 399-404.

Heitzer, J. and Mattheck, C. 1989, FEM calculation of the stress intensity factors of a circular arc crack under uniaxial tension, *Eng. Fracture Mechanics*, 33, 91-104.

Hills, D.A. and Nowell, D. 1989, Stress intensity calculations for closed cracks, *J. Strain Analysis*, 24, 37-43.

HKS, ABAQUS Users Manual version 5.4, 1995, HKS Inc., Rhode Island.

Kant, R. and Bogy, D.B. 1980, The elastostatic axisymmetric problem of a cracked sphere embedded in a dissimilar matrix, *J. Applied Mechanics*, 47, 545-550.

Karami G. and Fenner, R.T. 1987, A two dimensional boundary element method for thermo-elastic body forces contact problems, *Boundary Elements IX*, edited by C. A. Brebbia, W.L. Wendland and G. Kuhn, pp. 417-437, Springer-Verlag, Berlin.

Ito, Y.M. and Nelson, R.B. 1983, Numerical modelling of microcracking in two-phase ceramics, in *Fracture Mechanics of Ceramics - Volume 5*, edited by R.C. Bradt, et al., pp. 479-493, Plenum Press, New York.

Ito, Y.M., Rosenblatt, M., Cheng, L.Y., Lange, F.F. and Evans, A.G. 1981, Cracking in particulate composites due to thermal-mechanical stress, *Int. J. Fracture*, 17, 483-491.

Olukoko, O.A., Becker, A.A. and Fenner, R.T. 1993, A new boundary element approach for contact problems with friction, *Int. J. Numerical Methods in Eng.*, 36, 2625-2642.

Olukoko, O.A., Becker, A.A. and Fenner, R.T. 1994, A review of three alternative approaches to modelling frictional contact problems using the boundary element method, *Proc. Royal Society of London*, Series A, 444, 37-51.

Paris, P.C. and Tada, H. 1975, The stress intensity factor for cyclic reversed bending of a single edge cracked strip including crack surface interference, *Int. J. Fracture*, 11, 1070-1072.

Park, Y.H., Plant, R.C.A. and Becker, A.A. 1992, Separation and fracture of inclusions in multi-phase materials and ceramics, *Structural Integrity Assessment*, edited by P. Stanley, pp. 326-335, Elsevier Applied Science.

Park Y.H., Becker, A.A. and Fenner, R.T. 1994, Fracture mechanics study of microcracks in and around inclusions, *Proc. Second ASME Conference on Engineering Systems Design*, edited by A. Ertas, I.I. Esat and S. Peker, PD-Vol. 64, 95-105, ASME, New York.

Park Y.H., Becker, A.A. and Fenner, R.T. 1995, Contact mechanics analysis of frictional slip of inclusions embedded in dissimilar matrices using the boundary element method, *J. Strain Analysis*, 30, 245-255.

Parker, A.P. 1981, *The Mechanics of Fracture and Fatigue*, E & FN Spon, London.

Tan, C.L. and Gao, Y.L. 1990, Stress intensity factors for cracks at spherical inclusions by the boundary integral equation method, *J. Strain Analysis*, 25, 197-206.

Wilson, H.B. and Goree, J.G. 1967, Axisymmetric contact stresses about a smooth elastic sphere in an infinite solid stressed uniformly at infinity, *J. Applied Mechanics*, 34, 960-966.

Modern Practice in Stress and Vibration Analysis, Gilchrist (ed.) © *1997 Balkema, Rotterdam, ISBN 90 5410 896 7*

Linear structured uncertain system reduction using stability equation method

O. Ismail

Department of Electrical Engineering, Indian Institute of Technology, Bombay, India (On leave from: University of Aleppo, Syria)

ABSTRACT: This paper presents a method of reduction for linear structured uncertain system using stability equation. The denominator $d_m(s)$ of the reduced model is obtained by discarding the poles or zeros with larger magnitudes from stability equation polnomials of the four Kharitonov's polynomials associated with the denominator $d_s(s)$ of the original uncertain system. The numerator $n_m(s)$ of the reduced model is obtained by matching the first k_1 interval time moments and k_2 interval Markov parameters of the reduced model with that of the original uncertain system such that $k_1 + k_2 = r$. A numerical example illustrates the proposed procedure.

1 INTRODUCTION

The approximation of high order plant and controller models by models of lower order is an integral part of control system design. Until relatively recently model reduction was often based on physical intuition. For example, chemical engineers often assume that mixing is instantaneous and that packed distillation columns may be modelled using discrete trays. Electrical engineers represent transmission lines and the eddy currents in the rotor cage of induction motor by lumped circuits. Mechanical engineers remove high frequency vibration modes from models of aircraft wings, turbine shafts and flexible structures. It may also be possible to replace high order controllers by low order approximations with little sacrifice in performance.

Model reduction of continuous and discrete systems have been extensively studied [1-2]. Typical methods are: Aggregation method [3], Moment matching technique [4], Padé approximation [5], Routh approximation [6], and recently L_∞ optimization technique [7].
Recently model reduction for continuous and discrete linear structured uncertain systems have been extensively studied [8-18].

In this paper a method of reduction for linear structured uncertain system using stability equation is presented. The denominator $d_m(s)$ of the reduced model is obtained by discarding the poles or zeros with larger magnitudes from stability equation polnomials of the four Kharitonov's polynomials associated with the denominator $d_s(s)$ of the original uncertain system. The numerator $n_m(s)$ of the reduced model is obtained by matching the first k_1 interval time moments and k_2 interval Markov parameters of the reduced model with that of the original uncertain system such that $k_1 + k_2 = r$. The brief outline of this paper is as follows: Section 2 contains the basic results where as Section 3 includes a numerical example followed by the concluding Section.

2 THE BASIC RESULTS

Let the transfer function of a high order linear structured uncertain system be represented by:

$$G_s(s) = \frac{[a_0^-, a_0^+] + [a_1^-, a_1^+]s + \cdots + [a_{n-1}^-, a_{n-1}^+]s^{n-1}}{[b_0^-, b_0^+] + [b_1^-, b_1^+]s + \cdots + [b_n^-, b_n^+]s^n} \equiv \frac{n_s(s)}{d_s(s)} \quad (1)$$

where, $[a_i^-, a_i^+]$ for $(i = 0, 1, \ldots, n-1)$ and $[b_j^-, b_j^+]$ for $(j = 0, 1, \ldots, n)$ are known interval parameters, (interval parameters mean parameters a_i and b_j which can take independently any values in respective intervals $[a_i^-, a_i^+]$ and $[b_j^-, b_j^+]$), which is to be reduced to the $r^{th} (r < n)$ order model:

$$G_m(s) = \frac{[c_0^-, c_0^+] + [c_1^-, c_1^+]s + \cdots + [c_{r-1}^-, c_{r-1}^+]s^{r-1}}{[d_0^-, d_0^+] + [d_1^-, d_1^+]s + \cdots + [d_r^-, d_r^+]s^r} \equiv \frac{n_m(s)}{d_m(s)} \quad (2)$$

where, $[c_i^-, c_i^+]$ for $(i = 0, 1, \ldots, r-1)$ and $[d_j^-, d_j^+]$ for $(j = 0, 1, \ldots, r)$ are unknown interval parameters.

The four Kharitonov's polynomials [19] associated with the denominator $d_s(s)$ of the original uncertain system are:

$$\begin{aligned}
\mathbf{k}_s^1(s) &= b_0^- + b_1^- s + b_2^+ s^2 + b_3^+ s^3 + b_4^- s^4 + b_5^- s^5 + \cdots\cdots \\
&\equiv \alpha_0 + \alpha_1 s + \alpha_2 s^2 + \cdots + \alpha_n s^n \\
\mathbf{k}_s^2(s) &= b_0^- + b_1^+ s + b_2^+ s^2 + b_3^- s^3 + b_4^- s^4 + b_5^+ s^5 + \cdots\cdots \\
&\equiv \beta_0 + \beta_1 s + \beta_2 s^2 + \cdots + \beta_n s^n \\
\mathbf{k}_s^3(s) &= b_0^+ + b_1^+ s + b_2^- s^2 + b_3^- s^3 + b_4^+ s^4 + b_5^+ s^5 + \cdots\cdots \\
&\equiv \gamma_0 + \gamma_1 s + \gamma_2 s^2 + \cdots + \gamma_n s^n \\
\mathbf{k}_s^4(s) &= b_0^+ + b_1^- s + b_2^- s^2 + b_3^+ s^3 + b_4^+ s^4 + b_5^- s^5 + \cdots\cdots \\
&\equiv \delta_0 + \delta_1 s + \delta_2 s^2 + \cdots + \delta_n s^n
\end{aligned} \quad (3)$$

The polynomials $\mathbf{k}_s^j(s)$ for $(j = 1, 2, 3, 4)$ can be decomposed into even and odd components as follows:

$$\begin{aligned}
\mathbf{k}_{se}^1(s) &= \alpha_0 + \alpha_2 s^2 + \alpha_4 s^4 + \alpha_6 s^6 + \cdots\cdots \\
\mathbf{k}_{so}^1(s) &= \alpha_1 s + \alpha_3 s^3 + \alpha_5 s^5 + \alpha_7 s^7 + \cdots\cdots \\
\mathbf{k}_{se}^2(s) &= \beta_0 + \beta_2 s^2 + \beta_4 s^4 + \beta_6 s^6 + \cdots\cdots \\
\mathbf{k}_{so}^2(s) &= \beta_1 s + \beta_3 s^3 + \beta_5 s^5 + \beta_7 s^7 + \cdots\cdots \\
\mathbf{k}_{se}^3(s) &= \gamma_0 + \gamma_2 s^2 + \gamma_4 s^4 + \gamma_6 s^6 + \cdots\cdots \\
\mathbf{k}_{so}^3(s) &= \gamma_1 s + \gamma_3 s^3 + \gamma_5 s^5 + \gamma_7 s^7 + \cdots\cdots \\
\mathbf{k}_{se}^4(s) &= \delta_0 + \delta_2 s^2 + \delta_4 s^4 + \delta_6 s^6 + \cdots\cdots \\
\mathbf{k}_{so}^4(s) &= \delta_1 s + \delta_3 s^3 + \delta_5 s^5 + \delta_7 s^7 + \cdots\cdots
\end{aligned} \quad (4)$$

The roots of $\mathbf{k}_{se}^j(s)$ and $\mathbf{k}_{so}^j(s)$ for $(j = 1, 2, 3, 4)$, are called zeros z_i and poles p_i, respectively. The polynomials $\mathbf{k}_{se}^j(s)$ and $\mathbf{k}_{so}^j(s)$ for $(j = 1, 2, 3, 4)$ are called the stability equations of the four Kharitonov's polynomials $\mathbf{k}_s^j(s)$ for $(j = 1, 2, 3, 4)$.

For an asymptotically stable, non-minimum phase system equation (4) can be factored as:

$$\mathbf{k}_{se}^1(s) = \prod_{i=1}^{l_1}(s^2 + z_i^2), \qquad \mathbf{k}_{so}^1(s) = s\prod_{i=1}^{l_2}(s^2 + p_i^2)$$

$$\mathbf{k}_{se}^2(s) = \prod_{i=1}^{l_1}(s^2 + z_i^2), \qquad \mathbf{k}_{so}^2(s) = s\prod_{i=1}^{l_2}(s^2 + p_i^2)$$

$$\mathbf{k}_{se}^3(s) = \prod_{i=1}^{l_1}(s^2 + z_i^2), \qquad \mathbf{k}_{so}^3(s) = s\prod_{i=1}^{l_2}(s^2 + p_i^2)$$

$$\mathbf{k}_{se}^4(s) = \prod_{i=1}^{l_1}(s^2 + z_i^2), \qquad \mathbf{k}_{so}^4(s) = s\prod_{i=1}^{l_2}(s^2 + p_i^2) \quad (5)$$

where,

$$\begin{aligned}
l_1 &= n/2, & &if \quad n \quad is \quad even. \\
l_2 &= (n-1)/2, & &if \quad n \quad is \quad odd.
\end{aligned}$$

and

$$\begin{aligned}
p_1^2 &< p_2^2 < p_3^3 < \cdots\cdots \\
z_1^2 &< z_2^2 < z_3^3 < \cdots\cdots
\end{aligned}$$

Introducing the concept of 'dominant quantities' for the poles or zeros with smaller magnitudes, and discarding the poles or zeros with larger magnitudes from polynomials $\mathbf{k}_{se}^j(s)$ and $\mathbf{k}_{so}^j(s)$ for $(j = 1, 2, 3, 4)$, we get:

$$\begin{aligned}
\mathbf{k}_m^1(s) &= \tilde{\alpha}_0 + \tilde{\alpha}_1 s + \tilde{\alpha}_2 s^2 + \cdots + \tilde{\alpha}_r s^r \\
\tilde{\mathbf{k}}_m^2(s) &= \tilde{\beta}_0 + \tilde{\beta}_1 s + \tilde{\beta}_2 s^2 + \cdots + \tilde{\beta}_r s^r \\
\tilde{\mathbf{k}}_m^3(s) &= \tilde{\gamma}_0 + \tilde{\gamma}_1 s + \tilde{\gamma}_2 s^2 + \cdots + \tilde{\gamma}_r s^r \\
\tilde{\mathbf{k}}_m^4(s) &= \tilde{\delta}_0 + \tilde{\delta}_1 s + \tilde{\delta}_2 s^2 + \cdots + \tilde{\delta}_r s^r
\end{aligned} \quad (6)$$

The denominator $d_m(s)$ of the reduced model can be obtained as:

$$d_m(s) = \sum_{i=0}^{r}[\min(\tilde{\alpha}_i, \tilde{\beta}_i, \tilde{\gamma}_i, \tilde{\delta}_i), \max(\tilde{\alpha}_i, \tilde{\beta}_i, \tilde{\gamma}_i, \tilde{\delta}_i)]s^i$$

$$\equiv [d_0^-, d_0^+] + [d_1^-, d_1^+]s + \cdots + [d_r^-, d_r^+]s^r \quad (7)$$

The numerator $n_m(s)$ of the reduced model is obtained by matching the first k_1 interval time moments and k_2 interval Markov parameters of the reduced model with that of the original uncertain system such that $k_1 + k_2 = r$, where the interval time moments and interval Markov parameters of the reduced model and original uncertain system can be obtained as explained in [12]. (i. e., the interval time moments and interval Markov parameters of the reduced model and original uncertain system are obtained by solving a set of linear interval equations $Q^I.x = p^I$. The solution x^I will be understood to mean [20-22],

$$x^I = \{x : Q.x = p,\ Q \in Q^I,\ p \in p^I\}$$

There are several methods for computing x^I and we have applied the method defined by [20-22], the solution x^I makes $Q^I.x^I \supseteq p^I$. Consequently, the interval time moments and interval Markov parameters of the original uncertain system are a subset or equal to that of its uncertain model.

The advantages of this method are:
(i) The stability of the reduced order model is preserved.
(ii) The roots of the stability equations can easily be obtained.

3 NUMERICAL EXAMPLE

Consider a third order linear SISO structured uncertain system with a transfer function:

$$G_s(s) = \frac{[18,25]+[30,34]s+[6,9]s^2}{[8,10]+[12,15]s+[24,28]s^2+[5,6]s^3} \equiv \frac{n_s(s)}{d_s(s)}$$

It is required to find a second order reduced model:

$$G_m(s) = \frac{[c_0^-,c_0^+]+[c_1^-,c_1^+]s}{[d_0^-,d_0^+]+[d_1^-,d_1^+]s+[d_2^-,d_2^+]s^2} \equiv \frac{n_m(s)}{d_m(s)}$$

for this system using the proposed method.
The four Kharitonov's polynomials $\mathbf{k}_s^j(s)$ for $(j = 1,2,3,4)$, associated with the denominator $d_s(s)$ of the original uncertain system are obtained. The stability equations $\mathbf{k}_{se}^j(s)$ and $\mathbf{k}_{so}^j(s)$ for $(j = 1,2,3,4)$ of the four Kharitonov's polynomials associated with the denominator $d_s(s)$ of the original uncertain system are obtained as follows:

$$\begin{aligned}
&\mathbf{k}_{se}^1(s) = 8+28s^2, && \mathbf{k}_{so}^1(s) = 12s+6s^3\\
&\mathbf{k}_{se}^2(s) = 8+28s^2, && \mathbf{k}_{so}^2(s) = 15s+5s^3\\
&\mathbf{k}_{se}^3(s) = 10+24s^2, && \mathbf{k}_{so}^3(s) = 15s+5s^3\\
&\mathbf{k}_{se}^4(s) = 10+24s^2, && \mathbf{k}_{so}^4(s) = 12s+6s^3
\end{aligned}$$

The denominator $d_m(s)$ of the reduced model is formed by discarding the poles with larger magnitudes from polynomials $\mathbf{k}_{so}^j(s)$ for $(j = 1,2,3,4)$, as explained in Section 2:

$$d_m(s) = [8,10]+[12,15]s+[24,28]s^2$$

The numerator $n_m(s)$ of the reduced model is obtained by matching the first interval time moment and first interval Markov parameter of the reduced model with that of the original uncertain system.
(i. e., $m_{m(0)}^I = m_{s(0)}^I$ and $y_{m(0)}^I = y_{s(0)}^I$). Thus, the second order model is obtained as:

$$G_m(s) = \frac{[18,25]+[28.8,42]s}{[8,10]+[12,15]s+[24,28]s^2}$$

Simulation results in Figures (1) and (2) show the step response envelopes and the step responses of the nominal system and reduced model, respectively.

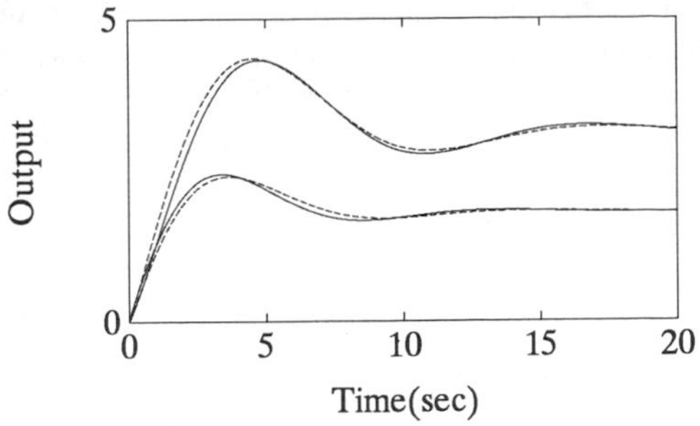

Fig.1:Step response envelopes.

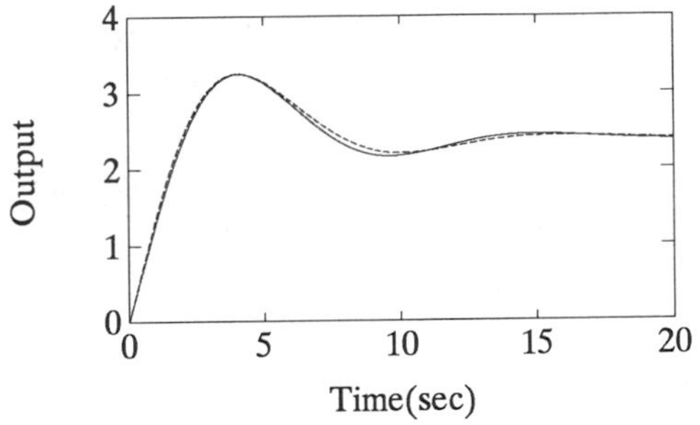

Fig.2:Step responses.

4 CONCLUSIONS

A method of reduction for linear structured uncertain system using stability equation is presented in this paper. The denominator $d_m(s)$ of the reduced model is obtained by discarding the poles or zeros with larger magnitudes from stability equation polnomials of the four Kharitonov's polynomials associated with the denominator $d_s(s)$ of the original uncertain system. The numerator $n_m(s)$ of the reduced model is obtained by matching the first k_1 interval time moments and k_2 interval Markov parameters of the reduced model with that of the original uncertain system such that $k_1 + k_2 = r$. The interval time moments and interval Markov parameters of the original uncertain system are obtained by solving a set of linear interval equations $Q^I.x = p^I$. The solution x^I makes $Q^I.x^I \supseteq p^I$. Consequently, the interval time moments and interval Markov parameters of the original uncertain system are a subset or equal to that of its uncertain reduced model.

References

[1] Jamshidi, M. 1983. *Large scale systems modeling and control.* North-Holland: Elsevier Science Publishing Co., Inc.

[2] Fortuna, L., G. Nunnari & A. Gallo 1992. *Model order reduction techniques with applications in electrical engineering.* London: Springer-Verlag.

[3] Aoki, M. 1968. *Control of large scale dynamic systems by aggregation.* IEEE Trans. Auto. Contr., 13: 246-253.

[4] Sinha, N. K. & B. Kuszta 1983. *Modelling and identification of dynamic systems.* New York: van Nostrand Reinhold. Ch. 8: 133-163.

[5] Shamash, Y. 1974. *Stable reduced order models using padé type approximation.* IEEE Trans. Auto. Contr. 19: 615-616.

[6] Hutton M. F. & B. Friedland 1975. *Routh Approximation for reducing order of linear time invariant system.* IEEE Trans. Auto. Contr., 20: 329-337.

[7] Glover, K. 1984. *All optimal hankel norm stem and their L^∞ error bounds.* Int. J. Contr. 39(6): 1115-1193.

[8] Argoun, M. B. 1993. *Model reduction with preserved uncertainty bounds.* Proc. IEEE Conf. on Decision and Control, San Antonio, Texas: 3355-3356.

[9] Bandyopadhyay, B., O. Ismail & R. Gorez. 1994 *Routh-padé approximation for interval systems.* IEEE Trans. Auto. Contr. 39(12): 2454-2456.

[10] Bandyopadhyay, B., R. Gorez & O. Ismail 1994. *Routh-padé approximation for discrete interval systems.* Proc. 14th IMACS World congress on computation and applied mathematics, Atlanta, Georgia, U.S.A.

[11] Ismail, O. & B. Bandyopadhyay 1995. *On padé approximation for scalar and multivariable interval systems.* Proc. IEEE Int. Conf. On systems, man and cybernetics, Canada.

[12] Ismail, O. & B. Bandyopadhyay 1995. *Model reduction of linear interval systems using padé approximation.* Proc. IEEE Int. Symposium on circuits and systems, Seattle, Washington, U.S.A.

[13] Ismail, O. 1995. *Model reduction of discrete interval systems using padé approximation.* Proc. IASTED Int. Conf. On modelling and simulation, Colombo, Sri Lanka.

[14] Ismail, O. 1996. *Model reduction for structured linear uncertain systems.* Proc. Int. Conf. On identification in engineering systems, Swansea, U.K.

[15] Ismail, O. & J. M. Jahabar 1996. *Structured linear uncertain systems reduction.* Proc. IEEE Southeastern symposium on system theory, Baton Rouge, Louisiana, U.S.A.

[16] Ismail, O. 1996. *On multipoint padé approximation for discrete interval systems.* Proc. IEEE Southeastern symposium on system theory, Baton Rouge, Louisiana, U.S.A.

[17] Ismail, O. & J. M. Jahabar 1997. *Model order reduction for linear structured uncertain systems.* Proc. Modern practice in stress and vibration analysis. Dublin, Ireland.

[18] Ismail, O. 1997. *On multipoint padé approximations for linear structured uncertain systems.* Proc. IASTED Int. Conf. On modelling and simulation, Pittsburgh, Pennsylvania, U.S.A.

[19] Kharitonov, V. L. 1978. *Asymptotic stability of an equilibrium position of a family of system of linear differential equations.* Differential 'nye Urauneniya.14(11): 2086-2088.

[20] Hansen, E. 1969. *On linear algebraic equations with interval coefficients. In E. Hansen (ed) Topics in interval analysis.* Oxford: Clarendon press.

[21] Moore, R. E. 1988. *Reliability in computing: The role of interval methods in scientific computing.* London: Academic press, Inc.

[22] Neumaier, A. 1990. *Interval methods for systems of equations.* Cambridge University Press.

Modern Practice in Stress and Vibration Analysis, Gilchrist (ed.) © *1997 Balkema, Rotterdam, ISBN 90 5410 896 7*

Model order reduction for linear structured uncertain systems

O. Ismail
Department of Electrical Engineering, Indian Institute of Technology, Bombay, India (On leave from: University of Aleppo, Syria)

J. M. Jahabar
Department of Electrical Engineering, Indian Institute of Technology, Bombay, India

ABSTRACT: This paper presents a method for reducing the high order linear structured uncertain system described by differential equation, with a set of specified initial conditions, to a lower order one. The coefficients of the reduced model are obtained by matching the integral of the square of the output and its first $r-1$ derivatives of the reduced model with that of the original uncertain system. A numerical example illustrates the procedure.

1 INTRODUCTION

The modelling problem of complex dynamic systems is one of the most important subjects in engineering. Moreover, a model is too complicated to be used in real problems, so approximation procedures based on physical considerations or using mathematical approaches must be used to achieve simpler models than the original one. Model reduction of continuous and discrete systems have been extensively studied [1-2]. Typical methods are: Aggregation method [3], Moment matching technique [4], Padé approximation [5], Routh approximation [6], and recently L_∞ optimization technique [7].
Recently model reduction for continuous and discrete linear structured uncertain systems have been extensively studied [8-19].

In this paper a method for reducing the high order linear structured uncertain system described by differential equation, with a set of specified initial conditons, to a lower order one is presented. Where the coefficients of the reduced model are obtained by matching the integral of the square of the output and its first $r-1$ derivatives of the reduced model with that of the original uncertain system. The brief outline of this paper is as follows: Sections 2 contains the basic results where as Section 3 includes a numerical example followed by the concluding Section.

2 THE BASIC RESULTS

Let the high order linear structured uncertain system be described by the differential equation:

$$[a_n^-, a_n^+]y_s^{(n)} + [a_{n-1}^-, a_{n-1}^+]y_s^{(n-1)} + \cdots + [a_0^-, a_0^+]y_s = 0 \tag{1}$$

with initial conditions $y_s^{(i)} = c_i$ for $(i = 0, 1, \ldots, n-1)$ where, $y_s^{(i)} \equiv \frac{d^i y_s}{dt^i}$ and $[a_i^-, a_i^+]$ for $(i = 0, 1, \ldots, n-1)$ are known interval parameters, with $[a_n^-, a_n^+] = [1, 1]$, (interval parameter means parameter a_i which can take independently any values in the interval $[a_i^-, a_i^+]$) .

The integral of the square of the output and its first $n-1$ derivatives of the uncertain system $I_{s(i)} = (-1)^i 2\int_0^\infty [y_s^{(i)}]^2 dt$ for $(i = 0, 1, \ldots, n-1)$ can be obtained in the following way:
Equation (1) is to be multiplied in succession by $y_s, y_s^{(1)}, \ldots\ldots, y_s^{(n-1)}$ and the resulting set of interval equations are to be integrated from $(0 \longrightarrow \infty)$ using integration by parts and the known boundry conditions. A general formula for $2\int_0^\infty y_s^{(i)} y_s^{(j)} dt$ where, $i > j$ can be obtained as given below:

$$2\int_0^\infty y_s^{(i)} y_s^{(j)} dt = p_{ij} \qquad if \quad k \quad is \quad odd. \tag{2}$$

$$2\int_0^\infty y_s^{(i)} y_s^{(j)} dt = p_{ij} + (-1)^j I_{(i+j)/2}$$

$$if \quad k \quad is \quad even. \tag{3}$$

where, $k = i - j$ and

$$p_{ij} = (-1)^{(k+1)/2} c^2_{(i+j-1)/2} + 2\sum_{l=1}^{(k-1)/2} (-1)^l c_{i-l} c_{j+l-1}$$

$$if \quad k \quad is \quad odd. \tag{4}$$

$$p_{ij} = 2\sum_{l=1}^{k/2} (-1)^l c_{i-l} c_{j+l-1}$$

$$if \quad k \quad is \quad even. \tag{5}$$

The matrix equation to evaluate I_s^I 's can be written as:

$$A^I \cdot I_s^I = P \cdot a^I \tag{6}$$

where,

$$A^I = \begin{bmatrix} [a_0^-, a_0^+] & [a_2^-, a_2^+] & [a_4^-, a_4^+] & \cdots & \cdots \\ [0,0] & [a_1^-, a_1^+] & [a_3^-, a_3^+] & \cdots & \cdots \\ \vdots & \vdots & \vdots & \vdots & \vdots \\ [0,0] & [0,0] & [0,0] & \cdots & [a_{n-1}^-, a_{n-1}^+] \end{bmatrix}$$

$$I_s^I = \begin{bmatrix} [I_{s(0)}^-, I_{s(0)}^+] \\ [I_{s(1)}^-, I_{s(1)}^+] \\ \vdots \\ [I_{s(n-1)}^-, I_{s(n-1)}^+] \end{bmatrix}$$

and

$$P = \begin{bmatrix} 0 & -p_{10} & -p_{20} & \cdots & \cdots & -p_{n,0} \\ p_{01} & 0 & p_{21} & \cdots & \cdots & p_{n,1} \\ \vdots & \vdots & \vdots & \vdots & \vdots & \\ (-1)^n p_{0,n-1} & \cdots & \cdots & \cdots & \cdots & (-1)^n p_{n,n-1} \end{bmatrix}$$

$$a^I = \begin{bmatrix} [a_0^-, a_0^+] \\ [a_1^-, a_1^+] \\ \vdots \\ [a_n^-, a_n^+] \end{bmatrix}$$

The problem is to find the coefficients of the reduced model described by the differential equation:

$$[b_r^-, b_r^+] y_m^{(r)} + [b_{r-1}^-, b_{r-1}^+] y_m^{(r-1)} + \cdots + [b_0^-, b_0^+] y_m = 0 \tag{7}$$

with a set of specified initial conditions, $y_m^{(i)} = d_i$ for $(i = 0, 1, \ldots, r-1)$ where, $[b_i^-, b_i^+]$ for $(i = 0, 1, \ldots, r-1)$ are unknown interval parameters and $[b_r^-, b_r^+] = [1, 1]$, such that the output $y_m(t)$ of the reduced model will be close as possible to the output of system $y_s(t)$.

The integral of the square of the output and its first $r-1$ derivatives of the reduced model $I_{m(i)} = (-1)^i 2\int_0^\infty [y_m^{(i)}]^2 dt$ for $(i = 0, 1, \ldots, r-1)$ can be obtained in similar way as for the original uncertain system. This will lead to:

$$B^I \cdot I_m^I = Q \cdot b^I \tag{8}$$

where,

$$B^I = \begin{bmatrix} [b_0^-, b_0^+] & [b_2^-, b_2^+] & [b_4^-, b_4^+] & \cdots & \cdots \\ [0,0] & [b_1^-, b_1^+] & [b_3^-, b_3^+] & \cdots & \cdots \\ \vdots & \vdots & \vdots & \vdots & \vdots \\ [0,0] & [0,0] & [0,0] & \cdots & [b_{r-1}^-, b_{r-1}^+] \end{bmatrix}$$

$$I_m^I = \begin{bmatrix} [I_{m(0)}^-, I_{m(0)}^+] \\ [I_{m(1)}^-, I_{m(1)}^+] \\ \vdots \\ [I_{m(r-1)}^-, I_{m(r-1)}^+] \end{bmatrix}$$

and

$$Q = \begin{bmatrix} 0 & -q_{10} & -q_{20} & \cdots & \cdots & -q_{r,0} \\ q_{01} & 0 & q_{21} & \cdots & \cdots & q_{r,1} \\ \vdots & \vdots & \vdots & \vdots & \vdots & \\ (-1)^r q_{0,r-1} & \cdots & \cdots & \cdots & \cdots & (-1)^r q_{r,r-1} \end{bmatrix}$$

$$b^I = \begin{bmatrix} [b_0^-, b_0^+] \\ [b_1^-, b_1^+] \\ \vdots \\ [b_r^-, b_r^+] \end{bmatrix}$$

The coefficients of reduced model described by the differential equation (7), with a set of specified initial conditions $y_m^{(i)} = d_i$ for $(i = 0, 1, \ldots, r-1)$ can be obtained by matching the integral of the square of the output and its first $r-1$ derivatives of the reduced model with that of the uncertain system. This will lead to a set of r linear interval equations involving the unknown interval parameters of the reduced model $[b_i^-, b_i^+]$ for $(i = 0, 1, \ldots, r-1)$, i. e., $(W^I \cdot x = v^I)$ and this set of r linear interval equations can be solved to obtain $[b_i^-, b_i^+]$ for $(i = 0, 1, \ldots, r-1)$. The solution x^I will be understood to mean [19] and [20-22], $x^I = \{x : W \cdot x = v, W \in W^I, v \in v^I\}$, where the solution x^I makes $W^I \cdot x^I \supseteq v^I$, therefore the integral of the square of the output and its first $r-1$ derivatives of the original uncertain system are a subset or equal to that of its uncertain reduced model.

3 NUMERICAL EXAMPLE

Consider a third order linear SISO structured uncertain system described by the differrential equation:

$$y_s^{(3)} + [24,25]y_s^{(2)} + [35,36]y_s^{(1)} + [4,5]y_s = 0$$

with initial conditions $y_s(0) = 0$, $y_s^{(1)}(0) = 1$ and $y_s^{(2)}(0) = 0$.
It is required to find a second order reduced model:

$$y_m^{(2)} + [b_1^-, b_1^+]y_m^{(1)} + [b_0^-, b_0^+]y_m = 0$$

with initial conditions $y_m(0) = 0$ and $y_m^{(1)}(0) = 1$, for this system using the proposed method.

The integral of the square of the output and its first and second derivatives of the original uncertain system $I_{s(i)} = (-1)^i 2 \int_0^\infty [y_s^{(i)}]^2 dt$ for $(i = 0, 1, 2)$ are obtained as:

$$[I_{s(0)}^-, I_{s(0)}^+] = [3.3483, 4.3137]$$
$$[I_{s(1)}^-, I_{s(1)}^+] = [-0.7392, -0.7302]$$
$$[I_{s(2)}^-, I_{s(2)}^+] = [1.5568, 1.6123]$$

The integral of the square of the output and its first derivative of the reduced model $I_{m(i)} = (-1)^i 2 \int_0^\infty [y_m^{(i)}]^2 dt$ for $(i = 0, 1)$ are obtained in similar way as for the original uncertain system. The coefficients of the reduced model are obtained by matching the integral of the square of the output and its first derivative of the reduced model with that of the original system. This will lead to:

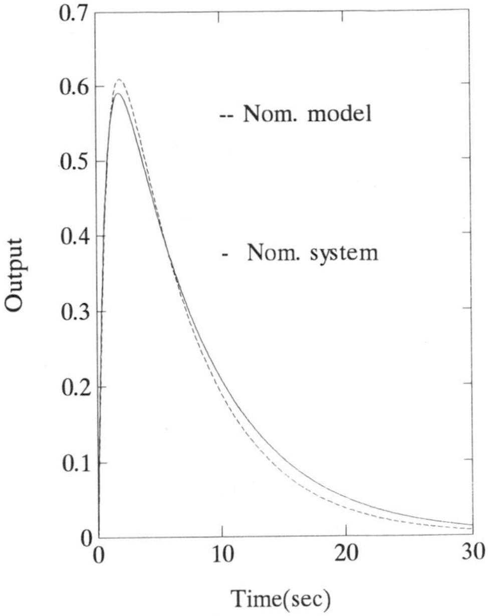

Fig.2:Responses.

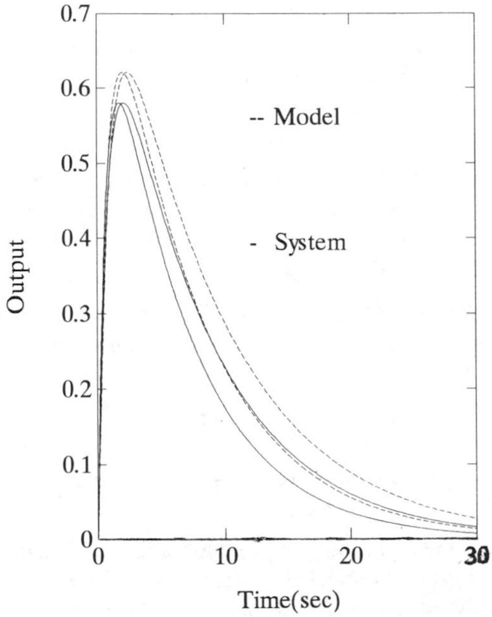

Fig.1:Responses.

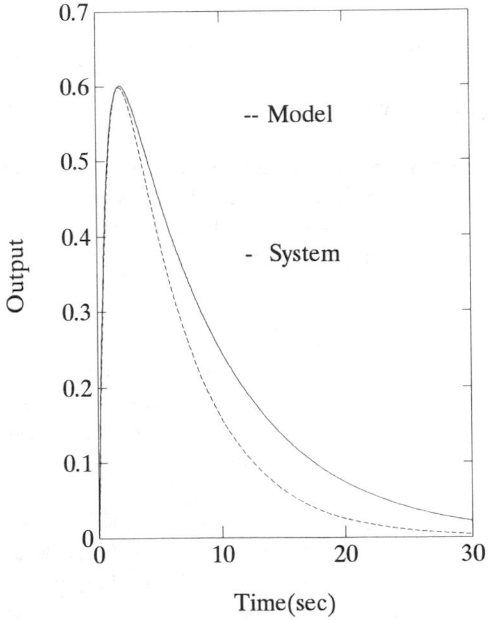

Fig.3:Responses.

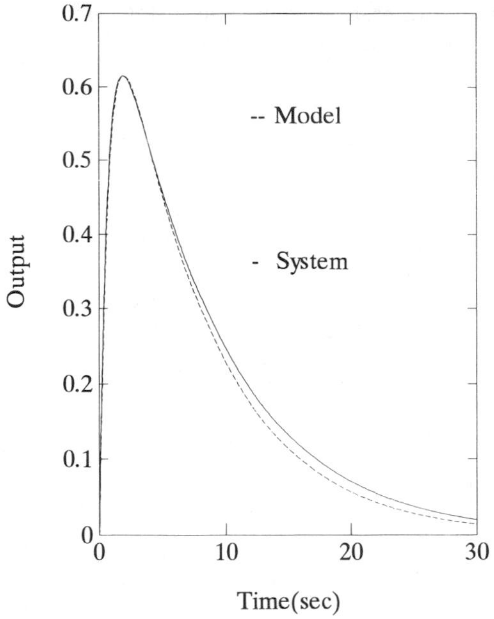

Fig.4:Responses.

$$\begin{bmatrix} [3.3483, 4.3137] & [0, 0] \\ [0, 0] & [-0.7392, -0.7302] \end{bmatrix} \begin{bmatrix} [b_0^-, b_0^+] \\ [b_1^-, b_1^+] \end{bmatrix} = \begin{bmatrix} [0.7302, 0.7392] \\ [-1, -1] \end{bmatrix}$$

This set of linear interval equations is solved and the reduced model described by the differential equation is obtained as:

$$y_m^{(2)} + [1.3528, 1.3695] y_m^{(1)} + [0.1714, 0.2181] y_m = 0$$

with initial conditions $y_m(0) = 0$ and $y_m^{(1)}(0) = 1$.

Simulation results in Figures (1),(2),(3) and (4) show the responses of some of the systems and reduced models constructed from the original uncertain system and its uncertain reduced model.

4 CONCLUSIONS

A method for reducing the high order linear structured uncertain system described by differential equation, with a set of specified initial conditons, to a lower order one is presented in this paper. The coefficients of the reduced model are obtained by matching the integral of the square of the output and its first $r-1$ derivatives of the reduced model with that of the original uncertain system. This will lead to a set of r linear interval equations $W^I.x = v^I$. This set of r linear interval equations can be solved to obtain the coefficients of the reduced model $[b_i^-, b_i^+]$ for $(i = 0, 1, \ldots, r-1)$. The solution x^I makes $W^I.x \supseteq v^I$. Consequently, the integral of the square of the output and its first $r-1$ derivatives of the original uncertain system are a subset or equal to that of its uncertain reduced model.

References

[1] Jamshidi, M. 1983. *Large scale systems modeling and control.* North-Holland: Elsevier Science Publishing Co., Inc.

[2] Fortuna, L., G. Nunnari & A. Gallo 1992. *Model order reduction techniques with applications in electrical engineering.* London: Springer-Verlag.

[3] Aoki, M. 1968. *Control of large scale dynamic systems by aggregation.* IEEE Trans. Auto. Contr., 13: 246-253.

[4] Sinha, N. K. & B. Kuszta 1983. *Modelling and identification of dynamic systems.* New York: van Nostrand Reinhold. Ch. 8: 133-163.

[5] Shamash, Y. 1974. *Stable reduced order models using padé type approximation.* IEEE Trans. Auto. Contr. 19: 615-616.

[6] Hutton M. F. & B. Friedland 1975. *Routh Approximation for reducing order of linear time invariant system.* IEEE Trans. Auto. Contr., 20: 329-337.

[7] Glover, K. 1984. *All optimal hankel norm approximation of linear multivariable system and their L^∞ error bounds.* Int. J. Contr. 39(6): 1115-1193.

[8] Argoun, M. B. 1993. *Model reduction with* Conf. on Decision and Control, San Antonio, Texas: 3355-3356.

[9] Bandyopadhyay, B., O. Ismail & R. Gorez. 1994 *Routh-padé approximation for interval systems.* IEEE Trans. Auto. Contr. 39(12): 2454-2456.

[10] Bandyopadhyay, B., R. Gorez & O. Ismail 1994. *Routh-padé approximation for discrete interval systems.* Proc. 14th IMACS World congress on computation and applied mathematics, Atlanta, Georgia, U.S.A.

[11] Ismail, O. & B. Bandyopadhyay 1995. *On padé approximation for scalar and multivariable interval systems.* Proc. IEEE Int. Conf. On systems, man and cybernetics, Canada.

[12] Ismail, O. & B. Bandyopadhyay 1995. *Model reduction of linear interval systems using padé approximation.* Proc. IEEE Int. Symposium on circuits and systems, Seattle, Washington, U.S.A.

[13] Ismail, O. 1995. *Model reduction of discrete interval systems using padé approximation.* Proc. IASTED Int. Conf. On modelling and simulation, Colombo, Sri Lanka.

[14] Ismail, O. 1996. *Model reduction for structured linear uncertain systems.* Proc. Int. Conf. On identification in engineering systems, Swansea, U.K.

[15] Ismail, O. & J. M. Jahabar 1996. *Structured linear uncertain systems reduction.* Proc. IEEE Southeastern symposium on system theory, Baton Rouge, Louisiana, U.S.A.

[16] Ismail, O. 1996. *On multipoint padé approximation for discrete interval systems.* Proc. IEEE Southeastern symposium on system theory, Baton Rouge, Louisiana, U.S.A.

[17] Ismail, O. & J. M. Jahabar 1997. *Model order reduction for linear structured uncertain systems.* Proc. Modern practice in stress and vibration analysis. Dublin, Ireland.

[18] Ismail, O. 1997. *On multipoint padé approximations for linear structured uncertain systems.* Proc. IASTED Int. Conf. On modelling and simulation, Pittsburgh, Pennsylvania, U.S.A.

[19] Ismail, O. 1997. *Linear structured uncertain system reduction using stability equation method.* Proc. Modern practice in stress and vibration analysis. Dublin, Ireland.

[20] Hansen, E. 1969. *On linear algebraic equations with interval coefficients. In E. Hansen (ed) Topics in interval analysis.* Oxford: Clarendon press.

[21] Moore, R. E. 1988. *Reliability in computing: The role of interval methods in scientific computing.* London: Academic press, Inc.

[22] Neumaier, A. 1990. *Interval methods for systems of equations.* Cambridge University Press.

Modern Practice in Stress and Vibration Analysis, Gilchrist (ed.) © 1997 Balkema, Rotterdam, ISBN 90 5410 896 7

Dynamic analysis of laminated orthotropic hemispherical shells

M. Shakeri, M. R. Eslami, M. H. Yas & M. Kashani
Department of Mechanical Engineering, University of Amirkabir, Iran

ABSTRACT: This paper deals with the dynamic analysis of multi−layered composite hemispherical shells.Each layer of the shell is assumed to be made of orthotropic materials, and second order shear deformation theory is used. The in−plane displacement components of the shell are assumed to vary linearly through the shell thickness, while the lateral displacement is assumed to be of second order polynomial which allow both lateral normal stressand shear deformation to be reduced in the governing equations. The variatonal principle is employed to derive the general governing equations of the shells of revolution. The resulting governing equations are then reduced for spherical shells and the Galerkin finite element is used to analyse the problem.
To show the degree of accurary of this solution, the results are compared with the recent work of the authors, based on the three−dimensional elasticity solution [1]. It is shown that the higher order theories can be used as a suitable alternative to rigorous three dimensional analysis.

1-INTRODUTION

The free vibration and dynamic analysis of spherical shells have been under investigation for many years, due to their important role in structual engineering applications. In a number of cases, closed and open spherical shells have been analysed by closed form solution, using Legendre polynomials and Legendre functions, as given by kraus [2] and Sodel [3]. In these analysis, traditional membrane and bending theories of thin shells were employed, and the Kirchhoff−Love simplified hypothesis were used. For the case of moderately thick shells, the effect of transverse shear and rotary inertia were taken into account by kalnins and kraus [4], to study the vibration of spherical shells.
Free vibration of orthotropic spherical shells, and the dynamic response of laminated orthotropic spherical shells, were studied respectively by chao[5] and Narasimhan [6]. In both analysis first order shear deformation theory was used. Recently the authors have considered the free vibrations of laminated orthotropic cross−ply hemispherical shells [7] and cross−ply laminated cylindricul shells [8], using elasticity solution.

2- THEORITICAL FORMULATION

The strain−displacement relations in general curvilinear coordinate are [2]

$$\varepsilon_i=\frac{\partial}{\partial\alpha_i}\left(\frac{u_i}{\sqrt{g_{ij}}}\right)+\frac{1}{2g_{ii}}\sum_{S=1}^{3}\frac{\partial g_{ii}}{\partial\alpha_s}\frac{u_s}{\sqrt{g_{ss}}}\qquad i=1,2,3$$

$$\gamma_{ij}=\frac{1}{\sqrt{g_{ii}g_{jj}}}\left[g_{ii}\frac{\partial}{\partial\alpha_j}\left(\frac{u_i}{\sqrt{g_{ii}}}\right)+g_{jj}\frac{\partial}{\partial\alpha_i}\left(\frac{u_j}{\sqrt{g_{jj}}}\right)\right]\qquad i,j=1,2,3\quad i\neq j\quad (1)$$

where g_{11},g_{22} and g_{33} are the principle values of reference metric tensor in the directions of orthogonal curvilinear coordinates α_1,α_2 and z respeetively.
For the shell of revolution, where $u_1=u$, $u_2=v$, $u_3=w$ the magnitude of base vectors are relaled to the Lame parameters A_1 and A_2 and the principle radius of curvatures of the middle plane of shell R_1 and R_2 as

$$g_{11}=[A_1.(1+\eta/R_1)]^2, g_{22}=[A_2(1+\eta/R_2)]^2, g_{33}=1 \quad (2)$$

upon substituation Eqs(2) into Eqs.(1), the general strain−displacement relations for shell of revolution is obtained as

$$\varepsilon_1=\frac{1}{A_1(1+z/R_1)}\left(\frac{\partial u}{\partial\alpha_1}+\frac{v}{A_2}\frac{\partial A_1}{\partial\alpha_2}+\frac{A_1 w}{R_1}\right)$$

$$\varepsilon_2=\frac{1}{A_2(1+z/R_2)}\left(\frac{\partial v}{\partial\alpha_2}+\frac{u}{A_1}\frac{\partial A_2}{\partial\alpha_1}+\frac{A_2 w}{R_2}\right)$$

$$\varepsilon n=\frac{\partial w}{\partial z}$$

$$\gamma_{12}=\frac{A_2(1+z/R_2)}{A_1(1+z/R_1)}\frac{\partial}{\partial\alpha_1}\left[\frac{v}{A_2(1+z/R_2)}\right]+\frac{A_1(1+z/R_1)}{A_2(1+z/R_2)}\frac{\partial}{\partial\alpha_2}\left[\frac{u}{A_1(1+z/R_1)}\right] \quad (3)$$

$$\gamma_{1n}=\frac{1}{A_1(1+z/R_1)}\frac{\partial w}{\partial\alpha_1}+A_1\left(1+\frac{z}{R_1}\right)\frac{\partial}{\partial z}\left[\frac{u}{A_1(1+z/R_1)}\right]$$

According to second order shell theory , the displacements are defined as [2].

$$u(\alpha_1,\alpha_2,z,t)=u_\circ(\alpha_1,\alpha_2,t)+z\psi_1(\alpha_1,\alpha_2,t)$$

$$v(\alpha_1,\alpha_2,z,t)=v_\circ(\alpha_1,\alpha_2,t)+z\psi_2(\alpha_1,\alpha_2,t)$$

$$w(\alpha_1,\alpha_2,z,t)=w_\circ(\alpha_1,\alpha_2,t)+z\omega_1(\alpha_1,\alpha_2,t)+\frac{z^2}{2}w_2(\alpha_1,\alpha_2,t) \quad (4)$$

where $u_\circ$,$v_\circ$ and $w_\circ$ denote the midplane displacements along the α_1 , α_2 and z directions respectively , while ψ_1 , ψ_2 denoting rotations of a transverse normal about the α_1 , α_2 axes respectively. Upon inserting Eqs. (4) into (3) one obtains [2].

$$\varepsilon_1=\frac{1}{1+z/R_1}\left(\varepsilon^\circ_1+z\varepsilon'_1+\frac{z^2}{2}\varepsilon''_1\right)$$

$$\varepsilon_2=\frac{1}{1+z/R_2}\left(\varepsilon^\circ_2+z\varepsilon'_2+\frac{z^2}{2}\varepsilon''_2\right)$$

$$\varepsilon_n=w_1+zw_2$$

$$\gamma_{12}=\frac{1}{1+z/R_1}(\beta^\circ_1+z\beta'_1)+\frac{1}{1+z/R_2}(\beta^\circ_2+z\beta'_2)$$

$$\gamma_{2n}=\frac{1}{1+z/R_2}\left(\mu^\circ_2+z\mu'_2+\frac{z^2}{2}\mu''_2\right)$$

$$\gamma_{1n}=\frac{1}{1+z/R_1}\left(\mu^\circ_1+z\mu'_1+\frac{z^2}{2}\mu''_1\right) \quad (5)$$

where the terms ε°_1 , ε°_2 , ε'_1 , ε''_2 , ε''_1 , ... μ°_2 , μ'_1, μ'_1 , μ'_2 , μ''_1 , μ''_2 are given in term of displacement components in Ref [2].
Based on the second shell theory , the stresses and forces and moments per unit length are related as

$$\begin{Bmatrix}N_1\\N_{12}\\Q_1\end{Bmatrix}=\int_z\begin{Bmatrix}\sigma_1\\\tau_{12}\\\tau_{1n}\end{Bmatrix}(1+z/R_2)dz$$

$$\begin{Bmatrix}N_2\\N_{21}\\Q_2\end{Bmatrix}=\int_z\begin{Bmatrix}\sigma_2\\\tau_{21}\\\tau_{2n}\end{Bmatrix}(1+z/R_1)dz$$

$$\begin{Bmatrix}M_1\\M_{12}\end{Bmatrix}=\int_z\begin{Bmatrix}\sigma_1\\\tau_{12}\end{Bmatrix}z\left(1+\frac{z}{R_2}\right)dz$$

$$\begin{Bmatrix}M_2\\M_{21}\end{Bmatrix}=\int_z\begin{Bmatrix}\sigma_2\\\tau_{21}\end{Bmatrix}z\left(1+\frac{z}{R_1}\right)dz$$

$$\begin{Bmatrix}S_i\\P_i\\T_i\\A\\B\end{Bmatrix}=\int_z\begin{Bmatrix}\tau_{in}.z.(1+\frac{z}{R_j})\\\sigma_i(\frac{z^2}{2})(1+\frac{z}{R_j})\\\tau_{in}(\frac{z^2}{2})(1+\frac{z}{R_j})\\\sigma_n(1+\frac{z}{R_1})(1+\frac{z}{R_2})\\\sigma_n(1+\frac{z}{R_1})(1+\frac{z}{R_2})\end{Bmatrix}d_z \qquad i,j=1,2 \quad i\neq j \quad (6)$$

The equilibrium equations describing the dynamic behavior of a composite shell based on the terms of Eqs.(6) , using the variational methods are

$$\frac{\partial(A_2N_1)}{\partial\alpha_1}+\frac{\partial(A_1N_{21})}{\partial\alpha_2}-N_2\frac{\partial A_2}{\partial\alpha_1}+N_{12}\frac{\partial A_1}{\partial\alpha_2}+A_1A_2\left(\frac{Q_1}{R_1}+q_1\right)=A_1A_2(I_1\ddot{u}_\circ+I_2\ddot{\psi}_1)$$

$$\frac{\partial(A_2N_{12})}{\partial\alpha_1}+\frac{\partial(A_1N_2)}{\partial\alpha_2}-N_1\frac{\partial A_1}{\partial\alpha_2}+N_{21}\frac{\partial A_2}{\partial\alpha_1}+A_1A_2\left(\frac{Q_2}{R_2}+q_2\right)=A_1A_2(I_1\ddot{v}_\circ+I_2\ddot{\psi}_1)$$

$$\frac{\partial(A_2Q_1)}{\partial\alpha_1}+\frac{\partial(A_1Q_2)}{\partial\alpha_2}-A_1A_2\left(\frac{N_1}{R_1}+\frac{N_2}{R_2}-q_n\right)=A_1A_2\left(I_1\ddot{w}_\circ+I_2\ddot{w}_1+\frac{I_3}{2}\ddot{w}_2\right)$$

$$\frac{\partial(A_2M_1)}{\partial\alpha_1}+\frac{\partial(A_1M_{21})}{\partial\alpha_2}-M_2\frac{\partial A_2}{\partial\alpha_1}+M_{12}\frac{\partial A_1}{\partial\alpha_2}-A_1A_2(Q_1-m_1)=A_1A_2(I_2\ddot{u}_\circ+I_3\ddot{\psi}_1)$$

$$\frac{\partial(A_2M_{12})}{\partial\alpha_1}+\frac{\partial(A_1M_2)}{\partial\alpha_2}-M_1\frac{\partial A_1}{\partial\alpha_2}+M_{21}\frac{\partial A_2}{\partial\alpha_1}-A_1A_2(Q_2-m_2)=A_1A_2(I_2\ddot{v}_\circ+I_3\ddot{\psi}_2)$$

$$\frac{\partial(A_2S_1)}{\partial\alpha_1}+\frac{\partial(A_1S_2)}{\partial\alpha_2}-\left(\frac{M_1}{R_1}+\frac{M_2}{R_2}\right)A_1A_2-(A-m_n)A_1A_2=A_1A_2(I_2\ddot{w}_\circ+I_3\ddot{w}_1+I_4/2\,\ddot{w}_2)$$

$$\frac{\partial(A_2T_1)}{\partial\alpha_1}+\frac{\partial(A_1T_2)}{\partial\alpha_2}-A_1A_2\left(\frac{P_1}{R_1}+\frac{P_2}{R_2}\right)-A_1A_2\left(B-\frac{h^2}{8}q_n\right)=A_1A_2\left(\frac{I_3}{2}\ddot{w}_\circ+\frac{I_4}{2}\ddot{w}_1+I_5/4\ddot{w}_2\right) \quad (7)$$

where $I_i=\int_{-h/2}^{h/2}\rho\left(1+\frac{z}{R_1}\right)\left(1+\frac{z}{R_2}\right)z^{(i-1)}dz,\quad i=1,2,..5$

$$q_{ni}=q_i^+\left(1+\frac{h}{2R_1}\right)\left(1+\frac{h}{2R_2}\right)+\bar{q}_i\left(1-\frac{h}{2R_1}\right)\left(1-\frac{h}{2R_2}\right)$$

$$m_i=\frac{h}{2}[q_i^+(1+\frac{h}{2R_1})(1+\frac{h}{2R_2})-\bar{q}_i(1-\frac{h}{2R_1})(1-\frac{h}{2R_2}) \quad (8)$$

the inside and outside surfaces respectively.
It is assumed that the shell is made of laminate with M layers, where the material of each layer is orthotropic. The constitutive law of the kth layer in the reference coordinate system is written as

$$\begin{bmatrix}\sigma_1\\ \sigma_2\\ \sigma_n\\ \tau_{2n}\\ \tau_{1n}\\ \tau_{12}\end{bmatrix}=\begin{bmatrix}\overline{Q}_{11} & \overline{Q}_{12} & \overline{Q}_{13} & 0 & 0 & \overline{Q}_{16}\\ \overline{Q}_{12} & \overline{Q}_{22} & \overline{Q}_{23} & 0 & 0 & \overline{Q}_{26}\\ \overline{Q}_{13} & \overline{Q}_{23} & \overline{Q}_{33} & 0 & 0 & \overline{Q}_{36}\\ 0 & 0 & 0 & \overline{Q}_{44} & \overline{Q}_{45} & 0\\ 0 & 0 & 0 & \overline{Q}_{45} & \overline{Q}_{55} & 0\\ 0 & \overline{Q}_{26} & \overline{Q}_{36} & 0 & 0 & \overline{Q}_{66}\end{bmatrix}\begin{bmatrix}\varepsilon_1\\ \varepsilon_2\\ \varepsilon_n\\ \gamma_{2n}\\ \gamma_{1n}\\ \gamma_{12}\end{bmatrix} \quad (9)$$

Substituating for strain matrix from Eqs.(5) and (9) in Eq.(6) and then into Eqs.(7) results into the matrix equation for forces and moments in terms of the strains and rotation compnents of the middle plane of shell.

2.1 spherical shell

consider a composite multilayer spherical shell of mean radius R. The general governing Eq.(7) reduce to the following equilibrium equations, describing the dynamic behavior of the spherical shell.

$$N_{\phi,\varphi}+N_\varphi Cot\varphi-N_\theta Cot\varphi+Q_\varphi+Rq_\varphi=R(I_1\ddot{u}_\circ+I_2\ddot{\psi}_\varphi)$$

$$N_{\phi\theta,\varphi}+2N_{\varphi\theta}Cot\varphi+Q_\theta+Rq_\theta=R(I_1\ddot{v}_\circ+I_2\ddot{\psi}_\theta)$$

$$Q_{\phi,\varphi}+Q_\varphi Cot\varphi-N_\varphi-N_\theta+Rq_2=$$
$$R(I_1\ddot{w}_\circ+I_2\ddot{w}_1+\frac{I_3}{2}\ddot{w}_2)$$

$$M_{\phi,\varphi}+M_\varphi Cot\varphi-M_\theta Cot\varphi-RCos\varphi+Rm_\varphi=$$
$$R(I_2\ddot{u}_\circ+I_3\ddot{\psi}_\varphi)$$

$$M_{\varphi\theta,\varphi}+2M_{\theta\varphi}Cot\varphi-RQ_\theta+Rm_\theta=R(I_2\ddot{v}_\circ+I_3\ddot{\psi}_\varphi)$$

$$S_{\phi,\varphi}+S_\varphi Cot\varphi-M_\varphi-M_\theta-RA+Rm_z=$$
$$R(I_2\ddot{w}_\circ+I_3\ddot{w}_1+\frac{I_q}{2}\ddot{w}_2)$$

$$T_{\phi,\varphi}+T_\varphi Cot\varphi-P_\varphi-P_\theta+RB-R\frac{h^2}{8}q_2=$$
$$R(\frac{I_3}{2}\ddot{w}_\circ+\frac{I_q}{2}\ddot{w}_1+\frac{I_5}{q}\ddot{w}_2) \quad (10)$$

Where

$$I^i=\sum_{k=1}^{N}(\rho)_k\int_{Z_k-1}^{Z_k}(1+\frac{Z}{R})^2 Z^{(i-1)}dz \qquad i=1,2,...5$$

The strain – displacement relations reduce to

$$\varepsilon^\circ_\varphi=\frac{1}{R}(u_{\circ,\varphi}+w_\circ) \qquad \varepsilon^\circ_\theta=\frac{1}{R}(u_\circ Cot_\varphi+w_\circ)$$

$$\varepsilon'_\varphi=\frac{1}{R}(\psi_{\phi,\varphi}+w_1) \qquad \varepsilon'_\theta=\frac{1}{R}(\psi_\varphi Cot_\varphi+w_1)$$

$$\varepsilon''_\varphi=\frac{w_2}{R} \qquad \varepsilon''_\theta=\frac{w_2}{R}$$

$$\beta^\circ_\varphi=\frac{1}{R}v_{\circ,\varphi} \quad \beta^\circ_\theta=\frac{1}{R}v_\circ Cot_\varphi \quad \beta'_\varphi=\frac{1}{R}\psi_{\theta,\varphi}$$

$$\beta'_\theta=-\frac{1}{R}\psi_\theta Cot\varphi \qquad \mu^\circ\varphi=\frac{1}{R}(w_{\circ,\varphi}-u_\circ)+\psi_\varphi$$

$$\mu^\circ\theta=-\frac{v_\circ}{R}+\psi_\theta \qquad \mu'_\varphi=\frac{1}{R}w_{1,\varphi}$$

$$\mu''_\varphi=\frac{1}{R}w_{2,\varphi} \qquad \mu''_\theta=\circ \qquad \mu'=\circ \quad (11)$$

Substituating Eqs. (11) into the constitutive law (9) and finally into the equilibrium Eqs. (10) results into the dynamic governing equations of the spherical shell in terms of the displacement components. The boundary conditions may be either kinematical or forced, depending on whether the displacements and rotations are specified on the boundary or forces and moments. These conditions are in general as follows

$u_\circ=u_\circ^*$	or	$N_\phi=N_\varphi^*$
$v_\circ=v_\circ^*$	or	$N_{\theta\varphi}=N_{\varphi\theta}^*$
$w_\circ=w_\circ^*$	or	$Q_\phi=Q_\varphi^*$
$\psi_\varphi=\psi_\varphi^*$	or	$M_\phi=M_\varphi^*$
$\psi_\theta=\psi_\theta^*$	or	$M_{\phi\theta}=M_{\varphi\theta}^*$ (12)

where $(\ )^*$ denotes the known values on the boundary

2.2 Galerkin Finite Element Formulation

while the variational formulation of shell problems require higher order field approximations for the finite element modeling of shell structures, the Galerkin method provides acceptable accuracy with lower field approximations [9] . On this basis, the field variable are approximated by linear shape functions. The nodal degrees of freedom, according to the equilibrium equations are $u_\circ$,$v_\circ$, $w_\circ$, ψ_φ , ψ_θ , w_1 and w_2 . Approximating these functions with linear shape for the base element equilbrium equations yields.

$$\oint_{v(e)}^{i} \phi_i R_1^{(e)} N_s dv = \circ \qquad i=1,2,...,7 \quad S=i,j \qquad (13)$$

where R_1 is the residue of equilibrium equations and N_s is the shape function. The process of weak formutation should be carried out for the terms with the second and first order derivatives such that the resulting weak forms are compatible with one of the essential kinematical and forced boundary conditions.

The final matrix form of the finite element equilibrium equation reduce to the follwing

$$[M]\{\ddot{X}\}+[K]\{X\}=\{F\} \qquad (14)$$

The unkown matrix {x} is

$<x>=<u_{oi}, v_{oi}, w_{oi}, \psi_{\varphi i}, \psi_{\theta i}, w_{1i}, w_{2i}>$

The equilibrium equation (14) is solved by time marching technique and the results are discussed in the following.

3 RESULTS

A two–layer cross–ply hemispherical shell (0/90 from outer – ply) composed of graphite–ep oxy is considered with the following propreties

$E22=E33=4.8\times10^9$ Pa . $\frac{E_{11}}{E_{22}}=25$ $\quad \nu_{12}=\nu_{23}=0.25$

$\frac{G_{12}}{E_{22}}=0.5$ $\quad$ h=10 mm $\quad$ R=100 mm

The blast load is taken as Fig.(1).
$P_\circ$= 1 MPa , T_1=4 MSec.

The hemispherical shell is clamped at the edge. The radial displacement on the middle surface and at the crown which is obtained are compared with elasticity solution [1] in table (1) As it is obsereved there is good agreement between the results and the maximum difference is about 7% .

This difference is underestood to be due to the approximate nature of solution which is based on the single equivalent layer. It is noticed that the difference and can be used as a suitable alternative to rigorous three dimensional analysis.

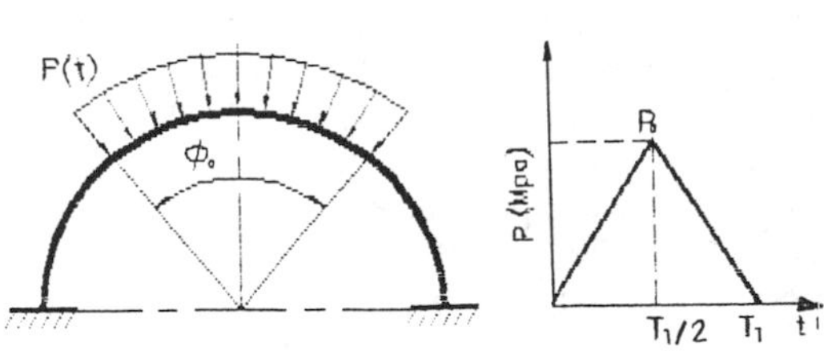

Fig.(1) Dynamic Lood

Table (1). Comparision of the results

msec	mm	Ref(1) mm
2	-2.94	-3.18
5	2.96	3.18
8	-3.33	-3.61
11	2.92	3.18

The time history of radial displaeement on the middle surface and at crown for $\varphi_\circ=90^\circ$ are shown in Fig.(2).

In Fig.(3) a comparision is made between two different orientation of layers (0/90) and (90/0) for radial displacement versus time. As it is noticed, theradial displacement for (90/0) is more than (0,90). In other words the hemisphrical shell is stiffer when the outer layer is reinforced.

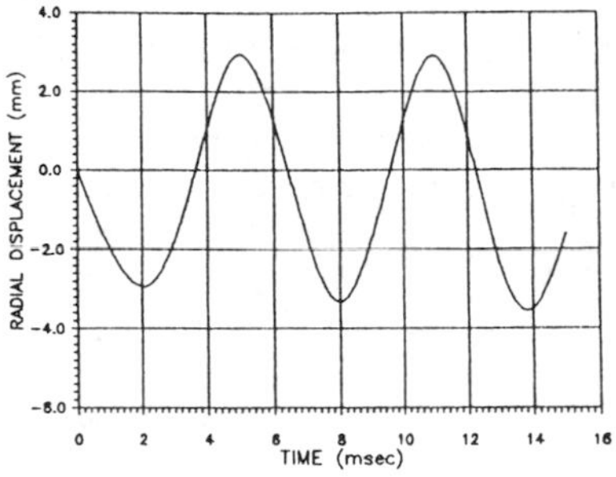

Fig(2). Time history of radial displacement

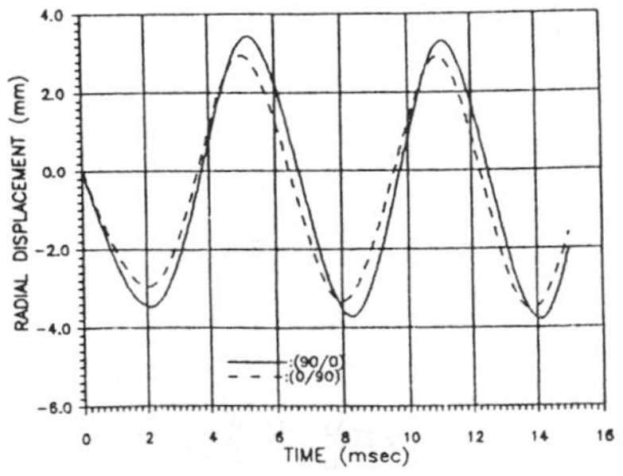

Fig (3). comparision of two different arientation of layers

REFERENCES

Yas, M.H. 1996. Three–dimensional dynamic revolation Ph.D thesis
kraus, H., 1967. Thin elastic shells, John wily and Sons.
Sodel, w. 1981. vibrations of shells and plates, Marrel Dekker.
Kalnins, A. 1966. Effect of transverse shear and rotary inertia on vibration of spherical shells. proceeding of 5 th U.S. Nat'l congress of Appl. Mech 134–138
Chao, c.c. 1988. Axisymmetric free vibration of orthotropic complete spherical shell. Composite Meterial J. 22: 1116–1130
Narasimhan, M.C. 1992. Dynamic response of laminated orthotropic spherical shells, J. Acoustical society of America, 22(5) : 719–720
Shakeri, M. Yas, M.H. 1995.
Three–dimensional axisymmetric vibrations of orthortropic cross– ply laminated hemispherical shells. Proceeding of 32 th SES Conference New–orlean Shakeri, M. Yas, M.H. 1996, Three– dimensional vibration of orthotropic and cross–ply laminated cylindrical shells, proceeding of 1st. Conference, ACAM 389–393. Melbourne
Sshakeri, M., Eslami, M.R. Yas, M.H. 1994. Galerkin finite element analysi of shells of revolution under blast load. proceeding of second ASMA/ESDA Conference london.

Modern Practice in Stress and Vibration Analysis, Gilchrist (ed.) © *1997 Balkema, Rotterdam, ISBN 90 5410 896 7*

Compression strength of composite suspension push-rods for Formula 1 racing cars

L.Curley, J.Mallon & M.D.Gilchrist
Mechanical Engineering Department, University College Dublin, Ireland

ABSTRACT: Advanced composite materials are extensively used in the construction of a contemporary Formula 1 racing car. This paper describes the manufacture and ultimate mechanical performance under compression of composite suspension push-rods that could typically be used in a Grand Prix racing car. An aerofoil-type cross-section was used with different lay-ups of unidirectional and woven cross-ply carbon/epoxy composite. Failure mechanisms including compression and buckling were observed and the ultimate strength of the component under compression was significantly less than that of the material.

1 INTRODUCTION

A central load-bearing structure in a modern F1 car connects the front and rear suspension systems; this load-bearing structure consists of the monocoque, the engine and the gearbox casing. The driver, fuel tank and front suspension dampers are housed within the monocoque whilst the engine is jointed to the back of the monocoque on four studs. The gearbox casing is attached to the rear face of the engine. This three-piece box-beam structure carries the inertial loads to the four corners of the car. Various wing structures, underbodies, cooler ducting and bodywork are attached to and around this box-beam.

More than 80% of a modern Formula 1 car is made from some form of composite material, with the majority being based on carbon/epoxy systems. Such extensive use originates back to the mid-1970s when the "wing-car", developed by Lotus, created large downforce by using the underneath of the car. This required large wing-shaped underbodies to be attached to a chassis of reduced width, the torsional rigidity of which could only be maintained efficiently by use of composite materials. Additionally, turbochargers emerged in the late 1970s and, producing in excess of 1400 bhp, these led to severe loads being applied to the chassis. Composite materials offered greater specific stiffnesses and greater flexibility in design than the aluminium alloys that had been used previously.

In 1981 the monocoque of the McLaren MP4 F1 car was first moulded from a carbon fibre reinforced epoxy polymer. The monocoque was moulded over a machined aluminium tool which was subsequently removed in sections through the cockpit opening. Unidirectional carbon/epoxy was used for the skins whilst aluminium honeycomb was used for the core. This design was used virtually unchanged for six racing seasons, so successful was the one-piece construction. A two-piece construction was pioneered by Gustav Brunner in 1983 for his ATS F1 car by moulding the monocoque as top and bottom halves in a female mould. This had advantages of providing greater flexibility in the geometry and size of the monocoque over the one-piece construction.

More recently, however, composites have begun to be used in manufacturing components other than primary structural parts, such as, for example, high-strength components, the gearbox casing, where torsional rigidity is crucial, and suspension components, which require high stiffness. Traditional metal suspension components are being replaced by composites in order to increase the stiffness of the individual suspension members, and thereby give the designer more control over the overall stiffness of the suspension system. It is the push-rod which has the single major influence on the stiffness of the suspension system. However, the change from metal to composite components has not been without problems for many F1 teams. Williams, for example, replaced the metal push-rod by a composite push-rod but had to revert to the metal component due to a series of rear suspension failures in testing.

This particular paper aims to investigate the performance of a composite push-rod under compression which should consequently assist design engineers to predict the ultimate limits to which a composite push-rod can be used. The geometry and stacking sequence that were used to manufacture the push-rods are discussed, as is the experimental test setup that was used to apply direct compression to the components. The performance of the push-rods under compression, the manner of failure and the fracture mechanisms that were observed are discussed.

2 PUSH-ROD DESIGN & MANUFACTURE

2.1 *Geometry and stacking sequence*

In order to minimise the effects of wind-drag around the push-rods it was decided to utilise an aerofoil cross-section instead of a circular cross-section. Uniform and tapered layups were used, the purpose of the taper being to increase the equivalent modulus along the critical section of the push-rod and consequently, to increase the load at which buckling would occur. Since the end sections of both the tapered and uniform layups were identical it was anticipated that the load at which compression failure should occur would be identical for both types of push-rod. The push-rod was 650mm long whilst the nominal wall thickness was 1.825mm for the uniform push-rod and varied between 1.825-2.450mm for the tapered push-rods. The external major and minor dimensions of the airfoil axes were nominally specified at 38mm x 18mm.

One objective of this project was to investigate the influence of the lay-up on the possible buckling response of the push-rod. Since the principal in-service mechanical load on the push-rod was uniaxial compression, it was necessary to maximise the number of 0° plies within the stacking sequence in order to provide maximum uniaxial stiffness. A number of cross-plies were necessary, however, to prevent longitudinal splitting of the push-rod. The first stacking sequence that was considered was a uniform layup of ($0°/90°,0°_9,0°/90°$), i.e., two external 0°/90° cross-plies of woven prepreg surrounding nine unidirectional 0° plies. This layup differs from that which is typically used in current F1 design only in that there are no tapered plies within the stacking sequence. As such, it was anticipated that the ultimate mechanical response of this push-rod design would be a lower bound limit and failure would be due to buckling.

The remaining two push-rods were identically tapered centrally along their mid-lengths and the particular stacking sequence that was used was ($0°/90°,0°_7,90°,0°_6,0°/90°$), i.e., two outer 0°/90° cross-plies of woven prepreg surrounding seven 0° plies, one 90° ply and six 0° plies. The taper was obtained by only placing some of the 0° and 90° plies along part of the 650mm length of the push-rods.

Commercially available laminate analysis software was used (LAP, 1991) to estimate the equivalent laminate properties (shown in Table 1) from the precise ply properties of Table 2.

2.2 *Manufacture of Push-rods*

Three separate carbon/epoxy push-rods were manufactured by wrapping the various plies of

Table 1. Equivalent mechanical properties of uniform and tapered layups used to manufacture the different push-rods.

Equivalent laminate property	Uniform Layup	Tapered Layup
E_{xx}, [GPa]	211	221
E_{yy}, [GPa]	23.7	34.7
ν_{xy}	0.115	0.068
ν_{yx}	0.013	0.011
G_{xy}, [MPa]	11.6	11.7

prepreg around a hollow elliptical silicone mandrel. This was then placed within an elliptical two-part cavity mould and cured in an autoclave. The hollow mandrel acted as an expandable bladder during the curing cycle, thereby pressing the prepreg firmly against the walls of the mould and ensuring that a uniform wall thickness was produced along the length of the push-rod.

The autoclave curing cycle for the woven and unidirectional carbon/epoxy prepreg involved a 90 minutes cure at 125°C and 700kPa with a heat-up and cool-down rate of 2.3°C per minute. When the temperature reached 125°C the vacuum was vented to atmosphere. Pressure was then introduced and ramped at 50kPa per minute to 700kPa. When the pressure cycle was completed the pressure was ramped down at 50kPa per minute to 0kPa, at which stage the vacuum was reintroduced.

Both the mould and the silicone mandrel were reused when manufacturing all three push-rods and these were cleaned and degreased before being coated with release agent (Freekote) prior to the plies of carbon/epoxy prepreg being wrapped around the mandrel and placed within the mould. The complete assembly was vacuum bagged to evacuate air, solvents and entrapped volatiles from the laminate and to allow the positive autoclave pressure to consolidate the laminate against the mould surface. A breather cloth bagging assembly was used to absorb any excess resin flow and also to smooth out the sharp corners of the mould, which could cause the vacuum bag to rupture under the high autoclave pressures. A solid release film was placed against the mould walls to prevent the breather cloth from sticking to the mould surfaces.

Upon completion of the curing cycle, the vacuum bag assembly was removed from the autoclave. The bag and breather were discarded and the end plates were removed prior to the mould being opened. The composite push-rod was then taken from the mould and the silicone mandrel removed from the centre of the push-rod.

Before the actual carbon/epoxy push-rods could be manufactured, it was necessary to manufacture a suitable elliptical mould and elliptical silicone mandrel so that the finished push-rods would be of the required thickness and cross-section. The mould was machined from aluminium whilst the silicone mandrel was manufactured using GFRP slips, an elliptical copper pipe and the mould. The

Table 2 Mechanical properties of the 0°/90° woven and unidirectional carbon/epoxy material systems used to manufacture the composite push-rods.

Mechanical property	0°/90° woven ply	unidirectional ply
Thickness	0.35mm	0.125mm
Longitudinal stiffness	53GPa	310GPa
Transverse stiffness	52GPa	5.9GPa
Shear modulus	0.011GPa	0.012GPa
Poisson's ratio	0.1	0.2
Longitudinal tensile strength	690MPa	1960MPa
Longitudinal compressive strength	59MPa	700MPa
Transverse tensile strength	690MPa	354MPa
Transverse compressive strength	59MPa	354MPa
Shear strength	80MPa	100MPa

copper pipe was located centrally within the mould cavity through an aluminium end-plate. The end-plate was subsequently bolted to the mould and the mould was inverted. The GFRP slips were placed against the mould walls and de-aerated liquid silicone rubber was poured into the space between the GFRP slips and the copper pipe in the mould. This assembly was left under room conditions for fourteen hours to allow the rubber compound to cure and was then placed in an oven at 120°C for 90 minutes to complete the curing process. The hollow silicone mandrel was then removed from the mould and the copper pipe was extracted from the mandrel. No significant air bubbles or voids, which would have made the mandrel unsuitable for manufacturing the push-rods, were detected visually.

The GFRP slips were fabricated using the two halves of the mould. After spraying release agent on both halves of the mould, six plies of GFRP were stacked in each half of the mould. The two halves of the mould were covered in a release ply, covered with a bleeder cloth and placed in a vacuum bag, which was then sealed. The assembly was cured in the autoclave using an appropriate cycle.

A fourth carbon/epoxy push-rod was manufactured using a sand-bag technique instead of the silicone mandrel, which ruptured when being removed from the third push-rod. The procedure involved in making this core used a cylindrical nylon

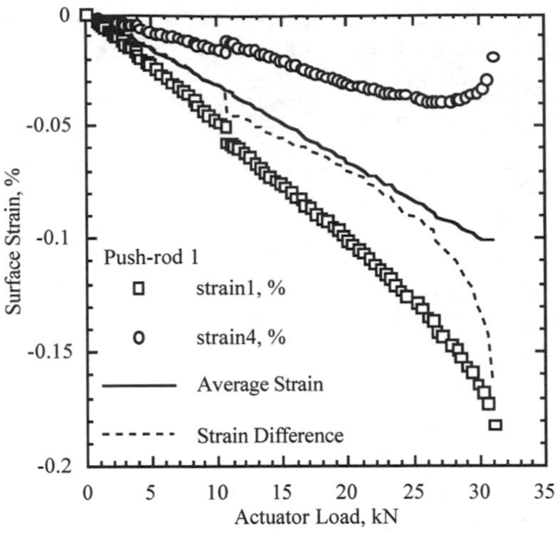

Figure 1. Variation of surface strains with actuator load during testing of push-rod 1. Buckling is identified by the difference between the values of the two surface strain readings and begins with the onset of actuator load. Incipient fracture begins at approximately 26kN.

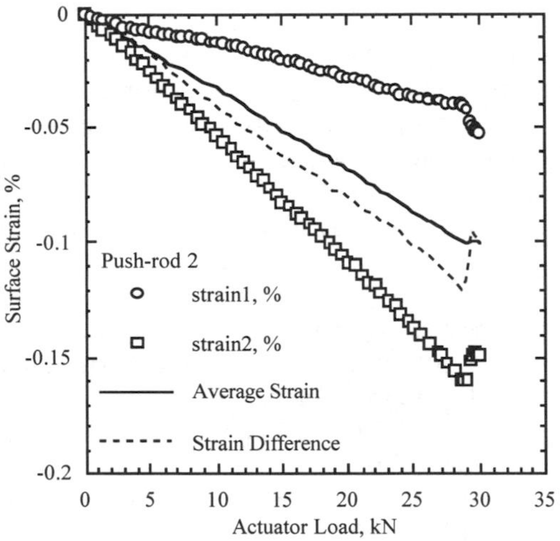

Figure 2. Variation of surface strains with actuator load during testing of push-rod 2. Buckling is identified by the difference between the values of the two surface strain readings and begins with the onset of actuator load. Incipient fracture begins at approximately28kN.

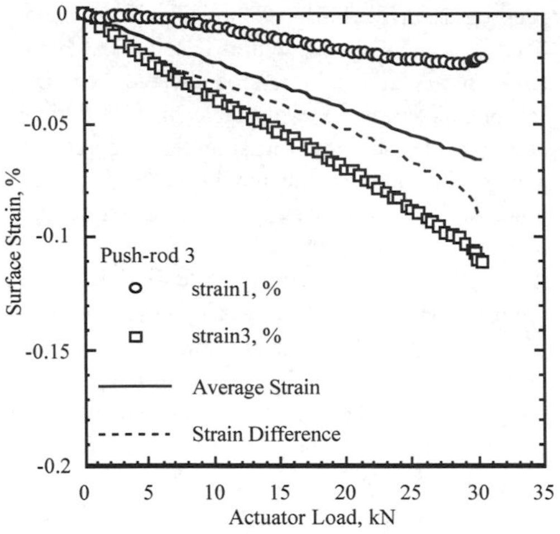

Figure 3. Variation of surface strains with actuator load during testing of push-rod 3. Buckling is identified by the difference between the values of the two surface strain readings and begins with the onset of actuator load. Incipient fracture begins at approximately 29kN.

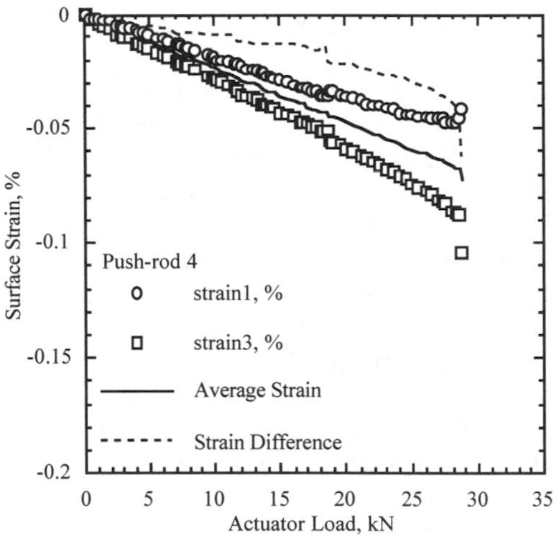

Figure 4. Variation of surface strains with actuator load during testing of push-rod 4. Buckling is identified by the difference between the values of the two surface strain readings and begins with the onset of actuator load. Incipient fracture begins at approximately 28kN.

tube (thermally stable, thin and impermeable). The mould, with the GFRP slips, was then bolted together and the nylon tube was inserted into the mould cavity. The tube was sealed at one end using sealant tape and dry sand was then added to the tube and compacted by means of a vacuum pump. The subsequent procedure for manufacturing this fourth push-rod was identical to that based on using the silicone mandrel.

However, this fourth push-rod was laid up incorrectly with a slight overlap in the first ply, which prevented the first ply from expanding and consolidating against the other plies in the mould during the autoclaving process. This prevented resin from flowing to the surface and consequently exposed fibres were detected at the outer surface of the push-rod after manufacture. In normal operating conditions such a component would be scrapped. Nevertheless, this push-rod was tested in the same manner as the other three push-rods and the results of this test is also discussed in the following sections.

3 EXPERIMENTAL TEST PROCEDURE

Four push-rods have been tested statically to failure under compression using a displacement mode of control on a 100kN uniaxial servohydraulic fatigue machine (Series 8501 Instron). The loading was introduced at both ends of a push-rod using female end-fixtures which had been designed to provide boundary conditions that were pin-jointed at the bottom and cantilevered at the top in order to simulate in-sevice support conditions.

Surface strains, from gauges at three different positions on the push-rods, were recorded using a data acquisition system which operated on a keypress sequence. Strain gauges were aligned longitudinally and transversely along the length of the push-rods to measure the performance under compressive load. Two gauges were aligned axially at the midlength and opposite faces of the push-rods: these provided information on the presence of buckling, the deviation from linearity in the mechanical response of the push-rod and fracture strains. A third strain gauge was aligned normal to the first two gauges and was used to calculate values of Poisson's ratio.

Table 3. Summary of ultimate actuator loads, surface strains and compressive failure locations.

Specimen	Load	Strain	Location
Push-rod 1	31.00kN	0.182%	40mm from centre
Push-rod 2	30.00kN	0.160%	pin-joint end
Push-rod 3	30.25kN	0.111%	pin-joint end
Push-rod 4	28.75kN	0.104%	pin-joint end

4 BEHAVIOUR OF PUSH-RODS UNDER COMPRESSIVE LOADING

All push-rods were loaded statically to failure by means of a displacement mode of control. Load, displacement and strain values were collated at increments of actuator load. As the applied load was increased from zero, the response of the push-rods was initially linear elastic. Figures 1-4 detail the variation of compressive surface strains at the mid-length position on opposite sides of the four push-rods with increasing actuator load. The strain responses deviated from linearity at approximately 90% of the final failure load although minor fracture events occurred before this deviation from linearity in the first and fourth push-rod tests (at 10.8kN in Figure 1 and 18.75kN in Figure 4, respectively). This deviation of the strain difference (i.e., magnitude of strain difference between front and back faces of the push-rods) from linearity, which occurred at approximately 90% of the final failure load identified the onset of catastrophic fracture. Ultimate failure of the first uniformly laid-up push-rod occurred some 40mm from the centre of the specimen whilst failure of the remaining three push-rods was concentrated around the the pin-jointed end-fixture of the testing machine. Table 3 summarises the loads and strains at which ultimate failure ocurred during the four tests.

The measured actuator loads and surface strains are presented in Figures 1-4 for the four push-rods. The average load-strain relationship for all the push-rods is essentially linear almost until fracture. However, the individual strain-gauge readings deviate from linearity immediately with the application of load and this deviation continues to increase directly with applied load up until failure.

Table 3 identifies the maximum direct compressive strains which were measured during each test and may be compared against the failure strains of the carbon-fibres of 1.5% (Lovell, 1991).

Figure 5. Compressive fracture of push-rod 4 as identified visually. The outer 0°/90° woven ply is clearly visible. The damage mechanism that initiated failure was due to compressive stress (near side in photograph).

While the maximum direct strain reading at failure of push-rod 1 (i.e., 0.18%) is greater than those recorded during the other three tests (this is to be expected since the strain gauge position of this push-rod was closer to the failure site than in all other tests), this is considerably less than the fibre failure strain. Consequently, failure of these components is due to geometric and manufacturing limitations rather than material limitations.

4.1 *Buckling Behaviour*

A simple first mode of buckling was apparent along the length of the four test specimens, with maximum lateral deformation (i.e., crest of the buckle) occurring close to the mid-length of the push-rods. Buckling initiated with the application of load in all push-rod tests, as can be seen from the deviation of the two sets of surface strain gauge readings (Figures 1-4) from the average compressive strain. The amplitude of the buckle increased linearly in size with actuator load until failure. No dial gauges were used during the tests to quantify the amplitude of the buckle although this could be estimated from the degree of curvature and bending that has been measured by the surface strain gauges.

Figure 6. Micrograph of compressive fracture of push-rod.

4.2 *Damage in Push-rods*

The compressive failure mechanisms that occurred in all four push-rods were similar although failure of push-rod 1 occurred at a position close to the mid-length of the component whereas failure was close to the pin-jointed end for the other three push-rods. The reason for this different failure site is due to the fact that push-rod 1 was manufactured without any tapered region in its mid-section, unlike the other three push-rods. The general appearance of the fracture associated with push-rod 4 is shown in Figure 5. The appearance of the fracture surface is different both around the perimeter of the push-rod and through the thickness of the push-rod. The fracture is not uniformly compressive around the perimeter and this is due to the differential buckling strains that existed on opposite sides of the push-rod. The lack of similarity of the through-thickness fracture features is partly due to the variation of compressive strains and partly due to the different ply orientations through the thickness of the push-rod.

Figure 7. Detailed micrograph of fractured fibres due to compressive failure of push-rod.

Figures 6 and 7 detail the initiating compressive failure sites that led to ultimate fracture of the push-rods using scanning electron microscopy. Many broken fibres are evident in Figure 6 and the manner in which these fibres fractured is characteristic of compressive failure, i.e., fibre microbuckling and localised fibre fracture (Gilchrist et al., 1996a, b).

5 CONCLUSIONS

Unidirectional and woven cross-ply carbon/epoxy composites were used to manufacture suspension push-rods that could typically be used in a Formula 1 racing car. These were subsequently loaded to failure under compression using cantilevered and pin-jointed end supports. Three push-rods had a tapered mid section consisting of 0° and 90° plies whilst an initial trial specimen was of constant thickness along its length. Fracture of the trial specimen (push-rod 1) occurred close to the mid-length whilst fracture in all other cases was close to the pin-jointed support in the loading frame. Buckling occurred in all cases and this increased directly with the application of load. The ultimate performance of these particular push-rods was limited by geometric, manufacturing and support parameters and was not close to the ultimate fibre failure strains of the materials that were used.

ACKNOWLEDGEMENTS

Financial support in the form of a President's Research Award from University College Dublin is gratefully acknowledged. The authors are happy to express their gratitude to Jordan's Formula 1 for providing composite prepreg materials.

REFERENCES

Gilchrist, M. D., Kinloch, A. J., Matthews, F. L. & Osiyemi, S. O., 1996a, Mechanical performance of carbon fibre and glass fibre-reinforced epoxy I-beams: I - Mechanical behaviour. *Composites Science & Technology*, 56, pp. 37-53.

Gilchrist, M. D., Kinloch, A. J. & Matthews, F. L., 1996b, Mechanical performance of carbon-fibre and glass-fibre reinforced epoxy I-beams: II - Fractographic failure observations. *Composites Science & Technology*, 56, pp. 1031-45.

LAP, 1991, *Laminate Analysis Program*, Centre for Composite Materials, Imperial College, London SW7 2BY, UK.

D. R. Lovell, 1991, *Carbon & high performance fibres directory*, Edition 5, Chapman & Hall.

Modern Practice in Stress and Vibration Analysis, Gilchrist (ed.) © *1997 Balkema, Rotterdam, ISBN 90 5410 896 7*

Fundamental solutions for equations of harmonic vibrations in the theory of elasticity

S.V. Kuznetsov
Institute for Problems in Mechanics, Moscow; Russia

ABSTRACT: Fundamental solutions for equations of harmonic vibrations in anisotropic media with arbitrary elastic anisotropy are constructed by the multipolar and analytic expansion method. Properties of the series representing fundamental solutions are investigated. For isotropic medium fundamental solution is obtained in a closed form, valid for any real frequencies.

INTRODUCTION

Fundamental solution in a closed form for the equations of harmonic vibrations of isotropic elastic media in R^3 for the first time was derived (Kupradze 1953). The form of Kupradze's solution is such, that at the zero frequency it becomes undefined and hence does not coincide with the Kelvin's fundamental solution which corresponds to the equations of equilibrium. Meanwhile in the similar situation of the Helmholtz equation its fundamental solution remains valid for any real frequencies and reduces to the fundamental solution of the classical potential theory at $\omega = 0$.

For an arbitrary elastic anisotropic medium in R^3 closed expressions for fundamental solutions are not known. By using Fourier transform (Natroshvili 1979) an improper volume integral representation was obtained for fundamental solutions of vibration problems and pseudo vibrations. But, there is no effective analytical or numerical procedures for computing that improper integral.

For non-stationary dynamics, the problem of fundamental solution construction can be solved by the use of the Radon transformation (John 1955), (Synge 1957), which appears at the stage of Fourier inversion. In the case of arbitrary elastic anisotropy this leads to multiple computations of successive line integrals and integrals on diametrical circles on the unit sphere (Burchuladze & Gegelia 1985). Some times this method is referred to as "plane-wave decomposition method".

The plane-wave decomposition method can also be applied to the discussed elliptic problem of harmonic vibrations in anisotropic media, but the form of the fundamental solution becomes even more complicated in comparison with the problems of non-stationary dynamics. This is because of non-homogeneity of a polynomial which represents the symbol of the operator of harmonic vibrations. In a more general situation of arbitrary elliptic equations with constant coefficients, presumably the first attempt to construct fundamental solution by the use of the plane-wave decomposition method is due to (Herglotz 1928).

Numerical estimates (Wilson & Cruse 1978) and in a later work (Deb, Henry & Wilson 1991) show, that in a simpler case of equilibrium equations for an anisotropic medium, the Radon transform method is highly time consuming in numerical algorithms.

The present paper gives an alternative method for the construction of the fundamental solution based on multipolar and analytical expansions. For the static problems multipolar expansions were used (Kuznetsov 1989, 1995). Some theoretical estimates show (Kuznetsov 1989), that for these problems the multipolar expansion method can have some advantage in comparison with the methods based on the Radon transformn.

In some cases of harmonic vibrations, the developed method allows to obtain closed expressions for fundamental solutions. In this respect, solutions to the Helmholtz equation and equations of the harmonic vibrations in isotropic medium demonstrate the validity of the method. It should be noted that in contrast to the Kupradze's fundamental solution, the present fundamental solution for an isotropic medium remains valid for any real frequency and transforms to Kelvin's fundamental solution at the zero frequency.

1 BASIC OPERATORS

A uniform anisotropic elastic medium in R^3 is considered with the equation of harmonic vibrations written in the form

$$\mathbf{A}_\omega(\partial_\mathbf{x})\,\mathbf{u}(\mathbf{x}) \equiv \equiv [\mathbf{A}_0(\partial_\mathbf{x}) - \rho\omega^2\mathbf{I}]\mathbf{u}(\mathbf{x}) = \mathbf{g}(\mathbf{x}) \tag{1.1}$$

where u is the displacement field, ρ is the density, ω is the frequency of vibrations, I is unit diagonal matrix, and g is the field of body forces. In the equation (1.1) A_0 is matrix differential operator of the equation of equilibrium

$$\mathbf{A}_0(\partial_\mathbf{x})\mathbf{u}(\mathbf{x}) \equiv -\mathrm{div}_\mathbf{x}\mathbf{C}\cdot\cdot\nabla_\mathbf{x}\mathbf{u}(\mathbf{x}) \tag{1.2}$$

The fourth order elasticity tensor C in (1.2) is assumed to be strongly elliptic and symmetric, provided it is regarded as a linear operator in the space of the second order tensors, so $C^{ijkl} = C^{klij}$.

The Fourier transform

$$f^\wedge(\xi) = \int_{R^3} f(\mathbf{x})\exp(2\pi i\,\xi\cdot\mathbf{x})dx, \qquad \xi \in R^3$$

applied to the operator A_ω gives corresponding symbol

$$\mathbf{A}^\wedge_\omega(\xi) \equiv \equiv \mathbf{A}^\wedge_0(\xi) - \rho\omega^2\mathbf{I} = (2\pi)^2\xi\cdot\mathbf{C}\cdot\xi - \rho\omega^2\mathbf{I} \tag{1.3}$$

Definition of the fundamental solution for equation (1.1) provides

$$\mathbf{A}_\omega(\partial_\mathbf{x})\mathbf{E}_\omega(\mathbf{x}) = \delta(\mathbf{x})\mathbf{I} \tag{1.4}$$

The Fourier transform of (1.4) gives following expression for the symbol of the fundamental solution

$$\mathbf{E}^\wedge_\omega(\xi) = \mathbf{A}^{\wedge\,-1}_\omega(\xi) \tag{1.5}$$

Expressions (1.3), (1.5) show that the symbols A_0^, and E_0^ corresponding to the static equations are positively homogeneous with respect to ξ of degree 2 and -2 respectively.

2 PROPERTIES OF THE SYMBOLS

Spectral decomposition of the symbols A_0^, and E_0^ gives

$$\begin{gathered}\mathbf{A}^\wedge_0(\xi) = \mathbf{Q}(\xi')\cdot\mathbf{D}_0(\xi)\cdot\mathbf{Q}^t(\xi')\\ \mathbf{E}^\wedge_0(\xi) = \mathbf{Q}(\xi')\cdot\mathbf{D}_0^{-1}(\xi)\cdot\mathbf{Q}^t(\xi'),\\ \xi' = \xi/|\xi|\end{gathered} \tag{2.1}$$

where Q is an orthogonal matrix and D_0 is a diagonal matrix with eigenvalues $\lambda_i(\xi), i = 1,2,3$ of the symbol $A^\wedge_0$ on its diagonal. Formulas (1.3), and (1.4) show that spectral decompositions of the symbols A_ω^, and E_ω^ are similar to (2.1):

$$\begin{gathered}\mathbf{A}^\wedge_\omega(\xi) = \mathbf{Q}(\xi')\cdot\mathbf{D}_\omega(\xi)\cdot\mathbf{Q}^t(\xi'),\\ \mathbf{E}^\wedge_\omega(\xi) = \mathbf{Q}(\xi')\cdot\mathbf{D}_\omega^{-1}(\xi)\cdot\mathbf{Q}^t(\xi'),\\ \mathbf{D}_\omega(\xi) = \mathbf{D}_0(\xi) - \rho\omega^2\mathbf{I}, \qquad \xi' = \xi/|\xi|\end{gathered} \tag{2.2}$$

It is essential that orthogonal tensors Q in (2.1) and (2.2) are identical. That can be easily proved for symbols $A^\wedge_0$ and A_ω^ and then proved for E_0^, and E_ω^ by (1.5).

The following identity is a direct consequence of (2.1)

$$(\mathbf{E}^\wedge_0)^n = \mathbf{Q}\cdot(\mathbf{D}_0)^{-n}\cdot\mathbf{Q}^t \tag{2.3}$$

THEOREM 2.1. *a) Symbol* E_ω^ *is analytic with respect to* ξ *and* ω *in* $R^3\times R^1\setminus\mathrm{Con}_d$, *where* Con_d *is the characteristic cone of the operator* $\mathbf{A}(\partial_\mathbf{x},\partial_t)$; b) *For any real* $\xi,\omega \notin \mathrm{Con}_d$ *the symbol* E_ω^ *admits Taylor series expansion with respect to* ω:

$$\mathbf{E}^\wedge_\omega(\xi) = \sum_{n=1}^{\infty}(\mathbf{E}^\wedge_0(\xi))^n\rho^{n-1}\omega^{2(n-1)} \tag{2.4}$$

c) *The series in the right side of* (2.4) *is absolutely convergent in the topology of compact convergence in* $R^3\times R^1\setminus\mathrm{Con}_d$; *d) For any* $\omega\neq 0$ $\mathbf{E}^\wedge_\omega(\xi)\to \frac{1}{\rho\omega^2}\mathbf{I}$ *for* $\xi\to 0$; e) *For any* ω *the series in the right side of* (2.4) *converges weakly in* $\mathcal{S}'$, *where* $\mathcal{S}'$ *is space of tempered tensor distributions in* R^3.

Remark 2.1. For arbitrary anisotropy Con_d consists of three cones with common apex, whereas for some particular kinds of anisotropy there could be other points of intersection. In particular for any isotropic medium two of the cones coincide.

Proof. Point *a)* follows from the analyticity of the symbol A_ω^ in $R^3 \times R^1$. The determinant of this symbol vanishes only on the characteristic cone, and due to (1.5) characteristic cone represents polar set for the symbol E_ω^.

To prove *b)* it is needed to remark that according to (1.3), and (1.5) the symbol E_ω^ is the resolvent operator for symbol $\mathbf{A}_\omega\hat{}(\xi)$ in the normed algebra of symmetric tensors of the second rank in R^3. Now, we expand the symbol $\mathbf{E}_\omega\hat{}(\xi)$, which is analytic for $\xi, \omega \notin \mathrm{Con}_d$, in Taylor series with respect to ω. This gives:

$$\mathbf{E}\hat{}_\omega(\xi) \equiv (\mathbf{A}\hat{}_0(\xi) - \rho\omega^2 \mathbf{I})^{-1} = \sum_{n=1}^{\infty} \mathbf{A}\hat{}_0(\xi)^{-n} \rho^{n-1} \omega^{2(n-1)} \tag{2.5}$$

In the expression (2.5) the formula for successive derivatives of the resolvent operator (Bourbaki 1967) was used:

$$\frac{d^n}{d\lambda^n} R(\lambda) = n! R^{n+1}(\lambda) \tag{2.6}$$

where $R(\lambda) = \mathbf{E}_\omega\hat{}(\xi)$, $\lambda = \rho\omega^2$, so $R(0) = \mathbf{E}_0\hat{}(\xi)$. But series in the right side of (2.5) obviously coincides with (2.4).

The point *c)* follows from the analyticity of the symbol E_ω^ in $R^3 \times R^1 \setminus \mathrm{Con}_d$, which ensures absolute and hence uniform convergence on any compact subsets of $R^3 \times R^1 \setminus \mathrm{Con}_d$.

To prove *d),* we remark that $\forall\omega$ $\mathbf{A}_\omega\hat{}(\xi) \to \rho\omega^2 \mathbf{I}$ for $\xi \to 0$. This together with the results of the general theory of normed algebras provides convergence $\mathbf{E}_\omega\hat{}(\xi) \to \frac{1}{\rho\omega^2}\mathrm{I}$, if $\xi \to 0$ for $\omega \neq 0$.

By virtue of (2.5) the proof of point *e)* can be done for an arbitrary tensor function $\mathbf{g} \in C^\infty$ with compact support containing a vicinity of the origin (in the ξ-space), where all singularities of the terms in (2.5) are located. But for $\omega \neq 0$ point d) of the theorem shows that in a small ε-vicinity of the origin in the ξ-space, the series (2.5) can be replaced by $\frac{1}{\rho\omega^2}\mathbf{I}$ with the asymptotic error $O(\varepsilon)$. While for $\omega = 0$, the proof of *e)* becomes trivial since the series in (2.5) is reduced to the first term.

3 FUNDAMENTAL SOLUTION E_ω.

THEOREM 3.1. *a) The fundamental solution* E_ω *is real-analytic with respect to* x *and* ω *everywhere in* $R^3 \setminus 0 \times R^1$; *b) The following decomposition is valid*

$$\mathbf{E}_\omega(\mathbf{x}) = \sum_{n=1}^{\infty} \mathbf{E}_n(\mathbf{x}) \rho^{n-1} \omega^{2(n-1)}, \quad (\mathbf{x}, \omega) \in R^3 \setminus 0 \times R^1 \tag{3.1}$$

where E_n *denotes the inverse Fourier transform of the symbol* $(\mathbf{E}\hat{}_0)^n$; *c) The series in the right side of* (3.1) *is absolutely convergent in the topology of compact convergence in* $R^3 \setminus 0 \times R^1$.

Point *a)* flows out from the known results on the analyticity of solutions of differential equations with analytic coefficients (Treves 1982).

The application of inverse Fourier transform to the series (2.4) and the use of the topological isomorphism of the space $\mathcal{S}'$ on itself under Fourier transform proves point *b).*

To prove *c)* we remark that tensorial functions E_n in (3.1) are positively homogeneous with respect to $|\mathbf{x}|$ of order $2n$-3 (Gel'fand & Shilov 1964), so the singularity at $\mathbf{x} = 0$ is only in the first term $\mathbf{E}_1 \equiv \mathbf{E}_0$. Then, under Fourier inversion of the symbols $(\mathbf{E}\hat{}_0)^n$ there arise Bochner's multipliers γ which depend on n and another parameter, such that

$$|\gamma| = O\left(\frac{\pi^{2n}}{(n-1)!\,\Gamma(-1/2+n)}\right), \; n \to \infty$$

Expressions for Bochner's multipliers will be given explicitly further. So, for any ω and $\mathbf{x} \neq 0$ we have

$$|\mathbf{E}_n(\mathbf{x})\rho^{n-1}\omega^{2(n-1)}| = O\left(\frac{a^n}{(n-1)!\Gamma(-1/2+n)}\right), \quad (3.2)$$

$$n \to \infty, \quad a > 0$$

But the series with asymptotic estimate (3.2) is absolutely convergent. This ensures uniform convergence on compact subsets in $R^3 \setminus 0 \times R^1$.

In conclusion it should be noted that Fourier inversion improves the domain of convergence of the series (3.1) in comparison with the initial series (2.4).

To construct kernels E_n in (3.1) we expand corresponding symbols $(\mathrm{E}\hat{}_0)^n$ in multipole series

$$(\mathbf{E}\hat{}_0(\xi))^n = |\xi|^{-2n} \sum_{k=0,2,4,\ldots} \sum_{m=1}^{2k+1} (\mathbf{E}_0)^n_{km} Y^k_m(\xi'), \quad (3.3)$$

$$\xi' = \xi / |\xi|$$

where Y^k_m are spherical harmonics of degree k and index m. Matrix coefficients $(\mathbf{E}_0)^n_{km}$ in (3.3) are determined by integration over the unit sphere S in R^3:

$$(\mathbf{E}_0)^n_{km} = \int_S (\mathbf{E}\hat{}_0(\xi'))^n Y^k_m(\xi')\, d\xi'$$

Presence of the harmonics of even order in (3.3) is due to the positive homogeneity with respect to $|\xi|$ of the symbols $(\mathbf{E}\hat{}_0)^n$.

Inverse Fourier transform applied to $(\mathbf{E}\hat{}_0)^n$ gives (Gel'fand & Shilov 1964):

$$\mathbf{E}_n(\mathbf{x}) = = |\mathbf{x}|^{2n-3} \sum_{k=0,2,4,\ldots}^{\infty} \gamma_{nk} \sum_{m=1}^{2k+1} (\mathbf{E}_0)^n_{km} Y^k_m(\mathbf{x}') \quad (3.4)$$

$$\gamma_{nk} = (-1)^{k/2} \pi^{2n-3/2} \Gamma(\tfrac{(3+k)}{2} - n) / \Gamma(\tfrac{k}{2} + n)$$

Here γ_{nk} are Bochner's multipliers.

COROLLARY. *a)* $\mathbf{E}_\omega(\mathbf{x}) \to \mathbf{E}_0(\mathbf{x})$ *if* $\omega \to 0$ *uniformly (with respect to* x*) on any compact subsets in* $R^3 \setminus 0$; *b) For any* ω

$$|\mathbf{E}_\omega(\mathbf{x})| = O(|\mathbf{x}|^{-1}), \ |\mathbf{x}| \to 0.$$

REMARK 3.1. Preceding corollary reveals the possibility to construct solution of a boundary value problem for the theory of harmonic vibrations by perturbations of the static solution, provided frequency ω is sufficiently small.

In some cases series (3.1) can be summed analytically. Next sections give examples of fundamental solutions constructed in closed forms: these are fundamental solutions to Helmholtz equation and equations of harmonic vibrations of isotropic elastic medium.

4 HELMHOLTZ EQUATION

Fundamental solution of the Helmholtz equation must satisfy identity

$$(-\Delta - \omega^2)\, E_\omega(\mathbf{x}) = \delta(\mathbf{x}) \quad (4.1)$$

Applying expansion (2.4) to the symbol of the fundamental solution E_ω, we get

$$E_\omega(\xi) = \sum_{n=1}^{\infty} (2\pi|\xi|)^{-2n} \ \omega^{2(n-1)} \quad (4.2)$$

A characteristic cone of the equation (4.1) is determined by the equation $(2\pi|\xi|)^2 = \omega^2$.

Fourier inversion of the expression (4.2) on account of formulas (3.1), (3.4) produces

$$E_\omega(\mathbf{x}) = = \pi^{-3/2} \sum_{n=1}^{\infty} 2^{-2n} |\mathbf{x}|^{2n-3} \Gamma(3/2-n)/\Gamma(n)\, \omega^{2(n-1)} = = (4|\mathbf{x}|)^{-1} \pi^{-1/2} \sum_{n=1}^{\infty} \frac{(-1)^n}{(n-1)!\,\Gamma(-1/2+n)} \left(\frac{\omega|\mathbf{x}|}{2}\right)^{2(n-1)} \quad (4.3)$$

But the last series up to a multiplier gives Taylor expansion of the Neumann's function $N_{1/2}$. So, (4.3) can be rewritten in the form

$$E_\omega(\mathbf{x}) = -1/4\, \omega^{1/2} (2\pi|\mathbf{x}|)^{-1/2} N_{1/2}(\omega|\mathbf{x}|) \quad (4.4)$$

which coincides with the formula, obtained by the plane-wave decomposition method (John 1955).

Sometimes fundamental solution of the Helmholtz equation is written in the form (Sanchez-Palencia 1980)

$$E_\omega(\mathbf{x}) = -(4\pi|\mathbf{x}|)^{-1}\exp(\pm i\omega|\mathbf{x}|) \tag{4.5}$$

It is easy to see that its real part coincides with (4.4) and hence with (4.3).

5 FUNDAMENTAL SOLUTION FOR ISOTROPIC ELASTIC MEDIUM

An isotropic elastic medium is considered with elasticity tensor of the form

$$C^{ijmn} = \lambda\delta^{ij}\delta^{mn} + \mu(\delta^{im}\delta^{jn} + \delta^{in}\delta^{jm}) \tag{5.1}$$

where λ, μ are Lamé's constants, and δ^{ij} is the Kronecker symbol.

On account of (5.1) symbol of the operator of harmonic vibrations can be represented by the following expression

$$\begin{aligned}&\mathbf{A}\hat{}_\omega(\xi) = \\ &= (2\pi)^2(\mu|\xi|\mathbf{I} + (\lambda+\mu)\xi\otimes\xi) - \rho\omega^2\mathbf{I}\end{aligned} \tag{5.2}$$

With the use of (1.4), (5.2), the symbol of the fundamental solution has the form

$$\begin{aligned}&\mathbf{E}\hat{}_0(\xi) = (2\pi)^{-2}(\mu(\lambda+2\mu))^{-1}\times \\ &\left[(\lambda+2\mu)|\xi|^{-2}\mathbf{I} - (\lambda+\mu)|\xi|^{-4}\xi\otimes\xi\right]\end{aligned} \tag{5.3}$$

The following identity, which is flowing out from (5.3), can be easily proved by induction on n

$$\begin{aligned}&(\mathbf{E}\hat{}_0(\xi))^n = (2\pi)^{-2n}\times \\ &\left[\mu^{-n}|\xi|^{-2n}\mathbf{I} - (\mu^{-n} - (\lambda+2\mu)^{-n})|\xi|^{-2n-2}\xi\otimes\xi\right]\end{aligned} \tag{5.4}$$

Formulas (2.4), and (5.4) give Taylor's expansion of $\mathbf{E}\hat{}_\omega$ with respect to ω:

$$\begin{aligned}&\mathbf{E}\hat{}_\omega(\xi) = \sum_{n=1}^{\infty}(2\pi)^{-2n}\times \\ &\times\left(\mu^{-n}|\xi|^{-2n}\mathbf{I} - (\mu^{-n} - (\lambda+2\mu)^{-n})|\xi|^{-2n-2}\xi\otimes\xi\right)\times \\ &\times\rho^{n-1}\omega^{2(n-1)}\end{aligned} \tag{5.5}$$

Fourier inversion of (5.5) with the use of expressions (4.2) - (4.5) gives the fundamental solution we are looking for

$$\begin{aligned}&\mathbf{E}_\omega(\mathbf{x}) = -(4\pi\mu)^{-1}\Big\{|\mathbf{x}|^{-1}\cos(\alpha\omega|\mathbf{x}|)\mathbf{I} - \\ &-2^{-1}\nabla_{\mathbf{x}}\nabla_{\mathbf{x}}\left(|\mathbf{x}|\left(\cos(\alpha\omega|\mathbf{x}|) - \frac{\mu}{\lambda+2\mu}\cos(\beta\omega|\mathbf{x}|)\right)\right)\Big\}\end{aligned} \tag{5.6}$$

where

$$\alpha = (\rho/\mu)^{1/2}, \quad \beta = (\rho/(\lambda+2\mu))^{1/2}$$

A straightforward verification shows that if $\omega \to 0$, fundamental solution (5.6) transforms into Kelvin's fundamental solution.

ACKNOWLEDGEMENT

Author thanks International Science Foundation (Grant M7Y000) for financial support.

REFERENCES

Bourbaki, N. 1967. *Théories spectrales.* Ch. 1. *Algèbres normées.* Paris: Hermann.

Burchuladze, T.V. & T.G.Gegelia 1985. Development of the potential method in the theory of elasticity (in Russian). *Trudy Tbil. matem. in-ta.* 79: 5-226.

Deb, A., D.P.Henry Jr., & R.B.Wilson 1991. Alternate BEM formulation for 2- and 3D anisotropic thermoelasticity. *Int J. Solids Struct.* 27: 1721-1738.

Gel'fand, I.M. & G.E.Shilov 1964. *Generalized functions. Vol.* 1. *Properties and Operations.* New York: Academic Press.

Herglotz, G. 1928. Über die Integration linearer partieller Differentialgleichungen mit Konstanten Koeffizienten. *Abh. Math. Sem. Hamburg*: 189-197.

John, F. 1955. *Plane waves and spherical means: Applied to partial differential equations.* N.Y.: Springer.

Kupradze, V.D. 1953. Boundary-value problems of free elastic vibrations (in Russian). *Uspech. Matem Nauk.* 8(3): 21-74.

Kuznetsov, S.V. 1989. Fundamental solutions for Lame equations in anisotropic elasticity (in Russian). *Izv. AN SSR. Mech. Tverdogo Tela.* 4: 50-54.

Kuznetsov, S.V. 1995. Direct boundary integral equation method in the theory of elasticity. *Quart. Appl. Math.* 53: 1-8.

Natroshvili, D.G. 1979. On fundamental matrices of the equations of steady-state oscillations and pseudooscillations in the theory of anisotropic elasticity (in Russian). *Soobsch. AN Gruz. SSR.* 96(1): 49-52.

Sanchez-Palencia, E. 1980. *Non-homogeneous media and vibration theory.* Berlin: Springer.

Synge, J.L. 1957. *The hypercircle in mathematical physics.* Cambridge: Cambridge Univ.Press.

Treves, F. 1982. *Introduction to pseudodifferential and Fourier integral operators, Vol.*1. *Pseudodifferential operators.* New York & London: Plenum Press.

Wilson, R.B. & T.A.Cruse 1978. Efficient implementation of anisotropic three dimensional boundary-integral equation stress analysis. *Int. J. Num. Meth. Eng.* 12: 1383-1397.

Modern Practice in Stress and Vibration Analysis, Gilchrist (ed.) © *1997 Balkema, Rotterdam, ISBN 90 5410 896 7*

Finite element analysis of wheelchair structure

J.Gu
RSEL, Middlesex University, London, UK

ABSTRACT: This paper presents a static and dynamic force analysis of the structures of both the surrogate wheelchair and the production wheelchair. A combination of the multibody techniques with finite element analysis, validated by dynamic sled tests, is employed to allow a more detailed description of the crash performance of the wheelchairs. The correlation of a static load to the actual dynamic load in a given crash severity was developed. The correlation between the computer models and experiments was also presented. The loading analysis of the production wheelchair was based on dynamic test results. The dynamic analysis was used to investigate the crash performance of the surrogate wheelchair.

1 INTRODUCTION

Under impact condition, the loads transferred to the wheelchair are of sufficient magnitude to cause its deformation, and even collapse of the joints in the wheelchair frame and hence injury to the disabled occupant. In order to strengthen the joints of the wheelchair's tubular structure, it is necessary to determine the values and directions of the forces and moments acting on the individual joints and tubes.

The sled impact testing of Wheelchair Tiedown and Occupant Restraint System (WTORS) was carried out in Middlesex University Road Safety Engineering Laboratory (MURSEL). It was shown that the wheelchair frame itself was the limiting factor in the frontal impact.

The Finite Element Analysis (FEA) method facilitated the designers in getting quick solutions to the force analysis of the mechanism (Hoffmann R. 1990). In this paper, the wheelchair was modelled using the FEA method to gather data, such as, the wheelchair centre of gravity (CG) and the mass moment of inertia of the wheelchair in the three principal direction, Ixx, Iyy, Izz, for the construction of Crash Victim Simulation (CVS) model. All components placed in the wheelchair, such as the tube structure of the wheelchair, the battery and other parts, are incorporated into the model by scaling their masses and distributing them about the places where they are attached to the sled.

The solid modelling program, PIG and finite element modelling program, PAFEC (level 8.1) were used running on the VAX cluster. The CVS programs, DYNAMAN and MADYMO were also used.

2 WHEELCHAIR STRUCTURES

Wheelchair is a 'seating system composed of a frame, a seat, and wheels that is designed to provide support and mobility for persons with physical disability. (ISO/CD 10542-1, 1996, ISO WD 7176-1995E).

2.1 Production wheelchair

Many types of production wheelchairs have been used in the WTORS crash testing in the previous researches (Gu J. et al. 1995), such as standard manual wheelchairs, power wheelchairs, scooter-type wheelchairs, and specialised seating base wheelchairs. As a result of these tests, the strong and weak points of standard wheelchairs have been determined. The performance of specific tiedown systems with certain chairs has been observed.

Power wheelchairs are typically used in WTORS tests because their weight (approximately 85 kg) makes them a worse case loading severance. Testing has revealed that for the some parts, welded frame manual and power wheelchairs could not withstand the forces of a 30 mph (48 km/h), 20g frontal crash with disaster although sufficiently restrained (Fig. 1).

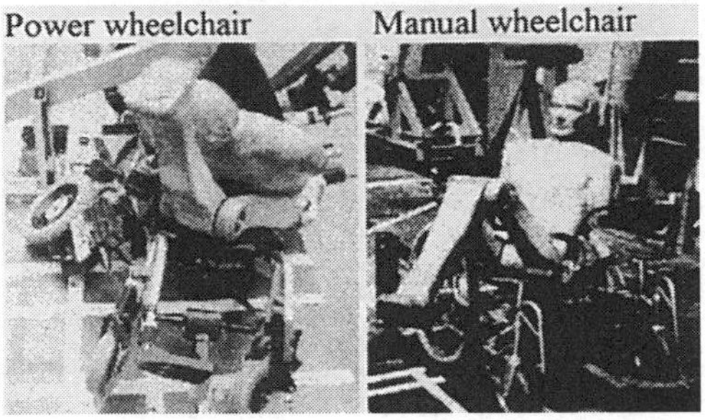

Fig.1 Production wheelchair collapsed at 30 mph, 20g

2.2 Surrogate wheelchair

The extreme crash always results in a large amount of chair deformation and involves the high cost of replacing damaged parts. The use of these chairs is not feasible to evaluate the WTORS in the frontal impact.

A solution to the problems in WTORS testing is the use of a reusable surrogate wheelchair. A surrogate wheelchair is a chair having the general dimensions, shape, and geometry of a standard production wheelchair, but suitably reinforced to insure that the chair will not deform in a 30 mph crash. It facilitates a standardised test, ensuring that all WTORS can be tested repeatable within-lab consistency and reproducibly between-lab consistency (Shaw G. S. Etc. 1994). In essence, the surrogate wheelchair provides a level playing field for all WTORS and presents a worse case loading severance for systems. As this chair does not deform, that is, dissipate energy, the entire energy management of the crash could be done by the restraint systems. Surrogate wheelchairs are suggested to be used for dynamic testing of different restraint systems.

One of the surrogate wheelchair designed by the Transport Research Laboratory (TRL) in UK was selected in this research program (Fig. 2). The design for the chair is based on standard power based wheelchairs in UK. The extra masses are attached to the chair so that the total chair mass is 85 kg.

Fig. 2 TRL Surrogate wheelchair and its webbing tiedown restraint system

2.3 Wheelchair tiedown restraint system

The tiedown hardware attachment is another important issue. Wheelchair can not be secured properly because of lack of attachment points. Securement of structurally weak places on the wheelchair, like the wheels, armrests and footrests are inadequate. It was found that different tiedown in buses with different attachment points on the wheelchair lead to a lot of problem and put a burden on the wheelchair manufacturers. The solid attachments or opening points have been suggested on the wheelchair front and rear independently.

In this paper reference have been made to the surrogate wheelchair and one of the production wheelchair (DHSS, Model 8L, specified in BS 5568 1978). The dimensions of the wheelchair were measured in order to be able to reproduce a wheelchair.

3 STATIC AND DYNAMIC TESTING OF WHEELCHAIRS

Static testing involves applying a constant force to a structural member at a very low rate over a fairly long period of time. Dynamic testing involves a force application of short duration (approximately 100 ms in actual crash) at a high rate.

3.1 Static tests

Manufacturers designed their wheelchair tiedown restraint system usually based on simple static calculations: the restraint force = mass of chair times peaks' vehicle deceleration, and then tested their systems based on these force levels. Unfortunately, a static analysis invariably underestimates the loads generated for given crash conditions. This is due to the fact that a static test oversimplifies the crash, and cannot account for phenomena unique to the dynamic environment, such as the amplification of vehicle deceleration effect. A wheelchair tiedown restraint system undergoing a 20g vehicle deceleration seems not a horizontal force of 17 kN. In fact, the restraint experiences a greater load. The shortcomings of static tests involved the load's point of application. Ideally, the force point of application should be at the weakest point in the system, however this location was never obvious due to the complexity of the entire system. Another problem of static testing involved how to determine the direction of load paths. During the actual crash, load paths in three directions are created. It is impossible to predict all of the load paths analytically using the inherent over-simplifications of static testing.

3.2 Dynamic tests

Dynamic testing exposes systems to real-world crash environments. It also reveals modes of hardware failure, like cutting of webbing at areas of high stress concentration. The system failed the dynamic test because stress concentrations were created due to inertial loading of the chair and its resulting deformation during impact. There are a lot of important performance parameters, such as, wheelchair and dummy excursions, chair and dummy interactions, the secondary collision, which can be evaluated only in dynamic crash. It is widely recognised that the secondary collision, the impact between the occupant and the interior structures of the vehicle resulting from occupant excursions, is the primary cause of injury and death in an accident.

3.3 Correction between static and dynamic tests

In order to eliminate the need of dynamic testing of the WTORS, attempts have been made to find a correlation between static and dynamic testing. Through the use of computer simulations, a simple relation coefficient could be determined. How to find a correlation of a static load to the actual dynamic loads in a given crash severity? In this paper, this effort has proved successful. The loading analysis method based on dynamic test's results was used to analysis the crash performance of the manual wheelchairs. The tiedown restraint forces are estimated as following formula:

Restraint force = mass of chair times resultant dummy deceleration.

4 DYNAMICS ANALYSIS OF SURROGATE WHEELCHAIR

Since paramedic sled tests are costly and time consuming to perform, computer simulation has been popular to emulate the crash environment. It has been used as a tool for examining the effects of crash pulse variations. How to find the correlation between the computer models and experimental work is another subject in this research programme.

The most straightforward type of dynamic analysis is the determination of natural frequencies and mode shapes. This type of calculation gives considerable insight into the dynamic behaviour of a structure. The natural frequencies and model shape are calculated using a limited number of dynamic freedoms called masters.

All the separate elements of the surrogate wheelchair, such as tubes were built with simple beam element (34000) (Fig. 3). The surrogate wheelchair was simplified as simple beam elements without wheels as these wheels had no constructive part to play in the structural strength of the wheelchair.

The five modes were selected in order to know the complete mode shape. The system has to perform an operation known as back-substitution in order to find the complete mode shape. The mode shapes at the master degrees of freedom are printed out for the first twenty frequencies irrespective of the number modes for which full back-substitution is required.

The following example (Fig. 4) demonstrates the determination of natural frequencies in a restrained three dimensional surrogate wheelchair structures.

From Fig. 4, these deformations and distortion were compatible with those observed during the actual dynamic impact testing.

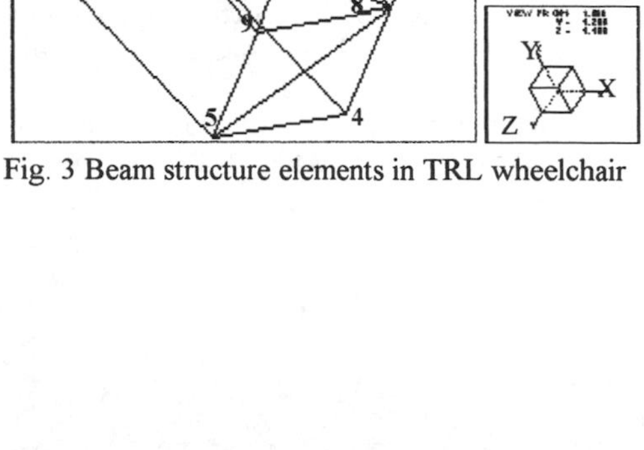

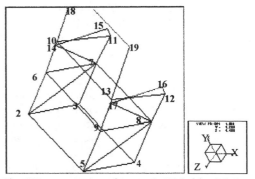

Fig. 3 Beam structure elements in TRL wheelchair

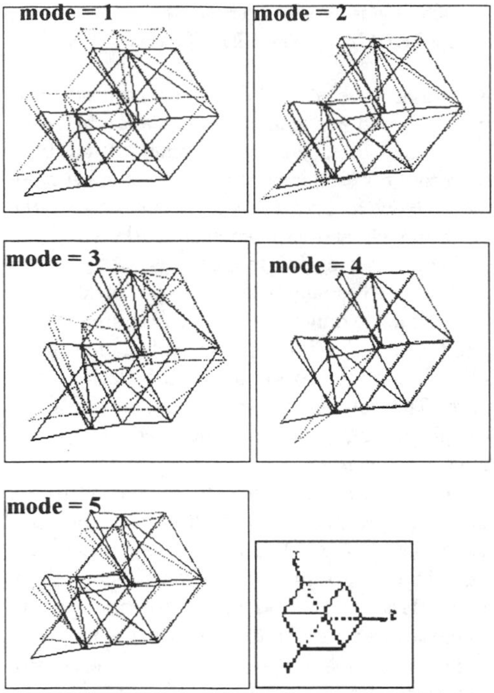

Fig. 4 Five modes analysis of TRL wheelchair

The investigation has shown that a twisting of the wheelchair about the Y-axis occurred causing uneven forces in the two halves of the wheelchair frame.

5 LOADING ANALYSIS OF MANUAL WHEELCHAIR

Sled tests implied that the production wheelchair frame becomes severely distorted and that the tubes at certain load-bearing joints were pulled apart, such as, the caster tube (Fig. 5). The forces causing the deformation are not pure tension or compression. It is the combined effect of tension and compression with large bending moments.

Dynamic sled tests also show that it is difficult to decide the magnitude and directions of loads acting on the wheelchair frame from the deceleration force of sled during impact. This is because the combination movements of dummy sitting on the wheelchair and wheelchair floating on the sled. After examination of the testing high speed video it was found that the wheelchair was progressively loaded by the deceleration and the loads transferred to the wheelchair moved forward to the front caster wheels.

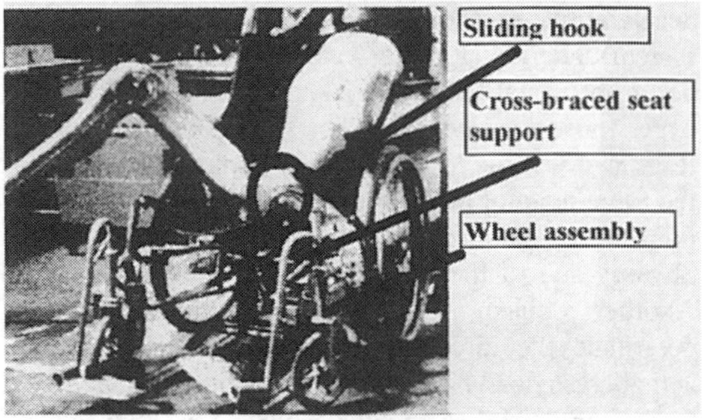

Fig. 5 A manual wheelchair frame distorted after frontal impact

The impact tested wheelchairs were then examined to determine the areas of maximum deformation and gain an overall picture of the impact forces effect on the wheelchair. All load cases are the result of initial assumptions of the force distribution in the frame of the wheelchair. The two halves of the wheelchair frame are initially assumed as equal force allocation due to the co-ordinate structure. The front edge of the seat is the worst possible position. If the seat belts restraining the movements of the dummy are loose the dummy body could slide forward until the CG lies directly over the front edge of the seat. The load would be increased on the front caster wheels and be increased to bend or even to be cracked.

The wheelchair was progressively loaded by dummy resultant deceleration (about 40g) at the sled severity of 20g, 30 mph. The force acting through the CG of the dummy (mass 75 kg) was 30 kN. This load transferred to the seat cushion would be assumed in the order of two thirds in the front and one third to the rear. Furthermore, the loads transferred to the seat belt restraints would be assumed in the order of half the force to waist's restraints and half the force to the shoulder restraint.

The detailed real deformed parts were modelled as follows:

- The sliding hooks

As these hooks become distorted and do not restrain the movements of the seat member in the z-plane or y-plane in the impact situation. The locating hook was modelled as the seat member is free to slide in the joint in the x-plane but is fixed and unable to rise in the y-plane.

- Cross-braced seat supporting member

The positioning of the cross-braced members is behind the centre of the seat, which the occupant's centre of gravity lies directly overhead in normal use, thus causing maximum bending moment at the cross braced joint.

- Wheel assembly

The rear wheel axles were bent. They have been modelled passing through the rear axle joint and the front caster-wheel axle passing through front axle joint. The frame is able to pivot about the rear axle, thus the axle cannot move in the Y-axis and Z-axis but is free to move in the x-axis. The same boundary conditions are applied to the front caster wheel axle.

All removable parts, such as armrests and foot rests have been removed. It was also assumed that all these parts had no constructive part to play in the structural strength of the standard wheelchair.

A combination of the following four element types were used in FEA manual wheelchair model:

Type 1 Simple beam element (34000)

A straight uniform beam element with two nodes.

Type 2 Shear deformation and rotary inertia beam element (34100)

A straight uniform beam element with shear deformation included. There are six degrees of freedom (u_x, u_y, u_z, ϕ_x, ϕ_y, ϕ_z) at each of the two nodes. This element was applied in seat hook structure and wheel axis.

Type 3 Curved beam element (34300)

This element is part of a circle. Two node numbers are given in the topology and these are positioned at the centres of area of the cross-section at the two ends of the element. Shear deformation and rotary inertia are included. This element was applied all round corners in the manual wheelchair.

Type 4 Tension Bar Element (34400)

A straight uniform element that carries end load only.

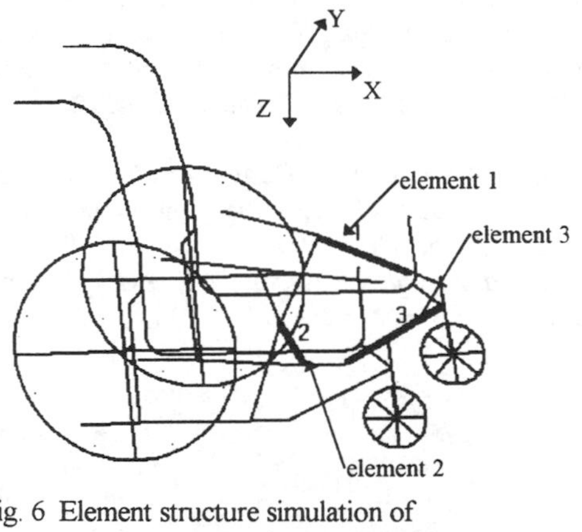

Fig. 6 Element structure simulation of Manual wheelchair (DHSS, 8L)

As the larger deformation was found in this model, the static loading analysis was used to investigate the positions of higher loading. The higher load positions in the DHSS wheelchair (marked as 1,2,3 in Fig 6) were found from the model. The highlighted element 1 posses higher shear force (z) (range from 820 N to 1000 N) and higher bending moment (y) (range from 211 Nm to 251 Nm); The highlighted element 2 posses higher shear force (y) (range from 580 N to 770 N) and higher bending moment (x) (range from 680 Nm to 990 Nm); The highlighted element 3 posses higher shear force (x) (range from 500 N to 760 N).

6 CVS MODELLING OF A SURROGATE WHEELCHAIR-OCCUPANT SYSTEM

Considering the data supplied from the above FEA model, a CVS model was initially written within DYNAMAN program and then modified with MADYMO package.

6.1 TRL wheelchair model

All segments placed in the wheelchair were incorporated in the model by scaling their mass and moment of inertia, distributing them about the places. The following two systems were built in this model (Fig. 7 and Fig. 8):

Inertial system: Sled and bus stop target

System 1: Wheelchair and tiedown securement points

6.2 A wheelchair-dummy model

The system 2 (TNO10 dummy data) was added in the above TRL wheelchair model. After simulation, the complete wheelchair-dummy model was created in Fig. 9.

The surrogates WTORS system was modelled as linear segments whose stiffness properties were

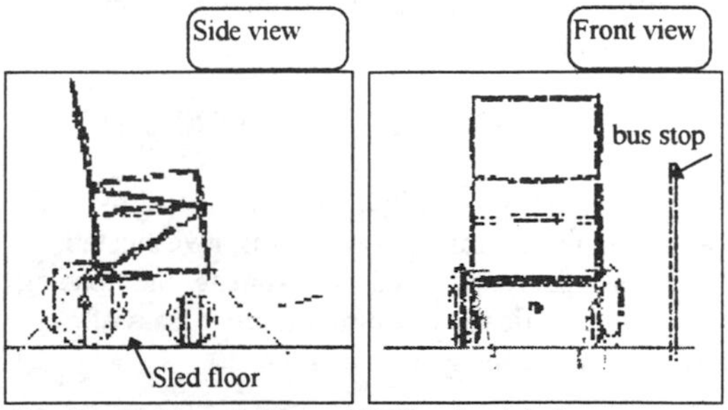

Fig. 7. CVS modelling of TRL wheelchair

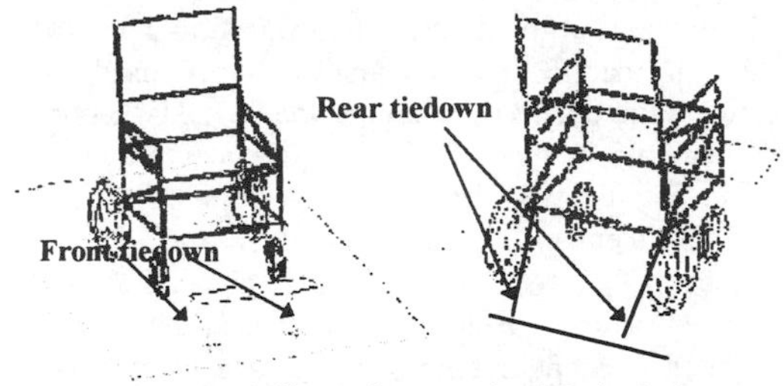

Fig. 8 CVS modelling of TRL wheelchair tiedown restraint system

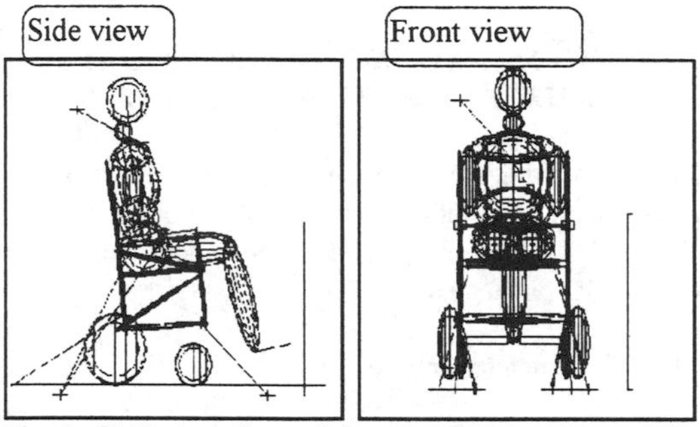

Fig. 9 CVS modelling of the WTORS system

initially determined experimentally from static testing of the belts. Finally, they were validated and adjusted by dynamic sled test results. Damping and permanent deformation properties of the system were also accounted for.

Interactions between the dummy, chair, and floor were modelled using Force-deflection function (FDF), Inertial spike function (I), energy absorption factor function (R), permanent deflection factor (G), and frictional properties of the various contacts. Contacts between the wheelchair wheels and vehicle floor, the wheelchair and dummy, and seat belts and dummy torso were emulated.

The crash was emulated by specifying the crash pulse and determining system responses at predetermined time intervals.

7 CONCLUSION AND FURTHER STUDY

The correlation of a static load to the actual dynamic load in a given crash severity was investigated. It could reasonably be used to estimate the loading conditions: the tiedown restraint force = mass of chair times resultant dummy deceleration.

The simulation produced by the FEA method compared reasonably well to the actual test results from the full scale sled tests.

A computer model of crash performance of wheelchair (CVS' model) was constructed and validated by the crash tests. A close approximation was made more accurate adjusting the various stiffness functions, friction penetration factors, and correction factors although the uncontrollable factors during crash are difficult to model.

Further study of impact properties is needed to get a better correlation between the models and experiments. One of the areas in which an improvement is required is a means of modelling of the contact and friction forces exerted between the ground and the pneumatic tyres. Another area for improvement is the modelling of different restraint systems.

REFERENCE

Hoffmann R. (1990). Finite Element Analysis of Occupant Restraint System Interaction with PAM-CRASH. *34th Stapp Car Crash Conference Proceedings*, Paper No. 902325.

ISO/CD 10542-1: 1994E, 1995E, and 1996 Wheelchairs Tiedown and occupant restraint systems for motor vehicles.

ISO WD 7176-1995E. Wheelchair- wheeled mobility devises for use in motor vehicles- Requirements and Test Methods.

Shaw G.S. et al. (UVA), Schneider L.W. (UMTRI), Roy A.P. (MURSEL)(1994). Interlaboratory Study of Proposed Compliance Test Protocol for Wheelchair Tiedown and Occupant Restraint Systems. *Proc. 38th Stapp Car Crash Conference*, Fort Lauderdale. October 1994, also published as SAE Paper No. 942229

Gu J., Roy A.P. (1995). Current Research to Evaluate the Performance of Wheelchairs in Frontal Impacts (Evaluation of Floor Reaction Forces). *Proc 23rd Transport Forum, University of Warwick, September 1995, PTRC (0 86050 285 6).*

Modern Practice in Stress and Vibration Analysis, Gilchrist (ed.) © *1997 Balkema, Rotterdam, ISBN 90 5410 896 7*

Modelling of fracture for composites in compression along layers

I.A.Guz
Institute of Mechanics, Kiev, Ukraine

ABSTRACT: The problem of fracture caused by instability in a composite of an arbitrary laminated structure with interlaminar defects is discussed. The exact statement of the problem is formulated for cases of compressible, elastic and elastic-plastic, orthotropic and isotropic layers in compression along cracks. This exact statement is based on the model of piecewise-homogeneous medium and equations of the three-dimensional linearized theory of deformable bodies stability (TLTDBS) (A.N. Guz 1986). The bounds for critical loading parameters are suggested and substantiated. Numerical results are obtained for particular cases of real composites for typical dispositions of cracks. Besides that, it's shown that approximative approaches to so fine and complex phenomenon in the material structure cannot describe it even on the qualitative level.

1 INTRODUCTION

The modern level of engineering and technology demands new materials to be designed which are simultaneously lightweight, reliable, durable, resistant to the ambient influence, but efficient and cost-effective in manufacturing. Composite materials which can display such various properties have being widely utilized in many branches of industry, and, for example, production of metal matrix composites is expected to exceed in 2010 the production of steel.

One of the most interesting, peculiar and inadequately investigated phenomena in mechanics of composites (mechanics of non-homogeneous media) is fracture in compression, in which mechanisms of fracture not specific for homogeneous media but for composites (non-homogeneous media) only are revealed. One of the mechanisms referred to is fracture owing to the loss of stability in the composite structure. This instability is so called the internal instability (Biot 1965) and is derived not by dimensions and the shape of the specimen (constructive element) but by relationships between physical and geometrical parameters of the composite components and various heterogeneities.

At the present time there is a large amount of studies devoted to the stability problems of composites and in the most of them it is assumed that structural elements of the material are rigidly attached. The abundant structure of composites is a laminated one. Therewith, intercomponent defects of various nature (separations onto layers, cracks, exfoliations, zones of slacked adhesion etc) may occur in real laminated composites due to the fabrication technology or operating conditions. In the case of compression of laminated composites along layers and, therefore, along the mentioned defects the classical Griffith-Irwin criterion of fracture or its generalizations are inapplicable, since in such situation all stress intensity factors and crack opening displacements are equal to zero. This fact emphasizes the importance of the investigation of fracture just owing to the loss of stability in the composite structure.

Until now the mentioned investigations of composites with interlaminar cracks were carried out only within the scope of applied design schemes (for example, bar and shell design schemes) and approximative theories (for instance, the continuum theory of composites). The application of so approximative approaches to rather fine and complex phenomenon in the material structure furnishes the suitable identification of results obtained within the scope of these theories. The most exact results may be obtained only using the model of piecewise-homogeneous medium and equations of the three-dimensional linearized theory of deformable bodies stability (TLTDBS) (A.N. Guz 1986).

The present investigation was fulfilled just within the scope of this exact approach.

2 PROBLEM STATEMENT

2.1 Composites with interlaminar cracks

Let us consider a composite with an arbitrary laminated structure, which may consist of an arbitrary combination of layers and half-spaces. Feasible kinds of such structures had been described earlier (I.A. Guz 1992, 1993c). Suppose that the composite is situated in conditions of the plane strain state in compression along layers by "dead" loads applied at infinity in such a manner that equal deformations along all layers is provided in direction of loading. This is the uniform precritical state (Figure 1) with

$$\epsilon_{11}^{0(i)} = \epsilon_{11}^{0(j)} \tag{1}$$

Hereafter index "(i)" means number of the layer, index "0" means that this value concerns the precritical state.

Layers of the investigated structure may have various mechanical characteristics and thicknesses (the thickness of the i-th layer is $2h^{(i)}$). They are simulated by compressible, elastic or elastic-plastic, isotropic or orthotropic (with elastically equivalent directions which are parallel and perpendicular to interfaces) bodies. In the case of elastic-plastic layers we will utilize the generalized conception of the continuous loading which allows to neglect the changing of loading and offloading zones during the stability loss. All investigations will be fulfilled with the Lagrangian coordinate system which is Cartesian one in the non-deformed state (Figure 1).

Problems of internal and local instability for such structures had been solved (I.A. Guz 1989a, 1989b, 1990, 1991, 1992, 1993b) for the case of absence of interlaminar cracks (rigid-connected or sliding without friction layers) within the scope of the above-mentioned exact approach. Now assume that there are N cracks on the boundary between

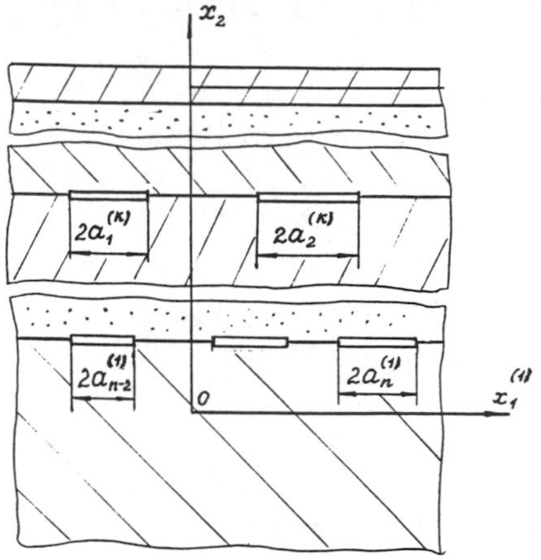

Figure 1: Macrocracks.

layers. We'll simulate them by mathematical sections regardless of reasons of their occurrence and denote their lengths as $2a_i$ ($i = 1, \ldots, N$). Owing to the kind of applied loads and (1) the static (Euler) method of investigation of static problems of TLTDBS may be used in this case (A.N. Guz 1986). The formulation of the stability problem will be as follows.

Main equations of TLTDBS for every layer

$$\frac{\partial}{\partial x_i}\left(\omega_{ij\alpha\beta}^{(i)} \frac{\partial u_\alpha^{(i)}}{\partial x_\beta}\right) = 0 \tag{2}$$

and the uniform boundary conditions for perturbations of displacements (u_i) and stresses (t_{ij}) on the interfaces with rigid contact of layers S_r

$$t_{2\beta}^{(i)} \mid_{S_r} = t_{2\beta}^{(i+1)} \mid_{S_r}, u_\beta^{(i)} \mid_{S_r} = u_\beta^{(i+1)} \mid_{S_r}; \ \beta = 1,2 \tag{3}$$

on the crack surfaces S_c

$$t_{2\beta}^{(i)} \mid_{S_c} = 0 \quad , \quad t_{2\beta}^{(i+1)} \mid_{S_c} = 0; \quad \beta = 1,2 \tag{4}$$

on the free surface S_s

$$t_{2\beta}^{(j)} \mid_{S_s} = 0; \quad \beta = 1,2 \tag{5}$$

and conditions of the attenuation of perturbations in moving from the cracks

$$\vec{u}^{(k)} \mid_{x_1 \to \infty} \longrightarrow 0 \quad , \quad \vec{u}^{(k)} \mid_{x_2 \to \infty} \longrightarrow 0 \tag{6}$$

Relations (2)-(6) are the eigenvalue problem with respect to the loading parameter which is included in components of the tensor $\tilde{\omega}$. This tensor also

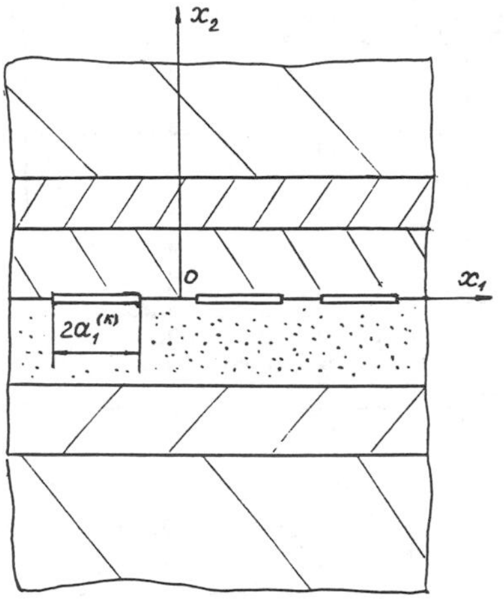

Figure 2: Structural cracks.

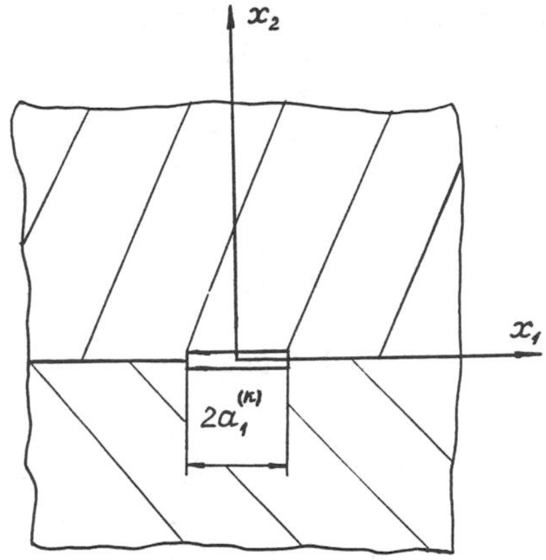

Figure 3: Microcrack.

depends on the constitutive equation for the corresponding layer (A.N. Guz 1986).

2.2 Classification of interlaminar defects

The above-stated formulation of the problem is the most general and does not impose any limitations on the number of cracks and the ratio of their lengths and thicknesses of layers. However, the solution in such formulation may be extremely complicated. In this connection, the necessity in certain indiscrepant simplifications, connected not with the changing of equations, but only with geometrical characteristics of the composite and the cracks, appears. For this purpose we introduce the classification of cracks.

Macrocracks, structural cracks and microcracks are discerned in this classification. So, the most general case is the case of macrocracks. Namely, cracks, which lengths are commensurate or exceed thicknesses of layers, are denoted as macrocracks (Figure 1). But, may be, there are layers laying above and below the cracks such as their thicknesses considerably exceed lengths of cracks

$$\max_{1\leq n\leq N} a_n \ll \min\{h^{(i)}, h^{(j)}\} \tag{7}$$

We'll denote cracks satisfying to (7) as structural cracks. In this case the investigation of stability of the composite is reduced to the investigation of the structure like as on Figure 2 - several layers with cracks between two half-planes instead of thick layers (i) and (j). And, finally, lengths of cracks may be considerably less of thicknesses of layers, between which they are located

$$\max_{1\leq n\leq N} a_n \ll \min\{h^{(i)}, h^{(i+1)}\} \tag{8}$$

These are the microcracks. In the case of microcracks we may restrict ourself by the investigation of stability of two half-planes with the same cracks between them (the particular case of structural cracks – Figure 3).

The problem statement indicated above was formulated for the crack models which are ideal in a certain sense. Such models (the free of stresses crack surfaces) are used in classical fracture mechanics. However, other types of interlayer adhesion breakdown can also occur in composite materials. For example, a change in the nature of contact of layers is possible, when their interaction is implemented by friction forces. In this case, naturally, there are no gaps between the layers and the continuity of normal components of stresses and displacements at interfaces is retained. If the friction force between the layers is sufficiently large, it can be assumed that rigid contact does not break down. If the friction force is very small, zones of similar defects can be assumed to be slippage zones without friction with boundary conditions for the stress and displacement perturbations in the form

$$t_{21}^{(i)}\,|_{S_c} = 0 \quad , \quad t_{21}^{(i+1)}\,|_{S_c} = 0 \tag{9}$$

$$t_{22}^{(i)}\,|_{S_c} = t_{22}^{(i+1)}\,|_{S_c} \quad , \quad u_2^{(i)}\,|_{S_c} = u_2^{(i+1)}\,|_{S_c} \tag{10}$$

The problem statement (2)-(6) remains in force also for the defects described above, which will be called below "defects with connected edges". It is only necessary to replace everywhere the boundary conditions (4) by the boundary conditions (9).

3 NUMERICAL RESULTS

3.1 Using of the finite difference method

Let us consider linear-elastic layers of a composite. The solving of the above-formulated problem will be carried out utilizing the method of finite differences. For this purpose the difference scheme having the first order of approximation is proposed. The difference eigenvalue problem is formulated in accordance with the differential problem of stability (2)–(6) by the variational-difference method. This problem is a generalized completely determined eigenvalue problem, because a uniform precritical state (1) is realised in a composite. The convergence of the mentioned difference scheme was proved in (Guz and Kokhanenko 1993), in which the gradient iterative process was used for the finding of the least eigenvalue of the difference eigenvalue problem. This iterative process is also applied in the present paper.

3.2 One microcrack between orthotropic layers

Numerical results had been obtained by the above method for isotropic layers (Guz and Kokhanenko 1993, I.A. Guz 1993a). Below we dwell on the investigation of stability of composites with orthotropic layers.

Let us consider composite, each layer of which consists of a matrix reinforced by continuous parallel fibers. Therewith, directions of the reinforcement are mutually perpendicular in adjacent layers. This is so called composite of the lateral-torsional packing. Assume that there is one microcrack between two adjacent layers (8). In this case the upper layer is reinforced by fibers, which are perpendicular to the plane x_1ox_2, the lower one – by parallel to the axis ox_1 fibers, and both may be replaced by half-planes due to the small length of the crack in comparison with thicknesses of them. Let us suppose also that geometrical dimensions of structural elements inside of each half-plane are such that these half-planes are orthotropic homo-

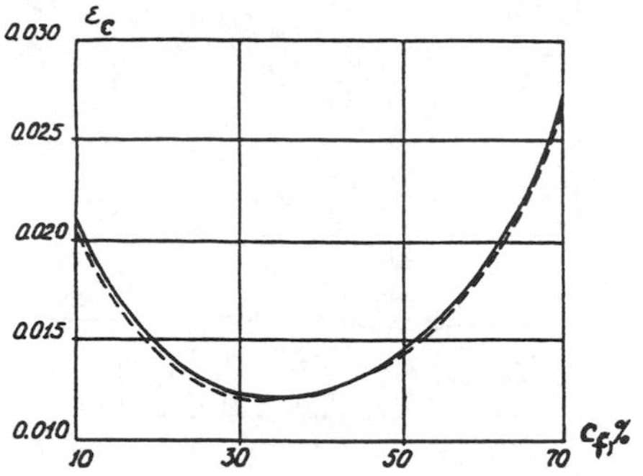

Figure 4: Critical strains for the case of microcracks between boron/epoxy layers ($E_f = 400GPa$, $\nu_f = 0.21$, $E_m = 2GPa$, $\nu_m = 0.4$); c_f – the volume concentration of fibers in layers.

geneous in continuum approximation. In so doing, values of Young moduli and Poisson ratios for fibers (E_f, ν_f) and matrix (E_m, ν_m) refer to real constructional materials of a similar structure (*"Handbook ..."* 1978). Effective constants for upper and lower half-planes may be calculated from E_f, ν_f, E_m, ν_m according to the corresponding formulae (*"Mechanics ..."* 1982). These effective constants are used in defining of critical strains of the problem under consideration. The results of calculations for one microcrack are presented in Figure 4.

3.3 Two microcracks on one interface

The above-stated problem was solved also for the case of two microcracks with the distance l between them. Critical strains were obtained for numerous values of the ratio l/a for considered composites of the lateral-torsional packing. The following conclusion was formulated after the analysis of these results.

In the case of compression along interlaminar microcracks the critical strain does not depend on the number of microcracks and the distance between them. This conclusion for laminated composites (piecewise-homogeneous bodies) is in agreement with the exact solution (A.N. Guz 1990) for cracks in homogeneous anisotropic materials. (Hatched curve in Figure 4 is calculated using the mentioned exact solution as applied to compression of the unidirectional fibrous composite along cracks. The direction of compression coincides

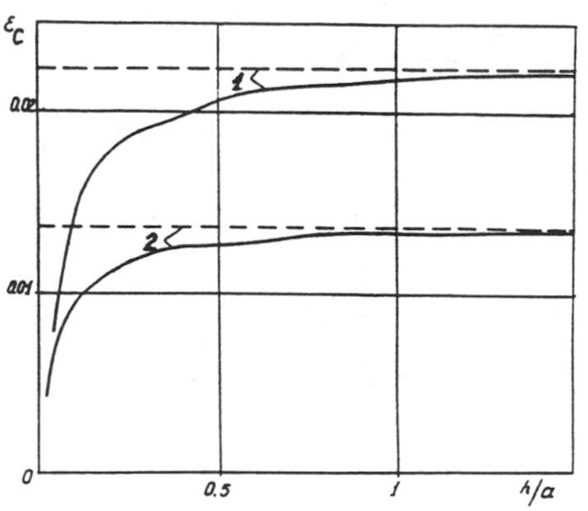

Figure 5: Critical strains for cases of Tornel-300/epoxy (curves 1: $E_f = 239GPa$, $\nu_f = 0.2$, $E_m = 2GPa$, $\nu_m = 0.4$) and boron/epoxy (curves 2: $E_f = 400GPa$, $\nu_f = 0.21$, $E_m = 2GPa$, $\nu_m = 0.4$) layers; the volume concentration of fibers in layers $c_f = 50\%$.

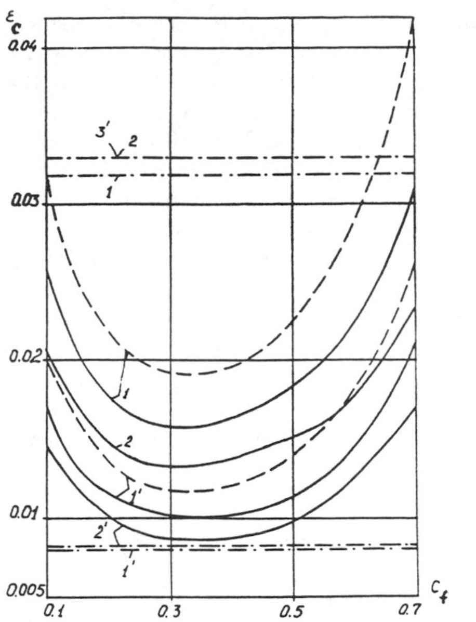

Figure 6: Critical strains for cases of Tornel-300/epoxy (curves 1,2: $E_f = 239GPa$, $\nu_f = 0.2$, $E_m = 2GPa$, $\nu_m = 0.4$) and boron/epoxy (curves 1',2',3': $E_f = 400GPa$, $\nu_f = 0.21$, $E_m = 2GPa$, $\nu_m = 0.4$) layers; c_f – the volume concentration of fibers in layers.

with the direction of reinforcement; properties of fibers and matrix for continuous and hatched curves are the same.) Values of critical strains for compression along microcracks are presented also in Table 1 as ϵ_c^m for the cases of E-glass fibers (the first row), Kevlar-49 fibers (the second row), Tornel-300 fibers (the third row) and boron fibers (all other rows).

3.4 Two parallel structural cracks

Let us suppose that the considered composite of the lateral-torsional packing has two parallel structural cracks with the distance h between them. The upper and the lower layers are replaced by half-planes due to the small lengths of cracks in comparison with thicknesses of layers (7). These half-planes are reinforced in the perpendicular to the plane x_1ox_2 direction, the middle layer – in the direction of axis ox_1. Calculating the effective constants as for the above case of microcracks the following results were obtained.

Values of critical strains (ϵ_c) are presented in Figures 5,6 for the concerned problem of compression along two parallel cracks (continuous curves) in comparison with critical strains (ϵ_c^m) for the problem of compression along microcracks (hatched curves). In so doing, values of the ratio h/a were put equal to 0.2 (curves 1,1') and 0.1 (curves 2,2') for Figure 6. Table 1 also contains values of critical strains ϵ_c and ϵ_c^m for various properties of layers. Figures 5,6 and Table 1 demonstrate that the interaction of two parallel interlaminar cracks may cause the decrease of the critical strains in several times in comparison with the case of interlaminar microcracks. But when the distance between parallel cracks more than their length, the influence of this interaction is not more than some percents.

As it was noted above, the formulation of the problem utilized in the present paper — the model of piecewise-homogeneous medium and equations of TLTDBS (A.N. Guz 1986) — is of the most strict nowadays. The stability problem for the case of compression along two parallel cracks had been solved by various authors within the scope of approximative approaches based on calculation of critical strains for an infinite in one direction plate with properties of the middle layer and different conditions of the end fixation (rigid or hinge restrain). Results of such approaches are presented by hatched-dotted curves (Figure 6). Namely, lines 1 and 1' correspond to the case of critical strains for hinge-restrained ends calculated using the three-dimensional theory (A.N. Guz 1971). Lines 2 and 2' correspond to the Euler critical strains for rigid-restrained ends. Evidently, the approximative approaches do not describe the phenomenon under consideration even on the qualitative level. Sure, hatched-dotted lines (Figure 6) do not depend on properties of materials (c_f) and their difference from the results of the exact approach (continuous curves) may be more than 10 times.

Table 1. Values of critical strains for microcracks (ϵ_c^m) and two parallel structural cracks (ϵ_c) in the case of $\nu_m = 0.4$.

h/a	c_f, %	E_f, GPa	ν_f	E_m, GPa	ϵ_c, %	ϵ_c^m, %
0.2	35	73.5	0.22	2	4.05	5.59
0.2	50	131	0.2	2	2.99	3.92
0.2	50	239	0.2	2	1.81	2.26
0.2	35	400	0.21	2	1.00	1.16
0.2	50	400	0.21	2	1.13	1.37
0.2	50	403	0.21	2.8	1.53	1.89
0.2	50	403	0.21	4	2.11	2.62
0.1	50	400	0.21	4	1.72	2.72
0.5	50	400	0.21	4	2.46	2.72

4 BOUNDS FOR THE CASE OF VARIOUS INTERLAMINAR IMPERFECTIONS

4.1 Substantiation of the bounds

To estimate the critical loading parameters in composite material with various imperfections (9) between layers the following bounds can be proposed basing on the well-known principle of mechanics, namely, stating that release from some connections cannot increase the value of the critical load, under which stability loss of the system takes place.

According to the above-said, the critical parameter f^+, under which stability loss takes place in a structure without imperfections (i.e. with rigid contact of layers) must be larger than the critical loading parameter f for the same structure with imperfections (9). In other words, it can be said that a structure with imperfections is obtained from a structure without imperfections by releasing from connections those parts of the boundary

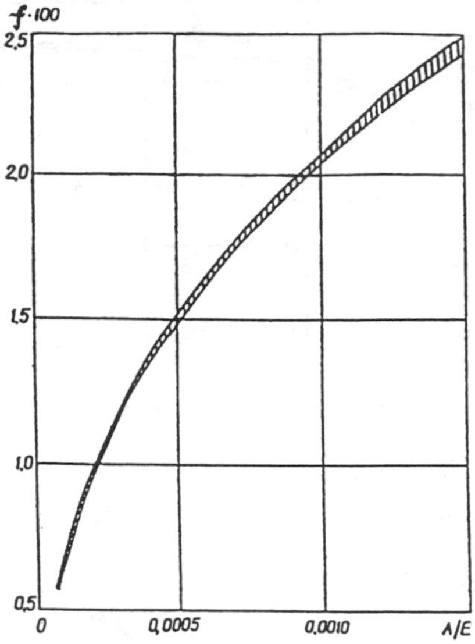

Figure 7: Bounds for critical loading parameters for metal matrix composites with interlaminar defects ($k = 0.34$, $\nu = 0.2$, $h^{(1)}/h^{(2)} = 0.015$).

where imperfections must be located, i.e.

$$f < f^+ \tag{11}$$

At the same time, if in a structure with imperfections all remaining connections (zones of the rigid connection) are released, structure of the same type with ideal slippage (sliding without friction) between all the layers and half-planes is obtained. Applying the same principle, it can be said that the critical loading parameter f^- for a structure with sliding layers must be smaller than the critical loading parameter f for a structure with imperfections (9)

$$f^- < f \tag{12}$$

The parameters f^+ are determined from the solution of equations (2) with boundary conditions (3), (5), (6), where in this case

$$S = S_r \cup S_s \quad , \quad S_c = 0 \tag{13}$$

The parameters f^- are determined from the solution of equations (2) with boundary conditions (5), (6), (9), where in this case

$$S = S_c \cup S_s \quad , \quad S_r = 0 \tag{14}$$

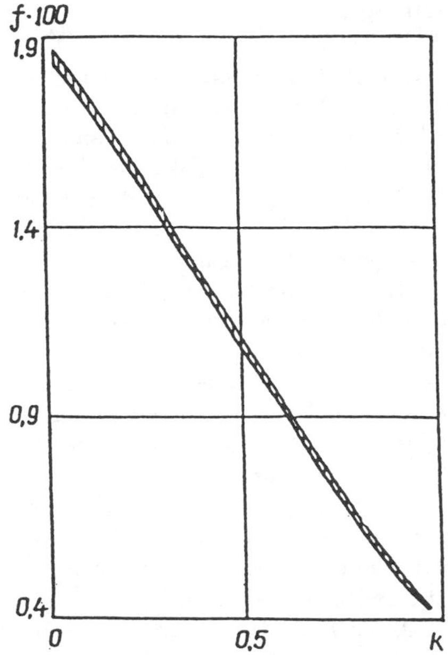

Figure 8: Bounds for critical loading parameters for metal matrix composites with interlaminar defects ($k = 0.34$, $A/E = 0.0003$, $h^{(1)}/h^{(2)} = 0.015$).

When the f^+ and f^- are found, the bounds required follow from (10), (11)

$$f^- < f < f^+ \tag{15}$$

An example of such bounds for one of the models of laminated composite containing "defects with connected edges" will be given below.

4.2 Example of the bounds for the particular model

Let us consider a composite consisting of alternating layers of a linear-elastic isotropic compressible filler (of thickness $2h^{(1)}$)

$$\sigma_{ij}^0 = \delta_{ij}\frac{E\nu}{(1+\nu)(1-2\nu)}\epsilon_{nn}^0 + \frac{E}{1+\nu}\epsilon_{ij}^0 \tag{16}$$

and an elastic-plastic incompressible matrix (of thickness $2h^{(2)}$) with power-mode dependence between intensities of stresses (σ_I^0) and strains (ϵ_I^0) in the form

$$\sigma_I^0 = A(\epsilon_I^0)^k \tag{17}$$

Non-axisymmetrical problems for the composites with the purpose of determining f^+ and f^- for

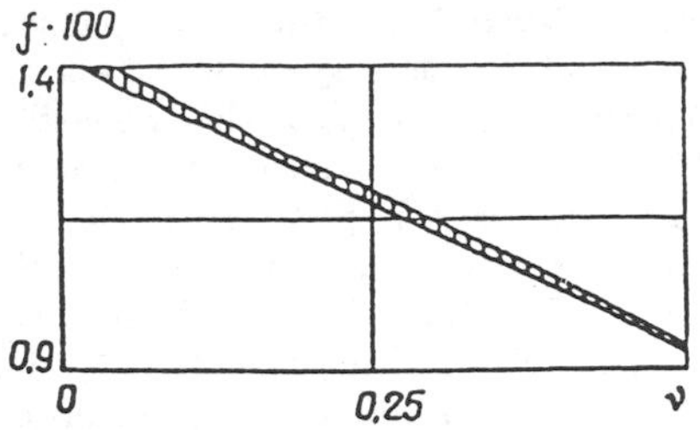

Figure 9: Bounds for critical loading parameters for metal matrix composites with interlaminar defects ($\nu = 0.2$, $A/E = 0.0003$, $h^{(1)}/h^{(2)} = 0.015$).

various types of layer models, including the abovementioned one, have been considered (I.A. Guz 1991). The bounds obtained similarly for the case of three-dimensional problem are shown in Figures 7-9. The shaded region is located between f^+ and f^-. According to the above-said, the critical loading parameters for composite with interlaminar cracks are located precisely in this region. Figure 7 provides the critical loading parameters as a function of A/E where E is Young modulus for filler. Figure 8 shows the dependence of critical loading parameters on the Poisson ratio ν, Figure 9 – on k.

As one sees, for several parameter combinations the suggested bounds provide quite accurate result for f, particularly for engineering (the strain value ϵ_{11}^0 is used here as f).

5 CONCLUSIONS

The problem of instability of laminated composite materials in compression along interlaminar cracks was considered in this paper within the scope of the exact statement using the model of piecewise-homogeneous medium and equations of TLTDBS (A.N. Guz 1986). The eigenvalue problem with respect to the loading parameters was formulated for cases of compressible, elastic and elastic-plastic, orthotropic and isotropic layers.

The classification of cracks was introduced. It is based on certain indiscrepant simplifications, connected not with the changing of equations, but only with geometrical characteristics of the composite and the cracks. Macrocracks, structural cracks and microcracks were discerned in this classification.

Basing on the results for composites with rigid-connected layers, the estimation of critical loading parameters was suggested. This estimation establishes the upper and the lower bounds of critical loads for laminated structures in compression along the interlayer imperfections from the results for rigid-connected and sliding layers. Substantiation of the bounds is based on the one of the general principles of mechanics on the influence of liberation from a part of connections on value of critical loads for the mechanical system. Numerical investigations for composites with elastic-plastic matrix had shown that suggested bounds present reasonable fair results for particular cases of composites.

Numerical results were obtained for various sorts of orthotropic layers with the particular interlaminar cracks utilizing the method of finite differences. In so doing, the difference scheme having the first order of approximation was proposed. The following conclusions were formulated after the analysis of these results.

In the case of compression along interlaminar microcracks the critical strain does not depend on the number of microcracks and the distance between them.

The interaction of the two parallel interlaminar cracks may cause the decrease of the critical strains in several times in comparison with the case of interlaminar microcracks. But when the distance between parallel cracks is more than their length, the influence of this interaction is not more than some percents.

Besides that, it was shown that approximative approaches to so fine and complex phenomenon in the material structure cannot describe it even on the qualitative level and their difference from the results of the exact approach may be more than 10 times.

6 ACKNOWLEDGEMENT

This research was partially fulfilled within the fellowship of the Royal Society (London, UK). Author is very grateful to Dr. W.J. Stronge (Department of Engineering, University of Cambridge, UK) for his much appreciated attention and support of the investigation.

REFERENCES

Biot, M.A. 1965. *Mechanics of incremental deformations.* New York: Wiley.

Guz, A.N. 1971. *Stability of three-dimensional bodies.* Kiev: Naukova Dumka. (In Russian)

Guz, A.N. 1986. *Foundations of three-dimensional theory of deformable bodies stability.* Kiev: Vyshcha Shkola. (In Russian)

Guz, A.N. 1990. *Mechanics of fracture of composite materials in compression.* Kiev: Naukova Dumka. (In Russian)

Guz, I.A. 1989a. Spatial nonaxisymmetric problems of the theory of stability of laminar highly elastic composite materials. *Soviet Appl. Mech.* 25: 1080-1085.

Guz, I.A. 1989b. Three-dimensional nonaxisymmetric problems of the theory of stability of composite materials with metallic matrix. *Soviet Appl. Mech.* 25: 1196-1201.

Guz, I.A. 1990. Continuum approximation in three-dimensional nonaxisymmetric problems of the stability theory of laminar compressible composite materials. *Soviet Appl. Mech.* 26: 233-236.

Guz, I.A. 1991. Asymptotic accuracy of the continuum theory of the internal instability of laminar composites with an incompressible matrix. *Soviet Appl. Mech.* 27: 680-685.

Guz, I.A. 1992. Estimation of critical loading parameters for composites with imperfect layer contact. *Int. Appl. Mech.* 28: 291-296.

Guz, I.A. 1993a. Computational schemes in three-dimensional stability theory (the piecewise-homogeneous model of a medium) for composites with cracks between layers. *Int. Appl. Mech.* 29: 274-280.

Guz, I.A. 1993b. Investigation of local forms of stability loss in laminated composites (three-dimensional problem). *Proc. of the Ninth Int. Conf. on Composite Materials.* 6: 377-383.

Guz, I.A. & Yu.V. Kokhanenko 1993. Stability of the laminated composite material in compression along the microcrack. *Int. Appl. Mech.* 29: 702-708.

Handbook of fillers and reinforcements for plastics (Edited by Katz, H.S. & V. Milevski). 1978. New York: Van Nostrand Reinhold Company.

Mechanics of composite materials and constructive elements (Edited by Guz, A.N.). Vol.1. 1982. Kiev: Naukova Dumka. (In Russian)

Modern Practice in Stress and Vibration Analysis, Gilchrist (ed.) © *1997 Balkema, Rotterdam, ISBN 90 5410 896 7*

An elasto-plastic finite element method analysis on the crack growth of super-strength-steel material

D. Sun
Basics Department, Transportation Engineering Institute, Tianjin, People's Republic of China

X. Ma
China Aeronautics & Space Technology Institution, Beijing, People's Republic of China

ABSTRACT: This paper has studied non-self-similar stable crack growth of a cracked body in case of plane stress which was analyzed by Finite Element Method (FEM) based on linear constitutive equations and incremental theory. An alternative method, the effective maximum stress criterion, has been established in FEM analysis. Several parameters, such as the stress near crack tip and plastic zone, have been prezented and compared with the experimental results.

1 INTRODUCTION

Several scholars [1-4] have studied self-similar stable crack growth (Mode I) in case of plane stress. Complex-mode (Mode II or III) crack problems are fairly complicated, for the direction and critical condition of the crack propagation must be considered in the selection of fracture criterions. The different researchers have their own criterions in control of the crack growth, thus the parameters are quite different as they are related not only to materials concerned but also to the configuration of cracked body, therefore the selection of criterions is rather controversial.

The COD criterion put forward in [5] assumes that the direction of the crack growth is perpendicular to that of crack opening displacement and the effect of bilateral stress on it is considered, and as a result the fracture stress increases with the increasing of Mode II.The analysis of non-self-similar crack growth is conducted by the experiment and the finite element method based on elasto-plasticity of small displacement in [6] which points out that fracture stress increases with the increasing of b/W,where 2W is the width of the cracked plate,2b the subtraction of the project of the crack length on x-axis from the plate width. In case of plane stress the fracture stress is no longer related with the thickness and the direction of the crack growth is vertical with the apllied stress. The paper [7] in analysing the non-self-similar stable crack growth demonstrates the stress near crack tip, plasic energy and growth path during the process of crack propagation but no experimental results.

In the paper non-self-similar stable crack growth of the cracked body in case of plane stress has been studied. An experiment shows that there even exits stable crack growth in the super-strength-steel, 37SiMnCrM.V, subjected to plane stress. In FEM analysis,the curve of applied stress and crack length is used for governing the crack growth, and the criterion of effective maximum stress has been established to determine the direction of the crack growth. The parameters of fracture mechanics such as crack tip stress, plastic zone during the the process of crack growing have been analyzed by the FEM based on incremental theory and compared with the experimental results.

2 THE COMPUTATIONAL ANALYSIS OF CRACKED PLATE

The focus of the article is on super-strength-steel cracked plate shown in Figure 1, the width of which is 2W=50mm, length 2L=100mm, thickness t=1.2mm, the original crack length $2a_0$. Only half of the plate is neccerary because of the symmetry of its geometry and loading. The boundaries are shown in Figure 2.

Using the maximum stress criterion or s criterion

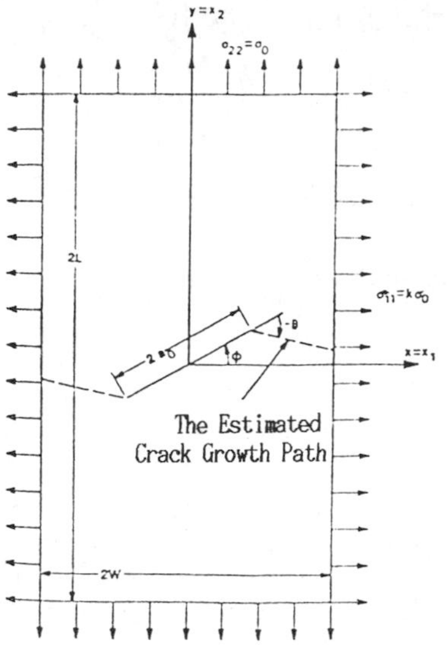

Figure 1. The geometry and applied stress of cracked plate.

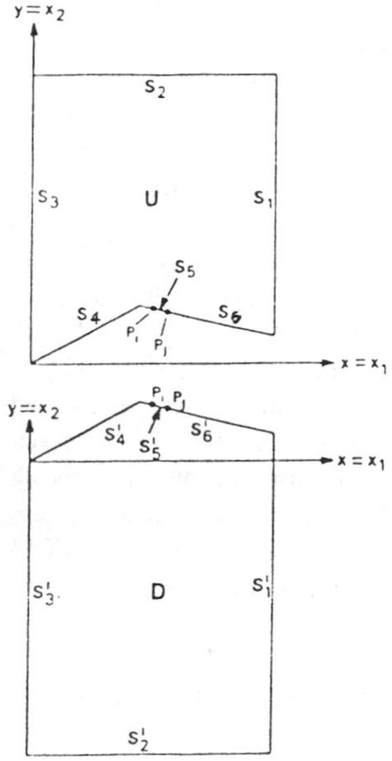

Figure 2. FEM boundaries.

as the parameter to govern the direction of the crack growth is not reasonable hereby because it is based on linear elastic fracture mechanics and meanwhile stress intensity factor here is difficulty to calculate.

It is interesting that two additional equations must be set up as both length and direction of the crack

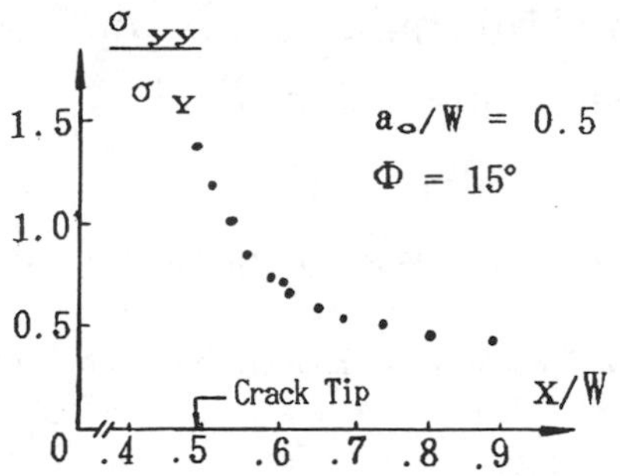

Figure 3. The distribution of stress near crack tip

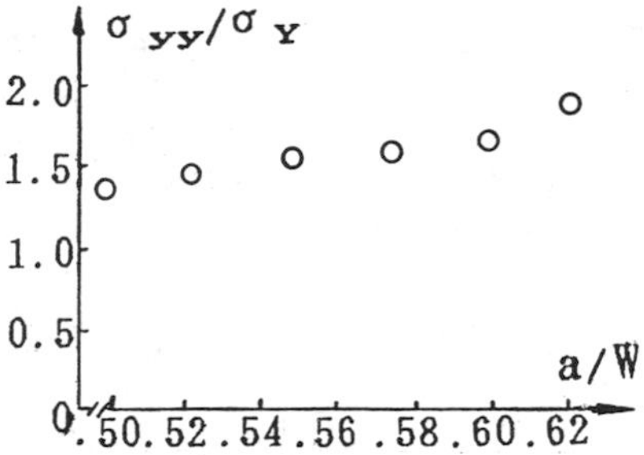

Figure 4. The variation of stress in crack tip during the crack growth.

growth are unknown. A curve of applied stress and crack length derived from the experiment is to be used as the governing equation of crack propagation.

Therefore an alternative method, the effective maximum stress criterion, has been initiated to do the non-self-similar crack growth, meaningly its direction is perpendicular to that of the force of the nodal point when the crack tip moves from one nodal point to the other. Suppose the angle between the direction of the crack growth and x-axis as θ_m, thus

$$\theta_m = \text{arctg}\,(F_x / F_y)$$

where F_x and F_y correspond the forces of the nodal points on crack tip in x and y axises respectively. While the growing direction is to be varied after every step of the computation, the mesh modification of the FEM programme should be required before the next step starts up.

3 THE RESULTS OF CALCULATION AND EXPERIMENT

Two kinds of specimens, $\Phi = 15°$ and $\Phi = 30°$ are performed in the experiment and calculated in FEM analysis.

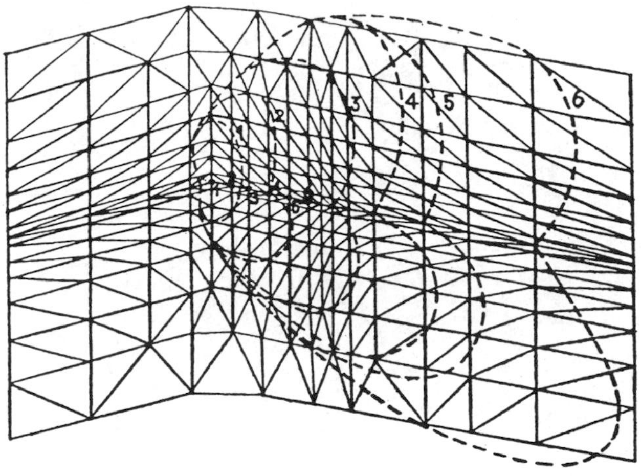

Figure 5. The changes of plastic zone. 1,2,··· ··· ,6 and dash lines refer to the positions of the boundaries of the zone.

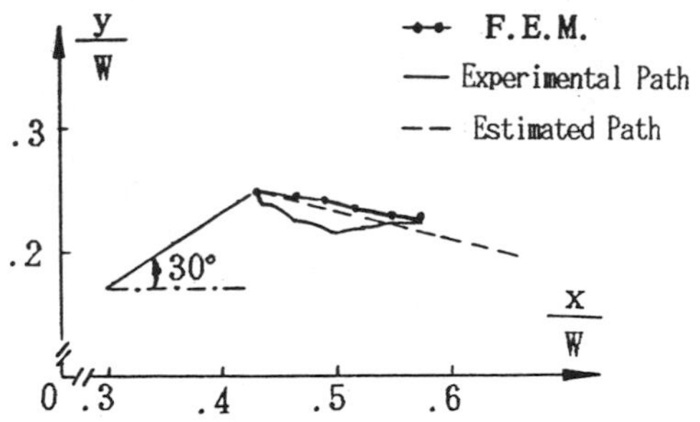

Figure 6. The path of the crack growth ($\Phi = 30°$).

From the FEM analysis and experiment, we may see as follows:

The distribution of stress in the crack tip is given in Figure 3, while Figure 4 stands for the variation of stress near crack tip. Figure 5 refers to the plastic zone and Figure 6 the path of the crack growth.

4 CONCLUSIONS

1. The stress variation of non-self-similar crack growth is similar with that of self-similar one, but the values are a little bit higher than the latter one, meaningly the resistant energy against crack growth is increasing with the increasing of the non-self-similar crack growth, in other words the energy comsumption during crack growth is growing up.

2. The plastic zone is different from that of the self-similar one and the contour of that both bigger and asymmetrical.

3. The experimental path of crack growth is some different from that of the computation, but little change at the later stage of the growth. That shows the reasonability of the effective maximum stress put forward in the paper.

REFERENCES

1.Du Shanyi and Lee , J . D . , *Engng. Fract. Mech.* , 17, p.173(1983)

2. G. B. May and A. S. Kobayashi, Int. *Journal of Solids and Struct.*,Vol.32,p.857-881(1995)

3.Du Shanyi,*Proceedings of ICF*,Beijing,p.18,(1983)

4.Ahoor,A.and Abou-Sayed,I.S.,*Computers and Structures*,13, p.137(1981)

5.Ku Jilin et al.,*Proceedings of ICF*, Beijing, p.267(1983)

6.Ueda,Y., *Proceedings of ICF,Beijing*, p.225(1983)

7.Lee, J. D. , et al., *Presented at Int.Conf.on Fract. Mech.* , Australia,August 10-13,1983

Modern Practice in Stress and Vibration Analysis, Gilchrist (ed.) © *1997 Balkema, Rotterdam, ISBN 90 5410 896 7*

Thermo mechanical analysis of rail wheel

L. Ramanan & R. Krishna Kumar
Indian Institute of Technology, Madras, India

R. Sriraman
Integral Coach Factory, Madras, India

ABSTRACT : The combined effect of stresses arising in a railway wheel is due to the mechanical and thermal loads. This paper analyses these stresses in the railway wheel with a three dimensional finite element model using the commercial finite element codes. The effect of assembling the axle into the railway wheel has also been analyzed and validated with that of the results obtained from the manufacturing process.

1. INTRODUCTION

The various terminology used in expressing the portions of the railway wheel are as shown (Figure 1).The stresses arising in a railway wheel due to mechanical and / or thermal loads strongly depends upon the geometrical configuration of the wheel disc (Fermer 1994).It has been found that railway wheels which are subjected to tread braking experiences two different types of failures. They are

1. Failure in the tread portion;
2. Failure in the disc portion.

On an average in European railways every solid wheel breakages results in an accident (Edel & Schaper 1992).Thermal cracks are produced easily and propagates in the radial direction as deep fatigue cracks leading to fracture (Wise 1987). These two types of failures are reported in the Indian Railways as well as in British and American Railways (Edel & Schaper 1992). Constraint of thermal expansion at the rail wheel contact area can cause stresses which may eventually initiate and propagate as fatigue cracks (Michael & Sehitoglu 1985).

Lunden (1991) analyzed the railway wheel under combined mechanical and thermal loads using an axisymmetric model and elasto-plastic constitutive equations.

Johan Jergeus (1991) considered a small segment of the railway wheel to analyze the wheel rail contact problem. No thermal loads were considered in this analysis.

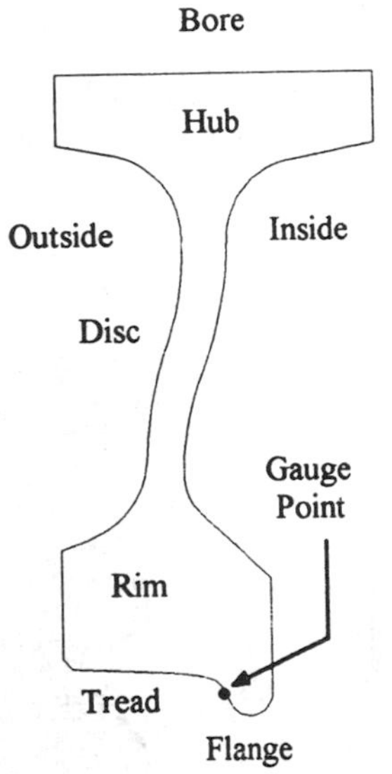

Figure 1. Section of Railway wheel

Even the report of railway engineering research (1992) does not consider thermal loads. Mechanical loads alone have been considered for the analysis

2. FINITE ELEMENT ANALYSIS

A coupled thermo-mechanical analysis has been performed to determine the total stress distribution in the wheel. As can be clearly seen from the literature, not much work has been done to calculate the stress during the application of brakes. The mechanical analysis is elasto-plastic in nature. Further, two types of models have been compared. The first model considered in this study is as per the American association of rail roads standards (S-660-83 1984). The second model, which includes the axle also is according to the Chalmers university report (1992). These models are as shown in Figure 2 & 3.

Both the models use 8 noded hexa elements predominantly (with a few wedge elements at the axle).For *AAR* model 5496 elements and 6978 nodes have been used. The second model consists of 10032 elements and 11550 nodes. The geometry of the wheel and the material is as per *RDSO* standards (R-19 1993) The properties of the materials are summarized in Table 1

2.1 Loading and boundary conditions

The energy imparted during braking raises the temperature levels in the wheel. The total energy during braking has been calculated and applied as uniformly distributed heat flux on the tread portion of the wheel this is according to the recommendations of the *AAR* standards.(S-660-83,1984). The heat transfer coefficient for the side of the wheel which is exposed to the atmosphere is taken to be 22.7128 J/sq.m °K (S-660-83 ,1984) with a surrounding temperature of 40 °C. Symmetry boundary conditions has been applied in the 1-2 plane for the wheel and 1-2 as well as 1-3 plane for the axle. The temperature thus calculated has been used for the subsequent mechanical analysis.

The major mechanical load is the service load, including the self weight of the coach and the payload. The vertical load considered are with 40% dynamic conditions on the axle load (Lunden 1992) (Fermer 1994). Apart from the vertical load, a horizontal load acts in the wheel due to swaying of the coach. This swaying motion is due to track irregularities and perturbation. As suggested by Fermer (1994) and *AAR* (S-660-83, 1984), these loads are 50 % of the vertical loads.

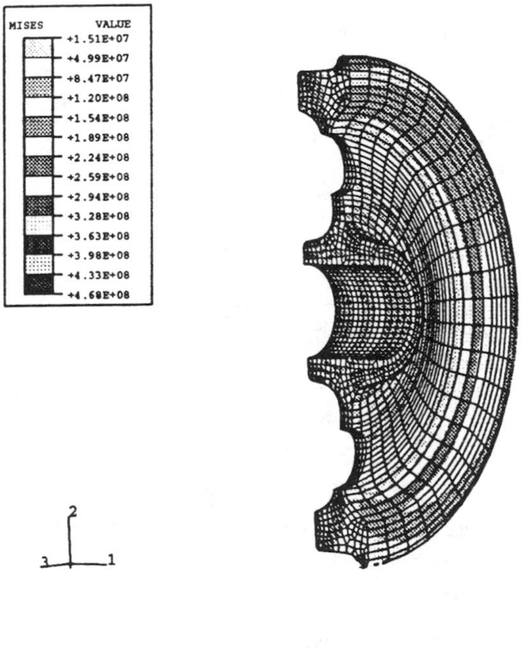

Figure 2. AAR model with vonmises stress results of the coupled analysis.

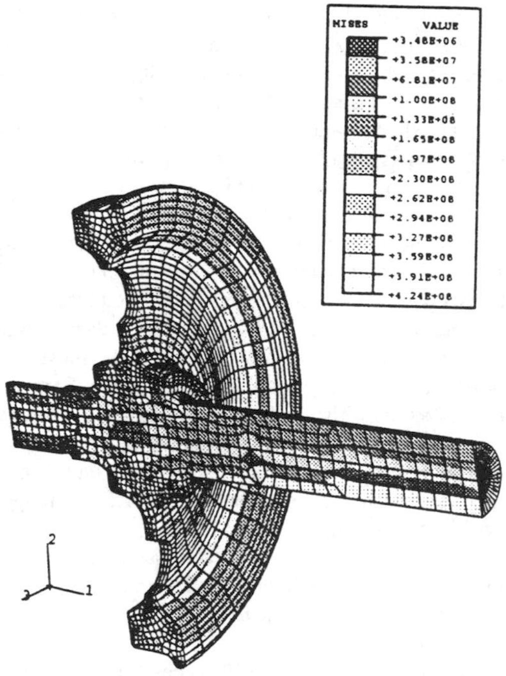

Figure 3.Chalmers University model with vonmises stress results of the coupled analysis.

Table-1 Material properties

Modulus of Elasticity	2.10E+11 pascals
Poissons ratio	0.29
Thermal Conductivity	48.6336 J/m°K sec
Specific heat at constant pressure	427.054 J/kg °K
Yield Stress at log strain of 0	4.10E+08 pascals
at log strain of 0.1484	7.60E+08 pascals
Isotropic hardening	
Mass density	7820 kg/m^3
Speed of the train	120 km/h
Braking distance	800 m
Convection Coefficient	22.7128 J/m^2°K
Surrounding temperature	40 °C
Axle Load	16.25 tonnes
Wheel dia	0.915 m

The point of application of these loads differ in both models. Since the *AAR* model does not consider the axle, the loads are applied as point loads at the gauge point (Figure 1). The bore nodes are fixed in both radial and axial directions of the wheel.

On the other hand, in the second model, the vertical load is applied at the axle position and the reactions are taken at the part of contact of the tread and the rail. Horizontal load is applied at the contact position. This model is similar to that recommended in Chalmers university report (1992). Symmetry boundary conditions has been applied at appropriate planes (Figure 1 & 2). Commercial software package *ABAQUS* has been used for the analysis and *SDRC I-DEAS* has been used for the model.

2.2 Model for axle push-in

An axisymmetric model has been used for this analysis. A total number of 277 nodes and 205 elements have been used. The model is as shown in Figure 4. A nonlinear analysis with contact has been performed to study the load required to assemble the axle. A contact algorthim based on perturbed lagrangian formulations (Pavan 1996) has been used for the analysis.

The top nodes of the hub are restrained in the 2 direction. The end of the axle has been effected a displacement boundary condition in the 2 direction.

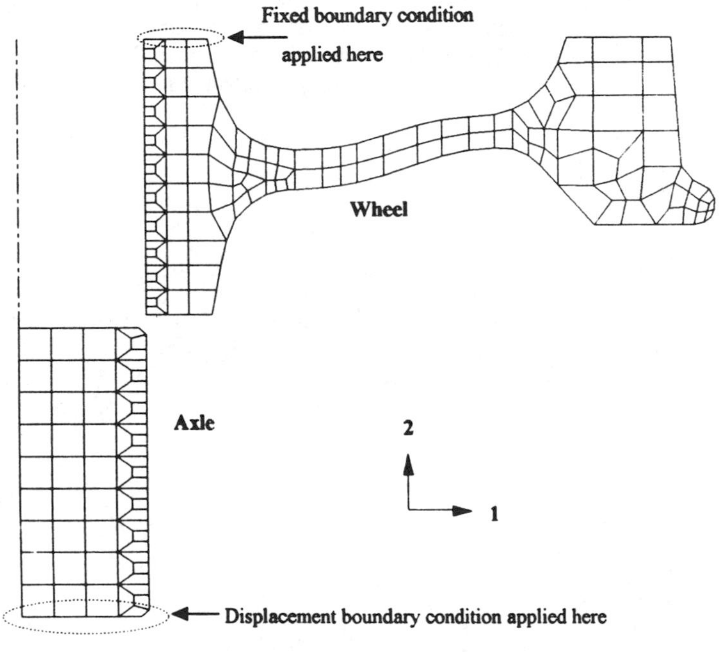

Figure 4 Axi-symmetric model

Table 2. Results of the analysis.

Analysis	Model	Region	Vonmises stress in pascals
Coupled	*AAR*	Tread	1.54 E08
		Disc	4.68 E08
	Chalmer's	Tread	3.19 E08
		Disc	4.24 E08
Mechanical	*AAR*	Flange	1.88 E08
		Disc	1.01 E08
		Tread	5.79 E09
	Chalmer's	Tread	4.10 E08
		Disc	4.40 E04
Thermal	*AAR*	Tread	1.273E08
		Disc	4.63 E08
	Chalmer's	Tread	3.84 E07
		Disc	4.24 E08

3. RESULTS AND DISCUSSIONS

The results of the coupled thermo elasto-plastic analysis is shown in Table-2. From the stress distribution as can be seen from the table there are two regions at which stresses are high. They are the flange and the tread region of the wheel.

The Coupled analysis (Figure 2 & 3) shows that the disc part of the wheel has very high stresses. This is the case with both the models. The stresses are in to the plastic region due to braking and interestingly a number of failures have been reported in the disc region (Edel 1992).On the other hand the stresses at the tread portion of the wheel are high in the Chalmers university model (391 N/sq.mm) and is low with *AAR* model (154 N/sq.mm).But a number of wheel failures have been reported in this area aswell (Wise 1987). In order to understand these results two separate finite element analysis was carried out. The first analysis considered only the mechanical part of the load. The second analysis considered loading due to braking (temperature effects alone). The results clearly show that the high stresses in the disc region is solely due to braking. This effect requires serious consideration in design.

The radial stress distribution due to axle push - in alone is given in Figure 5. As can be seen from this figure, the residual stresses in the disc region is as high as 50 N/sq.mm. This is the region at which stresses there are high stresses in the previous analysis. Hence it is advised that assembly stresses may also be important and should be considered.

It is found during this analysis that the bore nodes expand in the radial direction by a value equal to 80% of the interference. In practice, while assembling the axle and wheel, solid lubricants like graphite are applied. It is reported (Martin) that by applying solid lubricants it is possible to reduce the

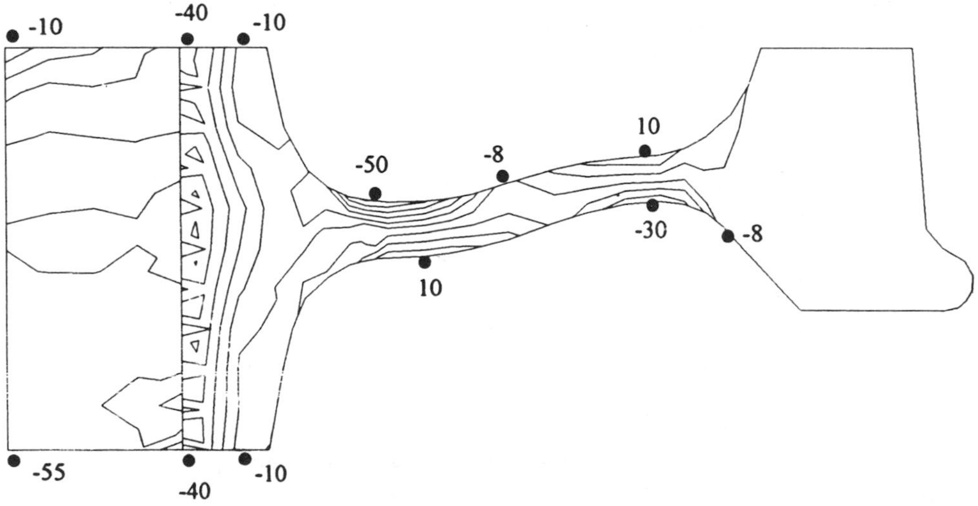

Figure 5. Radial stress contour plot due to axle push-in (in N/mm^2)

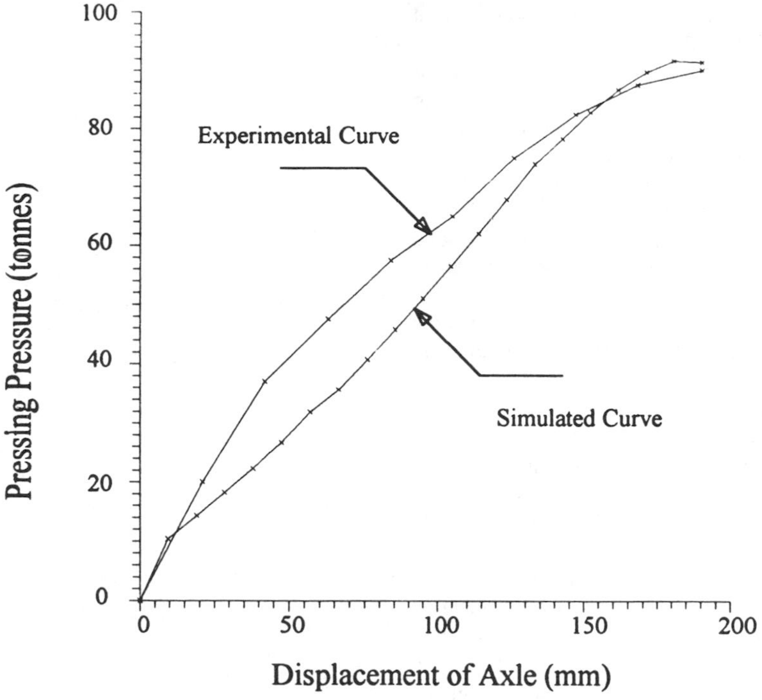

Figure 6. Comparison between experimental and simulated axle pressing pressure

coefficient of friction to as low as 0.05. However it has been found during this analysis that the coefficient of friction may vary between 0.08 - 0.11. The applied pressure for assembling the wheel and axle in practice and as well as of the analysis coincides with each other, as shown in the Figure 6.

4.CONCLUSIONS

As found in the analysis,the disc region is highly stressed ,hence it is not advisable to have any holes in disc portion.

The Chalmers university model predicts both the failures in the tread and disc region.Hence this model can be used effectively for design.

The stresses due to mechanical loads is found high at the contact region. Hence it requires contact analysis with full wheel set and rail inorder to understand more clearly about these contact stresses.A three dimensional analysis is being carried out for this purpose.

The results of the axi-symmetric analysis brings out that the stresses due to the axle push-in gets extended in the disc region also. Hence it is essential to consider the displacement of the bore nodes in the radial direction by a value of 80 % of the interference for the coupled and mechanical analysis of the wheel.

5.ACKNOWLEDGEMENTS

This paper is the part of research work for the M.S. thesis of the first author. He expresses his sincere gratitude to the General Manager of Integral Coach Factory, Madras, India for his constant encouragement and for the necessary permission to publish this research work.

REFERENCES

Edel, K.O & Schaper,M 1992. Fracture mechanics fatigue resistance analysis of the crack damaged tread of over braked solid railway wheels. Rail International J(1992)

Fermer,M.1994. Optimization of railway freight car wheel by use of fractional factorial design method. Proceedings of the Institution of Mechanical Engineers. Vol 208

Johan Jergeus 1991. Wheel rail contact problems modeled with three dimensional finite element in a super computer. Report T125 Solid Mechanics division of the Charmers University of Technology

Kenneth Anderson, Krister Johnson, Herik Simonsen and Eric Wang 1991. Martensite formation in railway wheel treads. Report T126 Solid Mechanics division of the Chalmers University of Technology.

Lunden,R 1991. Contact region fatigue of railway wheels under combined mechanical rolling pressure and thermal brake loading. Wear 102,31-42

Lunden,R 1992. Cracks in railway wheels under rolling contact. International. Wheelset Congress 1991

Martin J.Sigel, Valdimir L Maleev & James B Hartman. Text book on Mechanical design of Machines. pp 368-374

Michael C-Fec & Huseyin Sehtoglu. 1985. Thermal mechanical damage in railroad wheels due to hot spotting. Wear 102:31-42

Moyar,G.J.& Stone,D.H. 1991. An analysis of thermal contributions to railway wheel shelling. Wear 144:117-138

Pavan Chaand,Ch. 1996, Yield function for metal powder compaction based on micromechanics of particle deformation. Ph.D thesis, IIT Madras.

Report of railway engineering research at Chalmers University in Gothenburg 1992 Proceedings of the Institution of Mechanical Engineers. 206:145

S-660-83 Procedure for the analytical evaluation of locomotive and freight car wheel designs. Association of American Railroads (*AAR*) Mechanical division, Manual of standards and recommended practices. 1984

Wise,S.1987.Railway wheelsets-a critical review. Proceedings of the Institution of Mechanical Engineers 201:D4

Modern Practice in Stress and Vibration Analysis, Gilchrist (ed.) © *1997 Balkema, Rotterdam, ISBN 90 5410 896 7*

Integrated design and FEA of composite materials

T.J.Adams, V.Middleton & I.A.Jones
Department of Mechanical Engineering, University of Nottingham, UK

ABSTRACT: Recent improvements in composites manufacturing have not been matched by improvements in structural analysis methods, in particular the time taken to perform a finite element analysis. The analysis of composites often requires the material properties to be defined for each element. Systems allowing automatic meshing exist; however, these are somewhat limited when analysing composites. Manual data entry is time consuming and error-prone. A need exists for a system which can import the design geometry, together with fibre architecture data, and create a representative FE model with minimal operator effort. A two stage approach has been adopted; i) component geometry is meshed using a pre-processor, ii) the composite material properties are assigned to the elements. This is implemented by means of a program which uses the element's positions to establish the properties from a representation of the internal architecture. Composite properties data relates to isoparametric blocks covering sections of the component with properties defined at nodal points. The system is implemented for 2D structures using axisymmetric quadrilateral elements and has been used to obtain results comparable with a previously verified structure.

INTRODUCTION

Composites can offer mass, strength and cost advantages compared with metals in aero-engine applications. Aerospace companies are in the process of substituting numerous metal parts in their engines with composite parts. These include guide vanes, ducting, nozzles, casings and front end spinners. Increasingly components are being used in load bearing applications. By integrating the function of several metal parts into one, the composite components offer lower complexity in the final construction.

However a major deterrent in the greater usage of polymer matrix composites is the lack of validated design and analysis methods, particularly in the area of Finite Element Methods (FEM). Work is in progress in the aerospace industry to validate these methods within design and manufacturing systems. An outstanding problem is the time taken to perform an FE analysis of a composite part in comparison to a metal part. The FE systems in use require the material architecture (laminate structure etc.), fibre directions and material properties to be defined manually element by element (or, small region by small region if the shape's geometry is very simple).

This is: a) time consuming in all but very simple models, which are rarely realistic and b) requires high levels of skill in understanding the FE system and the model under construction. This state of affairs clearly must be improved if polymer composite materials are to continue to find increasing usage in aero-engines and other aerospace applications and so contribute to cost and mass savings in the future.

The objective of the work being undertaken by the authors is to establish methods and systems that enable the creation of detailed and accurate finite element models from existing definitions of external geometry and existing information concerning the design data and/or design rules pertaining to the internal laminate architecture.

The aerospace industry is a traditional user of composites, but other users of composites technology will encounter similar analysis problems. For example, in the water treatment/sewage industry, cast iron is currently used for the column situated at the centre of a biological filter and used to support the radial arms which distribute the effluent over the filter media. Previous work (Corden 1997) has studied the feasibility of replacing this heavy component with a lighter, corrosion-resistant alternative made from glass-reinforced polyester, and various analyses of it have already been undertaken. The component is essentially a tapered tube with a flange and lends itself (with some simplification) to being modelled as an axisymmetric object. Since finite element results (and corresponding experimental results) for static deflection were available for this component, it was used here in preference to an aerospace component as a test case.

BACKGROUND

The aerospace industry, in particular aeroengine manufacturers, has provided the impetus for much of the innovative structural materials developed since the last war. These include the ever expanding group of polymer, metallic and ceramic matrix composites. Modern aero-engines are also using new manufacturing techniques such as Resin Transfer Moulding (RTM) for the construction of large multi-ply composite components (Backman 1992).

Polymer and composite usage in gas-turbines is growing at an increasing rate as polymers with better stability at raised temperatures become available (Ecklund 1994). Reduction in weight is the main reason for the use of composites, but other properties have made composites popular as metal substitutes: inherent damping thus reducing engine vibration and high strengths possible with suitably orientated fibres.

With the exception of the fan and nozzle most of the modern aeroengine seen from the outside is composite. Engines such as the Rolls-Royce Trent extend the use of composites into the core and in smaller engines, typified by the Tay and the BMW-RR 700, the bypass duct is a complex composite part supporting the engine in the airframe so showing that composites can be structural parts (Dominy 1994).

Though lauded as highly attractive for use in many fields, the fact that composites can be constructed using an infinitely large combination of resins, fibres, orientations and sequences leads to the design process being more complex than first appears. This points towards FEA being useful in composite design through the reduced need for physical prototypes. Analysis of composite components begins at an earlier stage as manufacturing methodology has to be considered when finalising architecture as well as required strength etc. (Mills 1991).

Applying material properties to the model along with orientations and stack sequences is a problem encountered in the FEA of composites (Mills 1991). This becomes more complex the larger the composite component modelled. Selection of elements is also a problem. Specialist elements reduce expensive computing time and produce results not achievable with regular elements e.g. interlaminar shear. Most laminates are used in shell-like structures so shell elements are used; however solid or brick elements exist for more complex shapes.

This shows that composite analysis is feasible but is carried out in a rather ad hoc way. Most FEA software producers now include some type of composite element but the methods of attaching the correct data to these elements are not fully developed. This is borne out by the situation with ABAQUS™. HKS have recently introduced a composite brick (to complement their composite shell) but a considerable volume of information is required to define the properties of each layer with respect to global directions.

Both CAD and FEA are widely used but the programs rarely exchange information well, resulting in analysts having to recreate geometry and designers revising component designs to fit optimised FEA models. This is costly in time and money and the problems of integration become even more of a concern in high technology industries such as aerospace, where a company may have many software packages involved in the system.

There are three general approaches to solving the problem of integration (Rankin 1992). These are:

a) *Direct approach*: a custom designed interface translates model geometry information from one database to a format that is readable by another.

b) *Integrated approach*: a comprehensive set of CAD/CAE/CAM packages is developed to run from one database; this eliminates the need to interface with other packages.

c) *Open approach*: data is read from and written to a neutral format that has its specification controlled by an independent organisation, allowing transfer of data from package to package.

Within the last two decades the field of numerical techniques has expanded vastly with applications relating to many fields such as aerospace, mechanical and civil engineering (Basu and Kumar 1995). The two most useful are considered to be Finite Element Methods (FEM) and Boundary Element Methods (BEM). It is thought that it may be easier to pre-process a FEM or BEM model in a CAD system with meshing capabilities and then move the model to other systems for analysis.

Spaindour et al (Spaindour 1991) emphasise in their discussion on the integration of composite data into FEA that the key is getting complete and accurate data to predict the component behaviour and that composite design is a repetitive process suited to automation. However the optimum design may not be possible as it may include geometries or lay-ups that cannot be manufactured.

Translating a CAD model to FEA has yet to become a simple operation (Finkel 1988). A large amount of effort still goes into the decision concerning the path taken (Rankin 1992). On top of this the amount of data required for accurate analysis must be decided upon. The ease of revising these decisions to ensure accurate modelling has a major part in choosing the FEA code to be used with a CAD system and the method of the data transfer.

It can be seen that the integration of FEA with existing design systems has been examined in respect to many industries, with an aim to integrate the two functions as a unit or to build them both into an overall CAE (Computer Aided Engineering) system.

Integration, where the object being analysed is constructed from an isotropic material, appears to have be partly successful. Integration involving composites is far more complex due to the almost infinite range of possible laminate properties and consequently only limited success has occurred.

The ideal way forward would be a truly open approach in which design data is read and written in a neutral format so allowing any system to link with any other. However this results in huge and often unwieldy files as all eventualities must be covered. The approach of this work is to create a methodology for use, initially, between two systems (in this case FemGen™ and ABAQUS™) and then move it towards a more open format.

APPROACH

The information generated by FemGen™ (or other pre-processors) describes the physical shape of an object under investigation and generates a suitable mesh with the associated nodal and elemental information. However the material information that can be defined in FemGen™ is that for homogeneous isotropic materials such as aluminium or steel. This makes it impossible to create a model of a composite component without excessive simplification which will result in the model being unrealistic.

The component's material properties are dependent on fibre direction within each lamina and also the laminate stacking arrangement i.e. how many laminae are arranged at what angle to the orthogonal axes. Information such as E_1, E_2, E_3, υ_{12}, υ_{13}, υ_{23}, for each lamina are available from manufacturers or from simple laboratory tests and the local lamina axes (1,2,3) can be defined by use of the angle at which the fibres lie in relation to the global (x,y,z) axes.

The physical properties can be represented in a *MATERIAL module, the stacking arrangement in the *SOLID/SHELL SECTION module and the angle of the fibres by a *ORIENTATION module. The *MATERIAL module is easily added to the ABAQUS™ input file, but the fibre angle varies throughout the object and is needed for the other two modules.

The fibre angle can be defined relatively easily at various points such as edges and corners. These angles can be used to calculate the fibre angle at other points away from these references.

To evaluate the fibre angle at any point (in this case in each element) an interpolation method must be used. For the purposes of this work shape functions are being used by extending their traditional role of interpolating a field variable within an element to the new and novel role of interpolating material properties over a region of elements.

Shape functions are a convenient way of describing the behaviour of an element in terms of the variable value at the nodes of the element. It may seem easy to use global co-ordinates (x,y) to define the shape functions, but the expressions evolved can soon become unwieldy and complex to change for other element shapes. It is therefore easier to define a new (curvilinear) co-ordinate system (ξ,η) local to the element under investigation. This is illustrated in Figure 1.

Each side of the element is a quadratic parametric curve with nodes at each end and the mid point. The local axes are positioned so that the values of ξ and η vary from -1 to +1 from one side of the element to the other.

The geometry of the element can be described as follows by use of the nodal co-ordinates and the shape functions:

$$x(\xi,\eta) = \sum_{c=1}^{8} N_c(\xi,\eta)x_c = N_1(\xi,\eta)x_1 + N_2(\xi,\eta)x_2 + \ldots N_8(\xi,\eta)x_8 \quad (1)$$

$$y(\xi,\eta) = \sum_{c=1}^{8} N_c(\xi,\eta)y_c = N_1(\xi,\eta)y_1 + N_2(\xi,\eta)y_2 + \ldots + N_8(\xi,\eta)y_8 \quad (2)$$

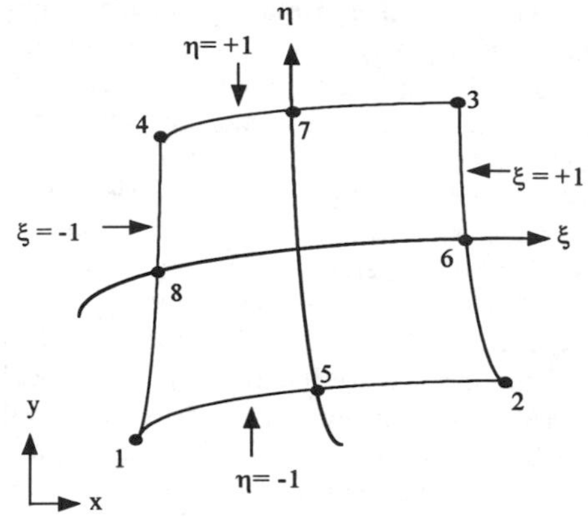

Figure 1. Element topology and axes used to describe quadrilateral elements.

Similarly other variables can be described the same way, such as for the purpose of this work fibre angle (α):

$$\alpha(\xi,\eta) = \sum_{c=1}^{8} N_c(\xi,\eta)(\alpha)_c \qquad (3)$$

where $N_c(\xi,\eta)$ are quadratic equations that must satisfy two criteria:

1) $N_c(\xi,\eta) = 1$ at c.
2) $N_c(\xi,\eta) = 0$ at all other nodes.

To obtain explicit expressions for the shape functions a typical quadratic expression is assumed:

$$N_c(\xi,\eta) = a_1 + a_2\xi + a_3\eta + a_4\xi^2 + a_5\eta^2 + a_6\xi\eta + a_7\xi\eta^2 + a_8\xi^2\eta \qquad (4)$$

From this eight equations can be written to calculate the unknowns so giving the following shape functions:

$$N_1 = \tfrac{1}{4}(1-\xi)(1-\eta)(-\xi-\eta-1) \qquad (5)$$
$$N_2 = \tfrac{1}{4}(1+\xi)(1-\eta)(\xi-\eta-1) \qquad (6)$$
$$N_3 = \tfrac{1}{4}(1+\xi)(1+\eta)(\xi+\eta-1) \qquad (7)$$
$$N_4 = \tfrac{1}{4}(1-\xi)(1+\eta)(-\xi+\eta-1) \qquad (8)$$
$$N_5 = \tfrac{1}{2}(1-\xi^2)(1-\eta) \qquad (9)$$
$$N_6 = \tfrac{1}{2}(1+\xi)(1-\eta^2) \qquad (10)$$
$$N_7 = \tfrac{1}{2}(1-\xi^2)(1+\eta) \qquad (11)$$
$$N_8 = \tfrac{1}{2}(1-\xi)(1-\eta^2) \qquad (12)$$

All this allows the interpolation of a variable within the boundaries of an element which is the traditional role of shape functions. However by using the corners and midsides of a region as the nodes the shape functions can be used to interpolate a variable across a region consisting of as many elements as required.

Equations (5) through to (12) (and adaptations of them) have been used to build the interpolation routines that allow the angle (and other variable) information to be assigned correctly to the relevant elements.

PROGRAM TRANSANGLES

This program aims to provide a system that creates an ABAQUS™ input file that has the correct composite information for the mesh generated over the object under analysis. This information includes lamina fibre directions and orientations, lamina thicknesses and material properties all referenced to the correct ELSETS, *SOLID/SHELL SECTION modules and *ORIENTATION modules.

The geometry of the model is created using FemGen™ and meshed using suitable element types and biasing the mesh in areas of particular concern. This meshed model is then outputted as a skeletal ABAQUS™ input file. The same geometry is used to create an outline of the object being analysed and this is divided up into regions (termed ParaBlocks) along lines of sub-assemblies or areas of similar composite lay-up.

In its simplest terms a ParaBlock can be thought of as a element set within which the lamina lay-up does not vary too greatly. The ParaBlock covers an area of the object being analysed where the fibre angles are known at certain points.

Two differing subroutines can used to interpolate the fibre angle, the choice being made by the operator. One routine, the nodal interpolation method, interpolates from the angles defined at the corners and mid-side points of the ParaBlock arranged in the same manner as an eight noded quadrilateral element. At these points fibre angles and the lamina material names are defined. The other routine, the parallel edge method, works on the assumption that the fibres lie parallel to the edges of the ParaBlocks and estimates the orientation of the

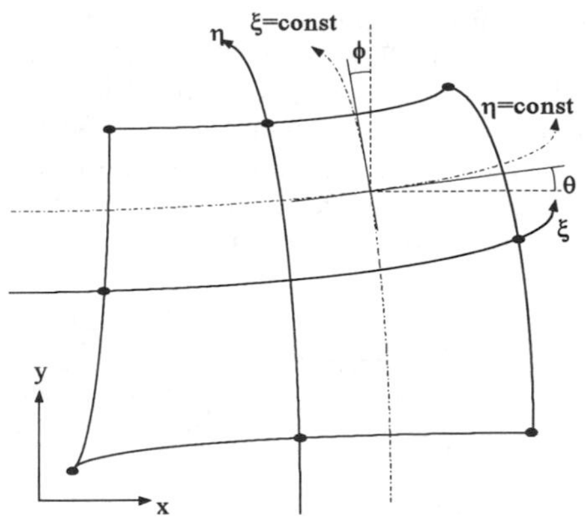

Figure 2. Angles calculated in the parallel edge interpolation method.

fibres at the element centres to be the angle between the ξ and x axes or the η and y axes, for example,

$$\theta = \tan^{-1}\left(\frac{\partial y / \partial \xi}{\partial x / \partial \xi}\right) \quad \text{and} \quad \phi = \tan^{-1}\left(\frac{-\partial x / \partial \eta}{\partial y / \partial \eta}\right)$$

These angles are shown in Figure 2.

Using the translation program the fibre direction information is interpolated from the ParaBlocks to the centre point of each element and assigned to the relevant *SOLID/SHELL SECTION and *ORIENTATION modules. This allows the physical properties to be applied in the correct direction for that point in the model and an accurate analysis to be carried out. Figure 3 shows a flow chart the program known as TRANSANGLES.

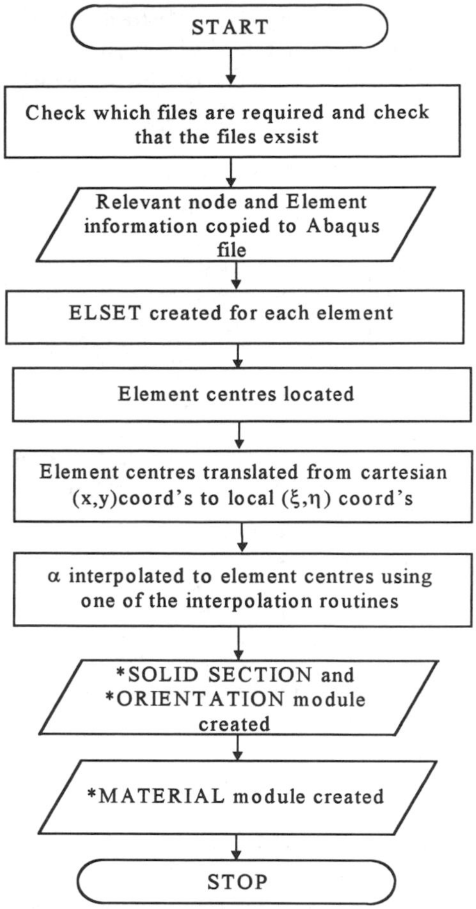

Figure 3: Stages of the program TRANSANGLES.

VERIFICATION

To verify the interpolation methods and the program TRANSANGLES work has been carried to compare the results obtained from a model that has had its material data added by the program TRANSANGLES via the nodal interpolation method and the parallel edge method.

In addition to this a comparison has been carried out between the FEA results obtained from a model of a sewage distributor column created by FemGen™ and TRANSANGLES and those generated in a previous analysis of this structure (Corden 1997).

The column geometry was used to generate a two dimensional axisymmetric model in FemGen™ which was then meshed. This model was then used as the input for the program TRANSANGLES, which was run using both interpolation methods.

The ParaBlocks were set up to divide the model into five areas; three of these have almost constant fibre angles and the remaining two represent the transition between the previous three.

The same geometry that was used to create the model was used to generate the nodal information for the ParaBlocks and the same ParaBlocks were used in both interpolation methods. Figure 4 shows a detail of the base of the column

Table 1 shows a comparison of the fibre angles generated via the two interpolation methods used in the program TRANSANGLES for a selection of the element in the meshed column model (figure 4).

Generally the correlation between the angles is very good. In ParaBlocks 1, 3 and 5 the correlation is particularly close with angles agreeing to within

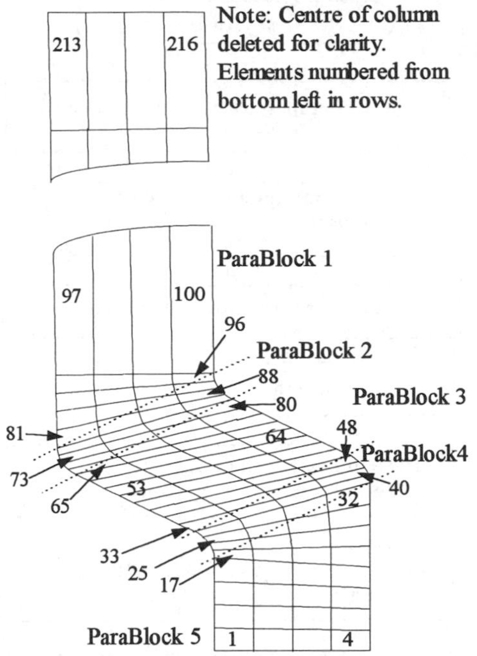

Figure 4. Detail of the base of Sewage Distributor Column, showing mesh.

1.09° (1.81 %). This should be expected as these blocks are very regular and the parametric description of the edges matches the geometry closely.

In ParaBlocks 2 and 4 the correlation is not as close, agreeing to within 3.85° (10.28 %). This is thought to be due to the fact the parametric description of the Blocks does not match exactly to the mesh geometry (parametric description uses only three points to describe a curve) so leading to interpolation from a curve that does not exactly match the edges of the area. A method of reducing this error would be to use a higher order of ParaBlock, such as a 12 noded cubic ParaBlock, but this would increase the complexity of the program and lead to higher computing cost. However this error is considered to be acceptable as the angle actually described by the fibre in the component can vary +/- 5° from that designated in the manufacturing instructions.

Table 2 shows a comparison between the tip deflections generated through the use of differing methods when the structure is loaded laterally at the tip with a load of 2.2 kN. It can be seen that the deflection produced from the analyses agree very closely to the experimental (-0.62%, +0.31%). This suggests that the interpolation methods used in the program TRANSANGLES produces models that are acceptable for the analysis of two dimensional axisymmetric structures. It should be noted that the simple beam theory result used a simplified structure and was carried out to obtain a guide result to check the acceptability of the other results.

Table 1. Comparison of fibre angles assigned to elements by interpolation methods.

Element	Nodal method (degrees)	Parallel edge method (degrees)
1	0.00	0.00
4	0.00	0.00
17	0.00	0.00
25	26.86	26.59
32	0.00	0.00
33	60.00	59.40
40	34.00	32.64
48	60.00	58.97
53	60.00	59.32
64	60.00	58.91
65	60.00	59.29
73	33.60	37.45
80	60.00	58.85
81	0.91	0.91
88	28.66	31.52
96	0.91	0.94
97	0.91	0.91
100	0.91	0.94
213	0.91	0.91
216	0.91	0.94

Table 2. Comparison of column tip deflections.

Analysis Method	Tip Deflection (mm)	Percent Difference
ABAQUS (nodal method)	3.19	-0.62
ABAQUS (parallel edge method)	3.19	-0.62
Patran (Corden 1997)	3.22	+0.31
Simple Beam theory	3.63	+13.1
Experimental (Reference)	3.21	+/-0.0

PROGRAM ASSIGN.

As an alternative to the approach described in Program TRANSANGLES the following system has been developed. Rather than define properties at all the nodes of each ParaBlock it has been decided to define the laminate properties at a single normal in

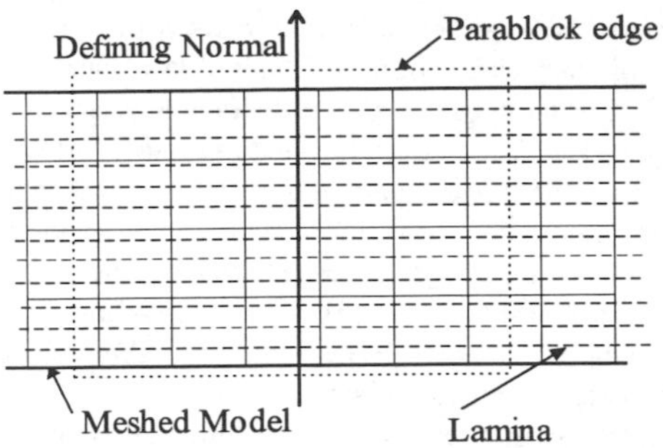

Figure 5. Example of Defining Normal in the program ASSIGN.

each ParaBlock (termed the defining normal) and interpolate the information from this to each element in the ParaBlock abiding by the design rules used in the components construction. Figure 5 shows a simple example of how the defining normal is set up.

The system starts with a meshed component as in the program TRANSANGLES and the ParaBlocks are set up to represent areas of similar composite construction. The centre line of each block is defined as a surface normal and this is taken to be the Defining Normal (DN) for that block with respect to composite properties. The elements through which the DN passes are then located and the points of intersection between the DN and element edges are calculated. This allows the thickness of the material to be calculated at the DN. By defining the materials present at the DN a description of the material present in each ParaBlock is now available.

By locating each element in relation to its parent ParaBlock (the ParaBlock in which the element is located) and calculating the surface normal of the element it is possible to evaluate the component thickness at the location of the element and the element thickness.

By comparing the defining normal thickness and the component thickness at the element under question and using composite design rules the composition at the element normal can be interpolated. By dividing up the normal's composition pro rata between the elements upon which it lies the composition of the element in terms of what laminae (or packs of laminae) lie within it can be evaluated.

The ABAQUS™ two dimensional axisymmetric elements have no facility for the definition of composite laminae, so the combined laminae properties have to be "smeared" over the element. This smearing will result in approximate properties being assigned, however, these should be sufficient to give adequate results with respect to global behaviour such as stiffness of a structure, provided the mesh is reasonably fine. Another limitation of this smearing is that it ignores factors such as interlaminar shear which laminated elements can account for.

To enable the "smearing" to be carried out the individual lamina properties have to be resolved into a single laminate via laminate analysis theory so giving E_1, E_2, υ_{12} etc. for the laminate. This provides the engineering constants required for the *SOLID SECTION and *MATERIAL modules but does not resolve in which direction these constants should be applied.

The fibre angle is assigned using a similar routine to that used in TRANSANGLES, working on the assumption that the fibres lie parallel to the edges of the ParaBlocks, it estimates the angle of the fibres at the element centres to be the angle subtended by the x and ξ axis or the angle subtended by the y and η axis (see figure 2). Figure 6 shows the sections of program ASSIGN.

PRESENT STATUS

At the present time the program TRANSANGLES interpolates composite information into an ABAQUS™ input file which is used to run analyses of 2D axisymmetric structures. The program may use either of the interpolation methods: interpolating from values assigned at node or interpolating from the fibres lying parallel to the sides of the regions.

The program ASSIGN which interpolates laminate architecture information from the central normal of the ParaBlocks to the centre point of each element is at present in the developmental stage and tenable results are not available.

FURTHER WORK

The next stage of the work under way is to complete the development of the program ASSIGN and then carry out testing of the program.

Though useful results have been obtained from the two dimensional models that the programs can create, composite components are rarely simple axisymmetric structure such a pipes etc. Therefore, the development of systems that can cope with truly three dimensional structures is to be undertaken. The first step will be to move the system on to 3D shells. These should present fewer problems than bricks as

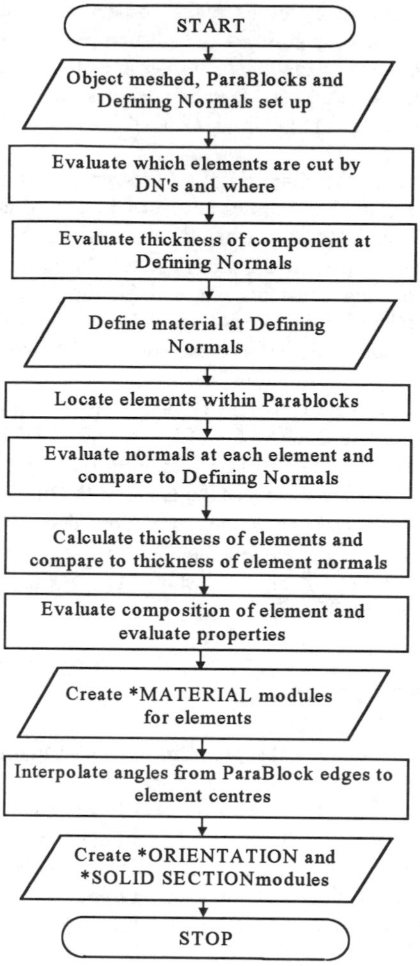
START
Object meshed, ParaBlocks and Defining Normals set up
Evaluate which elements are cut by DN's and where
Evaluate thickness of component at Defining Normals
Define material at Defining Normals
Locate elements within Parablocks
Evaluate normals at each element and compare to Defining Normals
Calculate thickness of elements and compare to thickness of element normals
Evaluate composition of element and evaluate properties
Create *MATERIAL modules for elements
Interpolate angles from ParaBlock edges to element centres
Create *ORIENTATION and *SOLID SECTIONmodules
STOP

Figure 6. Stages of the program ASSIGN.

the laminae are constrained to lie in the plane of the element so reducing the information required to define the orientation. Also, as a further initial simplification the properties of the laminate will be approximated as an equivalent set of properties covering the whole element rather than defining each individual lamina (a facility that exists in ABAQUS^{TM}), although definition of each lamina is another aim of the continuation of this work. The work then will continue towards creating a system that can define fully 3D laminated bricks.

CONCLUSIONS

It has been shown that the integration of design information into FE analysis is an area that has advanced considerably in recent years with respect to isotropic materials, but is still limited in its success in the field of composites. Methods have been proposed and developed to integrate the required information into an FEA model which has given encouraging results comparing favourably with those previously produced. The programs developed by the authors can be used to produce two-dimensional axisymmetric models containing all the relevant composite laminate data required to analyse applicable structures. There is still further work to be carried out to expand the methods into a fully three dimensional system capable of generating realistic models of complex components.

ACKNOWLEDGEMENTS

The authors gratefully acknowledge the support to one of the authors (TJA) from the University of Nottingham via a University Studentship and the guidance given by Dr. R Lewin and Mr. R. Bailey of Rolls-Royce Plc.

REFERENCES

Basu, D; Kumar, S.S. (1995) "Importing mesh entities through IGES/PDES" *Advances in Engineering Software*. vol. 23, no. 3. pp. 151-161.

Corden, T.J. (1997) "Development of Design and Manufacturing Techniques for Glass Reinforced Plastic Waste Water Treatment Equipment", PhD thesis, University of Nottingham, UK.

Dominy, J. (1994) "Structural composites in civil gas turbine aero engines", *Composites Manufacturing*. vol. 5, no. 2. pp. 69-72.

Ecklund, R. (1994) "Polymer and Composite use in Gas Turbine Engines", *Flight-Vehicle Materials, Structures and Dynamics - Assessment and Future Directions*. vol. 2, chap. 3, section 1. pp. 385-394.

Finkel, J.I. (1988) "Appropriate levels of CAD Translators for Finite Element Analysis", *1988 ASME Pressure Vessels and Piping Conference, Pittsburgh PA, June 1988*. pp. 7-12.

Mills, R. (1991) "FEA for Composites", *Computer-Aided Engineering*. vol. 10, no. 8. pp. 30-36.

Rankin, J.J; Ott, D.A. (1992) "The Open Approach to FEA Integration in the Design Process", *Mechanical Engineering*. vol. 114, no. 9. pp. 70-75.

Spaindour, L.K.; Rasdorf, W.J; Patton, E.M; Burns, B.P; Collier, C.S. "A Computer-Aided Analysis System with DBMS Support for Fiber-Reinforced Thick Composite Materials", *Engineering Databases: An Engineering Resource*. ASME 1991, pp. 37-48

Modern Practice in Stress and Vibration Analysis, Gilchrist (ed.) © *1997 Balkema, Rotterdam, ISBN 90 5410 896 7*

Identification of hydrodynamic torque converter controlled by physical properties of working fluid

Z. Kęsy, A. Kęsy & J. Madeja
Department of Mechanical Engineering, Technical University of Radom, Poland

ABSTRACT: This paper presents the results of identification of a hydrodynamic torque converter controlled by change of density and viscosity of working fluid. Three first-order nonlinear, nonstationary differential equations describing a hydraulic torque converter working in a power transmission system are used. This mathematical model is verified by experimental tests. Next, the equations are linearized and considered as a multiple degree of freedom lumped parameter system. Division of signals into inputs and outputs is shown. Linear model equations are formed as transfer functions. Finally, the coefficients of the transfer functions are identified.

1 INTRODUCTION

Automatic transmissions are very popular in many types of vechicles. A hydrodynamic torque converter (HTC) is its unavoidable integral part of automatic transmissions. Its chief advantage is an ample capacity for an automatic stepless change of the torque multiplication according to a change of load. In addition, the HTC provides vibrational isolation by absence of mechanical connection between input and output shafts.

The HTC characteristics describing the automatic change for typical design solutions are stable. This disadvantage reduces the benefits from hydrodynamic torque converter application in power transmission systems. There are a few ways of changing HTC's characteristics, e.g. by adjustment of guide blades, by exchanging whole impellers, by throttling fluid flow (Voith 1990), but the results are still rather poor because of a decrease in efficiency of the transmission system and more complicated HTC construction.

Because HTC impellers are not connected mechanically, but with the fluid flow in the flow area, so it seems that the optimal way of the change in hydrodynamic torque converter characteristics is control of working fluid properties (density and viscosity) (Kęsy 1993).

The present investigation on the development of controllable mechanical devices focuses on the utilization of electromagnetic and electrorheological fluids. The fluids change their viscosity under the electric or magnetic field. In addition, electromagnetic fluid can change its apparent weight under the magnetic field gradient.

Application of electromagnetic and electrorheological fluids as an HTC current-varying working fluid could allow to control torque transmission or angle velocities of shafts by an active closed-loop system (Kęsy 1995).

The first step for the design of such a control system is to describe the influence of density and viscosity change on HTC characteristics by some mathematical equations. Next, the mathematical model should be identified by comparing the model solution to experimental data.

The purpose of this paper is to present an HTC with working fluid changing its density and viscosity in time as a typical element of a control system by using identification methods.

2 MATHEMATICAL MODEL

A mathematical model of a power transmission system with an HTC including density and viscosity of working fluid has been derived based on a model of the one-dimensional flow in an HTC flow area. The following additional assumptions were added to typical ones: density and viscosity of a working fluid do not depend on pressure, are functions of time and are the same in the whole HTC flow area, rotating disc drag torques are considered. In order to derive mathematical equations the transmission system is broken down

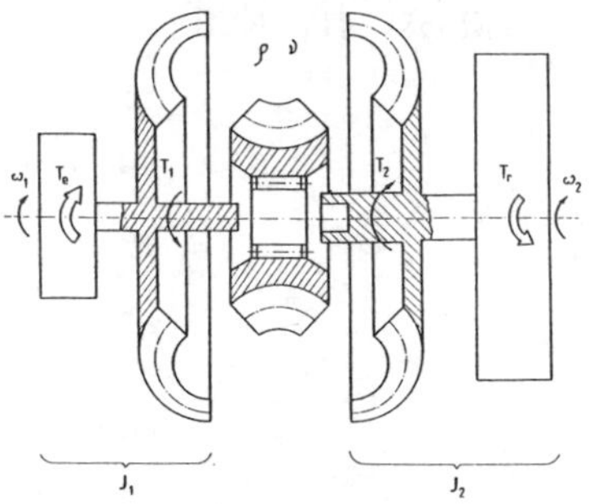

Fig.1. Scheme of a transmission system with a 3-member HTC relevant to the derivation of the dynamic equations of motion.

into two parts - the engine with an HTC pump, and components joined with the HTC turbine. The two parts are connected by means of fluid rotating in the flow area of an HTC which impresses torques equal T_1 and T_2 on the pump and the turbine, respectively. The equation of the non-steady state motion of the transmission system schematically shown in Fig.1 can be written as:

$$T_e = J_1 \dot{\omega}_1 + T_1 \quad (1)$$

$$T_r = - J_2 \dot{\omega}_2 + T_2 \quad (2)$$

where T_e = engine torque; T_r = resistance torque; ω_1= angular velocity of pump; ω_2 = angular velocity of turbine; J_1 = inertia moment of pump; J_2 = inertia moment of turbine.

The T_1, T_2 -torques described by n-index for n = 1, 2 may be divided into the static part T_{no} and the dynamic part T_{nd} (including derivatives):

$$T_n = T_{no} (\omega_1, \omega_2, Q, \rho) +$$

$$T_{nd} (\omega_1, \omega_2, Q, \rho, \dot{\omega}_1, \dot{\omega}_2, \dot{Q}, \dot{\rho}) \quad (3)$$

where Q = axial torus volume flow; ρ = working fluid density.

On the basis of physical interpretation of motion, density of working fluid ρ and its derivative ρ are independent variables and volume flow Q is the third dependent variable. The third equation follows from the power balance of the entire HTC: the net power supplied by the shafts is equal to the total power dissipated. There are two major contributors to the total power dissipation i.e., the flow losses and the shock losses. For a non-steady state motion the total power dissipated includes, additionally, the power wasted on acceleration of the fluid in the HTC impellers and in the gaps between the impellers. The equation is described as follows:

$$P_o (\omega_1, \omega_2, Q, \rho, \nu) +$$

$$P_d (\omega_1, \omega_2, Q, \rho, \dot{\omega}_1, \dot{\omega}_2, \dot{Q}, \dot{\rho}) = 0 \quad (4)$$

where ν = working fluid viscosity.

The form of equations (3), (4) depends on formulae describing the rotating disc torque and the power losses. Finally, the mathematical model as the set of non-linear differential equations (1), (2), (3), (4) can be written in matrix notation as follows:

$$\mathbf{V} \dot{\mathbf{x}} = \mathbf{z} \quad (5)$$

where **x** = dependent variable vector, **V** = matrix of inertia, **z** = vector of the other quantities. The numerical coefficients in equation (5) can be obtained based on technical documentation of the HTC impellers and the transmission system.

At first stage of the analysis of the influence of density and viscosity of a working fluid on HTC characteristics the mathematical model (5) is too complicated. The model can be simplified by neglecting: gaps between impellers, inertia of working fluid moving in relative direction and the density derivative. When the simplifications are taken into consideration, the number of model equations and the number of variables in the equations are reduced. The simplified equations describing torques T_1 and T_2 are static and can be written as :

$$T_1 = \lambda \rho D^5 \omega_1^2 \quad (6)$$

$$T_2 = k \lambda \rho D^5 \omega_1^2 \quad (7)$$

where λ = torque coefficient; D = external pump diameter; k = torque ratio.

Now, the mathematical model of a power transmission system with an HTC including density and viscosity of working fluid is the set of equations (1), (6), (7). The mathematical model is still non-linear. Analysis of linear models is very important from the point of view of the design of the control system. The main advantage of linear models is the fact that they can be solved algebraically. Thus, the simplified mathematical model was linearized for small perturbations around a certain steady-state operating point. Therefore equations (6) and (7) were written in the symbolic form convenient for linearization:

$$f_n = T_n(\omega_1, \omega_2, \rho, \nu) \quad (8)$$

The f_n- function was expanded by using Taylor's series around the steady state operating point denoted here by index " o ":

$$f_n = f(z_o) + \sum_{j=1}^{4} \left(\frac{\partial f_n}{\partial z_j} \right) \Delta z + r_n \quad (9)$$

where $z = \omega_1, \omega_2, \rho, \nu$ for $j = 1 - 4$, respectively; $\Delta z = z - z_o$; r = nonlinear rest. For a steady-state operating point with assumption that $r = 0$ equation (9) gives:

$$\Delta T_n = L_{n1}\Delta\omega_1 + L_{n2}\Delta\omega_2 + L_{n3}\Delta\rho + L_{n4}\Delta\nu \quad (10)$$

where Δ = perturbation; L_{nj} = partial derivatives.
Finally, combining linearized equations (1), (2) and equation (10) linear model of a power transmission system with HTC is obtained as follows:

$$(J_1 p + L_{11})\Delta\omega_1 + L_{12}\Delta\omega_2 =$$

$$\Delta T_e - L_{13}\Delta\rho - L_{14}\Delta\nu \quad (11)$$

$$- L_{21}\Delta\omega_1 + (J_2 p - L_{22})\Delta\omega_2 =$$

$$- \Delta T_r + L_{23}\Delta\rho + L_{24}\Delta\nu \quad (12)$$

where $p = d/dt$ = differential operator.

3 TRANSFER FUNCTIONS

Because the number of output quantities has to be equal to the number of equations it was assumed here that $\Delta\omega_1$, $\Delta\omega_2$ are the output quantities. Now, solution of equations (11), (12) under condition $\Delta\rho = 0$, $\Delta\nu = 0$ can be written in matrix form as:

$$\boldsymbol{\omega} = \mathbf{G}_{nj}\,\mathbf{t} \quad (13)$$

where $\mathbf{G}_{nj}$ = transformation matrix (n = 1, 2 - the number of output quantities; j = 1, 2 - the number of input quantities); $\boldsymbol{\omega} = [\Delta\omega_1 \;\; \Delta\omega_2]^T$; $\mathbf{t} = [T_e \;\; T_r]^T$. Elements of $\mathbf{G}_{nj}$ transformation matrix were expressed as:

$$G_{11} = (J_2 p - L_{22})/M, \quad G_{12} = L_{12}/M$$

$$G_{21} = L_{21}/M, \quad G_{22} = (-J_1 p - L_{11})/M \quad (14)$$

where $M = J_1 J_2 p^2 + (J_2 L_{11} - J_1 L_{22})p + (L_{12} L_{21} - L_{11} L_{22})$ = the characteristic equation.
It should be noted here that G_{11}, G_{22} -transfer functions describe the differentiating elements and G_{12}, G_{21} -transfer functions describe the non-periodic elements of a linear second-order system.
Solutions of equations (11), (12) for ΔT_e, $\Delta T_r = 0$ can be presented with the help of $\mathbf{G}_{nj}$-transformation matrix as:

$$\boldsymbol{\omega} = \mathbf{G}_{nj}\,\mathbf{L}_{ni}\,\mathbf{u} \quad (15)$$

where $\mathbf{L}_{ni}$ = transformation matrix (n = 1, 2, i = 3, 4); $\mathbf{u} = [\Delta\rho \;\; \Delta\nu]^T$.
Elemets of $\mathbf{L}_{nj}$ transformation matrix are equal to patrial derivatives L_{nj}, respectively.
Solutions of equations (11), (12) for all input quantities can be easilly obtained from equations (13), (15) with the help of the superposition method.

4 IDENTIFICATION

The identification study has been undertaken based on experimental investigations with a test rig for modelling a power transmission system with PH1 280 1 HTC. This was made by comparing experimental and theoretical step function responses.

4.1 *Partial derivatives*

Dimensionaless characteristics of the HTC are shown in Fig. 2. PH1 280 1 has external pump diameter D equal to 0.28 m.

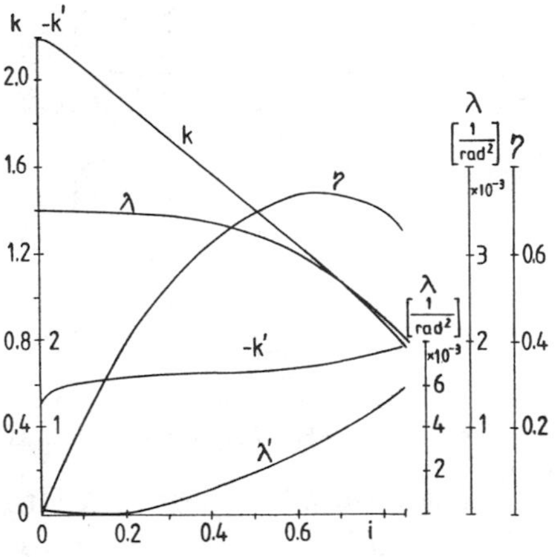

Fig. 2. Diagram of experimental dimensionless characteristics of PH1 280 1 HTC.

Partial derivatives L_{nj} were calculated for an established steady state operatin point. The operating point is determined by speed ratio $i = \omega_{2o} / \omega_{1o}$ and T_{eo} .

Values of L_{nj} for $n = 1,2$ and $j = 1,2,3$ were calculated with values k, k', λ, λ', η read out for a chosen i from diagram (Kęsy 1988) shown in Fig.2 as:

$$L_{11} = P (2- i \lambda' / \lambda), \qquad L_{12} = P (\lambda' / \lambda)$$

$$L_{21} = P (2 k -i k' - \eta \lambda' / \lambda)$$

$$L_{22} = P (k' + k \lambda' / \lambda) \tag{16}$$

$$L_{13} = T_{eo} / \rho_o, \qquad L_{23} = k T_{eo} / \rho_o$$

where $P = \sqrt{T_{eo} \lambda \rho_o D^5}$; $\lambda' = \partial\lambda / \partial i$; $k' = \partial k / \partial i$.

Values of L_{nj} for $n = 1,2$ and $j = 4$ cannot be calculated as the other values, because there is no ν in equations (6). In order to calculate the derivatives theoretical considerations and experiments were performed on the influence of working fluid viscosity on PH1 280 1 dimensionless characteristics. Some examples of the results are presented in Fig. 3.

It follows from this figure that the λ-curves practically do not depend on ν, especially for $i < 0.7$.

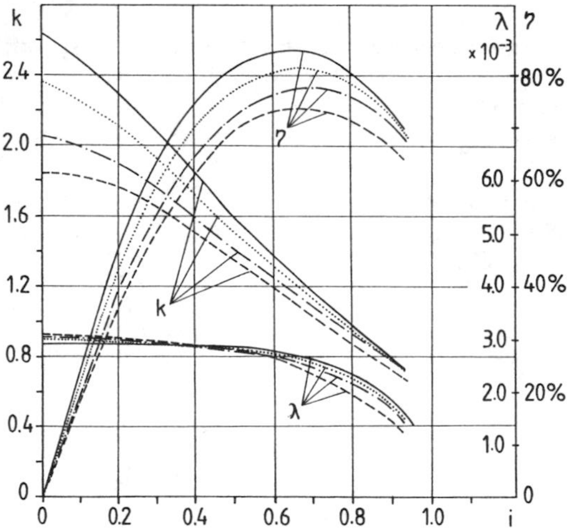

Fig.3. Influence of ν on PH1 280 1 dimensionless characteristics;

________ - $\nu = 0.8 \times 10^{-6}$ m^2/s,

.... - $\nu = 4 \times 10^{-6}$ m^2/s,

-.-.-. - $\nu = 20 \times 10^{-6}$ m^2/s,

------ - $\nu = 100 \times 10^{-6}$ m^2/s.

Because of that, it was assumed in the study that $L_{14} = (\partial T_1/\partial\nu)_o = 0$. Partial derivative L_{24} can be determined under this condition as:

$$L_{24} = (\partial T_2 / \partial\nu)_o = (\partial k T_1 / \partial\nu)_o =$$

$$T_{1o} (\partial k / \partial\nu)_o \tag{17}$$

Partial derivative $(\partial k/\partial\nu)_o$ was calculated for i = constant with k-curves obtained for different values of ν_o. See Fig. 4.

4.2 *Identification results*

In order to identify the linear mathematical model (15), first the non-linear mathematical model was verified by comparing the solutions of equation (5) obtained by means of the fourth order Runge-Kutta method with the experimental investigation results. An example of the verification results is presented in Fig. 5.

Generally, the relative errors, defined as the ratios of the differences between values ω_1 and ω_2 for the mathematical model and for the test, calculated for each time instant, were less than 20%. The result can be considered as good in HTC practice.

Next, step function responses obtained by algebraic solution of equation (15) have been compared with step function responses calculated numerically from equation (5).

The reason for this was a dificulty in keeping T_e and

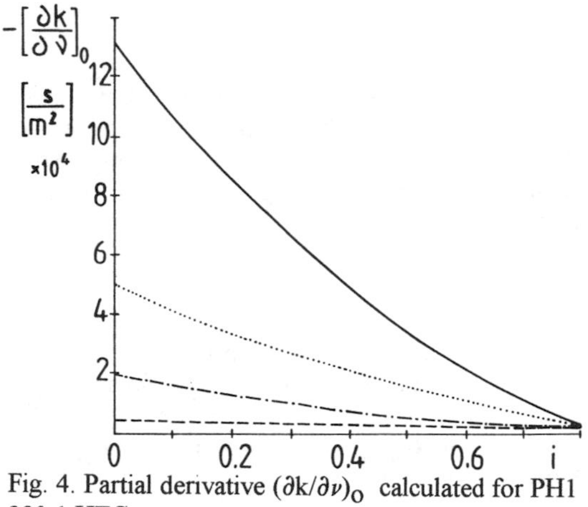

Fig. 4. Partial derivative $(\partial k/\partial\nu)_o$ calculated for PH1 280 1 HTC;

________ - $\nu_o = 0.8 \times 10^{-6}$ m^2/s,

.... - $\nu_o = 4 \times 10^{-6}$ m^2/s,

-.-.-.- - $\nu_o = 20 \times 10^{-6}$ m^2/s,

-------- - $\nu_o = 100 \times 10^{-6}$ m^2/s.

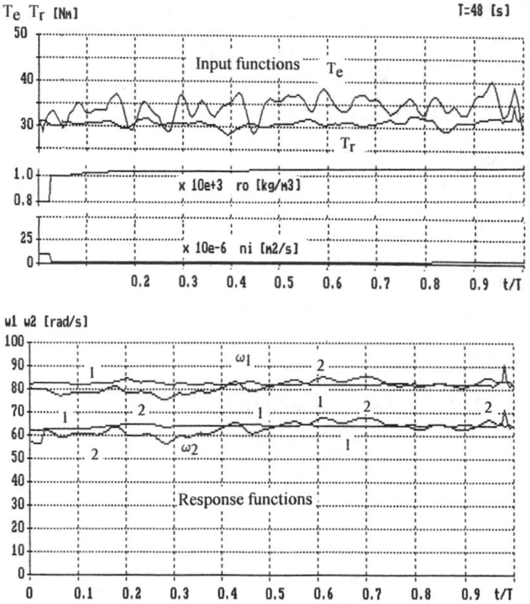

Fig. 5. Results of verification for step change of ρ and ν; 1 - test results, 2 - solutions of non-linear mathematical model (5).

T_r -torques constant during the test. Some examples of the identification results for $J_1 = 0.2$ kgm^2, $J_2 = 3.5$ kgm^2 are presented in Fig. 6 - Fig. 9.

4.3 *Discusion*

It can be concluded from these results that in spite of a big step change in ρ and ν values, the step function responses do not vary much.

Differences between values of the pump angular speeds (as seen in Fig. 9, ω_1 decreases monotonically for solution of equation (5) and increase monotonically for solution of equations (15)) are caused by assumption λ = constant for $i > 0.7$.

Generally, for practical application the ρ-change could be less than 20% and the ν-change could be less than 50 times.

It is readily obtained from equations (15) and (16) that G_{13} , G_{23} -transfer functions positioned by $\Delta\rho$ can be written symbolically as:

$$G = I + k D \tag{18}$$

where I = transfer function of a non-periodic element; D = transfer function of a differentiating element.

This equation shows that the influence of differentiation rises proportionally to k (inversely proportionally to i). The influence is stronger for the turbine angular speeds, thus in Fig. 6 and Fig. 7 angular speed ω_2 first drops and then rises. The speed drop is bigger in Fig. 6 than in Fig. 7 because i-value is smaller. Generally, it was found that the speed drop increases if J_2 / J_1 -ratio is decreased.

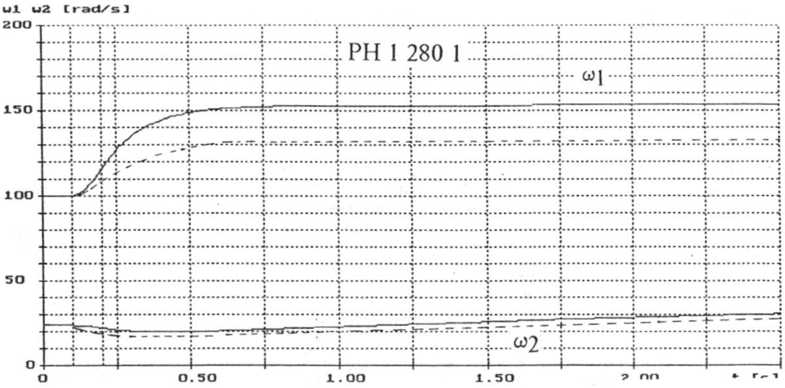

Fig. 6. Identification results for step change of ρ from 2000 kg/m^3 to 850 kg/m^3 for i = 0.30 and T_{eo} = 120 Nm; ________ - solution of equation (5), ------ - solution of equation (15).

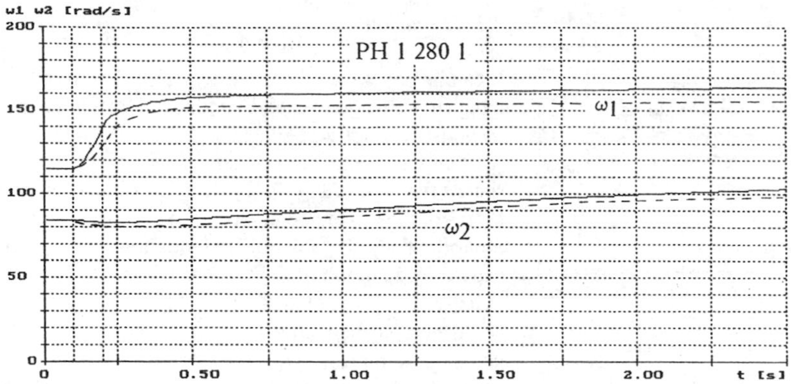

Fig. 7. Identification results for step change of ρ from 2000 kg/m^3 to 850 kg/m^3 for i = 0.75 and T_{eo} = 120 Nm; ________ - solution of equation (5), - - - - - - - solution of equation (15).

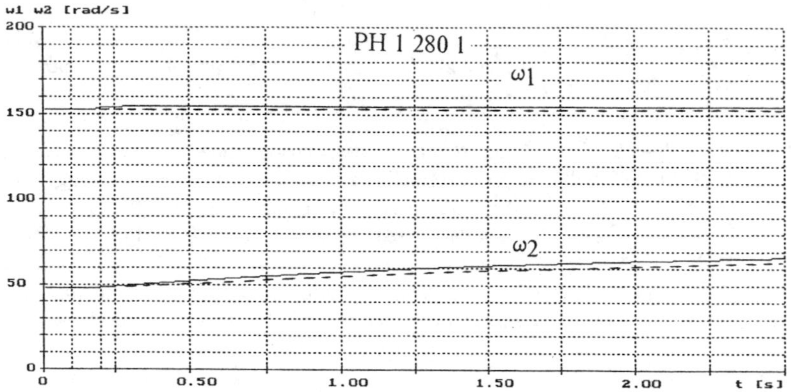

Fig. 8. Identification results for step change of ν from 120×10^{-6} m^2/s to 0.5×10^{-6} m^2/s for i = 0.30 and T_{eo} = 120 Nm; ________ - solution of equation (5), - - - - - - - solution of equation (15).

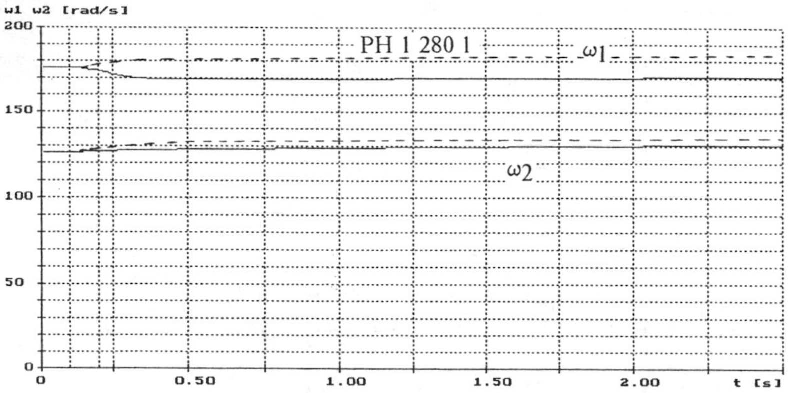

Fig. 9. Identification results for step change of ν from 120×10^{-6} m^2/s to 0.5×10^{-6} m^2/s for i = 0.75 and T_{eo} = 120 Nm; ________ - solution of equation (5), - - - - - - - solution of equation (15).

5 CONCLUSIONS

A power transmission system including an HTC with working fluid changing its density and viscosity in time has been presented as a typical element of a control system by using methods of identification in order to simplify a control system analysis.

It was found that the trasfer functions corresponding to $\Delta\rho$ and $\Delta\nu$ -input quantities can be obtained with the transfer functions corresponding to ΔT_e and ΔT_r -input quantities. The conclusion is very beneficial to the analysis of transient phenomena because there have been a number of theoretical and experimental studies on the non-stationary motion of a power transmission system with an HTC caused by torque changes. Now, the results of those studies can be used.

Also, damping effects caused by density and viscosity changes can be studied in the same way in order to define damping properties of an HTC working in a power transmission system.

Full identification of the mathematical models considered should be undertaken in theoretical and experimental investigations for sinusoidal input signals that occur in control practice.

REFERENCES

Kęsy, Z. & Kęsy, A. 1988. *Modelling of a torque converter used in automotive transmission.* Dissertation. Technical University of Łódź. (In Polish).

Kęsy, Z. & Kęsy, A. 1993. *Ferrofluid as a working fluid of a hydrodynamic torque converter.* XVI Symposion of Machine Elements Design. Szczyrk. Poland. (in Polish).

Kęsy, Z. 1995. *Prospects for control of torque converter using magnetic fluid. Colloquium on Innovative actuators for mechatronic systems.* IEE Grup 16. Londyn.

Voith GmbH. 1990. *Hydrodynamic power transmission.* Brochure.

Modern Practice in Stress and Vibration Analysis, Gilchrist (ed.) © *1997 Balkema, Rotterdam, ISBN 90 5410 896 7*

Experimental and numerical studies of creep crack growth in filled adhesive systems

A. D.Crocombe
Department of Mechanical Engineering, University of Surrey, Guildford, UK
G.Wang
Department of Mechanical Engineering, East China Institute of Metallurgy, Anhui, People's Republic of China

ABSTRACT: Bulk adhesive compact tension specimens have been subjected to sub-critical creep loads. Measurements of crack growth, load-line displacement and crack tip profile have been taken. Tensile creep tests have been carried out on the same material and the results used to generate models for both steady state power law (SSPL) creep and finite element analyses. Standard solutions for SSPL creep of compact tension specimens have been used to calculate fundamental parameters such as C* and transition time. From these data it would seem likely that the crack growth is controlled by creep, rather than elastic, singular fields. Finite element analyses have been carried out for both a stationary crack and a propagating crack. The former are relevant to the incubation period and allow the redistribution of stress and accumulation of creep strain to be assessed. The latter enable the predicted conditions in the domain surrounding the propagating crack to be investigated for appropriate crack driving parameters.

1 INTRODUCTION

In previous work one of the authors Xu et al (1996) carried out fatigue tests at various frequencies on adhesively bonded structures. Paris law plots for the cyclic fatigue crack propagation rates (CFCPR) are shown in fig. 1. It can be seen that for a given frequency the CFCPR follows the conventional linear plot (with some evidence of a threshold). However this relationship varies with frequency; at a given ΔG the CFCPR increases as the test frequency decreases. These data can be considerably rationalised by plotting the temporal rather than the cyclic FCPR. This then indicates that creep can be a significant factor in the frequency dependent FCPR exhibited by some adhesives. This will be important for bonded structures subjected to low frequency loading such as the operational cycle of aircraft etc. It is important to be able to identify relevant crack driving parameters for such adhesives and this paper outlines both experimental and analytical studies carried out to achieve this for a particular adhesive.

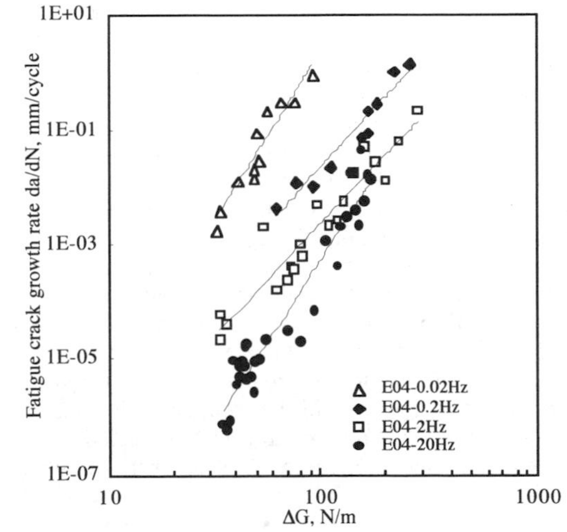

Fig 1 - Paris law plots for adhesively bonded DCB fatigued at various frequencies

Although a significant amount of work (Webster, 1994) has been carried out on the creep behaviour of metals at elevated temperatures comparatively little has been undertaken in the field of polymers in general and adhesives in particular.

Various constitutive models for the rate dependent response of adhesives have been considered including power law creep (Althof 1982) and visco-plasticity (Kitagawa, 1989). Only the first of these approaches is currently viable for finite element solutions covering long time periods and complex finite element meshes. With regard adhesive joint response to sustained loads Allen (1976) tested lap joints at various load levels and suggested that the observed secondary creep rate was proportional to applied load and that there was some evidence for failure at a critical creep strain. Wake (1989) is one of a number that report that the time to failure is an exponential function of the applied stress.

Almost no numerical modelling work has been carried out to support these experimental studies in creep response. The purpose of this work is both to assess the most relevant creep crack driving parameters for the adhesive used in the fatigue studies referred to earlier and to develop modelling procedures that enable the state of the adhesive near the tip of a creeping crack to be investigated. To carry out this work it has been necessary to obtain relevant adhesive creep material properties for use in analyses, to carry out creep crack growth tests and to analyses the resulting data. These aspects are all discussed in more detail below.

2 BULK ADHESIVE TENSILE CREEP TESTS

The adhesive used in this study is the same adhesive that was used in the fatigue tests described above. This is a two part, cold cure epoxy with mineral filler and toughened with rubber. Specimens are manufactured by injecting the mixed adhesive into closed steel and perpsex moulds which have been coated in mould release to facilitate specimen removal. The filled moulds are left to cure at 40°C for 48 hours. Flat strip dumbbell tensile test specimens, conforming to BS 2782, 5mm thick, were made in this way. These were then tested within a temperature cabinet on a 6025 Instron servo-mechanical testing machine at three different levels of creep load 10MPa, 7.5MPa and 5MPa. The test machine was run in "hold" mode under load control which continually adjusts the crosshead to maintain a constant load. Extensometers were attached to the specimen and gave a continual output of strain throughout the test. The resulting creep strains can be seen as the thicker solid lines in fig. 2 where the highly non-linear nature of the response (invalidating the use of conventional rheological visco-elastic models) can be seen. The 10MPa test failed after about 60 minutes at a creep strain of 37%. The tests at the two lower levels of stress had not reached this level of strain when they were halted after 120 minutes.

Analytical solutions are available for materials that conform to steady state (secondary) power law (SSPL) creep i.e.:

$$\frac{\varepsilon_c'}{\varepsilon_o'} = \left[\frac{\sigma}{\sigma_o}\right]^n \qquad (1)$$

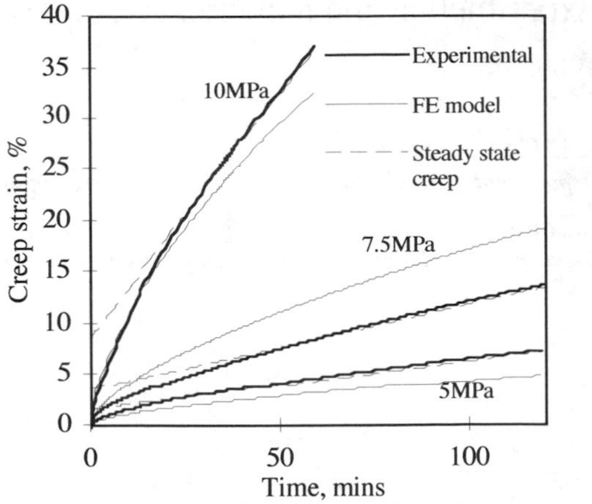

Fig 2 - Tensile creep strains at three different levels of creep load

Table 1 - Constants for the SSPL creep strain rate equation (1)

ε_o'	min^{-1}	$3.76x10^{-4}$
σ_o	MPa	5.0
n		3.2

In order to utilise these solutions it is necessary to determine the material constants in eqn. 1 (i.e. ε_o', σ_o' and n) that best fit the experimental tensile secondary creep rate data shown as dotted lines in fig. 2.

The experimental creep do not take an exact power law distribution but a "best fit" has been determined. The values of the relevant constants in eqn 1 are given in Table 1. Table 2 shows the nature of this best fit by comparing the SSPL creep rates from the experimental data and from the application of eqn 1 at 5, 7.5 and 10 MPa. It can be seen that eqn 1 underestimates the strain rate slightly at 10MPa and overestimates it at 7.5MPa. Nevertheless it should enable reasonable analytical creep solutions to be obtained for later compact tension tests.

To carry out more detailed numerical analyses a fit was obtained for the experimental data to the following finite element creep strain rate model which has been expressed both in terms of time and strain hardening:

$$\varepsilon_c' = A\,\sigma^N t^M = [A\sigma^N]^{1/(M+1)}\,[\varepsilon_c(M+1)]^{M/(M+1)} \qquad (2)$$

Table 2 - SSPL rates from experiment and eqn 1

Stress (MPa)	Rate-exp. (min^{-1})	Rate-eqn 1 (min^{-1})
10	4.75E-03	3.46E-03
7.5	8.21E-04	1.38E-03
5	4.71E-04	3.76E-04

Table 3 - Constants for the finite element creep model eqn (2)

A	N	M
$6.28x10^{-6}$	3.4	-0.372

As with the SSPL creep this model cannot be made to fit the experimental data exactly and the "best fit" chosen can be seen in fig 2 as thin solid lines. It slightly underestimates the data for 10 and 5MPa whilst overestimating the data at 7.5MPa. The values of the constants used in eqn. 2 are given in Table 3 where stress is in MPa, time in minutes and strain is non-dimensional.

3 BULK ADHESIVE COMPACT TENSION SPECIMENS

Compact tension specimens were manufactured in the same way as the bulk tensile specimens. The geometry of the specimens essentially conforms to ASTM E 399/78 with a thickness of 6mm and a width (W, measured from the load line) of 40mm. A sharp pre-crack was generated in all specimens propagating away from the tip of a tapped razor blade. By pre-compressing the specimen at an appropriate position the length of this pre-crack could be controlled fairly accurately, to about $15^{+/-1}$ mm. These specimens were then subjected to various levels of sub-static failure load in the Instron 6025 using the same procedures outlined in the previous section. The crack propagation was monitored as a function of time through the use of a travelling microscope and paper scale with 100μm gradations attached to the specimen adjacent to the anticipated crack path. Also recorded as a function of time was the separation of the loading points. These data are shown in fig. 3 for creep loads of 150N and 180N and it can be seen that the tests lasted about 20 and 4 hrs respectively. Further it can be seen that there was a similar amount of crack growth in each specimen and that final failure occurred at about the same load-line displacement in each case. In fact both were slightly lower for the

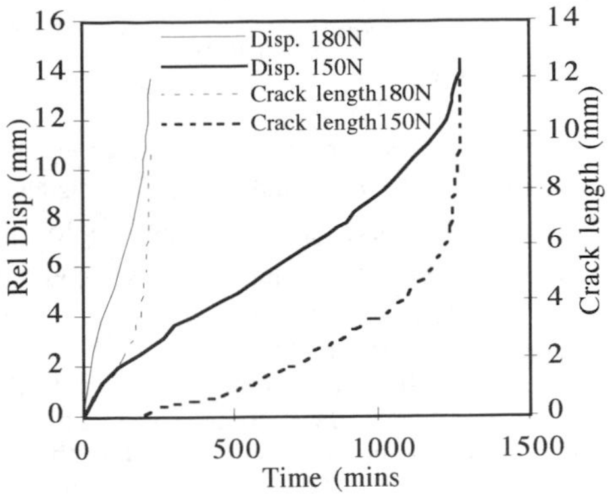

Fig 3 - Details of the temporal crack growth and load-line displacement in the compact tension specimens at two levels of loading

higher load (faster) test which is consistent with its more highly strained state. Both load-line displacements display the conventional stages of creep response i.e. primary (hardening), secondary (steady state) and tertiary (softening) regimes. The crack growth exhibits a period of very limited growth or incubation period (particularly the lower load) followed by accelerating growth caused by the increased intensity of the load with crack length. From these data the creep crack growth rate (CCGR) can be determined as a function of crack length and a CCGR law determined. There are two commonly used crack driving parameters in these CCGR laws, one is the stress intensity factor K and the other the C* integral. This latter parameter is essentially the rate equivalent to the J integral and will be discussed in more detail in the next section.

4 STEADY STATE POWER LAW (SSPL) CREEP ANALYSIS OF THE COMPACT TENSION SPECIMEN

Under the assumptions of power law creep (eqn. 1) it can be shown that the steady state creep solution to a problem is identical to the plasticity solution where the stress-strain data takes the same form as eqn. 1 with strains replacing strain rates. Using this analogy the SSPL creep stress and strain fields near the crack tip are characterised by the parameter C* introduced in the previous section just as the fields are characterised by K for linear elastic behaviour, [3]:

$$\sigma_{ij} = \left[\frac{C^*}{\varepsilon_o' \sigma_o I_n r}\right]^{1/(n+1)} \sigma^*_{ij}(n,\theta) \qquad (3a)$$

$$\varepsilon'_{ij} = [\frac{C^*}{\varepsilon_o' \, \sigma_o I_n r}]^{n/(n+1)} \, \varepsilon'^*_{ij}(n,\theta) \qquad (3b)$$

Here ε_o', σ_o and n are the SSPL creep material constants (Table 1), r and q are the distance and angle from the crack tip and expressions for I_n, σ^*_{ij} and ε^*_{ij} are available in the literature. Note that the strain singularity is stronger than in LEFM (with our adhesive 0.76, compared to 0.5).

Stress intensity K is used as a crack driving parameter when the strain field around the propagating crack is dominated by the elastic singularity whilst C* is used when the field is dominated by the steady state creep singularity. The elastic parameter K is relevant when crack propagation is much quicker than the relaxation time of the material whilst C* is relevant when crack growth is much slower. In order to determine which these is most pertinent to this work a redistribution time t_r can be calculated by equating the transient creep and elastic strains at a 'skeletal' point where the elastic and SSPL creep stresses are the same. After some manipulation it can be shown that t_r is given as

$$t_r = \frac{K^2}{EC^*} \qquad (4)$$

In order to evaluate the redistribution time and also to use as a crack driving parameter it is necessary to calculate C*. As C* is analogous to J for a plastic solution where the elastic strains are negligible it is possible to use existing tables for J_{pl} to determine C*. Such data is commonly expressed as:

$$C^* = \sigma_o \,.\, \varepsilon_o' \,.\, [w-a] \,.\, h_1(a/w, n) \,.\, [P/P_o]^{n+1} \qquad (5)$$

Here P_o is the limit load which has been expressed in terms of the geometry and crack length for various configurations. Thus by use of equations (4) and (5) the redistribution time at various crack lengths for three different creep loads have been calculated. The redistribution times for the initial crack lengths are shown in Table 4.

The redistribution time decreases steadily with in increasing load and crack length. This is because the relative amount of creep deformation increases, i.e. C* in eqn. 4 increases more quickly than K. What is also significant is that the redistribution times are lower than the duration of the complete creep tests and of the same order as the time required to propagate a crack through the creep zone and thus it is likely that the CCGR will be governed by C*.

Table 4 - Redistribution times for the initial crack lengths

Load (N)	180	150
Redistribution time (min)	13	22

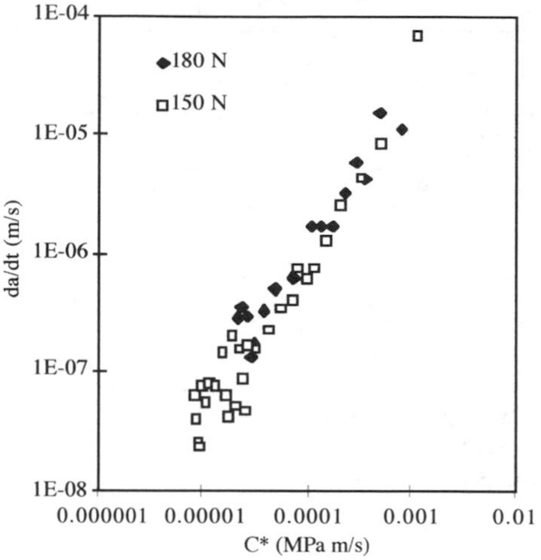

Fig 4 - Variation of crack growth rate in the CT specimens with C*

The relevance of C* as creep crack driving parameters can be seen in fig 4. There is good overlap between data from both levels of loading in the plots. Indeed supplementary tests not reported here at even lower levels of load were also consistent with these data. Further, from the straight line nature of the data when plotted on logarithmic axes, it is apparent that a power law in C* would correlate the crack growth rate very well. A semi-experimental check can be made on the analytical value for C* using the standard formula for power law creeping materials involving the steady state load line displacement rate Δ' i.e.

$$C^* = \frac{n}{n+1} \, \frac{h(a)}{w-a} \frac{P}{B} \, \Delta' \qquad (6)$$

Here h is a function of crack length that varies with specimen type. The steady state load-line displacement rate can be found from the experimental data in fig. 2. There is good correlation between the two approaches and for both load levels the value from each approach agree at a crack length that lies within the domain of steady state crack growth as outlined in Fig 2. This implies that the SSPL creep model used in this section is a reasonable representation of the material behaviour.

5 FINITE ELEMENT ANALYSIS OF THE COMPACT TENSION SPECIMEN WITH A STATIONARY CRACK

To obtain a better prediction of the strain field in the compact tension specimens finite element analyses were undertaken. The FE material model illustrated in fig 1 was used which, unlike the model used in the last section, approximates the whole of the material response not just the secondary, constant strain rate portion. Further, the SSPL solutions reported in the previous sections assume that the creep strains dominate the elastic strains whilst solutions using this FE modelling approach include both elastic and creep strains. Analyses have been carried out assuming strain hardening rather than time hardening and the resultant creep strain increment (Δe_{cr}) is determined from eqn. 2 from the current stress and creep strain level and time increment. The component creep strain increments ($\Delta\varepsilon_{crij}$) are then determined using the von Mises associated flow rule i.e. in terms of the deviatoric stresses (S_{ij}) and the equivalent stress (σ_{eq}) as shown in eqn 7

$$\Delta\varepsilon_{crij} = \Delta\varepsilon_{cr}\,\frac{S_{ij}}{\sigma_{eq}} \qquad (7)$$

Two different types of analyses have been undertaken. The first used a refined material model only considering the initial crack length and can be considered to be relevant to the incubation period seen in the experimental data before significant crack growth occurred. Analyses have only be carried out for fractions of the total test times, these being the times over which sub-millimetre crack growth occurred. The second type of analyses model the propagating crack and thus are carried out over the entire test duration. For this reason a less refined model is used in this work. The results of the stationary crack are contained in this section whilst the propagating crack analyses are discussed in the following section.

The mesh used for the stationary crack analyses is shown in Fig 5. This consists of 348 predominantly eight noded plane strain quadrilateral elements. Due to symmetry only half the compact tension specimen needs to be modelled. All nodes on the crack surface are free to move whilst those on the line of symmetry ahead of the crack tip are constrained against movement normal to the plane of symmetry. The smallest elements have been placed in the crack tip region and have a length of 0.375 along the crack front. The load has been applied as a pressure over two elements which span the actual point of loading.

As well as the creep material properties discussed in an earlier section an elastic modulus and Poisson's ratio of 800MPa and 0.4 respectively have been used. These values being derived from separate tensile tests carried out on the bulk adhesive not described here. Analyses have been carried out for the two levels of load reported in the section outlining the creep tests of the compact tension specimen. A summary of the loads and the corresponding times periods over which the analyses have been carried out (which are typical times for sub-millimetre crack growth) are shown in Table 5. Automatic time stepping within the finite element package (ABAQUS) was used, the time steps were determined so that the predicted creep strain rates at the beginning and end of the step were the same to a specified tolerance. This resulted in very small time steps initially which increased with solution time. The initial time step size was $1.0\mathrm{x}10^{-17}$ min. and the total number of time steps used within each analysis are shown in Table 5.

The results from the analyses can be compared with some of the SSPL creep parameters derived in the last section. The transient (as opposed to steady state) stress and strain distribution in a creep crack problem are given by equations similar to eqn 3 but with C* replaced by a parameter C(t). At long times, under SSPL creep C(t) should tend to C*. The variation of C(t)/C* with time has been investigated for each of the load levels considered. At early times C(t) has been found to be many times larger than C* implying some considerable error in the corresponding SSPL stress and strain fields. However it has been found that the value tends to unity as the time approaches the redistribution time calculated earlier. The value drops below unity with increasing time because the finite element material model used is not a true SSPL model.

As the time exposed to loading increases so does the size of the creep zone. This is illustrated in fig 5 which shows the increase with time of the size of zone that contains creep strains that are higher than the elastic strains for the 180 N creep test. It can be seen that by the time the crack begins to show significant growth around (60min.) the creep zone

Table 5 - Time incrementation details of the stationary crack finite element analyses undertaken

Load (N)	Total time (min.)	Total number of steps
180	60	853
150	125	784

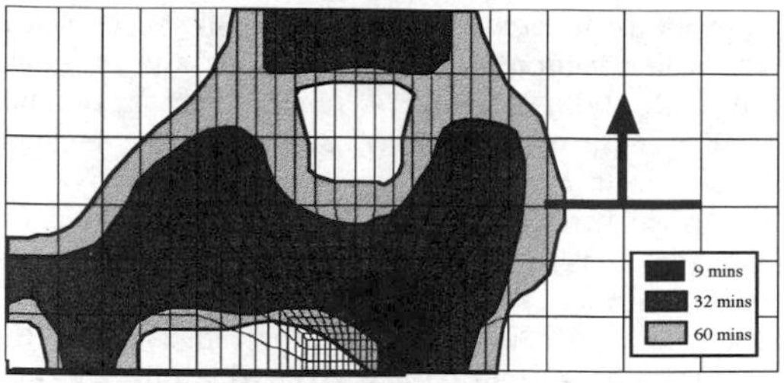

Fig 5 - Growth of the predicted creep zone size with time for the 180N creep test

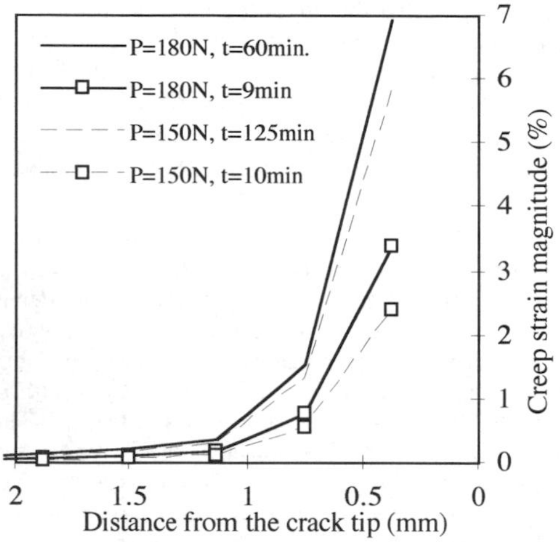

Fig 6 - Distribution of resultant creep strain magnitude on the centre-line at various times in all three creep analyses

size is very extensive. Similar plots for the other creep loads, although not shown here, show a very similar trend. Clearly for the initial crack growth at least it would be valid to assume that C* is an appropriate crack driving parameter. As the crack growth accelerates, material ahead of the crack tip that initially experiences low creep strains will see an increase in the level of these strains. However it is not possible to determine the state of strain (and hence whether C* will be a valid parameter) for other than the initial crack length and this is the reason for propagating crack analyses reported in the next section.

Probably the most relevant results from these stationary crack analyses are the redistribution of stress and strain distributions. As expected the stresses near the crack tip reduce with time and a significant redistribution takes place over the time period of the analyses. Note that the equivalent stresses will be lower than these normal stresses and that even with the mesh refinement used in this model the maximum stresses are not significantly outside the range of experimental data that is used to generate the material model. The resultant creep strain magnitude on the centre line of the specimen ahead of the crack tip are shown in fig 6 where appropriate data for each of the loads are illustrated at three times pertinent to crack initiation; initial, intermediate and final. When considering the strains it can be seen that there is a build up of creep strain magnitude at the crack tip. It is possible to speculate that significant creep crack growth only begins once some level of creep strain has been accumulated. This thesis is supported by these numerical results which suggest that crack propagation might begin at a creep strain magnitude of about 6%.

6 FINITE ELEMENT ANALYSIS OF THE COMPACT TENSION SPECIMEN WITH A PROPAGATING CRACK

These full life analyses have been carried out on the less refined mesh shown in fig 7 because of the extensive computational effort that is involved. The elements are the same as the previous analysis except that they have only four nodes in order to be compatible with the interface elements used to implement the crack propagation behaviour. Crack front location is prescribed as a function of time and is thus made to follow the experimental compact tension behaviour illustrated in fig 2. Crack growth is achieved by replacing a constrained node on the centre line with its corresponding force and then gradually reducing this force to zero at the time at which the next node on the crack path begins its release. In this way a quasi-continuous debonding process is achieved in a discrete modelling procedure, namely finite element analysis. It is now possible to correlate the numerical and experimental data.

The actual and simulated load-line data correlates reasonably well. Discrepancies are attributed to the coarseness of the finite element model and the mismatch between the actual and finite element material models. Indeed current work using the same refined mesh as used in the stationary crack analyses is showing an even better correlation.

The experimental and numerical crack tip profiles are compared in fig 7. The former were obtained photographically during testing and are shown in fig 7 for both sub-millimetre and extensive crack growth. In the case of sub-millimetre crack growth the rounding at the crack tip caused by blunting can be seen quite clearly whilst the sharp crack tip of the propagating crack and the excessive deformation around the original crack tip can be seen in the case of the extensive crack growth. The numerical

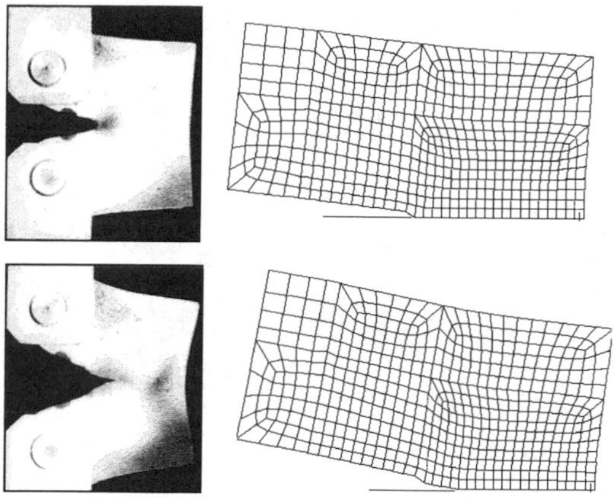

Fig 7 - Crack tip profiles: experimental a=16mm and 19.5mm, numerical a = 16mm and 21mm

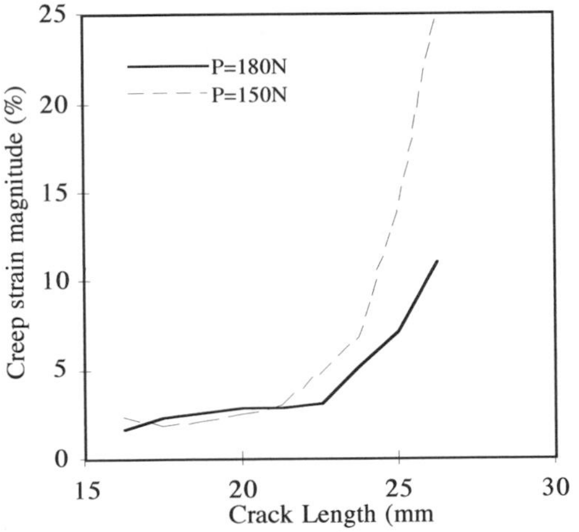

Fig 8 - The value of the maximum resultant creep strain magnitude during crack propagation for both creep loads

deformation for similar crack lengths (i.e. sub-millimetre and extensive) can also be seen in fig 7 and exactly the same features of a blunt crack in the sub-millimetre growth and sharp crack tip and original extensive crack tip deformation in the extensive crack growth can be observed. This is further evidence of the validity of the propagating analyses discussed in this section.

Finally we consider the validity of various possible controlling parameters for creep crack growth. In Fig 8 we show, for both levels of creep load, the predicted value of the resultant creep strain magnitude averaged over the element immediately ahead of the crack tip just prior to the release of the crack tip node (i.e. at the point of crack growth). The incubation period has not been included in these data and it can be seen that apart from long crack lengths the creep strain magnitude does not vary significantly during crack growth. Further the level of this critical strain is similar for all three levels of creep load suggesting that it would be an appropriate parameter to use in subsequent analyses to predict rather than model the point of rupture at the creep crack growth tip. At long crack lengths the observed crack growth becomes unstable and the crack tip stresses lie outside the range defined by the material behaviour in fig 1.

Another potentially useful parameter to consider is C(t) as this defines the strain during transient power law creep. As with the resultant creep strain magnitude, its value at the point of crack growth has been evaluated for the range of crack lengths. Evaluation of C(t) is carried out around contours surrounding the crack tip, the first contour around the first layer of elements, is rather inconsistent but the values at other contours are similar and these values have been used. It was found that, unlike the resultant creep strain magnitude, the value of C(t) varies significantly with crack length and even for a given crack length is not similar for the different levels of applied load. Thus it would appear to be inappropriate to use this parameter to predict rupture conditions in a creep crack growth situation for the adhesive in question.

7 CONCLUSIONS

Creep tensile data has provided a useful basis for various material models based on power laws in stress and/or strain (time) and although these models are not an exact fit to the experimental data when used in conjunction with analytical or numerical approaches they provide a reasonable fit to the observed experimental behaviour.

Creep tests have been carried out on compact tension specimens and load-line displacement exhibits the characteristic three stages of creep response. Crack growth generally did not occur immediately but after some incubation period. Thence crack growth accelerated until it became catastrophic. Crack tip profiles showed significant blunting associated with the incubation period with much unrecoverable deformation showing as a ridge on the extended crack tip profile.

Steady state power law creep models in conjunction with the experimental data have shown that the time periods over which crack growth occurs are such that one would expect creep rather than elastic parameters to control crack propagation. Subsequent crack propagation rate plots showed that growth laws could be formulated as a power functions of C* or K. Analytical and semi-experimental values for C* showed reasonable agreement.

Finite element creep analyses of a stationary crack in compact tension specimens correlated well with the analytical SSPL creep solutions. Significant stress redistribution was found to occur. Extensive creep zones developed over time periods similar to that required for sub-millimetre crack growth and it would appear that macro-crack propagation began when the resultant creep strain magnitude reached about 6%.

Finite element analyses with propagating cracks gave reasonable correlation with observed displacement-time and crack tip profile plots. The value of the creep strain has been shown to remain sensibly constant over a large period of the creep crack growth and it is suggested that the creep strain would make a good criterion for crack tip rupture in subsequent creep analyses of other structures bonded with this adhesive. Values of C(t) have shown significantly more variation over the period of crack growth and appear less appropriate as an indicator of crack tip rupture.

REFERENCES

Althof W, 1982, Creep recovery and relaxation of shear loaded bond lines, *J Rein Plastics and Comp*, 1, 29-39

Allen KW and Shanahan MER, 1976, *J Adhesion*, 8-1, 43 (1976)

Kitagawa M et al, 1989, Rate dependent nonlinear constitutive equation of polypropelyne, *J Polymer Sci, pt B Polymer Physics*, 27, 85-95

Wake WC,1989, *Elastomers; criteria for engineering design*, ed Hepburn and Reynolds, Applied Sci Publ

Webster GA and Ainsworth RA, 1994, *High temperature component life assessment*, Chapman and Hall

Xu XX, Crocombe AD and Smith PA, 1996, Fatigue crack propagation of adhesive joints tested at different rates, *J Adhesion*, 58, 191-204

Free vibration analysis of clamped-free circular cylindrical shell with end cap by receptance method

J.S.Yim & D.S.Sohn
Department of Future Fuel Design Development, Korea Atomic Energy Research Institute, Yusong, Taejon, Korea
Y.S.Lee
Department of Mechanical Design Engineering, ChungNam National University, Yusong, Taejon, Korea

Abstract: An analysis of free vibration of cantilevered circular cylindrical shell with end cap is presented using receptance method. The natural frequencies and mode shapes of the combined system were calculated and they were compared with those of commercial finite element code, ANSYS, as well as those from vibration test. The results focused on the natural frequencies at present showed good agreement with those from ANSYS fairly well and with test results.

1. Introduction

Analyses of free vibration for the combined structures such as a shell with a plate welded at the shell have been studied by various authors. Suzuki et al.(1983) analyzed the free vibration of cantilevered shell with a circular plate attached at top by applying Mindlin theory on the plate. Nagamatsu et al.(1983,1984) analyzed the vibration of machine center using substructure synthesis. Irie et al.(1984) presented an analyses for the free vibration of joined conical-cylindrical shell by way of transfer matrix technique while Tavakoli et al.(1989) presented the eigensolution of joined/hermitic shell structures using the state space method.

One of the most representative study on this subject seemed to be the receptance method which was first introduced by Faulkner(1969). Huang and Soedel(1993) applied the method to the analysis of free vibration for the circular cylindrical shells with plate attached at arbitrary axial position for the boundary condition of simply supported at both ends.

However the analysis of the shells with a plate for non-symmetric boundary condition such as cantilevered boundary condition using receptance method was not found.

Here, in order to analyze the free vibration of cantilevered cylindrical shells with a plate attached at arbitrary axial position(s), the receptance method was used. The modal displacements of the shell were assumed as beam functions and through applying Rayleigh-Ritz method the frequencies of the shell were calculated from the cubic frequency equation.

The modal displacement of the shell was expressed by way of the mode superposition(Soedel, 1993). The analytic results were compared with those of the commercial finite element code ANSYS (Swanson, 1992) and free vibration test.

Detailed results and discussions of the frequencies are concentrated on the low axial half waves.

2. Formulation

2.1 Displacement Function due to Dynamic Loading

Figure 1 represents one of the case of a shell with end cap and the coordinates of the structure and

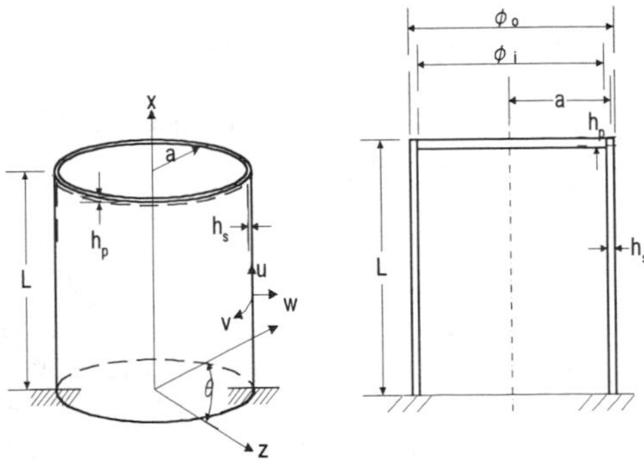

Figure 1 Coordinates System and Notation

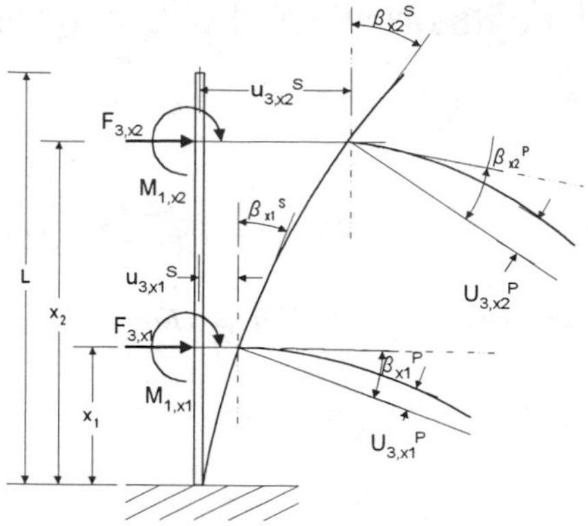

Figure 2 Displacement due to Loads for Shell with Plate at Top and Middle

displacements. Figure 2 shows the cross sectional view of the displacements and the slopes at joining points due to the dynamic transverse line loads and line moments around the shell exerted by the motion of vibration. The displacements of the shell subjected to a dynamic loading can be expressed by the modal displacement and mode participation factor (Soedel,1993) as equation (1).

$$u_i(x,\theta,t) = \sum_{k=1}^{\infty} \eta_k(t) U_{ik}(x,\theta), i = 1,2,3 \tag{1}$$

In equation (1) the mode participation factor η_k is the root of the following modal equation.

$$\ddot{\eta}_k + 2\varsigma_k \omega_k \dot{\eta}_k + \omega_k^2 \eta_k = F_k^* e^{j\omega t} \tag{2}$$

where

$$F_k^* = 1/\rho h N_k \int_x \int_\theta (f_1 U_{1k} + f_2 U_{2k} + f_3 U_{3k})\ a dx d\theta \tag{3}$$

$$N_k = \int_x \int_\theta (U_{1k}{}^2 + U_{2k}{}^2 + U_{3k}{}^2) a dx d\theta \tag{4}$$

Here, the input forcing functions in equation (3) may have the following form.

$$f_i(x,\theta,t) = F_i^*(x,\theta) e^{j\omega t}, \quad i = 1,2,3 \tag{5}$$

The displacement components of the shell in equation (3) can be represented by beam functions as below.

$$U_{1k}(x,\theta) = A\phi'(x)\cos n\theta \tag{6}$$

$$U_{2k}(x,\theta) = B\phi(x)\sin n\theta \tag{7}$$

$$U_{3k}(x,\theta) = C\phi(x)\cos n\theta \tag{8}$$

Where $\phi(x)$ is a beam function and $\phi'(x)$ is derivative of the function which is expressed as below.

$$\phi(x) = \cosh p_r x - \cos p_r x - C_r(\sinh p_r x - \sin p_r x) \tag{9}$$

$$\phi'(x) = \sinh p_r x + \sin p_r x - C_r(\cosh p_r x + \cos p_r x) \tag{10}$$

C_r and p_r can be obtained from following equations.

$$C_r = \frac{\sinh p_r L - \sin p_r L}{\cosh p_r + \cos p_r L} \tag{11}$$

$$\cosh p_r L \bullet \cos p_r L + 1 = 0 \tag{12}$$

From equation (2) the mode participation factor of k-th mode can be obtained as below.

$$\eta_k = \frac{F_k^*}{\omega_k^2 \sqrt{\left(\left[1 - \left(\frac{\omega}{\omega_k}\right)^2\right]^2 + 4\zeta_k^2 \left(\frac{\omega}{\omega_k}\right)^2\right)}} e^{j\omega t} \tag{13}$$

Neglecting damping of the system, the displacements of the structure can be expressed as equation (14).

$$u_i(x,\theta,t) = \sum_{k=1}^{\infty} \frac{F_k^*}{\left(\omega_k^2 - \omega^2\right)} e^{j\omega t} U_{ik}(x,\theta) \tag{14}$$

When a circular plate attached at axial position of the circular cylindrical shell, the transverse dynamic excitation exerted at the junction due to the constraint of the displacements of the shell by the plate can be assumed as equation (15) from the assumption of neglecting the other components of displacements except transverse normal displacement of the shell u_3.

$$F_3(x^*,\theta,t) = F_3^S \cos n\theta \delta(x - x^*) e^{j\omega t} \tag{15}$$

From equation (3) with the excitation of equation (15) the forcing term in equation (2) can be obtained as below. If we let

$$Denom = (\frac{A}{C})^2_{imn} \int_0^L \phi'^2(x)dx + (\frac{B}{C})^2_{imn} \int_0^L \phi^2(x)dx + \int_0^L \phi^2(x)dx \quad (16)$$

then

$$F_k^* = \frac{F_3^S \phi(x^*)}{\rho h \bullet Denom} \quad (17)$$

Equation (4) results in the following equation (18).

$$N_{imn} = \pi a \bullet Denom, \quad \text{for } n \neq 0 \quad (18)$$

Using the results of equation (17) and in relation with equation (14) the dynamic displacement of the shell can be obtained using mode summation. Neglecting the displacements $u_1(=u)$, $u_2(=v)$ of the shell, the only displacement to be considered can be expressed as following form.

$$u_3^S(x,\theta,t) = \sum_{m=1}^{\infty} \sum_{n=0}^{\infty} \sum_{i=1}^{3} \frac{1}{\left(\omega_{imn}^2 - \omega^2\right)} \bullet \frac{\left[F_3^S \phi(x^*)\phi(x)\right] \cos n\theta e^{j\omega t}}{\rho h \bullet Denom} \quad (19)$$

The slope of shell at the junction in the axial direction can be obtained from equation (19) by differentiation with respect to the axial coordinate x.

$$\beta_{x1}^S(x,\theta,t) = -\frac{\partial u_3^S}{\partial x} = -\sum_{m=1}^{\infty} \sum_{n=0}^{\infty} \sum_{i=1}^{3} \frac{1}{\left(\omega_{imn}^2 - \omega^2\right)} \bullet \frac{\left[F_3^S \phi(x^*) p_r \phi'(x)\right] \cos n\theta e^{j\omega t}}{\rho h \bullet Denom} \quad (20)$$

Next, the dynamic moment loading exerted at the junction due to the constraint by the plate can be expressed as equation (21).

$$M_1(x^*,\theta,t) = M_1^P \cos n\theta \delta(x - x^*) e^{j\omega t} \quad (21)$$

With this moment loading the forcing function of the equation (3) can be obtained(Soedel,1976).

$$F_k^* = \frac{1}{\rho h N_k} \int_x \int_\theta U_{3k} \left[\frac{1}{a}\left(\frac{\partial(M_1)}{\partial x}\right)\right] a dx d\theta \quad (22)$$

Here, N_k is the same as in equation (18). Neglecting the damping of the structure the final displacement of the shell due to the moment at junction can be obtained from equation (14) and (22).

$$u_3^S(x,\theta,t) = -\sum_{m=1}^{\infty} \sum_{n=0}^{\infty} \sum_{i=1}^{3} \frac{M_1^S}{\left(\omega_{imn}^2 - \omega^2\right)} \frac{p_r \phi'(x^*)\phi(x) \cos n\theta e^{j\omega t}}{\rho h \bullet Denom} \quad (24)$$

The slope of the shell by the moment at the junction can be calculated as equation (25).

$$\beta_{x1}^S(x,\theta,t) = \sum_{m=1}^{\infty} \sum_{n=0}^{\infty} \sum_{i=1}^{3} \frac{M_1^S}{\left(\omega_{imn}^2 - \omega^2\right)} \bullet \frac{p_r{}^2 \phi'(x^*)\phi'(x) \cos n\theta e^{j\omega t}}{\rho h \bullet Denom} \quad (25)$$

The dynamic transverse loading and moment loading at the junction yield the translational rigid displacement u_3^P and the slope of the plate β_{x1}^P as below.

$$u_3^P = -\frac{F_3 \cos\theta e^{j\omega t}}{a \rho h_P \omega^2} \quad (26)$$

$$\beta_{x_1}^P(r,\theta,t) = -\sum_{m=1}^{\infty} \frac{\lambda^2 a \pi M_1{}^P}{\rho h_P N_{mn}(\omega^2{}_{mn} - \omega^2)} \bullet \left[J_{n+1}(\lambda a) - \frac{J_n(\lambda a)}{I_n(\lambda a)} I_{n+1}(\lambda a)\right]^2 \cos n\theta e^{i\omega t} \quad (27)$$

2.2 Frequency Equation for Shell with End Cap

When a plate is attached at arbitrary axial position of the shell, the frequency equation can be deduced by considering the continuity condition at the shell/plate joining point. Here, only the transverse normal displacement of the shell and the transverse rigid body motion of the plate due to the transverse dynamic loading at the junction and the slope in the axial direction of the shell and the slope in the radial direction of the plate due to dynamic moment were taken into considered because the other components of displacement could be ignored for negligibly small compared with the two. By applying the continuity condition at the junction, equation (28) can be deduced.

$$\begin{bmatrix} \alpha_{11} + \beta_{11} & \alpha_{12} \\ \alpha_{21} & \alpha_{22} + \beta_{22} \end{bmatrix} \begin{Bmatrix} F_3 \\ M_1 \end{Bmatrix} = 0 \quad (28)$$

From the condition of having non-trivial solution of equation (28), the frequency equation of the combined system can be deduced as.

$$\begin{bmatrix} \alpha_{11}+\beta_{11} & \alpha_{12} \\ \alpha_{21} & \alpha_{22}+\beta_{22} \end{bmatrix} = 0 \tag{29}$$

where

$$\alpha_{11} = \frac{u_3^S(x^*,\theta,t)}{F_3(x^*,\theta,t)} = \sum_{m=1}^{\infty}\sum_{i=1}^{3}\frac{1}{\left(\omega^2{}_{imn}-\omega^2\right)} \bullet \frac{\phi(x^*)\phi(x^*)}{\rho h \bullet Denom} \tag{30}$$

$$\alpha_{21} = \frac{\beta_{x1}^S(x^*,\theta,t)}{F_3(x^*,\theta,t)} = -\sum_{m=1}^{\infty}\sum_{i=1}^{3}\frac{1}{\left(\omega^2{}_{imn}-\omega^2\right)} \bullet \frac{p_r\phi(x^*)\phi'(x^*)}{\rho h \bullet Denom} \tag{31}$$

$$\alpha_{22} = \frac{\beta_{x1}{}^S(x^*,\theta,t)}{M_1(x^*,\theta,t)} = \sum_{m=1}^{\infty}\sum_{i=1}^{3}\frac{1}{\left(\omega^2{}_{imn}-\omega^2\right)} \bullet \frac{-p_r{}^2\phi'(x^*)\phi'(x^*)}{\rho h \bullet Denom} \tag{32}$$

$$\alpha_{12} = \frac{u_3^S(x^*,\theta,t)}{M_1(x^*,\theta,t)} = \sum_{m=1}^{\infty}\sum_{i=1}^{3}\frac{1}{\left(\omega^2{}_{imn}-\omega^2\right)} \bullet \frac{-p_r\phi(x^*)\phi'(x^*)}{\rho h \bullet Denom} \tag{33}$$

While the receptances of the circular plate are the following forms(Soedel,1993).

$$\beta_{11} = \frac{u_3^P(a,\theta,t)}{F_3^P(a,\theta,t)} = \begin{cases} \dfrac{1}{a\rho h_P\omega^2} & for \quad n=1 \\ 0 & \text{for} \quad n\neq 1 \end{cases} \tag{34}$$

$$\beta_{22} = -\frac{\beta_{x1}^P}{M_1(x^*,\theta,t)} = \sum_{m=1}^{\infty}\frac{\pi\lambda^2 a}{\rho h_P}\frac{\left[J_{n+1}(\lambda a)-\dfrac{J_n(\lambda a)}{I_n(\lambda a)}I_{n+1}(\lambda a)\right]^2}{N_{mn}(\omega_{mn}^2-\omega^2)} \tag{35}$$

Here,

$$N_{mn} = \pi\int_0^{r=a}[J_n(\lambda r)-\frac{J_n(\lambda a)}{I_n(\lambda a)}I_n(\lambda r)]^2 r dr \tag{36}$$

Using equation (28), the moment to lateral force ratio can be calculated and consequently the mode shapes of the plate and the shell will be the following forms.

$$u_3{}^P(r,\theta) = -F_3{}^S\sum_{m=1}^{\infty}\frac{\left(M_1^S/F_3^S\right)\pi\lambda a}{\omega^2{}_{mn}-\omega^2}\frac{\left[J_n(\lambda r)-\dfrac{J_n(\lambda a)}{I_n(\lambda a)}I_n(\lambda r)\right]}{\rho h_P N_{mn}} \bullet \left[J_{n+1}(\lambda a)-\frac{J_n(\lambda a)}{I_n(\lambda a)}I_{n+1}(\lambda a)\right] * \cos n\theta \tag{37}$$

$$u_3{}^S(x,\theta,t) = \sum_{m=1}^{\infty}\sum_{i=1}^{3}\frac{F_3{}^S\phi(x)}{(\omega^2{}_{imn}-\omega^2)} \bullet \frac{\{\phi(x^*)-(M_1^S/F_3^S)p_r\phi'(x^*)\}\cos n\theta}{\rho h \bullet Denom} \tag{38}$$

2.3 Frequency Equation for Two Plates Attachment

If two plates are attached at arbitrary axial positions of the shell separately, the frequency equation which is deduced under the conditions of the continuity condition at the junctions will be of the form

$$\begin{bmatrix} \alpha_{11}+\beta_{11} & \alpha_{12}+\beta_{12} & \alpha_{13} & \alpha_{14} \\ \alpha_{21}+\beta_{21} & \alpha_{22}+\beta_{22} & \alpha_{23} & \alpha_{24} \\ \alpha_{31} & \alpha_{32} & \alpha_{33}+\beta_{33} & \alpha_{34}+\beta_{34} \\ \alpha_{41} & \alpha_{42} & \alpha_{43}+\beta_{43} & \alpha_{44}+\beta_{44} \end{bmatrix}\begin{Bmatrix} F_{3,X1} \\ M_{1,X1} \\ F_{3,X2} \\ M_{1,X2} \end{Bmatrix} = 0 \tag{39}$$

From equation (39) the frequency equation for the combined system will be

$$\begin{bmatrix} \alpha_{11}+\beta_{11} & \alpha_{12}+\beta_{12} & \alpha_{13} & \alpha_{14} \\ \alpha_{21}+\beta_{21} & \alpha_{22}+\beta_{22} & \alpha_{23} & \alpha_{24} \\ \alpha_{31} & \alpha_{32} & \alpha_{33}+\beta_{33} & \alpha_{34}+\beta_{34} \\ \alpha_{41} & \alpha_{42} & \alpha_{43}+\beta_{43} & \alpha_{44}+\beta_{44} \end{bmatrix} = 0 \tag{40}$$

where the plate receptances β_{ij} are the same as in equation (34), (35) and α_{ij} are the receptances of shell which are listed in Appendix I.

3. Numerical Results and Discussions

The theoretical formulation was programmed which can be executed by Lahey FORTRAN(Lahey Computers Inc.,1994). Individual receptances of the plate and shell were calculated and using these values frequencies of the system were obtained for each modes using numerical method. The incremental root search finding and bisection method was employed to get more accurate results and fast iteration. Then, the mode shapes can be calculated with the calculated frequencies. Convergence was examined by comparing the results as a function of number of terms of mode summation in the displacement expression. The outer normal displacement of the shell was taken into considered to be converged if the displacement was represented by mode summation of up to 40-th mode.

Material properties and dimensions of the shell and plate used were: L=500mm, a=104.5mm, $h_s=h_p=3$mm, E=20.6x10^4 N/mm^2, $\rho_P = \rho_S = 7.85\text{x}10^3$ Ns^2/mm^4 and ν=0.3.

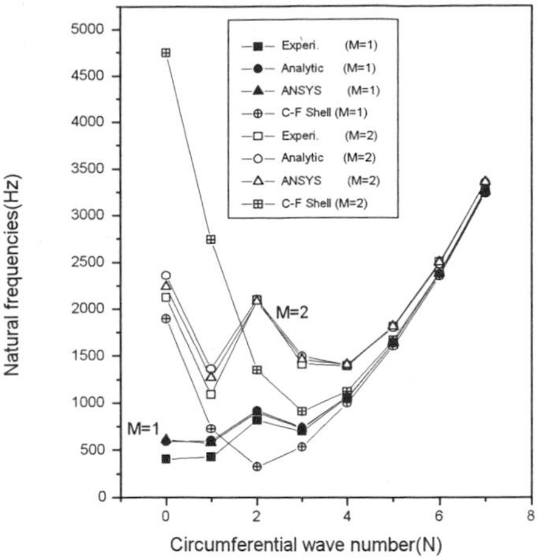

Figure 3 Natural Frequencies of Shell with End Cap

3.1 Natural Frequencies of Cantilevered Shell with End Cap

Table 1 shows the frequencies from analytic, ANSYS and test. The frequencies as a function of circumferential wave number are plotted in Figure 3 where the analytic results show good agreement with those of ANSYS and test in values while the values from test mark always lower than those of the two especially in the lower circumferential wave range. It is shown that at higher circumferential wave number they are well coincident with each other.

The frequencies at circumferential wave number 0 of the cantilevered shell with plate mark far below from those of shell without plate. This is attributed to the fact that the effect of the plate attachment and for this mode of vibration, as a matter of fact, the mode was proven as plate dominant mode. Thus the frequency at this mode highly depends on the frequency of the plate itself that makes the frequencies low for the combined system at axial half wave number M=1 and M=2. The effect of the plate on the frequencies shows negligibly small for the higher circumferential wave numbers mainly because of the high frequencies of the plate.

3.2 Natural Frequencies of Cantilevered Shell with Plates at Top and Middle

The natural frequencies of the shell with plates attached at top and middle of the shell obtained from analytic and ANSYS are listed in Table 2 and it is plotted in Figure 4 where the two low frequencies, which are not much different, are seen at the circumferential wave number 0 and 1.

In this case there might occur two plate dominant mode of vibration with almost the same frequency at independent mode number for the top and the middle plate. Therefore the lowest two frequencies seem to be the plate dominant mode of vibration and it was

Table 1. Natural Frequencies from Analytic, ANSYS and Test for Shell with End Cap (Hz)

N	M=1			M=2		
	Analytic	ANSYS	Test	Analytic	ANSYS	Test
0	589	609	400	2360	2235	2125
1	603	575	425	1365	1271	1088
2	918	904	812	2103	2079	2100
3	739	728	700	1493	1464	1413
4	1060	1059	1050	1400	1400	1388
5	1636	1637	1637	1802	1813	1812
6	2371	2373	-	2477	2494	2500
7	3249	3248	-	3344	3349	-

Table 2. Natural Frequencies from Analytic and ANSYS for Shell with Plate at Top and Middle (Hz)

N	M=1		M=2	
	Present	ANSYS	Present	ANSYS
0	586	609	590	646
1	594	562	1365	1271
2	1667	1947	2103	2079
3	1295	1395	1493	1464
4	1334	1372	1400	1400
5	1772	1805	1802	1813
6	2457	2492	2477	2494
7	3314	3349	3344	3349

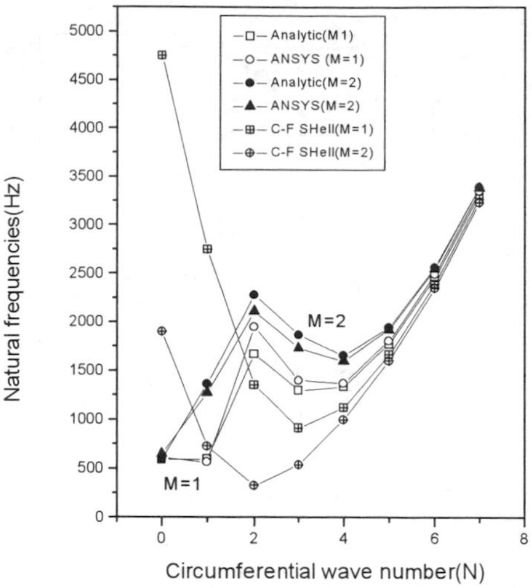

Figure 4 Natural Frequencies of Shell with Plate at Top and Middle

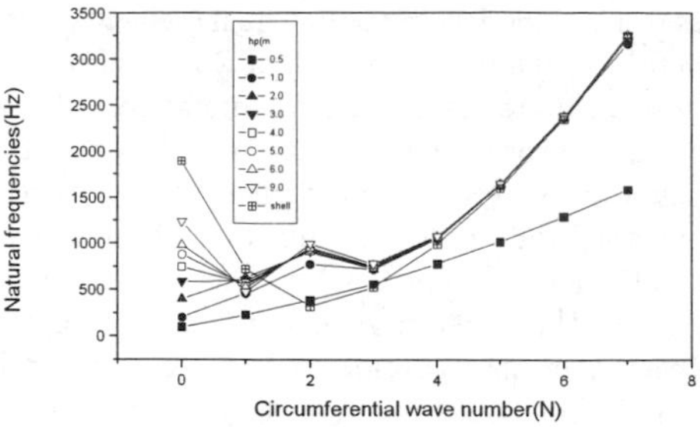

Figure 5 Natural Frequencies as a Function of Plate Thickness for Axial Half Wave Number One(M=1)

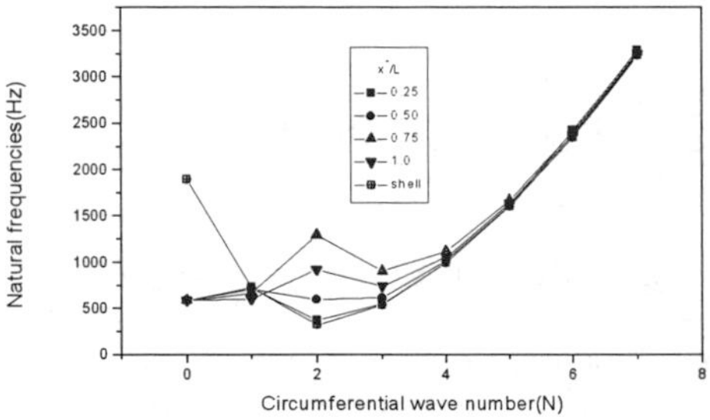

Figure 6 Natural Frequencies as a Function of Plate Location for Axial Half Wave Number one(M=1)

confirmed by ANSYS. At circumferential wave number 2(N=2) much higher frequency increment than that of one plate attachment are shown due to the more constraint of the shell oval motion by the two plates attached at top and middle. Generally the frequencies of the cantilevered shell only are shown much increased by the plate attachment for the modes higher than circumferential wave number 2. The fundamental frequency of the shell only is no longer the fundamental of the combined system by the plate.

3.3 Influence of the Plate Thickness on the Natural Frequencies of the Combined System

The frequencies of the combined system with variation of the thickness of the plate are calculated and is shown in Figure 5. In this calculation the thickness of the plate were changed from 0.5mm up to 9mm. For circumferential wave number 0 whose mode is a plate dominant mode the frequencies of the combined system also shown to be significantly affected by the plate and the system frequencies depend on the frequency of the plate itself. Again the thicker the thickness of the plate is, the higher the frequencies are shown due to the increment of the frequencies of the thicker plate. However at circumferential wave number 1(N=1), i.e. swaying mode of the shell, the thicker the plate is, the lower the frequencies are shown owing to the effect of mass of the plate on the frequencies. At circumferential wave number 2, where the fundamental frequency of the shell without plate appeared, the significant increase of the frequencies are shown due to the constraint of the shell oval mode of vibration by the plate. As was already discussed the effect of the plate on the frequencies of the combined system fade out for higher range of circumferential wave numbers.

3.4 Influence of the Plate Position on the Natural Frequencies

The frequencies of the combined system with varying the location of the plate attachment at 0.25L, 0.5L, 0.75L and 1.0L are plotted in Figure 6 for M=1.

In Figure 6, at circumferential wave number 0 the frequencies of the combined system decrease significantly and it seems like to be independent on the location of the plate. This is stemmed from the truth that this mode is a plate dominant mode thus it does not affect the system frequencies.

At circumferential wave number 1(N=1) whose

mode is a beam-like swaying mode, the plate location also does not affect on the frequencies of the combined system. However at circumferential wave number 2 which is a shell oval mode, the effect of the location of the plate on the frequencies differ from the location of the plate. The most influential location of the plate on the frequencies is 0.75L among the four locations. Here also the influence of the plate on the natural frequencies is shown to be negligibly small as the
circumferential wave number goes higher mode.

4. Conclusions

An analysis of free vibration of cantilevered circular cylindrical shell with end cap is presented using receptance method. The natural frequencies of the combined system were calculated and they were compared with those of ANSYS as well as those from vibration test. The results showed good agreement with those from ANSYS well and with test results.
The process developed herein can be used to determine the dimensioning of the cantilevered circular cylindrical shell with plate attached at arbitrary axial position(s) in order to avoid resonance of the structure in low modes of vibration.

References

Faulkner L. 1969, Ph. D. *Thesis*, Purdue University, Vibration analysis of shell structures using receptances.

Huang D. T. and Soedel W., 1993, Natural frequencies and modes of a circular plate welded to a circular cylindrical shell at arbitrary axial positions, *Journal of sound and vibration* Vol. 162, No. 3, pp. 403-427.

Huang D. T. and Soedel W., 1993, On the free vibration of multiple plates welded to a cylindrical shell with special attention to mode pairs, *Journal of sound and vibration*, Vol. 166, No. 2, pp. 315-339.

Irie T., Yamada G. and Muramoto Y., 1984, Free vibration of joined conical-cylindrical shells, *Journal of sound and vibration*, Vol. 95, No. 1, pp.31-39.

Lehey Computers Inc., 1994, Lahey Fortran User's manual.

Nagamatsu A., Iwamoto T. and Fujita Y., 1983, Vibration analysis using substructure synthesis method(written in Japanese), *Journal of Japanese Mechanical Society*, Div. C, Vol.49, No. 439, pp. 314-322.

Nagamatsu A., Ookuma M., Yamada T. and Ueno H., 1984, Analysis of vibration for machine center(written in Japanese), *Journal of Japanese Mechanical Society*, Div. C, Vol. 50, No. 460, pp. 2276-2282.

Soedel W., 1976, Shells and plates loaded by dynamic moments with special attention to rotating point moments, *Journal of sound and vibration*, Vol. 48, No.2, pp. 179-188.

Soedel W., 1993, *Vibration of shells and plates*, Marcel Dekker, Inc., Hong Kong.

Suzuki K., Takahashi S., Anzai E. and Kosawada T., 1983, Vibration of cylindrical shell with circular plate varying the thickness(written in Japanese), *Journal of Japanese Mechanical Society*, Div. C, Vol. 49, No. 438, pp. 154-164.

Swanson analysis systems Inc., 1992, ANSYS User's manual.

Tavakoli M. S. and Singh R., 1989, Eigensolution of joined/hermitic shell structures using the state space method, *Journal of sound and vibration*, Vol. 100, No. 1, pp. 97-123.

Yamada G., Irie T. and Tamiya T., 1986, Free vibration of a circular cylindrical double-shell system closed by end plates, *Journal of sound and vibration*, Vol. 108, No. 2, pp. 297-304.

Nomenclature

A, B, C	unknown coefficient of displacement in x,r, θ direction of shell
A_1, A_2	generalized coordinates of shell
a	radius of shell or plate
$F^S_{3,x1}, F^S_{3,x2}$	transverse dynamic force of shell at position x1 and x2
f_i	forces at shell/plate junction
h, h_S, h_P	thickness of shell, plate
k	mode number
L	length of shell
m ,M	axial half wave number
$M_{1,x1}, M_{1,x2}$	moment at point x_1,x_2
M_P	mass of plate
N, n	circumferential mode number
p_r	Eigenvalue of beam function
r	radial coordinate of plate
t	time
u_3^P	transverse normal displacement of plate
u_3^S	transverse normal displacement of shell
u_i	displacement of shell in i direction

U_i	displacement of shell in i direction
x_1, x_2	axial coordinates of shell where plates are welded
x^*_1, x^*_2	axial coordinates where plate attached at shell
$\alpha_{ij}, i,j = 1,2,3,4$	receptances of shell
$\beta_{ij}, i,j = 1,2,3,4$	receptances of plate
$\beta^P_{x_1}, \beta^S_{x_1}$	slope of plate and shell at axial coordinates x_1
δ	Dirac Delta function
λ	eigenvalue of plate
η_k	mode participation factor of k-th mode
ς	damping of structure
ϕ	beam function
θ	coordinates in the circumferential direction
$\omega, \omega_k, \omega_{imn}$	natural frequency, of k-th mode, of *imn* mode

Appendix I. Receptances of shell for the case of two plates attachment

$$Denom = \left[(A/C)^2_{imn}\int_0^L \phi'^2(x)dx + (B/C)^2_{imn}\int_0^L \phi^2(x) + \int_0^L \phi^2(x)dx\right]$$

$$\alpha_{11} = \sum_{m=1}^{\infty}\sum_{i=1}^{3}\frac{1}{\omega^2_{imn}-\omega^2}\frac{\phi'(x_1^*)\phi(x_1^*)}{\rho h \bullet Denom}$$

$$\alpha_{12} = \sum_{m=1}^{\infty}\sum_{i=1}^{3}\frac{1}{\omega^2_{imn}-\omega^2}\frac{-p_r\phi'(x_1^*)\phi(x_1^*)}{\rho h \bullet Denom}$$

$$\alpha_{13} = \sum_{m=1}^{\infty}\sum_{i=1}^{3}\frac{1}{\omega^2_{imn}-\omega^2}\frac{\phi(x_1^*)\phi(x_1^*)}{\rho h \bullet Denom}$$

$$\alpha_{14} = \sum_{m=1}^{\infty}\sum_{i=1}^{3}\frac{-1}{\omega^2_{imn}-\omega^2}\frac{p_r\phi'(x_1^*)\phi(x_1^*)}{\rho h \bullet Denom}$$

$$\alpha_{21} = \sum_{m=1}^{\infty}\sum_{i=1}^{3}\frac{1}{\omega^2_{imn}-\omega^2}\frac{-p_r\phi(x_1^*)\phi'(x_1^*)}{\rho h \bullet Denom}$$

$$\alpha_{22} = \sum_{m=1}^{\infty}\sum_{i=1}^{3}\frac{1}{\omega^2_{imn}-\omega^2}\frac{p_r^2\phi'(x_1^*)\phi'(x_1^*)}{\rho h \bullet Denom}$$

$$\alpha_{23} = \sum_{m=1}^{\infty}\sum_{i=1}^{3}\frac{1}{\omega^2_{imn}-\omega^2}\frac{-p_r\phi(x_2^*)\phi'(x_1^*)}{\rho h \bullet Denom}$$

$$\alpha_{24} = \sum_{m=1}^{\infty}\sum_{i=1}^{3}\frac{-1}{\omega^2_{imn}-\omega^2}\frac{p^2{}_r\phi(x_1^*)\phi'(x_2^*)}{\rho h \bullet Denom}$$

$$\alpha_{31} = \sum_{m=1}^{\infty}\sum_{i=1}^{3}\frac{1}{\omega^2_{imn}-\omega^2}\frac{\phi(x_1^*)\phi(x_2^*)}{\rho h \bullet Denom}$$

$$\alpha_{32} = \sum_{m=1}^{\infty}\sum_{i=1}^{3}\frac{-1}{\omega^2_{imn}-\omega^2}\frac{p_r\phi'(x_1^*)\phi(x_2^*)}{\rho h \bullet Denom}$$

$$\alpha_{33} = \sum_{m=1}^{\infty}\sum_{i=1}^{3}\frac{1}{\omega^2_{imn}-\omega^2}\frac{\phi(x_2^*)\phi(x_2^*)}{\rho h \bullet Denom}$$

$$\alpha_{34} = \sum_{m=1}^{\infty}\sum_{i=1}^{3}\frac{-1}{\omega^2_{imn}-\omega^2}\frac{p_r\phi(x_2^*)\phi'(x_2^*)}{\rho h \bullet Denom}$$

$$\alpha_{41} = \sum_{m=1}^{\infty}\sum_{i=1}^{3}\frac{-1}{\omega^2_{imn}-\omega^2}\frac{p_r\phi(x_1^*)\phi'(x_2^*)}{\rho h \bullet Denom}$$

$$\alpha_{42} = \sum_{m=1}^{\infty}\sum_{i=1}^{3}\frac{1}{\omega^2_{imn}-\omega^2}\frac{p_r^2\phi'(x_1^*)\phi'(x_2^*)}{\rho h \bullet Denom}$$

$$\alpha_{43} = \sum_{m=1}^{\infty}\sum_{i=1}^{3}\frac{-1}{\omega^2_{imn}-\omega^2}\frac{p_r\phi(x_2^*)\phi'(x_2^*)}{\rho h \bullet Denom}$$

$$\alpha_{44} = \sum_{m=1}^{\infty}\sum_{i=1}^{3}\frac{1}{\omega^2_{imn}-\omega^2}\frac{p_r^2\phi'(x_2^*)\phi'(x_2^*)}{\rho h \bullet Denom}$$

$$\alpha_{31} = \alpha_{13}$$

$$\alpha_{32} = \alpha_{23}$$

Modern Practice in Stress and Vibration Analysis, Gilchrist (ed.) © *1997 Balkema, Rotterdam, ISBN 90 5410 896 7*

Energy of plane elastic waves in anisotropic media

A.S.Grishin & A.R.Loshitskiy
Moscow Institute of Civil Engineering, Russia

ABSTRACT: Numerical data for organic wooden materials showed that the direction of the minimal energy flux is not orthogonal to the front of plane wave. In respect to the maximal energy flux, we observed that it coincides with the direction of the longitudinal plane wave.

1. INTRODUCTION

The equation for determination of the speeds propagation of plane elastic waves in anisotropic medium with arbitrary elastic anisotropy was proposed (Fedorov 1964). The precise description of propagation longitudinal and transverse waves in anisotropic medium in a closed analytical form was obtained (Truesdell 1966) with generalisation of the previous results (Fedorov 1964). Further investigations were mainly devoted to the analysis of the interface waves, namely Rayleigh's waves which arise on the plane interface between solid and gas, and Lamb's waves which arise on the interface between two solids.

The present analysis is concerned with determination of the energy transferred by progressive plane waves propagating in linearly elastic anisotropic medium with arbitrary anisotropy. The main concept of the analysis is based on the decomposition of the energy flux in orthogonal directions, which generally speaking do not coincide with the directions of wave propagation and the transverse to it.

2. BASIC OPERATORS

A homogeneous anisotropic medium is considered with the equation of wave propagation written in the form:

$$-\operatorname{div}\overset{4}{\mathbf{C}}\cdot\cdot\nabla\mathbf{u} = \rho\cdot\ddot{\mathbf{u}} \quad \text{or}$$

$$A(\partial_x)\mathbf{u} - \rho\cdot\ddot{\mathbf{u}} = 0 \qquad (2.1)$$

where $\overset{4}{C}$ = fourth-order tensor of elasticity; $\mathbf{u}$ = the displacement vector; ρ = medium density.

The plane progressive elastic waves are considered (Gurtin 1972)

$$\mathbf{u}(\mathbf{x},t) = \mathbf{m}\psi\ (\mathbf{n}\cdot\mathbf{x} - c\cdot t) \qquad (2.2)$$

where $\mathbf{u}(\mathbf{x},t)$ = the displacement field on the front of a shock wave; $\mathbf{m}$ = a unit vector called the amplitude of the plane wave. It characters the direction of displacement field for the wave ($\mathbf{m} \in R^3$); $\mathbf{n}$ = a unit vector which is orthogonal to the front of wave. It shows the direction of wave propagation ($\mathbf{n} \in R^3$); c = speed of the wave propagation; ψ - any function which is: $\psi \in C^2(R)$; $\psi'' \neq 0$.

For the further analysis the symbol of the differential operator of the equilibrium equations is required. Fourier's transform applied to eq. (2.1) gives

$$A^V(\mathbf{x}) = (2\pi)^2\cdot\mathbf{x}\cdot\overset{4}{C}\cdot\mathbf{x} \qquad (2.3)$$

Substituting formula (2.2) in (2.1), We obtain:

$$A(\partial_x)\mathbf{u} = -C_{ijpq}\cdot\mathbf{u}_{p,q,j} = \mathbf{n}\cdot\overset{4}{C}\cdot\mathbf{m}\otimes\mathbf{n}\cdot\psi''_S \qquad (2.4)$$

where:

$$\nabla \mathbf{u} = \mathbf{m} \cdot \frac{d\psi}{dS} \nabla S = \mathbf{m} \otimes \mathbf{n} \cdot \psi_S{}' \tag{2.5}$$

$$\ddot{\mathbf{u}} = \mathbf{m} \cdot \psi_S'' \frac{dS}{dt} \cdot \frac{dS}{dt} = \mathbf{m} \cdot c^2 \cdot \psi_S'' \tag{2.6}$$

Dividing (2.4) by ψ_S'' as the latter is not equal to zero, We arrive to the equation of wave propagation in anisotropic medium

$$\mathbf{n} \cdot \overset{4}{\mathrm{C}} \cdot \cdot \mathbf{n} \otimes \mathbf{m} = \rho \cdot c^2 \cdot \mathbf{m} \tag{2.7}$$

Taking into consideration (2.3), formula (2.7) can be rewritten in the form:

$$\frac{1}{(2\pi)^2} \mathrm{A}^V(\mathbf{n}) \cdot \mathbf{m} = \rho \cdot c^2 \cdot \mathbf{m} \tag{2.8}$$

From (2.8) it is easy to obtain a formula for speeds of wave propagation. To achieve it, We multiply the both left parts of (2.8) by $\mathbf{m}$-vector and obtain (as $\mathbf{m} \cdot \mathbf{m} = 1$)

$$\frac{1}{(2\pi)^2} \mathbf{m} \cdot \mathrm{A}^V(\mathbf{n}) \cdot \mathbf{m} = \rho \cdot c^2 \cdot \mathbf{m} \cdot \mathbf{m} \Rightarrow$$

$$c = \sqrt{\frac{\mathbf{m} \cdot \mathrm{A}^V(\mathbf{n}) \cdot \mathbf{m}}{(2\pi)^2 \cdot \rho}} \tag{2.9}$$

3. ENERGY OF PLANE ELASTIC WAVE

The formula for energy can be written as full convolution of an elasticity tensor and a stress tensor:

$$\mathrm{W} = \frac{1}{2} \overset{2}{\varepsilon} \cdot \cdot \overset{2}{\sigma} \tag{3.1}$$

Where Hooke's law takes the form:

$$\overset{2}{\sigma} = \overset{4}{\mathrm{C}} \cdot \cdot \overset{2}{\varepsilon} \tag{3.2}$$

In accordance to (3.2) formula (3.1) can be rewritten in the following way:

$$\mathrm{W} = \frac{1}{2} \overset{2}{\varepsilon} \cdot \cdot \overset{4}{\mathrm{C}} \cdot \cdot \overset{2}{\varepsilon} \tag{3.3}$$

From Cauchy's identity

$$\overset{2}{\varepsilon} = \frac{1}{2}\left(\nabla \mathbf{u} + (\nabla \mathbf{u})^T\right) \tag{3.4}$$

According to (2.2) formula (3.4) is written as

$$\overset{2}{\varepsilon} = \mathrm{sym}(\mathbf{m} \otimes \mathbf{n}) \cdot \psi_S' \tag{3.5}$$

Substituting (3.5) in (3.3) We obtain

$$\mathrm{W} = \frac{1}{2} (\mathbf{m} \otimes \mathbf{n}) \cdot \cdot \overset{4}{\mathrm{C}} \cdot \cdot (\mathbf{n} \otimes \mathbf{m}) \tag{3.6}$$

Expression for energy can be rewritten in terms of speed for which it is necessary to consider formula (2.9). According to formula (3.6) the expression for energy in terms of speed is the form

$$\mathrm{W} = \frac{1}{2} \rho c^2 \tag{3.7}$$

The expression (2.8) can be represented in the form:

$$\mathrm{A}^{\mathrm{V}}(\mathbf{n})\mathbf{m} = \lambda \mathbf{m} \tag{3.8}$$

Multiplying formula (3.8) by $\mathbf{m}$-vector from leftside, We obtain

$$\mathbf{m} \cdot \mathrm{A}^{\mathrm{V}}(\mathbf{n}) \cdot \mathbf{m} = \lambda\, \mathbf{m} \cdot \mathbf{m} \qquad \Rightarrow$$

$$\lambda = \mathbf{m} \cdot \mathrm{A}^{\mathrm{V}}(\mathbf{n}) \cdot \mathbf{m}, \text{ but}$$

$$\mathbf{m} \otimes \mathbf{n} \cdot \cdot \overset{4}{\mathrm{C}} \cdot \cdot \mathbf{n} \otimes \mathbf{m} = \frac{1}{(2\pi)^2} \mathbf{m} \cdot \mathrm{A}^{\mathrm{V}}(\mathbf{n}) \cdot \mathbf{m} = \frac{\lambda}{(2\pi)^2} \tag{3.9}$$

The following expression is the formula We look for:

$$\mathrm{W} = \frac{1}{2} \frac{\lambda}{(2\pi)^2} = \frac{1}{8} \frac{\lambda}{\pi^2} \tag{3.10}$$

4. RESULTS OF NUMERICAL ANALYSIS

After the numerical analysis of different anisotropic media having been done, the results which are shown in Table 1, based on the experimental data (Fedorov 1964, Ashkenazi 1972), were obtained.

Table 1. Results of numerical analysis.

Material Name	Energy Emax/ Emin	Angle at Emax/ Emin
Hexagonal	crystals	
Mg	5.75 / 1.17	0.02 / 90.00
SiO_2	6.17 / 1.80	0.08 / 89.99
H_2O (crystal)	0.75 / 0.16	0.00 / 90.00
Co	17.90 / 3.55	0.00 / 90.00
Cd	8.65 /2.46	0.00 / 90.00
Zn	8.14 / 1.14	0.08 / 71.93
Cubic	crystals	
Al	5.75 / 1.17	0.02 / 90.00
Ni	17.33 / 2.48	0.97 / 90.00
Pb	3.04 / 0.19	0.69 / 90.00
Mo	23.00 / 5.50	0.00 / 90.00
Tetragonal	crystals	
Sn	4.35 / 1.10	0.00 / 90.00
In	2.73 / 0.10	0.03 / 89.97
$BaTiO_3$	15.41 / 1.52	0.13 / 81.54
$ZrSiO_4$	3.67 / 0.69	0.00 / 90.00
Wooden	materials	
Oak-tree	13.07 / 0.73	0.21 / 56.02
Beech	11.28 / 0.34	0.59 / 34.01
Maple	18.27 / 0.47	0.37 / 33.97
Birch	27.19 / 0.29	0.37 / 21.99
Ash-tree	18.46 / 0.30	0.51 / 26.00
Silver fir	34.72 / 0.38	0.21 / 22.00
Fir	16.73 / 0.30	0.48 / 17.99
Nut-tree	22.35 / 0.42	0.61 / 26.00

For all media with crystallic structure, the direction of maximum energy flux coincides with the direction of corresponding plane wave propagation while the direction of minimum energy flux is orthogonal to the front of wave propagation. The only one exception has been discovered, it is for Zn crystal, where the angle between $\mathbf{E}_{min}$ and $\mathbf{n}$ was 72^0. Possibly, the constants of elasticity for Zn were not correct.

At the same time analysis which has performed for organic media showed a greater deflexion from normal between vectors $\mathbf{E}_{min}$ and $\mathbf{n}$. In this kind of media, there are no media (amongst those tested) for which the flux of minimum energy should be orthogonal to wave propagation. So, the minimum angle for fir is about 18^0. It should be pointed that the fact showed above is common for all tested sorts of wood. These examples show, as well as the previous one, that there might have been mistakes in determination of elastic constants, or the minimum energy flux is not orthogonal to the direction of a wave propagation.

REFERENCES

Fedorov F.I. 1964. *To theory of elastic waves in crystals.* Moscow: Vestn. MGU . Ser. fisika, astronomia, №6. P.36-40.

Truesdell Cl. A. 1966. *Existence of longitudinal waves* . J.Acoust. Soc. Amer. V.40. No.3. P.729-730.

Gurtin M.E. 1972. *The linear theory of elasticity*. Handbuch der Physik. Bd. 6a/2. Berlin: Springer, P.1-295.

Ashkenazi E.K. 1972. *Anisotropy of constructive materials.* Spravochnik, Leningrad: Mashinostroenie.

Modern Practice in Stress and Vibration Analysis, Gilchrist (ed.) © *1997 Balkema, Rotterdam, ISBN 90 5410 896 7*

A numerical technique for simulating the dynamic stresses in piping systems with limited vibration measurements

W.A. Moussa
Mechanical Engineering Department, The American University in Cairo, Egypt

A.N. AbdelHamid
Science and Engineering Department, The American University in Cairo, Egypt

ABSTRACT: A practical procedure is investigated for the determination of dynamic stresses in pipelines through the use of Finite Element Method (FEM) in dynamic analysis and field measurement vibrations at selected points. Numerical simulation of a dynamically loaded pipeline is used to establish the validity of the procedure. The analysis was carried out in the frequency domain with harmonic excitation of the selected pipeline structure. The results show that if the location(s) of the actual exciting force(s) or the translation and rotational "measured" displacements at the ends of a force free pipe segment are completely specified the stress picture resulting from the application of the procedure is exact. Alternatively, lack of coincidence between the vibration measurement points and the input force(s), or the use of only translational vibration inputs results in an approximate stress picture. The extent of the "error" in these cases is found to depend on the spectrum of the vibration frequency, the proximity between the vibration measurement points and the input forces, the density of these points and the type of vibratory-motions they measure.

1 INTRODUCTION

The safety-related piping in industrial sites such as nuclear power plants and offshore oil platforms may be subjected to pump- or fluid- induced vibrations that, in general, affect only local areas of the piping systems around the vibration source (Huang 1991). Pump- or fluid-induced vibrations are typically characterized by low levels of amplitudes and high number of cycles over the lifetime of plant operation. Thus, the resulting fatigue damage to the piping systems could be an important safety concern (Goyder 1980).

The complexity of a typical built-up pipeline often prevents its detailed vibrational characteristics being predicted theoretically. Alternatively, numerical techniques were used to analyze an existing structure by employing only measured vibrations, with no assumptions being made about the form of possible governing differential equations (Coockey 1992). On one hand, however, measured vibrations may be limited because of instrumentation lackness in critical piping locations and/or because of the difficulty in obtaining data in inaccessible areas. On the other hand, the characterization of the vibration source/s represents another important measurement that may be missing from the acquired vibrations. In such cases, it is difficult to correlate the measured data with the pipeline vibration characteristics because the necessary information is either insufficient or missing (Huang 1991).

2 PROPOSED SIMULATION TECHNIQUE

The main theme of this work is to establish a technique that uses measured vibrations at selected points on a pipeline structure, as seen in Figure (1), along with the dynamic Finite Element Analysis (FEA) to approximately predict the structure's dynamic behavior under unknown "real" loading conditions, (Moussa 1993). The question which arises at this point is : what happens if the vibratory-motion at a point other than the "real" loading points is used as an input "load" ? The answer depends on whether or not the vibration inputs account for all the actual loading points on the structure. If they do, then the added input is superfluous and will not introduce an "error" in the result. On the other hand, if one or more of the "real" loading points are not accounted for, then some "error" is introduced in the results. This "error" is expected to depend not only on the number and location of missed "real" loading points but also on the frequency content of the structure vibration. The simulation procedure proposed is summarized in the following points :

1. A Finite Element (FE) model of the pipeline structure is generated and tested using a commercially available FE package [ANSYS IV].

2. The pipeline structure is assumed to be excited by a single concentrated harmonic force at its mid span. The resulting vibratory-motions, such as the structure's internal forces and stress spectrum, are determined using dynamic FEA. This load case is termed the "real" load case.

3. The vibratory-motions at some points on the structure, as determined from the previous step, are used selectively to simulate the vibrations "measured" in a "real" field case.

4. The simulated vibration measurements are then used individually or in groups as displacement loading to the structure, where the resulting stress spectrum for each verification case is then determined. The range of cases considered is wide and addresses the question raised in the last paragraph.

5. The results for each of the simulated load cases are then compared with those of the "real" load case to determine the extent of the "error" in the dynamic stresses in the pipe structure and to establish a criterion for the validity of the proposed simulation technique.

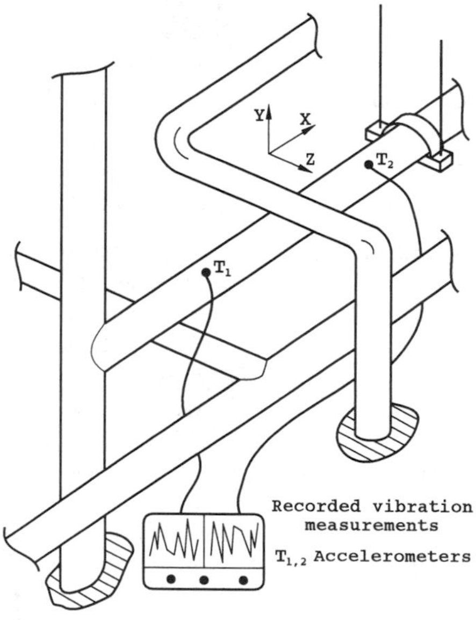

Figure 1: A typical vibration measurement setup of a complex piping system seen in industrial planes.

Material & Geometric Properties

- Clamped-Hinged Tubular Beam Structure
- Total Length = 240 in
- Distance Between Measuring Points = 12 in
- Outer Diameter = 4.5 in
- Wall Thickness = 0.75 in
- Modulus of Elasticity = 3000 Ksi
- Mass Density = 0.28299 lb/in^3

(b)

FE Model

- Element Type : Pipe Element
- Number of Elements = 20 (STIF16)

n	f_n	$f_{FEM}-f_n$
1	11.159	0.157
2	36.014	0.260
3	74.698	0.308

(c)

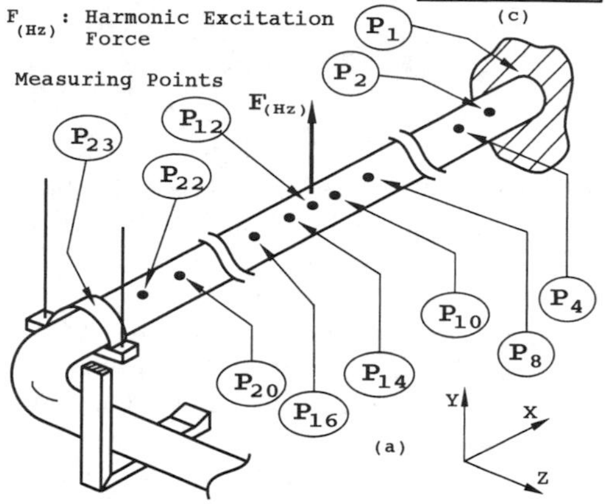

Figure 2: Clamped-Hinged Pipe model (a) Vibration measurements layout (b) Material & geometric properties (c) Natural frequencies [first 3].

3 NUMERICAL MODEL

The modeled pipe is a Clamped-Hinged beam structure with a total length of (20) feet and a uniform tubular cross section. The asymmetry of the ends conditions, as seen in Figure (2-a), is used to include possible ends effects on the results and to avoid symmetric mode shapes. The geometric and material properties of the pipe structure used are described in Figure (2-b). The dimensions of the FE pipe model were obtained from a typical pipeline system in an offshore oil platform, (AbdelHamid 1992).

The first step in the dynamic FEA is to verify the basic dynamic properties of the pipe model used versus the analytical closed-form solution

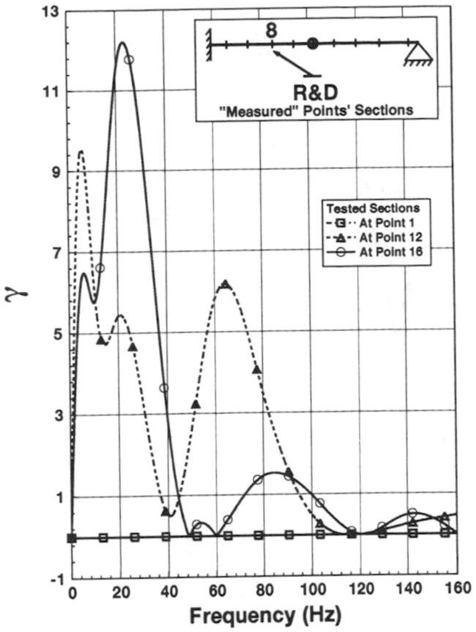

Figure 3: Distribution of γ versus frequency at different sections of the pipeline FE model for translational and rotational vibratory-motion at point (8).

found in literature, (McConnell 1995),

$$f_n = \frac{\lambda_n^2}{2\pi L_p^2}\sqrt{\frac{E_x I_z}{m}} = \frac{\lambda_n^2}{8\pi L_p^2}\sqrt{\frac{E_x(D_o^2 + D_i^2)}{\rho}} \quad (1)$$

where f_n is the natural frequency for mode n, m is the pipe mass per unit length, L_p is the pipe length, E_x is the modulus of elasticity in the x-axis direction, ρ is the mass density of the pipe material, I_z is the modulus of inertia about the z-axis, λ_n is a mode constant, D_o and D_i are the outer and inner diameters of the pipe, respectively. Neglecting the mass of the fluid inside the pipe, the first three natural frequencies of the pipe structure obtained using FE modal analysis were compared against the values from Equation (1). The negligible difference in the natural frequencies, as seen in Figure (2-c) validates the dynamic FE solution of the pipe model.

4 VERIFICATION CASES

The simulation process was used in the frequency domain to verify the use of FE dynamic analysis with input displacement (translational and/or rotational). As a starting point, the FE pipe model was harmonically excited with a single external force of constant amplitude in the y-direction within a frequency range of (5, 10, 20,..., 160) Hz. In reality, such force could result from parametric resonance situations. In such cases, The forcing function in the axial direction of the pipe, caused by a checking valve, for example, may induce parametric resonance in the lateral (i.e. y-axis) direction, which is a secondary mode of vibration for the pipe structure.

Applying the "auxiliary nodes" technique discussed in an earlier work (Moussa 1996), the resulting vibratory-motion was used as "measured" vibrations at different points along the pipe x-axis and was applied to the FE model to simulate the effect of the original exciting force F_y.

4.1 *The effect of lack of coincidence between the position of the input force and the input vibratory-motion*

In the first verification case, the resultant vibratory-motion at node (8), where ($\frac{L}{D_o} = 8$), was used as the simulating loading source. The percentage difference in the stress spectrum, γ, was

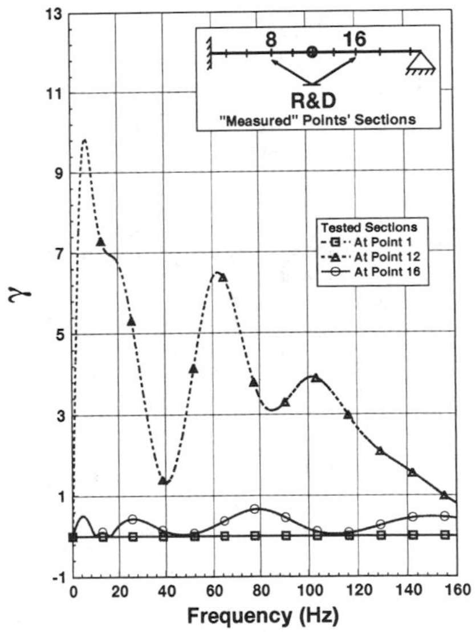

Figure 4: Distribution of γ versus frequency at different sections of the pipeline FE model for translational and rotational vibratory-motion at points (8 & 16).

calculated from Equation (2). In this equation, σ_F and σ_D are the maximum Von-Misses stresses in the tested sections for the force excited "real" case and the displacement excited "simulated" case, respectively. While the γ value at the fixed root, section (1), was (0%) for the simulated case, the maximum values of γ at sections (12) and (16) were calculated as (9.4%) and (12.2%), respectively.

In these two sections, as seen in Figure (3), the maximum values of γ were found at low frequency levels (ie. $\leq$ 30 Hz).

$$\gamma \; = \; \left| \frac{\sigma_F - \sigma_D}{\sigma_F} \right| \; \times \; 100 \qquad (2)$$

4.2 *The effect of having dual rotation and translational vibratory-motion inputs with distance in between*

In case (2), dual displacement inputs have been investigated in the frequency domain for both translational and rotational vibratory-motions. In this case, the resultant vibratory-motion at points (8) and (16) with ($\frac{L}{D_o} = 8$) was used as the simulating loading source, see Figure (4). Compared to the "real" case, the maximum value of γ was (9.85%) at the position of the exciting force, section (12). However, in the hinged section, (16), the maxi-

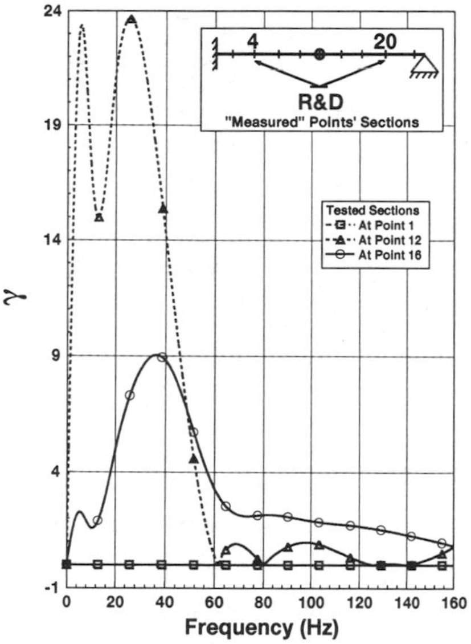

Figure 5: Distribution of γ versus frequency at different sections of the pipeline FE model for translational and rotational vibratory-motion at points (4 & 20).

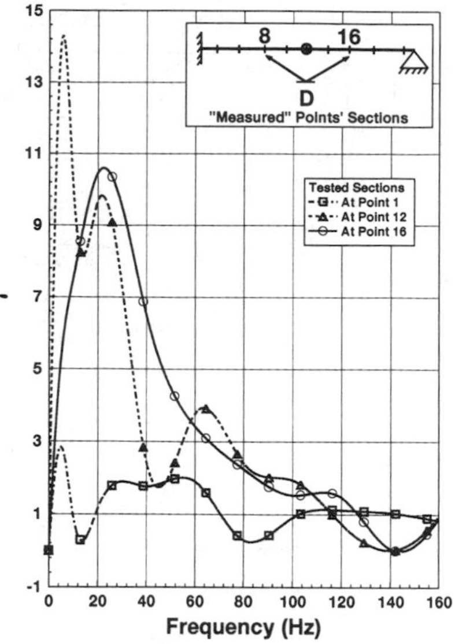

Figure 6: Distribution of γ versus frequency at different sections of the pipeline FE model for translational vibratory-motion at points (8 & 16).

mum γ decreased to (0.66%). Although there was no noticeable change in the γ value in section (12) compared with case (1), the significant decrease of γ in section (16) emphasize the need for having a balance in the measuring points surounding a "suspected" exciting-force position.

In case (3), the resultant translational and rotational vibratory-motion at points (4) and (20) with ($\frac{L}{D_o} = 18.66$) was used as the simulating loading source, as seen in Figure (5). By increasing the ($\frac{L}{D_o}$) value, the maximum γ values in sections (12) and (16) have raised to (23.66%) and (9.03%) respectively, indicating the proportional relation between ($\frac{L}{D_o}$) and γ. As for section (1), γ remained at the zero level.

In case (2) and (3), the variation of γ remained either zero or relatively stable within the frequency range tested for sections (1) and [(16) in case (2)]. In contrast, γ decreased for high ranges of excitation frequency at section (12) and [(16) in case (3)] as ($\frac{L}{D_o}$) increased.

4.3 *The effect of having dual translational vibratory-motion inputs with distance in between*

To study the lack of measurements effect for rotational displacement along the z-axis, dual dis-

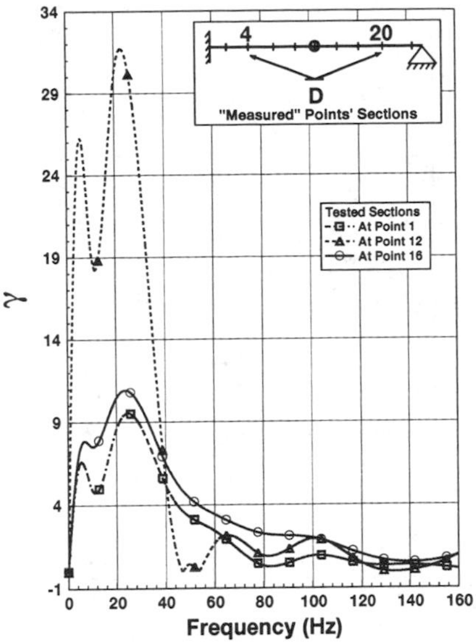

Figure 7: Distribution of γ versus frequency at different sections of the pipeline FE model for translational vibratory-motion at points (4 & 20).

placement inputs have been investigated for only translational measurements in the same frequency range studied above.

In case (4), the resultant translational vibratory-motion at points (8) and (16) with ($\frac{L}{D_o} = 8$) was used as the simulating loading source, as seen in Figure (6). Compared to case (2), the maximum value of γ raised slightly to (14.05%) at the position of the exciting force, section (12). However, in the hinged section, (16), dramatic increase of γ to (10.6%) was noticed. Also, the maximum γ value at the fixed end increased to (2.85%) for the first time. The last two results indicate the significant impact of the rotation motion absence on the accuracy of the predicted stress spectrum in the FE pipe model.

In case (5), the resultant translational vibratory-motion at points (4) and (20) with ($\frac{L}{D_o} = 18.66$) was used as the simulating loading source, as seen in Figure (7). As expected, γ increased further to its highest value through out the presented work. At section (12), for example, the maximum γ value was (31.72%). Also, compared with case (3), higher γ values of (9.53%) and (10.92%) were obtained at sections (1) and (16), respectively.

4.4 *The effect of multiple input points with translational Vibratory-Motion Only*

To investigate the effect of substituting rotational vibratory-motion with a higher number of translational motion measuring-point, the following two cases were applied.

In case (6), the resultant translational vibratory-motion at points (2,4,20, and 22) with (minimum $\frac{L}{D_o} = 18.66$) was used as the simulating loading source, as seen in Figure (8). Compared to case (5), the maximum value of γ slightly decreased to (28.45%) at the position of the exciting force, section (12). While the maximum γ remained stable at (10.6%) in the hinged section, its value at the fixed end considerably decreased to (1.68%).

The resultant translational vibratory-motion at points (2,10,14, and 22) with (minimum $\frac{L}{D_o} = 2.66$) was used as the simulating loading source in case (7). As seen in Figure (9), the maximum value of γ has dropped to (0%) at both ends of the pipe model. As for sections (12), the maximum γ value was (5.35%), which is the lowest in all seven cases studied. The last result is encouraging since it shows the positive effect of increasing the number of measured translational vibratory-motion on reducing the maximum γ value throughout the pipe model.

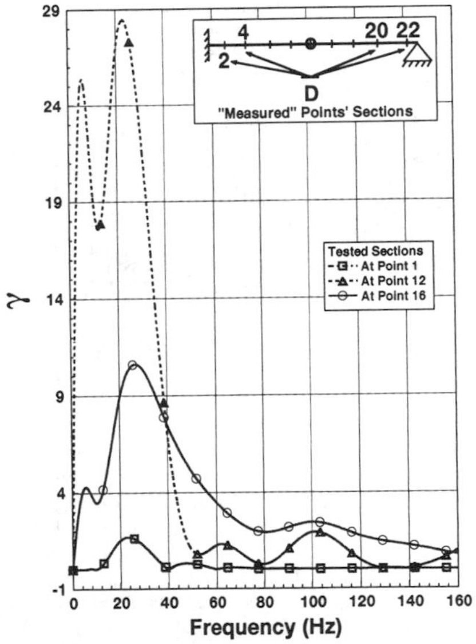

Figure 8: Distribution of γ versus frequency at different sections of the pipeline FE model for translational vibratory-motion at points (2,4,20 & 22).

5 DISCUSSION

Examining the modeled cases above, the overall efficiency of the proposed dynamic simulation technique could be improved based on the following guidelines:

1. Vibration measurement points balancing is required in the vicinity of a "suspected" exciting-force position.

2. Lower values of γ could be obtained by reducing the minimum $(\frac{L}{D_o})$ between the measuring points and the "suspected" exciting force position in the tested pipeline, as seen in Figure (10) .

3. Dual translational vibration measurements could be used to predict the stress spectrum in pipelines at regions where exciting forces are not likely to fall within, for example, sections (1) and (16) in both cases (4) and (5).

4. Vibration measurements of rotational motion could be substituted with a higher number of translational measurements, as seen in cases (6) and (7).

5. The highest value of γ is expected to be at the position of the exciting force, which may not necessary be the same section with maximum stress level.

6. In all cases discussed above, the optimum measurement setup was used in case (7), as seen in Figures (10) and (11). This result emphasis the significance of prior knowledge of the exciting force "suspected" position in simulating the stress spectrum in a pipeline with limited vibration measurements.

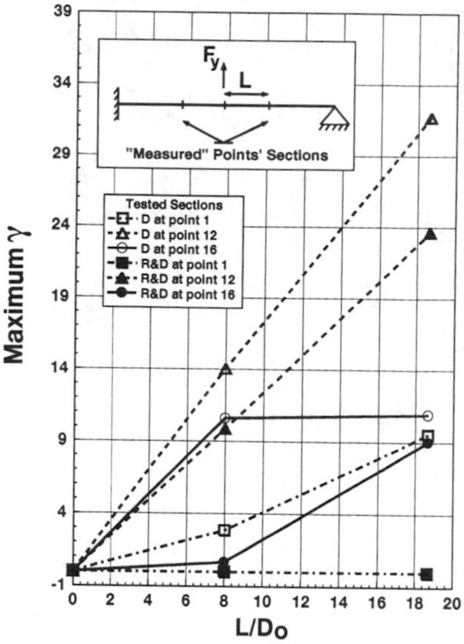

Figure 10: Maximum γ values versus $\frac{L}{D_o}$ at different sections of the pipeline FE model for rotational and/or translational vibratory-motion cases.

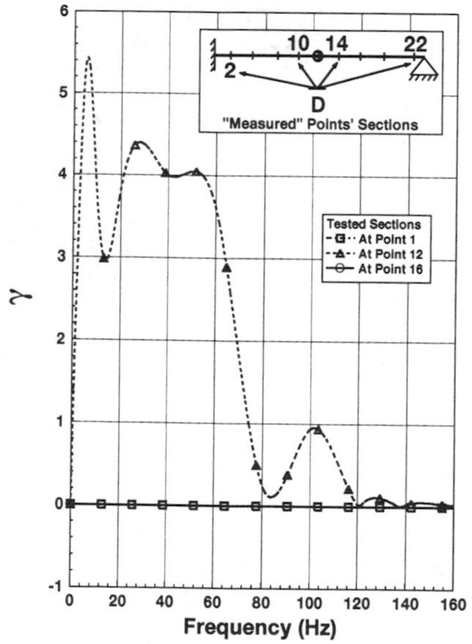

Figure 9: Distribution of γ versus frequency at different sections of the pipeline FE model for translational vibratory-motion at points (2,10,14 & 22).

6 CONCLUSIONS

A practical procedure is investigated for the determination of dynamic stresses in pipelines through the use of FEM in dynamic analysis and field measurement vibrations at selected points. Numerical simulation of a dynamically loaded pipeline is used to establish the validity of the procedure. The analysis was carried out in the frequency domain with harmonic excitation of the selected pipeline structure.

In the studied cases, the highest and lowest percentage difference between the "real" and "simulated" stress spectrum at different section of the pipe model were (31.72%) and (5.35%), respectively. These results show that if the location(s) of the actual exciting force(s) or the translation

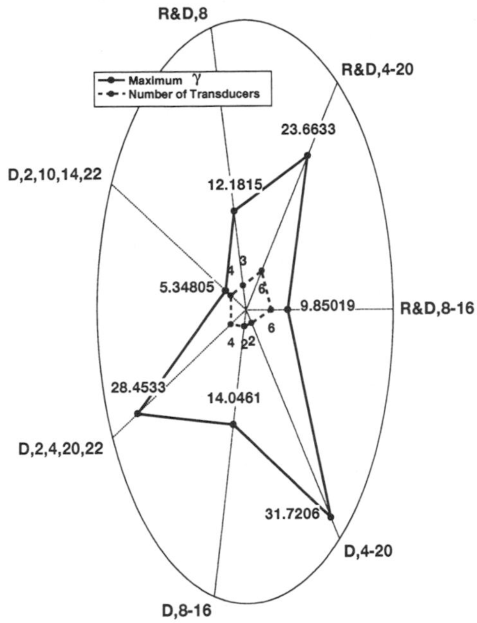

Figure 11: Maximum γ values versus the number of vibration transducers for each measurement setup simulated.

and rotational "measured" displacements at the ends of a force free pipe segment are completely specified the stress picture resulting from the application of the procedure is exact. Alternatively, lack of coincidence between the vibration measurement points and the input force(s), or the use of only translational vibration inputs results in an approximate stress picture. The extent of the "error" in these cases is found to depend on the spectrum of the vibration frequency, the proximity between the vibration measurement points and the input forces, the density of these points and the type of vibratory-motions they measure.

7 ACKNOWLEDGEMENT

The support of Esso Sues Inc. throughout contract 22-838 is gratefully acknowledged.

REFERENCES

AbdelHamid, A. N. 1992. Static and dynamic analysis of stresses in flow lines 7 and 12 on the offshore platform of Esso Sues Inc. Technical Report 22-838, The American University in Cairo, Cairo.

Coockey, W. M. & Butcher, N. A. 1992. Development of a technique of analyzing the durability aspects of automotive structure by computational methods. *Fatigue Analysis*, 15–18.

Goyder, H. G. D. 1980. Methods and application of structural modelling from measured structural frequency response data. *Journal of Sound and Vibration 68*(2), 209–230.

Huang, S. N. 1991. Fatigue evaluation of piping systems with limited vibration test data. In *Proceedings of (ASME) Pressure Vessels and Piping Conference, San Diego.*, Volume 220, San Diego, pp. 261–265.

McConnell, K. G. 1995. *Vibration Testing.* New York: John Wiley & Sons.

Moussa, W. A. 1993. Evaluation of dynamic stresses and fatigue life predictions for a randomly loaded structure. Master's thesis, The American University in Cairo, Cairo.

Moussa, W. A. 1996. Evaluation of dynamic stresses for a randomly loaded structure using a new numerical simulation technique. In *Proceedings of Mechanics in Design*, Volume 2, Toronto, pp. 785–793.

Modern Practice in Stress and Vibration Analysis, Gilchrist (ed.) © 1997 Balkema, Rotterdam, ISBN 90 5410 896 7

Author index